Worlds Before Adam

Martin J. S. Rudwick

WORLDS *Before* ADAM

The Reconstruction of Geohistory in the Age of Reform

The University of Chicago Press *Chicago & London*

The University of Chicago Press, Chicago 60637
The University of Chicago Press, Ltd., London

Paperback edition 2010
Printed in the United States of America

19 18 17 16 15 14 13 12 11 10   2 3 4 5 6

ISBN-13: 978-0-226-73129-2 (paper)
ISBN-10: 0-226-73129-4 (paper)

Library of Congress Cataloging-in-Publication Data

Rudwick, M. J. S.
Worlds before Adam : the reconstruction of geohistory in the age of reform / by Martin J.S. Ruwick.
p. cm.
Includes bibliographical references and index.
ISBN-13: 978-0-226-73128-5 (cloth : alk. paper)
ISBN-10: 0-226-73128-6 (cloth : alk. paper) 1. Geology—Europe—History—19th century. 2. Science—Europe—History—19th century. I. Title.
QE13. E85R83 2008
551.7094'09034—dc22

2007041676

♾ The paper used in this publication meets the minimum requirements of the American National Standard for Information Sciences—Permanence of Paper for Printed Library Materials, ANSI Z39.48-1992.

As we explore this magnificent field of [geological] inquiry, the sentiment of a great historian of our times may continually be present to our minds, that "he who calls what has vanished back again into being, enjoys a bliss like that of creating".

Charles Lyell, *Principles of Geology*, 1830,
quoting Barthold Georg Niebuhr,
Römische Geschichte [*Roman History*], 1811

CONTENTS

ILLUSTRATIONS

ACKNOWLEDGMENTS

As in *Bursting the Limits of Time*, to which this volume is the sequel, my first duty is to thank the Master and Fellows of Trinity College, Cambridge, for appointing me Tarner Lecturer for 1994–97. This first gave me the incentive and opportunity to put my ideas into some kind of order, which I have in fact retained in these two volumes although my historical evidence has of course been hugely expanded. My second obligation, no less pleasurable, is to thank the members and associates of the Department of History and Philosophy of Science at Cambridge for giving me a congenial and stimulating intellectual home after my return to England from California. Though they have probably been quite unaware of it, their lively arguments in several seminars, and particularly Nick Jardine's Cabinet of Natural History, have helped keep my brain alive, and my knowledge of the field reasonably up-to-date, while I was working the rest of my time, somewhat eremitically (I've long wanted to find a use for that lovely word), deep in the Fens or away on the Marches bordering wild Wales.

I will only mention one specific occasion that was particularly helpful in the writing of this second volume, although it took place long ago while I was still teaching in La Jolla. In 1991 I offered a seminar on "Catastrophisms Old and New" at the Scripps Institution of Oceanography. It considered the relation between the classic early nineteenth-century debates covered in the present volume, and those that erupted in the late twentieth century around claims that there had been a massive extraterrestrial impact event (popularly believed to have knocked out the last dinosaurs) at the "K/T boundary" in geohistory. My seminar was attended not only by graduate students in Science Studies at UCSD but also by a formidable group of senior earth scientists at SIO, some of whom had been personally involved in the debates that were then still very recent history. Their firsthand comments on those debates helped to bring to life, in my mind, the parallel arguments among an equally impressive cast of geologists a century and a half earlier. The similarities did much to convince me that in writing what has become the present volume I would need to extricate those earlier debates from the incubus of a largely illusory conflict between "Science" and "Religion", and to show how on the contrary they had dealt with legitimate and perennial problems at the heart of mainstream science.

I will not repeat here the list of friends and colleagues whom I thanked in my previous volume for their help and support. I hope and believe that they know how deeply I appreciate all they have done to sustain this project, and my sometimes fragile self-confidence about it, over many years. I have however thanked some of them again in certain footnotes, for enlightening me on specific points on which their knowledge and understanding was and is far greater than mine. Among others whom I did not name before, but who have also helped me greatly with the present volume, are Mike Bassett (Cardiff), Ernie Hamm (Toronto), Jean-Paul Schaer (Neuchâtel), and Philippe Taquet (Paris). I am especially grateful to my fellow hillwalker and long-standing arguing partner Jack Morrell (Leeds) for wading through the entire draft text of the book and giving me a multitude of suggestions for its improvement.

Again as before, I am grateful to those in charge of all the museums, libraries, and archives to which I have had access, and in some cases for permission to quote or reproduce images from books and documents in their care. The use of images is acknowledged specifically in the relevant captions; illustrations without such a note have been reproduced from materials in my own collection, or are my own drawings, or maps of my own design.

Last but far from least, I am immensely grateful to Christie Henry and all her colleagues at the University of Chicago Press, and to their former colleague Catherine Rice, for their outstandingly fine work with this book and its predecessor. The Press has made them volumes that are, quite apart from their content, a delight to read. If they are rather a load on one's lap—but a burden worth bearing, since the high-quality paper makes it possible to appreciate the illustrations in the detail they deserve—they can always double as doorstops.

A NOTE ON FOOTNOTES, REFERENCES, AND QUOTATIONS

The footnotes give the detailed information on sources that some readers will find essential, and suggestions for further reading on specific topics (whether or not I myself have derived my ideas from these particular sources). I hope these references will encourage some readers to use my work as a starting point for further research, some of which will, I hope, be in directions that I cannot imagine. The size of this volume and its predecessor may suggest otherwise, but in fact they should be regarded not as the last word on their subject but as one of the first.

The footnotes are also used to add ancillary points that might disrupt the main text, for example to mention where certain historically important sites and specimens can now be seen. I hope these notes too will encourage some readers to follow not so much in my footsteps but rather in the footsteps of the nineteenth-century geologists themselves, and to try looking at their evidence through their eyes. The notes also explain scientific and historical terms that may be unfamiliar to some readers.

Sources in the form of books and articles, archives and museums, are cited in the footnotes in abbreviated form, and listed in full in the Sources sections at the end of the volume.

Quotations from sources not originally in English are my own translations (unless stated otherwise in the footnotes). Words that are particularly significant in the original language are noted in square brackets after the relevant translated word or words, for the benefit of readers familiar with the language concerned. All words and phrases printed in italics were printed in that way in the original (or, in manuscript letters and notebooks, underlined for emphasis); punctuation is mainly original, but I have modified it occasionally in the interests of intelligibility.

In the main text, cross-references in the form "(Chap. *)" or "(§*.*)" refer to relevant chapters or sections in this volume; those in the form "(*BLT* Chap. *)" or "(*BLT* §*.*)" refer back to chapters or sections in Rudwick, *Bursting the Limits of Time* (2005), to which the present volume is the sequel.

William Buckland crawling into Kirkdale Cave in 1821, and finding his extinct "antediluvial" hyenas alive and well: a contemporary caricature depicting Buckland's cave research as time-travel into a "world before Adam".

INTRODUCTION

In February 1822, at three successive weekly meetings of the Royal Society in London, a large audience of its Fellows and their guests listened with fascination as William Buckland, the first reader in geology (and self-styled professor) at Oxford, described how he had reconstructed a den of large extinct hyenas, and their meals of mammoth steak, water rats, and other delicacies, in what had since become a quiet corner of rural Yorkshire. The Society's president, the chemist Humphry Davy, told Buckland he could not recall any paper that had aroused so much interest. Later in the year it was published in the Society's *Philosophical Transactions*, the world's oldest and most prestigious scientific periodical; and the Council decided to give Buckland the Copley Medal, the Society's highest award, in recognition of his paper's outstanding value and importance.

Buckland's "hyaena story" about Kirkdale Cave was used in my book *Bursting the Limits of Time* as the culmination of a narrative that spanned several previous decades of research in the sciences of the earth. Buckland's reconstruction of a vanished ecosystem in recent geohistory (to borrow useful terms from modern biologists and geologists respectively) was seen at the time as a sensational foretaste of what the great Parisian naturalist Georges Cuvier had eloquently set out, just ten years earlier, as a worthy and attainable goal for the new science of geology. "Would it not be glorious", Cuvier had asked rhetorically, for geologists to "burst the limits of time", just as astronomers had "burst the limits of space"? The allusion was not to the expansion of traditional conceptions of the magnitude of either time or space, but to the extension of reliable human knowledge in both dimensions, beyond the limits of direct human experience. Astronomers, although physically confined to one small planet, had already learned to calculate the motions of the whole solar system, and were even reaching beyond it to the stars in the inconceivable vastness of deep space. In the same way, geologists, although physically confined to the present moment, could hope to penetrate back into the inconceivable vastness of deep time, and to reconstruct the earth's own history—geohistory—with a similar degree of confidence and reliability. As portrayed in a celebrated caricature, Buckland, by crawling into Kirkdale Cave with the light of science in hand, had burst the temporal limits of recorded human history and penetrated back into the "antediluvial" world of the extinct cave hyenas (see figure opposite page 1). And this, as Davy recognized, was just a beginning: Buckland had shown by his example that geologists had now developed a conceptual time-machine, which could also be used to explore far earlier periods of geohistory.

The present book is a sequel to *Bursting the Limits of Time*, but it is self-contained and should be wholly intelligible to readers who have not seen the earlier volume. All the main characters carried over from the earlier narrative are introduced again, however briefly, for the sake of those who have not yet encountered them; so are the main empirical issues that were the stuff of the earlier debates. I have tried to make this book, like the earlier one, accessible both to readers who are unfamiliar with the historical period it covers, and to those who know little about geology. So I hope it will also be found intelligible by those who have no great background knowledge in either direction, but who are curious to know how the main outlines of the earth's own complex history were first reconstructed. And, as I emphasized before, no one should be put off by the small print of the footnotes: they give the detailed information that some readers will need, particularly on sources, and they add ancillary points that would disrupt the main text, but the book should be wholly comprehensible without looking at any of them.

In the earlier volume I traced the gradual development of the practice of geohistory within the sciences of the earth. In the late eighteenth century it was an infrequent and marginal feature of scientific research, but within a few decades it became a defining element of a newly named science of "geology". I argued that geology became the first truly *historical* natural science; and that it did so as a result of deliberately transposing methods and concepts from the human sciences of history itself. The earth's own immensely lengthy history, almost all of it apparently prehuman, then began to seem reliably knowable. By the time Buckland reconstructed his antediluvial hyena den, geohistorical ways of thinking were becoming so deeply embedded in the practice of geology that the explicit analogies with human history—the telling metaphors of *nature's* monuments and documents, *nature's* coins and inscriptions, *nature's* archives and chronologies—were beginning to drop out of scientific discourse, though they remained valuable as illustrative analogies whenever geologists sought to reach a wider audience.

The present volume takes as its starting point this sense that the earth's deep or prehuman geohistory could in principle be reconstructed almost as reliably as, say, the history of the ancient Greeks and Romans. The narrative takes a new tack, however, in that its focus is now on the character of the geohistory that this novel perspective disclosed; or, in other words, on what geologists discovered as they deployed their virtual time-machine. Of course, the research described in my earlier volume did often suggest what the course of geohistory had been like; but generally it did so either with a very broad brush, covering the whole globe in outline, or else in a detailed local vignette such as that of antediluvial Yorkshire. Furthermore, priority was quite properly given to establishing the sheer historical reality of the events of the deep past, shelving if necessary the problems of discovering their causal explanation. It seemed more important, for example, to establish that in recent geohistory there had indeed been some kind of geological "revolution"—which had apparently wiped out the cave hyenas and mammoths—and to determine its approximate date and physical character, rather than speculating how exactly it might have been caused. However, during the years covered in the present volume the geohistorical and the causal were increasingly brought together and integrated. By the end of this period—that is, well before the middle of the nineteenth century—geologists had reconstructed a consistent outline of geohistory with plausible causal explanations of many of its main features; and this geohistory was at least congruent with, and in some respects foreshadowed quite precisely, what is accepted as valid by geologists in the twenty-first century.

The narrative in this volume, then, covers a later period than its predecessor (with a little overlap to ensure continuity). It spans the quarter-century from about 1820 to about 1845. Put another way, it starts in the aftermath of the Napoleonic era in European history, and it extends into the earliest years of Britain's Victorian age and equivalent phases in the history of other nations. As a convenient label for this period, my subtitle borrows from political historians the phrase "the Age of Reform": the Great Reform Bill in Britain and the July Revolution in France, for instance, lie neatly near the middle of my chosen quarter-century. As in my previous volume, however, I do not intend this definition to imply that the political was necessarily the most important element in the context of the scientific work: the period covered by the present volume might also be defined—perhaps with no less relevance—as stretching from Schubert to early Verdi, from Scott to early Dickens, or as that phase of the Industrial Revolution that saw the coming of railways, steamships, and the electric telegraph. Although geology was already firmly established by the start of this quarter-century, it was a period marked by exceptionally intensive and innovative research, which was of outstanding importance for all the later history of the science. This justifies devoting a substantial volume to its tangled debates.

Apart from the period it covers, however, this book adopts much the same parameters as *Bursting the Limits of Time*, and they can be summarized here quite briefly. First, and somewhat negatively, I should emphasize again that I do not claim to offer a comprehensive history of geology, even within my chosen period. I make no attempt to describe or analyze the whole range of geological research; many important topics are mentioned only in passing or not at all, simply because they did not impinge directly or substantially on the reconstruction of geohistory. Even with that restriction, it would not be possible to cover all the relevant work, much of which—in ever increasing volume—was reported in a burgeoning range of scientific periodicals published throughout Europe (and a few beyond it, in Russia, the United States, and elsewhere). So my policy has once again been to trace and analyze those topics that were either *innovative* or *exemplary*, and to let them represent a far larger body of relevant research.

Second, I hope that this volume, like its predecessor, will help to correct an overwhelmingly anglocentric and anglophone bias in much of the historiography of geology. I try to replicate the highly international outlook and interactions of the leading scientific figures, by describing research that was being done in all the main scientific nations, which at this period meant primarily those in Europe (including Britain, of course, as a constituent part of Europe). It was France, however, that remained dominant in all the sciences, as in many other aspects of cultural life, despite the military defeat of its Napoleonic regime; and French remained the preeminent scientific language worldwide. France and the French language had, respectively, much the same positions as the United States and the English language in today's scientific world. But their primacy was just beginning to be challenged, not so much by Britain but rather by the still politically fragmented German-speaking lands; by the end of the century it was German rather than English that had overtaken French to become the dominant language for serious scientific research. Those who practiced any of the natural sciences outside Europe—for example, in Russia to the east or the United States to the west, let alone in still more remote parts of the world—were, and knew themselves to be, very much on the periphery; their work was valued primarily as factual evidence for the theoretical or interpretative debates that were taking place in Europe.

Third and last, as this implies, my focus once again is on the work and interactions of the leading scientific figures—those doing original research of international significance—rather than, say, amateur naturalists or the general reading public, let alone still wider classes of society. This book is, without apology, an elitist account of certain aspects of the science of geology at one of its most innovative periods; but this particular elite was based not on birth or wealth or social class, but primarily on intellectual effort and originality. Studies of popular science have their rightful place in scholarly work on the history of the sciences, but they are no substitute for the study of the original scientific research that gave the popularizers most of their raw material. Scientific popularization is no longer treated by historians simply as a story of diffusion *de haut en bas*: popularizers made creative use of what they found available, and refashioned it—sometimes in directions opposed to the original authors' intentions—for their own varied purposes. But in doing so their work was derivative from, and therefore secondary to, the original research that they sought to make accessible and meaningful to a wider audience.

There is therefore a strong case for giving as much historical attention to elite science as to popular science, if not more. For the topic of the present volume, as for that of its predecessor, this is an urgent necessity, because historical research on the earth sciences in the crucial period of the late eighteenth and early nineteenth centuries is still quite patchy and often seriously defective. This is in striking contrast to our historical understanding of other sciences in the same period. To mention just one example, scholars in the so-called "Darwin industry" have reconstructed in impressively thorough detail, and in its full intellectual and social context, the circumstances in which Darwin's evolutionary theory first took shape (also largely in the Age of Reform). This research has provided reliable foundations for valuable studies of the popular presentation, reception, and application of evolutionary ideas, both then and later in the nineteenth century, far beyond the circles of the scientific elite. There is much comparably fine research on the physical sciences in the same period. Geology, however, remains a Cinderella in the historical study of the sciences.

To mention Darwin and evolution in a context of the history of geology is to evoke, with seeming inevitability, images of "conflict" between geology and Genesis, or more generally between allegedly monolithic entities labeled "Science" and "Religion". In the introduction to *Bursting the Limits of Time* I mentioned briefly why this kind of historiography has long been abandoned by historians, although it remains popular in the modern media, and above all in the rhetoric of those self-appointed spokespersons for "Science" who are, in effect, atheistic fundamentalists. Here it is necessary only to emphasize that *all* the geologists with whom this book is concerned were convinced that geohistory had been played out on a timescale of humanly inconceivable magnitude (though they had as yet no means to put reliable figures to it). The many who were also religious believers saw no conflict between their geology and their understanding of the Creation stories in Genesis; they had long since learned that it was a *religious* mistake to treat biblical texts as if they were scientific sources, because an inappropriate literalism deflected attention away from religious meaning. It is true that some of these geologists, particularly in England, had to confront vocal critics—the self-styled "scriptural" writers—who relentlessly pursued a literalist line on matters of "geology and Genesis". But this, like the modern and peculiarly American phenomenon of creationism and other forms of religious fundamentalism, was a contingent feature of a particular time, place, and, above all, social location. In keeping with the focus adopted for this book, these issues will not be described or analyzed here, except when and where they impinged directly on the work of the scientific elite, rather than on its more general handling of its relations with the wider public; the latter is an important topic in its own right, but it is a different topic and it deserves separate treatment.

Toward the end of this book, Darwin himself comes on stage, but as a young and promising *geologist*, the role in which he first made his name and earned respect as a "man of science". As in *Bursting the Limits of Time*, I want to claim that the reconstruction of geohistory—within which the history of life or "biohistory" was just one major theme among several—was an outstanding intellectual achievement in its own right, and one that was only loosely linked to the search for a causal explanation of the diversity of life or of the origin of the human species. The establishment of a reliable geohistory was a part of the necessary infrastructure for any adequate evolutionary theory, but the latter was not its only outcome. All roads did not lead to Darwin. Appropriately—though without its having been my intention when first planning the book—my narrative in this volume reaches its culmination with a major phase of geological research in which Darwin's contribution was one that he himself later repudiated as "a great failure" and even as "one long gigantic blunder". My purpose in describing this episode is not to topple Darwin from his well-deserved historical pedestal, but simply to situate him in the scientific debates of his time. Like many outstanding scientists in the modern world, he did not always get everything right.

Darwin's failure when tackling one specific puzzle was in fact a minor episode within a much larger and more important (and international) story. This volume culminates in an account of research that vividly suggested how the earth's own history—no less than the human history that had once been its conceptual template—was intrinsically unpredictable (or, more correctly, unretrodictable), because it was ineluctably *contingent.* It became clear that at every turn, geohistory could conceivably have taken a different course, without impairing in any way the uniform action of the underlying "laws of nature". In consequence—and again as in human historiography—it could be reconstructed only bottom-up from a detailed study of surviving evidence of what had *in fact* happened, rather than top-down by the application of fixed laws of nature to determine what supposedly "must" or "should" have happened. As one of the most distinguished polymaths of the time recognized explicitly—and this forms the finale of the volume—geology had become the outstanding exemplar of a new *kind* of natural science, in which the historical dimension was central and constitutive. That *nature has had its own history* was a lesson already well learned by the young Darwin, when he transposed that insight from geology, his first love, to zoology and botany. Animals and plants, no less than mountains and volcanoes, glaciers and oceanic islands, could be fully understood only by taking into account the histories built into them. But Darwin's is another story.

To summarize: the scientific research described in my previous volume demonstrated that it was feasible in principle to gain reliable knowledge of the earth's history long before the earliest human records, and thereby to "burst the limits of time". The geologists whose research is described and analyzed in the present volume, spanning a subsequent quarter-century of exceptional fertility and importance for the future of the earth sciences, were able to take this approach for granted. They went on from there to reconstruct systematically and in detail what course geohistory had in fact taken, to reach the earth's present state. On a timescale that they agreed had been of inconceivable (though unquantifiable) magnitude, they plotted a sequence of distinctive periods and eras in geohistory, most strikingly distinct in the kinds of animals and plants that had flourished successively on earth. Many other long-term changes, for example in global geography and climate, were more problematic and controversial, as was the reality or otherwise of occasional episodes of rapid or even catastrophic change. However, what seemed a cumulatively reliable conclusion was that, in one way or another, the earth had had a highly eventful *history*, passing through many different and distinctive phases, long before the present world came into existence. And almost all this complex and eventful geohistory had preceded not only the earliest written records of human history but the human species itself. In terms of the traditional archetypal figure of *ha'adam*, The Man, it was a story of a long and complex sequence of Worlds before Adam.

~ ~ ~

The first half of *Bursting the Limits of Time* was devoted to a survey of the sciences of the earth as they were being practiced around the end of the Old Regime in Europe, or on the eve of the French Revolution; this acted as a baseline for the narrative in the second half of the book. That narrative is continued in the present volume, but is arranged this time in a larger number of much shorter chapters. As before, but to an even greater extent, the narrative has to sustain several subplots running in parallel, tracing the course of different loosely linked "focal problems" within what was defined by this time as the science of geology. But I hope I have provided sufficient flashbacks, recaps, and cross-references to enable the reader to follow the intricate interweaving of the various themes and their evolving interactions: classic Victorian novels such as Charles Dickens's *Bleak House* have been my inspiration and model, though I cannot begin to match their narrative flair.

The chapters are grouped loosely into four Parts, in chronological sequence though with some overlaps. Part One describes the first postwar years (that is, post-Waterloo and post-Napoleonic) and the early 1820s, when the great Georges Cuvier was still a dominant presence in geology. Part Two deals with the immensely fruitful period of the later 1820s and earliest 1830s, during which much (though not all) of the most innovative research was shared between France and England. Part Three is focused on the pivotal figure of Charles Lyell, and describes the gestation, publication, and immediate reception of his *Principles of Geology*, in the years around 1830. Finally, Part Four follows the development of geohistorical themes through the 1830s and into the early 1840s, partly (but not wholly) in response to Lyell's provocative synthesis. My substantial focus on Lyell will surprise no one familiar with the existing historiography of geology; but I try to show how his contemporaries were, at one and the same time, profoundly critical of some aspects of his work while being profoundly impressed by others.

The narrative as a whole traces the reconstruction—at first tentative, but gradually more confident—of an eventful geohistory, which in fact is congruent with what geologists in the twenty-first century accept as valid (at least in outline, and for the Phanerozoic or post-Precambrian portion of geohistory). More specifically, the story culminates in the formulation of the glacial theory, and the utterly unexpected inference of an exceptional and drastic "Ice Age" in the geologically recent past. It was this, more than any other single development, that forced geologists to recognize the contingent character of geohistory as a whole. The Concluding (Un)Scientific Postscript (with apologies to Kierkegaard), which lies outside this narrative, considers briefly some of the broader issues that the whole work has addressed.

I should here explain the relation between this volume (and its immediate predecessor) and my earlier work on the history of the earth sciences. My first historical book, *The Meaning of Fossils* (1972), set out a first sketch of some of the debates that I now see as central to the reconstruction of geohistory, but only insofar as they affected what came to be called paleontology (the science

I had been practicing professionally when that book was conceived). Much of the present volume is founded on my subsequent research on the early work of Charles Lyell, and that of his contemporaries and critics, centered on the meaning of the "uniformity" of nature and the legitimacy of "catastrophist" explanations in geology. This research was published in a string of papers, long and short, many of them now reprinted in *The New Science of Geology* (2004) and *Lyell and Darwin, Geologists* (2005). I make no apology for citing these papers rather fully in the present volume, because they contain far more detailed documentation than can be repeated here. Three other books are also relevant, more or less tangentially: *The Great Devonian Controversy* (1985) is primarily about stratigraphy, which was (and is) a major source of evidence for geohistory, but not coincident with it; *Scenes from Deep Time* (1992) deals primarily with the wider public's apprehension of the new geohistory, and it extends into later decades; conversely, *Georges Cuvier* (1997) focuses on the earlier work of the outstanding naturalist who provides the present volume with its starting point. However, I should emphasize that I have here used many new sources (new at least to me), and have re-studied the others; and I have interpreted them all within a largely new conceptual framework, namely the *reconstruction of geohistory* mentioned in my subtitle. So the present volume is much more than just a restatement or even a synthesis of my earlier publications.

As in *Bursting the Limits of Time*, readers who feel daunted by the scale and density of this narrative can get a brief overview of its argument by reading the concluding section of each chapter, which is in fact a summary of its contents; or by looking at the illustrations and their captions, which collectively embody much of the argument in visual form; or, paradoxically, by going first to the brief Postscript at the end of the volume.

I hope that the overall narrative framework I have adopted, in preference to any division of the material into separate topics, will convey the strong sense of unity of purpose and scientific progress that the participants experienced (albeit with plenty of false starts and dead ends). Contrary to what many modern scientists wrongly imagine—that the lives of their intellectual ancestors were altogether more leisurely than their own—early nineteenth-century geologists felt in fact that they were living in stirring times, and that the science to which they were contributing was changing and developing at almost breakneck speed. We in the modern world are the beneficiaries of what they first pieced together with such imaginative scientific skill.

Part One

Fig. 1.1 A portrait of Georges Cuvier at the height of his power and fame: the frontispiece to the first version of his famous *Discourse on the Earth's Revolutions* to be issued separately in book form in its original French (1826). A prominent member of the scientific elite in France, he wears the ceremonial uniform of the Académie Royale des Sciences, of which he was one of the two "permanent secretaries". His stepdaughter criticized this print for failing to reproduce the smile that the painter had faithfully recorded.

CHAPTER 1

Cuvier's model for geohistory (1817–25)

1.1 CUVIER'S *FOSSIL BONES*

In the peacetime era after the fall of Napoleon, the Muséum d'Histoire Naturelle in Paris (hereafter just "Muséum") remained the world's most prominent institution for research in all the sciences—animal, vegetable, and mineral—that were grouped together as natural history. A quarter-century earlier, when the Muséum was founded during the Revolution (or rather, when it was "democratized" from the old royal museum and botanic garden), it had been the first institution anywhere to give formal recognition to what was then the newly named science of "*géologie*" (*BLT* §6.3). But in the event, the Muséum's greatest impact on the sciences of the earth had come not from Barthélemy Faujas de St Fond (1741–1819), its first and rather undistinguished professor of geology, but from the much younger naturalist who became its professor of comparative anatomy, Georges Cuvier (1769–1832), Napoleon's exact contemporary. In post-Napoleonic Europe, Cuvier was the Muséum's most powerful figure and one of the most famous savants in the world (Fig. 1.1).[1]

Back in 1812, while Napoleon, the self-styled Emperor of the French, held sway over most of mainland Europe, Cuvier had produced four volumes of *Researches*

1. Fig. 1.1 is reproduced from Bultingaire, "Iconographie de Cuvier" (1932), pl. 4, reproduced in turn from Cuvier, *Discours sur les révolutions* (1826), quarto issue, frontispiece, lithographed by Charles-Louis Constans after a painting by Nicolas Jacques. The Académie was the royalist successor of the natural-scientific "First Class" of the Institut de France set up during the Revolution, at which Cuvier had presented most of his earlier research: see Crosland, *Science under control* (1992). The word "*Muséum*", with its French accent, will here serve to distinguish the great Parisian institution from any other museum or *musée*. The word "*savant*", which at the time was widely used in English and almost invariably in French, denotes all who did original work in *any* of the sciences, natural or human; its use avoids the anachronistic connotations of the word "scientist", which was first coined during the 1830s (see §25.3) but failed to gain wide acceptance until the twentieth century.

on Fossil Bones. This massive work reprinted the many specialized papers he had published in still earlier years in the Muséum's in-house periodical, the *Annales du Muséum* (*BLT* §9.3). But it was prefaced by a new, long, and eloquent "Preliminary Discourse", adapted from some earlier lectures for the general educated public in Paris, which set out his ideas not only on fossil bones and but also more broadly on geology as a whole (*BLT* §10.3). It was in this celebrated essay that he had suggested that geologists could and should aspire to "burst the limits of time" by reconstructing in reliable detail what he was convinced was a vast expanse of *prehuman* geohistory, dwarfing the totality of subsequent recorded human history.[2]

The main outlines of Cuvier's ideas, as set out in his Discourse and substantiated in the rest of his *Fossil Bones*, will be recalled here quite briefly. First, his work was indeed geohistory. He presented himself as "a new species of antiquarian", a *historian* of the earth, who was using fossil bones instead of human artifacts to recover the distant past (*BLT* §7.2, §9.3). He repudiated a project that was still going strong among his contemporaries, that of devising a high-level causal theory that would explain all the major features of the earth: the genre of "theory of the earth" (or *geotheory*, as I have named it: *BLT* §3.1) was rejected as premature (*BLT* §8.4). In its place, Cuvier proposed more limited enquiries that had some hope of being resolved; in particular, he gave priority to work that might establish the geohistorical reality of specific events in the deep past, while deferring if necessary the more knotty problems of finding their causal explanation.

Like all other scientific savants at this time, Cuvier rejected emphatically the traditional *short timescale* for the earth, which had been derived from the purely textual evidence assembled by the venerable science of "*chronology*" (*BLT* §2.5); its most famous product, James Ussher's date of 4004 B.C. for the primal Creation, had been proposed more than a century and a half earlier and was now treated as utterly outdated. But Cuvier also rejected the unlimited time, indeed the *eternity*, imputed to the world by influential *philosophes* in the Enlightenment and by some of his own contemporaries. These positions—*both* of them profoundly unmodern—were being used to support political agendas, respectively religious and anti-religious in intent. In contrast, Cuvier adopted a third (and modern) alternative, arguing that the earth was of inconceivably vast antiquity *yet not eternal* (*BLT* §8.3). He claimed that it had truly had a *history*, a sequence of distinctive events and unrepeated eras, rather than the endless repetition or recycling of more or less similar physical states that was entailed by any kind of eternalism. But he insisted that this lengthy geohistory was not co-extensive with human history; on the contrary, almost all of it had been prehuman (*BLT* §10.3).

Cuvier, like almost all savants working actively in the sciences of the earth around this time, regarded this vastly extended but finite geohistory as one of gradual directional change towards the earth's present physical state. It was generally conceived as being marked by a progressive lowering of global sea level and a consequent gradual reduction in the extent of the world's oceans; this had long been geotheory's, and geohistory's, "*standard model*" (as I have termed it: *BLT* §3.5). But in the course of this directional change, long periods of calm conditions, recorded for example by thick beds of rock containing well-preserved shells and other marine fossils, seemed to have been punctuated by occasional episodes of rapid and perhaps violent change, recorded by broken and distorted strata or by thick beds of coarse gravel (often consolidated into "pudding-stone" or conglomerate). Like many other savants, Cuvier defined these episodes as the earth's natural "*revolutions*" (*BLT* §6.1); they were often treated as analogous to the traumatic political Revolution in France, which was still fresh in the collective memory of all Europeans.

The most recent of all these physical revolutions was of decisive importance in Cuvier's understanding of geohistory. Not only were its traces the clearest to see and the easiest to interpret, because the least effaced by the passage of time. It was also of unique significance because it seemed to separate the "present world" of human civilizations from the vastly more extended "ancient" or "former world" [*ancien monde*] before the first appearance of the human species. For by this time the few alleged finds of human bones or artifacts in ancient rocks were all regarded—with good reason—as spurious or at best dubious (*BLT* §5.4); and Cuvier, adopting the role of a textual critic, also rejected the historical records that were alleged to prove a vast antiquity for certain human cultures and were even taken as evidence for the eternity of the human species itself (*BLT* §10.1).

Cuvier claimed that the most important result of his detailed work on fossil bones was to show that a whole mammalian fauna had suffered a mass extinction. On the basis of his functional analysis of the anatomy of the extinct species, he argued that they had all been well adapted to particular modes of life. So their disappearance could be explained, in his view, only as the consequence of a sudden and overwhelming revolution, and a quite recent one at that. He remained uncertain about the physical character of this decisive event; unsurprisingly so, since he was an indoor museum naturalist with little firsthand experience of the relevant field evidence. At different times he provisionally adopted either of two current alternatives. Before the turn of the century the Parisian naturalist Déodat de Dolomieu (1750–1801) had suggested that the event might have been, as it were, a "*mega-tsunami*", a huge transient wave analogous to the humanly catastrophic tsunamis that often accompany submarine earthquakes, but on a far larger scale (*BLT* §6.3). An alternative explanation, suggested by the Anglo-Genevan naturalist Jean-André de Luc (1727–1817) in publications spread over many years of his long life, was that it might have been due to the sudden collapse of parts of the earth's crust, such that the continents of "the former world" had sunk below sea level, while the former ocean floors had been left high and dry to form the present continents: in effect, a massive interchange between continents and oceans (*BLT* §3.3, §6.2). However, Cuvier's uncertainty on this point was evidently tolerable to him, because he regarded it as even more important to establish the sheer geohistorical reality of the event, one that had been catastrophic enough to wipe out a whole fauna of apparently well-adapted animal species.[3]

2. Cuvier, *Ossemens fossiles* (1812) 1, "Discours préliminaire", translated in Rudwick, *Georges Cuvier* (1997), 165–252.

3. Gaudant, *Dolomieu* (2005), which was published just too late to be cited in *BLT*, does not include among its otherwise valuable essays any assessment of Dolomieu's influential mega-tsunami theory.

Still less was Cuvier concerned to determine the physical cause of this event, though he took it for granted that it must have a natural explanation of some kind. But he rejected the idea put forward by some of his contemporaries, that there had never been any such revolution or special event, and that everything could be explained by the gradual action of ordinary processes observably at work at the present day (*BLT* §6.1, §8.3). On the contrary, he agreed with de Luc that while these processes—which de Luc had termed "*actual causes*"—were real enough in themselves, they were utterly inadequate to explain the observable features, most of all the mass extinction for which Cuvier himself had presented so much new evidence based on fossil bones (*BLT* §10.3). Here he was simply following the prescriptive rules of the great Isaac Newton: any causes invoked as explanations should be proportionate to their observed effects.[4]

So the physical character and cause of the earth's most recent major revolution remained enigmatic. But its date was another matter. Cuvier argued that it could be dated reliably, albeit approximately, by collating the earliest *human* records. Although in his opinion these were all more or less garbled and often shrouded in myth and legend, when analyzed critically they provided multicultural evidence for a global catastrophe. This had apparently been a huge aqueous "deluge" of some kind, at or near the dawn of human history; Cuvier was no biblical literalist, and he treated the Flood story in Genesis on a par with similar stories from other ancient cultures, extending as far away as China. So the venerable science of chronology, when pursued in this multicultural fashion with the latest tools of critical textual analysis, did after all have a place in geohistorical practice. Cuvier argued that the last revolution could be dated, not to the very year (as Ussher and other chronologers had claimed long before, for Noah's Flood), but at least approximately, to no more than a few millennia in the past. In effect, therefore, the last revolution was the unique boundary event that tied human history back into geohistory, because the evidence for its historical reality came both from human records, however obscure, and from natural evidence such as that of fossil bones.

Cuvier's "Preliminary Discourse" was not confined to his analysis of the earth's last revolution; that event was unique only in its date, and hence in its link to human history. As already mentioned, he argued that it was just the most recent in a long succession of physically similar natural events, separated by lengthy periods of tranquil conditions.

Cuvier's best evidence for this picture of geohistory had come from his joint fieldwork with his near-contemporary the mineralogist Alexandre Brongniart (1770–1841), the director of the state porcelain factory at Sèvres just outside Paris. Together they had surveyed what they termed the "Paris Basin"—it was almost the only geological fieldwork that Cuvier ever did—and had described a complex sequence of rock formations, some of them rich in fossil shells. Their methods had been somewhat similar to what the English surveyor William Smith (1769–1839) was doing around the same time (*BLT* §8.2); and Brongniart may have heard about Smith's work while he was visiting London during the brief Peace of Amiens. But he and Cuvier had radically transformed what Smith later called the "stratigraphical" use of fossils (*BLT* §9.5). The Frenchmen had treated fossils not just as empirically "characteristic" of particular formations, but as diagnostic of specific environmental conditions. They had interpreted the pile of formations in the region around Paris in geohistorical terms: they claimed that tranquil seas had alternated with periods of equally tranquil freshwater lakes, lagoons, or marshes. Occasional sudden changes in the environment were then held responsible for the occasional abrupt changes in the observable rocks and fossils (*BLT* §9.1). In his Discourse, Cuvier had cited this work with Brongniart as the best exemplar of how geohistory could be reconstructed, reliably and in detail. He reprinted their joint monograph in his own work; indeed he had it bound into his first volume, following his introductory essay and preceding any of his detailed studies of fossil bones.

Cuvier and Brongniart had recognized that their research on the Paris Basin could also play a crucial role in the reconstruction of the whole sweep of global geohistory. The rock formations around Paris were clearly older than the "*Superficial*" deposits, such as the river gravels that contained the bones of most of Cuvier's extinct mammals; for the solid rocks had evidently been excavated to form the valleys in which the gravels lay. But these rocks in turn overlay the highly distinctive and widely distributed Chalk formation, which rose to the surface around the Paris region. The Chalk had previously been regarded as the uppermost (and therefore youngest) of all the thick and varied "*Secondary*" formations, those that often contained abundant fossils. Far lower in the pile, and therefore far older still, were the so-called "*Transition*" rocks, mostly slates and "greywacke" with only a few obscure traces of life. Lowest of all were the enigmatic "*Primary*" rocks such as schist, gneiss, and granite; having no trace of fossils of any kind, these were generally regarded as dating from the very earliest phases of geohistory, probably before the origin of life itself (*BLT* §2.3, §4.4). So the formations in the Paris Basin and others of the same age elsewhere—which were soon being distinguished as "*Tertiary*"—could act as the link or cognitive gateway, as it were, between the present world and the even stranger worlds of the still deeper past. One could use these formations as the first stage, and the least unfamiliar, on the way towards deciphering the more obscure periods of even earlier geohistory. Cuvier therefore urged geologists to make the close study of these youngest or Tertiary formations their highest priority (*BLT* §9.3).[5]

4. De Luc's actual causes [*causes actuelles*] were "actual" not in the modern anglophone sense of real and not imaginary, but in the older sense still retained in other European languages, meaning current or of the present day (the news bulletins on French television, for example, are the day's *actualités*). Hence the analytical term *actualism*, applied to the earth sciences, denotes the methodological strategy of using a comparison with observable present features, processes, or phenomena as the basis for inferences about the unobservable deep past: in epigrammatic form, "the present is the key to the past". As the following chapters will indicate, the heuristic value of this strategy was taken for granted by *all* the geologists with whom this book is concerned: it was *not*—as modern historical myth would have it—first proposed by Charles Lyell in 1830. The arguments were about its *adequacy* for causal explanation, not about its validity or its value.

5. In modern terms the Superficial deposits were roughly equivalent to the Quaternary, and the "regular" or Secondary formations to the rest of the Phanerozoic (where not altered by metamorphism). However, soon after Cuvier's work was published the younger Secondary formations (those lying above the Chalk, in the Paris Basin and elsewhere) began to be distinguished as Tertiary—a term still used by modern geologists to denote all the Cenozoic apart from the Quaternary—so that the word "Secondary"

In much of this, Cuvier was simply borrowing from the work of others in the well-established science of *geognosy*, the science of rock structures and rock formations (*BLT* §2.3), parts of which were changing into what would later be called "*stratigraphy*" (see §3.1 and Fig. 9.1). But his own work on fossil bones added a geohistorical gloss to the accepted geognostic picture. This will be clear from a summary of his evidence, plotted back in time from the present world. From the Superficial deposits he described a whole fauna of fossil mammals, many of spectacular size (in modern terms, the Pleistocene *megafauna*). He claimed that all of them were of species unknown alive, though most belonged to known genera: the mammoth, for example, belonged to the same genus as the living elephants but was distinct from either the Indian or the African species. He interpreted all the fossil species as truly extinct, the victims of the earth's most recent revolution. Moving back in time to the Tertiary formations around Paris, he described another mammalian fauna, much more distinct from that of the present world, consisting wholly of unknown genera (*BLT* §7.5). Still further back, in the Secondary rocks proper (the Chalk and the formations below it), he found no trace of any mammals, but instead a wide variety of reptiles. Some of these were relatively familiar, such as crocodiles and turtles, but others were extremely strange. For example he identified the already famous "Maastricht animal", found in the Chalk near the Dutch city, as a huge marine lizard. Sensationally, he interpreted a much smaller but equally striking fossil, from a well-known rock at Solnhofen in Bavaria, as a *flying* reptile, which he named the "*ptéro-dactyle*" or wing-fingered animal (*BLT* §9.3). So Cuvier suggested that there might have been an age of reptiles long before there were any mammals, let alone any humans. Although highly tentative, this reconstruction of the history of the "quadrupeds" (in modern terms, the mammals, reptiles, and amphibians) accentuated the directional character of geohistory as a whole.[6]

Cuvier could and did claim credit for much of this picture of geohistory, though for certain aspects he duly acknowledged his intellectual debts to other savants such as de Luc and Dolomieu, and above all to Brongniart. But his "Preliminary Discourse" was far more eloquent and readable than the published work of his colleagues, and it became widely known throughout the savant world and indeed among the educated public generally, in part through its translation into other European languages (*BLT* §9.3, §10.3). Together with the detailed studies that it served to introduce, it became in effect the baseline or starting point for the work of other and often younger savants, as they sought to take Cuvier's research in new directions. They might try to extend and confirm his inferences, or use new evidence to try to refute them, but in either case Cuvier could not be ignored.

1.2 THE *FOSSIL BONES* REVISED

Once Cuvier's *Fossil Bones* was in the public realm, he had resumed work on what, at the start of his career before the turn of the century, he had regarded as his main field of research. In 1817, two years after Napoleon's final defeat at Waterloo, Cuvier had published his great *Animal Kingdom*, also in four volumes, using rigorous comparative anatomy to set out a radically revised classification of the whole range of living animals. This work was recognized immediately as being of equally decisive importance in its own field. In effect, it invalidated the traditional "scale of beings" [*échelle des êtres*], which had supposedly linked all animals in a single linear chain or "series" from the simplest to the most complex. Therefore it also undermined the "*transformist*" (in modern terms, evolutionary) interpretation of that linearity, which had been proposed most forcefully by Cuvier's older colleague Jean-Baptiste de Lamarck (1744–1829), the professor of invertebrate animals at the Muséum (*BLT* §7.4, §8.3, §10.1).[7]

Cuvier the indefatigable workaholic then turned back to fossil bones, while also shouldering ever-increasing responsibilities in the educational administration of the restored Bourbon monarchy of Louis XVIII (who in 1820 awarded him the title of *baron*). His *Fossil Bones* had been in effect a massive progress report. He had hoped and intended that it would stimulate naturalists everywhere to study the bones found in their own areas and conserved in their cabinets and museums; and that they would send him further specimens, or at least accurate pictures of them, to supplement those he had already used, and thereby enable him to correct or extend the interpretations he had proposed (*BLT* §7.3, §7.5). His hopes had been amply fulfilled, for other naturalists were keen to be associated—with due acknowledgment, of course—in a work of such outstanding importance. Specimens, and *proxy* specimens in the form of accurate drawings and paintings, accompanied by manuscript letters and offprints of published articles, converged on Paris from around the world, in even greater abundance than in earlier years.[8]

In one important case, however, Mahomet went to the mountain. Cuvier visited England for the first time in 1818, with his wife and daughters and his

---

then denoted, in modern terms, only the Mesozoic and the Paleozoic. The Transition rocks were mostly Paleozoic in age, but usually somewhat affected by metamorphism; the Primaries were, in modern terms, mostly "basement" rocks of igneous or highly metamorphic origin and Precambrian age (it was not yet recognized that in some regions they might be much more recent in origin).

6. Unlike the celebrated "Ohio animal", which Cuvier had named the *mastodon*, the "Maastricht animal" remained without a distinctive label until Conybeare (see §2.2) named it the *mosasaur*, or lizard of the Maas or Meuse, the river on which the Dutch city lies: see Parkinson, *Oryctology* (1822), 298, and Bardet and Jagt, "*Mosasaurus hoffmani*" (1996). Throughout the present volume, fossil animals and plants are referred to by the names given them at the time, not their modern equivalents; Cuvier's "*ptéro-dactyle*", for example, was in modern terms a pterosaur. And as was then customary, such names are printed here in their informal English style, without italics or initial capitals, unless there is reason to use their formal Latin names.

7. Cuvier, *Règne animal* (1817), like its brief trailer "Nouveau rapprochment" (1812), defined four "*embranchements*" of animals with sharply distinct basic anatomies; it was the origin of the modern concept of sharply distinct *phyla* such as the chordates (including all the vertebrates from fish to humans) and the arthropods (including insects, spiders, and crustaceans). See the classic accounts in Daudin, *Cuvier et Lamarck* (1926), and Coleman, *Georges Cuvier* (1964), chap. 4.

8. The flow of material to Paris can be assessed from Cuvier's massive research files in Paris-MHN and from his incoming correspondence in Paris-IF: the latter summarized chronologically in Dehérain, *Manuscrits du fonds Cuvier* 1 (1908). The term "proxy", first introduced in this context in Hineline, *Visual culture* (1993), denotes accurate—and sometimes strikingly trompe l'oeil—pictorial representations of particular specimens, which made them "mobile" across the international network of naturalists (*BLT* §2.1): see Rudwick, "Picturing nature" (2005). On the role of this "noble commerce", as Cuvier termed it, in the production of the first edition of *Ossemens fossiles*, see Rudwick, "Alliés internationaux" (1997) and "Cuvier's paper museum" (2000).

research assistant Charles Laurillard (1783–1853). As on his earlier wartime tours on the Continent, he traveled in an official capacity, this time representing the French universities and the Académie des Sciences; he was presented at court, watched the House of Commons in session, and was entertained by Sir Joseph Banks (1743–1820), the long-standing president of the Royal Society, and other leading savants. But he also found time to study fossil bones and other specimens in the major English museums, which had been frustratingly inaccessible to him during the years of war. In particular, he studied the outstanding collections at the Royal College of Surgeons, amassed in the previous century by John Hunter (1728–93). At the British Museum, he confirmed the opinion of Charles König (1774–1851), the German-born curator of its natural history collections, that a famous human skeleton from Guadaloupe in the West Indies was not truly fossilized and probably only a few centuries old, although it was embedded in solid (and therefore ancient-looking) limestone; so it gave no support to claims that the human species had existed far back in geohistory (*BLT* §10.3). He also visited Oxford to meet William Buckland (1784–1856), the university's most prominent geologist, and to inspect his museum collections (see §5.1). But Cuvier's overall opinion of the scientific scene in England was probably echoed faithfully in what his assistant reported to a friend after their return to France:

> The English anatomists, like the naturalists, are not strong in zoology or comparative anatomy, and do not even realize the value of their own riches [i.e., museum collections]. In general, the scientific institutions in England are almost nothing, the government favoring only the art of making money, which is brought to perfection in this country. Money is made on everything: art collections, natural history cabinets, seeing [historical] monuments; in England everything pays for itself, in effect, everything relates to money and money relates to everything.[9]

Cuvier's visit to England filled the last major gaps in his firsthand study of relevant specimens, and in 1821 a new edition of his *Fossil Bones* began to appear in print. Over the next four years the four original volumes were replaced by seven larger ones, even more fully illustrated with newly engraved pictures of a vast range of specimens.[10]

Cuvier's second edition opened with a revised version of his "Preliminary Discourse" (*BLT* §10.5). With one important exception, however, the revisions were slight. Cuvier was tacitly leaving to others, and particularly to Brongniart, the further development of his geological ideas. He himself was concentrating instead on the improvement of his evidence for the range and distinctiveness of the fossil species that he claimed had gone extinct. Only at one point was the text of his essay greatly enlarged with new material. Significantly, it concerned the relation of geohistory to *human* history. Cuvier cited massive further evidence from the textual records of ancient cultures, to refute the renewed claims being made at this time for a vast antiquity for early civilizations. His own view conceded that there must have been some kind of human presence even before the geologically recent "revolution": without it there could have been no later records of the event. But he insisted again that it was only after that event that humankind had developed literate civilizations, which had so far lasted no more than a few millennia. So nature's most recent revolution did still form, in effect, the boundary between the brief era of recorded human history and the vast spans of almost entirely prehuman geohistory.[11]

In the rest of the first volume and the whole of the next there were enlarged special studies of the fossil "pachyderms" (such as mammoths and mastodons, rhinoceros and hippopotamus), together with fossil horses, pigs, tapirs, and so on, all of them preserved in the Superficial deposits. These comprised much of the mammalian megafauna that Cuvier claimed had gone extinct in the earth's most recent major revolution (the roster was completed in later volumes with studies of ruminants, carnivores, rodents, and edentates from the same deposits).[12]

The following volume included a greatly enlarged version of his and Brongniart's study of the geology of the Paris Basin; the revisions were in fact mostly Brongniart's work. It was still further enlarged by Brongniart's account of similar rocks in other parts of western Europe, the fruits of his extensive fieldwork since the end of the wars (*BLT* §9.6). All this stratigraphical detail demonstrated the importance and wide distribution of these formations (above the Chalk and its equivalents), which other geologists were beginning to distinguish as "Tertiary". It was therefore becoming clear that the corresponding Tertiary era in geohistory could indeed act as a gateway leading from the familiar present world (and the still relatively familiar world of the extinct megafauna wiped out by the last revolution), back in deep time towards the much stranger worlds represented by the very thick Secondary formations that underlay and were therefore still older than the Tertiaries (see Chap. 10).[13]

Brongniart's work on the Tertiary formations now provided, even more firmly than before, a geohistorical context for Cuvier's detailed studies of the most important vertebrates of Tertiary age. These were the fossil animals from

9. Laurillard to Georges Louis Duvernoy, 30 September "1817" [printed in Mathiot and Duvernoy (M.), "Lettres inédits de Laurillard" (1940), 10–11, and in Duvernoy, "Sophie Duvaucel" (1939), 54]; the year must have been transcribed or printed in error, for there seems to be no other evidence that Cuvier visited England in 1817 rather than 1818. On his clan-like research team, see Outram, "Le Muséum après 1793" (1997). His stepdaughter Sophie Duvaucel (1789–1867) and his natural daughter Clémentine (at this point only a thirteen-year-old) were both intelligent, well educated, and fluent in English. This made them valuable to Cuvier, who spoke little English: see Orr, "Cuvier's daughters" (2007). Taquet, "Reptiles marins anglais" (2003), reproduces some of the drawings of fossil bones that Cuvier made at London-RCS in 1818, and later (in 1824) annotated as having been those of the plesiosaur (see §2.3).

10. Cuvier, *Ossemens fossiles*, 2nd ed. (1821–24), 5 vols. in 7.

11. Cuvier, *Ossemens fossiles*, 2nd ed., 1 (1821), Discours (i–clxiv); the thirty-six pages (lxxix–cxv) on ancient human records—over a fifth of the whole essay—were enlarged from twelve pages (in larger print) in the first edition. Those whom Cuvier was criticizing were claiming—speculatively—an antiquity for *literate* civilizations of, in some cases, more than a hundred thousand years. Cuvier's own dating was of course much closer to that assigned to the same kind of evidence by modern archaeological research.

12. The pachyderms are in Cuvier, *Ossemens fossiles*, 2nd ed., 1 (1821) and 2(1) (1822); the other mammals, including Buckland's cave hyenas (*BLT* §10.6), are in *ibid.* 4 and 5(1) (1823).

13. Cuvier, *Ossemens fossiles*, 2nd ed., 3 (1822). The term "Secondary" will be used henceforth in its then newly restricted sense, i.e., excluding the Tertiaries and therefore roughly equivalent, in modern terms, to the Mesozoic and Paleozoic combined.

the Gypsum formation that outcropped widely around Paris (where it was quarried to make "plaster of Paris"). Cuvier's analysis of them occupied the whole of the next volume. His earlier descriptions of several species of the mammals that he called "palaeotherium" and "anoplotherium" (*BLT* §7.5)—anatomically much stranger than anything in the Superficial deposits—as well as fossil birds and other less important forms, were greatly enlarged, utilizing the many further specimens that had been found by quarrymen in the intervening years or newly identified in older collections.[14]

Finally, two years later, the last volume of Cuvier's great work presented his revised and enlarged studies of fossil reptiles. He dealt with those of all geological ages, among them the celebrated specimen from a Tertiary formation at Oeningen near Konstanz—first found a century earlier and named at that time the "man a witness of the deluge"—which in the first edition Cuvier had famously debunked as being in fact a giant salamander (*BLT* §9.3). But by far the most important were the fossils from the Secondary formations, on which he had based his hunch that an age of reptiles might have preceded the age of mammals. As before, they included not only crocodiles and turtles, but also the huge marine lizard from Maastricht and the bizarre flying pterodactyl from Solnhofen. However, some sensational new discoveries, of an even wider range of peculiar fossil reptiles, were just in time to be included; but they were so important in their own right that they will be described separately, later in this narrative (see Chaps. 2, 5).[15]

1.3 CUVIER'S SECULAR RESURRECTION

In Cuvier's massively impressive new edition of *Fossil Bones*, only one other new feature need be noted here. In his earlier research on the Tertiary fossils found in the Gypsum formation around Paris, he had used hundreds of disarticulated bones (and, very rarely, more complete assemblages) to make careful reconstructions of the skeletons of the strange mammals; and he had published his drawings of these skeletons in lively and lifelike poses (*BLT* §7.5). His generation would not have missed the biblical overtones of his dramatic claim that his work was "almost a resurrection in miniature, and I did not have the almighty trumpet at my disposal . . . [but] at the voice of comparative anatomy each bone, each fragment of bone, took its place again".[16]

However, Cuvier had kept to himself his astonishing pictorial reconstructions of their whole bodies—the skeletons reclothed in muscles and skin, with ears, eyes, and all—which would have made these long-extinct animals as vividly real as the living ones in the Muséum's *ménagerie* or zoo just round the corner from his house (*BLT* §7.5). He may have feared that such pictures would be criticized as merely fanciful, and thereby detract from the scientific authority of his research. Whatever the reason, in his new edition he did publish some pictorial reconstructions, though they were far inferior to his private ones. It may be significant that he delegated to Laurillard the task of drawing this new set, as if to keep himself a little aloof from them. Laurillard's drawings were much less lively than Cuvier's, and showed only the external form rather than the internal anatomy. Nonethe-

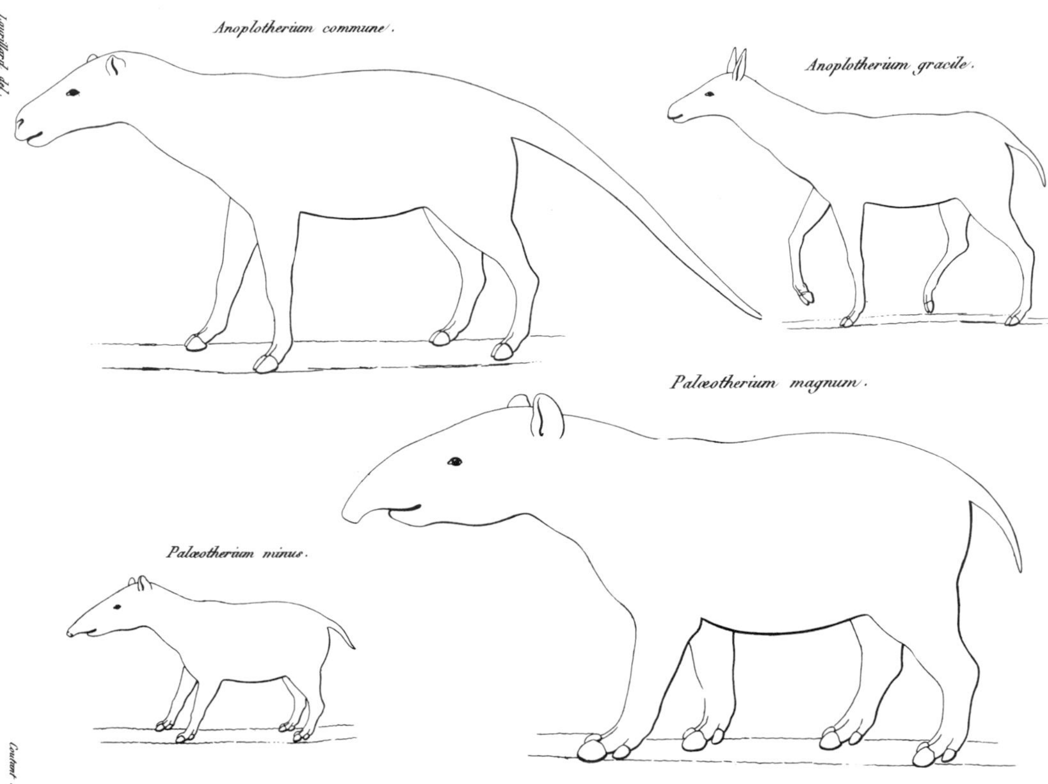

Fig. 1.2 Reconstructions of the body outlines of four of the strange extinct Tertiary mammals from the Gypsum formation around Paris, as published in 1822 in the second edition of Cuvier's *Fossil Bones*. The drawings were not by Cuvier himself but by his assistant Charles Laurillard. Their curiously stilted style, with rigorously lateral profiles, recalled the ancient Egyptian friezes that were attracting intense antiquarian interest at this time.

less, for readers (or rather, viewers) of the published work they did enhance the sense that Cuvier had successfully brought these extinct mammals back to life, at least in the mind's eye (Fig. 1.2).[17]

This represented a significant further step towards turning mere fossil bones into geohistory. Cuvier took yet another step when he described not only what these animals might have looked like but also how they might have lived in their environment. More fully than before, the valley of dry bones was becoming a

14. Cuvier, *Ossemens fossiles*, 2nd ed., 3 (1822).

15. Cuvier, *Ossemens fossiles*, 2nd ed., 5(2) (1824); the nominal fifth and final volume may have been split in two, and this its second half postponed, in order to allow these most recent discoveries to be included. At this time the class Reptilia was so defined as to include the amphibians (and among them the salamanders) as a subclass.

16. Cuvier, *Ossemens fossiles* (1812) 3, Introduction: 3; alluding to the prophetic vision in Ezekiel 37:1–10 and the eschatological one in 1 Corinthians 15:52. Contrary to later myths about him, Cuvier did not claim to be able to *reconstruct* a fossil animal on the basis of a single bone, but only (in favorable cases) to *identify* the kind of animal from which it had come.

17. Fig. 1.2 is reproduced from Cuvier, *Ossemens fossiles*, 2nd ed., 3 (1822), pl. 66, engraved by Jean Louis Denis Coutant. Laurillard, who was not unskillful as a scientific artist, also made many of the new drawings of specific bones for the second edition.

scene from deep time. The example of *Anoplotherium commune* (Fig. 1.2, top left) will make the point:

> What distinguished it most was its long tail. This gave it something of the appearance of an otter, and it is very probable that like that carnivore it was often on or under the water, above all in marshy places; but this was certainly not in order to fish there. Like the water rat, the hippopotamus, and all kinds of boar and rhinoceros, our Anoplotherium was a herbivore; so it went in search of the succulent roots and stems of aquatic plants. With its habits as a swimmer and diver, it would have the smooth skin of the otter.[18]

The second edition of Cuvier's *Fossil Bones*, including these reconstructions, completed his survey of fossil quadrupeds, but not the demand for it. In 1825 he reissued all the volumes as a third edition, little changed from the second. Then, with the work securely in print in its definitive form, he had its famous "Preliminary Discourse" reprinted as a separate small volume. As was common practice at this time, it was decorated with a portrait of the author, to enable a wider public to see what manner of man had achieved this celebrated resurrection of a "former world" of animal life (Fig. 1.1).[19]

Cuvier's essay had in fact been available in this form from the start, but only in English. When *Fossil Bones* first appeared, Robert Jameson (1774–1854), the professor of natural history at Edinburgh, had promptly commissioned a translation. But he had entitled it an *Essay on the Theory of the Earth*; and successive editions of this work, to which Jameson added more and more material of his own, had reinforced the impression among anglophone readers that Cuvier was contributing to the genre of geotheory (*BLT* §10.3). Significantly, Cuvier himself gave his text a quite different title: it was a "Discourse on the revolutions at the earth's surface, and on the changes that they have produced in the animal kingdom". The work was not on geotheory but on *geohistory*, and with the implication that the history of life was causally connected with the history of its physical environment. In this new form Cuvier's ideas became even more widely known, for his *Discourse* was soon translated into German, Italian, and Swedish.[20]

1.4 CONCLUSION

Cuvier's *Researches on Fossil Bones* was a seminal work: in the years that followed its first publication, it provided the baseline and suggested a starting point for many separate lines of research by other savants. Whether or not they agreed with his conclusions, his work could not be ignored.

Cuvier presented a forceful and eloquent case for treating geology as a *historical* science, and for hitching the short span of recorded human history on to the tail end of an inconceivably longer span of prehuman geohistory. Penetrating geohistory backwards from the known present—"bursting the limits of time"—Cuvier claimed, first, that around the dawn of human history there had been the most recent of the earth's physical "revolutions"; these, he argued, were natural events that had occasionally interrupted far longer periods of relative tranquility. A watery catastrophe of some kind was obscurely recorded in several ancient literate cultures (the biblical story of Noah's Flood being just one of them); but it was also recorded more clearly, in his opinion, in the apparent mass extinction of the diverse mammalian megafauna that had been revealed by his own careful analysis of fossil bones. Second, his joint research with Brongniart on the rocks of the Paris Basin showed that, long before that last revolution, there had been a succession of similar sudden events, which had several times altered the environment abruptly from marine to freshwater and back again. These Tertiary rocks (as they were soon being called) contained another mammalian fauna, but one much less like the present fauna of living species. Third, the underlying Secondary rocks recorded an even earlier era in geohistory, which seemed to have been an age of reptiles, with no mammals at all. Fourth and last, the Transition rocks, with very few fossils of any kind, were a transition to the Primary rocks, which formed a basement to the whole geognostic (or stratigraphical) pile; the Primaries were evidently the oldest, and they contained no fossils at all and were assumed to be likely to date from before the origin of life itself.

This picture of a broadly directional geohistory—on a vast timescale, but not an eternity—was far from being original to Cuvier. But he provided it with a striking new dimension, by analyzing fossil bones and interpreting them in terms of a *history* of quadruped life. In the particular case of the Tertiary mammals from around Paris, he showed by example how a vanished fauna could be brought back to life, at least in the mind's eye, in a feat of secular resurrection: the valley of dry bones could become a scene from deep time.

Even as Cuvier was revising his great work for its enlarged second edition, soon reissued as its third, other naturalists were taking up various aspects of the agenda he had proposed, either to reinforce his conclusions or to try to refute them. Some of these diverse lines of research will be described in the rest of Part One, in narratives that run in parallel over much the same years.

18. Cuvier, *Ossemens fossiles*, 2nd ed., 3 (1822), 247–48; all the verbal reconstructions are translated in Rudwick, *Scenes from deep time* (1992), 34–36. It should be obvious that Cuvier's paleoecology, no less than his functional anatomy, was underlain by the standard actualistic method: in both cases he was using living animals as the key to their extinct analogues.

19. Cuvier, *Ossemens fossiles*, 3rd ed. (1825). The 4th ed. (1834–36) was posthumous.

20. It is just possible that Cuvier considered issuing the second edition of his essay with *Jameson*'s title, and then thought better of it. A version entitled "Discours sur la théorie de la terre" and dated 1821 is so rare that it was probably only a trial printing: Smith (J. C.), *Georges Cuvier* (1993), no. 665, records a copy in Washington-SI, and there is another in Paris-MHN [Ch. 334]; apart from the title page it appears to be identical to the text in *Ossemens fossiles*, 2nd ed., 1 (1821). Smith also lists the early translations (nos. 662, 664, 688).

CHAPTER 2

Monsters from deep time (1819–24)

2.1 THE STRANGE ICHTHYOSAUR

In the first edition of his *Fossil Bones* Cuvier had described the fossil reptiles found in Secondary formations in various parts of Europe, and had conjectured that they might represent an ancient age of reptiles, in which mammals had not yet "begun to exist". And if the huge marine lizard from Maastricht and the flying pterodactyl from Solnhofen were representative members of that fauna, the deep past must have been strange indeed, and profoundly unfamiliar. Nothing, therefore, could better reinforce the *geohistorical* character of Cuvier's study of fossil bones than for the fossil reptiles from the Secondary formations to be studied more thoroughly. Fortunately for him, an astonishing sequence of newly discovered specimens began to throw further light on this most ancient era of quadruped life, just at the time when he was working on fossil reptiles for the final volume of the new edition of his great work.

The new discoveries were not just the result of fortunate accidents. The exciting work of geologists since the turn of the century had generated a heightened awareness of fossils, among both savants and amateur naturalists. In Britain, for example, the *Organic Remains* (1804–11) published by the London physician James Parkinson (1755–1824) had catered to this interest among amateurs (*BLT* §8.2); and it was in effect continued in the superbly illustrated *Mineral Conchology* published in installments (from 1812) by the commercial naturalist James Sowerby (1757–1822) and his family, also in London (see Fig. 4.1). There was a rising market for fine specimens. The demand was met by the assiduous work not only of quarrymen but also of lower-class "*fossilists*", collectors who combed the fields and sea shores for specimens that they could sell. They all became increasingly knowledgeable about what their middle- and upper-class customers most desired and would pay them for. This widespread trade in fossils flourished throughout western Europe, but in the event it was in England that the most significant new specimens of fossil reptiles were found.[1]

Cuvier had known of a few specimens that he identified as the bones of crocodiles, from both sides of the Channel; but for the English ones he had had to depend on earlier published accounts. However, some much better specimens had been found during the war years, although at the time Cuvier had no access to them. Most of them came from the Secondary formation that William Smith called the "Lias", which he had mapped right across England. At both ends of its outcrop, at Whitby (Yorkshire) in the northeast and Lyme Regis (Dorset) in the southwest, fine coastal exposures of the Lias rocks were well known for their fossil "crocodile" bones, notably some large vertebrae. When, at the first sign of peace in 1814, Buckland had told a friend that he hoped the latter might see "Cuviers and crocodiles" on his forthcoming visit to Paris (*BLT* §9.5), he was alluding to what had become a exciting topic among naturalists, on which Cuvier would of course be treated as the authority to whom they would all defer.

The finest and most decisive new specimen of a crocodile-like animal had been found at Lyme by two children of the Anning family. Richard Anning had already been supplementing his modest income as a cabinet maker by working also as a fossilist, collecting and selling common Lias fossils such as ammonites to the gentry who visited the seaside resort after the annual social season at the fashionable spa town of Bath. After his death in 1810 his widow Mary had an

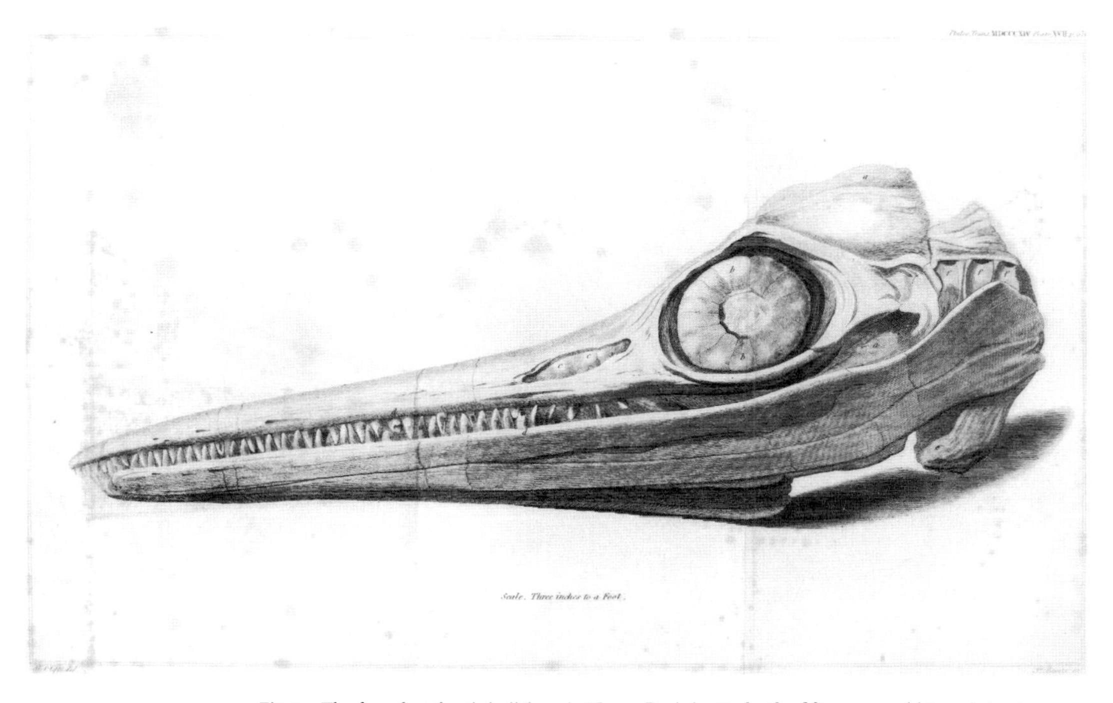

Fig. 2.1 The four-foot fossil skull found at Lyme Regis in 1811 by the fifteen-year-old Joseph Anning, as published in 1814 in the Royal Society's *Philosophical Transactions* by the anatomist Everard Home; Joseph's sister Mary had found some of the rest of the huge skeleton in 1812. Home suggested at first that it was a fish, but it was later assigned to a new reptilian genus, the ichthyosaur. The ring of bony plates surrounding the eye is characteristic. (By permission of the Syndics of Cambridge University Library)

even greater need to maintain this business, and as usual in their social class her children were expected to help. Two of them had made a sensational discovery: in 1811 Joseph (1796–1849) had discovered a huge skull, and the following year his sister Mary (1799–1847) had found some of the rest of the skeleton. The boy was already apprenticed and had little time for further fossil hunting, but the girl proved to have an outstanding knack for finding good specimens; as she grew up she acquired a formidable reputation among the family's fossil-collecting customers.[2]

The Anning children's first great find had been purchased by a local landowner who was also a keen collector—the payment would have been a major windfall for the family—and was exhibited at a commercial museum in London. In 1814 it had been described at the Royal Society by the surgeon Everard Home (1756–1832), who much earlier had been Cuvier's contact in London (*BLT* §7.5) and was still one of the few comparative anatomists in England. As a result of the fame generated by this and Home's later papers, the "crocodile in a fossil state" fetched double the price paid to the Annings, when in 1819 it was sold at auction to König, who was bidding for the British Museum (Fig. 2.1).[3]

Home had suggested that the affinities of this animal lay with fish, although it was unlike any known alive. Other specimens had revealed that it had paddle-like limbs, so it was certainly not a crocodile. Two years later, having seen these further specimens, Home had retracted his earlier opinion and suggested an affinity to the Australian duckbilled platypus that he had described at the turn

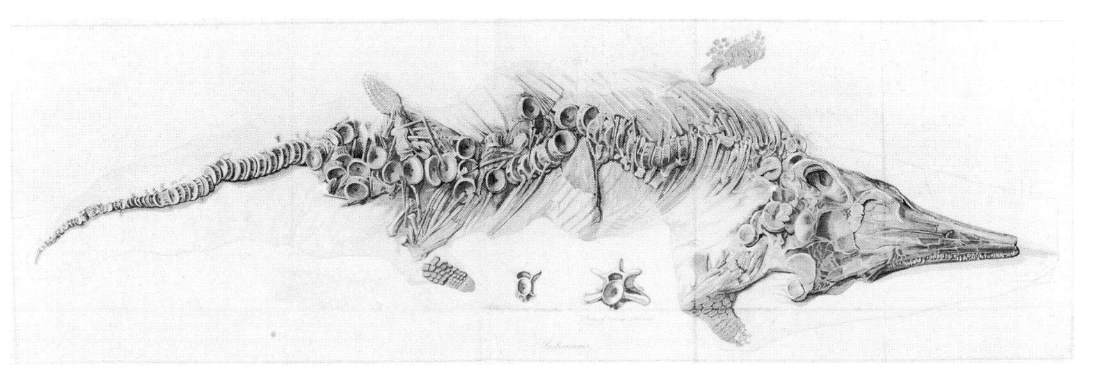

Fig. 2.2 A small but almost complete ichthyosaur specimen (90cm in length) from Lyme Regis, used by Home in 1819 to illustrate one of his papers to the Royal Society. It clearly showed that both pairs of limbs were paddles and the general form like that of a dolphin. (By permission of the National Museum of Wales)

1. Cleevely, "The Sowerbys" (1974) and "Bibliography of Sowerby family" (1974). Knell, *Culture of English geology* (2000), is a fine social and cultural history of fossil collecting in England in the postwar years; comparable historical research on the rest of Europe is as yet lacking.

2. Torrens, "Mary Anning" (1995), records valuable biographical information and traces some of the luxuriant myths that have grown up around her (primarily because she was a woman); on the early discoveries, see also Howe et al., *Ichthyosaurs* (1981), 7–20; and Taylor, "Before the dinosaur" (1997).

3. Fig. 2.1 is reproduced from Home, "Fossil remains of an animal" (1814) [Cambridge-UL: T340:1.b.85.103] , pl. 17; this skull, joined to part of the backbone found subsequently by Mary Anning, is now on display in London-NHM (the successor to the former Natural History department of the British Museum). Home did not mention the Annings, but it was not usual for naturalists to record by name the fossilists—of either sex—who had found or sold them their specimens.

of the century. But in 1819 he claimed that the fossil animal was anatomically intermediate between salamanders [*Proteus*] and lizards, and he therefore named it "*Proteo-saurus*". His shifting opinions reflect his bewilderment at such a strange creature, but also some lack of competence (he must have been one of the English anatomists who evoked Laurillard's, and probably Cuvier's, scorn). Anyway, by this time König had already proposed that it be called "*Ichthyosaurus*" [fish-lizard], and this was the name that was soon adopted generally. Whatever it was to be called, it was obviously very strange indeed (Fig. 2.2).[4]

2.2 THE GEOLOGICAL SOCIETY

By the time Home published his last paper on these fossils, other and more competent naturalists were becoming involved. They were to move the center of English debate on fossil bones from the Royal Society to the Geological Society. The latter had been founded back in 1807, but it was several years before some of its members had begun to shift the focus of their attention from the hard old Primary rocks to the overlying Secondary formations, or, in other words, from rocks without fossils to those in which fossils were often abundant (*BLT* §8.4, §9.4). In the last years of war and the first of peace, a new generation of lively young savants joined the Geological Society, and soon confounded Laurillard's (and probably Cuvier's) earlier scorn for the English scene by making London a powerhouse of geological research to rival Paris. When the wars finally ended in 1815, the Society promptly signaled its international ambitions by creating a new category of Foreign Members; the first batch of seven included, unsurprisingly, both Cuvier and Brongniart, as well as the great Saxon geognost Abraham Gottlob Werner (1749–1817).[5]

In the postwar years the Geological Society grew rapidly in numbers and in social and scientific prestige. Yet its core of highly active members was small enough for it to retain the informality with which it had been founded. The character of its meeting room helped to sustain this atmosphere. The members sat facing each other as in the House of Commons, with the president in the Speaker's position, acting as an impartial chairman. This was in striking contrast to the awe-inspiring formality of the Royal Society's premises, where the president and secretaries faced serried ranks of Fellows as in a lecture room. Furthermore, at the Geological the reading of papers was followed on the spot by discussions that were often as lively as any in Parliament: "though I don't much care for geology", one distinguished visitor famously remarked, "I do like to see the fellows fight". This was an innovative custom that contrasted again with the Royal Society's prohibition on any such ungentlemanly action (at least until its Fellows had adjourned for coffee). There was nothing in Paris—except the *salons* at Cuvier's home and those of other leading savants—to match the Geological Society as a site for the fruitfully informal communication of ideas, arguments, and conclusions in the new science (Fig. 2.3).[6]

One of the most prominent among these lively newer members was William Daniel Conybeare (1787–1857), who, had he not chosen to get married, would have occupied Buckland's teaching position at Oxford (*BLT* §9.5); it was he who

Fig. 2.3 A meeting of the Geological Society, with an ichthyosaur skull displayed on the central table and a large geological section hung on the far wall. The sketch was made from behind the president, who, flanked by the two secretaries, occupies a position similar to that of the (non-partisan) Speaker in the House of Commons. The leading members—Henry De la Beche (see below) is recognizable by his spectacles—face each other on the front benches, while the less important "back-benchers" are crowded behind them and scarcely sketched in. The parliamentary arrangement of the meeting room (which is lit by coal gas, the latest technology) facilitated some famously lively discussions. This sketch was probably made around 1830, when ichthyosaur specimens were still much under discussion. (By permission of the Geological Society of London)

Buckland had hoped would be able to meet "Cuviers and crocodiles" on his first visit to Paris. While the two men were jointly making a stratigraphical survey of the Bristol region (*BLT* §9.5), Conybeare—who was then living there, working as an Anglican parson—also began a thorough study of the reptiles from the Lias. In this latter project he collaborated with a younger and more recent recruit to

4. Fig. 2.2 is reproduced from Home, "Proteo-saurus" (1819), pl. 15 [Cardiff-NMW]; the specimen was purchased in 1820 for London-RCS when the fossil collection of an amateur naturalist was sold for the benefit of the Annings, who by then were almost destitute; it was destroyed by bombing during the Second World War. See also Home, "Farther account" (1816) and "Additional facts" (1818).

5. Geological Society, 5, 19 May 1815 (London-GS: OM/1). The very first, preceding these seven, was the leading Prussian geognost von Buch (see §3.1), who was already in Britain and had been elected at the previous meeting; it may have been his presence that spurred the English geologists into taking this step.

6. Fig. 2.3 is reproduced from an anonymous sketch in London-GS; Herries Davies, *Under the earth* (2007), 59–63, dates it (by the gas lighting) to 1828–32, after the Society has moved into Somerset House. Thackray, *Fellows fight* (2003), i, suggests it may be by De la Beche, but the style is quite unlike other drawings that are certainly his: see Figs. 13.2, 22.7, and others reproduced in McCartney, *De la Beche* (1977). The undated "fighting" remark by John Lockhart (1794–1854), who edited the influential Tory *Quarterly review* from 1825 (see §14.4), is quoted in Allen, *Naturalist in Britain* (1976), 70. On the Geological Society at this period, see also Rudwick, *Devonian controversy* (1985), 18–27. The Society of Antiquaries had a larger meeting room with a similar layout, which could have served as a model; but many early members of the Geological Society, including its first president George Greenough (see below), were also Members of Parliament and therefore familiar with the arrangement they adopted. The meeting room of the Geological Society retained its Parliamentary plan, in different premises, until

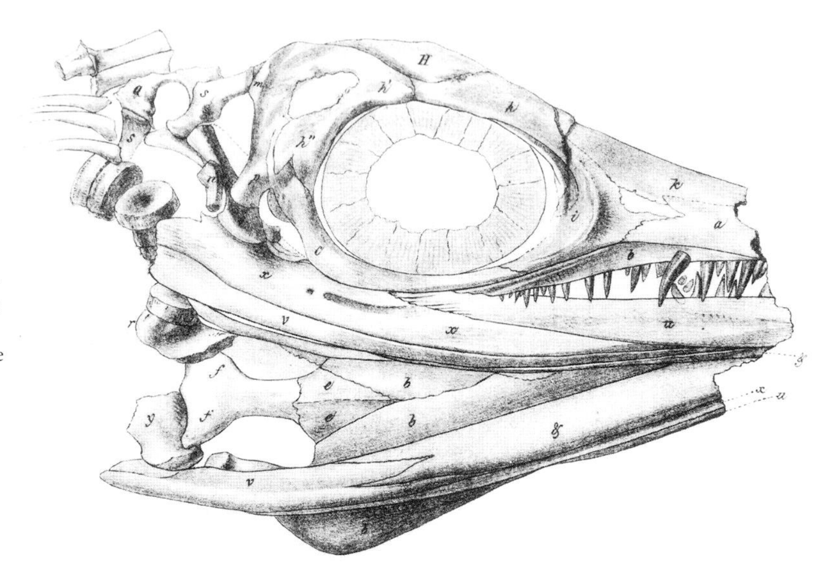

Fig. 2.4 A lithograph of a "matchless specimen" of the skull of an ichthyosaur from Lyme Regis, drawn by De la Beche, who owned it at the time. Conybeare skillfully identified its component bones—in this specimen the palate was crushed flat into the same plane as the side of the skull—and analyzed it in Cuvierian style in a supplement to their joint paper, which was read at the Geological Society in 1821. (By permission of the Syndics of Cambridge University Library)

the science. Henry Thomas De la Beche (1796–1855) was an Englishman—despite his Frenchified name—who as a child had inherited sugar plantations in Jamaica. In 1812 he and his widowed mother had settled in Lyme Regis, where he met the Annings and soon had a reputation as an assiduous and wealthy young collector and a fine scientific artist. He had been encouraged in his geological interests by George Bellas Greenough (1778–1855), the first president of the Geological Society, and by Buckland, whose family home at Axminster was only a few miles from Lyme. De la Beche joined the Geological Society in 1817 and soon began presenting specimens and drawings of the fossils from Lyme and other parts of Dorset. Once he began to collaborate with Conybeare, he concentrated on doing the necessary fieldwork, studying the specimens scattered in public and private collections, and making superb drawings of them, while leaving Conybeare to analyze their anatomy in Cuvierian style (Fig. 2.4).[7]

2.3 CONYBEARE'S PLESIOSAUR

Meanwhile Conybeare had recognized among the Lias fossils some that clearly belonged to a different animal. As he told Buckland, its anatomy was strangely intermediate between the ordinary lizards and the ichthyosaur, "furnishing a beautiful series of links between known & unknown Saurii in all its structure"; he sent a copy of this letter to Cuvier's young Irish assistant Joseph Pentland (1797–1873), to make sure his discovery was known in Paris without delay. Like the ichthyosaur the new animal had a streamlined body and four paddles, but its long snake-like neck contained far more cervical vertebrae than any known quadruped, living or fossil; the skull was as yet unknown. The cautious Cuvier found this anatomy so anomalous that he advised Conybeare to check very carefully that he had not been duped, before announcing it in public: there were plenty of earlier cases (as there still are today) in which naturalists had been misled by specimens cunningly "enhanced" by unscrupulous fossilists to increase

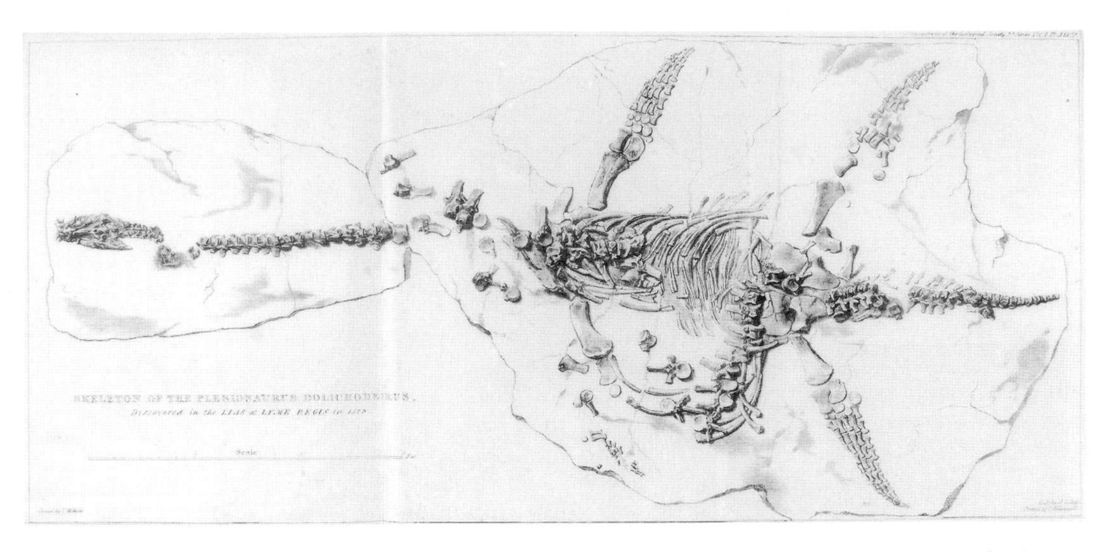

Fig. 2.5 The "almost perfect skeleton of the Plesiosaurus" found by Mary Anning in 1823 in the Lias formation at Lyme Regis, as depicted by Conybeare in his analysis of it for the Geological Society (1824). The four paddles were rather like those of the ichthyosaur, but the long neck was without parallel and made the plesiosaur the most bizarre of all known fossil quadrupeds. Conybeare estimated that in life this specimen had been almost ten feet (about 3m) from head to tail, but other separate bones already known showed that some individuals must have been far larger. (By permission of the Syndics of Cambridge University Library)

their commercial value. But Conybeare was convinced that his specimen had not been fraudulently fabricated from bits of different fossils. Reporting on it to the Geological Society, he proposed that it be called the "*Plesio-saurus*" [almost-lizard]; at the next meeting his paper on it described it as "an entirely new animal". He took the opportunity to compare it closely with the ichthyosaur, stressing that "the method of Mr Cuvier is followed throughout these descriptions".[8]

1975, when it was radically remodeled into the present modern but characterless auditorium: see Herries Davies, 60–62, 132–43, 278–80, and Le Bas, *Milestones in geology* (1995), frontispiece.

7. Fig. 2.4 is reproduced from Conybeare, "Additional notices" (1822) [Cambridge-UL: Q365.b.12.6], pl. 17, lithographed by the German-born topographical artist and scene-painter George Scharf (1788–1860): the paper was a supplement to De la Beche and Conybeare, "New fossil animal" (1821), which in fact was written by the latter. Geological Society, 2 May, 21 November 1817, 5 June, 6 November 1818, 5 March, 18 June 1819 [London-GS: OM/1]; Buckland to his father, 25 May 1818 [Exeter-DRO, F24]. On Conybeare, see Taylor, "Plesiosaur's birthplace" (1994); his osteological analysis should leave no doubt about the remarkable skill of this versatile and highly intelligent parson-naturalist. De la Beche's father Thomas Beach had changed his name to one with a Norman flavor, perhaps to suggest the higher social status to which he hoped his wealth would entitle him; McCartney, *De la Beche* (1977), gives valuable biographical information.

8. Conybeare to Buckland, 21 April 1821 [copy to Pentland in Paris-MHN, MS 629, no. 26]; a later but undated letter to Pentland [same archive] described its osteology in detail, denied that he had ever thought it a crocodile, and insisted that "I distinctly regarded it as an animal sui generis". De la Beche and Conybeare, "New fossil animal" (1821), read at the Geological Society 16 March, 6 April. Since the two fossil genera came from the same formation and were of the same geological age, the putative "links" represented by the plesiosaur were (in modern terms) those of a *morphological* series, not a phyletic or evolutionary one: Conybeare explicitly repudiated Lamarck's "monstrous" transformist idea (see §17.1) of "real [*sic*] transitions from one branch to another of the animal kingdom" (561n). Sarjeant and Delair, "Irish naturalist" (1980), prints many of Pentland's letters to Buckland in 1820–22 (letters II–XXXI); they record the intense interactions between Cuvier and the English geologists, for whom Pentland acted as a convenient anglophone intermediary.

When at last the missing skull of the plesiosaur was found, it turned out to be quite small and, as expected, somewhat like a lizard's. Even more important, however, was that at the end of that year Mary Anning discovered an almost complete skeleton. It was sold for the princely sum of £100 to a wealthy collector, the duke of Buckingham, who had the huge slab of rock shipped to London for display at the Geological Society. Ten workmen struggled to get it upstairs, but it proved too large to put into the meeting room and had to be parked in a passage outside. Still, it allowed Conybeare to give a full description of the extraordinary animal (Fig. 2.5).[9]

Cuvier was informed of the discovery without delay, and was of course greatly excited by it: as he told Conybeare, "one shouldn't anticipate anything more monstrous to emerge from the Lias quarries". For the plesiosaur was a monster not only in the colloquial but also in the scientific sense: as Buckland put it years later, in characteristically dramatic language, "to the head of the Lizard, it unites the teeth of the Crocodile; a neck of enormous length, resembling the body of a Serpent; a trunk and tail having the proportions of an ordinary quadruped, the ribs of a Chameleon, and the paddles of a Whale". But Conybeare also analyzed its strange anatomy in functional terms, reconstructing its likely mode of life in a resurrection scene that outdid Cuvier's, in that the animal was even more remote from any living analogue:

> May it not therefore be concluded that it swam upon or near the surface, arching back its long neck like the swan, and occasionally darting it down at the fish which happened to float within its reach? It may perhaps have lurked in shoal water along the coast, concealed among the sea-weed, and raising its nostrils to a level with the

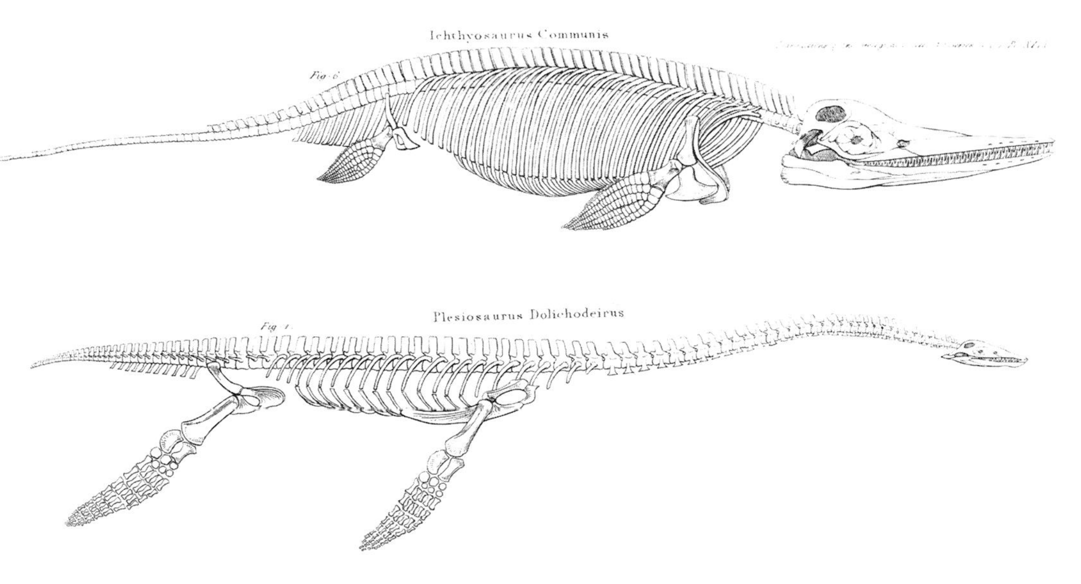

Fig. 2.6 Reconstructions of the skeletons of the ichthyosaur and the plesiosaur, based on exceptionally complete specimens, drawn by Conybeare in 1824 to illustrate his analysis of their comparative anatomy. The bizarre character of both genera, as aquatic reptiles unlike any known in the present world, greatly accentuated the strangeness of the "ancient world" represented by the Secondary formation (Lias) in which they were found. (By permission of the Syndics of Cambridge University Library)

surface from a considerable depth, may have found a secure retreat from the assaults of dangerous enemies; while the length and flexibility of its neck may have compensated for the want of strength in its jaws and its incapacity for swift motion through the water, by the suddenness and agility of the attack which they enabled it to make on every animal fitted for its prey, which came within its extensive sweep.[10]

De la Beche was not at the meeting, being away in Jamaica inspecting his patrimony (and witnessing for the first time the slavery on which his wealth was based). But as Conybeare jokingly told him, "I made my beast roar as loud as Buckland's Hyaenas". His was a verbal reconstruction quite as striking as Buckland's scene of the antediluvial hyena den at Kirkdale (figure opposite page 1, and *BLT* §10.6), and equally a demonstration of the practicality of Cuvier's aspiration "to burst the limits of time". And it was based, like Cuvier's examples, on a rigorous anatomical analysis, culminating in a careful reconstruction of the skeleton in a lifelike pose, in his case with the ichthyosaur for comparison (Fig. 2.6).[11]

The Geological Society received its final social accolade in 1825, during Buckland's presidency and under his astute leadership, when King George IV granted it a Royal Charter. Its members were then restyled "Fellows", like those of the Royal Society; and it was considered necessary, or at least desirable, to adopt an official seal and armorial bearings. Significantly, one of the designs suggested by De la Beche used the ichthyosaur and the plesiosaur as heraldic supporters, their skeletons posed like a rampant lion and unicorn. Buckland told him that his designs were considered the best, though in the end the council settled for a more sober one, fearing that anything so "emblematical" was too easily "convertible to caricature". Nonetheless, the proposal shows how the two reptiles were now being treated—and not only by De la Beche—as icons of the new triumphs of English geology.[12]

9. Fig. 2.5 is reproduced from Conybeare, "Plesiosaurus" (1824) [Cambridge-UL: Q365.b.12.6], pl. 48, drawn by Thomas Webster; the paper was read 20 February. Conybeare, "Additional notices" (1822), read 3 May, had described the first specimen of the skull (pl. 19). Conybeare to Mantell, 4 March 1824 [Cardiff-NMW: no. 302, printed in part in Taylor, "Before the dinosaur" (1997), xxiii–xxiv]. Norman, "De la Beche and plesiosaur's neck" (2000), suggests that De la Beche interpreted the scattered bones at the base of the neck of this specimen as the result of its being fatally bitten by an ichthyosaur, and later depicted that moment in his famous scenic reconstruction *Duria antiquior* (see Fig. 11.3). The specimen is now on display in London-NHM.

10. Conybeare, "Plesiosaurus" (1824), 389; the later quotation is in Buckland, *Geology and mineralogy* (1836), 1: 202 (where "whale" was used in the colloquial sense that included dolphins and other cetaceans). Cuvier to Conybeare, 20 March 1824 [Wellington-ATL]; Buckland to Cuvier, 3 April 1824 [Paris-MHN, MS 629, no. 22], enclosing a superb drawing of the specimen (closely similar to the published one reproduced in Fig. 2.5) by Mary Morland, soon to be Buckland's wife: see Taquet, "Reptiles marins anglais" (2003), 49, 52. Like Cuvier, Conybeare took it for granted that his fossil animals had been well adapted to a specific mode of life; but when—like modern biologists—he used "design language" to describe this kind of functional inference, its explicit reference was to *nature's* design rather than the divine design of traditional natural theology (see §29.1).

11. Fig. 2.6 is reproduced from Conybeare, "Plesiosaurus" (1824), [Cambridge-UL: Q365.b.12.6], pl. 49, lithographed by Scharf. Geological Society, 20 February 1824 [London-GS: OM/1]; Conybeare to De la Beche, 4 March 1824 [Cardiff-NMW, no. 302]. De la Beche, *Negroes in Jamaica* (1825), set out his relatively liberal, but not abolitionist, views on slavery.

12. De la Beche to Buckland, 13 June 1825 [Exeter-DRO, 138M/f730]; Buckland to De la Beche, 7 July 1825 [Cardiff-NMW, no. 173]. De la Beche's design is reproduced in McCartney, *De la Beche* (1977), 58; the coat-of-arms itself featured ammonites, a bone cave, and a geological section.

However, the two strange marine reptiles from the Lias were no merely chauvinistic emblems. Conybeare's impressive analysis of the anatomy of the ichthyosaur and the plesiosaur reached Cuvier just in time for the Frenchman to summarize it—with full approval and ungrudging praise—in the final chapter of the last volume of his revised *Fossil Bones* (§1.2). Conybeare's careful descriptions enabled Cuvier to recognize that similar but fragmentary specimens had also been found on the Continent. And so, finally, Cuvier predicted that further discoveries—perhaps within the next few years—would show that his own massive and laborious work on other extinct animals had been "only a slight glimpse, a first quick glance over these immense creations of ancient times". It was a prescient remark.[13]

2.4 CONCLUSION

The strange new fossil reptiles from the Lias rocks of England were found by fossilists such as Anning, described and named by geologists such as Conybeare, and analyzed authoritatively by Cuvier himself. But the significance of the ichthyosaur and plesiosaur, and the other fossil reptiles, could be appreciated fully only if they could each be assigned to their correct relative dates within geohistory. In other words, their value as traces of the long history of life depended on an accurate assessment of their positions within the pile of rock formations. Cuvier's comments show that in general terms he was well aware of the relative ages of, say, the new reptiles from the English Lias and the fossil mammals that he himself had described from the Gypsum formation around Paris. What he and others needed, however, was more precise knowledge of the order, and hence the relative age, of *all* the formations that were yielding important fossils. The next chapter reviews the kind of geological research—soon to be defined as *stratigraphy*—that was clarifying the order of the formations in just this way. Stratigraphy was not itself geohistorical, but it did provide the indispensable foundations for any reliable reconstruction of geohistory.

13. Cuvier, *Ossemens fossiles*, 2nd ed., 5(2) (1824), 487, the conclusion of the main text of the whole work. The complete plesiosaur specimen, and Conybeare's reconstructions, were redrawn in pls. 31, 32. Cuvier's pl. 29, fig. 1, depicts the skull from Lyme—already published in Home, "Proteo-saurus" (1819), pl. 13—that Cuvier had purchased for the Muséum in 1820 and improved by carefully exposing the reptile's external nares: see Taquet, "Reptiles marins anglais" (2003), 42–43. Conybeare to Cuvier, (undated but on paper watermarked 1821) [Paris-MHN, MS 629, no. 20], had given Cuvier a detailed summary, with pictures, of English finds of fossil reptiles of various ages from the Lias to the London Clay (Tertiary), for comparison with French finds.

CHAPTER 3

The new stratigraphy (1817–25)

3.1 THE PRACTICE OF GEOGNOSY

In 1817, William Smith issued a brief explanation of the methods he had used in compiling his great geological map of England and Wales, which after long delays had been published in 1815, the year of Waterloo (*BLT* §8.2, §9.5). His text was entitled *Stratigraphical System of Organized Fossils.* This brought a new word into the English language, which, in the form of *stratigraphy*, was soon adopted internationally. Smith chose his word carefully: it denoted the *description* of what he always called "Strata", not their causal explanation, nor their place in geohistory. His map was far from being the first to depict the outcrops of what others called *formations, terrains*, or *Gebirge*—the terminology was fluid and often contentious—and it did not change the world, even the world of geology, as later and historically ignorant mythmaking has often claimed. Nonetheless it was a remarkable achievement: it covered a wider area in greater detail than any of its predecessors, and it was the almost unaided work of just one man. Above all, it demonstrated the value of paying close attention to what Smith called the "characteristic fossils" found in many (though not all) formations; again, the perception that some rocks had distinctive fossils was not new, but it had never previously been explored so consistently or in such detail (*BLT* §8.2, §9.5).[1]

Smith's contemporaries recognized his work as an enriched variant of one branch of what had long been practiced under the name of *geognosy.* The science

1. Smith (W.), *Stratigraphical system* (1817). The term "organized fossils" in his title was antiquated: by this time, qualifiers such as "organized", "extraneous", and "accidental" (*BLT* §2.1) were dropping out of general use, because the distinction between objects of organic and inorganic origin had been clarified in all but a few rather unimportant cases, and the bare word "fossils" therefore referred unambiguously to the former.

of geognosy aimed to give accurate descriptions of the three-dimensional *structure* of rock masses of all kinds, independently of any theoretical conjectures about their causal origin or place in geohistory (*BLT* §2.3). For the Primary rocks such as granite and gneiss, schist and marble, it was often difficult anyway to infer any temporal order, except, for example, where mineral veins cut through other rocks and were obviously later in origin; and even in this case the causal problems were highly contentious. For the Secondary rocks, on the other hand, the elucidation of structure was more straightforward. It involved above all the determination of the *order* in which the formations were lying one above another (often more or less tilted or "dipping" in one direction), which in favorable cases could be observed directly in coastal cliffs or on a mountainside. But here too the emphasis was on observable structural order, often described from top to bottom. Any claim about the temporal or geohistorical order of deposition—normally presumed to be from bottom to top—was rightly treated as an inference derived from what could be observed directly.[2]

Problems arose for geognosts when they tried to "*correlate*" the rocks of one region with similar ones in another. It was important to do so, not least in view of the economic value of identifying particular rock masses such as those associated with mineral veins and coal seams. Formations were generally correlated between distant regions by their similar rock types. But distinctive rock types were found to recur at different positions in the pile, making correlations uncertain. In England, for example, major formations of red sandstone were found both above and below the uniquely distinctive and economically important Coal formation; so if similar red sandstones were found in some new region, there was no obvious way of knowing whether they might be underlain by valuable coal seams, or whether that enticing possibility was out of the question.

It was therefore in the matter of correlation that Smith's "characteristic fossils" came to the rescue, because they turned out to be more reliable than other criteria. Smith's methods were, in effect, those of a geognosy *enriched* by the use of the fossil criterion (*BLT* §9.5). But his practice remained unmistakably that of a geognost, however much he himself, as a robustly insular Englishman, would have repudiated that "foreign" label. Like other geognosts, he was concerned primarily with the *order*—his own favorite word—of the formations that he mapped so effectively. When, four years *before* Smith's map was published, Brongniart and Cuvier had produced their closely similar map of the Parisian formations, they had, significantly, called it a "*carte géognostique*" (*BLT* §9.1).

However, Smith's new "stratigraphy" could not simply replace the older practice of geognosy, for the obvious reason that it could be applied only in areas of Secondary and Tertiary formations, where at least some of the rocks contained plenty of fossils. So the science of geognosy continued to flourish in its own right as a practice of wider scope, within which the enriched variety that came to be called stratigraphy was adopted wherever it was feasible to do so (other parts of geognosy were eventually transformed into such modern practices as tectonic, igneous, and metamorphic geology).

In the early post-Napoleonic years, geognosy as a basically structural science received its most penetrating analysis from the great Prussian naturalist Alexander von Humboldt (1769–1859), the almost exact contemporary of both Cuvier and Smith (and Napoleon). When Greenough visited Paris in 1814, just after Napoleon's first defeat, Humboldt, who had already settled in the world center of the sciences, had given him a draft of his forthcoming "work on the *stratification of rocks and the identity of formations*", explaining that since before the turn of the century this had been "the goal of all my fieldwork [*courses*]". His celebrated expedition to Latin America and other extensive travels had made him as well qualified as anyone to compare the formations on a global scale or, as he put it, "in the two hemispheres" of the Old and New Worlds. His conclusions were published in 1822 as a very long article in the multi-volume *Dictionary of Natural Sciences* edited in Paris by the professors at the Muséum; and the following year this was reprinted in book form as his *Geognostic Essay* and also published in German and English translations. It was by any measure one of the most important geological works of its time.[3]

Humboldt's work was first printed under the title "Independence of Formations". That phrase expressed the concept of "formation" that had emerged from geognostic practice in the years since Werner's early classification of *Gebirge* (*BLT* §2.3). Referring to a famous lecture "On the Concept of a Formation", given in Berlin back in 1809 by his colleague the Prussian geognost Leopold von Buch (1774–1853), Humboldt stated that "the essential character of an independent formation is its spatial relation [*rapport de position*], the place that it occupies in the general series of strata [*terrains*]". A formation was "independent" if it could be recognized in different circumstances and in different regions, but always in the same *structural* position relative to other formations with the same status. This kind of language gave the status of "formation" to all the main sets of rock masses, which could be defined primarily by their unique places in the pile, but also by their rock types and in some cases by their fossils. But in practice Humboldt gave priority to "mineral character", because, as he pointed out, fossils could not be deployed except among the more recent formations, such as those described by Brongniart and Cuvier around Paris. Given Humboldt's definition, it was not unrealistic, and certainly not foolish, to hope in due course to identify the *same* "formations" on opposite sides of the world.[4]

2. The historiography of geology has been seriously distorted by the failure of many historians to recognize the marginality of geohistory in scientific practice at this period. This is shown by the anachronistic use of the modern term "historical geology" to denote a stratigraphy that until around the 1820s was generally *not* geohistorical at all, or only marginally so.

3. Humboldt to Greenough, 11 December 1814 [London-UCL]; Humboldt, "Indépendance des formations" (1822) and *Essai géognostique* (1823); as usual he wrote in French, the international scientific language, not in his native German. Since the theme of Humboldt's book is relatively peripheral to that of the present narrative, it must be described here far more briefly than it would deserve in another context.

4. Humboldt, "Independence des formations" (1822), 60–62; *Essai géognostique* (1823), 5–7; he wrote his essay too soon to be aware of Smith's work as well as that of Brongniart and Cuvier. Buch, "Ueber dem Gabbro" (1810). Some historians and later geologists, assuming erroneously that the word "formation" here carried its modern meaning, have ridiculed the geognosts' ambition to identify the same formations globally or "universally". They fail to notice that this is just what has been done with great success, with the modern "geological column" of globally valid "systems" such as Cambrian and Jurassic, and the corresponding relative timescale of "periods" bearing the same names. The only difference lies in the criteria that are regarded as most reliable for correlation, but this is just what geognosts such as Humboldt were trying to discover.

Voici quelques exemples de l'emploi de ces douze signes pasigraphiques des roches :

α, $\gamma + \pi$, $\delta\tau'$, κ', π, σ, α.

Le terrain de transition commence après $\gamma + \pi$ (le micaschiste avec des bancs de porphyre primitif). C'est presque la suite des formations de Norwége (page 148). On voit suivre une formation complexe de thonschiefer et de calcaire (noir) avec débris de coquilles, du grauwacke, un porphyre, de la syénite et du granite. Les termes $\delta\tau'$ et κ', qui précèdent π, ς, α, caractérisent ces trois roches comme des roches de transition. En Angleterre, où le terrain intermédiaire offre deux formations calcaires bien distinctes (celle de Dudley et du Derbyshire), on voit se succéder:

β, $\sigma\pi$, δ', κ^{g}, τ', κ^{b}, τ', ξ, κ^{a}, τ^{a}, $\kappa^{n} + \vartheta$, τ^{o}, κ^{l}. τ^{c}, κ^{2l}. . . .

Le terrain de transition commence avec la formation de syénite et porphyre (Snowdon) placée sur un gneis qu'on croit primitif; puis se suivent : un thonschiefer avec trilobites, le grauwacke de May-Hill, le calcaire de transition de Longhope, le old red sandstone de Mitchel Dean, le mountain-limestone du Derbyshire, la grande formation de houille, le new red conglomerate qui représente le grès rouge, le calcaire magnésifère, le red marl avec sel gemme, le calcaire oolithique, le grès secondaire à lignites (greensand), la craie, le grès tertiaire à lignites ou argile plastique, etc. Sur le continent, les formations secondaires, si elles s'étoient toutes développées, se succéderoient de la manière suivante :

τ', κ^{b} || $\pi\kappa^{a} + \xi$, $\tau^{a} + \vartheta$, κ^{n}, τ^{m}, κ^{q}, τ^{o}, κ^{l}, τ^{c} || κ^{2l}. . . .

Fig. 3.1 A typical passage in Humboldt's essay (1822) on his "pasigraphical method" or universal algebraic language for correlating the geognostic sequence of rock formations. Each truly "independent" formation was assigned a Greek letter (their constituent parts being further qualified with superscripts): among the Primary rocks, for example, α denoted granite and β denoted gneiss. The line of symbols in the middle of this passage summarized the English sequence from Primary gneiss at the base (left) to Tertiary Plastic Clay at the top (right); the line of symbols at the foot of the page showed the full sequence of Secondary formations on the Continent, separated (by double vertical lines) from the top of the Transitions (left) and the base of the Tertiaries (right). Representing the rocks as a horizontal line of symbols recalled Leopold von Buch's earlier analogy between a pile of formations and a row of distinctive houses in a street: "order" in the sense of structural or spatial *sequence*—not geohistorical *age*—was what mattered in geognosy.

Humboldt distilled his concept of formations into a "pasigraphy" or universal language, specifically adapted to the needs of geognosy. At its simplest, Humboldt explained, "it offers an easy way to indicate *geognostic equivalents* or *parallel rock masses*, even in the case where, by the local suppression [i.e., absence] of formation β, formation α directly underlies formation γ". In other words, it was designed to express, in a kind of algebraic notation, the structural sequence of formations observed in any given region; and this could then serve to denote its correlation with other regions, and ultimately display a unitary global sequence in a highly condensed form (Fig. 3.1).[5]

Although Humboldt's notation was never widely adopted by other geognosts, it expressed exceptionally clearly their shared conception of their science: geognosy dealt primarily with the structural *order* of rock masses and their correlation into an international and even global sequence. Humboldt was of course well aware that a three-dimensional pile of solid rock formations usually (though not always) represented a temporal sequence in their deposition. But he was careful to treat their relative age as a property inferred from their geognostic position, not the other way round. The concrete certainty of observable superposition was, he suggested, like that of the anatomy of plants, the heights of mountains, and so on. He contrasted these with the uncertainties of such questions as "the former state of our planet", the origin of its atmosphere, and the origins of the present distribution of plant species and human races. He rightly put superposition into epistemic priority, although in practice this had the effect of sidelining questions of geohistory. To state the point once more, geognosy was not a geohistorical science, even in this postwar period and in the hands of a savant as outstanding and polymathic as Humboldt, any more than it had been for the young Werner thirty years earlier (*BLT* §2.3).

3.2 "CONYBEARE AND PHILLIPS"

In the same year that Humboldt in Paris first published his essay on geognosy, with its aspiration to define a complete sequence of "independent formations" of global validity, another compilation was published in London with more limited goals but, in the event, an even wider impact. This was the *Outlines of the Geology of England and Wales* (1822), compiled largely by Conybeare but based in part on earlier work by William Phillips (1775–1828), a London publisher and one of the eleven founding members of the Geological Society (*BLT* §9.5). "Conybeare and Phillips", as their book was soon known, summarized the sequence of British formations. It was designed as a synthesis of all that had recently come to be known about the stratigraphy of Britain, as a result of intensive geological fieldwork not only by Smith but also by Greenough and many other members of the Geological Society; local surveys by the latter were in many cases condensed onto Greenough's map.

The formations were described from the top down—in the reverse of any inferred geohistorical order—from the "Superior" (Tertiary) formations overlying the distinctive Chalk, as far down as the base of the "Carboniferous" formations (Conybeare's new name for the group that included the industrially important Coal formation itself). At the top end (overlying even the Tertiaries), the thin, loose, and variable Superficial deposits were dismissed in a couple of pages. At the bottom end, any description of the Transition and Primary rocks, which contained few fossils or none at all, was deferred to a planned sequel (which in the event never appeared). Between these excluded limits, the book covered all the "regular", that is, Tertiary and Secondary formations, many of which contained more or less abundant fossils. These were the very formations that Smith

5. Fig. 3.1 is reproduced from Humboldt, "Independence des formations" (1822), appendix on pasigraphy (373–82), 380. In a linguistic context, pasigraphies were systems of symbols that (like numerals) would directly represent ideas rather than words, and therefore, it was hoped, transcend cultural differences and create a universal human language.

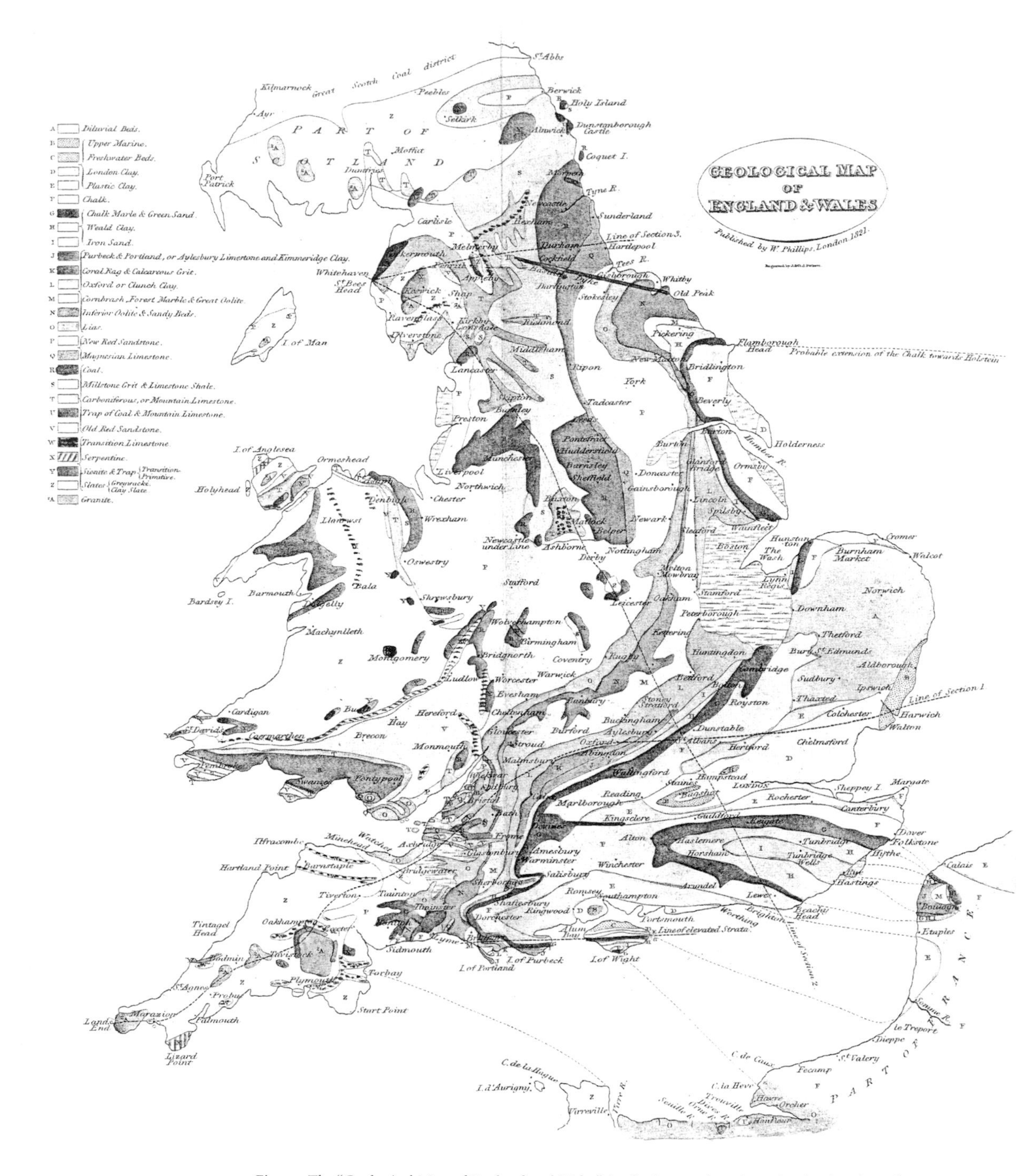

Fig. 3.2 The "Geological Map of England and Wales" (and of parts of southern Scotland and northern France), illustrating the famous summary of British stratigraphy published in 1822 by William Conybeare and William Phillips. Their text described the formations from the Tertiary "Upper Marine" (B) down to the "Old Red Sandstone" (V) at the base of the "Carboniferous" group; but the map also showed the Superficial "Diluvial Beds" (A) above that sequence, and the Transition and Primary rocks (W, X, Y, Z, and ²A) below it. The "Coal" (R), the formation of greatest economic importance, outcropped in the coalfields around or near such newly expanding industrial cities as Newcastle, Sheffield, Manchester, and Swansea. The inclusion of the north French coast, and the dashed correlation lines across the Channel (and across the North Sea towards northern Germany), reflected the authors' hope that their work might become a reference standard for Europe as a whole.

had mapped so effectively. Phillips's earlier and more modest compilation had included a small and simplified version of Smith's map (this had first made the surveyor's work more widely known than the large and expensive original). The greatly enlarged "Conybeare and Phillips" likewise had a small map that was based on Greenough's later and rival survey, though in fact the authors adopted most of Smith's names for the formations (Fig. 3.2).[6]

3.3 THE STRATIGRAPHY OF EUROPE

Conybeare's aspirations, unlike Smith's, were anything but narrowly nationalistic. He was able to benefit from the extensive travels on the Continent that Buckland had undertaken almost annually since the end of the wars (unlike his bachelor friend, Conybeare's own opportunities for fieldwork were now limited by his pastoral duties and a young family). Buckland's principal goal was to correlate Continental sequences of formations with the British. On his first trip he had met Werner in Freiberg, and had compared the Saxon's classic geognostic sequence with his own (*BLT* §10.4). In the summer of 1820, again with Greenough as a companion for part of the time, he made another extensive tour—starting as usual in Paris, to pick up the best advice on where to go and what to see—and ranged this time as far east as Prague. The following year he synthesized his conclusions from all this fieldwork in an important paper on the sequences north and south of the Alps and "their relation to the Secondary and Transition rocks of England"; it was published in the *Annals of Philosophy*, one of the rapidly expanding range of British scientific periodicals, and promptly translated in the Parisian *Journal de Physique*, which made it known throughout Europe (*BLT* §9.5). Buckland's remarkably effective correlations—using fossils in the Smithian manner that was already becoming routine—contributed to the widespread sense that the English sequence was indeed worth treating as a standard of reference.[7]

So it is not surprising that the map in "Conybeare and Phillips" included the nearest part of the Continent, nor that their text summarized what had been published about similar and perhaps equivalent formations throughout Europe. The authors—or at least Conybeare—intended the work to be a standard against which the stratigraphy of the rest of Europe, and ultimately the rest of the world, could be compared. This was no crass scientific imperialism. Conybeare had good reason to appreciate, on the basis of his own extensive fieldwork as well as his wide reading of geological publications, that his own country happened to be exceptionally suitable as a stratigraphical standard, at least for the Secondary formations. In England these rocks outcropped in wide variety across a relatively small landmass; in general they were little disturbed by later folding and

6. Fig. 3.2 is reproduced from Conybeare and Phillips, *Geology of England and Wales* (1822), frontispiece, dated 1821 and engraved by J. and G. J. Pickett, and based on Greenough, *Geological map* (1820).

7. Buckland, "Structure of the Alps" (1821). In the titles of the periodicals mentioned, "philosophy" and "physics" were both used in their broad older senses, to mean respectively all the natural sciences and all those with causal rather than descriptive goals (*BLT* §1.4); the long-running French periodical included "natural history" and the "arts" (or technologies) in its full title, so it too covered much the same wide range.

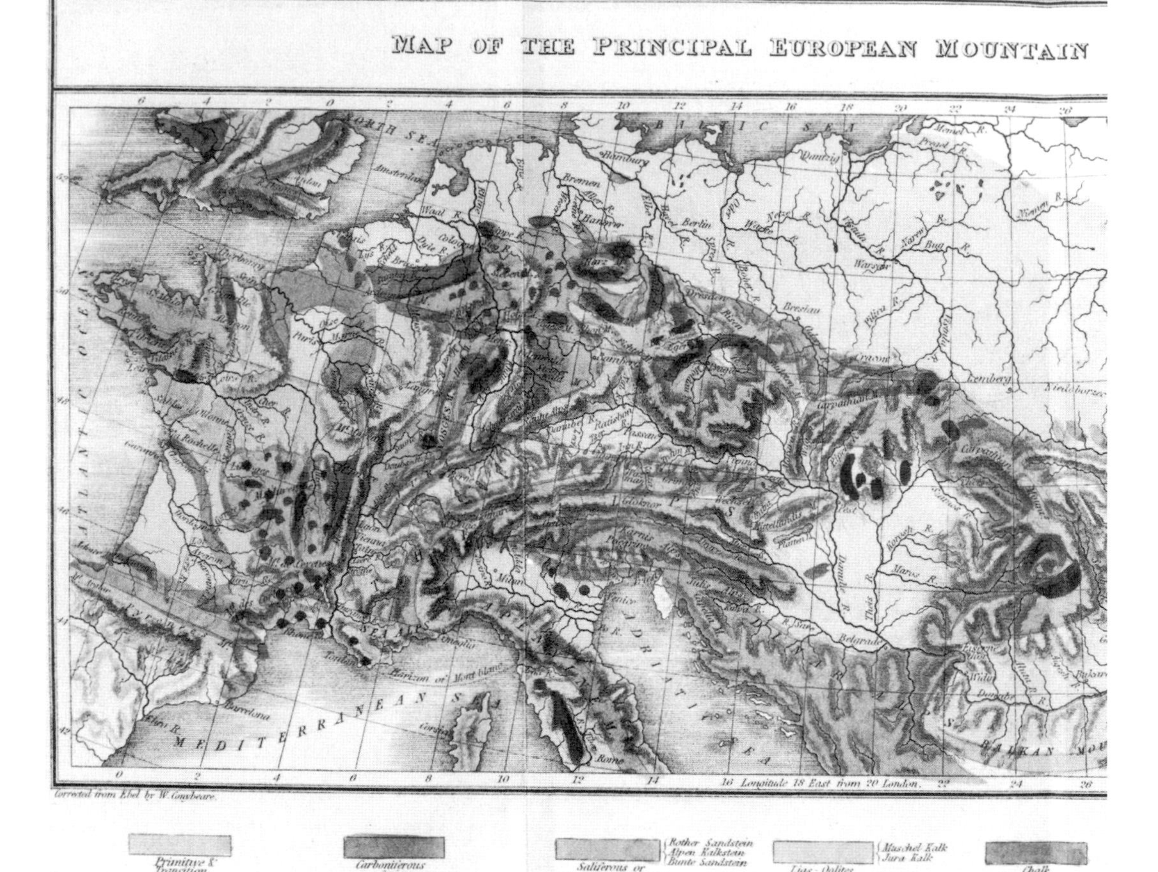

Fig. 3.3 Conybeare's "Map of the principal European mountain chains" (1823), compiled from the detailed geological surveys already available for most parts of Europe; only Spain and the more remote parts of Russia (mostly omitted here) had to be left blank. Even this uncolored reproduction shows (in the palest tone) the main areas of Tertiary and Superficial deposits: the Paris Basin and southwest France; London, East Anglia, and Hampshire; the plain of the Po in northern Italy; the Swiss plain; the Vienna and lower Danube basins; and a vast area stretching from the Low Countries through northern Germany and Poland into western Russia. Also clear are mountain chains such as the Alps, Pyrenees, Apennines, and Carpathians, mainly composed of "Primitive and Transition" rocks; but the key also distinguishes the Carboniferous or coal rocks, the Chalk, and two broad categories in between (spanning, in modern terms, roughly from Permian through Jurassic). Conybeare's native country, although treated by him and others as a useful standard of reference for the new stratigraphy, is here peripheral. (By permission of the Syndics of Cambridge University Library)

faulting, and were now well mapped; and they often contained abundant fossils, which were being assiduously collected by a small army of amateur naturalists and working fossilists.

The year after "Conybeare and Phillips" appeared, Conybeare—the senior author in geological credentials though not in years—published a sketch map of the geology of almost the whole of Europe, based on his wide reading of the massive body of published work already available. It epitomized the way that stratigraphy was already a fully international research project (Fig. 3.3).[8]

In this postwar era, then, stratigraphy was becoming more international than

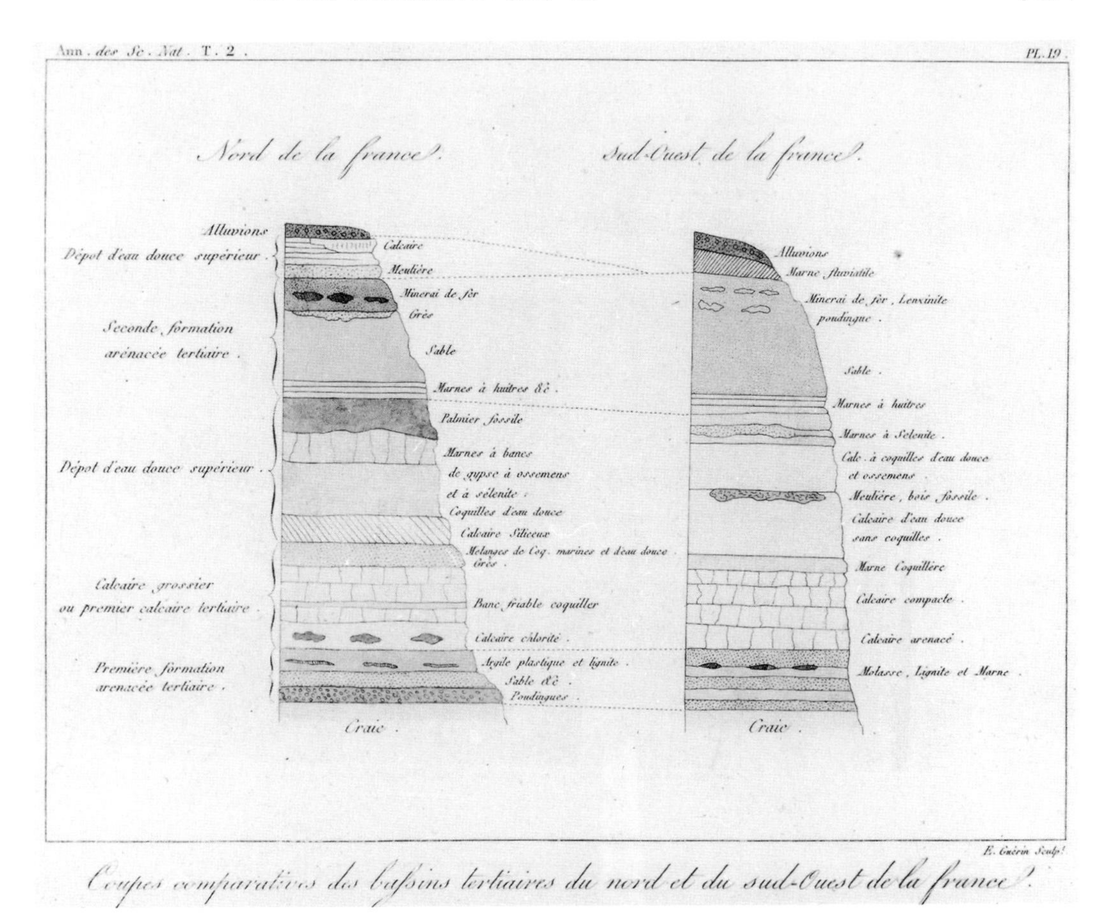

Fig. 3.4 Ami Boué's "comparative sections of the Tertiary basins of the north and southwest of France" (1824), showing his correlation between Brongniart's Parisian stratigraphy (left) and the comparable but different sequence in the southwest (right). The dotted lines linking the two sections indicate Boué's suggested correlations of the boundaries between five major formations (named on the left). Such detailed comparisons—and these innovative linking lines—soon became standard practice in stratigraphy. (By permission of the Syndics of Cambridge University Library)

ever, and not only from the English side of the Channel. Ami Boué (1794–1881) is a good francophone example of the trend. Boué had a cosmopolitan background that made him particularly well qualified to compare the geology of different regions right across Europe. The son of a Swiss businessman based in Berlin, Boué had studied first under Jameson in Edinburgh, and then, after the coming of peace, in Paris, Berlin, and Vienna. This made him an exemplary multilingual European; and when he came of age and inherited his father's assets he chose, like Humboldt, to settle in Paris and become a full-time savant. He first became known among geologists on the Continent for his book in French on the geol-

8. Fig. 3.3 is reproduced from Conybeare, "Geological map of Europe" (1823) [Cambridge-UL: Q340:1.c.32.21], part of pl. 19, omitting the eastern portion (Ukraine and the Black Sea) for which his information was scanty. Many fine geological maps were already available as a basis for such a compilation: see for example Keferstein, *General Charte von Teutschland* (1821), a small colored map covering much of central Europe, and, on a larger scale, those in Ebel, *Alpengebirge* (1808), and in Keferstein, *Teutschland geognostisch-geologisch dargestellt* 1 (1821).

ogy of Scotland (1820), and in Britain the same year for his comparison of the volcanic rocks of France and Scotland; the latter was published in Jameson's *Edinburgh Philosophical Journal*, which was becoming one of the leading scientific periodicals in English. Over the next few years Boué traveled extensively, examining the sequences of formations in the field and mastering the growing body of published material. His study of southwest France, for example, described its stratigraphy in detail, suggesting correlations with other and better known regions. For the Tertiary formations the correlation was not only verbal but also visual (Fig. 3.4).[9]

In 1825 Boué synthesized all his detailed local studies, offering a tabular summary of "the principal geognostic facts" about all the formations from Primary to Tertiary throughout Europe (and even, more sketchily, beyond it), listed in true *geohistorical* order. In effect, Boué was following Buckland's example, while boldly challenging him on specific points. His "synoptical table" was an astonishing achievement for a geologist still in his twenties; others quarreled with its details, but none could ignore it. And Boué chose to publish it in English, in Jameson's periodical. This not only highlighted his challenge to Buckland but was also a sign that he recognized the growing importance of British geologists on the international stage, particularly in this Smithian kind of stratigraphy.

Boué boasted in conventional style that he had drawn up his summary "taking nature, and not theoretical ideas, for the basis of my description". But in fact

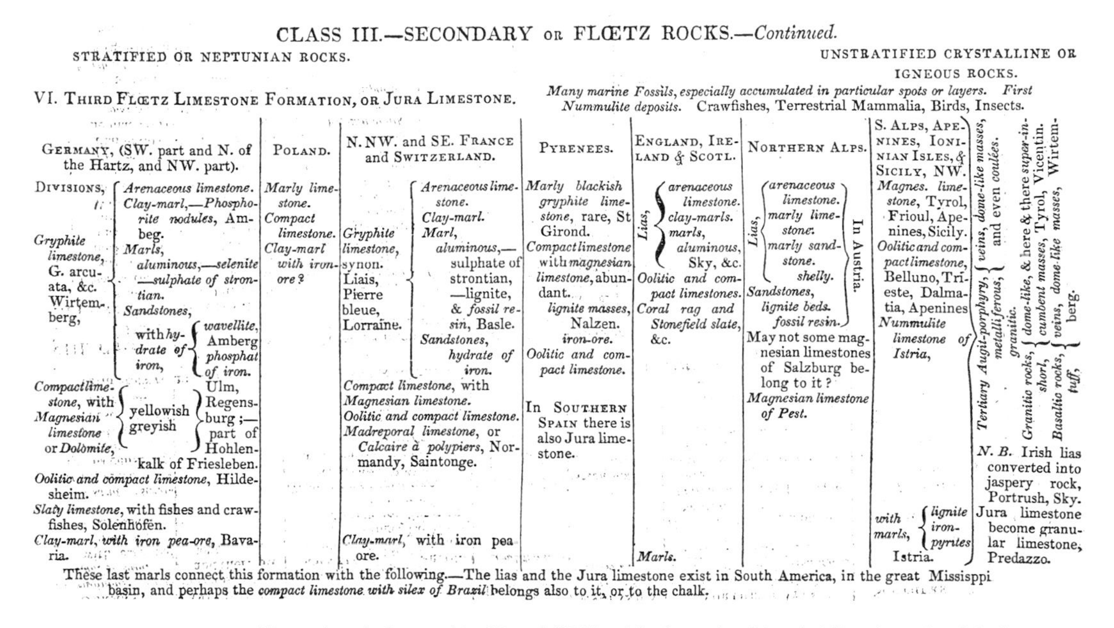

CLASS III.—SECONDARY OR FLŒTZ ROCKS.—*Continued.*

STRATIFIED OR NEPTUNIAN ROCKS. UNSTRATIFIED CRYSTALLINE OR IGNEOUS ROCKS.

VI. THIRD FLŒTZ LIMESTONE FORMATION, OR JURA LIMESTONE.

Many marine Fossils, especially accumulated in particular spots or layers. First Nummulite deposits. Crawfishes, Terrestrial Mammalia, Birds, Insects.

GERMANY, (SW. part and N. of the Hartz, and NW. part).	POLAND.	N. NW. and SE. FRANCE and SWITZERLAND.	PYRENEES.	ENGLAND, IRELAND & SCOTL.	NORTHERN ALPS.	S. ALPS, APENINES, IONIAN ISLES, & SICILY, NW.	
DIVISIONS, { *Arenaceous limestone.* *Clay-marl,—Phosphorite nodules,* Am-beg. *Marls, aluminous,—selenite —sulphate of strontian. Sandstones,* with *hydrate of iron,* { *wavellite,* Amberg *phosphat of iron.* *Gryphite limestone,* G. arcuata, &c. Wirtemberg, *Compact limestone,* with *Magnesian limestone* or *Dolomite,* { yellowish greyish } Ulm, Regensburg;—part of Hohlenkalk of Friesleben. *Oolitic and compact limestone,* Hildesheim. *Slaty limestone,* with fishes and crawfishes, Solenhofen. *Clay-marl, with iron pea-ore,* Bavaria.	*Marly limestone. Compact limestone. Clay-marl with iron-ore?*	*Gryphite limestone,* synon. Liais, Pierre bleue, Lorraine. { *Arenaceous limestone. Clay-marl. Marl, aluminous,*— sulphate of strontian, —lignite, & *fossil resin,* Basle. *Sandstones, hydrate of iron.* *Compact limestone,* with *Magnesian limestone.* *Oolitic and compact limestone.* *Madreporal limestone,* or *Calcaire à polypiers,* Normandy, Saintonge. *Clay-marl,* with iron pea ore.	*Marly blackish gryphite limestone,* rare, St Girond. *Compact limestone* with *magnesian limestone,* abundant. *lignite masses,* Nalzen. *iron-ore.* *Oolitic and compact limestone.* In SOUTHERN SPAIN there is also Jura limestone.	*Lias,* { *arenaceous limestone. clay-marls. marls, aluminous,* Sky, &c. *Oolitic and compact limestones.* *Coral rag and Stonefield slate,* &c. *Marls.*	*Lias,* { *arenaceous limestone. marly limestone. marly sandstone. shelly.* *Sandstones, lignite beds. fossil resin.* } In Austria. May not some magnesian limestones of Salzburg belong to it? *Magnesian limestone of Pest.*	*Magnes. limestone,* Tyrol, Frioul, Apenines, Sicily. *Oolitic and compact limestone,* Belluno, Trieste, Dalmatia, Apenines *Nummulite limestone of Istria,* *with marls,* { *lignite iron-pyrites* Istria.	*Tertiary Augit-porphyry,* { *veins, dome-like masses, metalliferous,* and even *coulées. granitic.* *Granitic rocks,* { *dome-like,* & here & there *super-incumbent masses,* Tyrol, Vicentin. *shorl,* *Basaltic rocks,* { *veins, dome-like masses,* Wirtemberg. *tuff,* *N. B.* Irish lias converted into jaspery rock, Portrush, Sky. Jura limestone become granular limestone, Predazzo.

These last marls connect this formation with the following.—The lias and the Jura limestone exist in South America, in the great Missisppi basin, and perhaps the *compact limestone with silex of Brazil* belongs also to it, or to the chalk.

Fig. 3.5 A typical page of Ami Boué's "Table of the formations" (1825): it lists the rocks of the "Jura Limestone" (in modern terms, the Jurassic system) throughout Europe. The sequences of sedimentary ("stratified or Neptunian") rocks are listed, in *geohistorical* order, for each of seven geographical regions: in the British sequence, for example, the Lias, the lowest and oldest formation in this group, is listed at the top, and the somewhat younger Stonesfield "slate" further down. The eighth and last column (right) lists the "unstratified crystalline or igneous rocks" judged to be of the same general age, and notes cases of sedimentary rocks changed (in modern terms, metamorphosed) by contact with them. At the foot of the table are brief notes on rocks possibly of the same age in the Americas. (By permission of the Syndics of Cambridge University Library)

one of its innovations embodied an important tacit claim for geohistory, and indeed for causal theorizing. In each category, from Primaries through to the youngest of the Tertiaries, the "Stratified or Neptunian rocks" (in modern terms, sedimentary formations) were listed in parallel columns for different regions. But in another column he listed the "Unstratified crystalline or Igneous rocks" that appeared to have been formed at the *same* periods. This made explicit what other geologists had long accepted implicitly, namely that volcanic rocks, and even perhaps rocks such as granite, might have been formed at *any* period in the earth's history, not just in the remote era of the Primaries: volcanic activity, and even the formation of "igneous" rocks at high temperatures in the depths of the earth, might have been constant features of geohistory, or at least repeated events, no less than the formation of layered sedimentary rocks at its surface (Fig. 3.5).[10]

3.4 CONCLUSION

Smith's stratigraphical method, as embodied in the geological map of England and Wales that he published right at the start of the postwar era, rapidly became a routine practice among geologists everywhere. However, as Humboldt's great synthesis showed, stratigraphy was just a variant—enriched by the use of fossils—of the much older and already well-established science of geognosy, which focused not only on the sequential order of the "regular" formations, but more generally on the spatial relations of rock masses of all kinds. All such work was primarily structural in orientation; questions of causal origin and geohistorical reconstruction were either disregarded, or treated as secondary and derivative from what was directly observable.

Smith's map was soon superseded for most purposes by Greenough's rival compilation, which incorporated the detailed local work of many others in the Geological Society. But the practice of using "characteristic" fossils to help distinguish otherwise similar rock formations was increasingly adopted, wherever possible; that is, it was used in the study of the "regular" Secondary and Tertiary formations, many of which contained abundant fossils. This was precisely the scope of the handbook soon dubbed "Conybeare and Phillips", which summarized current knowledge of the sequence of Secondary and Tertiary formations in England and Wales. Conybeare and his friend Buckland helped ensure that

9. Fig. 3.4 is reproduced from Boué, "Sud-ouest de la France" (1824) [Cambridge-UL: T382.b.8.1], pl. 19; "Comparative view" (1823), an abridged translation (without illustrations), had already been published in Jameson's *Journal*; the paper subsequently became famous for initiating the modern concept of regional metamorphism (see §27.4). See also Boué, *Essai géologique sur l'Écosse* (1820) and "Comparison of the volcanic rocks" (1820). Laurent, "Ami Boué" (1993), gives biographical information.

10. Fig. 3.5 is reproduced from Boué, "Table of the formations" (1825) [Cambridge-UL: Q340:1.c.6.13], 140; on this sample page, his inclusion of "first nummulite deposits" suggests that—in modern terms—some Cenozoic (Tertiary) formations were being misidentified as Jurassic. Boué, "Synoptische Darstellung" (1827), was a later and much fuller version in German, broadening still further the impact of his correlations; similar tables had already appeared in Keferstein, *Vergleichende Geognosie* (1825). Nitsch, "Keuper, 1820–34" (1996), is a fine study of the history of stratigraphical research on one specific formation, the Keuper being a major part of what in 1834 was first named the "Trias" (in turn representing, in modern terms, the early part of the Mesozoic era: see §36.2).

the British sequence, which was already well known and seemed to be exceptionally complete, came to be treated as a useful reference standard for the whole of Europe and even, more tentatively, for the rest of the world. And younger geologists such as Boué were already extending such work by proposing correlations between the sequences in all parts of Europe.

Yet all this was still stratigraphy, or enriched geognosy; it was not geohistory. It was confined to the description of three-dimensional rock masses or formations, to the elucidation of their spatial or structural relations in some kind of solid sequence, and to their correlation from one region to another by means of fossils and other criteria. It was not primarily concerned with the causal origins of the rocks and the fossils they contained, nor with the sequence of conditions and events that might have led to their formation and preservation. The next chapter describes work that did use this new stratigraphy, in the same years, as a foundation on which to build just such geohistorical inferences.

CHAPTER 4

Outlines of life's history (1818–27)

4.1 "PALEONTOLOGY" DEFINED

During the first few postwar years, the rapid adoption of William Smith's fieldwork methods, by geologists throughout the scientific world, transformed much of the older geognosy into the newer practice of stratigraphy. A close attention to fossils enriched geognosy by making it possible to trace much the same sequence of Secondary formations throughout Europe and, more provisionally, eastwards into Russia and westwards across the Atlantic to North America. Even the fossils sent back from still more remote parts of the world could often be assigned to the same sequence. Generally, however, this new stratigraphy remained on the same *structural* level as the geognosy from which it had developed: fossils were used primarily to enrich geognosy by clarifying the sequence of three-dimensional rock masses or formations, and their correlations from one region to another. Of course, geologists were well aware that this spatial sequence of rocks also represented a sequence of events in time; but the temporal dimension usually remained in the background, the subject of an occasional passing remark about which rocks were older or younger, but often not even that. Only a few geologists were beginning to enrich their stratigraphy still further, or in a double sense, by using fossils to give their work a primarily *geohistorical* focus. Rather than being treated just as objects useful for recognizing specific formations, fossils could then be imagined in the mind's eye as fragmentary traces of organisms that had truly been alive at specific periods of geohistory: Cuvier rather than Smith could be taken as the model.

However, the relative novelty of the Cuvierian approach was reflected in the lack of any general term to denote the scientific study of these relics of past life. Not by coincidence, the gap was filled at just this time by the editor of what was still one of Europe's most important scientific periodicals. The Parisian monthly *Journal de Physique*, which has been mentioned already, had a distinguished record stretching well back into the eighteenth century. Its pages had seen the emergence of the scientific research paper in its modern form; but it was not tied to any specific institution and was open to authors of all nationalities; and its editor's annual reviews of all the natural sciences had provided savants with useful summaries of the latest research (*BLT* §1.3).[1]

In the postwar world the *Journal de Physique* remained influential, though its position was being challenged by an increasing range of similar periodicals throughout Europe. When the chemist Jean-Claude de La Métherie, its editor since well before the Revolution, died in 1817, his assistant Henri Marie Ducrotay de Blainville (1777–1850), a comparative anatomist and one of Cuvier's former students, had taken his place. He continued La Métherie's tradition of reviewing annually the latest research in all the natural sciences. In his first such essay, Blainville coined the word "*paléozoologie*" to characterize the study of fossil animals, which his mentor's *Fossil Bones* had made well known and in which he himself was working. A more inclusive term was desirable, however, to cover the work of other naturalists who were giving increasing attention to fossil plants (see §4.4). So Blainville proposed "*paléosomiologie*", to denote the science of the ancient "bodies" of all kinds of organisms (in French, *corps organisés*). This cumbersome word evidently gained no positive welcome from other naturalists, for in 1822 he replaced it with "*paléontologie*", the science of ancient "beings", the organisms themselves; and this was the neologism that was soon adopted by naturalists everywhere. Its English form can therefore be used from this point in the present narrative, without jarring anachronism.[2]

4.2 LIFE'S OWN HISTORY

Blainville's new word reflected the growing value of fossils in their own right, as evidence for the history of life itself. Rather than being merely ancillary aids, as they were in Smithian stratigraphy, fossils could become the focus of attention in a new way, because stratigraphy could be used to assign them relative dates in geohistory. In the cognitive relation between fossils and rocks, the tables could be turned.

An agreed and reliable framework of stratigraphy, as summarized for example in Buckland's and Boué's correlations, enabled earlier speculations about the broader features of the history of life to be set on much firmer foundations. The study of vertebrate animals had already led the way. Cuvier, with his usual strategic sense, had recognized that terrestrial vertebrates—his own favorite subjects—were the most effective for this purpose (*BLT* §9.3). Early in his research on fossil bones, he had boldly guessed from the fossil record, as then understood, that reptiles had preceded mammals and that non-human mammals had preceded humans. As already recalled (§1.2), that hunch had been confirmed by more than two decades of further research, clarified above all by the results of the new stratigraphy. The human species still appeared to have no authentic fossil record at all. Other mammals were limited to the Superficial deposits and the younger formations that were now defined as Tertiaries. Reptiles extended—in newly discovered diversity—well back into the true (pre-Tertiary) Secondary formations, most notably in the Lias of England and its Continental equivalents (§2.1, §2.3), but also as far back as the German *Kupferschiefer* or copper shale (in modern terms, of Permian age). Since reptiles were, of course, important components of *living* faunas, alongside mammals, this suggested that vertebrate life had become progressively more diverse in the course of geohistory, by the successive addition of animals with arguably "higher" kinds of organization. It implied an overall directionality, or even progress in the history of the quadrupeds, which might also apply to the history of life as a whole.

The fossil record of other organisms, however, was more difficult to interpret, either because they were usually preserved in a highly fragmentary condition (as in the case of plants), or because they might still be flourishing in remote places and not be extinct at all (as in the case of marine invertebrates). The "living fossil" explanation (as I have termed the latter argument: *BLT* §5.2) had not been disproved; indeed it could not be, since it depended on negative evidence. But it had been tacitly sidelined, and with good reason. In the 1820s it would have been no great surprise if some further "living fossils", like the Australian *Trigonia* shells that had so delighted Lamarck many years earlier (*BLT* §8.3), had continued to turn up occasionally in the course of expeditions and voyages. But naturalists no longer expected to be able to conclude—as Lamarck and others around the turn of the century had confidently claimed—that no organisms had ever gone extinct (unless by human agency), or that all fossil species might still be alive somewhere on earth (unless in the meantime they had been transformed, or in modern terms evolved, into something else).

During the early years of peace in Europe, the massive expansion of research on fossils of all kinds, combined with the equally rapid development of stratigraphy, confirmed earlier hunches about the directional character of the fossil record—and not just that of the quadrupeds—suggesting that it was a broadly reliable record of the history of life as a whole. It now seemed almost certain, for example, that major groups of apparently marine fossils, such as the distinctive *ammonites*, were completely extinct: the intense exploration of the globe, by all the scientific nations, consistently failed to discover any trace of a single living species that could be assigned to this group of mollusks (Fig. 4.1).[3]

1. At the height of the Revolution, *Observations sur la physique*, as it was then called, had briefly suspended publication, and then reappeared with its new title; but its format and broad scope—covering all the natural sciences and the related technologies—were unchanged (*BLT* §6.5).

2. Blainville, "Resumé" (1818), 71; "Analyse" (1820), 80; "Analyse" (1822), liv. In the German-speaking world the terminological gap had already been recognized, as shown for example by the title of Schlotheim, *Petrefaktenkunde* [fossil science] (1820), a wide-ranging survey of fossils of all kinds; but it was the French word that eventually became the international standard, even in German. The *Journal de physique* ceased publication in 1823, perhaps as a result of increasing competition from other and newer scientific periodicals in Paris or because Blainville was not an effective editor; its final issues were rather thin by comparison with its distinguished past. Blainville's penchant for neologisms can be judged from his use of "*bélemnitologie*" in his monograph on those fossils (see §4.3).

3. Fig. 4.1 is reproduced from Sowerby, *Mineral conchology* 2 (1818) [Cambridge-UL: S365.b.81.2], pl. 130, described on 69; although the fossil was 21 inches (54cm) in diameter, the size of the central hole must, sadly, throw some doubt on the Buckland anecdote. The next two plates (pls. 131, 132) depicted other ammonite species named in honor of Conybeare and Greenough.

Fig. 4.1 A colored engraving of *Ammonites Bucklandi*, a large ammonite species from the Lias formation in England, named by James Sowerby in honor of William Buckland and published in 1818 in Sowerby's long-running series of illustrations of fossils. The inner (and therefore juvenile) whorls of the plane-spiral shell were missing in this specimen, giving it a hollow center. According to Sowerby, when it was first collected Buckland had carried it from the field on horseback, slung around his neck, causing his companions—with the punning humor so popular at the time—to dub him an "Ammon-Knight". The repeated wavy lines running across the whorls are the traces of the frilly edges of the shelly "septa" or partitions across the original shell. Although analogous to those separating the gas-filled buoyancy chambers of the living pearly nautilus, the far more complex septa of ammonites marked them out as a distinct group, which had been varied and prolific in the era of the Secondary formations but appeared to be wholly extinct. (By permission of the Syndics of Cambridge University Library)

This growing confidence in the fossil record as a reliable trace of the true history of life was reflected in changing forms of publication. There had earlier been broad inventories of fossils of all kinds, for example Knorr and Walch's great German paper museum (*BLT* §2.1) and Parkinson's later British volumes (*BLT* §8.2). There had also been many local monographs, such as Burtin's description of all the fossils found around Brussels (*BLT* §4.2) and Volta's magnificent paper museum of fossil fish from Monte Bolca in the Alpine foothills behind Verona (*BLT* §5.2). But these traditional kinds of publication were beginning to be supplemented by two novel genres: descriptions of all the fossils in a specific rock formation, wherever it might be found; and descriptions of all the fossils of a particular kind, in whatever formation they were found.[4]

In the present context the latter genre, focusing on a particular group of fossil organisms, is the more significant. Cuvier's *Fossil Bones* was of course an early example, and a massive and inspiring one at that, but in the 1820s its new and enlarged edition was joined by several analogous studies by other naturalists. A few will be described briefly in the rest of this chapter.

4.3 THE LIFE OF ANCIENT SEAS

One of the earliest, unsurprisingly, was by Cuvier's collaborator Brongniart. His research dealt with the *trilobites*, distinctive fossils that were particularly attractive to amateur collectors. Brongniart helped a younger and less well-established naturalist, Anselme-Gaëtan Desmarest (1784–1838), by combining his own work on trilobites with the latter's parallel account of other fossil "crustaceans". Brongniart claimed explicitly that their pair of monographs was innovative in having both zoological and geological dimensions. Zoologically, Desmarest's fossils were recognizably similar to living crabs and lobsters, whereas Brongniart's trilobites were much more peculiar in appearance. And the geological relations of the two groups illustrated the same effect that Cuvier had claimed for his quadrupeds. Like them the "crustaceans" became increasingly strange the further back in time they were traced: Desmarest's familiar crabs and lobsters came from the Tertiary and Secondary formations, whereas most of Brongniart's strange trilobites were from the still older Transition rocks.[5]

Werner had inserted the "Transition" category between the Secondaries and the Primaries, to allow for the recognition of ancient rocks (mostly slates and "greywacke") that did contain a few fossils, the very earliest known. Even with the more recent development of stratigraphy, however, the Transition rocks remained obscure, because their fossils were usually rare and poorly preserved, the rocks often highly folded, and the formations therefore difficult to unravel. But there were exceptions. In a few localities, rocks that underlay and were clearly older than the Carboniferous (as Conybeare called it) and therefore by definition pre-Secondary, were relatively undisturbed and yielded fossils that were as abundant and well preserved as those in much younger formations. Such were the trilobites that Brongniart had described at the Institut (soon to be refounded as the Académie), back in the first months of peace. But the preparation of the illustrations that his fine specimens required and deserved delayed his publication by several years. However, it was worth waiting: his trilobites established the distinctive character of the geohistorical period represented by the Transition rocks, as surely as the better known ammonites helped define that of the subsequent Secondaries (Fig. 4.2).[6]

A second example of this kind of detailed study of a particular group of fossil invertebrates was the work of the naturalist John Samuel Miller (1779–1830). Born Johann Müller, he was a native of Danzig in Prussia (now Gdansk in Poland) but had settled in Bristol in the west of England, where he worked as an accountant. He amassed, or had access to, superb fossil collections of what he called "*Crinoidea*", a Classical adaptation of their vernacular English name

4. Both new genres demanded the internationalism that continued to characterize the science: it would have made little sense, for example, to describe all the fossils found in the Chalk in England while turning a blind eye to those in the very same formation just across the Channel in France (exposed respectively in the white cliffs of Dover and the Pas de Calais, each visible from the other on a clear day).

5. Brongniart and Desmarest, *Crustacés fossiles* (1822), divided into Brongniart's "Trilobites" and Desmarest's "Crustacés proprement dits": as that phrase shows, the main title used the word crustaceans in a broader sense than the modern one, and Brongniart was well aware that his trilobites were significantly different from Desmarest's crabs and lobsters, and indeed from any other living arthropods. Anselme-Gaëtan was the son of the much more distinguished naturalist Nicolas Desmarest (1725–1815), who had been famous for his pioneer geohistorical research on the extinct volcanoes of central France (*BLT* §4.3), and whose long life had ended just as the peace began.

6. Fig. 4.2 is reproduced from Brongniart and Desmarest, *Crustacés fossiles* (1822) [London-BL: 443.f.19], part of pl. 1. Brongniart's portion of the book (1–65, pls. 1–4) was based on a paper read at the Institut in October 1815; as he noted, the delay in publication had allowed him to include many more specimens, particularly those in collections that had been inaccessible to him during wartime. For example, some of the trilobites illustrated here came from the famous Wenlock Limestone locality at Dudley near Birmingham (in modern terms, of Silurian age: see §30.3). A few trilobites were known from Carboniferous formations, but many more from the underlying Transitions.

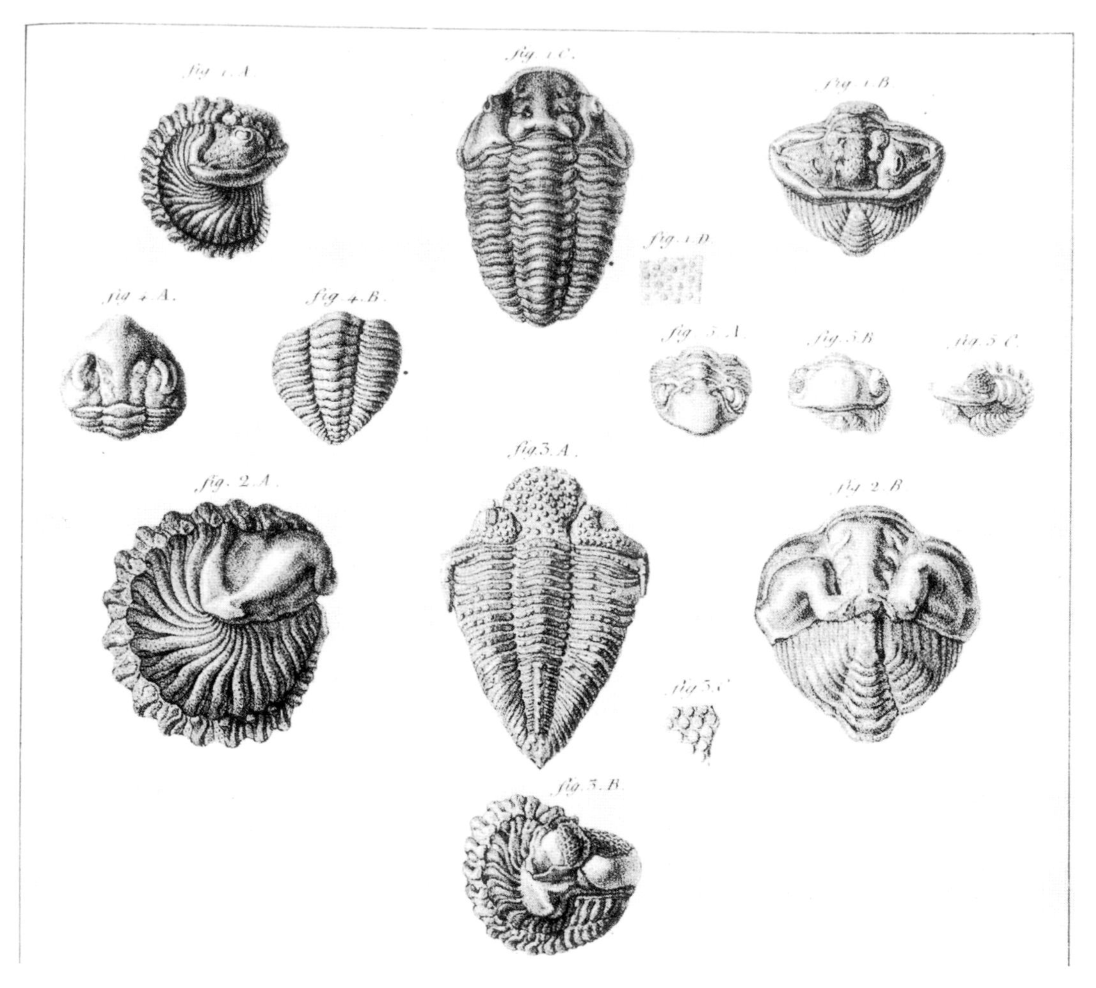

Fig. 4.2 Fossil trilobites of four species of the genus *Calymene*, as illustrated in Brongniart's monograph on the group (1822); he claimed that trilobites collectively characterized the Transition formations, representing the earliest known periods in the history of life. Some of the trilobites shown here came from a well-known fossil locality in the English Midlands, some had been sent to the Muséum as drawings of specimens in English collections, and one had been brought to Paris from the United States; the geographical spread reflected the internationalism that Brongniart took for granted in his research. One species (top row, fig. 1) had been described by, and was named after, Blumenbach of Göttingen (see below). All these specimens showed that *Calymene* had had large compound eyes and had been able to curl into a ball—presumably for protection against predators—in a way that was strikingly analogous (on quite different scales) to living woodlice and armadillos. (By permission of the British Library)

"sea lilies". Although the crinoids were somewhat plantlike in form, naturalists had long recognized that their striking fivefold symmetry and plated structure aligned them with the much better-known sea-urchins, brittle-stars, and starfish (all of which were soon to be given their modern collective name as the *echinoderms*). Crinoids were well known as a spectacular example of the "living fossil" effect. The fossils were quite common, and some Secondary limestones were largely composed of their fragmentary remains. But crinoids were still alive in the present world, for extremely rare specimens were brought up occasionally from great depths at sea (*BLT* §5.2). However, the far wider variety of fossil forms suggested that the great majority were extinct. Miller described both living and fossil species in detail, illustrating his specimens with his own drawings reproduced by lithography (a new technique that was as yet rather crude in effect but far less expensive and time-consuming than the usual engravings on copper). Although Miller was a provincial naturalist, he was well known to the leading metropolitan geologists: his handsome volume, published in Bristol at his own expense, counted among its subscribers such prominent figures as Buckland, Conybeare, De la Beche, and Greenough, and also foreigners such as the great Göttingen naturalist Johann Friedrich Blumenbach (1752–1840). Like Brongniart, Miller recognized that his work was relevant to both zoology and geology, for he dedicated it tactfully to the members of both the Linnean and the Geological Societies in London.[7]

Encouraged by the success of his study of crinoids, Miller then tackled another distinctive group of Secondary fossils, the *belemnites*. These solid bullet-shaped objects, with a crystalline internal structure, had been among the most puzzling of all fossils, and their organic origin had only been established when, in the course of the eighteenth century, rare but better-preserved specimens were found. It then became clear that they were simply the most robust and readily fossilized part of some peculiar mollusks with chambered shells, roughly intermediate in structure between the pearly nautilus and the cuttlefish, both of which belonged to the group of living mollusks that Cuvier had defined as "*cephalopods*". In a paper read at the Geological Society in 1823, Miller described the structure of belemnites in exemplary fashion. He tentatively reconstructed the whole animal, taking the living cuttlefish as a model for its likely anatomy. He showed that the solid belemnite itself must have been like the "bone" of a cuttlefish, embedded in the animal's body, not an external shell like that of the nautilus. And he suggested how it might have functioned to counterbalance the buoyancy provided by the putatively gas-filled shelly chambers, to make possible a free-swimming mode of life much like that of the cuttlefish. And Miller's Cuvierian resurrection was of animals that had evidently flourished only during a specific part of the Secondary era. He claimed that a variety of species had existed from the time of the Lias through that of the Chalk, but he knew of no trace of belemnites either before or after that specific span of deep time (Fig. 4.3).[8]

It was a sign of the accelerating pace of research in "paleontology", and its international character, that much of Miller's work was unknowingly duplicated by the author of that neologism. Blainville had almost completed his own study of belemnites before Miller's paper reached Paris; there had been the usual long

7. Miller, *Crinoidea* (1821); his drawings included innovative "exploded" diagrams of the detailed plate structure of some of his species. Lithography had been invented in Germany at the turn of the century, and was used initially for the inexpensive reproduction of musical scores: see Twyman, *Lithography* (1970). It was only around 1820 that in an improved form it began to be exploited for purposes of natural history, Miller's book being one of the earliest examples.

8. Fig. 4.3 is reproduced from Miller, "Observations on belemnites" (1826) [Cambridge-UL: Q365.b.12.7], pl. 9, read at the Geological Society on 4 April 1823, when it was presented by his Bristol neighbor Conybeare, since Miller himself was not a member; in the same year, and again through Conybeare's patronage, his competence as a naturalist was rewarded by appointment as curator of the museum at the Bristol Institution: Neve, "Science in a commercial city" (1983). Miller's functional reconstruction was quite similar to the modern interpretation of these extinct pelagic mollusks.

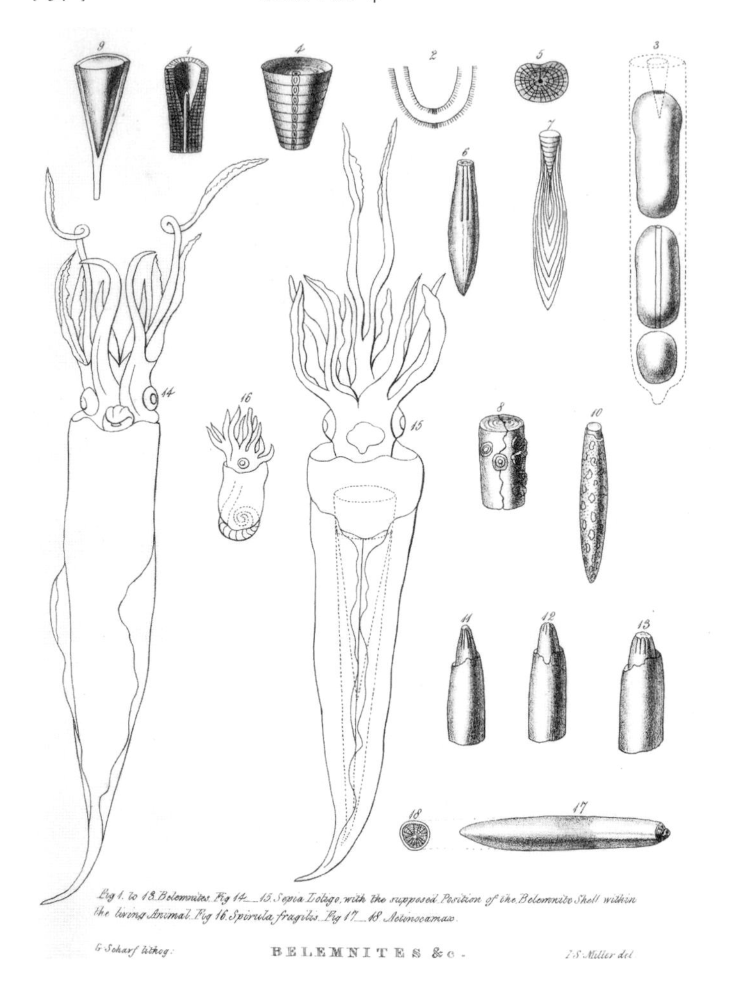

Fig. 4.3 John Miller's reconstruction of the functional anatomy of a belemnite, one of an abundant group of fossil animals that had evidently flourished during a long period of geohistory, but which appeared to have been completely extinct since the end of the Secondary era. Miller used the living cuttlefish *Sepia* (fig. 14) as a model for a tentative reconstruction of the fossil animal (fig. 15), with the conical belemnite (dotted) enclosed by the animal's tissues; the living *Spirula* (fig. 16), with its half-internal spiral shell of gas-filled chambers, suggested another partial analogue. He conjectured that the solid part of the belemnite could have provided a counterweight to the head and tentacles, between which the putatively gas-filled shelly chambers could have provided buoyancy and enabled the animal to maintain a horizontal position. Among other drawings that fill the space on the plate, two belemnites (figs. 6, 17) illustrate the variations in shape and size that distinguish different species; sections across them (figs. 5, 18) show the concentric rings that record the growth stages; a section along the length of one of them (fig. 7) also shows the apex of the conical series of chambers, which are seen more clearly in another specimen (fig. 4). Other specimens illustrate problems of preservation that had misled earlier naturalists. (By permission of the Syndics of Cambridge University Library)

delay between the reading of the paper in London and its publication in the Geological Society's *Transactions*. Blainville published his own work in Paris the following year, acknowledging Miller's priority. Like Brongniart's monograph on trilobites, the full title of Blainville's made explicit reference to its twofold character: belemnites would be "considered zoologically and geologically". On the anatomy and mode of life of the animals, Blainville had come to much the same conclusions as Miller. But his study was more detailed, and he described some fifty different species, many of them confined to particular formations and therefore potentially useful for Smithian correlation (see Fig. 17.2). He also extended their total stratigraphical range: he knew of belemnites from as far down in the Secondary sequence as the German *Muschelkalk* (in modern terms a limestone of Triassic age), which was not represented among the British formations. Translated into geohistorical terms, belemnites had flourished for a somewhat longer period than Miller supposed, but they still seemed to have become extinct by the end of the Secondary era. Blainville's work thus confirmed and extended Miller's; jointly, their research demonstrated that invertebrate animal life had indeed had a *history*, and one that could be known with confidence and in some detail.[9]

4.4 ANCIENT PLANT LIFE

The term "paleontology" had in effect become necessary, as already mentioned, because the study of fossil plants was beginning to emerge from its somewhat Cinderella-like earlier status. Cuvier and his fossil bones hardly constituted an Ugly Sister, but his and others' studies in paleozoology had certainly held center stage. It was difficult to emulate Cuvier with fossil plants, because their leaves, stems, and roots were usually preserved in isolation and in a fragmentary state, and could not be reassembled with any confidence; and the reproductive organs (flowers, fruits, and seeds) that were most important in the Linnaean classification of living plants were extremely rare as fossils (*BLT* §5.3). Nonetheless, fossil botany was, in effect, a field just waiting to be cultivated.

Twenty years earlier, the civil servant (and part-time naturalist) Ernst Friedrich von Schlotheim (1764–1832) of Gotha had published a preliminary study of the flora of the "primitive world" [*Urwelt*], or more precisely of the Coal formation (*BLT* §8.1). But he had not followed it up with the promised fuller account, though his massive *Petrefaktenkunde* (1820), with systematic descriptions of fossils of all kinds, kept his name in the forefront of the science. Schlotheim's near contemporary, count Kaspar Maria von Sternberg (1761–1838), an Austrian naturalist in Prague, had ambitions to fill the gap: he had begun to publish a major illustrated work on the flora of the "former world" [*Vorwelt*]—the title echoed Schlotheim's key word—emphasizing its double reference by calling it a "geognostic-botanical" study. But large and expensive monographs rarely made rapid progress, and this one was no exception.[10]

Alexandre Brongniart's son Adolphe Théodore (1801–76) therefore saw an opportunity to make his name as a naturalist (he hedged his bets by also qualifying as a physician). In focusing on what would later be termed paleobotany, he had paternal encouragement, and doubtless influence too. On his twenty-first birthday he made his debut at the Académie with a precocious paper "On the classification and [stratigraphical] distribution of fossil plants in general". Following his father's lead, he stressed the need to study the fossils under the "*double rapport*" of both botany and geology; but for the latter he used unambiguously geohistorical language, making it a matter of "epochs" in nature's history. He focused on the plants of the Tertiary formations, the nearest to the present both in botanical character and in geohistory: emulating his father's research (*BLT* §9.6) he recognized that the Tertiaries were the gateway to the deeper past and a good place to start. But he did also describe the more puzzling fossils from the far older Coal formation. He conceded that they might indicate a formerly tropical climate, but pointed out their differences from any living tropical flora; similar fossils found in coal mines in India and Australia suggested there might have

9. Blainville, *Bélemnites* (1827). His history of earlier *bélemnitologie* (2–25), as he called this special branch of *paléontologie*, remains, even today, a useful source.

10. Sternberg, *Flora der Vorwelt* (1820–38). Schlotheim, *Petrefaktenkunde* (1820), covered the entire range of fossils in his vast collection, though all its illustrations were of fossil plants, not animals. Cleal *et al.*, "'Golden Age' of palaeobotany" (2005), describes their illustrations and those of the younger Brongniart (see below).

been a global climate far more uniform than in the present world (these mines were just beginning to become strategically important for fueling the worldwide voyages of the new steamships). But for the time being these were no more than hints of a possible geohistorical interpretation of fossil plants.[11]

Adolphe Brongniart's approach to fossil plants was directly analogous to Cuvier's exemplary work on fossil quadrupeds: it was primarily an indoor study of museum specimens, aimed at classifying them and reconstructing their place in geohistory. But it could be complemented by field studies that might serve to reconstruct their mode of life and, in modern terms, their paleoecology. Alexandre Brongniart had in effect already done just that, for the case of freshwater and marine mollusks, in his collaboration with Cuvier on the Tertiary formations of the Paris Basin (*BLT* §9.1). Even before Adolphe's debut as a paleobotanist, Alexandre made at least a small contribution towards a similar geohistorical reconstruction of the flora of the Coal formation, suggesting in effect how his son's research might develop.

The origin of coal had long been a matter of controversy. That its substance was derived from ancient land plants was not in doubt. But it was quite uncer-

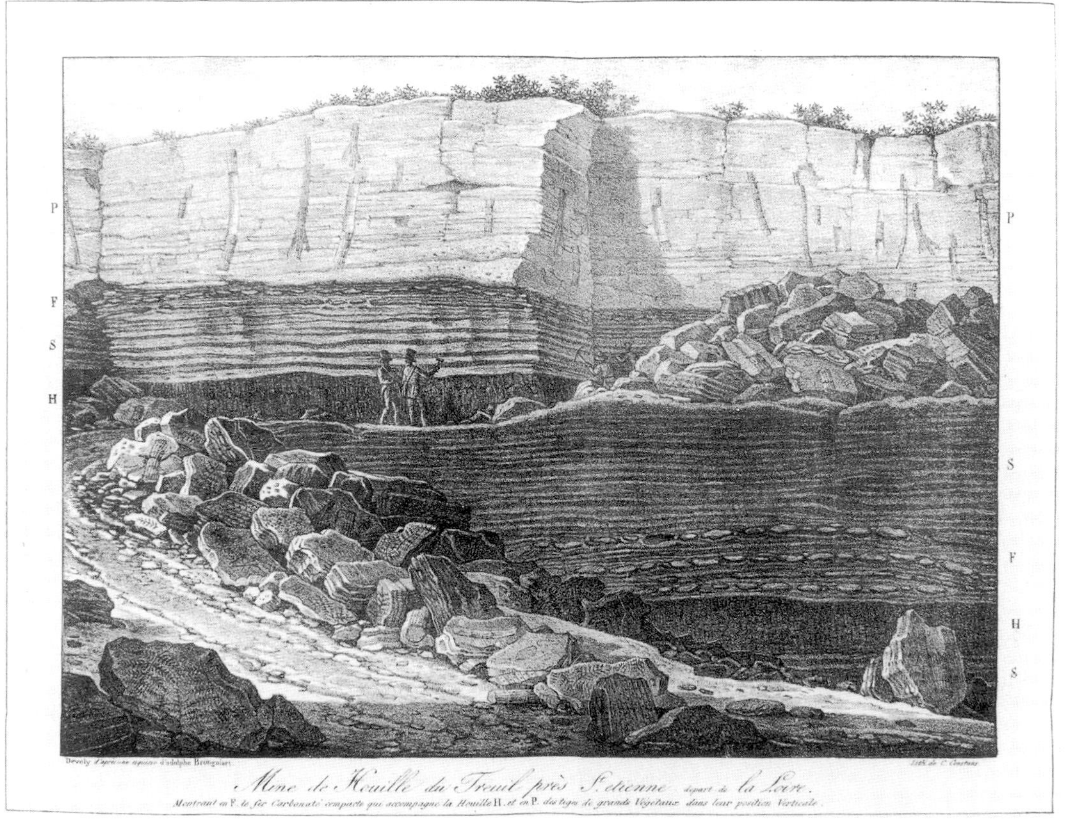

Fig. 4.4 Tree trunks preserved upright in strata overlying a coal seam (the quarrymen show the scale): Adolphe Brongniart's sketch of an opencast coal mine near Saint Étienne in central France, reproduced as a lithograph to illustrate a paper published by his father Alexandre in 1821. The latter concluded that "this is a true fossil forest of monocotyledonous plants, like bamboos or giant *Equisetum* [horsetails], petrified in situ". But the case was inconclusive, because the trunks appeared to be broken off at their base, and the sandstone in which they were embedded (P) was separated from the coal seam (H) by some two meters of other strata (F, S). (By permission of the Syndics of Cambridge University Library)

tain whether the material had been deposited on the seafloor after drifting out to sea, or whether coal seams represented the products of plant decay on land (or at least in freshwater swamps of some kind). The attractions of the standard model of geotheory (*BLT* §3.5), which implied that all sedimentary formations were of marine origin, inclined geologists to favor the former interpretation. But Brongniart had already rejected that presumption, by reconstructing the geohistory of the Paris Basin (with Cuvier's encouragement) as a story of repeated alternations between marine and freshwater conditions. This implied that the region had for long periods been at or above sea level. During the quite different (and far earlier) period represented by the Coal formation, similar conditions might therefore have generated swampy forests and future coal seams. So Brongniart, while doing fieldwork with his son in central France, was not surprised to find a striking new case of tree trunks preserved *upright* in strata overlying a coal seam: to him they suggested that an ancient forest had indeed been buried in situ. Though the case was somewhat ambiguous, it did in his opinion invalidate the earlier inference that Coal plants had somehow been swept all the way from the tropics. If the plants had indeed been tropical in habitat, then the climate of Europe at that time must also have been tropical (Fig. 4.4).[12]

4.5 CONCLUSION

Blainville's new word "paleontology" signaled the growing attention being given to fossils in their own right, rather than just as useful tools in stratigraphy. Following Cuvier's example rather than Smith's, fossils could be reconstituted as organisms that had once been truly alive. They then became evidence for the history of life itself, which in turn was an important component of geohistory as a whole. In the early 1820s Cuvier's enlarged new edition of *Fossil Bones* provided the outstanding exemplar that other naturalists could and did emulate. His early collaborator Brongniart, for example, made a special study of trilobites, distinctive fossils that were almost confined to the Transition formations. Thus they represented the oldest era in the known history of life, and were ancient relatives of the more recent fossil crustaceans described in the same volume by the younger Desmarest.

Miller focused on the "sea lilies", or what he called crinoids, which were almost confined to the Secondary formations that overlay—and were therefore younger

11. Brongniart (Ad.), "Végétaux fossiles" (1822), read at the Académie des Sciences on 14 January; on the Coal flora, see 333–46; he was scornful about Schlotheim but tactfully acknowledged the value of Sternberg's work. Launay, *Les Brongniart* (1940), although focused on Alexandre, is a valuable source on all the family.

12. Fig. 4.4 is reproduced from Brongniart (Alex.), "Végétaux fossiles" (1821) [Cambridge-UL: CP432.c.16.44], part of pl. 3, based on a sketch by Adolphe Brongniart and lithographed by C. Constans; quotation on 362. The site of the quarry ("Mine de houille du Treuil") now lies within the industrial city of Saint Étienne (Loire). The frontispiece of Sternberg, *Flora der Vorwelt* (1820–38), showed a gallery in an underground coal mine, with fossil trunks preserved upright in the strata in much the same way; but the Brongniarts' proxy, published in the widely distributed *Annales des mines*, made the phenomenon better known and more persuasive. De la Beche, *Geological memoirs* (1824), included this paper (with a redrawn version of this plate) among those that he translated from the *Annales des mines* for the benefit of his linguistically challenged compatriots.

than—the Transition formations. However, unlike the totally extinct trilobites a few crinoids had straggled into the present world as "living fossils", since some extremely rare specimens proved that they still survived in the ocean depths. Nonetheless, apart from these rare exceptions crinoids had evidently flourished only during a specific portion of geohistory. Miller and Blainville, at first unaware of each other's work, both tackled another group of distinctive fossils, the belemnites. These were confined without exception to the Secondary formations, and therefore to that era in geohistory. Added to the already well-known case of the ammonites, which had much the same stratigraphical range, it became clear that the marine invertebrates of the Secondary era had been highly distinctive. They were significantly different, both from those of earlier times (the older Secondary and Transition formations), and from those of later times (the Tertiaries), let alone the present world. And the younger Brongniart made his debut with a preliminary study of the fossil record of land plants; it pointed to a similar history, with fairly familiar plants in the Tertiary formations and some very strange ones in the far older Coal formation. Finally, fieldwork by both Brongniarts turned up a striking new case of a forest of tree trunks preserved upright and almost in situ in the Coal formation; and since the Coal plants included such things as tree-ferns, it was plausible to suppose that Europe's climate at this remote period might have been tropical.

All these indications of a true *history* of life reinforced Cuvier's early hunch that the era of the Secondary formations had been an age of reptiles, without mammals, let alone humans. The diversity of the early reptilian fauna was being strikingly demonstrated by the new studies of ichthyosaurs and plesiosaurs (Chap. 2). The next chapter picks up that thread of the story, by describing discoveries that heightened still further the sense of the sheer strangeness of the living world of the deep past.

CHAPTER 5

Ancient monsters on land (1818–25)

5.1 BUCKLAND'S MEGALOSAUR

When Cuvier visited Oxford in 1818, during his first trip to England (§1.2), Buckland showed him a remarkable fossil. It came from the underground quarries at Stonesfield, where a seam of limestone that could be split into very thin slabs was worked extensively to provide roofing for Oxfordshire houses. From its field context Buckland and other English geologists judged that the Stonesfield "slate" clearly belonged among the "Oolite" limestone formations that Smith and Greenough had both mapped above the Lias; that is, it was definitely right in the middle of the Secondaries. Like the other Oolites it was well known for its abundant fossil mollusks, but the Stonesfield rock also contained sharks' teeth and—very rarely—bits of large bones. One of the latter, which had long been in the Oxford museum, was a fragment of a lower jaw bearing some distinctive flattened conical teeth. Cuvier recognized at once, from the character of the teeth, that it was reptilian, which in view of its stratigraphical position was no surprise. He told Buckland that its affinities seemed to be with the lizards, and that it might have come from an enormous monitor. Since he had already interpreted the animal from the Chalk formation at Maastricht as a huge marine lizard, this conclusion was not unexpected, but it was certainly of great interest.[1]

1. The underground quarries at Stonesfield (16km northwest of Oxford) have long been abandoned, though some of the deep shafts are still accessible. Stonesfield "slates" can still be seen on the roofs of many of the older buildings in Oxford and the surrounding villages; they are slates in the vernacular, not the scientific sense (i.e., the rock splits along the original bedding, not along a true "slaty cleavage" imposed subsequently: see Fig. 27.2). The "Oolites" were so called because many of their constituent strata are limestones with a fine structure of tiny spheres with a fortuitous resemblance to hard roe (fish ovaries); they include many famous building stones such as those of Bath (Somerset) and Portland (Dorset).

Buckland did not follow up at once on Cuvier's lead, perhaps because he hoped the Stonesfield quarries might yield further specimens to make the identification more definite. Anyway he was diverted soon afterwards onto the quite different problems of the Superficial deposits, and the "geological deluge" of

Fig. 5.1 A part of one side of the lower jaw of the fossil *Megalosaurus* ["giant lizard"] from Stonesfield near Oxford, and separate teeth: drawn by the young amateur naturalist Mary Morland (soon to become Buckland's wife) and lithographed for his paper (1824) on this sensational addition—terrestrial rather than marine—to the strange reptilian fauna from the Secondary formations. The jaw fragment, which alone is one foot (30cm) long, is shown in internal (fig. 1) and external (fig. 2) views, with one large backward-curving tooth in place. In the lower row, one tooth is shown end-on embedded in a jaw fragment (fig. 3), and two other three-inch (8cm) teeth (figs. 4, 5) help to suggest the monstrous size of the whole animal. (By permission of the Syndics of Cambridge University Library)

which he believed they were the traces (§6.1; *BLT* §10.5). However, Conybeare mentioned the Stonesfield specimen in his paper on the plesiosaur (§2.3), and in 1822 Buckland told Cuvier—through Pentland, as usual—that they were working jointly on the "Stonesfield monitor". The following summer he sent Cuvier proof copies of the plates for a paper on "the great Animal of Stonesfield", assuring him that, like the one on the plesiosaur, it would be published in time to be incorporated in Cuvier's second edition. Meanwhile Parkinson had jumped the gun by giving it, in his introductory book on fossils, the appropriate name *Megalosaurus* [huge lizard].[2]

After many delays, the paper on the megalosaur was read at the Geological Society early in 1824, just after Buckland started a two-year stint as president. In the event it was by Buckland alone, but it shared the meeting at which Conybeare exhibited the duke of Buckingham's (or Mary Anning's) complete plesiosaur (§2.3); it was an impressive double bill. Buckland claimed—with Cuvier's approval, of course—that the teeth suggested an affinity with the "saurian" reptiles (thereby justifying Parkinson's name for it). Comparing the size of its teeth with those of living lizards, he followed Cuvier in estimating that it might have been more than 40 feet (12m) in length. Unlike the ichthyosaur and plesiosaur, of which almost complete skeletons had been found, Buckland's specimens were mere scraps. Even so, the megalosaur was an impressive further addition to the increasingly strange fauna being recovered from the Secondary formations. And if the animal was indeed related to living monitors, it suggested that unlike the other Secondary reptiles it might be a relic of an ancient *terrestrial* fauna (Fig. 5.1).[3]

5.2 MANTELL'S GIANT HERBIVORE

By the time Buckland read his paper on the megalosaur, he knew that somewhat similar fragments of a huge fossil reptile were turning up in another part of England and another Secondary formation. They had been found by Gideon Algernon Mantell (1790–1852) of Lewes in Sussex, a surgeon who had completed his training in London before returning to practice in his native town. Like many other medical men in rural England, Mantell took a keen interest in the antiquities and natural history of his local area, and particularly in its rocks and fossils; even before the end of the wars he had become one of the many local informants who were helping Greenough to compile his map. Greenough had encouraged Mantell, just as he had helped Conybeare, Buckland, and many other geologists in earlier years. They exchanged specimens, and in 1817 Mantell told the older geologist about his plans to publish a book in a familiar genre, on the geology and fossils of his home area. Most of the fossils in his growing collection were from

2. Buckland to Pentland, 11 February 1822; Buckland to Cuvier, 30 March, 10 May 1822, 9 July 1823 [Paris-MHN, MS 634, nos. 77–78, 81–82, 91–92; MS 629, no. 7]. Parkinson, *Oryctology* (1822), 298. Delair and Sarjeant, "Discoveries of dinosaurs" (1975) and "Dinosaur bones" (2002).

3. Fig. 5.1 is reproduced from Buckland, "Megalosaurus" (1824) [Cambridge-UL: Q365.b.12.6], pl. 41, lithographed by Henry Perry; Geological Society, 20 February 1824 [London-GS: OM/1]. Buckland inferred a "probably amphibious" mode of life, i.e., crocodile-like and partly terrestrial (392): hence their preservation in a deposit full of obviously marine mollusks. The fossils are now finely displayed in Oxford-MNH, the successor to the collection for which Buckland was responsible.

the Chalk, which forms the grassy hills of the South Downs rising steeply above the town of Lewes. But in a quarry in the then densely wooded Weald region to the north of the Downs, in one of several formations that were known to dip below the Chalk and were therefore older, a particular bed of sandstone was yielding fragments of large bones and bits of exotic plants. Mantell commissioned the quarrymen to send him further specimens, and these fossils soon became the focus of his research, and of the attention of other geologists (Fig. 5.2).[4]

Mantell was at no great disadvantage for being a provincial naturalist; the Geological Society had welcomed such recruits from the outset as valued local informants (*BLT* §8.4). He was proposed for membership in 1818, at the very next meeting after he asked Greenough how to join. Later, having read Greenough's

Fig. 5.2 A quarry near the village of Cuckfield, in the Weald region of southeast England, exposing the rocks from which, in 1817, Gideon Mantell began to collect fossil teeth and fragments of large bones and exotic plants. By the time this scene was drawn for his *Geology of Sussex* (1827), he had identified some of the fossils as belonging to Buckland's megalosaur, but defined other teeth and bones as those of another huge fossil reptile, which he named the "*Iguanodon*". The picture served to show that the fossils—a large plant stem being hammered, and a fern-like frond in front of the figure on the right—had been found in solid rock (C), part of a Secondary formation: Buckland and others had suspected at first that some of the teeth might have fallen from or been washed out of the loose Superficial "diluvial loam" (A) at the top of the quarry. Of the three gentlemen (wearing top hats), Mantell is probably the one joining the quarrymen (bare-headed) in wielding a hammer and chisel; the others, looking on, may be Buckland, revising his opinion on being shown the place, and his former student Charles Lyell. The church spire, a familiar feature of the English rural landscape, hints at the contrast with the strangeness of the fossils being discovered beneath it. (By permission of the Syndics of Cambridge University Library)

book on geology (1819), with its extreme skepticism about theorizing, Mantell told him that "I have given up *theory* and shall content myself hereafter with matters of fact"; he could be trusted to be a reliable country member who would not venture into geotheory. In 1821, at Greenough's suggestion, he sent Cuvier advance copies of the illustrations for his forthcoming book; they had been lithographed from drawings by his wife Mary, who, like Buckland's friend and future wife Mary Morland, was a fine artist. He also sent fossils from the Chalk, the local formation in which they were most abundant, probably hoping that they would be passed on to Brongniart, who was known to be working on the Chalk fauna (Mantell apologized for being unable to write in French, but assured Cuvier he could read it). He wanted to initiate a regular exchange of specimens with the Muséum, and was clearly trying to establish his credentials as a geologist who deserved to be taken seriously not only in England but also internationally. In addition to the Chalk fossils, however, Mantell sent Cuvier what he described as "a few bones from a stratum of sandstone that occurs in the Oak-tree-clay, beneath the Chalk". These were to dominate later exchanges across the Channel.[5]

Mantell's *Fossils of the South Downs* was published the following year. Buckland, Conybeare, and Greenough were among its geological subscribers, and further copies went to Sussex landowners and other local gentry, though there were too few of them for Mantell to cover his costs. The book included a preliminary description of what he called the "Tilgate beds" (named after an area to the north of Cuckfield), and of the fossil teeth and bones that had been found in them. Although they were fragmentary and poorly preserved, Mantell assigned most of these specimens to "an animal of the Lizard tribe of enormous magnitude"; sharp teeth like those from Stonesfield suggested a similar huge carnivore. But he had also found some teeth of a different kind, with flattened crowns rather like those of some herbivorous mammal. Mantell noted that the latter were "of a very singular character and differ from any previously known". Prudently he just ascribed them to "unknown animals"; but since he concluded that the Tilgate beds contained "one or more gigantic animals of the Lizard Tribe", he implied that the more puzzling teeth might be those of *another* huge reptile, and a herbivorous one at that. He took some specimens to London to show around at the Geological Society; but the opinion there—Buckland's among others—was that they were more likely to belong either to a large fish or to a herbivorous mammal

4. Fig. 5.2 is reproduced from Mantell, *Geology of Sussex* (1827) [Cambridge-UL: MF.41.9], frontispiece, drawn by F. Pollard. This is the earliest of four versions of a scene that was probably idealized and composite (in later ones the fossil plant being extracted was replaced by iguanodon bones, and the human figures were much altered); all the prints are reproduced in Cleevely and Chapman, "Mantell's *Geology of Sussex*" (2000). Mantell to Greenough, 1 February 1815, 20, 25 November 1817, 5 January, 1 April 1818 [London-UCL]. Dean, *Mantell* (1999), chaps. 2–4, gives a detailed biographical narrative, and states (65) that this scene depicts Mantell showing the place to Buckland and Lyell on 7 March 1824 (see below). Cuckfield is 18km northwest of Lewes.

5. Mantell to Greenough, 1 April 1818, 2 August 1819 [London-UCL]; Geological Society, 17 April 1818 [London-GS: OM/1]; Mantell to Cuvier, 28 July 1821 [Paris-MHN, MS 627, no. 69, printed in Taquet, "Cuvier, Buckland, Mantell" (1982), 482]. Mantell pointedly gave his book (see below) an epigraph *in French*, from Burtin, *Oryctographie de Bruxelles* (1784), a pioneer work on fossils (*BLT* §4.2). When, later, the Geological Society was considering designs for its corporate seal (§2.3), Mantell offered one with the motto "Facts not Theories" (Mantell to Buckland, 16 June 1825 [London-RS: MS 251, no. 22]).

from some Superficial deposit. The latter was not an unreasonable inference: if, like many of Mantell's specimens, the teeth had been found loose on the floor of the quarry, they might well have fallen out of the overlying Superficial deposit, rather than coming from any of the solid rocks (see Fig. 5.2). But such skepticism about his finds hardly accorded with Mantell's own conception of his rising status as a geologist.[6]

5.3 WEALDEN STRATIGRAPHY

Mantell's published sections show that he was in no doubt that what he called the Tilgate beds were part of the thick sequence of Secondary formations beneath the Chalk. On this point other geologists agreed: the general structure of the Weald region (in modern terms, a broad and gentle anticline, deeply eroded), lying between the North and South Downs composed of the Chalk, was well understood from Smith's and Greenough's maps and from subsequent fieldwork by others (see Fig. 3.2). But it was not clear in any detail how the formations in the Weald were to be correlated with those in other parts of England. Mantell himself was inclined to equate his Tilgate beds with Buckland's at Stonesfield, mainly on the Smithian argument that both contained what might be the same megalosaur. On the other hand, Buckland told Cuvier that in his opinion the strata with the new fossils were of about the same age as the Purbeck formation on the south coast of England: the Sussex sequence included some thin beds of shelly limestone not unlike the "Purbeck marble" that had long been quarried on the Isle of Purbeck (in fact part of the mainland). Buckland's correlation would put Mantell's fossils at a higher level in the stratigraphical pile than his Stonesfield ones, but in either case both localities remained firmly within the Secondary sequence as a whole.[7]

Before Mantell's book was published, he gained a collaborator and friendly critic in the form of a somewhat younger geologist. Charles Lyell (1797–1875), whose family lived in Hampshire but owned an estate on the edge of the Scottish Highlands, had attended Buckland's courses in three successive years while he was an undergraduate at Oxford; his father, a reputable botanist, had conceded that an amateur interest in geology was at least "rational and gentlemanlike . . . though lighter than Aristotle or the Law". Buckland proposed the younger Lyell for membership in the Geological Society even before he graduated and moved to London to train as a barrister (a lawyer entitled to speak in court). Lyell had heard about Mantell as a good local naturalist, and being in another part of Sussex in the autumn of 1821 he rode over to meet him and see his collection. He was greatly impressed by both, and the two geologists began an intensive correspondence, particularly about the stratigraphical—and therefore geohistorical—position of Mantell's new fossils. Lyell followed Buckland in adopting standard Smithian reasoning, and thought Mantell was "asking much more Philosophically" in correlating the Tilgate beds with Stonesfield rather than Purbeck, because the former entailed even greater problems with "Geognostic position" and implied "an extraordinary chasm in the usual order of succession". Yet the following spring he told Mantell about a giant bone newly discovered at Stonesfield, which, being similar to what Mantell was finding, seemed to confirm the "wonderful coincidence" of the correlation that Mantell had favored.[8]

Clearly some further research was needed, to determine the stratigraphical position of Mantell's important fossils. In the light of joint fieldwork with Lyell, Mantell reported to the Geological Society that the Tilgate beds were an integral part of the "Iron-Sand formation" of the Weald. Like the two young geologists, more senior figures were in fact already trying to establish the sequence of formations lying between the Chalk and the Oolites. One was William Henry Fitton (1780–1861), a physician who, having found himself a wealthy wife, had settled in London as a gentlemanly savant; he was currently one of the secretaries of the Geological Society. Thomas Webster (1773–1844), the Society's salaried administrator, also contributed to the argument, on the basis of his earlier fieldwork in Hampshire and particularly on the Isle of Wight (*BLT* §9.4).[9]

There was the usual element of competitiveness—not always gentlemanly—about this research, but it was straightforward Smithian stratigraphy. Mantell's Tilgate beds were eventually agreed to be a part of the thick "Wealden" formation that outcropped in the central part of the Weald and reappeared in the same stratigraphical position on the Isle of Wight; as Buckland had suggested, they therefore lay above the Purbeck formation (and far above the Stonesfield beds). So Mantell's reptiles were substantially younger than Buckland's megalosaur from Stonesfield, and even more so than the marine reptiles from the Lias. On the other hand, they were substantially older than Cuvier's giant marine lizard from the Chalk at Maastricht across the North Sea. But anyway, they belonged unquestionably within the Secondary sequence, and therefore within Cuvier's putative age of reptiles.[10]

5.4 MANTELL'S IGUANODON

Until these stratigraphical uncertainties had been resolved, the significance of Mantell's finds remained unclear. His claim to have found the "teeth and bones

6. Mantell, *Fossils of the South Downs* (1822), 37–60; Dean, *Mantell* (1999), 73.

7. Mantell, *Fossils of the South Downs* (1822), sections on 58, 296. Buckland to Cuvier, 16 March 1824 [Paris-MHN: MS 627, no. 21], and Buckland, "Megalosaurus" (1824), 394. The highly prized "Purbeck marble" (a limestone taking a high polish, not a marble in the scientific sense) had been quarried for centuries, and can be seen in the interiors of many English medieval churches and cathedrals; "Sussex marble" was a local substitute; both derive their decorative quality from their embedded fossil shells, later recognized as those of freshwater mollusks.

8. Lyell to Mantell, 19 December 1821, 11 April 1822 [Wellington-ATL]; Wilson, *Lyell* (1972), 44, 92–94; Wilson, "Lyell: the man" (1998), is a brief biographical summary.

9. Lyell to Mantell, 6 (7?) March, 19 April, 6, 16 June, 4 July 1822, 12 February, 11 June 1823 [Wellington-ATL]; Mantell, "Iron-Sand formation" (1826), read at the Geological Society, 17 January 1823 [London-GS: OM/1]; Fitton, "Chalk and Purbeck Limestone" and "Additions to a paper" (both 1824); Webster, "Beds below the Chalk" (1824) and "Reply to Dr Fitton's paper" (1825); Wilson, *Lyell* (1972), 103–15.

10. This stratigraphical research, here summarized too briefly, was complicated by confusion between two similar formations, later named the Lower Greensand and Upper Greensand, at different positions in the stratigraphical pile, and by the absence of the Wealden formation from the sequence outside the southern counties. In modern terms the Wealden is of early Cretaceous age, the Purbeck is late Jurassic, the Stonesfield beds middle Jurassic, and the Lias early Jurassic.

of crocodiles and other Saurian animals of an enormous magnitude" was uncontentious, since it paralleled what had long been known from Maastricht and had been found more recently in the Lias and at Stonesfield. But his further claim to have found the "teeth of an unknown Herbivorous Reptile, differing from any hitherto discovered either in a recent [i.e., living] or fossil state" was much more controversial, precisely because it was unparalleled elsewhere. When, in the summer of 1823, Lyell went to Paris to meet French geologists and improve his command of the international scientific language—he had just become one of the secretaries of the Geological Society—he took with him one of Mantell's "herbivorous" teeth, to get Cuvier's opinion of it. When shown it briefly at his Saturday *salon*, Cuvier suggested it was that of a rhinoceros. Coming from the greatest authority on fossil bones, Cuvier's judgment was a blow to Mantell's ambitions for his finds in Sussex. After his return to England, Lyell visited Oxford and mentioned to Buckland the controversial "Rhinoceros tooth of the Tilgate beds". As he then reported to Mantell, Buckland "seemed as much inclined to believe it as if we had asserted that a [fossil] child's head had been discovered there". In his usual flamboyant style, Buckland expressed the general opinion that to find a large mammal well down among the Secondary formations was improbable in the extreme, and contrary to all previous reports. Buckland thought it much more likely that the tooth had been washed out of some Superficial deposit; Lyell sat on the fence but urged his Sussex friend to be cautious.[11]

Mantell was in the audience when Buckland's paper on the megalosaur was at last read at the Geological Society early in 1824. During the subsequent discussion, he capped Buckland's report by describing the similar but even larger bones and teeth that he had found in Sussex. Buckland went to Sussex as soon as he could, to get Mantell to show him the Cuckfield quarry (Fig. 5.2) and to see the specimens that had been found there. He promptly reported Mantell's finds to Cuvier and acknowledged them in the text of his own paper. However, these were the uncontroversial teeth and bones of the megalosaur, the provenance of which was not in doubt. Mantell would not be able to persuade his fellow geologists that the other specimens—possibly from an equally large but herbivorous animal—came from the same Secondary formation, until he had some much better specimens, securely located in the solid rock rather than being picked up loose at the surface. However, "by stimulating the diligent search of the workmen by suitable rewards" he did indeed collect a much more persuasive set of specimens. Specifically, he assembled a growth series of the enigmatic teeth, showing that the young ones had a sharply ridged form that had gradually been obliterated by wear until they acquired the flattened crown seen on the first specimens he had found. Equally important, he found evidence that the teeth had been replaced in the course of life in just the same way as those of the megalosaur and living reptiles, so the enigmatic animal was certainly not a mammal.[12]

With such specimens at last in hand, Mantell hoped that Cuvier might change his mind, so he sent a series of drawings of the teeth over to Paris. Having seen these proxy specimens, Cuvier agreed that the teeth were indeed reptilian, yet not those of any carnivorous form: "might we not have here a new animal, a herbivorous reptile?" That it had been enormous would then be no surprise, since the largest living terrestrial mammals were likewise herbivores. This was just what Mantell had hoped to hear. With Cuvier's support, he could regain confidence in his own earlier hunch, namely that there might have been two distinct reptiles in the Tilgate fauna, both huge but one a carnivore and the other a herbivore. The former seemed to be Buckland's megalosaur, although of an even more monstrous size than at Stonesfield. The latter was the greater novelty, and a sensational one at that, for it had no apparent analogue among either fossil or living reptiles.[13]

To clinch his argument, Mantell needed to find such an analogue. Being unable to visit Paris to see the outstanding collections at the Muséum—with his busy professional life he could probably not afford either the time or the expense of a trip to Paris—he settled for the best alternative. In the autumn of 1824 he took his fossil specimens to London, to compare them with John Hunter's great anatomical collection at the Royal College of Surgeons. Its curator William Clift (1775–1849) was, unlike Home, a highly competent anatomist; but he and Mantell searched in vain until Clift's young assistant Samuel Stutchbury (1798–1859) drew their attention to the skeleton of an iguana from Barbados. This three-foot lizard had teeth that closely matched Mantell's specimens, except that the fossil teeth were about twenty times as large. If the match was valid, it seemed that the fossil animal must have been an enormous herbivorous lizard—perhaps 60 feet (18m) in length—quite distinct from the carnivorous megalosaur, and an important further addition to the strange reptilian fauna of the Secondary era. Mantell's fame was now at last assured. He sent specimens to Cuvier, Brongniart, and other leading naturalists on the Continent, at their request. Cuvier's were just in time for the iguanodon, like the megalosaur, to be described in the final volume of his revised *Fossil Bones*. Here the French naturalist readily conceded that his own earlier opinion had been mistaken. The mistake was understandable, since it had been based on a single deceptive specimen: "it is only since Mr Mantell sent me a series of pristine and more or less worn [*usées*] teeth, that I have been entirely convinced of my error".[14]

Mantell told Cuvier that he proposed to call his animal the "*Iguana-saurus*" [iguana-lizard], to match the ichthyosaur and plesiosaur; but Conybeare, a punctilious Classical scholar, pointed out that this would be tautologous, and that "*Iguanodon*" [iguana-tooth] would be more appropriate. The latter was the name that Mantell duly gave his huge herbivorous reptile. Aware of the importance of

11. Lyell to Mantell, 11 June, 4 December 1823 [Wellington-ATL]; Dean, *Mantell* (1999), 73–76.

12. Buckland to Cuvier, 16 March 1824 [Paris-MHN: MS 627, no. 21]; Buckland, "Megalosaurus" (1824), 394; Dean, *Mantell* (1999), 77–79.

13. Cuvier to Mantell, 20 June 1824 [excerpt printed in Mantell, *Petrifactions* (1851), 231, and quoted in Dean, *Mantell* (1999), 81]; Mantell to Cuvier, 9 July 1824 [Paris-MHN, MS 629/2, no. 8]; Taquet, "Cuvier, Buckland, Mantell" (1982), 480–84.

14. Mantell to Cuvier, 12 November 1824 [Paris-MHN: MS 629, no. 6], enclosing an advance copy of his plate (Fig. 5.3); Cuvier, *Ossemens fossiles*, 2nd ed., 5(2) (1824), 351 and pl. 21; Dean, *Mantell* (1999), 81–85. Cuvier and other naturalists derived their estimates of the sizes of the vanished reptiles by simple linear extrapolation, being as yet unaware of the complexities of allometric growth, or that these fossil reptiles—still known only from a few fragmentary scraps—might not have been at all lizard-like in form.

his discovery, he followed Buckland's example with the Kirkdale Cave hyenas (*BLT* §10.6) and submitted his paper not to the Geological but to the Royal Society. Not being a Fellow of the Royal, it was in the form of a "letter" to the vice-president; on the strength of it, he was elected a Fellow soon afterwards, being

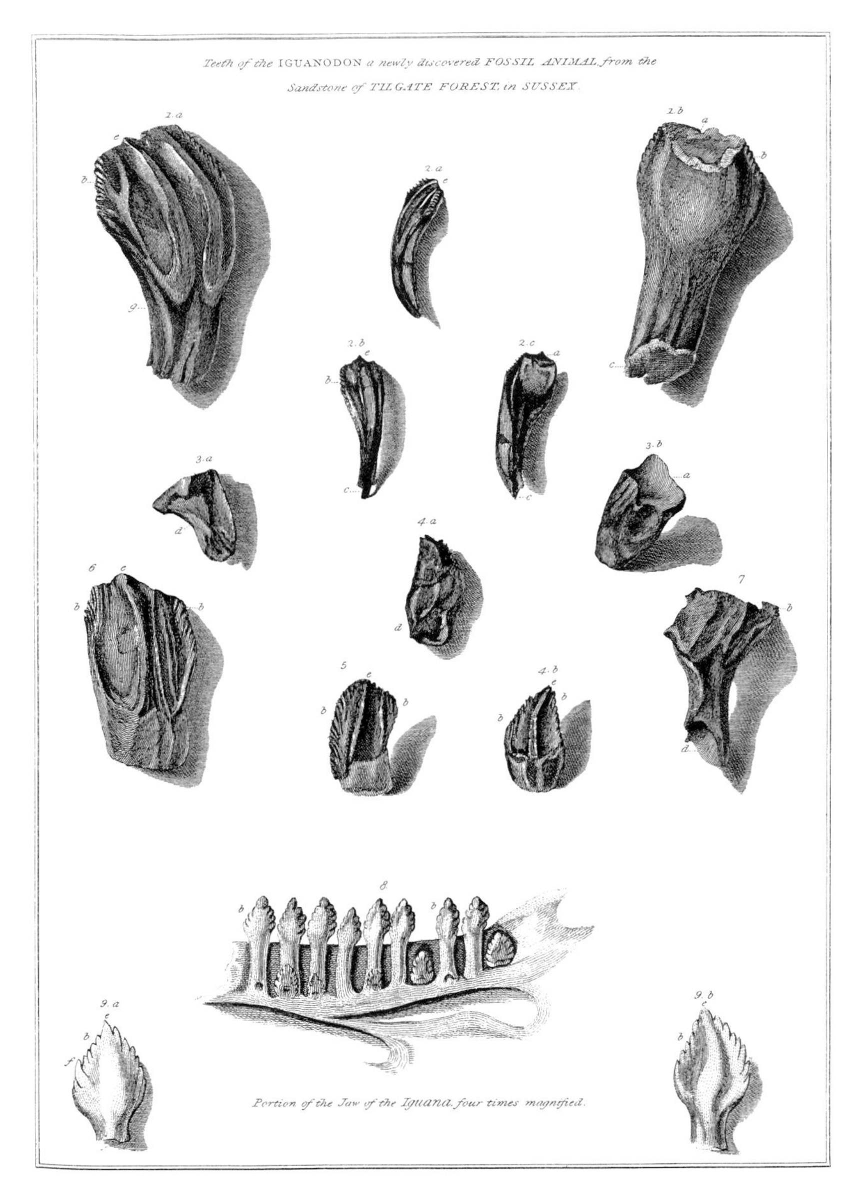

Fig. 5.3 Mantell's comparison between the teeth of his fossil *Iguanodon* (above) and the far smaller teeth of the living iguana (bottom row): engravings to illustrate the paper that he sent to the Royal Society early in 1825. It established the iguanodon as a huge herbivorous reptile and a contemporary of Buckland's carnivorous megalosaur (years later, both genera were assigned to the then newly defined group of the "dinosaurs"). Mantell's first fossil specimens had been dismissed by other naturalists as probably those of a rhinoceros from the Superficial deposits, but the full growth series shown here proved that that appearance had been deceptive. The mammal-like teeth were simply the oldest, worn down by a herbivorous diet (figs. 1, 6), whereas the youngest ones (figs. 2, 4) were similar—except in size—to those of the living lizard. The surface *a* was "worn by mastication" and a sign of a herbivorous diet; the cavity *d* is the inferred site of replacement teeth growing from below like those of the iguana and other living reptiles. The portion of the iguana's jaw (fig. 8) was magnified four times, and its individual tooth (fig. 9) even more, compared to the fossil teeth.

proposed by Buckland and a string of other leading geologists. He made his case in the most persuasive way possible, with a fine set of drawings to display the similarity between the living lizard and its huge fossil relative (Fig. 5.3).[15]

5.5 THE STONESFIELD MARSUPIALS

Cuvier's earlier hunch, that an age of mammals had been preceded by an age of reptiles, seemed to be confirmed by every strange new reptile to be discovered in the Secondary formations. It was hardly weakened by the discovery of a small exception. Years earlier, when Cuvier identified the celebrated "Maastricht animal" (the mosasaur) as a giant lizard rather than a whale, and dismissed some alleged deer antlers as fragments of a turtle's carapace, he had in effect eliminated *all* alleged mammals from the Chalk formation (*BLT* §7.2). Isolated bones from other Secondary formations in England had also been attributed to whales, but the new discoveries of giant fossil reptiles now made that identification equally dubious. However, two specimens of tiny fossil jaws, found at Stonesfield around 1812, were more difficult to dismiss. One had been purchased by Buckland, the other lent to him. When Cuvier visited Oxford in 1818 and saw them, he judged from the character of the teeth that they were mammalian, making them the first reliable indication of *any* mammals before the Tertiary era. Within a few years, however, all the new research on strange and varied Secondary reptiles gave this apparent exception much greater importance, and Cuvier was anxious to check it again before he completed the new edition of his *Fossil Bones*. Meanwhile, the better of the two specimens had somehow been mislaid, reducing the crucial evidence to a single piece.

Cuvier himself was far too busy to visit England again in person, but he soon found a reliable understudy. In 1824, Brongniart's former student Constant Prévost (1787–1856) was sent to England with government funding, to study the equivalents of the Secondary formations of northern France, and to bring back English fossils for the Muséum in Paris: this would help improve stratigraphical correlations across the Channel and thereby aid the new geological survey of the whole of France. Prévost was well qualified for such fieldwork. Living in Austria after the end of the wars, he had emulated Brongniart's example by exploring the Tertiary formations near Vienna. Later, settled back in Paris and married into enough money to support himself as a savant, he made his debut at the Académie with an important paper that compared the Austrian Tertiaries with Brongniart's around Paris and others elsewhere, making innovative suggestions about their geohistorical significance (§10.3; *BLT* §9.6). He also began research on the stratigraphy and fossils of the Secondary formations in Normandy, which

15. Fig. 5.3 is reproduced from Mantell, "Iguanodon" (1825), pl. 14; addressed to Davies Gilbert and read at the Royal Society on 10 February; Dean, *Mantell* (1999), 84–85. In 1842, Buckland's megalosaur and Mantell's iguanodon were the first two fossil reptiles to be assigned to the group that Richard Owen then distinguished as the "*Dinosauria*". A heroic mythology as luxuriant as Anning's and Smith's has therefore grown up around Mantell, usually contrasting him with alleged villains at the Geological Society such as Greenough and Buckland: see Torrens, "Dinosaurs and dinomania" (1996) and "Politics and paleontology" (1997).

clearly matched those on the opposite side of the Channel (Fig. 3.2) and were likewise yielding exciting new reptilian fossils.[16]

The previous year, when Lyell visited Paris (and first showed Cuvier one of Mantell's enigmatic fossil teeth), Prévost had taken him into the field to see the famous Parisian Tertiaries (§10.2). So Lyell now offered in return to act as the Frenchman's guide and companion in England and indeed as his interpreter, since Prévost, like Cuvier, spoke little English. Their tour took them first to Oxford (and Buckland) and later to Bristol (and Conybeare). Then they tackled the rocks and fossils of southwest England, before doubling back along the south coast by way of Lyme (and Anning), as far as Lyell's family home in Hampshire. By great good luck they stopped at Lyme just when a new plesiosaur specimen had been found, in quality second only to the duke of Buckingham's famous

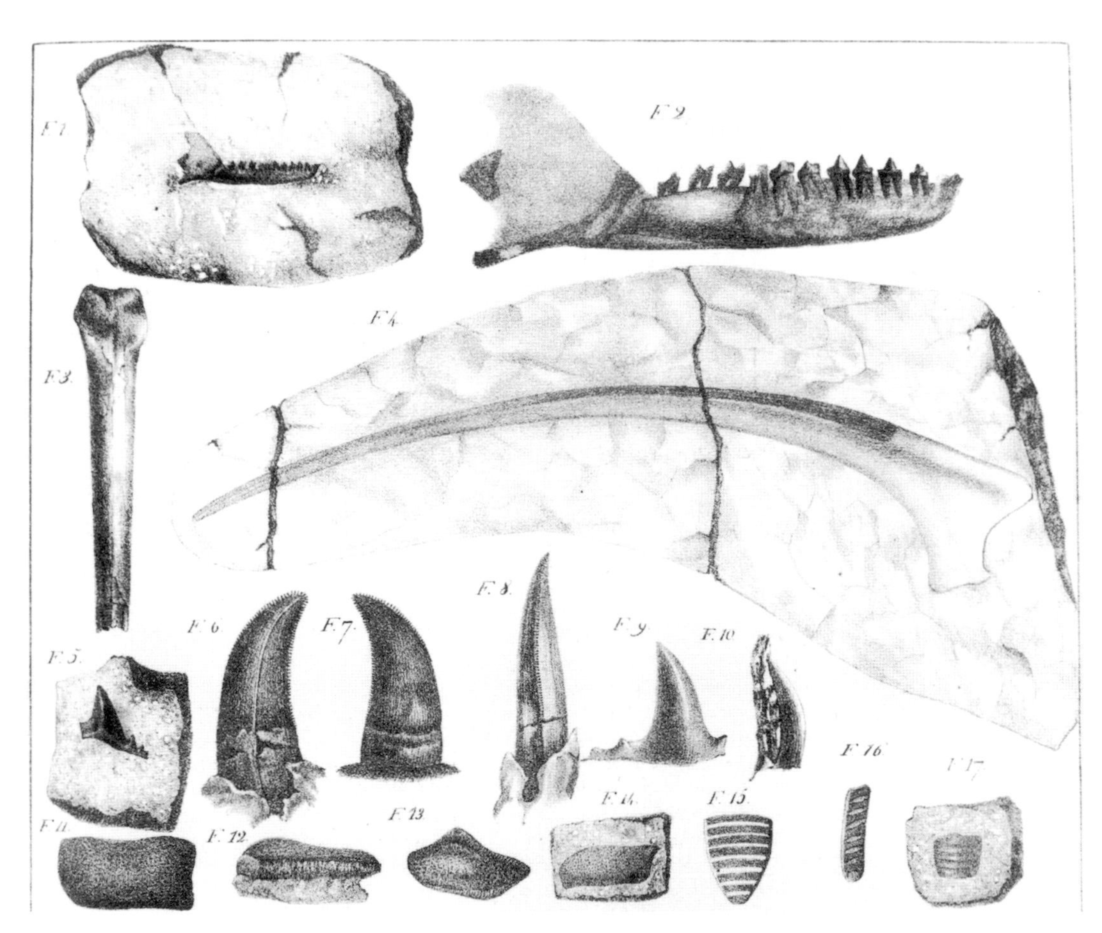

Fig. 5.4 Buckland's anomalous mammalian jaw from the Stonesfield "slate" near Oxford (fig. 1, enlarged in fig. 2), as depicted by Prévost (1825) in the context of a "bird" bone (fig. 3), a rib and some teeth of the megalosaur (figs. 4, 6-8), sharks' teeth (figs. 5, 9-10) and some palatal teeth of other fish (bottom row). Prévost accepted on Cuvier's authority that the jaw was mammalian, but questioned whether the Stonesfield beds were truly an Oolitic deposit and not an anomalous pocket of younger sediment. Other geologists, however, assimilated the case into the general rule that the quadruped fauna of the Secondary era, if not quite exclusively reptilian, had certainly been dominated by that group. This picture of the little marsupial jaw, published in a leading French periodical and displayed with the rest of the Stonesfield fauna and a report on its stratigraphical setting, made the English discovery well known to naturalists throughout Europe. (By permission of the Syndics of Cambridge University Library)

one; Prévost was able to acquire it for the Muséum. But earlier, while in Oxford, he had done Cuvier a service of scarcely less importance by making an accurate drawing of Buckland's mammalian jaw from Stonesfield; he was also taken to see the famous quarries, and he collected (or purchased) other fossils there (Fig. 5.4).[17]

Having scrutinized Prévost's proxy specimen, Cuvier confirmed to Buckland that it was indeed a mammal and more specifically a marsupial, a fact that he found "quite extraordinary"; it was not unlike the little fossil opossum that he had famously identified long ago from the Tertiary gypsum formation near Paris (*BLT* §7.5). However, it hardly dented his confidence in his broader inference about the history of life. Summarizing his new volume on the fossil bones of reptiles, he told Buckland that "the result—despite the two little jaws, which form only a very modest exception—is that at the time of the Secondary formations the earth had almost no mammals, but that it swarmed with reptiles that were as remarkable for their size as for the variety and singularity of their forms."[18]

Prévost found the anomaly harder to swallow. When Cuvier pronounced the little Stonesfield jaw to be mammalian, albeit of a kind often regarded as "lower" than ordinary placental mammals, Prévost was certainly not prepared to question the great naturalist's authority as an anatomist. But the stratigraphy was another matter. Having seen Stonesfield for himself, Prévost considered that the position of the beds within the sequence of the Oolitic formations was open to doubt: he thought they might in fact be a pocket of younger sediments, nestling deceptively in a hollow in older ones. Not surprisingly, this skepticism on a point of Oxfordshire stratigraphy was hardly welcomed by English geologists, who vigorously defended their own competence in the field. In the event, Prévost's doubts were soon forgotten, and other geologists adopted Cuvier's view that the anomaly was too small to disturb the much grander conclusion that the Secondaries represented a great age of reptiles.[19]

16. Humboldt to Cuvier, 1822, enclosing Prévost to Humboldt, 13 March 1822 [Paris-IF: MS 3244/41, 41bis], summarized the results of Prévost's earlier fieldwork in Normandy.

17. Fig. 5.4 is reproduced from Prévost, "Schistes de Stonesfield" (1825) [Cambridge-UL: T382.b.8.2], part of pl. 18, explained on 81 (Buckland later identified the "bird" bones as those of pterodactyls); his little map, pl. 17, fig. 1, shows his cross-Channel correlations; see also Prévost, "Gisement du *Mégalosaure*" (1825). Prévost to Cuvier, 28 May 1824 [Paris-MHN, MS 634(2), folder "Sarigue de Stonesfield"], enclosed his drawing of the mammalian jaw; Prévost to Webster, 8 July 1824 [printed in Challinor, "Correspondence of Webster" (1961–64), no. 56]. Taquet, "Reptiles marins anglais" (2003), describes Prévost's English fieldwork and reproduces Cuvier's engraving of the new plesiosaur specimen (still on display at Paris-MHN); it was inserted in his *Discours sur les révolutions* (1826), pl. 3, fig. 1, having been just too late for the last volume of the third edition of *Ossemens fossiles* (1825).

18. Cuvier to Buckland, 20 June 1824 [London-RS: MS Bu19], reporting that the jaws were those of a new *Didelphis*. Buckland added a note on the mammal in the published text of his "Megalosaurus" (1824), 391. The second, better, and previously mislaid specimen was later assigned to a different species of *Didelphis* on the basis of its dentition (see §17.2). The two specimens were, in modern terms, the first Mesozoic mammals ever found, and they long remained the only ones known; both are still classed as non-placentals, though no longer as marsupials.

19. Prévost, "Schistes de Stonesfield" (1825), pl. 17, figs. 3–4, show his interpretation of the Stonesfield slate as an outlier filling a hollow in the Oolitic strata (in modern terms, all are of middle Jurassic age). In his preliminary report, "Gisement du *Mégalosaure*" (1825), he suggested that the slate might be of the same age as Mantell's substantially younger "Weald" rocks (also with megalosaur bones), though this still kept them in the Secondary formations and alleviated the anomaly only slightly.

5.6 CONCLUSION

Of the new vertebrate fossils that English geologists—or the fossilists who supplied them with specimens—turned up during the first decade of peace, the ichthyosaurs and plesiosaurs, interpreted most competently by Conybeare, were the first and most sensational. Since almost complete skeletons were found, the animals could be confidently reconstructed as streamlined aquatic reptiles; and the associated fossils left no doubt that they had been marine. Buckland's megalosaur and Mantell's iguanodon, on the other hand, were based on far more fragmentary remains, and their reconstruction as gigantic lizards—one a carnivore and the other a herbivore—was much less certain. However, they did suggest that the ancient reptilian fauna had included not only marine but also terrestrial (or at least "amphibious") forms.

When the new fossil reptiles were added to the giant marine lizard and flying pterodactyl that Cuvier had earlier described from the Continent, they confirmed and reinforced the Frenchman's hunch that the Secondary formations represented a period in geohistory that was far stranger than that of his Parisian mammals from the Tertiaries. The discovery of very small and rare mammalian fossils in the same strata as Buckland's megalosaur did not seriously undermine confidence in the validity of Cuvier's generalization that the era of the Secondary formations had been predominantly an age of reptiles. Apart from some crocodiles and turtles, these reptiles were as peculiar in relation to living quadrupeds as their contemporaries the ammonites and belemnites were in relation to living mollusks. Together they implied that the "former world" represented by the Secondary formations had been strange indeed.

However, the very strangeness of this Secondary world was in effect a challenge. The fossils found in the overlying Tertiary formations might provide a kind of bridge between the Secondaries and the familiar present world, as Cuvier's and Brongniart's Tertiary research had already suggested (*BLT* §7.5, §9.6). But there was still an awkward disjunction between even the youngest of these "regular" strata and the present. This was the gap filled by the puzzling Superficial deposits and the major "revolution" that most geologists were convinced they represented. The next chapter describes the attempts that were being made, during these same years, to work out what this enigmatic revolution might have been like, and when, in relation to human history, it might have taken place.

CHAPTER 6

Geological deluge and biblical Flood (1819–24)

6.1 BUCKLAND'S "HYAENA STORY" AT KIRKDALE

In the first years of peace, Buckland had made his debut at the Geological Society with two quite modest papers: one on some hard old Transition rocks in the north of England, the other on one of the Tertiary formations in what Webster—emulating Brongniart's and Cuvier's classic work around Paris—had called the "London Basin" (*BLT* §9.4). Later, Buckland and Conybeare jointly described in detail the stratigraphy of the Secondary formations around Bristol, not far from Smith's home area (*BLT* §9.5). All this work established Buckland as a competent field geologist in the mainstream of current research; his work on the Stonesfield megalosaur (§5.1) fitted into this stratigraphical framework, while extending it in a zoological direction that depended crucially on Cuvier's expertise. On the other hand, Buckland's famous inaugural lecture at Oxford, focusing on the problems of the Superficial deposits and interpreting them as the physical traces of the "geological deluge" that he identified as the biblical Flood (*BLT* §10.5), represented a distinctly different line of research, carried on in parallel with his other work and only loosely linked to it.

Buckland's "diluvial" research was hugely enhanced by the chance discovery of Kirkdale Cave in Yorkshire, with its rich haul of fossil bones (*BLT* §10.6). These were identified by Clift in London, and confirmed by Cuvier in Paris; but this indoor museum work on comparative anatomy was complemented crucially by Buckland's own outdoor fieldwork on the geological context of the cave itself. Combining the two sources of evidence, he constructed the sensational "hyaena story" that he presented to the Royal Society in 1822, and for which he received the Copley medal. He interpreted the cave as a former den of large extinct hyenas, scavenging a varied diet of animals ranging in size from mammoths to water rats and including both carnivores and herbivores. He anchored this vanished ecosystem in geohistory, by using the stalagmitic layers above and below the silty bone deposit as "nature's chronometers" (as de Luc had called them). Buckland argued that they demonstrated that the den had been occupied at a geologically very recent period, and that the hyenas had been wiped out no more than a few millennia ago, a time compatible with any of the varied dates (Ussher's 2348

Fig. 6.1 William Buckland in academic dress, lecturing at Oxford in February 1823 and giving in effect a preview of his varied evidence for a geological deluge, which he published later that year in *Reliquiae Diluvianae* (see below). He is holding the fossil hyena jaw that helped prove that a large hyena species formerly lived in England, while the picture of spotted hyenas in a cave (middle right) is probably an imagined scene of the extinct animals at Kirkdale. Also from Kirkdale came the molars of mammoth, rhinoceros, and hippopotamus (lower right corner), while the large rhinoceros skull (immediately in front of him) came from the Wirksworth cave (Fig. 6.2). The skull of a modern bear (lower left corner) alludes to the fossils found in Gailenreuth cave in Bavaria, a section of which is on the left. Prints of the antler of an "Irish elk" (top left) and of a mammoth skeleton (top right) are among other exhibits illustrating the "antediluvial" fauna. The ammonite—the one Sowerby had named *Bucklandi* (Fig. 4.1)—and the ichthyosaur skull in front represent the far older and even stranger faunas of the Secondary formations, while his audience could locate the English finds, and grasp the relative ages of the formations, on Greenough's geological map in the background. (By permission of the National Portrait Gallery, London)

B.C. being just one) traditionally computed on textual grounds for the biblical Flood.

Buckland's interpretation of Kirkdale Cave as an "antediluvial" hyena den was buttressed by several other lines of research. In earlier fieldwork, for example, he had used telltale pebbles to track a "diluvial current" that, he claimed, had swept from the Midlands plain up over a watershed and down into the Thames valley, past Oxford and all the way to London, in a way that was quite inexplicable in terms of the ordinary action of the present rivers. Likewise he had interpreted the topography on the south coast as evidence that the valleys had been scoured out by a rapid current (*BLT* §10.5). But he also showed how the present could be the perfect key to the past, when he studied the eating and defecating habits of a living hyena and found that it left gnawed bones and fecal pellets exactly matching those at Kirkdale. Independent of Kirkdale, the discovery of a well-preserved and ungnawed hyena jaw in a Superficial gravel was welcome evidence that hyenas really had been living in England at the same time as mammoths, and that they were then wiped out by the same diluvial event. And Buckland revisited the famous bone caves near Muggendorf in Bavaria, confirming that they were closely similar to Kirkdale, except that in this case the dens had been those of extinct bears rather than extinct hyenas (*BLT* §10.6). All this research had reinforced his Cuvierian argument that a drastic and widespread diluvial event in the geologically recent past had been the last of the earth's occasional "revolutions".

Buckland's "hyaena story" based on Kirkdale made him not only well known in savant circles in London but also more prominent than ever in his university. In 1823 his new emphasis on geohistory was reflected in the title of his latest lecture course at Oxford: it was not only on "the Composition and Structure of the Earth" but also on "the Physical Revolutions that have affected its Surface, and the Changes in Animal and Vegetable Nature that have attended them". He attracted fifty-two subscribers, ranging in dignity from the bishop of Oxford and three other heads of colleges to nineteen undergraduates. His exposition of his new research was so striking that it was recorded in portraits showing him in action, surrounded by the visual aids for which his lectures were renowned (Fig. 6.1).[1]

6.2 BUCKLAND'S NEW "DILUVIAL" EVIDENCE

Even before he gave these lectures in Oxford, Buckland had begun to gather further evidence for a geologically recent deluge. For example, the ransacking of Kirkdale Cave on behalf of fossil collectors had unwittingly destroyed pre-

1. Fig. 6.1 is reproduced from a copy [London-NPG: no. 47981] of the lithograph drawn "from Nature" by George Rowe of Exeter; the title translates as "William Buckland, Bachelor of Divinity, Fellow of the Royal Society, Professor [*sic*] of Mineralogy and Geology at Oxford, 1823". This print, which is in a curiously archaic artistic style, is fully analyzed in Boylan, "Portrait of Buckland" (1970). Greenough's map, as shown, is much reduced in size, probably to make it fit into the design; some of the drawings and specimens depicted are reproduced in their own right in *BLT*, Figs. 10.2, 10.18, 10.20, 10.21, and 10.24. Another print (not reproduced here) of a lecture from the same course shows his specimens in less detail but includes his audience; it is described and illustrated in Edmonds and Douglas, "Oxford geological lecture" (1976), and is also reproduced in Rupke, *Great chain of history* (1983), 65, and elsewhere.

cious evidence by removing the bones from their material context in the field. Buckland's geohistorical analysis depended on their precise relation to the silt in which they were embedded and to the stalagmitic layers above and below; but he had first seen the cave too late to check this relation for himself. So when a similar cave was discovered nearby, some months after he read his paper at the Royal Society, Buckland returned to the area to study it before anything was disturbed; and he took two highly respectable witnesses with him, namely the chemist Humphry Davy (1778–1829), the Society's president, and the London geologist and businessman (and Member of Parliament) Henry Warburton (1784–1858), to add authority and veracity to whatever he might find. In the event this second cave proved to have no bones in it, but it confirmed what the stalagmite at Kirkdale had been like in its undisturbed state. Indirectly it also supported Buckland's reconstruction of Kirkdale as a hyena den; for if—on an alternative

Fig. 6.2 A section through Dream Cave near Wirksworth in central England: a drawing by Webster based on a sketch by Buckland, published in 1823. A skeleton of a rhinoceros (G) had been found buried in the debris below a "fissure" (D). Buckland argued that the carcass must have fallen into the cave while being swept along during a geologically recent deluge. The cave had been discovered by chance when it was penetrated by a shaft (A) exploiting a vein of lead ore (B) in the "Mountain" (Carboniferous) limestone (H). The natural opening nearby had been revealed by the collapse of the loose debris (E) formerly blocking it; the rhinoceros skeleton was shown as it might have been before that collapse disturbed it; there were also bones of oxen and deer (F). (By permission of the Syndics of Cambridge University Library)

interpretation—the bones had been swept in during the deluge, they should also have been preserved in this second cave and in similar ones elsewhere. Above all, however, the discipline of scrupulously careful excavation set a precedent and a new standard for cave research.[2]

At the end of the year, Buckland went to see another cave, newly discovered in Derbyshire. It turned out to be as instructive an example of a "fissure", into which animals had fallen, as Kirkdale was of a true "cavern" or den in which they had lived. At Wirksworth a fossil rhinoceros—an almost complete skeleton, in contrast to the fragmentary bones at Kirkdale—had been found buried in the debris inside the cave, along with other less spectacular animals. Buckland inferred that the carcass had been swept along in a violent mass of water during the deluge, and been preserved by chance after falling into the cave from above (Fig. 6.2).[3]

Around the same time, Buckland heard that fossil bones had just been discovered in a cave at the foot of a coastal cliff, directly below the prehistoric fort of Paviland on the Gower peninsula in south Wales. He wrote at once to his friend Lady Mary Cole, who lived nearby in Trerice Castle, asking for more information about the find; she and her daughter Mary Talbot (1795–1861) were competent amateur naturalists of a kind that was not uncommon at this time among upper-class women in Britain. Talbot visited the cave without delay, with two other local naturalists as suitable male companions. As she reported to Buckland, they collected masses of bones and teeth. He put to her some searching questions about the cave, which might help resolve its history; and to avoid a repetition of what had happened at Kirkdale, he asked that the site be sealed off until he could visit it and check the field context of the specimens for himself. As soon as he had seen the cave at Wirksworth he traveled to south Wales to do so.[4]

By far the most important discovery at Paviland was apparently made by Buckland himself. In addition to the usual "antediluvial" mammal bones, an incomplete human skeleton was found lying close to the surface of the deposit on the floor of the cave. Since Buckland identified the "geological deluge" as the biblical Flood, it might have been anticipated that he would be delighted—as Johann Esper had been long before, in the case of the Bavarian caves (*BLT* §5.4)—to find a human victim of that event and hence to have the date of the geological deluge confirmed as falling within the human period. But Buckland took his cue from Cuvier, who with good reason had been highly skeptical about all such alleged human fossils (*BLT* §10.3). Before he arrived, Buckland had set out for Talbot his own strict principles of careful excavation, which were designed to guard against false claims of this kind. These methods led him to dismiss the Paviland skeleton

2. On the barren cave, see Boylan, *Buckland* (1984), 120, 377. All parts of the present volume that deal with Buckland are deeply indebted to this fine and comprehensive work.

3. Fig. 6.2 is reproduced from Buckland, *Reliquiae* (1823) [Cambridge-UL: MF.40.10], pl. 20, explained on 273–74; Wirksworth is about 20km north of Derby.

4. Buckland to Cole, 24 December 1822; Buckland to Talbot, 31 December 1822 [Cardiff-NMW: MSS 165, 166]; Talbot was one of Cole's daughters by her first marriage. Paviland Cave is 3km west of Port-Eynon (and 7km west-southwest of Trerice Castle); the rocky entrance is inaccessible except at low tide, and even then is quite difficult to reach, which suggests that Talbot was no delicate blossom.

Fig. 6.3 Paviland Cave (also known as Goat Hole) on the Gower peninsula in south Wales: a lithograph after a drawing by Webster, based in turn on a sketch section made by Buckland when he studied the cave in 1823. The wide opening (A) in the coastal cliff of Carboniferous limestone was (and is) accessible only at low tide, and the outer part of the cave was floored with pebbles (G, H) washed in at high tide. Beyond a threshold of bedrock (I), and above high tide level, was "diluvial loam" (E, F), "much disturbed by ancient digging" but containing the usual fossil bones. Buckland argued that since the fissure or "chimney" (K, D) in the roof of the cave was too narrow for large carcasses to have fallen in from above (as one had at Wirksworth: Fig. 6.2), they must have been swept in during the deluge. A human skeleton was found close to a mammoth skull, but Buckland concluded that (as shown) it was lying in a shallow grave dug into the fossil deposit, and was therefore of a later and "postdiluvial" date; unlike the disarticulated animal bones, the skeleton was shown as almost complete. The ground plan (inset, right) shows the spatial relations of the "diluvial" deposit and the putative grave. (By permission of the Syndics of Cambridge University Library)

as irrelevant to his deluge story. For he judged that—as in the earlier cases that Cuvier had rejected—it was not embedded in the same material as the animal bones; and unlike them, the human ones were still articulated, and curiously stained red with ochre. Buckland concluded that the skeleton had been buried in a shallow grave, dug at a much later date into the truly antediluvial deposit (Fig. 6.3).[5]

Buckland assumed that the skeleton was male; and since smugglers often used coastal caves for storing their illicit liquor, he and his companions jokingly identified it as that of some ancient coastguard who had met his end after catching them in the act. However, writing after his return to Oxford, he told Talbot's mother that "the man whom we voted an Exciseman turns out to have been a woman". The wording might suggest that the sex change reflected the judgment of someone with more anatomical expertise than Buckland himself possessed. But in fact it seems to have been based—for anyone with the tacit assumptions of his generation—on the discovery of the remains of an ivory bracelet found with the skeleton, while a sheep's scapula found nearby suggested to him that she might have been practicing divination. Buckland joked that she would be a suitable subject for a historical novel entitled "The Red Woman, or the Witch of Paviland" (he might have had in mind the popular "Waverley" novelist, the still anonymous Walter Scott). More seriously, he adopted a suggestion by one of the local naturalists, that the woman might have been associated—in a manner left discreetly vague—with the prehistoric fortified site on the headland above the cave, which was "British" or pre-Roman, and certainly postdiluvial.[6]

However, the sex and occupation of the Paviland person—exciseman or Red Woman, witch or prostitute—were in the present context irrelevant. It was the *date* of this relic of humanity that was the crucial point at issue. Was he/she a contemporary of the typical antediluvial mammalian fauna preserved in the cave? Or was the skeleton no exception to the apparent rule, recently restated with renewed force in the new edition of Cuvier's *Fossil Bones*, that no genuine human fossils had yet been found anywhere? Buckland had good reason to stay on the side of caution and Cuvier, and to conclude that there was still no evidence that early humans had spread as far as Britain by the time that the deluge overwhelmed the country. For him, as for Cuvier, this did not weaken the case for identifying the geological deluge as the event recorded in Genesis. Both savants assumed that human life had been far more restricted geographically before the deluge than it became with the rise of civilizations after that event. It had long been conventional to locate the likely area of origin of the human species somewhere in Asia (a term that included the traditional site of Eden in Mesopotamia). So Buckland was not wriggling awkwardly out of a potential "conflict" between geology and Genesis, when he inferred that no humans had reached Britain at such an early date. And that inference was confirmed, in his opinion, by the wild and untamed character of the local antediluvial fauna, which he regarded as unlikely and unsuitable neighbors for any early human culture.

5. Fig. 6.3 is reproduced from Buckland, *Reliquiae* (1823) [Cambridge-UL: MF.40.10], pl. 21, lithographed by Scharf and explained on 274–76. The human skeleton was in fact more fragmentary than it was shown, and the skull was not preserved; the bones are now on display at Oxford-MNH. Modern research has confirmed that the skeleton was indeed later than the animal remains, though more ancient than Buckland supposed: Swainston and Brookes, "Paviland Cave" (2000). See also Boylan, *Buckland* (1984), 122, 385, 402; Grayson, *Human antiquity* (1983), 65–67.

6. Buckland to Cole, 24 December 1822, 15 February 1823 [Cardiff-NMW: nos. 165, 167]. Sommer, "Romantic cave" (2003), 184–97, sets Buckland's cave research in its Romantic context; her "Red Lady of Paviland" (2004) argues, in a detailed account of the episode, that the skeleton's sex was changed not on anatomical grounds but on those of gender stereotypes. The association of the skeleton with ivory ornaments, for which the ivory would obviously have come from the antediluvial mammoth tusks found on the spot, did not weaken the case for a postdiluvial date: it was well-known in Buckland's time that "fossil ivory" from mammoth tusks was an important item of commerce in Siberia. According to North, "Paviland cave" (1942), Buckland was not responsible for elevating the putative woman in social status to become the Red *Lady* of Paviland, though that has long been her title. Coastal erosion has greatly reduced the original extent of the prehistoric earthworks.

6.3 "RELICS OF THE DELUGE"

With discoveries of new caves giving cumulative support to his theorizing about a recent deluge, Buckland's fame was rising rapidly, and not only in Britain. The success of his paper at the Royal Society, and the acclaim it received on all sides, encouraged him to reprint it in book form, to reach a wider audience. The Latin title of *Reliquiae Diluvianae* [Relics of the Deluge], like that of his earlier inaugural lecture *Vindiciae Geologicae* (*BLT* §10.5), indicated both its academic respectability and its intended readers among the Classically literate classes, though in fact no knowledge of Latin was required in either case, beyond the title itself. However, although accessible to any educated person, Buckland's book was not a work of "popular" science, and he cut no corners in setting out the geological evidence in full and technical detail. He had declined Davy's invitation to lecture to the famously fashionable mixed audience at the Royal Institution in London, because he feared that it might "compromise the dignity of the university". In effect, he defined his proper audience as encompassing his academic colleagues, metropolitan savants, and serious amateur naturalists, but *not* the general public with a merely casual and superficial interest in the science.[7]

Buckland's book reprinted his now celebrated Kirkdale paper from the *Philosophical Transactions*, but also included material—and notably the illustrations—from his Geological Society papers on the "diluvial" gravels in the English Midlands and on the erosion of valleys on the south coast "by diluvial action" (*BLT* §10.5). It also expanded the part of his earlier text that had dealt with caves other than Kirkdale, at home and abroad, in the light of what he had done, for example at Wirksworth and Paviland, and in Bavaria, since the original paper was read. As before, he sent Cuvier an advance copy of this part of his text, to keep his hero informed in detail about his latest research.[8]

The "relics of the deluge" were in fact of many different kinds, some offering more persuasive evidence than others. Most important was that he reinforced Cuvier's case for a mass extinction of land mammals in the geologically recent past, by his detailed reconstruction of Kirkdale and other caves as samples of the living world that had apparently been wiped out at that time. His fieldwork around Oxford added to the evidence for specific "diluvial currents", traced in this case by their erratic pebbles. He particularly recommended those who doubted his interpretation to study the work of Hutton's friend Sir James Hall (1761–1832), who had reconstructed just such violent currents in the Edinburgh region and had explained them explicitly by analogy with modern tsunamis (*BLT* §10.2). Hall's research showed, as Buckland put it, that "the surface of the earth owes its last form not to the gradual action of existing causes, but to the excavating force of a suddenly overwhelming and transient mass of waters". For a major element of Buckland's synthesis was his claim that the whole observable topography had been radically reshaped by the erosive power of the deluge, clearly conceived as some kind of mega-tsunami; this explained, for example, why the Bavarian bone caves that had once been bear dens now had their openings high above the valley floors (*BLT* §10.4). He insisted that valleys could not

Fig. 6.4 "Large Block of Granite upon the Jura. Neuchâtel": De la Beche's drawing of the famous Pierre à Bot on the forested slopes of the Jura above the Swiss city, sketched in 1820 during his Continental travels. The granite of this huge erratic, resting on bedrock composed of Secondary limestone, had been traced by von Buch back to its source in the high Alps about 100km away. How this and hundreds of other smaller erratics had been transported remained extremely puzzling, but some kind of powerful "deluge"—perhaps a huge mega-tsunami—seemed the least unsatisfactory explanation.

be attributed to the gradual action of the streams and rivers that now flow in them, however long the time allowed.[9]

Buckland's argument was in fact even more powerfully supported by von Buch's already celebrated research on the erratic blocks found on the slopes of the Jura hills in Switzerland. The Prussian geologist had traced these back across the broad Swiss plain and all the way up the Rhône valley to their unquestioned source in the bedrock of the Alpine massif dominated by Mont Blanc (*BLT* §10.2). He had visited England soon after his paper was published (and just after the end of the wars), and was the very first geologist to be elected a Foreign Member of the Geological Society (§2.2). It is inconceivable that Buckland was unaware of his work, and his own tracing of a "diluvial current" from the Midlands via Oxford to London used the same method, albeit with little quartzite pebbles rather than large blocks of granite as the telltale evidence (*BLT* §10.5). Buckland did not mention von Buch's more spectacular case, probably because—unlike the Continental bone caves that he did cite—he had not seen it with his own eyes. Nonetheless, he must have heard reports from those such as De la Beche who

7. Buckland, *Reliquiae* (1823); on the Royal Institution proposal, see Rupke, *Great chain of history* (1983), 18.

8. Buckland to Cuvier, 24 February 1823 [Paris-MHN, MS 634(2), nos. 101–2]. He had requested the Council of the Royal Society (meeting of 14 November 1822) to allow him to republish his paper and especially its plates [London-RS: Council minutes 10: 12–13]. He read another paper there on 8 and 15 May 1823, on the bones from Continental caves, but it was never published in that form, because most of it was included in his book. Geologists on the Continent later expressed regret that Buckland's book was never translated into French and that his research was therefore less accessible beyond Britain.

9. Buckland, *Reliquiae* (1823), 203–5.

had seen the Swiss evidence, and above all the famous huge erratic that von Buch had cited at the very start of his argument. The Pierre à Bot above Neuchâtel was a "witness" fit to silence anyone who doubted the reality of a "geological deluge" *of some kind* (Fig. 6.4).[10]

To return to Buckland's book: it was, in sum, a compendium of his wide-ranging case for a geological deluge that had been not only recent enough to be identified with the biblical Flood, but also widespread enough to be regarded as probably global or "universal". Strictly speaking, the fossil bones of Kirkdale and other caves were not themselves "relics of the deluge", but traces of an earlier or "antediluvial" period that the deluge had brought to an end. Yet Buckland's chosen title was justified: the field context of the bones—their relation to the deposits left by the putative deluge and to the "postdiluvial" stalagmite—turned the caves into, as it were, nature's own reliquaries, which had preserved the precious material remains by which the reality of the historic event could be established. Dedicating his book to the elderly bishop of Durham—in whose cathedral was the famous reliquary of the early Celtic hero St. Cuthbert—Buckland was as explicit as he had been in his inaugural lecture about what he believed his research had achieved: "It has, already, produced conclusions, which throw new light on a period of much obscurity in the physical history of our globe; and, by affording the strongest evidence of an universal deluge, leads us to hope, that it will no longer be asserted, as it has been by high authorities, that geology supplies no proofs of an event in the reality of which the truth of the Mosaic records is so materially involved."[11]

For Buckland, as for de Luc before him, the religious value of the biblical writings was dependent on their reliability *as history*. Like many earlier savants (*BLT* §2.5), he had no hesitation in interpreting the Creation story in Genesis in a non-literal way, allowing for a vast prehuman antiquity of the earth. But once the biblical text became clearly historical in character, as he believed it did with the appearance of humankind, it became important to show that the geological evidence was compatible with the chronology derived from textual analysis. This is why Buckland was so delighted with the "natural chronometer" provided by the stalagmite at Kirkdale. For it showed not only that the "postdiluvial period" had been very brief by geological standards—as de Luc and Dolomieu had argued long before (*BLT* §6.2, §6.3)—but also that the "antediluvial period", tacitly defined as stretching from the deluge back to the creation of humankind, had covered a comparably brief span of time. This was not a matter of mindless biblical literalism. Buckland was concerned that the reliable results of careful scientific research should be shown to be compatible with what he believed to be the equally reliable results of careful textual research: both were historical, and at this point both involved the same periods of history.

6.4 CRITICS OF THE DELUGE

Buckland's analysis of bones caves was designed, then, to support an underlying *historical* claim: that the catastrophic "geological deluge" was so recent in date that it could be identified as the biblical Flood. This claim continued to have particular resonance in the British context, owing to its perceived implications for the reliability of the biblical documents as a whole, and hence for wider issues of the relation between church and state. On the Continent, where the political situation was quite different, the British preoccupation with issues of "geology and Genesis" left many savants amused, or at least bemused.

Buckland did not have it all his own way, even in Britain. But most of his critics distinguished carefully between two components of the diluvial theory that Buckland himself treated as indissoluble. The first was that there had been some kind of extraordinary event in the geologically recent past, probably a transient mega-tsunami, that had swept across the landmasses of Europe (and perhaps even more widely), eroding the surface and excavating or at least enlarging valleys, scratching the bedrock where it was hard enough to take scratches, and depositing the debris elsewhere as Superficial deposits of sand and gravel, "boulder clay", and large erratic blocks. The geohistorical reality of this "geological deluge" was widely accepted by geologists, however much they argued about its detailed character and effects. They were also agreed that it was very recent in date, by any geological standards: the Superficial deposits were clearly younger than even the youngest of the "regular" Tertiary formations, which in turn were obviously very recent in comparison with the huge pile of underlying Secondary formations and still older rocks (see §9.1).

Buckland's critics balked, however, at the second component of his diluvial theory. This was that the geological deluge was recent enough to be identified as none other than Noah's Flood, which in turn was dated by the traditional textual science of chronology at no more than about four or five millennia in the past. In claiming this equation between physical and textual evidence, Buckland was following the example of his hero Cuvier, who in turn had adopted the claims of earlier savants such as de Luc and Dolomieu. But Cuvier had set the argument in a much wider multicultural context, in which Genesis featured as just one ancient source among many, perhaps no less garbled than the rest (*BLT* §10.3). Buckland (and de Luc), in contrast, focused exclusively on the ancient textual testimony of Genesis. This served Buckland's purpose of giving the science of geology legitimacy and respectability in his local academic environment, while also bolstering the role of religious authority generally; but conversely it laid him open to criticism by those who had their own agendas for undermining that authority.

In the British context, these varied reactions to Buckland's book can be summarized briefly with three examples. First, it was welcomed enthusiastically in the heavyweight *Quarterly Review*, which articulated Tory opinion among the

10. Fig. 6.4 is reproduced from North, "Glacial theory" (1943), 6 [reproduced in turn from Cardiff-NMW: no. 347, dated 1820]. Buch, "Verbreitung grosser Alpengeschiebe" (1815). De la Beche was well aware of the enormity (in all senses) of the problem posed by such erratics, having explored the high Alps far away to the south on the same trip: see De Beer and North, "De la Beche on Mont Blanc" (1950), and §13.2 below.

11. Buckland, *Reliquiae* (1823), iii. The eighty-nine-year-old Shute Barrington (1734–1826) was known in the Church of England as an ally of William Wilberforce, an enlightened reformer, and a patron of scholars such as William Paley (see §29.1).

English social and intellectual elites. The review was anonymous, as was usual in this periodical, but everyone who mattered soon knew it had been written by Edward Copleston (1776–1849), the Provost of Oriel College at Oxford and a leading Anglican scholar. He singled out the uniquely *historical* character of the science of geology as the source of the arguments that surrounded Buckland's work. While rejecting biblical literalism, on the usual grounds that the purpose of scripture was not scientific but moral and religious, Copleston praised his colleague's efforts to show that the record of the Flood in biblical history was compatible with the natural evidence for some such drastic event in geohistory. In contrast, he noted that Cuvier, like other Continental naturalists, "betrays a morbid eagerness to separate his reasoning from scripture": a significant indication that Cuvier's work, despite the scriptural slant given to it in Britain by Jameson's editions (§1.3; *BLT* §10.3), was not always seen as friendly to the supposed interests of religion. In general, Copleston's review was strongly positive, which is hardly surprising since it was soon rumored—indeed it became notorious—that Buckland had coached him on how to handle the more technical issues.[12]

In the *Edinburgh Review*, the Whig counterpart of the *Quarterly*, Buckland's book received a more searching critique at the hands of a more knowledgeable reviewer. Fitton (also notionally anonymous) conceded that the new work was strictly scientific, and quite distinct from the increasingly popular genre of "*scriptural geology*". But since he and other geologists regarded the latter as scientifically worthless, he regretted that Buckland had not kept geology and Genesis more clearly separate. He accepted much of what the Oxford geologist had inferred about the caves, but claimed that he had not proved that the extinction of their inhabitants had been due to a deluge. Yet Fitton agreed that some kind of peculiar event or events had characterized the recent geological past, and he noted that, contrary to Hutton's and Playfair's much earlier claims (*BLT* §3.4, §8.5), "it is now almost universally admitted, that valleys have been excavated by causes no longer in action". And he agreed that the "diluvial" phenomena were extremely widespread, although he noted that Buckland had failed to prove that they were everywhere of the same date. In effect, therefore, Fitton accepted the reality of the diluvial effects, but questioned Buckland's conclusions about their date and duration. He suggested—with Hall's conjectural mega-tsunamis in mind (*BLT* §10.2)—that it was more likely that there had been *several* great "waves" over an extended period of time, rather than just a single unique event. And he urged that attempts to equate any physical events with the biblical Flood should at least be shelved until the strictly geological effects had been studied more thoroughly.[13]

Another well-informed critic, and a near contemporary, took issue with Buckland's assumption—in which he was again following Cuvier—that only a sudden and drastic physical event could have wiped out a whole fauna of well-adapted animals. John Fleming (1785–1857) was a highly competent naturalist who had been closely associated with Jameson in Edinburgh before becoming a Church of Scotland minister in the rural parish of Flisk in Fifeshire. In his two-volume *Philosophy of Zoology* (1822)—the title may have been a tacit challenge to Lamarck's famous transformist (evolutionary) treatise *Philosophie Zoologique* (*BLT*

§10.1)—Fleming had followed Cuvier's practice in animal classification, with a strong emphasis on Cuvierian "conditions of existence" (and, in modern terms, on animal ecology); he had treated the geographical distributions of species as largely determined by their life habits and environment. However, when reviewing Jameson's latest English edition of Cuvier's *Discourse*, Fleming had been highly critical of another aspect of the Frenchman's work. Cuvier was "a good anatomist, and a great naturalist", he wrote, "but we cannot consider him in the light of a geologist". In contrast to Cuvier's fine work on animal classification and on fossil bones, Fleming dismissed his geology as "always deficient, often wrong"; he touched a raw nerve when he added that Cuvier wrote as if Montmartre was the only mountain he had ever seen. Above all, Fleming criticized Cuvier's assumption that only a sudden inundation could have caused the apparent mass extinction: in his opinion more local factors might have been responsible.[14]

Fleming returned to this theme in 1824, in an article for Jameson's periodical. Modestly entitled "Remarks illustrative of the influence of [human] society on the distribution of British animals", it was in fact a powerful critique of the assumption that only a sudden event could have wiped out the fossil animals found in the Superficial deposits. In response to Buckland's *Reliquiae*, Fleming concluded his article with a vehement attack on the English geologist's "indiscreet union of Geology and Revelation", which in his opinion deserved Francis Bacon's classic censure for having led to "*Philosophia phantastica, Religio haeretica*". The criticism showed that not all ordained ministers of established churches approved of Buckland's deployment of Genesis in scientific debate.[15]

The core of Fleming's argument, however, was that Buckland's "Diluvium" did not deserve its name, because the material was local and variable, and lacked the marine fossils that might have proved that the alleged inundation had come from the sea. Fleming argued that the deposits were more likely to be due to "a cause, partial [i.e., local], sudden and transient, like the bursting of a lake": the

12. [Copleston], "*Reliquiae diluvianae* by Buckland" (1823); on Buckland's assistance, see Boylan, *Buckland* (1984), 127.

13. [Fitton], "*Reliquiae diluvianae* by Buckland" (1823). Penn, *Mineral and Mosaical geologies* (1822), was the model for a burgeoning genre of popular works. Millhauser's classic "Scriptural geologists" (1954) is still a useful survey of the genre; a representative selection of relevant texts is reprinted, with a historical introduction (1: ix–xx) by the editor, in Lynch, *Creationism and scriptural geology* (2002). Klaver, *Geology and religious sentiment* (1997), rightly distinguishes the "scriptural" writers from *all* the scientific geologists, whether religious or not; Roberts, "Geology and Genesis" (1998), explores the origins of the myth of "conflict". O'Connor, *Earth on show* (2007), an outstandingly important account of the popular impact of geology in early nineteenth-century Britain, includes (chap. 3) what is now the best historical analysis of the "scriptural" genre; this work reached me just too late to incorporate its insights into the present volume as thoroughly as they deserve.

14. Fleming, *Philosophy of zoology* (1822), 2: 88–105; "Theory of the earth, by Cuvier" (1823); see also Rehbock, "John Fleming" (1985). Gypsum quarries at Montmartre (a 100m-high hill then just outside Paris but long since absorbed within it) had yielded Cuvier's famous Tertiary fossil mammals (*BLT* §7.2, §7.5). Flisk is on the Firth of Tay 20km east of Perth: not a remote location, in view of the efficient coach and postal systems operating throughout Britain by this time.

15. Fleming, "Influence of society" (1824). The quotation was from Bacon, *Novum organum* (1620): "this unhealthy mixture of things divine and human begets not only fantastic philosophy but also heretical religion" (book 1, aphorism 65 [trans. *Oxford Francis Bacon* 11 (2004): 103]); the context was a general criticism of those who tried to build natural philosophy on biblical books such as Genesis and Job (I am grateful to Richard Sarjeantson for enlightening me about this). Fleming evidently thought that little had changed between Bacon's time and his own.

allusion was clearly to the notorious Alpine catastrophe in the Val de Bagnes in 1818, which had led von Buch to develop his concept of a mega-tsunami as something like a turbidity current (*BLT* §10.5). And Fleming claimed that the putative mass extinction of the "diluvial" megafauna was more likely to have been gradual and piecemeal, and due to human activities such as hunting and the clearing of forests.[16]

This suggestion pointed to what in the present context is most important about Fleming's position. He criticized geologists for being too fond of the alluringly remote eras of deep time: "were they to commence their investigations, with a knowledge of recent events, and proceed by degrees to those of remoter times, their conclusions would assume a more imposing character". In other words, Fleming advocated the heuristic method of penetrating back from the known present into the more obscure past, which Desmarest and other naturalists had pioneered so effectively in the previous century (*BLT* §4.3). He claimed that the historical records of the impact of human activities on animal life—he took his examples from Britain since the medieval period—could be extended much further back by using the *natural* "memorials" of still earlier times, and that this would show that no "physical revolution" or violent deluge need be invoked to explain the alleged mass extinction. It was a potentially lethal argument to deploy against Buckland's diluvial interpretation.[17]

However, Buckland was not greatly perturbed by the criticisms that were made of his work. The 1000 copies of *Reliquiae* had been sold within six months, which was a striking sign of its interest to a wide range of anglophone readers. An almost identical second edition appeared the next year; he added a brief postscript, giving the important news that Cuvier, after initial doubts, now agreed with his interpretation of the bone caves. In anticipation of finding further evidence in its support, and to give himself a chance to respond to his critics, he made plans for a second volume.[18]

6.5 CONCLUSION

Buckland interpreted Kirkdale Cave sensationally as a former hyena den, and its fossil bones as relics of a vanished "antediluvial" ecosystem; like Cuvier, he claimed that this megafauna had been wiped out by the sudden and geologically recent "deluge" that he equated with the biblical Flood. This diluvial theory was supported not only by his own earlier research on landforms and Superficial deposits, which he interpreted in terms of the same drastic event, but also by his exploration of new bone caves elsewhere in Britain. Paviland Cave in Wales was particularly important, because his discovery of a human skeleton impinged on the vexed question of the antiquity of the human species. But Buckland concluded from the field evidence that the "Red Lady" (as the skeleton was later dubbed) was in fact "postdiluvial", and that there was no good reason to abandon Cuvier's firm opinion that there was still no authentic case of any human remains from the diluvial or Superficial deposits. There was therefore no evidence, in Buckland's opinion, that early humans had spread from their putatively Asiatic point of origin as far as Europe, before the deluge struck; the apparent mass extinction had still to be attributed to that catastrophic natural event.

Buckland summarized his research and his conclusions in his book *Reliquiae Diluvianae* [Relics of the Deluge], which made his work much more widely known than his earlier papers to the Royal and Geological Societies in London. His critics distinguished sharply between his evidence for the reality of a sudden and drastic event at a geologically recent time, and his claim that it had been recent enough to be identified as none other than the biblical Flood: the former claim was widely accepted, but not the latter. As representative examples, Copleston welcomed his colleague's equation of the two, and recognized the issue as one of *history*, matching the human with the natural; Fitton accepted the reality of Buckland's "geological deluge" but not its identification as Noah's Flood; and the Scotsman Fleming, although like Buckland a minister of religion, deplored the Englishman's mixing of geology with Genesis and queried the reality of both the deluge event and the mass extinction that it had supposedly caused.

When Fleming suggested that the allegedly sudden mass extinction might in fact have been a gradual and piecemeal process without any drastic physical "revolution" at all, he was raising an important underlying issue of more general scope, namely the relation between the present world and even the most recent part of geohistory. It was a sign of a debate that was going on in parallel with Buckland's diluvial theorizing. The debate was concerned with the proper role of observable present processes (de Luc's "actual causes") as a means of explaining the deep past of geohistory. This is the topic of the next chapter.

16. On Fleming's geology, see Page, "Diluvialism and its critics" (1969), which summarizes his *Rise of the diluvial theory* (1963); also Burns, "John Fleming" (2007).

17. Fleming, "Influence of society" (1824); his argument for piecemeal extinction was one that he would soon develop further (see §14.3).

18. Buckland, *Reliquiae*, 2nd ed. (1824), 229–31; Buckland to Cuvier, 20 June 1824 [Paris-MHN: MS 629, no. 16]; his second volume never appeared. Cuvier had described the Kirkdale fauna in his *Ossemens fossiles*, 2nd ed., 4 (1823): 302–4, within a review of bone caves in general (291–309).

CHAPTER 7

The role of actual causes (1818–24)

7.1 THE ADEQUACY OF ACTUAL CAUSES

Cuvier's research, like that of other savants with interests in geohistory, had been based on the standard actualistic method of using the present as the primary key to the past. His identification of fossil bones as belonging to species distinct from any known alive, and hence his inference that there had been a mass extinction at the "last revolution", had been derived from his close and rigorous comparison between fossil bones and those of living animals (*BLT* §7.1, §7.5). Likewise Buckland's recognition of Kirkdale Cave as a hyena den was based on an equally rigorous comparison between what he had found there and what he saw living hyenas doing, eating and defecating (§6.1; *BLT* §10.6). His use of the stalagmite as a "natural chronometer" to estimate the lapse of time before and since their occupation of the cave was similarly based on well-known observations of the slow but perceptible rate of accumulation of that material. As has already been emphasized repeatedly in this narrative, an actualistic method was taken for granted on all sides as being obviously the best strategy for a science that was increasingly concerned with understanding events *in the past* as well as those directly observable in the present.

What was much more contentious was the extent to which the traces of events in the distant past could be explained *adequately* in terms of observable processes. The classic case was that of the origin of valleys, many of which had clearly been carved through otherwise undisturbed masses of solid rock. Ever since Desmarest's pioneer fieldwork on the extinct volcanoes in Auvergne (*BLT* §4.3), his inference that the valleys there had been excavated by the slow action of the streams and rivers that still flow in them had been countered by claims that, regardless of the magnitude of time allowed, such a puny process was quite inadequate to explain the scale of the phenomenon, which would therefore need to be attributed to some more powerful agent. Here it is sufficient to recall Cuvier's famous formulation of the problem in his "Preliminary Discourse" (*BLT* §10.3), which was reissued—unchanged on this point—in the new edition of his *Fossil Bones* (§1.2), just as Buckland was engaged in his research on diluvial phenomena and shortly before the discovery of Kirkdale Cave brought that work to its climax.

Cuvier was convinced that no known process could account for the apparently simultaneous disappearance of many mammalian species in the geologically recent past, or for the Superficial deposits in which their fossil bones were found. But he had extrapolated this specific point into a much more general claim, that the physical processes observably active in the present world were quite *inadequate* to explain all the equally observable effects that survived as the traces of past events (*BLT* §10.3). To justify this argument, Cuvier had reviewed currently "active causes", but in the event he had offered no more than a brief and perfunctory survey of the processes of subaerial erosion, marine erosion and sedimentation, and volcanic activity. Yet the inference he drew from it had been stated in characteristically confident and dramatic terms: "the thread of operations is broken; nature has changed course, and none of the agents she employs today would have been sufficient to produce her former works".[1]

Cuvier's claim that present processes were not "sufficient" for causal explanation of the deep past had not gone unchallenged. But in any case it remained unclear, for example, whether massive turbidity currents and mega-tsunamis—as invoked by von Buch and Hall as the possible cause of the earth's major "revolutions" (*BLT* §10.2)—should count as different in kind or merely different in degree from the well-attested effects of the lesser floods, mudslides, and tsunamis recorded in human history. If an actualistic method were to be pursued as the most effective strategy for a science dealing with unobservable past events, "actual causes" would need to be studied much more thoroughly. Despite Cuvier's bold rhetoric, they could not properly be pronounced inadequate—or adequate—to explain the "former world", until it was known just what they were capable of doing in the present world. A better understanding of the deep past depended critically on a fuller knowledge of the directly observable present (and of the recorded human past, which in this context counted as the present).

Two contrasting research projects had both attempted to tackle this issue. The first was derived from the kind of geotheory that had been formulated years earlier by Hutton (*BLT* §3.4) and championed after his death by Playfair (*BLT* §10.1). In order to avoid having to invoke any extraordinary or "accidental" effects such as the Flood or deluge (as traditionally understood), and conversely to demonstrate the total efficacy of the ordinary ahistorical "laws of nature", the Huttonians had insisted on the adequacy of present processes to explain *all* the traces of the deep past. Playfair, more clearly than Hutton himself, had emphasized the power of ordinary processes of erosion and sedimentation to produce even the largest of observed effects, provided they were allowed enough time in which to operate. Hutton's own work had almost dropped out of sight, its pervasive deistic metaphysics being hardly in tune with the empirical temper of the new century.

But Playfair's book had remained well known, at least in the anglophone world. It was respected, even by its critics, as offering a well-argued case not only for a sanitized version of Hutton's steady-state geotheory but also for the adequacy of actual causes in earth physics.

In the year of Waterloo, a complete translation of Playfair's book, balanced by a contemporary critique of Huttonian geotheory, had been published in Paris. The editor had rejected Cuvier's earlier and notorious jibe that such theories had reduced "geology" to a laughing matter (*BLT* §8.4); he had offered his edition as a way of promoting a discussion of rival theories that might eventually lead to a more satisfactory science of the earth. It had certainly made Huttonian geotheory much better known throughout Continental Europe than it had been during the years of war. But it also had the incidental effect of bringing the *causal* goals of earth physics (*BLT* §2.4) back into the international arena of scientific debate, by presenting persuasive arguments in favor of the explanatory use of actual causes.[2]

Ironically, the second project that promoted the discussion of actual causes was due to de Luc, Hutton's most persistent and formidable critic. For de Luc's dogged insistence on the reality of a sharp boundary at a "great revolution", and on its dating at the dawn of human history, had led him to study actual causes (as he was the first to call them) with greater thoroughness than any other savant of his generation. The ground of his persistent criticism of Hutton was that he concluded from his fieldwork that the effects of actual causes could *not* be traced back throughout geohistory, but only as far as his putative boundary event (*BLT* §6.2). Nonetheless, although his observations were all directed towards that geotheoretical objective, his steady stream of voluminous publications had served to keep alive a tradition of studying the changes of physical geography that had been recorded during human history.[3]

7.2 VON HOFF AND NATURE'S "STATISTICS"

De Luc had died at Windsor in 1817, at the ripe age of ninety, having retained his positions not only at the court of George III but also as the first professor of geology at Göttingen, in the King's Hanoverian territory recently recovered

1. Cuvier, *Ossemens fossiles* (1812), 1, "Discours préliminaire", 17; 2nd ed., 1 (1821), 14; trans. Rudwick, *Cuvier* (1997), 193.

2. Basset, *Explication de Playfair* (1815), "Avant-propos" (v–xv). César Auguste Basset (1760–1828) was the director of the École Normale in Paris; the volume contained his translations of Playfair, *Illustrations* (1802), and of the critique by Murray, *Comparative view* (1802). Significantly, its sole illustration (as a frontispiece) was Basset's re-drawing of Hutton's pictures of septarian nodules as evidence for the alleged high-temperature consolidation of *all* sedimentary rocks by fusion: this, together with the idea of the earth's dynamic interior and a consequent crustal mobility, continued to be regarded as the essential core of Huttonian theory (*BLT* §3.4). An indefinitely lengthy timescale, by contrast, had long been taken for granted by all serious naturalists, and was *not* regarded as a specifically Huttonian proposition.

3. Luc, *Traité élémentaire* (1809), contained the final statement of his geotheory, and his *Geological travels* (1810–11, 1813) made available a mass of earlier fieldwork observations that, he claimed, gave it further empirical support.

from Napoleonic occupation. It was probably no coincidence that in the following year the Royal Society of Sciences at Göttingen resolved to offer its next major prize for an essay on a topic that had been close to de Luc's heart. The prize question was proposed by Johann Friedrich Ludwig Hausmann (1782–1859), the university's professor of mineralogy and technology. It was duly approved by Blumenbach, the professor of medicine and the society's secretary, and by other members. Among the latter were the university's distinguished historians and philologists, since of course the Society interpreted the "sciences" [*Wissenschaften*] in its title with traditional Continental breadth (*BLT* §1.4); the great orientalist and biblical scholar Johann Gottfried Eichhorn (1752–1827), for example, approved the proposal "with enthusiastic pleasure". This consensus was appropriate, for the topic focused on ground that was common to the natural sciences and the humanities: it would require the critical use of human *historical* sources to investigate a fundamental issue in geology.[4]

The preamble to the Göttingen prize question summarized what could by now be taken for granted, at least among savants, about the course of geohistory. Echoing Blumenbach himself (*BLT* §6.1, §8.1), and of course Cuvier, it was now clear from the geognostic study of formations [*Lagen*] that they had been formed over vast spans of time, as the products of major "revolutions of the earth" [*Erdrevolutionen*]. But although these events could be placed in their correct temporal order, the absolute timescale of geohistory remained utterly uncertain. The Society therefore offered its prize for a careful study of the evidence for physical changes at the earth's surface within the span of "historical tradition" [*geschichtlicher Überlieferung*]. This was to provide a possible basis for estimating the timescale of deeper geohistory, or at least to throw further light on the changes or revolutions that lay "beyond the reach of [human] history". The prize question made no direct reference to de Luc, or even to his notion of "nature's chronometers"; but the suggested research was in fact grounded in what the deceased professor of geology had tried to do, in using actual causes to estimate the timescale of the most recent period of geohistory (*BLT* §6.2). At the same time, however, the prize question released such research from the limitations of de Luc's rather idiosyncratic goals, by assuming that the physical changes recorded in the span of human history might *not* be restricted to that period, and might instead be extrapolated backwards in time to form the basis for equally valid "chronometers" for much deeper geohistory.[5]

Three years later, in 1821, the prize was awarded to Karl Ernst Adolf von Hoff (1771–1837), a bureaucrat in the small ducal state of Gotha. There could hardly have been a more appropriate recipient, or a work that answered the question more effectively. Von Hoff had been a student at Göttingen, where he had covered a wide range of humanistic subjects relevant to his future career, but had also attended Blumenbach's famous lectures on natural history. Subsequently, his work as a civil servant and diplomat had limited his opportunities for scientific research; and unlike von Buch and Humboldt, for example, he had no private wealth. But Duke Ernst II of Gotha was an enlightened ruler, who encouraged scientific and scholarly work among his subjects, not least by subsidizing their publications. Early in his career, von Hoff had founded and edited a periodical for the earth sciences, though it had not lasted long; he had published articles on the basalt problem; he had made a careful barometric survey of the physical geography of Thuringia; and one brief article, more directly relevant to his later prize essay, had reported on the recent formation of a new island in a tributary of the Elbe. In parallel with all this work on the sciences of the earth, von Hoff had also published substantial historical research. For example he had written two volumes comparing the German lands before the start of the French Revolution in 1789 and after the treaty of Lunéville that in 1801 had brought peace (in the event, only temporarily) between France and Austria. In short, von Hoff was a savant equally at home in geology and in history.[6]

Von Hoff's prizewinning volumes were subtitled "a geographical-statistical parallel", which clearly located his research within the German tradition of *Statistik*. This had nothing to do with mathematical statistics in the modern sense. Instead it comprised the disciplined collection and classification of accurate factual information on all aspects of life relevant to the state—demography, education, commerce, religion, and so on—expressed where possible in quantitative terms and usually organized along geographical lines. *Statistik* was one of the bureaucratic sciences [*Kameralwissenschaften*] that served the "statist" or statesman in the governance of the many and varied independent states within the German lands. At Göttingen, however, it had taken a significant turn towards incorporating a historical dimension. In von Hoff's student days Johann Christoph Gatterer (1727–99), the then professor of history, had been famous for insisting that the scope of history should be enlarged to include "statistical" data; and from the other side his colleague August Ludwig Schlözer (1735–1809) had urged the importance of giving *Statistik* a historical perspective. In Schlözer's pithy aphorism, "history is ongoing statistics, statistics is stationary history". This was the kind of "statistics" that von Hoff had imbibed as a student; and thirty years later he transposed it from culture into nature, when he answered the Göttingen prize question by offering a historical *Statistik* of physical geography.[7]

Von Hoff's work was published at Gotha under the title *History of the Natural Changes at the Earth's Surface That Are Attested by Tradition*. The first volume (1822), which won him the prize, dealt with changes in the relation between land and sea; the second (1824), with changes due to volcanic activity and earthquakes. Like any work in human *Statistik*, the material was set out geographically. Von Hoff's coverage was of course best for Europe and the Mediterranean basin, but

4. Hamm, "Hoff's *History*" (1993), quotes Eichhorn's remark from the Society's archives; in the present section of this chapter I am greatly indebted to this important article and to Ernie Hamm himself.

5. The prize was offered, and the question announced with its preamble, in the Society's newsletter [*GGA* 1818: 2047–48], reprinted in the work that won the prize, Hoff, *Veränderungen der Erdoberfläche* 1 (1822), xiii–xiv (see below); it is translated in Hamm, "Hoff's *History*" (1993), 155–56.

6. Hoff, *Teutsche Reich* (1801–5); his scientific work is cited in Hamm, "Hoff's *History*" (1993); see also Mathé, "Karl von Hoff" (1985). Von Hoff himself assumed the noble prefix "von", although in fact this was only officially restored to his family in 1838, in posthumous recognition of his services to Gotha: see Reich, *Hoff der Bahnbrecher* (1905), 3n, a biography that is still valuable.

7. Hamm, "Hoff's *History*" (1993); on *Statistik* generally, see for example Hacking, *Taming of chance* (1990), 16–26. On Humboldt's and Candolle's "botanical arithmetic", an analogous project in natural history at this time, see Browne, *Secular ark* (1983), 58–64.

he also summarized what was known from more remote parts of the world, from Siberia to the Americas. All the changes in physical geography that he reported were arranged systematically: for example he grouped together his careful appraisals of cases where the sea had encroached on the land, and then those in which new land had extended across former areas of sea. His evidence drew on documents from many periods of human history, from the Classical literature to recent publications, from Herodotus to Humboldt. The older records demanded a critical treatment of material that might be merely fabulous or mythical. For example, von Hoff evaluated suggestions that the Classical story of Deucalion's Flood might be a faint trace of a real event that had breached the narrow isthmus at the Bosphorus, between the Mediterranean and the Black Sea. Like Cuvier in his review of the records of animal species in Antiquity (*BLT* §10.3), von Hoff

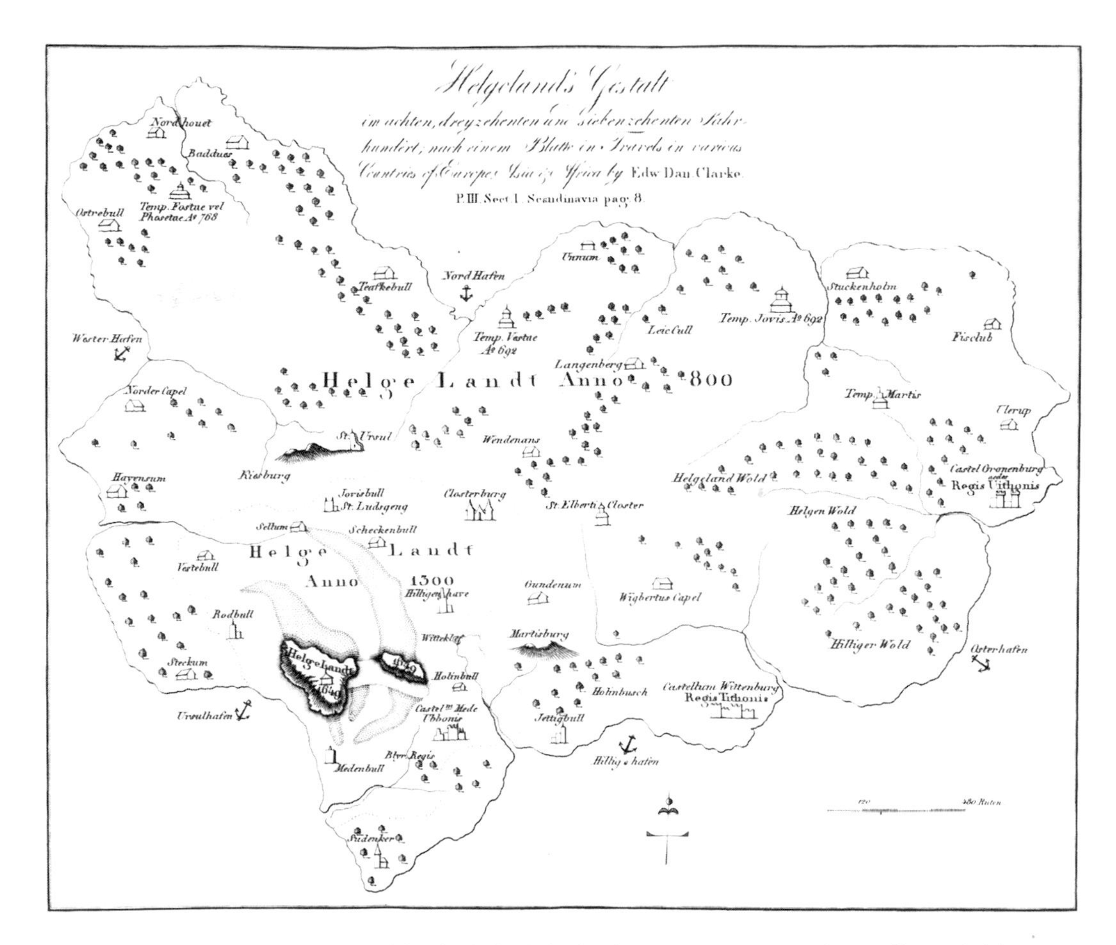

Fig. 7.1 A map of the island of Helgoland in the North Sea, showing its historically recorded diminution in size from around A.D. 800, through an intermediate state in about 1300 (outlined by a faint dotted line) to its small remnants in 1649 (the scale is of 480 *Ruten* [rods] or about 2.5km). This map, published by von Hoff in 1822, was the *only* illustration in his volumes describing "natural changes at the earth's surface attested by tradition". The example gave a vivid impression of the scale of geographical change that had taken place, at least locally, in the course of just one millennium of human history. Yet on a global scale even changes of this kind were almost negligible in relation to those attested by purely geological evidence, during the vast tracts of time "beyond the reach of [human] history". (By permission of the British Library)

was in effect applying to ancient records of physical geography the careful textual criticism that Eichhorn and his Göttingen colleagues were using on the Old Testament and other ancient texts.[8]

As von Hoff's own situation dictated, most of the research that lay behind what he himself described—too modestly—as "only a tedious compilation of many facts" had been library work rather than fieldwork; appropriately, he dedicated his volumes jointly to his scientific mentor Blumenbach and to the chief librarian at Gotha. His work was a descriptive inventory of physical geography with a historical dimension; it was not, and was not intended to be, a contribution to earth physics. Von Hoff was well aware that there were alternative causal explanations for the changes he described. For example, the land might have increased in area, either because sea level had fallen globally (in modern terms, eustatically) or because the earth itself had risen in that particular region (by crustal elevation), or because new sediment had been deposited in a delta or along a coastline. In the case of the well-known reports of continuing changes in sea level in the Baltic (*BLT* §3.5), Von Hoff reviewed impartially the evidence for and against the alleged effect; he left it to others to decide whether the sea was falling or the land rising, and to determine in either case what the physical cause might have been. He recognized that no conclusion could be reached on such matters from historical research alone, and he himself had no opportunity to investigate the phenomena at first hand in the field. His sole illustration reflected this limitation, in that it was borrowed from one of his published sources. But it exemplified with visual eloquence the kind of "change attested by tradition" that his documentation recorded (Fig. 7.1).[9]

Von Hoff's interpretative comments on his inventory were largely confined to his brief introduction to the work. He emphasized that the physical changes that he had been able to document historically extended back only to about 2000 B.C., and covered only a limited part of the globe. A total of some four millennia was as nothing to the vastness of geological time, as suggested for example by the thousands of feet of fossil-bearing Secondary strata exposed in some regions. Furthermore, although the changes "attested by tradition" were real enough, they were on such a small scale that they could be shown only on relatively large-scale

8. The "Deucalionische Fluth" was discussed, in a geographical context that extended to the Caspian Sea and even beyond it to the Aral, in Hoff, *Veränderungen der Erdoberfläche* 1 (1822), 105–44. Cuvier had cited Eichhorn in the course of assessing the historicity and date of the Flood recorded in Genesis (*BLT* §10.3).

9. Fig. 7.1 is reproduced from Hoff, *Veränderungen der Erdoberfläche* 1 (1822) [London-BL: 725.g.12], pl. at end of volume, explained on 56–57. It was redrawn from Clarke, *Travels in various countries* 5 (1819), pl. opp. 8. Edward Daniel Clarke (1769–1822), the professor of mineralogy at Cambridge, stated (8–9) that his engraving was based on an old map that he had been shown, but gave no further details; probably it was a map of about 1649, with the earlier topographies reconstructed from documents recording land tenure, etc. By his (and von Hoff's) time there had been further erosion; that Helgoland still exists at all is due to artificial sea defences (the island had been seized from Denmark during the wars, was formally ceded to Britain in 1814, and remained a British possession until it was transferred to Germany in 1890). Clarke's visit was just a brief stopover; on his extensive travels, see Dolan, *Exploring European frontiers* (2000). Von Hoff called his own work "nur ein mühsames Compilation" in a letter of 1 December 1824, quoted in Hamm, "Hoff's *History*" (1993), n108. He published a third volume in 1834, and two more appeared posthumously (1840–41), but these contributed to distinctly later debates (see §22.1).

maps (such as that of Helgoland). Even if inhabitants of the Moon—as he put it, in an imaginative thought experiment—had had access, throughout these millennia, to the best telescopes currently available to terrestrial astronomers, they would have been unable to detect any changes at all. Still, von Hoff was in no doubt that the effects he had catalogued would have been far more substantial in the course of the vast tracts of time "beyond the reach of history". So the prescriptive point was clear. Geological investigation should not begin with granite and suchlike Primary rocks, as was customary both in geognosy and in geotheory, for that path led only to "a chaos of hypotheses". Instead, the starting point should be the "newer and clearer appearances" for which there were or had been human witnesses. For von Hoff, the present world, including the recent human past recorded historically, was unquestionably the proper and most effective key to the far longer spans of geohistory.[10]

In line with the prize question to which he had responded, von Hoff did not attempt to penetrate beyond the span of recorded human history into those further reaches of geohistory: he limited himself explicitly to compiling the changes in physical geography for which there were reliable human records. He sidestepped the question of the historicity of a major "earth-revolution" in the geologically recent past, in the form of some kind of *global* deluge, by assuming tacitly that no such event was reliably "attested by tradition". He did not deny the possible historicity of the various reports on which the idea of a deluge had been based, but he implied that all of them, like Deucalion's Flood, were likely to refer to strictly local or regional events. In omitting all mention of the supposedly worldwide deluge of Genesis, it was tacitly sidelined as an event that could only have been "universal" in the limited perspective of the ancient tribes that had recorded it, just as Eichhorn had famously demythologized the "primal history" [*Urgeschichte*] of the Creation story at the start of the Genesis text. Von Hoff did not rule out the possibility that in the further reaches of geohistory there might have been far larger physical "revolutions", perhaps even ones with a truly global impact, but the evidence for them would be purely geological and therefore lay outside the terms of reference of the prize question.

At the start of his second volume (1824), von Hoff admitted that the historical evidence of volcanic eruptions and earthquakes was such that his compilation was unavoidably weighted much more heavily towards the geological description of the effects, with the historical material doing little more than supplying the dates. In other words, his work, however useful for documenting the incidence of eruptions and earthquakes within the past few millennia, became barely distinguishable from the long-established genre of descriptive physical geography (*BLT* §2.2). The practice of this science had continued unabated in the past decades—Humboldt was currently its most distinguished exponent—but it had remained almost independent of the development of the geohistorical perspective that is the focus of the present narrative. However, as von Hoff's work suggested, physical geography could now be seen more clearly to offer valuable raw material for geohistory. One example will here suffice to illustrate the kinds of local sources on which his great compilation was based (a second example will be described in detail in the next chapter).

7.3 ETNA: EUROPE'S GREATEST VOLCANO

Etna was not only the largest active volcano in Europe but also the most fully documented volcano anywhere in the world. Its records stretched far back into Antiquity; its eruptions had been observed long before Vesuvius sprang catastrophically into activity in A.D. 79, burying the towns of Herculaneum and Pompeii (*BLT* §4.1). But as Sir William Hamilton had noted—half a century before von Hoff studied his and other earlier accounts—some of the minor cones of volcanic ash on the flanks of Etna, and the lavas that had flowed from them, appeared to be much older than any human records, suggesting an unimaginable antiquity for the huge volcano as a whole. Hamilton's compatriot Patrick Brydone had gleefully trumpeted abroad the quandary in which the local naturalist (and priest) Giuseppe Recupero had found himself at that time, since any vast antiquity seemed incompatible with the traditional short timescale for the earth that was still taken for granted by Recupero's bishop, though not by savants (*BLT* §2.5). However, decades later, when Recupero's book on Etna was published at last, worries about the earth's timescale were much less of a problem, even in benighted Sicily, and the volcano was being studied more intensively than ever.

After the end of the wars, Francesco Ferrara (1767–1850), the professor of physics at Catania and the official in charge of Sicily's antiquities (and also, like

Fig. 7.2 "View from Catania of the 1787 eruption [of Etna] on the night of 18 July", showing explosive activity at the summit of the 3300m-high volcano, and a lava flow down its west (left) flank. This visual record was already a part of the *history* of the volcano, by the time it was republished by the local savant Francesco Ferrara in his *Description of Etna* (1818). (By permission of the Syndics of Cambridge University Library)

10. Hoff, *Veränderungen der Erdoberfläche* 1 (1822), "Einleitung" (1–20); see also his interpretative "Schlussbemerkungen" at the end of each chapter.

Recupero, a priest), reissued his own much earlier account of Etna, highlighting "the history of the eruptions" from Antiquity to his own time. Earlier descriptions of particular eruptions, like Hamilton's famous pictures of Vesuvius, now became *historical* evidence for an account of Etna that could trace the "actual cause" of volcanic activity through much of the timespan of "the modern world" (Fig. 7.2).[11]

Ferrara's work was complemented around the same time by a visual representation of the historical record of many more of Etna's eruptions. Mario Gemmellaro (1773–1839) was a local naturalist who had long been studying the volcano; he and his brothers had even built a refuge at the foot of the summit cone, to

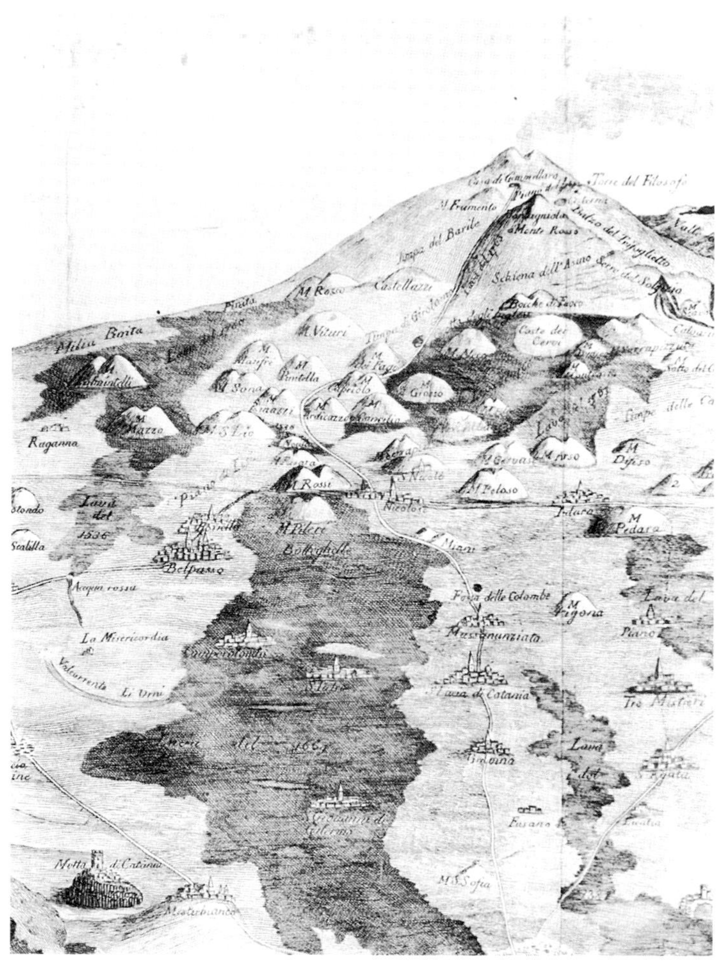

Fig. 7.3 A part of an imagined aerial view of the southern slopes of Etna, drawn by the Sicilian naturalist Mario Gemmellaro and published in 1820 or earlier; his refuge ("Casa di Gemmellaro") is marked just below the summit cone and crater. The engraving shows the many minor cones of volcanic ash dotted around the flanks of the main volcano, and those lava flows that could be identified from historical records (those shown here ranged in date from 1284 to 1792). The flow in the center was the largest ever recorded on Etna: it had erupted in 1669 from the double cone of Monti Rossi, and had breached the town walls of Catania on the coast many miles away. Etna was the most fully documented volcano in the world, so its humanly slow rate of accumulation was taken to be typical of volcanoes in general. Here the "actual cause" of volcanic eruptions could be traced back without interruption into the depths of geohistory, since the huge volcano seemed to be uniform in structure and many of the minor cones apparently dated from before the earliest human records.

which they regularly took the more energetic of visiting savants. Around 1820 he issued an engraving—probably for the benefit of such tourists—on which he plotted the minor cones on the flanks of Etna that faced towards Catania, together with those lava flows that could be dated from historical records. His drawing was crude in style but rich in scientific meaning. For it showed that although there had been eruptions of great magnitude in past centuries, they had added only a few new cones and lava flows to the vast bulk of the volcano. If they could be relied upon in Delucian manner to act as a natural chronometer, they implied that Etna as a whole must be of unimaginable antiquity. Even if the scale of its eruptions had been greater in the distant past, its total history would still have to stretch far further back than the earliest records. Hamilton had suspected this, but Gemmellaro's great local knowledge helped to put the inference on a sounder basis. What von Hoff's example of Helgoland showed for the timescale of marine erosion, Gemmellaro's view of Etna suggested for the timescale of volcanic activity (Fig. 7.3).[12]

7.4 ACTUAL CAUSES AND GLOBAL EXPLORATION

Finally, it should be noted that von Hoff was compiling his inventory of actual causes at a time of intense activity in geographical exploration by all the leading European nations. Although most voyages and expeditions had primarily strategic or commercial goals, they generated a wide range of scientific accounts of remote and exotic regions. These put into von Hoff's hands, and into those of other savants, a vast new store of information on physical geography worldwide. Much of this could not be squeezed into the parameters of the Göttingen prize question that von Hoff had set out to answer, because most expedition reports described only the *present* state of the lands that had been explored, not their mutations in the course of human history. Nonetheless, they suggested how the repertoire of "actual causes" might need to be expanded and enriched: not necessarily by any new *kinds* of physical process but by a new appreciation of the *scale*, and hence the explanatory power, of those already known. The geographical records of features in distant parts of the globe showed that even in the present world some actual causes were operating on a far larger scale than anything known in western Europe or around the Mediterranean.

11. Fig. 7.2 is reproduced from Ferrara, *Descrizione dell'Etna* (1818) [Cambridge-UL: MF.44.12/1], pl. 5, drawn by Nicola Bombara and engraved by Antonio Zacco, first published in Ferrara, *Storia generale dell'Etna* (1793), pl. 5; both titles include the crucial phrase "la storia delle eruzioni", where "history" includes the newer meaning of temporal narrative rather than only the older one of atemporal description. Hoff, *Veränderungen der Erdoberfläche* 2 (1824), 223–45, summarized data on Etna, and on the volcanic and seismic features of Sicily and Campania generally.

12. Fig. 7.3 is reproduced from part of the engraving, Gemmellaro (M.), *Prospetto meridionale dell'Etna* (n.d.). Alfieri, *De Aetna* (1981), pl. 22, reproduces what appears to be an earlier state (without decorative vignettes) of the same print, and dates it around 1820; but since the most recent lava shown is dated 1792 it may have been drawn (though perhaps not printed) before the turn of the century. A map of the whole volcano, of similar design but more diagrammatic in style, Gemmellaro (G.), *Quadro istorico dell'Etna* (1828), was later published in London with a bilingual text listing eruptions from 1226 B.C. to A.D. 1824; it is reproduced in part in Alfieri, colored pl. 24. Chester *et al.*, *Mount Etna* (1985), figs. 3–18 and table 3.3, are updated modern versions of the same kind of data.

The deltas of the Amazon, the Ganges, and the Mississippi, for example, showed that where large rivers reached the sea new land was being formed, by the deposition of muddy sediment, on a scale far beyond that of the Rhine, the Rhône, the Po, or even the Nile. Likewise, Humboldt's celebrated exploration of central America, and similar work on physical geography by other savants, had revealed the vast scale of the volcanoes strung out along the Pacific coast of the Americas, which dwarfed Etna in height, let alone Vesuvius and the other little volcanoes of Europe. This was neatly shown in a chart that the chemist Charles Giles Brindle Daubeny (1795–1867), a younger colleague of Buckland's, published to illustrate his Oxford lectures on volcanoes (Fig. 7.4).[13]

However, the most striking examples of newly appreciated physical features of the globe came from the polar latitudes. The Antarctic landmass had barely been glimpsed, and was known mainly at second hand by its drifting icebergs. But towards the other pole the commercial efforts of whalers at high latitudes, coupled with the many expeditions in search of the elusive Northwest Passage around the top of North America, were disclosing the physical geography of the Arctic as never before. Specifically, they revealed the sheer scale of the glaciers

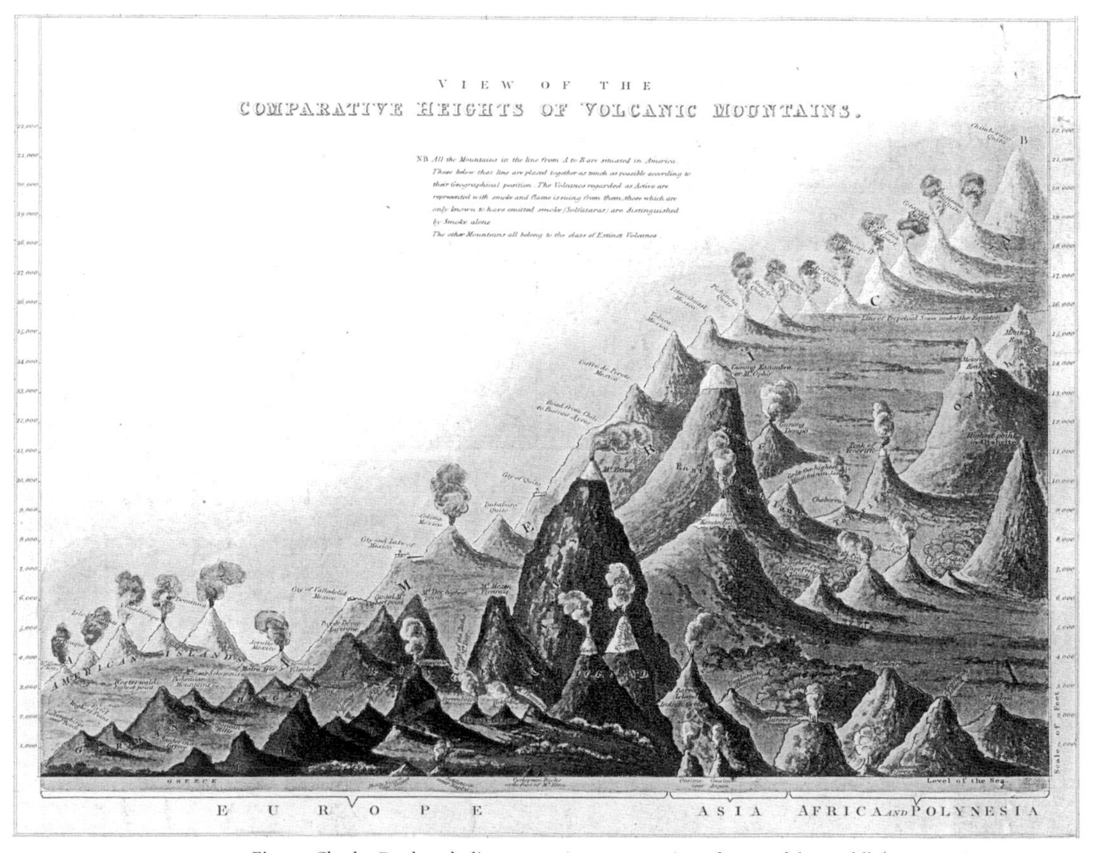

Fig. 7.4 Charles Daubeny's diagrammatic representation of some of the world's known volcanoes, arranged by height (scale in feet above sea level) and continent: those in Europe (center foreground, darkest tone) are dwarfed by those in the Americas (in the background), with Chimborazo in the Andes as the highest of all. This diagram was published to accompany Daubeny's book on volcanoes (1826); he had probably used a preliminary version to illustrate his earlier Oxford lectures.

around the coasts of Greenland, apparently derived from a vast ice-sheet in the interior. Spilling down to sea level, these glaciers generated huge icebergs, which drifted away into the north Atlantic, creating a serious hazard for flimsy wooden sailing ships. It was not clear what role, if any, great physical phenomena such as these might play in understanding the deep past of geohistory; but certainly they enlarged the range and scale of the actual causes that might in the future need to be taken into account (Figs. 7.5, 7.6).[14]

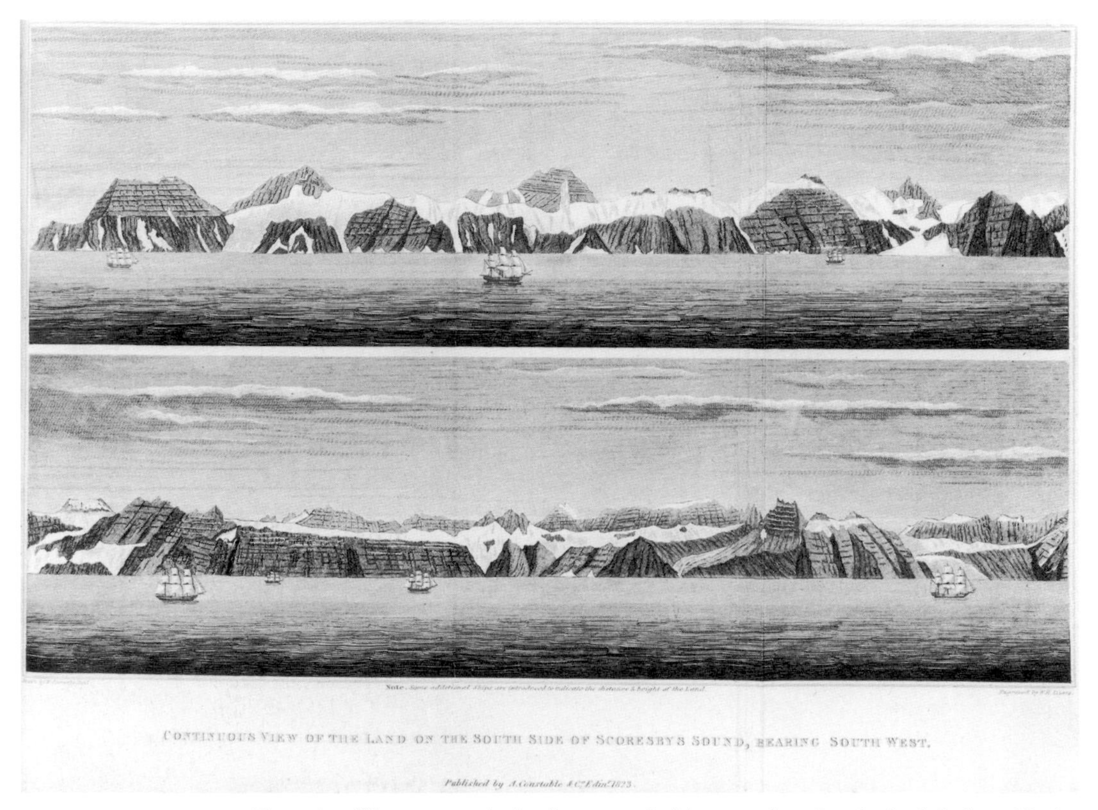

Fig. 7.5 An offshore panoramic view (in two parts) of the coast of east Greenland at latitude 70° North, sketched in 1822 on the younger William Scoresby's voyage to the whaling grounds in the far north Atlantic (and published in 1823). The artist inserted other—imaginary—ships, to help give a sense of the sheer scale of the coastal mountains and the ice-sheet capping them, although this gave a misleading impression of the vast emptiness of these Arctic waters. (By permission of the British Library).

13. Fig. 7.4 is reproduced from Daubeny, *Tabular view of volcanic phaenomena* (1826?), a large chart that listed the dates of eruptions of volcanoes around the world in a "statistical" style recalling von Hoff's work; it was published as a "companion" to Daubeny, *Active and extinct volcanos* (1826), which in turn was based on his lectures in previous years.

14. Fig. 7.5 is reproduced from Scoresby, *Arctic regions* (1823) 1 [London-BL: 792.e.1], pl. 7. The English explorer William Scoresby junior (1789–1857)—later a prominent Anglican clergyman—suspected that beneath its ice Greenland was not a single landmass and might even be an archipelago (328): the map of the coast surveyed (pl. 8) marks the incompletely explored Scoresby Sound as a channel that might cut right across to the west coast. Fig. 7.6 is reproduced from Parry, *Journal of a voyage* (1821–24), pl. opp. 17 [London-WL: V25136], described on 17–20 (see also the chart of their route, opp. 1). Beechey (whose father was the distinguished painter Sir William) had his sketch re-drawn for publication, and engraved, by the professional artist William Westall (1781–1850). Parry's expedition subsequently got as far west as Melville Island before having to turn back: Levere, *Science and the Canadian Arctic* (1993), 67–74.

Fig. 7.6 A huge iceberg sighted in 1819 by William Parry's Arctic expedition while on its way to search for a Northwest Passage. The iceberg, in Baffin Bay off the west coast of Greenland at latitude 73° North, towered 140 feet above sea level, but was aground in 120 fathoms and was therefore 860 feet (260m) in total height; in a near calm the ship had had to be towed clear to avoid drifting against it. It was probably derived from one of the large glaciers spilling off the ice sheet in the interior of Greenland. This engraving, published in 1821 in Parry's account of his voyage, was based on a sketch by Frederick William Beechey (1796-1856), then a young naval lieutenant with scientific interests and subsequently an Arctic explorer in his own right (see §14.2). (By permission of the Wellcome Institute Library, London)

7.5 CONCLUSION

The actualistic method of using the present as the most effective key to the deep past was taken for granted by all savants concerned with the newly *historical* science of geology. But Cuvier had famously claimed that the policy had limitations, because actual causes were *inadequate* to explain the earth's major "revolutions", and specifically the most recent one that Buckland called "diluvial". The contrary argument, that actual causes were adequate to explain *all* the preserved traces of all past events, had been championed by Hutton's follower Playfair, whereas Hutton's critic de Luc had tirelessly claimed that what he was the first to name "actual causes" could be traced back only as far as the revolution that allegedly separated the present world from the deep past. After de Luc's death, his colleagues at Göttingen offered a prize for research that might help resolve this issue, by discovering just what effects actual causes could be shown to have had during recorded human history.

The prize was won by the German savant von Hoff, for a detailed inventory—based on published sources rather than fieldwork—of historically recorded changes in physical geography, and volcanic eruptions and earthquakes. In effect, von Hoff transposed the methods of the social science of civil "statistics", which had already been given a temporal dimension by historians, from the human world into the natural. His work showed that although the recorded effects of actual causes such as erosion and sedimentation, volcanic and seismic action, were relatively modest on a global scale, their cumulative effects could have been substantial over the far greater spans of deep time that all geologists took for granted in geohistory.

The study of Etna—Europe's largest volcano and the world's most fully documented one—was an example of the kind of research that von Hoff was able to draw on; the local knowledge of local naturalists such as Ferrara and Gemmellaro made the *history* of the volcano well known, and suggested that its eruptive activity could be traced back, without major change of scale or frequency, into the deep past beyond even the earliest human record in Antiquity.

Finally, the many voyages and expeditions mounted around this time by all the leading Western nations were providing a mass of new information about the physical geography of remote parts of the world. The exploration of the huge volcanoes of the Andes, and the mapping of the high Arctic and its glaciers and icebergs, were just two striking examples. They may not have added to the range of known actual causes, but they certainly led to a better appreciation of their sheer scale, which in turn enlarged the explanatory potential of actual causes for understanding geohistory. Even if they had acted in the deep past at no more than their present intensity, they could still account for many of the ancient traces of change that geological research had revealed; if their intensity in the past had been greater, they might of course explain still more. Certainly a survey of their effects, such as the one for which von Hoff won his prize, put a large question mark against Cuvier's confident assertion that actual causes were quite inadequate to explain the deep past of geohistory. On that point the scientific jury was still out, but Cuvier's case could no longer be regarded as self-evidently justified.

Yet it remained uncertain whether actual causes could account for some of the most puzzling phenomena, such as those attributed to the "geological deluge" or most recent revolution (Chap. 6). The best explanation of the diluvial effects seemed to be in terms of large-scale versions of tsunamis, which in turn were usually attributed to sudden movements of the earth's crust. So the reality or otherwise of such crustal movements, not only in the deep past but also in the present, was of crucial importance, as the next chapter will explain.

CHAPTER 8

The dynamic earth (1818–24)

8.1 CRUSTAL ELEVATION

Von Hoff's prizewinning inventory of the effects of "actual causes" within the span of recorded human history was purely descriptive in scope. He did not aim to determine whether, for example, a recorded increase in land area on a certain coastline was due to sedimentation or to a fall in sea level or to a rise of the land itself; any such causal questions entailed fieldwork, which he was not in a position to undertake. But his compilation certainly provided raw material for debates on issues in earth physics, which in turn were being treated increasingly as germane to geohistory.

Prominent among such causal questions were those relating to putative movements of the earth's crust. Huttonian geotheory, for example, had demanded—a priori—huge movements of crustal elevation, in order to create new continents to replace those wasted by erosion, and to do so repeatedly, from and to eternity. But Hutton had not been able to point to any concrete evidence that such movements were currently in progress, and he had had to save the appearances by inferring that the movements were too slow to be perceptible within the timescale of human history (*BLT* §3.4). In his description of the Alps, the great Genevan physical geographer Horace-Bénédict de Saussure (1740–99) had famously described enormous folds in solid Secondary rocks, and had concluded—almost reluctantly, and contrary to his own earlier ideas—that they must have been caused by huge movements in the earth's crust in the distant past (*BLT* §2.4). This suggested that mountains had been formed not only in the earth's primal state, but also at much later periods and by crustal dislocation. But it was quite another matter to show that all this had been due to actual causes, and that the earth's interior was still dynamic and its crust still subject to major movements.

Earthquakes were notorious for shaking and destroying buildings and opening cracks in the ground, but it was not clear whether they ever resulted in permanent changes in the level of the land, let alone more dramatic effects in geography (*BLT* §2.2).

One of the few well-attested cases of relative sea level changing within human history came from the Swedish coast of the almost tideless Baltic. Marks inscribed on coastal rocks to record the level of the sea were being left progressively further out of the water as the years passed. But even here the significance of the change was unclear. It was usually attributed not to any movement of the earth itself but to a slow fall in the level of the sea worldwide (in modern terms, eustatically); the latter was what was anticipated on the widely accepted standard model of geotheory, which postulated a more or less steady fall in global sea level throughout geohistory (*BLT* §3.5). However, the Prussian savant von Buch, who had studied the evidence in the Baltic at first hand and was regarded as a highly reliable observer, had concluded that "the whole of Sweden is slowly heaving itself up", but particularly on the northern coastline around the Gulf of Bothnia. He reported this simply as a natural process genuinely operating in the present world. How any such crustal mobility was *caused* was a puzzle, but it was a separate question that belonged in earth physics.[1]

8.2 THE "TEMPLE OF SERAPIS"

In the previous chapter, Etna—the volcano for which human records were the longest and most complete—was cited as an example of the kind of local study that von Hoff was able to use in his great inventory of actual causes. A second example came from near Italy's (and Europe's) only other relatively large active volcano, Vesuvius. Apart from the Baltic coastline, this was almost the only well-attested case of substantial change in relative sea level—whether a genuine rise of the land or just a fall in the sea—within the span of human history.

Andrea di Jorio (1769–1851), an antiquarian living in Naples at the foot of Vesuvius, was studying a specific site that was also almost uniquely important to geologists. Jorio, who was in charge of Neapolitan education (and, like Ferrara and many other Italian savants, also a Catholic priest), had published a guidebook for the use of tourists: not to the celebrated sites of Herculaneum and Pompeii—which were already amply provided for—but to an almost equally interesting area around the town of Pozzuoli, to the west of Naples. The Phlegrean Fields [*Campi Flegrei*] was an area rich in antiquarian sites but also in hot springs and volcanic craters; it was therefore potentially important for any attempt to link actual causes in geology to the timescale of human history (Fig. 8.1).[2]

On the edge of Pozzuoli was a ruin of Roman date, known as the Temple of Serapis. The attribution to that Egyptian cult was based on an inscription that had been found when the site was excavated back in the 1750s in the hope of repeating the exciting discoveries at Herculaneum and Pompeii. Three limestone columns, which had previously stuck up out of the ground in the middle of a vineyard, had proved to be the upper portions of a ruined colonnade. This in turn was part of an extensive range of ruined buildings, surrounding a court-

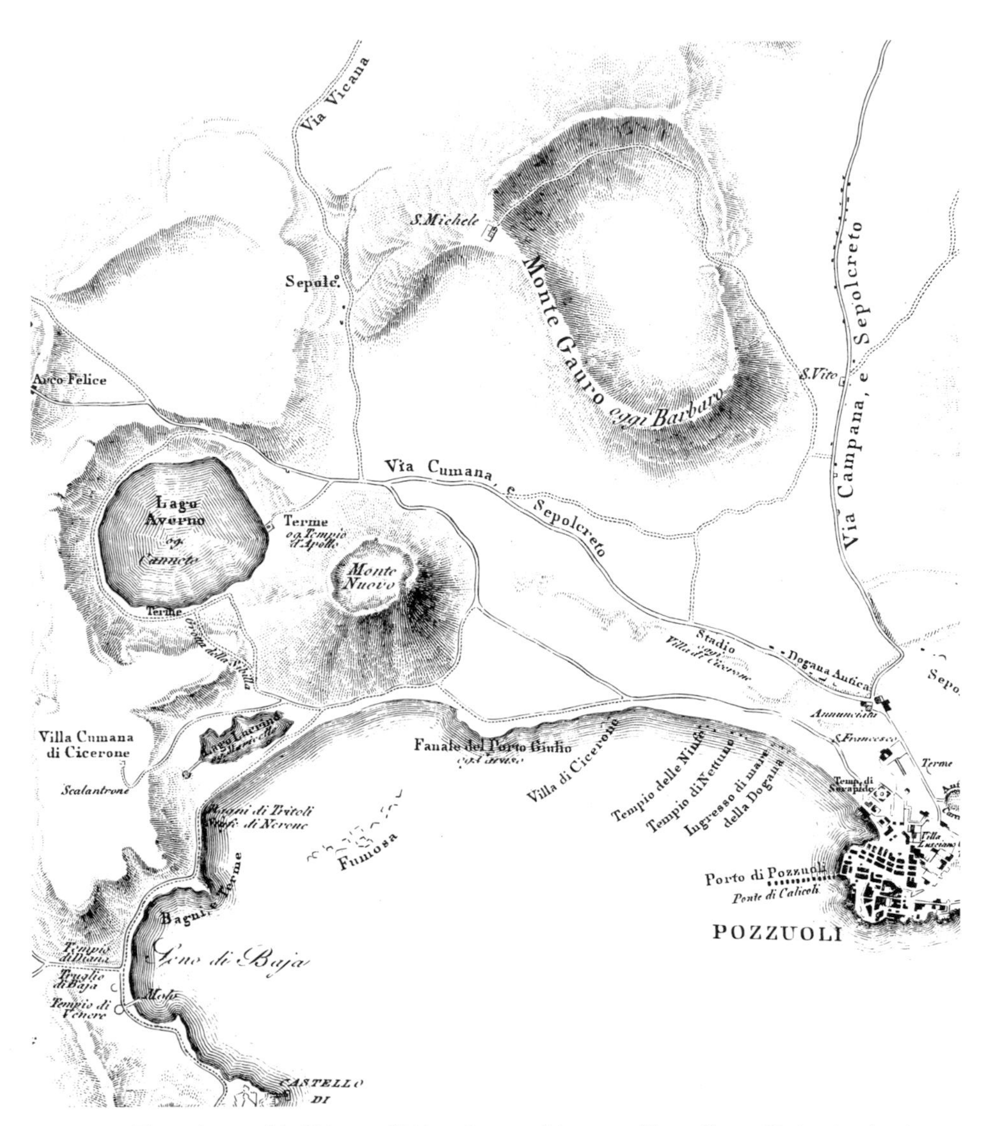

Fig. 8.1 A map of the Phlegrean Fields to the west of the town of Pozzuoli, near Naples, showing the sites of many temples [marked *tempio*] and other Classical ruins; it was first published by the Neapolitan savant Andrea di Jorio in his guidebook to the antiquities (1817). The map also showed hot springs [marked *terme*] and volcanic craters; among the latter was the large one occupied by Lake Avernus (in Classical mythology the site of an entrance to Hades), and also one at the top of Monte Nuovo ["new mountain"], a volcanic cone that had been formed rapidly during an eruption in September 1538. (Classical place names are in Roman lettering; modern ones [*oggi*, today] in italics.) The area was thus rich in material for studying geological "actual causes" within human history. The "Temple of Serapis", about which Jorio published a special study (see below), is shown on the northern fringe of Pozzuoli. This portion of his map is about 6km across. (By permission of the Syndics of Cambridge University Library)

1. Buch, *Norwegen und Lappland* (1810) 2, 289–92; Wegmann, "Déplacement des lignes de rivage" (1967), 133–39. The phrase "crustal mobility" is used in this chapter to denote *vertical* movements of the crust, not the horizontal ones that were invoked a century later in debates about continental displacement or "drift" (the forerunner of modern plate-tectonic theory).

2. Fig. 8.1 is reproduced from Jorio, *Tempio di Serapide* (1820) [Cambridge-UL: L.27.60/2], part of pl. 1; it was first published in his *Guida di Pozzuoli* (1817), pl. 1, a successful work that went through several editions. The map was drawn by Giosuè Russo and engraved by Antonio Rossi. Hamilton, *Campi Phlegraei* (1776), had used "Phlegrean Fields" in a broader sense to cover the whole volcanic region around Naples, including Vesuvius; it was also used, for example by Ferrara, *Campi Flegrei della Sicilia* (1810), to denote the similar region around Etna.

yard with a fine limestone pavement. Later in the century several antiquarians described and illustrated the ruins, and Hamilton included a distant view of it in his *Campi Phlegraei* (1776); they showed clearly that the Temple, although near the shore, had at that time been high and dry above sea level.[3]

By Jorio's time, however, the pavement of the Temple of Serapis was awash with seawater, which seeped in easily through porous subsoil from the nearby Mediterranean. This implied that the sea level must have risen a little, relative to the land, even in the decades since the ruins were excavated. Obviously, the earlier state of the building, with the paved area dry, would also have been what it was like in Roman times. Yet the sea level had not remained static during the many centuries between the building of Serapis and its excavation. For each of the three columns that had remained upright since the Roman period was marked by a zone full of borings made by marine mollusks, the shells of which could still be seen in their deep holes. This seemed to imply unambiguously that the sea had risen substantially—or the land had sunk—some time after the building was abandoned, and that at some later time that movement had been reversed, bringing the pavement back to much the same position it had had originally and in which it had first been excavated. The relative sea level had therefore changed

Fig. 8.2 "State of the Temple of Serapis at Pozzuoli in 1810", an engraving published by Jorio in his primarily antiquarian study of the ruins (1820). Since Hamilton's time the sea level had risen a little—or the land had sunk—so that the limestone pavement that had been excavated earlier was now just flooded by seawater. The excavation had also revealed that the columns still standing had been bored into, at about one-third of their height, by marine mollusks; this implied that at some earlier time the sea level had been much higher—and for an extended period—but that it had later returned almost to its original level. Jorio's engraving was a powerful visual proxy for what was already a celebrated object of discussion among geologists, for whom it was a vivid demonstration of the effects of actual causes within the span of human history. (By permission of the Syndics of Cambridge University Library)

substantially *in both directions* over the centuries, and had continued to change, though more slightly, over recent decades. The ruins offered almost uniquely clear evidence of a complex sequence of physical changes within the span of human history. When Jorio published a special study of Serapis, soon after his general guide to the Phlegrean Fields, he was well aware that its interest for savants extended far beyond his own antiquarian circles (Fig. 8.2).[4]

This complex physical *history*, within just the centuries since the Roman period, was confirmed by the topography of the immediate environs of the ruins. For Jorio published a large-scale map that showed several other Classical ruins lying offshore, partly submerged by the Mediterranean. Like the flooded pavement of Serapis, they showed that the present sea level was slightly but significantly higher than in Roman times. And behind the present beach and at a higher level was a sloping terrace, backed by a steep slope that looked like an ancient line of cliffs. Serapis itself was built on this terrace. If its columns had indeed been submerged for an extended period, during which they were bored by marine mollusks, the cliffs behind it might have been cut by wave action (or, at least, further eroded) at just the same time, while the terrace would have been a beach. And if at some later time the relative sea level had fallen again, the terrace might represent the "new" land that sixteenth-century documents recorded as having appeared at that time near Pozzuoli (Fig. 8.3).[5]

Jorio's detailed description of the Temple of Serapis was interesting in its own right to antiquarians. But his illustration of its columns was also a valuable proxy for geologists who had not visited the place, and his maps showed them the spatial context of the ruins. Blumenbach soon had a copy of Jorio's work, and told von Hoff about it in time for the latter to mention it in his second volume. In the first, von Hoff had already summarized the evidence around Pozzuoli, as a well-known example of recent changes in the relative positions of land and sea; in the second, he reviewed the possible role of earthquakes as the cause of these changes.[6]

3. Hamilton, *Campi Phlegraei* (1776), pl. 26; comparison with Jorio's accurate map, or with a modern one, shows how Fabris rearranged the topography (as was usual in landscape art) to make a more "picturesque" scene, but this did not detract from its scientific meaning. Other early illustrations are finely reproduced in Ciancio, *Teatro del mutamento* (2005), pls. 5–13. The later discovery that the ruins were those of a market building [*macellum*], though possibly on the site of an earlier temple, did not affect the scientific arguments that surrounded what continued to be called the "Temple of Serapis" or just "Serapis".

4. Fig. 8.2 is reproduced from Jorio, *Tempio di Serapide* (1820) [Cambridge-UL: L.27.60/2], pl. 7A, based on a drawing by John Izard Middleton. The evidence was unusually clear because the tidal fluctuation in the Mediterranean is almost as small as in the Gulf of Bothnia. When I visited Serapis in 1966, sea level had crept a little higher than it was in Jorio's (and Lyell's) time. But my photographs (like others from the same years) are now "historical", for that movement was reversed between 1968 and 1984, when the land rose again quite rapidly back to about where it was when the ruin was first excavated, leaving the pavement high and dry once more: see Dvorak and Mastrolorenzo, "Campi Flegrei caldera" (1991), figs. 19, 20, a fine study that makes good use of historical records to help analyze what have been, in modern terms, bradyseismic crustal movements.

5. Fig. 8.3 is reproduced from Jorio, *Tempio di Serapide* (1820) [Cambridge-UL: L.27.60/2], pl. 3. Dvorak and Mastrolorenzo, "Campi Flegrei caldera" (1991), summarizes modern research that dates the cliffs and raised beach to a prehistoric period, but also quotes from documents that suggest that the Starza was indeed flooded *again* in post-Roman times, until the uplift associated with the eruption of Monte Nuovo in 1538 (see below).

6. Hoff, *Veränderungen der Erdoberfläche* 1 (1822), 455–56; 2 (1824), xiii–xiv, 203–4.

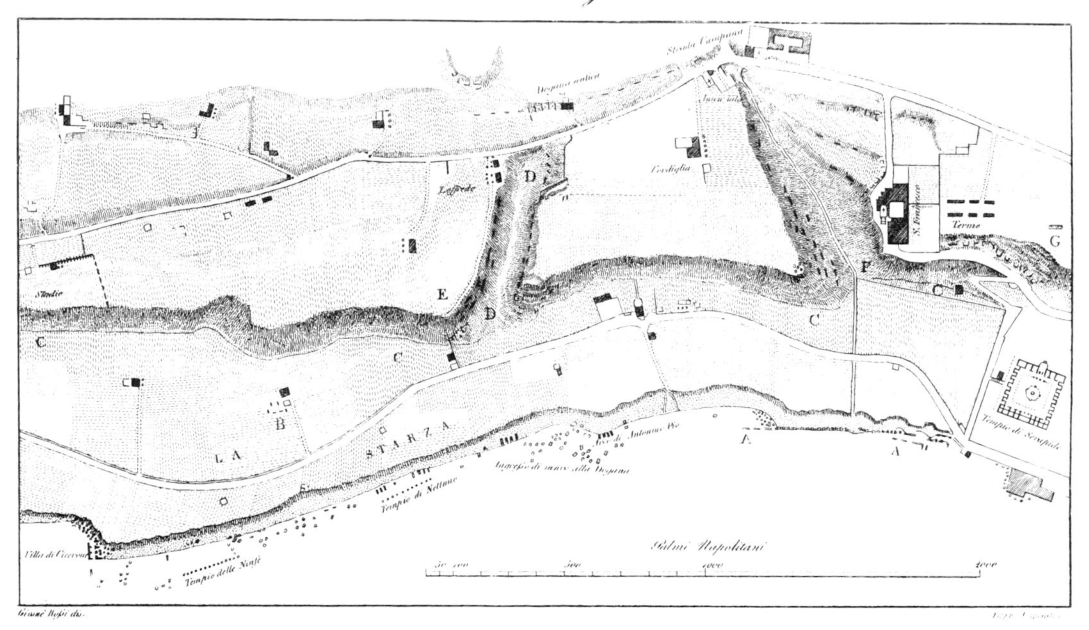

Fig. 8.3 Jorio's large-scale map of the coastline immediately west of the Temple of Serapis (right), showing other Classical ruins ("Cicero's villa", "Temple of the Nymphs", "Temple of Neptune", etc.) just offshore, indicating that sea level was lower in Roman times than in Jorio's. Serapis itself is shown on a broad terrace (*La Starza*) with an ancient beach (C) backed by a line of cliffs cut by gullies (D, F): this suggested an ancient period of higher sea level, perhaps the same period in which Serapis had been submerged and its columns bored by marine mollusks. The map shows about 1.5km of coastline. (By permission of the Syndics of Cambridge University Library)

There were two main alternative explanations. One was that the movements were causally related to the volcanic activity for which the region was famous, or rather with the seismic activity so often connected with it. One earthquake could have depressed the ruins under the sea; and at a later date, after the mollusks had had time to do their work, another could have raised the area back more or less to its original level. The other possibility was of course that the sea itself had changed in level. But there were problems with both explanations. Most earthquakes were notorious for shaking buildings into ruins, and it was difficult to imagine how three of the four large columns of Serapis could have remained upright if the site had suffered not just one but two major earthquakes. On the other hand, if it was the sea itself that had changed, it was puzzling that evidence for the same changes was not found around coastlines everywhere. So the importance of the problem went far beyond the immediate environs of Serapis, because it impinged on fundamental issues of the stability or mobility of the earth's crust and of its oceans.

Two contrasting assessments of the problem will illustrate how it was being tackled around the time that von Hoff reviewed it. In an earlier work on the physical geography of Campania, the Italian mineralogist and geognost Scipione Breislak (1748–1826) had carefully reconstructed the repeated movements in relative sea level in terms of four successive "epochs". In 1818 he incorporated this fieldwork into a major three-volume review of current geological theorizing (he had his work translated and published in French, explicitly to ensure it wider circulation). In the context of a critique of the standard model of geotheory (*BLT* §3.5), he reviewed the evidence for a falling sea level on a global scale, and argued that it was unsubstantiated. Yet he found the seismic explanation of the changes at Serapis and in the surrounding region equally unsatisfactory, for the reason just mentioned. He therefore concluded that the changes, both there and in the Baltic, must be due to *insensible* movements in the earth's crust, the level of which had changed so gently as to leave the Serapis columns in position throughout. As he must have recognized, however, this explanation might itself be thought unsatisfactory by other savants, since it attributed the effect to a "cause" that was *not* manifestly "actual" since it had not been directly witnessed.[7]

Among the skeptics was the veteran Weimar polymath Johann Wolfgang von Goethe (1749–1832), who in his lifetime was as famous for his scientific as for his literary achievements. The publication of von Hoff's work prompted "the great man" (as von Hoff called him) to recall his own explanation of this "architectural-natural-scientific problem", which he had devised when he visited the Phlegrean Fields during his celebrated Italian tour back in 1787. He regarded it as "a less desperate explanation" than that of major changes in sea level, for what he considered "such a minute purpose" [*einem so winzigen Zwecke*] as that of accounting for just one anomalous site. The Italian geognost Giovanni Battista Brocchi (1772–1826), who was already renowned for his superb description of the fossil mollusks in the Tertiary formations in the foothills of the Apennines (*BLT* §9.4), had recently given a plausible explanation of the curiously limited vertical range of the borings on the columns. Before the sea level rose, Brocchi suggested, the columns must already have been partially buried by alluvial or volcanic detritus, thereby protecting their lower part (and the rest of the ruins) from the mollusks' attentions. Goethe had in effect anticipated this idea, by suggesting that such detritus might also have formed a natural dam, which could have ponded back a small and purely local lagoon well above sea level. This could have partially submerged the upright columns and exposed them to the boring mollusks (Fig. 8.4).[8]

Goethe's was certainly an ingenious way of saving the appearances, but it was also a blatantly ad hoc explanation, and even more "desperate" than those he hoped it would replace. It got much wider attention when Jameson translated the article for his Edinburgh periodical, using it tacitly as a prestigious stick with which to beat Playfair and his Huttonian ideas of a dynamic earth. But in the event Goethe failed to persuade other naturalists, not least because the sea level *had* manifestly changed, even in the decades since Goethe visited the site; and

7. Breislak, *Topografia fisica della Campania* (1797), 300–7, and *Institutions géologiques* (1818) 1: xxviii–xxix, 74–86. He had a German father but was a Roman by birth: hence his hybrid name.

8. Fig. 8.4 is reproduced from Goethe, "Architektonisch-naturhistorisches Problem" (1823) [Cambridge-UL: F181.d.1.34], pl. opp. 79, published in what was in effect Goethe's personal scientific periodical *Zur Naturwissenschaft überhaupt*. See also Brocchi, "Tempio di Serapide" (1819); Goethe's original sketch of 1787 is reproduced in Ciancio, *Teatro del mutamento* (2005), pl. 14.

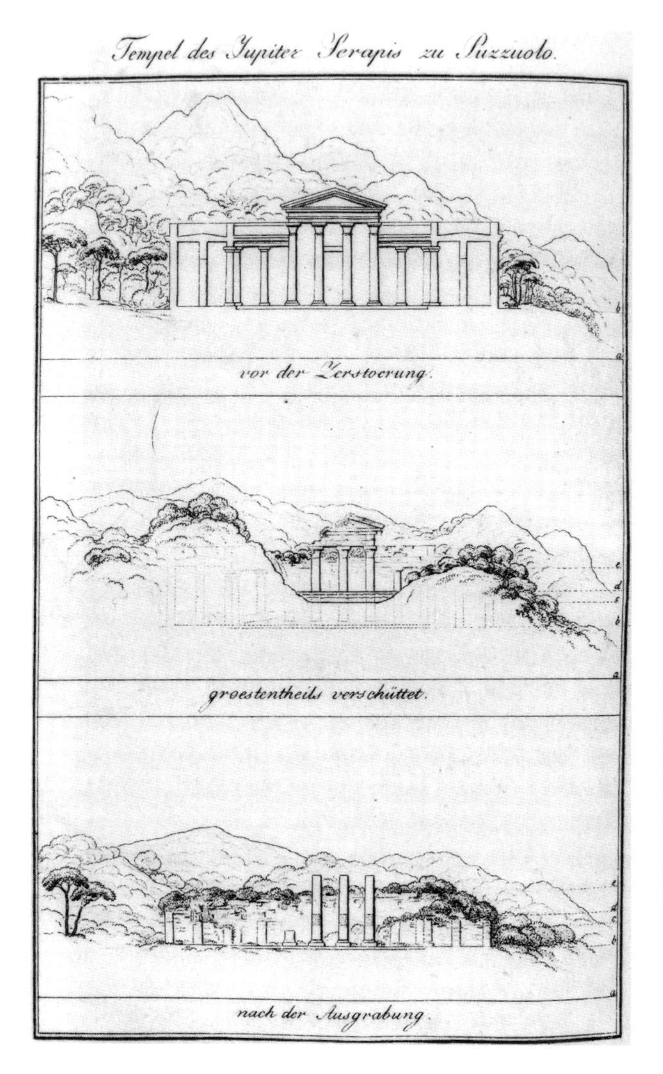

Fig. 8.4 The "Temple of Jupiter Serapis" in three successive historical states: engravings published by Goethe in 1823 to illustrate his explanation of how the columns could have been bored without any change in sea level, let alone any crustal movement. Between the original state "before the destruction" (top), and the present state "after the excavation" (bottom), is a hypothetical ("largely buried") state with the bases of the columns covered in detritus, and the middle parts submerged in an inferred lagoon well above sea level, ponded in by a barrier of more detritus (right). The bottom line (*a*) in each view represents sea level, which Goethe claimed had been *unchanged* throughout, below the level (*b*) of the paved courtyard at the base of the building. (By permission of the Syndics of Cambridge University Library)

anyway it was difficult to imagine that marine mollusks could have flourished well above sea level, in a conjectural lagoon that would surely have been filled with fresh, not salt water. Still, the intervention of such a distinguished savant does indicate how Serapis had become the subject of an international debate with far-reaching theoretical implications.[9]

Like much of von Hoff's material on the physical effects of volcanoes and earthquakes within the span of human history, the dramatic changes indicated by the famous columns of the Temple of Serapis were far clearer in the stones themselves than in any documentary records. However, there was also some textual evidence that substantial areas of new land appeared at Pozzuoli in the early sixteenth century. The records also suggested that relative sea level had fallen significantly, particularly during the days immediately before the dramatic formation of Monte Nuovo in September 1538. The coincidence in timing strengthened Breislak's hypothesis that the changes had been due to a gentle rise in the earth itself, probably confined to the region of the Phlegrean Fields. If so, the still earlier submergence of the columns might have been due to an equally gentle and local subsidence in the earth's crust; and the most recent change, just flooding the pavement in the courtyard, might be due to another slow downward movement. More generally, this would suggest that the earth's crust might indeed be dynamic, even within the span of human history: not only upwards, as Hutton's speculative geotheory had demanded, but *in both directions.*[10]

8.3 VON BUCH AND THE ORIGIN OF MOUNTAIN RANGES

Yet it was not clear whether a phenomenon apparently connected with local volcanic activity could be used more generally to infer a causal mechanism for movements of the earth's crust: either spatially, for movements in regions such

Fig. 8.5 Leopold von Buch's landscape view of the Peak of Tenerife [Pico de Teide] in the Canary Islands, published in 1825, showing the snowcapped volcanic cone (3718m) with the arcuate outer ridge (left) that he interpreted as an "elevation crater" formed by the forcible upward buckling of the earth's crust. It was this tectonic or crustal movement, he argued, that had created the conduit through which the truly volcanic materials had been able to rise to the surface and accumulate at the center of the buckled crust. (By permission of the British Library)

9. Playfair, *Illustrations* (1802), 450, and Basset, *Explication de Playfair* (1815), 360–62, had mentioned the Phlegrean Fields (and cited Breislak's earlier work) as a useful "illustration" of the truth of the Huttonian model of an internally dynamic earth. The highly international and multilingual argument about Serapis continued for many years; Ciancio, *Teatro del mutamento* (2005), gives a fine account, illustrated with superb reproductions of its iconography. The present narrative will return to it only at the point when one of Jorio's engravings (Fig. 8.2) was recycled by Lyell (see §21.1).

10. The reality of the movement as an actual cause was confirmed within the next few years, for the Neapolitan architect Antonio Niccolini (1772–1850) made accurate weekly measurements of the water level between 1822 and 1838, recording a further fall of about 10cm in the mean level of the pavement: Dvorak and Mastrolorenzo, "Campi Flegrei caldera" (1991), 16–17.

as the Baltic that were far from any volcanoes, or temporally, back into the deep past "beyond the reach of history".

In the case of the Baltic, von Buch himself had doubted whether the slow rise of the land had anything to do with the far larger crustal movements that had evidently operated in the deep past and were apparently responsible for the upheaval of whole mountain ranges. When the end of the wars made travel easier and less hazardous, he had sailed to the Canary Islands in the Atlantic to do extensive fieldwork in the volcanic terrain, and in 1820 he presented his conclusions in a major paper at the Academy of Sciences in Berlin. He made a sharp distinction between true volcanoes such as Vesuvius, which were manifestly built up by the accretion of lava flows and ash falls in the course of successive eruptions, and what he called "*elevation craters*" [*Erhebungs-Cratere*]. The latter, he claimed, had been raised by an upheaval of the earth's crust itself; a striking example from his own fieldwork in the Canaries was the arcuate ridge that half-encircles the Peak of Tenerife (Fig. 8.5).[11]

Von Buch described and illustrated many other examples of elevation craters: not only those he had seen for himself, such as the arcuate Monte Somma that half-encircles Vesuvius (see Fig. 22.1), but also those that were well documented from travel accounts by others. He claimed that they were all relics of episodes of sudden buckling, which had forced the earth's crust upwards at specific points, allowing the escape of magma (liquid rock) and other material from the deep interior and thereby generating "central volcanoes" [*Central-Volcane*] as a kind of secondary effect of the crustal movement. But active volcanoes were known to occur linearly in many parts of the world: von Buch published maps of these "linear volcanoes" [*Reihen-Volcane*] on archipelagoes such as the Kuriles and Aleutians, and those that are strung in a vast arc stretching through much of the East Indies (now in Indonesia). And other lines of volcanoes were known to lie along the axes of mountain ranges such as the Andes. So it seemed possible that elevation craters might be a special case of a much more general process of crustal elevation on a very large scale. The origin of mountain ranges might lie *not* in any observable "actual cause" but in large-scale natural processes that had operated intermittently in the deep past, and perhaps with sudden or catastrophic violence.[12]

However, nature itself obligingly provided a possible counter-example—a striking case of crustal mobility in action, albeit on a small scale—just when geologists were debating these matters with renewed vigor. On the night of 19 November 1822 the coast of Chile was struck by a major earthquake. Such events were (and are) not unusual along this coastal zone near the foothills of the Andes, and this one had the usual devastating impact on human settlements. Its effects were carefully recorded by a well-informed English observer who happened to be staying near Valparaiso, and whose account was later sent to Warburton and read at the Geological Society in London. By itself, this would simply have added a further firsthand account to the stock of earlier scientific descriptions of earthquakes, such as the famous one in Calabria in 1783, the effects of which had been thoroughly examined at the time by Hamilton and the Neapolitan savants (*BLT* §2.2). But the Chilean earthquake, unlike the Calabrian, was reported to have resulted in a permanent drop in relative sea level; and since the change was confined to the Chilean coast, it would have to be attributed to a genuine rise of the land in that region. This gave the report exceptional scientific importance. That it was by a woman lessened its credibility not at all. For the woman in question was the highly educated daughter of a British naval officer; she was well known for her intellectual tastes and abilities, and was already the author of several published works. Maria Graham (1785–1842) could clearly be trusted to be a reliable and accurate observer, no less than any man of similar background and achievements.

As a young woman Maria Dundas (as she then was) had sailed with her father to India, and had met and married Thomas Graham, a young naval officer, returning with him to live in London. Later, she had sailed with him to Naples—by then he had command of his own ship—and spent a year in Rome studying Italian art; a scholarly monograph on Poussin, the first in English, was the result. In 1821 she accompanied her husband to South America, again in his own ship, but tragically he died while they were rounding Cape Horn. On reaching Chile (which had recently gained independence from Spain), his widow stayed on in Valparaiso as the guest of Lord Cochrane, the British officer in command of the new Chilean navy. She was there when the earthquake struck. She kept a detailed record of its devastating effects and of its many aftershocks. After her return to England she prepared her lengthy Chilean journal for publication; but she also summarized what was of greatest geological interest, in the form of a letter that was immediately read at the Geological Society.[13]

Graham reported that "it appeared on the morning of the 20th [a few hours after the main shock] that the whole line of coast from north to south, to the distance of above 100 miles, had been raised above its former level". Though the horizontal extent was derived from reports by others, the vertical change was based on her own careful observations. At Quintero, where she was staying with Cochrane when the earthquake struck, the rise was about four feet: "although it was high water I found the ancient bed of the sea laid bare and dry, with beds of oysters, mussels, and other shells adhering to the rocks on which they grew, the [shell]fish being all dead and exhaling most offensive effluvia". It was an unnerving contrast to the peaceful Arcadian scene that she drew and published in her journal to accompany her record of the event (Fig. 8.6).[14]

11. Fig. 8.5 is reproduced from Buch, *Canarischen Inseln* (1825), Atlas [London-BL: 1259.d.27], pl. 6, fig. 1; it was complemented by an accurate topographical map of the island (pl. 1). In modern terms, many elevation craters were large calderas, formed by massive subsidence.

12. Buch, "Erhebungs-Cratere" (1820), read on 28 May 1818. Buch, *Canarischen Inseln* (1825), reported in full on his fieldwork in 1815–16; see esp. 323–28. The theory of elevation craters, being a part of earth physics rather than geohistory, is peripheral to the present narrative and cannot be dealt with here in the depth that it would deserve in another context.

13. Graham, "Earthquakes in Chili" (1824), dated 4 March 1824 and read at the Geological Society on 5 March [London-GS: OM/1]. Her sole (but serious) disadvantage as a woman was to be unable to attend this all-male gathering in person, even as a guest.

14. Fig. 8.6 is reproduced from Graham, *Residence in Chile* (1824) [Cambridge-UL: 8670.b.4], pl. opp. 329; the major quake (329–30) was followed by several weeks of aftershocks (330–40); extracts are printed in Mavor, *Captain's wife* (1993), 133–46. This drawing is a representative sample of her fine work as a landscape artist; she was equally skilled at botanical drawing, and equally valued by botanists. Quintero is about 30km north of Valparaiso.

Fig. 8.6 Maria Graham's drawing of Quintero Bay near Valparaiso, reproduced as an etching in her published account of her time in Chile (1824). This was the locality where she experienced the great earthquake of November 1822. The newly exposed shore, which proved that the land had risen abruptly by four feet in this area, is not, of course, visible in this distant view (which she may well have sketched before the event). She noted that in another direction "the snowy Andes" could be seen in the distance. (By permission of the Syndics of Cambridge University Library)

However, Graham was not content merely to report the details of the earthquake itself, the associated tsunami, and the permanent change in sea level. She concluded her letter to the Geological Society with a cogent inference of wider significance: "I found good reason to believe that the coast had been raised by earthquakes at former periods in a similar manner; several ancient lines of beach, consisting of shingle mixed with shells, extending in a parallel direction to the shore, to the height of 50 feet above the sea". It was obvious from this that even the highest of the Andes might owe their elevation to an extended sequence of similar small movements, rather than to a single huge event of the kind envisaged by von Buch. But the two explanations were not in fact exclusive alternatives. The recorded effect of the Chilean earthquake, and its inferred antecedents, might be small-scale versions—aftershocks, as it were—of an earlier and vastly more catastrophic mega-earthquake or episode of major crustal upheaval, just as the historic Lisbon tsunami might be analogous to occasional and far more devastating "deluges" or mega-tsunamis in the deep past.[15]

8.4 CONCLUSION

One of the most contentious issues for deciding the adequacy or inadequacy of actual causes, as a means of explaining all the events recorded in geohistory, was that of movements of the earth's crust. That there had been such movements, and on a huge scale, had long been clear from the folded strata seen in mountain regions. But it was another matter to decide whether this was due to occasional events of great magnitude in the deep past, or to the sum of far smaller movements that might still continue in the present world. In other words, even small movements, if genuinely recorded within the span of human history, might be the key to understanding far larger effects, even perhaps the origin of the highest mountain ranges. However, where relative sea level seemed to be changing on a human timescale, as on the coasts of the Baltic, it was difficult to determine whether the land was rising or the sea falling, and hence whether or not any crustal movement was taking place. In this case von Buch, who was no mean observer, concluded that the land was genuinely rising, but he recognized that it was quite another matter to propose any causal explanation for such crustal mobility.

By far the most intensely debated case was that of the so-called Temple of Serapis, in the volcanically active region of the Phlegrean Fields not far from Vesuvius. Here Jorio's antiquarian research provided geologists with a superbly detailed account of the Roman buildings, which had evidently witnessed substantial changes in the relative level of the nearby Mediterranean, during less than two millennia since the Roman period. Although there was much argument about the interpretation of the changes, they seemed to be due to crustal movement; they were confined to the Phlegrean Fields rather than being universal. Most importantly, they had been changes *in both directions*: the land seemed to have subsided substantially after the Roman period, and remained there for a very long time, but had then risen back to something like its original level. And the movement had continued within living memory, the land having slowly subsided once again. A savant as distinguished—as much in scientific as in literary affairs—as Goethe tried to save the appearances without conceding any change in sea level at all. But his hypothesis was judged inadequate and implausible, and the reality of crustal mobility in this case came to be generally accepted among geologists.

However, it was still far from clear that the Serapis case was relevant to the wider problem of accounting for crustal movements far from any volcanoes, as in the Baltic region, or for movements on the far larger scale seen in mountain regions. A more persuasive case derived from Maria Graham's report that in the terrible earthquake in Chile in 1822 the land had definitely risen along a long stretch of coastline. This provided a persuasive example of crustal mobility, for those geologists who argued that even the greatest mountain ranges—such as the Andes high above the affected Chilean coast—might have attained their present elevation by the summation of many similar movements, each relatively

15. Graham, "Earthquakes in Chili" (1824); see also Kölbl-Ebert, "Observing orogeny" (1999).

small. But the same case could equally well support the arguments of others, who claimed that the elevation of high mountain ranges must be due to proportionately large and rare events in the deep past, to which any historically witnessed events were merely small-scale analogues or aftershocks.

Like most of the other loosely linked issues described in the previous chapters, that of crustal mobility was the subject of intense research and often heated arguments. All these chapters (including this one) have described debates that were ongoing in the first postwar years and the first half of the 1820s, and that were still unresolved and open-ended around the middle of that decade. Part Two of this narrative picks up several of these focal problems and traces them, and in some cases their changing interactions or provisional resolution, through the rest of the 1820s and a little beyond.

Part Two

CHAPTER 9

The engine of geohistory (1824–29)

9.1 BRONGNIART'S GLOBAL STRATIGRAPHY

Any use of stratigraphy as a foundation for geohistory depended, of course, on a reliable understanding of the stratigraphy itself. Geologists in the 1820s were becoming increasingly confident that a generally valid sequence of formations was indeed being worked out, at least for the Secondary formations (Chap. 3). The English handbook dubbed "Conybeare and Phillips", for example, served as an effective model for detailed studies of the sequences of formations elsewhere in Europe and even, more tentatively, far away in Russia and across the Atlantic; and geologists such as Buckland and Boué were suggesting how these sequences could be correlated with one another internationally. Throughout the 1820s, the bulk of ordinary geological research, as reflected in the contents of the growing number of periodicals publishing such work, consisted of straightforward local or regional studies of stratigraphy, much of it carried out routinely along more or less Smithian lines.

At the end of the decade, all this research was summarized by the elder Brongniart in a major synthesis of stratigraphy. In one of the multi-volume "dictionaries"—in effect, encyclopedias—that were prominent in scientific publishing at this period, he wrote a book-length article on the "Theory of the structure of the earth's crust"; it was issued at the same time as a separate volume, entitled a "*Tableau* of formations" and an "Essay on the structure of the known part of the globe". Despite his use of the word "theory", the titles show that in fact Brongniart's goals were conventionally structural and geognostic, rather than those of geotheory or geohistory. Although he used an idiosyncratic set of names for the major groups of formations, which few other geologists adopted,

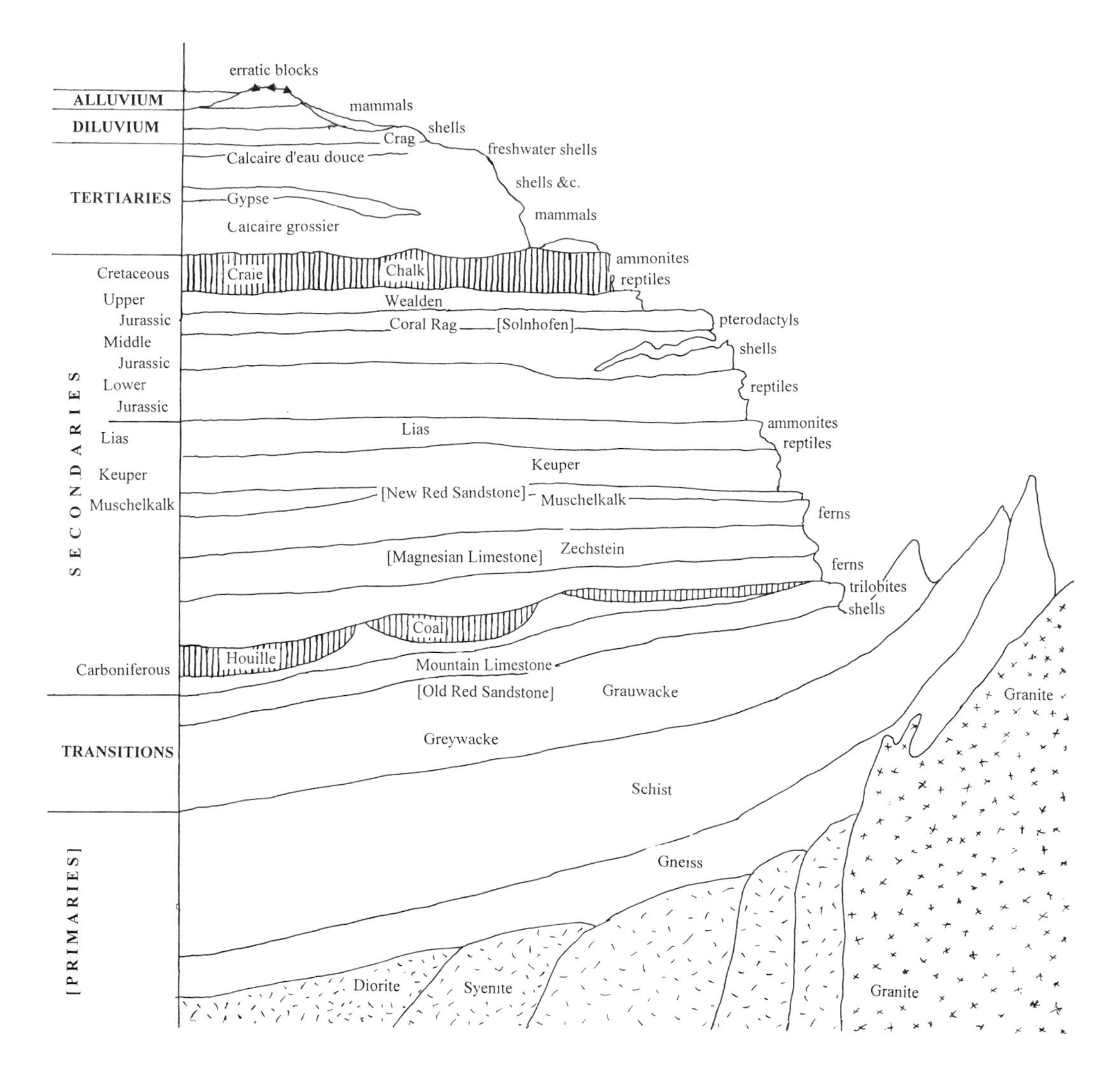

erratic blocks
ALLUVIUM
DILUVIUM
mammals
shells
Crag
freshwater shells
Calcaire d'eau douce
shells &c.
TERTIARIES
Gypse
Calcaire grossier
mammals
Cretaceous
Craie
Chalk
ammonites
reptiles
Upper
Jurassic
Wealden
Coral Rag
[Solnhofen]
pterodactyls
Middle
Jurassic
shells
Lower
Jurassic
reptiles
SECONDARIES
Lias
Lias
ammonites
reptiles
Keuper
Keuper
Muschelkalk
[New Red Sandstone]
Muschelkalk
ferns
[Magnesian Limestone]
Zechstein
ferns
trilobites
shells
Coal
Houille
Carboniferous
Mountain Limestone
[Old Red Sandstone]
Grauwacke
Granite
TRANSITIONS
Greywacke
Schist
[PRIMARIES]
Gneiss
Diorite
Syenite
Granite

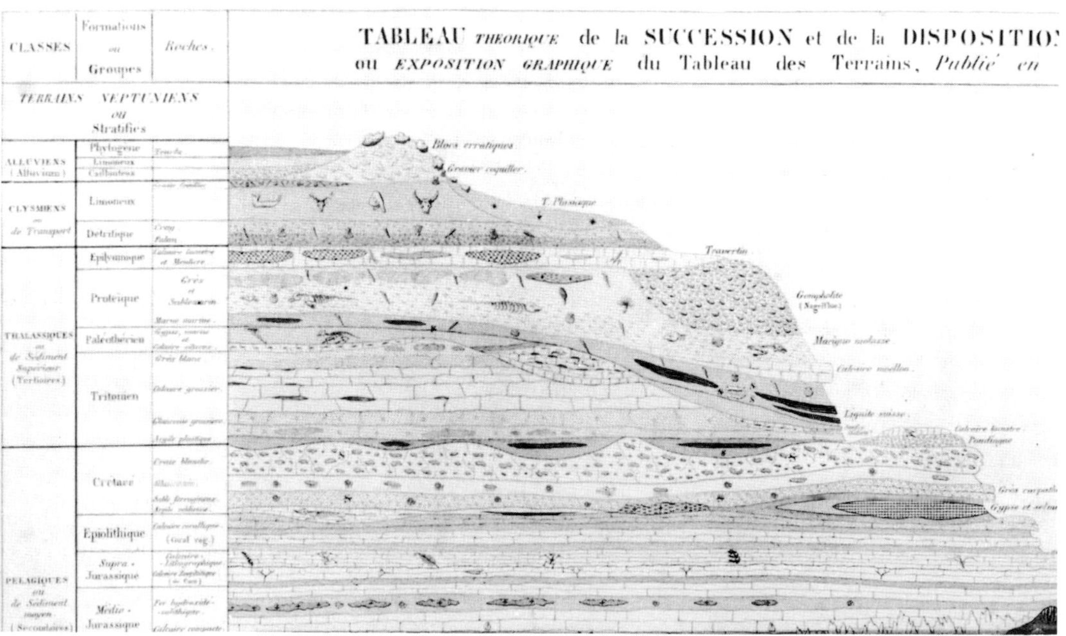

CLASSES
Formations ou Groupes
Roches
TABLEAU THEORIQUE de la SUCCESSION et de la DISPOSITIO
ou EXPOSITION GRAPHIQUE du Tableau des Terrains, Publié en
TERRAINS NEPTUNIENS ou Stratifiés
ALLUVIENS (Alluvium)
CLYSMIENS de Transport
Limoneux
Détritique
Blocs erratiques
THALASSIQUES
Epilymnique
Protéique
Paléothérien
Tritonien
Crétacé
Epiolithique
PELAGIQUES
Supra-Jurassique
Médio-Jurassique

his synthesis was a valuable review of current stratigraphical knowledge. It was not itself geohistorical, but it provided a reliable framework that could be used to explore further in that direction (Figs. 9.1, 9.2).[1]

With an agreed framework of stratigraphy, as summarized in Brongniart's work, earlier speculations about the broader features of the history of life could be put on much more reliable foundations. Cuvier's early conjecture, that reptiles had preceded mammals and that non-human mammals had preceded humans (*BLT* §9.3), had been strikingly confirmed by more than two decades of further research, clarified above all by the results of the new stratigraphy. The human species appeared to have no authentic fossil record at all; other mammals were virtually confined to the younger formations that were now distinguished as Tertiaries; while reptiles extended—in newly discovered diversity—well back into the true Secondaries (Chaps. 2, 5). Bringing still "lower" vertebrates into consideration, fossil fish were known from even older Secondary formations. Since fish and reptiles were of course important components of living faunas, alongside mammals, all this suggested that vertebrate life had become progressively more diverse in the course of geohistory, by the successive addition of groups with arguably "higher" kinds of organization. It implied an overall directionality, or even progress, in the history of the vertebrates; and there were increasing indications that this might also apply to the invertebrates and the plants, and hence to

Fig. 9.1 Brongniart's "Theoretical section [*tableau*] of the general European succession and disposition of the formations [*terrains*] and rocks of which the earth's crust is composed", published as a large chart to accompany his book-length article on "The structure of the earth's crust" (1829). This ideal section is redrawn here in outline (and greatly reduced in size), mainly to show the physical and conceptual centrality of the well-mapped Secondary formations. Two distinctive formations are accentuated here: the exceptionally widespread Chalk at the top; and the economically important Coal near the bottom, usually found in isolated "basins". Below all the "Carboniferous" are the still poorly understood Transition formations and "stratified" Primaries (depicted according to rock types rather than any well-established sequence); deepest of all are the puzzling unstratified Primaries (stippled) such as granite, some of which were apparently intruded into overlying rocks. In contrast, at the top, the Tertiaries are quite thin, and the Superficial deposits (Diluvium and Alluvium) even thinner. Thicknesses were tacitly treated by geologists as reflecting the likely relative periods of time that the formations represented; so the thin Alluvium, which was taken to correspond to most or all of human history, implied a very brief human epilogue to an immeasurably lengthy geohistory. Some of the formation names inscribed on the original chart are also shown here (in three columns for those in France, Britain, and Germany), and some of their distinctive fossils are named at the imaginary cliff face (right).

Fig. 9.2 The upper (and later) formations, as portrayed in Brongniart's ideal section (1829): a small portion of his large chart, to show the style in which the rocks were depicted. Most of his classification (in the two columns on the left) was not adopted by other geologists, but he did include some terms—e.g., *Cretacé, Jurassique*—that were already in use internationally (and have remained so ever since). The older formations are shown as relatively uniform, but among the younger ones there is much lateral variation. Some prominent fossils are depicted as miniature icons, for example the palaeotherium in the middle Tertiary (*Thalassique*) and other large mammals in the *Clysmien* deposits (Buckland's Diluvium).

1. Fig. 9.1 is traced, and Fig. 9.2 reproduced, from Brongniart, *Tableau théorique* (1829?), the large chart designed to accompany his "Structure de l'écorce" (1829), also published as a book. Following a long-standing geognostic convention (*BLT* §2.3), he depicted all the volcanic rocks separately and to one side, and therefore outside the temporal sequence implicit in the pile of formations (they are omitted altogether from Fig. 9.1); they were the subject of his "Volcans" (1829), another massive article also published in book form. The "known part of the globe" referred not to the limits of geographical knowledge but to the surface crust of the earth in contrast to its unknown deep interior.

the history of life as a whole (Chap. 4). This raised the question of the cause or causes that might underlie any such trend in geohistory. In just the years when the fossil evidence for directionality was becoming clearer than ever before, one possible cause emerged from an unrelated and unexpected direction.

9.2 FOURIER'S PHYSICS OF A COOLING EARTH

Back before the turn of the century, the great French mathematical astronomer Pierre Simon de Laplace (1749–1827) had put forward a "*nebular hypothesis*" to explain the origin of the whole solar system. Rather like Buffon before him (*BLT* §3.2), Laplace had conjectured that all the planets had condensed from a plume of gaseous "nebular" matter spun out from the sun; so the earth had initially been a spinning spheroid of hot fluid, and had been cooling ever since. In Napoleonic France, Laplace had been the patron of everyone who was anyone in the natural sciences (including Cuvier: *BLT* §9.3); even after the fall of Napoleon he remained one of the most powerful savants in France, and his nebular hypothesis remained a lively topic of debate among savants and the educated public.[2]

Around the time that Brongniart and Cuvier first presented their celebrated account of the Tertiary formations around Paris (*BLT* §9.1), the mathematician Jean-Baptiste Joseph Fourier (1768–1830)—who was acting at the time as Napoleon's *préfet* or governor in a part of the French Alps—had begun to send the Institut in Paris an equally celebrated series of highly mathematical researches on the nature of heat. When he returned to Paris after the Restoration, Fourier's Napoleonic past had only temporarily retarded his rapid rise to prominence in the scientific world. In 1822 he had become one of the permanent secretaries at the Institut's royalist successor the Académie des Sciences—complementing Cuvier by being responsible for the "exact" sciences—and his flow of important papers on his "analytical theory of heat" continued unabated. Although the core of his work was unmistakably a project in "general physics" (or just "physics" in the modern sense), his goal was to apply his theory of heat to *physique de la terre* or earth physics (*BLT* §2.4) and ultimately to the kind of cosmology represented by Laplace's nebular hypothesis.

As early as 1820, Fourier had published a paper on "The secular refrigeration of the terrestrial globe", which he described as an extract from a larger study in progress. It had appeared in the Parisian *Annales de Chimie et de Physique* (the title of which indicates how scientific periodicals were becoming more specialized). Immediately before Fourier's paper, the editors printed one by Laplace, who argued that the earth was now in a state of equilibrium: on astronomical grounds its period of rotation (or the length of the day) could be proved not to have changed perceptibly during the millennia since Antiquity, so that its size must also have remained unchanged. By itself, Laplace's argument might have been taken to imply that the earth had been stable throughout geohistory. But Fourier, taking his cue from Laplace's earlier work, interpreted this present stability as the end product of a long process of gradual cooling. Treating the mathematical principles of heat loss as already firmly established—mainly, of course, through his own work—Fourier argued that heat conduction through the earth's crust was so slow that the deep interior might still be incandescent although the surface was now cold. But he claimed that at earlier periods of geohistory the rate of cooling would have been rapid, and the earth's internal heat would have had a far greater effect at the surface. In effect, therefore, he offered geologists a physical model according to which surface conditions would have changed *directionally* and exponentially in the course of time, cooling rapidly at first, then more slowly, and eventually reaching its present state of stability and equilibrium, which Laplace had inferred on quite independent grounds.[3]

Four years later Fourier published another paper in the same periodical, offering a summary of his work "On the temperatures of the terrestrial globe and planetary spaces", shorn of its daunting mathematics and thereby made accessible to a wider range of savants. Again he argued that the earth's "central heat" was a *residual* heat. In earlier decades it had been easy to dismiss any narrative of global cooling as a mere "novel" [*roman*], the scornful epithet that had often been applied to conjectural geotheories such as that put forward by Buffon (*BLT* §3.2). But now a theory of long-term cooling was being advocated by one of the leading figures in the new world of mathematical physics. It could no longer be dismissed as a piece of science fiction.[4]

Fourier's mathematical arguments for a cooling earth were soon complemented by his geological colleague Pierre Louis Antoine Cordier (1777–1861). Like Fourier, Cordier had participated in Napoleon's expedition to Egypt before the turn of the century (*BLT* §7.2); he had then taught for many years at the Mining School in Paris (where he had earlier been one of the first students) while also surveying mineral resources for the Mining Corps. In 1819 he had become in addition the professor of *géologie*—by then no longer a contentious title—at the Muséum, after the death of Faujas, the first holder of that position. Cordier was a member of the Académie and a prominent and powerful figure in French geology. He has not been mentioned previously in this narrative, simply because he worked mainly on the "hard-rock" problems of Primary and volcanic rocks, and their mineralogy. But in 1827 he read a major paper at the Académie, "On the temperature of the earth's interior", which immediately made his work relevant to a wider range of geologists. It appeared promptly not only in the *Mémoires* of the Académie—which, not coincidentally, reprinted Fourier's matching paper almost immediately following it—but also in the *Mémoires* of the Muséum; a

2. Laplace, *Système du monde* (1st ed. 1796); see Brush, "Nebular hypothesis" (1987) and *Nebulous earth* (1996).

3. Fourier, "Refroidissement séculaire" (1820); Laplace, "Diminution du jour" (1820). In the event, Fourier's major work, *Théorie analytique* (1822), did not incorporate the theory's application to the earth itself.

4. Fourier, "Températures du globe" (1824). There were soon translations in German and Italian but none in English. Not until 1837 did Edward Hitchcock (1793–1864), the professor of natural history at Amherst College in Massachusetts, have it translated by a colleague; he then sent it to Benjamin Silliman (1779–1864) of Yale for his *American journal of science.* The time lag was symptomatic of the intellectual marginality of American savants, most of them even more linguistically challenged than Hitchcock. It explains why European geologists, including the British, tended at this period to give little attention to American contributions, unless they were purely factual.

translation in Jameson's Edinburgh periodical made it easily accessible to anglophone geologists the following year.[5]

Cordier duly acknowledged that it was Fourier who had rehabilitated the venerable theory of a cooling earth and given it firm scientific foundations for the first time. He himself focused on the vexed question of the alleged rise in temperature encountered in deep mines (in modern terms, the geothermal gradient), which had long been regarded as critical evidence for or against any such theory. On the basis of extensive and accurate measurements, including reliable reports from the deep mines in Cornwall, Cordier insisted on the reality of the phenomenon; and he reported that the rate of increase with depth varied in different regions by a factor of two or three, but was everywhere more rapid than had previously been recognized. Extrapolated inwards, the gradient implied that the deep interior must be extremely hot and therefore probably fluid; like Fourier, and Buffon long before him, Cordier attributed this to an *originally* fluid state, which had also been responsible for the shape or "figure" of the earth as an oblate spheroid. The regional differences in the gradient were then attributed to an uneven thickness of the solid crust; but even at its most substantial, Cordier envisaged it as a relatively thin skin, floating on a mainly fluid sphere. He also outlined how this physical model might help to give a causal explanation for such well-known phenomena as earthquakes and volcanoes, in terms of the shrinkage of a cooling crust lying on top of an hot interior of molten rock. These were observations and conjectures that would keep *physique de la terre* a lively field of debate for the rest of the century (during which this part of it evolved into the modern science of *geophysics*).[6]

9.3 SCROPE'S DIRECTIONAL GEOTHEORY

Even before Cordier's work in Paris lent it further support, Fourier's model of an exponentially cooling earth was adopted by an important new recruit to the Geological Society in London. George Poulett Scrope (1797–1876), whose father was a prosperous merchant in the Anglo-Russian trade, had begun his student career at Oxford but had migrated to Cambridge, where he attended the lectures given by the then fledgling geologist Sedgwick and the veteran mineralogist and traveler Edward Daniel Clarke (1769–1822). Even before he graduated, he had been taken by his parents on an extensive Grand Tour: seeing Vesuvius, Etna, and the eponymous Vulcano aroused what became a lifelong fascination with volcanoes. He had then married an heiress, the only child of a wealthy country landowner (who was also an able painter and a friend of Sir Walter Scott), adopting her family name in place of his own (Thomson) and thereby hoping to ensure the continuation of the Scropes, who traced their aristocratic line back to medieval times.[7]

A few weeks later, Scrope (as he was now known) had crossed to France with his wife and spent six months exploring the geology of the classic region of the Massif Central. As he recalled later, the local inhabitants, while friendly and helpful, expressed "unconcealed astonishment at the apparent objects of a geologist's researches"; and at one point he was arrested as a suspected spy. While still on the spot, or soon afterwards in Italy (where he witnessed the great 1822 eruption of Vesuvius), he had written a detailed description of the extinct volcanoes in France; but its publication was delayed for several years, because he wanted to illustrate it profusely, and therefore expensively, with his sketches of the volcanic landscapes (see §15.1). When he and his wife returned to England to settle on her father's estate in Wiltshire, Scrope—at the age of only twenty-six—probably had greater firsthand knowledge of volcanoes, active and extinct, than any other geologist in Britain. A few months later, in the spring of 1824, he was elected a member of the Geological Society, having been proposed by Webster and Lyell, the latter his almost exact contemporary. His paper on the volcanic region around Naples was read at the same meeting and effectively displayed his credentials. The following winter he was elected a secretary of the society, joining Lyell in that position.[8]

Later in 1825, Scrope's book entitled *Considerations on Volcanos* became his first major scientific publication. It combined a critical review of earlier research with material from his own extensive fieldwork. Scrope's primary goal was to understand volcanoes in terms of earth physics: as he put it in his lengthy title, it was to consider "the probable causes of their phenomena [and] the laws which determine their march". However, he also wanted to trace their geohistory, exploring "their connexion with the present state and past history of the globe". And finally, he promised—contrary to the current fashion among geologists—that all these investigations would lead to "the establishment of a new theory of the earth". Thus, while Scrope's book was welcome in other ways, it threatened to revive the genre of geotheory, which had been rejected and reviled by the leaders of the Geological Society ever since its foundation (*BLT* §8.4). In contrast, the book entitled *Active and Extinct Volcanoes*, by Buckland's younger Oxford colleague Daubeny (§7.4), which appeared the year after Scrope's, stayed safely

5. Cordier, "Intérieur de la terre" (1827), read on 4 June, 9, 23 July. The Muséum's in-house periodical had changed its title from *Annales* to *Mémoires* after the Restoration, to signal the political break with the Napoleonic past, but its format was unchanged.

6. Gay-Lussac and Arago, "Température de la terre" (1821), had reported Cornish measurements and related them to Fourier's heat theory. The possible influence of *pressure* on the inferred fluidity of the deep interior seems not to have been recognized by Cordier and his contemporaries: hence their conclusion that most of the globe was probably in a fluid state, rather than the relatively small fluid core envisaged (on far better evidential grounds, of course) by modern geophysicists.

7. Sturges, *Bibliography of Scrope* (1984), includes his writings on both economics and geology, and brief but invaluable biographical information. Scrope's wife was disabled in a riding accident soon after their marriage; possibly as a consequence, they had no children, although many years later, after her father's death, they adopted Scrope's illegitimate son by his long-term mistress, an actress in London. His name is confusing: not only did he change his family name on getting married, but while he was a student he had also replaced his baptismal middle name Julius with the Poulett that his father had recently adopted from a more aristocratic branch of the family: in the cause of social climbing George Julius Thomson became George Poulett Scrope. However, he kept the latter names for the rest of his life and always used them in his published work.

8. Scrope, *Central France* (1827), Preface dated Milan, 6 April 1822, and London, 1 January 1826; Geological Society, 19 March, 23 April, 4, 21 May 1824, 4 February 1825 [London-GS: OM/1]. In the event, he resigned from the position as secretary after only a year, being too rarely in London to be effective: Buckland to De la Beche, 4 February 1826 [Cardiff-NMW, no. 175].

within the limits of "a description" of them in the mode of physical geography; it was barely geohistorical, and its theoretical ambitions were limited to "inferences" about volcanic action.[9]

Scrope's book was notable for its trenchant insistence on the explanatory value of actual causes. With Buckland's geological "deluge" (Chap. 6) as his obvious if covert target, Scrope criticized those who speculated about "what *might be* rather than *what is*", and who invoked catastrophes without having first exhausted the explanatory potential of what they saw around them. He even alleged that such theorizing was harmful, on the grounds that it "stops further enquiry" by discouraging the search for observable causes that might in fact be adequate. Instead, he urged that these causes be studied minutely, applying them to the evidence of the deep past with "the most liberal allowances for all possible variations and an unlimited series of ages". There was to be no shortage of deep time in Scrope's theory. Although it was peripheral to his main theme of volcanic action, Scrope raised in this context the issue of the erosion of valleys, claiming in particular that valleys of sinuous form must have been eroded slowly, and that it was "idle to talk of sudden catastrophes, debacles or deluges" to account for them. In sum, he set out a policy for geological research closely similar to what had won von Hoff his prize (§7.2); it was also what Prévost was already pursuing, soon to be followed by Lyell (see §10.2).[10]

However, Scrope's insistence on the explanatory power of actual causes led him to a view of geohistory that was far removed from the Huttonian vision—by this time most familiar from Playfair's "illustrations" of it—of extremely slow processes producing an endless repetition of similar states (*BLT* §3.4). The geotheory that Scrope's book was designed to support was, on the contrary, strongly directional: dominated and in part determined by the slow cooling of the globe that Fourier was advocating with such authority. Nor did Scrope claim that sudden and violent events were always out of place in geological explanation. For example, although land might rise a little at a time as a result of ordinary earthquakes, the upheaval of mountains was "sometimes perhaps paroxysmal": he suggested that "the elevation of the Alps, and perhaps of the whole of Europe, [might be] attributable to such a catastrophe" (§8.3). In turn, such events might have caused mega-tsunamis—the 1755 Lisbon tsunami acted as usual as a small-scale analogue—which could have been responsible for the widespread diluvial effects: this, he claimed, was the "real character [of] deluges, cataclysms or debacles" in geohistory.

Above all, Scrope argued that volcanic action, and with it the forces of elevation and other processes internal to the earth, were likely to have been far more intense in early geohistory than they are now, simply as a consequence of the earth's progressive cooling from an originally hot and fluid state. In confirmation of this he mentioned in passing the tropical appearance of the fossils in the earlier formations (see §12.2). Geohistory as a whole, he concluded, was therefore a story of steady directional change, yet punctuated by occasional episodes of great intensity, without of course any abrogation of the ordinary laws of nature: "It has proceeded generally by a lent [i.e., slow] and uniform process, gradually diminishing in energy from the beginning to the present day; but occasionally presenting partial [i.e., localized] crises of excessive turbulence, resulting from accidental combinations of circumstances favourable to the maximum of violence."[11]

9.4 ÉLIE DE BEAUMONT'S SEQUENCE OF REVOLUTIONS

Another major theory in earth physics was taking shape around the same time, and might have seemed to be a rival, if not incompatible. But in the event the two were to merge, thereby forming the most powerful geotheory that had been seen for a generation or more, and one that might even help to rehabilitate that discredited genre.

The orientation of linear mountain ranges around the world had long been an important topic for the descriptive science of physical geography (*BLT* §2.2); Humboldt's early fieldwork in Latin America, expanded later into his transatlantic synthesis (§3.1), was particularly significant. But it was von Buch who effectively enlarged the context of this topic, both into earth physics and into geohistory. In parallel with his causal theory of "elevation craters" and the larger crustal movements to which they might be related (§8.3)—a topic clearly within earth physics—von Buch also tried to plot the *past* occurrence of such exceptional episodes of crustal disturbance, as a topic of geohistory. Out in the field, he looked to see which sedimentary formations had been affected by crustal movements in each mountain region, and which others lay undisturbed against or on top of the folded rocks. The "revolution" or episode responsible for the folding or buckling of the strata could then be given a relative or stratigraphical date, since obviously it must have taken place after the deposition of the youngest formation affected, but before the oldest formation unaffected and deposited on top. After extensive fieldwork throughout the German lands, von Buch concluded provisionally that in that part of Europe there had been four successive episodes of folding and faulting, each with a characteristic orientation.[12]

However, the further development of this idea was mainly the work of Léonce Élie de Beaumont (1798–1874), an up-and-coming French geologist a little younger than Scrope and a little older than Adolphe Brongniart. Élie de Beaumont, an outstanding student at the mathematically oriented Polytechnic School in Paris, had completed his education at the Mining School. In 1823 he had been the youngest of the three Mining Corps geologists who were sent to England

9. Scrope, *Considerations on volcanos* (1825); Daubeny, *Active and extinct volcanos* (1826), based on his lectures at Oxford (and illustrated by the diagram reproduced here as Fig. 7.4). Both books were published by William Phillips, the publisher (and founding member of the Geological Society) who had initiated what became "Conybeare and Phillips" (§3.2).

10. Scrope, *Considerations on volcanos* (1825), iv–v, 215, 242. It would be idle to claim priority for any one of these geologists rather than another, for their emphasis on actual causes; they were all reviving what had been commonplace half a century earlier in the time of Desmarest and Soulavie (*BLT* §4.3, §4.4).

11. Scrope, *Considerations on volcanos* (1825), 240. His characteristic use of unusual English words of French derivation (e.g., "lent" for slow, "march" for progress) reflects his fluency in the *lingua franca* of the time.

12. Buch, "Systeme von Deutschland" (1824), was a brief and provisional summary of his evidence; see also his superb *Geognostische Karte von Deutschland* (1826).

to study the fieldwork methods that lay behind Greenough's great geological map—regarded by that time as having superseded Smith's—in preparation for the state-funded geological survey of France. Élie de Beaumont had accumulated a wide range of fieldwork experience in the course of his official duties; in parallel with them, but not unrelated to their utilitarian goals, he was also pursuing his own research interests on the formation of mountain ranges and the folding and faulting of strata that they displayed. It was after one official field trip, for example, that he had reported to the elder Brongniart how he had seen spectacular evidence in the Swiss Alps (on the Diablerets, near the French border) that Secondary formations had been overturned on top of Tertiaries and even thrust over them (*BLT* §9.5). This had whetted his appetite, if it needed any whetting, for understanding how mountain ranges had been formed at different periods of geohistory.[13]

Élie de Beaumont aspired to extend von Buch's concept of repeated and localized crustal movements to the whole of Europe and even, by a critical study of published work, to the rest of the world. More importantly, however, he realized that the precision embodied in the new fossil-based or Smithian stratigraphy might allow an equal precision in the relative dating of the successive episodes of mountain-building movements in the earth's crust. It might therefore be possible to match the record of life with an equally reliable record of the successive physical "revolutions" that seemed to have punctuated geohistory. In 1829, after an immense research effort outdoors in the field and indoors in the Mining School's library, he tried out his conclusions among other younger savants and then in the august setting of the Académie; and he published a massive article in

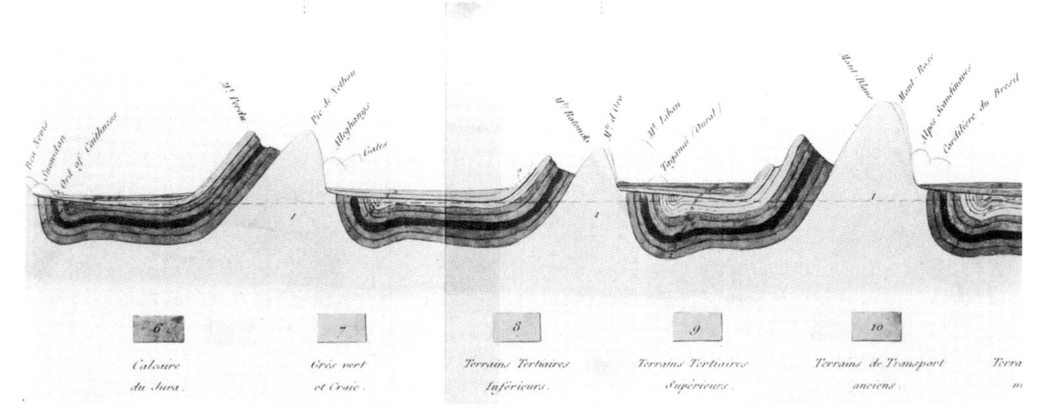

Fig. 9.3 Part of Élie de Beaumont's great theoretical section (1830) through the sedimentary formations, and the mountain ranges the upheaval of which had punctuated their accumulation. Time flows from left to right: the sudden elevation of one "system" of mountains after another (each with its own distinctive orientation, not shown on this section) had folded upwards all the formations previously deposited, while the subsequent formations had been laid down—after erosion—*unconformably* on their truncated edges. Each episode of crustal movement could therefore be dated to the interval between the youngest formation affected by its folding and the oldest unaffected. The three "systems" in this part of the section were, in geohistorical order from left to right, (a) at the end of the Cretaceous, elevating the Apennines and Pyrenees; (b) in the middle of the Tertiaries; and (c) at the end of the Tertiaries, elevating the western Alps. (The mountain ranges sketched only in outline indicate their provisional assignment to the same putatively worldwide episodes.) All the ranges are shown as having Primary rocks (1) at their core, as had long been observed in the field; but they are depicted as having buckled upwards at widely spaced intervals during geohistory.

four successive issues (1829–30) of the *Annales des Sciences Naturelles*, which had become the leading French periodical for geology. Like the elder Brongniart's massive synthesis of stratigraphy (published at almost the same time) it was also available in book form; extracts were soon translated into German and English, but in any case most geologists throughout Europe could read the original. Its lengthy title can serve as an effective summary of its author's claim to have correlated tectonic with stratigraphical events in geohistory: "Researches on some of the revolutions of the earth's surface, offering different examples of coincidence between the orientation [*redressement*] of the strata of certain mountain systems, and the sudden changes that have produced the lines of demarcation observed between certain successive stages of the sedimentary formations."[14]

Élie de Beaumont maintained that he had simply combined Cuvier's notion of a succession of sudden "*révolutions*", as witnessed by the fossil record, with von Buch's detection of sharp distinctions between the various systems of folded rocks in the mountain regions of Europe. But this was really too modest. At the start of his massive work, Élie de Beaumont, like von Buch, listed just four cases of coincidence: four sharp stratigraphical and paleontological breaks corresponded to four distinct sets of folding. From these he inferred four successive revolutions in the more recent history of the earth: at the end of the Jurassic formations; at the end of the Chalk formation (or of the Cretaceous group as a whole); at the end of the Tertiaries; and after most of the Superficial deposits (but before the Alluvium). However, by the conclusion of his work—most of which was devoted to a mass of local details supporting his argument—he had inserted a fifth case in the middle of the Tertiary and had added four more ancient ones reaching back to the Carboniferous, making in all a total of no fewer than nine revolutions. He summarized his geohistorical narrative with a theoretical diagram of astonishing originality: he called it "a kind of graphical recapitulation" of his evidence and his conclusions. It depicted the successive "epochs of elevation" or episodes of crustal disturbance as a single traverse section, in which the horizontal dimension was also a representation of time and geohistory (Fig. 9.3).[15]

13. Deville, *Travaux d'Élie de Beaumont* (1878), is a valuable source. On the origins of the French map, see Brochant, "Carte géologique" (1835), and Gaudant, "Carte géologique de France" (1991); also Gohau, "Élie de Beaumont" (1998), and, for the contrast with similar plans for Britain, Rudwick, *Devonian controversy* (1985), 90–91. The complete map was published in 1841, after less than twenty years of well-organized fieldwork, and was then by far the finest map of its kind in the world. Élie de Beaumont and his French colleagues used his double-barreled family name in full; his first or baptismal name was not Élie but Léonce.

14. Élie de Beaumont, "Révolutions du globe" (1829–30), title; it was expanded from a paper read at the Académie on 22 June 1829. The author sent an offprint to the Geological Society (the gift was recorded on 21 May 1830) so that his theory was promptly known to geologists in Britain.

15. Fig. 9.3 is reproduced from Élie de Beaumont, "Révolutions du globe" (1829–30), part of pl. 3, explained on 240. The other main illustration (pl. 1) is a map of the Alps and surrounding regions, showing (by colored straight lines) the precise orientations said to be characteristic of the different episodes of crustal movement or tectonic activity. The geometrical character of this was strongly criticized by other geologists at the time, as incompatible with the often curved and variable trends observed in the field; nonetheless he later pursued the idea with almost obsessive zeal, plotting a "pentagonal network" [*reseau pentagonal*] of fold lines right across the globe. However, his *geohistorical* conclusions were not dependent on this aspect of his theory. Lawrence, "Heaven and earth" (1977) and "Lyell versus central heat" (1978), rightly treat Élie de Beaumont's theory as the culmination of the research tradition represented by Fourier and Cordier. Greene, *Geology in the nineteenth century* (1982), chaps. 3, 4, sets his work in a broader context of tectonic theorizing.

The geohistory embodied in Élie de Beaumont's synthesis was not pinched for time. Each "epoch of elevation" might have been a "catastrophe"—to use one of Cuvier's favorite words—but they were certainly not being invoked to compensate for a failure to imagine the true magnitude of geological time. According to Élie de Beaumont, sudden and violent episodes of crustal deformation had only very occasionally punctuated the vast spans of time represented by the thick sedimentary formations. Whether what was sudden in geological terms would have been experienced as sudden by humans—had any been around—was left unclear; but there was no reason why they might not have been, in effect, *mega-earthquakes* of a magnitude far beyond anything recorded in the brief span of human history. On the other hand, the extremely long intervals of "tranquillité" *between* these events were described explicitly as having been much like the present world.

What was most innovative about Élie de Beaumont's synthesis was the precision, derived from the new fossil-based stratigraphy, with which the successive revolutions could be given relative dates. And the sharp changes in fossil faunas and floras between successive major formations—most strikingly, between the

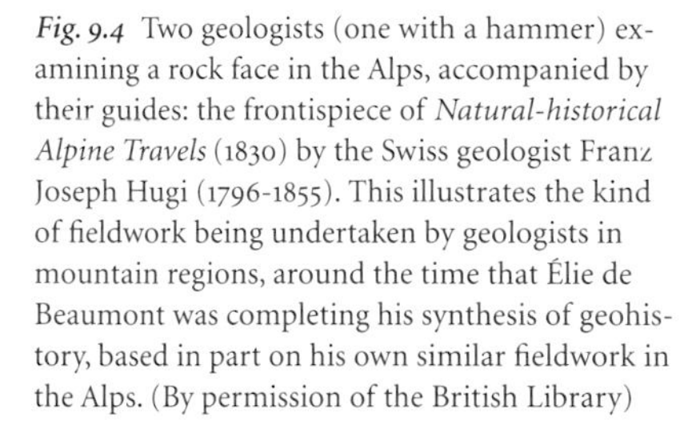

Fig. 9.4 Two geologists (one with a hammer) examining a rock face in the Alps, accompanied by their guides: the frontispiece of *Natural-historical Alpine Travels* (1830) by the Swiss geologist Franz Joseph Hugi (1796-1855). This illustrates the kind of fieldwork being undertaken by geologists in mountain regions, around the time that Élie de Beaumont was completing his synthesis of geohistory, based in part on his own similar fieldwork in the Alps. (By permission of the British Library)

Chalk and the earliest Tertiaries—now had a plausible *causal* explanation, in the drastic environmental disturbance that would have resulted from the sudden upheaval of a new mountain range, even in some distant part of the globe. This suggested how lines of evidence that had previously been almost independent—derived from stratigraphy, paleontology, and earth physics—might be integrated to form a single coherent narrative of geohistory.

In his printed account Élie de Beaumont disclaimed any ambition to offer a causal explanation, because he wanted his reconstruction to be judged first as geohistory. However, he had mentioned what he had in mind when presenting his work orally, and this later appeared in print. Picking up the hints that Cordier and the physicists had thrown out, he saw that a progressively cooling earth could have provided the necessary causal motor for a directional geohistory punctuated by occasional sudden revolutions. As the intensely hot fluid interior cooled down, it would necessarily have shrunk. So the even cooler solid crust might have become too large to fit, and the compressive strain might have been released by a sudden "epoch of elevation" or upward buckling along certain lines. Élie de Beaumont suggested that this had happened repeatedly, at widely spaced intervals in the course of geohistory, each time with a different orientation. Whatever his theory's defects or limitations in retrospect, it was an immensely fruitful perspective on geohistory, a wider-ranging synthesis than any hitherto. And, whatever its boldly hypothetical components, it was solidly based on his own extensive fieldwork, and that of his contemporaries, not least in the Alps and other mountain regions (Fig. 9.4).[16]

9.5 CONCLUSION

During the 1820s the rapid development of fossil-based or Smithian stratigraphy yielded an agreed framework for geohistorical interpretation, or at least for the portion of geohistory represented by the Secondary and Tertiary formations. This consensual stratigraphy was effectively summarized at the end of the decade by the elder Brongniart. It suggested a history of life marked by directional and even progressive change, with successively "higher" kinds of life making their appearance in the course of time. Other geologists also agreed with Brongniart's tacit assumption that the huge pile of formations represented a correspondingly vast span of time, in relation to which the whole of human history (represented by the Alluvial deposits) had been the merest sliver at the end.

The old idea of a gradually cooling earth made a dramatic comeback at just this time, because it was newly endorsed with all the authority and rigor of the

16. Fig. 9.4 is reproduced from Hugi, *Naturhistorische Alpenreise* (1830) [London-BL: 1429.d.2], frontispiece. Élie de Beaumont, "Histoire de l'Oisans" (1834), read on 7 and 20 March 1829 at the Société Philomathique and Société d'Histoire Naturelle respectively, was a detailed study of a part of the French Alps; but it included a long theoretical digression (15–19n) suggesting, in part by analogy with the Moon's surface features, that the earth's crust had buckled repeatedly during its "refroidissement séculaire". This paper was printed in 1829, and the author's ideas must have been well known at that time among Parisian savants, although publication of the volume was delayed for five years by the publisher's commercial problems. Meanwhile a revised version of his "Révolutions du globe" was printed in De la Beche, *Manuel géologique* (1833), 616–65: see especially "Remarques générales", 656–65.

latest physical reasoning. Fourier applied his mathematical theory of heat conduction to the specific question of the dissipation of the earth's internal store of heat. He inferred that the earth had cooled from an extremely hot initial state, rapidly at first but then more and more slowly, until it had reached its present state of stability and equilibrium; but even now a relatively thin solid crust was underlain by a hot and largely fluid interior. This model was promptly adopted by Scrope, who applied it to a theory centered on volcanic phenomena, which he had studied more extensively than anyone of his generation. Cordier then complemented Fourier's argument by establishing empirically the reality of the earth's internal heat, which he too interpreted as residual. This newly rehabilitated theory of long-term global cooling, although in itself a part of the causal science of earth physics, had clear implications for geohistory, and not least for the history of life.

The idea of occasional physical revolutions punctuating an extremely lengthy geohistory had been suggested by Cuvier and the elder Brongniart, in their study of the stratigraphy and fossils around Paris. Von Buch had then identified four successive episodes of folding and crustal disturbance in the German lands, each marked by a distinctive orientation. In the later 1820s Élie de Beaumont combined these approaches. He gave von Buch's revolutions greater precision when he recorded the youngest of the formations affected by the folding and the oldest unaffected, thereby locating each "epoch of elevation" in the geohistory established by fossil-based stratigraphy; he defined no fewer than nine such revolutions. Each set of fold structures had its own orientation, and he attributed them, in effect, to mega-earthquakes or sudden and violent episodes of crustal buckling. The mega-tsunamis thus generated might well have caused mass extinctions worldwide, and thereby account for the abrupt changes in fossil faunas and floras. And finally, he also suggested that the new theory of global cooling might explain the whole sequence, if the occasional buckling of the solid crust had accommodated a steadily shrinking core.

Élie de Beaumont's synthesis of stratigraphy, paleontology, and (in modern terms) tectonic or structural geology was to be profoundly influential, as later chapters will suggest. But first it is necessary to review what other geologists were doing during these same years, to clarify what the earth might have been like during the vast spans of time *between* the putative revolutions, not only in the strange and remote era of the Secondaries but also in the relatively recent era represented by the Tertiary formations. The latter is the subject of the next chapter.

CHAPTER 10

The Tertiary gateway (1824–27)

10.1 THE ADEQUACY OF ACTUAL CAUSES

Von Hoff's great prizewinning compilation summarized what was known about the effects of "actual causes" during the brief span of recorded human history (§7.2). Other geologists agreed that the "present world" offered the best possible key to understanding the "former world" of geohistory, or anyway the point at which investigations should begin. But there was no consensus about the adequacy of present processes to explain *all* the traces of events in the deep past, and there was much that made it seem, metaphorically, a profoundly foreign country. Indeed that sense of otherness, far from dissolving in the light of further research, was being deepened—just as von Hoff published his volumes—by the sensational discoveries of the reptilian monsters of the Secondary formations (Chaps. 2, 5). Brongniart's research on the Tertiary formations all over Europe was showing that in many respects they were quite like the Secondaries (*BLT* §9.6), but they were also characterized by much more familiar and modern-looking animals and plants. So if actual causes were to be used to explain the "former world" and to reconstruct its history, the Tertiaries were clearly the best place to start. If the Tertiaries could be understood in terms of processes that were observably active in the "present world", it would extend the scope of causal explanation in geology; and it would allow geologists to "burst the limits of time" not just at an isolated spy-hole such as Kirkdale Cave (*BLT* §10.6) but quite generally for this most recent era of geohistory. And if the Tertiaries could be linked onto the present in this way, then they in turn might act as a key for understanding the still deeper past of the Secondary formations.

In their already classic research on the Tertiaries of the Paris Basin (*BLT* §9.1), Brongniart and Cuvier had been ambivalent about the extent to which these formations could be understood in terms of the present world. On the one hand, Cuvier had reconstructed the fossil mammals of the Gypsum formation as being much like living mammals in their anatomy and physiology, although he had assigned them to rather strange and certainly extinct genera (*BLT* §7.5); and Brongniart had assumed that the fossil mollusks found in the various formations, when compared with living mollusks, could be used as reliable indicators of marine or freshwater conditions at different times (*BLT* §9.1, §9.2). On the other hand, they had claimed that some of the rock types, notably the solid freshwater limestones, were unlike anything forming at the present day; and they had left unexplained the "revolutions" or allegedly sudden changes that had apparently produced an alternation of marine and freshwater conditions. Indeed Brongniart's recent revision of their nominally joint work had accentuated the contrast with any comparable situations in the present world, for he had abandoned what had seemed the clearest case of two quite different kinds of sediment having been formed at the same time in different areas (*BLT* §9.6).

Whatever else was involved, it was clear that any further clarification of the Tertiaries would need to be based on the kind of detailed stratigraphical research that was in fact becoming standard practice among geologists. To mention just one small example, in 1824 the Parisian naturalist Jean-Jacques Nicolas Huot (1790–1845) published a paper devoted entirely to the famous little quarry at

Fig. 10.1 The quarry in the Coarse Limestone at Grignon in the Paris Basin, long famous for the abundance and superb preservation of its fossil shells of Tertiary age: a lithograph illustrating Huot's study of its stratigraphy (1824). His description of a dozen specific beds and their respective fossils was far more detailed than had been customary when, almost twenty years earlier, Lamarck had first described and named the fossils from Grignon and Brongniart had first mapped the Paris Basin. (By permission of the Syndics of Cambridge University Library)

Grignon, where the Coarse Limestone was exposed. This locality had been celebrated for the abundance and superb preservation of its fossil shells, ever since they were described and named by Lamarck in the early years of the century (*BLT* §7.4). Huot set out to amplify Brongniart's newly published revision of the stratigraphy of the Paris Basin, and on some points to correct it, just for this specific locality and in great detail (Fig. 10.1).[1]

10.2 INTERPRETING THE TERTIARY WORLD

It was Prévost (§5.5), however, who first seriously questioned his mentors' interpretation of the Parisian formations, and thereby opened up the problems of understanding the Tertiary era in geohistory. He had used his own fieldwork in the Vienna Basin as a peg on which to hang a significantly novel interpretation of the Tertiaries as a whole (*BLT* §9.6). He had suggested that during the Tertiary era the seas had gradually drained off the surface of stable continents, while sediments had been deposited in the various basins left in the irregular surface of the land. However, his adoption of this standard model of a steadily falling global sea level (*BLT* §3.5) had left him with a problem, namely that of accounting for the alternation of marine and freshwater formations that Brongniart and Cuvier claimed to have found in the Paris Basin (*BLT* §9.1). He was not content, as those older naturalists had been, just to ascribe the alternations to unspecified sudden "revolutions". Since similar alternations had also been found in other Tertiary basins—notably by Webster on the Isle of Wight (*BLT* §9.4)—they could not readily be explained by limited regional movements of the earth itself, for which the celebrated case of Serapis (§8.2) might have acted as a small-scale analogue. Yet on the other hand Prévost doubted whether global sea level could have fluctuated (in modern terms, eustatically) on the large scale required to generate the alleged alternations.[2]

Prévost had begun to tackle this problem by investigating the Parisian strata in much greater detail, collecting their fossils bed by bed: a method that, as just mentioned, was becoming standard practice among geologists. In a series of papers read and published in Paris, he had searched for an explanation of the stratigraphy that would avoid having to postulate "the repeated [*itératif*] lowering and raising of the oceanic waters", while retaining the now unquestioned distinction between the marine and freshwater fossils. The solution, he had claimed, lay in a close analysis of those few cases where there was a mixture of the two kinds of shells, at or near the boundaries between formations characterized by one or the other. He had argued that differences in preservation showed either that the freshwater shells had been swept into a marine environment, or that marine shells had been eroded out of an earlier deposit and redeposited in fresh water: in both cases some shells had been "reshuffled" [*remanié*], or in modern terms

1. Fig. 10.1 is reproduced from Huot, "Banc de Grignon" (1824) [Cambridge-UL: T382.b.8.1], pl. 1. The château de Grignon (which now houses a scientific research institute) is 30km west of the center of Paris and 14km west-northwest of Versailles.

2. Gosselet, "Prévost" (1896), is still a valuable source; see also Laurent, "Actualisme chez Prévost" (1976), Bork, "Constant Prévost" (1990), and especially Gohau, "Constant Prévost" (1995).

derived, and no longer represented their original environments. It was, he had claimed, a legitimate kind of "theoretical conjecture", which he contrasted with the "frivolous hypotheses" of the outmoded genre of geotheory. It was legitimate, because it was based on observable situations in the present world; that is, on actual causes.[3]

Prévost had therefore proposed a radical reinterpretation of the Parisian sequence, concluding that there was no true alternation between marine and freshwater episodes, and hence that no repeated revolutions or sudden changes in sea level need be invoked. The Plastic Clay formation overlying the Chalk, which Brongniart and Cuvier had interpreted as marking the first Tertiary freshwater episode, became, on Prévost's interpretation, a deposit of terrestrial material that had been swept by rivers in flood, down from the surrounding land areas into the sea that at that time still occupied the basin. The overlying Coarse Limestone, with the prolific marine fauna that Lamarck had described long before (particularly from Grignon), then marked a final and more tranquil period of truly marine deposition. After that, according to Prévost, the sea had withdrawn from the Paris Basin for the first and only time. *All* the overlying or subsequent formations, from the Gypsum onward, had accumulated under non-marine conditions; the marine fossils in the formation overlying the Gypsum were interpreted as being derived from the erosion of the Coarse Limestone elsewhere, and did not represent a true reversion to marine conditions.

Prévost's reinterpretation of the Paris Basin was certainly ingenious, though it might have seemed almost as "desperate" a way of saving the phenomena as Goethe's explanation of the bored columns of the Temple of Serapis (§8.2). Yet it did embody a new and explicit application of the principle that fossils needed interpretation in two distinct ways: not only as marking the relative *ages* of the formations in which they were found—which was now taken for granted as the basis for any effective stratigraphy—but also as witnessing to the *conditions* or circumstances of the time. And according to Prévost the latter might include not only the environmental conditions in which the organisms had once lived, but also the physical circumstances in which their remains had subsequently been preserved (in modern terms, their paleoecology and their taphonomy respectively). The following year, in his next paper, Prévost had made explicit this twofold way of using fossil evidence, as the necessary foundation for "inferring what has been *from what is*" or using the present world as a key to the deep past. As he put it in his title, this actualistic method underlined "the importance, for positive [i.e., fact-based] geology, of studying fossil organisms"; as he emphasized in his conclusion, it was making zoologists and anatomists as valuable to the science as mineralogists had always been.[4]

However, that coded compliment is unlikely to have appeased Cuvier for the rejection of his idea of successive sudden revolutions within the Tertiary era. For in 1825, in another paper published by the Société Philomathique, Prévost coupled it with a vigorous insistence that close comparisons with the present world—for which von Hoff had recently provided the indispensable database (§7.2)—had been insufficiently exploited by geologists, and that it was therefore *premature* to dismiss actual causes as inadequate:

> Taking the ground around Paris as an example, he [Prévost] has offered to show that suppositions conceivable in the present state of nature—and consequently having nothing contrary to the laws of general physics—suffice [*sic*] to explain the formation of the very different deposits of which the last [i.e., Tertiary] beds of the earth are composed . . . It was [only] before all the effects of causes still active had been observed, and compared step by step with the effects produced formerly, that some celebrated geologists could claim that everything in ancient nature happened otherwise than in present nature.[5]

This paper clearly revealed Prévost as being Cuvier's most serious critic on the question of the adequacy of actual causes in geological explanation. Only a few months later, however, geologists could see that he had already gained a useful ally across the Channel. Charles Lyell, primed by his youthful reading of Playfair's Huttonian *Illustrations*, had certainly been receptive to Prévost's ideas. When in 1823 he had visited Paris to make himself known to the leading savants and improve his fluency in the international scientific language (and incidentally to discover Cuvier's opinion on one of Mantell's fossils: §5.4), he had also spent much time with Prévost and evidently became quite a close friend. The Frenchman took him on several field trips to see the Tertiary formations and to collect their fossils, and had plenty of time to expound his ideas; since he spoke little or no English their conversations would have been good practice for Lyell. The following year they reversed their roles. Fulfilling a promise he had made in Paris, Lyell took Prévost on a long field trip to the furthest tip of Cornwall, and back along the south coast of England (§5.5). Apparently he did not cross to the Isle of Wight or see its Tertiary formations, as had probably been planned, but Lyell may at least have shown him those on the mainland coast near his family home in Hampshire.[6]

Later in the year, Buckland took Lyell on an extensive fieldwork tour in Scotland, and the younger man then spent several weeks on his family's estate near Forfar, on the edge of the Highlands. There he emulated Prévost's example by studying some small lakes nearby, finding in them an unexpectedly instructive analogue to the Tertiary freshwater limestones that he had been shown around Paris. Some of the lakes had recently been drained in order to exploit the marls that had been deposited in them, as lime to put on the fields. Lyell found in these marls not only the usual freshwater mollusks but also the tiny fossil *gyrogonites*,

3. Prévost, "Réunion de coquilles marines" (1821), 418–20, 427; and "Grès coquillers de Beau-Champ" (1822), read at the Société Philomathique on 28 July 1821; the localities were quarries at Bagneux and Beauchamp, respectively to the south and northwest of Paris (and both now in its suburbs). See also Gosselet, "Prévost" (1896), 68–77.

4. Prévost, "Corps organisés vivans" (1823), read at the Société d'Histoire Naturelle on 8 November 1822.

5. Prévost, "Formation des terrains" (1825), 74; the indirect prose, like the impersonal style used in modern scientific papers, does not imply that this report was written by someone else.

6. See Lyell's letters printed in Lyell (K.), *Life, letters and journals of Lyell* (1881) [hereafter "*LLJ*"] 1: 123–53; also Wilson, *Lyell* (1972), 115–29. I am indebted to Goulven Laurent for sending me his unpublished paper, given at a conference in Nantes in 1994, in which he rightly emphasized Lyell's debt to his French colleagues.

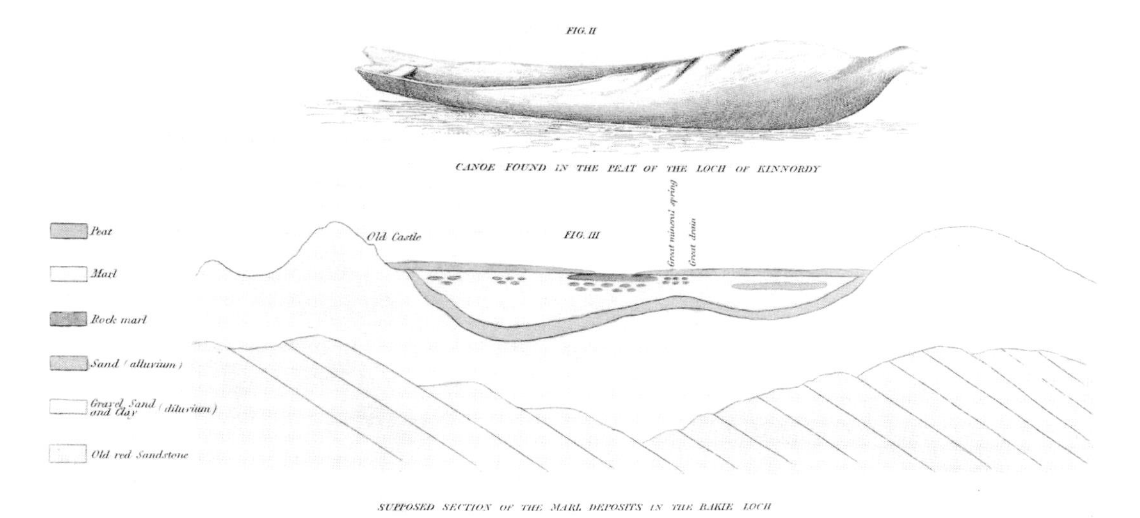

Fig. 10.2 Lyell's section through the recently drained Bakie Loch, near Kirriemuir in Scotland, published by the Geological Society in 1826. On a basement of ancient Secondary rock ("Old red sandstone") was an irregular Superficial deposit, for which Lyell adopted Buckland's term "diluvium". In a hollow in this deposit formerly occupied by the little lake, and above a layer of sandy "alluvium", was a deposit of calcareous marl up to ten feet thick; in one layer, and in smaller patches, this had been cemented into a hard rock (shown here by the darkest tint). Overlying the marl was a layer of peat dating from the human period, as shown by the primitive dug-out canoe (above) found in a similar deposit in another lake close to his family house of Kinnordy. Lyell argued that the marl was closely similar to the Tertiary freshwater limestones in the Paris Basin; and that even their hardest and oldest-looking varieties could be matched in these geologically very recent deposits. (By permission of the Syndics of Cambridge University Library)

or fruits of the freshwater plant *Chara*, which had been decisive in establishing the origin of the Tertiary freshwater limestones around Paris (*BLT* §9.2). It was equally important, however, that in places he found the marl cemented into a limestone just as hard and solid as any of the Tertiary ones. It was a striking proof that such solidity did *not* imply any disjunction of processes between that "former world" and the present world: it was a small but significant further demonstration that, as Prévost was claiming, everything about the Tertiary formations could be explained in terms of what could be seen happening at the present day (or at least in the geologically very recent past).

At the end of the year, the Forfarshire lake marls became the subject of Lyell's first paper to be published by the Geological Society. A sequel, read six months after the first, extended the analogy with the Parisian formations by considering the mammalian bones found in the lake marls. These papers were well received by those who mattered most, though Greenough, characteristically, expressed great skepticism about the way Lyell blurred the traditional sharp distinction between the alluvial and older deposits. Lyell's first paper was published the following year in the Society's now prestigious *Transactions* (Fig. 10.2).[7]

10.3 PRÉVOST'S REINTERPRETATION OF THE PARIS BASIN

In 1827, Prévost—fortified perhaps by the knowledge that he was not quite alone in his stand on the total adequacy of actual causes—took his earlier argument into the lions' den. He presented the Académie with a massive "geological dissertation" that set out his case in full. Moving from the present to the past, from the known to the unknown, Prévost intended to show how actualistic comparisons could explain much more than Brongniart and Cuvier, and perhaps others in his audience at the Académie, believed possible. Boldly, or rashly, Prévost challenged Cuvier even more explicitly than before:

> I have not been stopped at all in this attempt to link the past to the present, by what is called *a sharp limit between former nature and present nature*; on the contrary, I believe I have seen nuances, passages, everywhere, and I have been unable to convince myself that it is useless to search in the present order of things for the explanation of phenomena that have taken place on earth at remote times; my experience has prevented me from thinking "that the thread of operations is broken; nature has changed course, and none of the agents she employs today would have been sufficient to produce her former effects".[8]

Prévost's title asked, rhetorically, "Have the present continents been submerged by the sea several times?" Systematically and at length, he gave his reasons for answering that question in the negative. Citing a wide range of fieldwork, both his own and that of other geologists, he denied that there had been "repeated submersions" of the continents. In the second part of this massive work, in support of his general conclusion, Prévost set out in full his reinterpretation of the stratigraphy of the Paris Basin. Everything, he argued, could be explained naturally by inferring that the sea had retreated off the land surface just once and for all, turning an originally marine gulf into a brackish lagoon, then into a freshwater lake, and ultimately into dry land; but that local circumstances had led to a more complex sequence of formations, with either marine or freshwater shells, and in some places to anomalous mixtures or rapid alternations between them.[9]

7. Fig. 10.2 is reproduced from Lyell, "Freshwater limestone" (1826) [Cambridge-UL: Q365.b.12.7], part of pl. 10, read at the Geological Society on 17 December 1824 and 7 January 1825; Lyell to Mantell, 8 January 1825 [Wellington-ATL], describes its reception. His much less substantial "Remarks on quadrupeds imbedded in recent strata" was read on 3 June 1825 [London-GS: OM/1/3, 85–87]; like many other minor papers at this time, it was never published, but this hardly constituted the "humiliating failure" inferred by Dean, "Early career of Lyell" (1997), 222. Kinnordy House, Lyell's family home, with its lake (now a patch of marshy ground) just to the south, is 2km northwest of Kirriemuir; the former Bakie Loch is about 8km southwest of the town, which in turn is about 25km north of Dundee. Lyell's very first paper to be read at the Society, "Dike of serpentine" (1825), was later withdrawn and published in an Edinburgh periodical; this "hard-rock" topic (on a spot near the family estate) helped establish his credentials among geologists, and he gained reflected glory by persuading the already distinguished and polymathic John Herschel (1792–1871)—the son of the late and great William Herschel, de Luc's neighbor near Windsor long before—to add a note on a similar serpentine that he had seen in the Tyrol.

8. Prévost, "Continens actuels" (1828), 253; the famous quotation came from Cuvier, *Ossemens fossiles* (1812), 1, Discours préliminaire, 16–17 [trans. Rudwick, *Georges Cuvier* (1997), 193], repeated in *ibid.*, 2nd ed., 1 (1821), xiii. See also Gosselet, "Prévost" (1896), 35–59.

9. Prévost, "Continens actuels" (1828), read at the Académie on 18 June and 2 July 1827, used the phrase "submersions itératives" in its running head. Its sequel, enlarged from "Formation des terrains" (1825) and with a similar title, was read later in July 1827, but was not published until much later in *Candidature de Prévost* (1835), 93–126. See also Gosselet, "Prévost" (1896), 138–54.

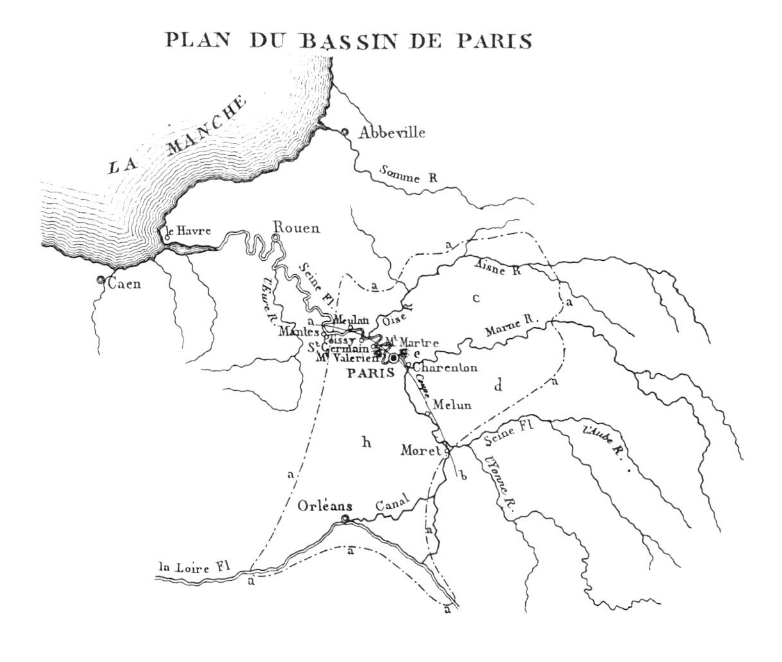

Fig. 10.3 Prévost's outline "Map of the Paris Basin", filled with Tertiary formations and surrounded by the underlying Chalk (*a*); showing also the present drainage basin of the Seine, which flows through Paris into the Channel [*La Manche*] beyond Rouen. (By permission of the Department of Earth Sciences, University of Cambridge)

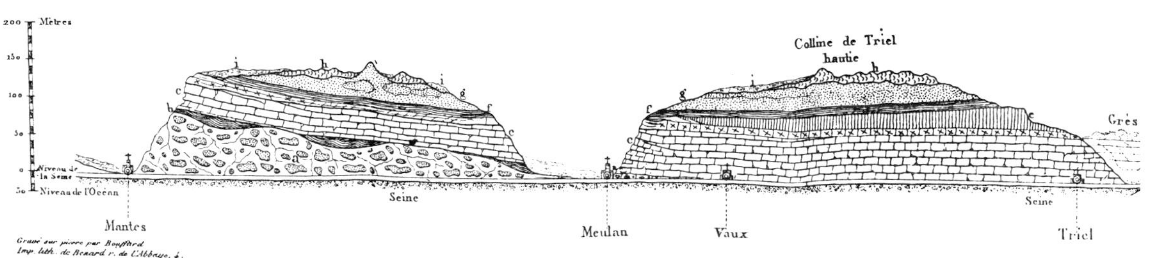

Fig. 10.4 One end of Prévost's very long and detailed section through the Paris Basin (its curved line through Paris is marked on Fig. 10.3 as *Coupe*). It shows the lowest formation, the Chalk, rising to the surface (left) from beneath the pile of Tertiary formations; the detailed stratigraphy formed the basis for Prévost's geohistorical interpretation, as had Brongniart's comparable sections a decade earlier. Two examples of the lateral variations that he emphasized (see below) can be seen here: the Plastic Clay (*b*) sandwiched discontinuously between the Chalk (*a*) and the Coarse Limestone (*c*); and the thinning of the Gypsum formation (*e*, with close vertical shading) in the Colline de Triel (right) and its total absence further west (left). (By permission of the Department of Earth Sciences, University of Cambridge)

Prévost's pursuit of "the analytical route [*marche*]" or "the philosophical route from the known to the unknown"—his wording echoed Desmarest's celebrated analysis of the Auvergne volcanoes long before (*BLT* §4.3)—led him to heuristic "conjectures" that may have been modeled on Cuvier's well-known thought experiment with the imaginable future of Australia (*BLT* §10.3). Prévost imagined a *future* for the Channel between France and England, but it was based on the *present* state of the region and was designed to illuminate the *past* geohistory of the nearby Paris Basin, as he had inferred it from his detailed fieldwork (Figs. 10.3, 10.4).[10]

Prévost claimed that the present state of the Channel illustrated what the Paris Basin of the Tertiary era might have been like in its earlier, marine phase. He reckoned that the waters of the Seine, particularly when muddy in times of flood, must now be depositing clay or silt on the floor of the Channel near the French coast, while on the opposite side the erosion of the Chalk cliffs of England must be producing gravels composed of flint pebbles (derived from the hard flint nodules embedded in the soft chalk). So these contrasting kinds of sediment must be accumulating *at the same time*, just as he suspected—and as Brongniart too had thought originally (*BLT* §9.1)—that one of the Parisian sandstone formations had been deposited at the same time that the Coarse Limestone was accumulating in other parts of the Paris Basin. Formations might therefore be limited in lateral extent because they were deposited under specific local conditions; and their fossils might reflect those circumstances, either of environments for life (in fresh- or seawater) or in conditions of preservation (in situ or derived). Prévost was in fact exploring how formations with quite different kinds of sediments and fossils—in modern terms, formations of contrasting *facies*—might have been deposited simultaneously in different areas (see §31.2). This heightened the doubly enriched or *geohistorical* character of his stratigraphy, for it emphasized the local and contingent circumstances that had been responsible for the accumulation of successive formations and for their fossil contents.

Prévost's inferences based on the present state of the Channel were then extended into an imaginable future, as a prelude to using them as a key to the deep past. Soundings recorded on hydrographic maps suggested that there were two basins within the Channel, an outer or western one open to the Atlantic, and an inner or eastern one enclosed by submarine ridges at both ends. Any modest future lowering of global sea level—whatever its physical cause—would expose the shallower ridge (between Calais and Dover), and turn the Channel into a deep gulf isolated from the North Sea; further lowering would then expose the other ridge (between Dieppe and Brighton), and turn the inner or eastern basin into an enclosed lake. If, subsequently, evaporation from this lake exceeded the inflow of water from rivers, its surface would fall below sea level (rather as the present Caspian was known to lie below the level of the Black Sea). If at some still later time one of the enclosing isthmuses was breached by coastal erosion, the lake would quite suddenly be flooded with seawater, and marine sediments would accumulate on top of freshwater ones; yet this would *not* have been caused by any rise in global sea level or local subsidence of the earth itself. And finally, the lake might become completely silted by freshwater sediments brought in by the rivers flowing into it, until eventually its surface was colonized by terrestrial plants and became habitable by terrestrial animals.[11]

This was roughly the sequence of events that Prévost believed had affected the Paris Basin in the course of Tertiary geohistory. His thought experiment was directed towards the causal explanation of the formations, in terms of their physical modes of origin and of the modes of life and subsequent preservation of their

10. Figs. 10.3 and 10.4 are reproduced from Prévost, *Candidature de Prévost* (1835) [Cambridge-ES: B.19.281], parts of folding pl. at end. This very large lithographed plate is marked "2me édition 1835"; Prévost probably used a first version when he read the paper at the Académie in 1827. The part of his section reproduced in Fig. 10.4 represents about 20km, and shows a total relief of about 150m (greatly exaggerated, as usual, in the interests of clarity).

11. Modern hydrographic maps do not show Prévost's western ridge, and hence no separate basin in the eastern part of the Channel, but this does not, of course, affect the validity of his thought experiment.

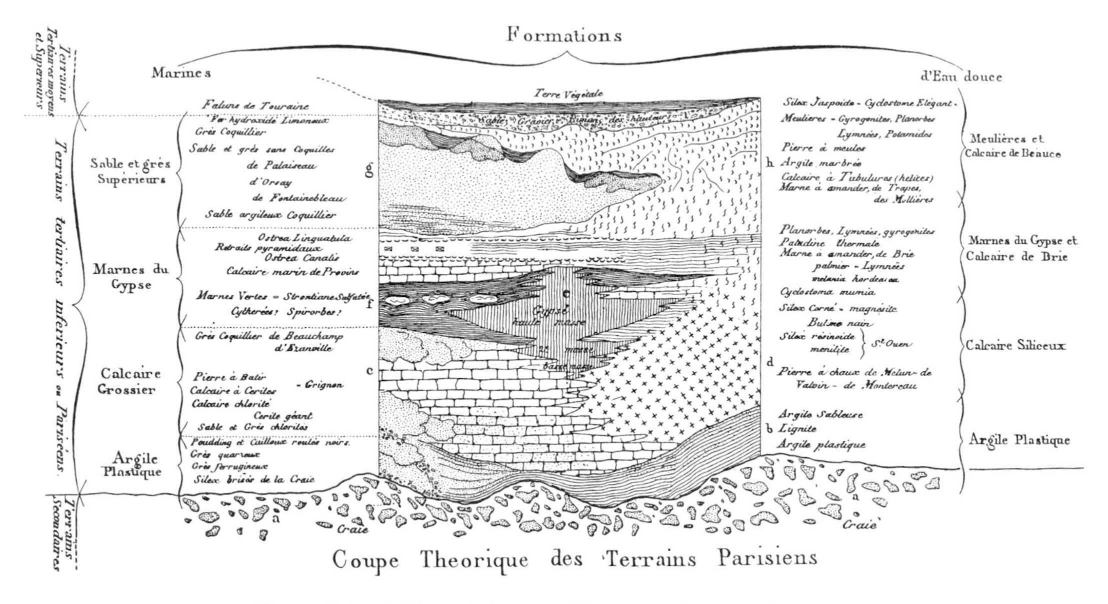

Fig. 10.5 Prévost's "theoretical section of the Parisian formations", illustrating the interpretation of Tertiary stratigraphy and geohistory that he presented to the Académie des Sciences in 1827 (though it was not published until 1835). The horizontal dimension represents space, from marine (left) to freshwater (right) environments; the vertical dimension represents the pile of formations, and therefore the sequence of conditions in time. Prévost depicted marine and freshwater formations interfingering in a highly complex manner: he claimed that they had been deposited *at the same time* in different environments, and hence with contrasting sets of fossils, on a uniform but irregular foundation of Chalk (*Craie*, *a*). He restored the lateral equivalence between the marine Coarse Limestone (*Calcaire Grossier*, *c*, left) and the freshwater Siliceous Limestone (*Calcaire Siliceux*, *d*, right) that Brongniart had originally depicted but later abandoned; but Prévost made it just one instance of a much more general phenomenon. The Gypsum formation (*Gypse*, *e*, center), which had yielded Cuvier's strange Tertiary mammals, was shown as a local lagoonal deposit, which had accumulated midway between marine and fully freshwater formations. (By permission of the Department of Earth Sciences, University of Cambridge)

fossil organisms. He summarized his interpretation of Parisian stratigraphy with a novel (and astonishingly modern-looking) visual representation of its variations across both time and space (Fig. 10.5).[12]

Prévost reconstructed not only the circumstances under which the Parisian formations might have been deposited but also the sources from which the sedimentary materials might have been derived. He suggested, for example, that the sandstones that are so prominent (and famously picturesque) a feature of the Forest of Fontainebleau to the south of Paris (Fig. 10.5, *g*, left) could have been formed from sand swept into the Paris Basin by rivers flowing at that time from the Primary rocks of the Massif Central to the south or from those of the Vosges to the east. In the present context, however, such details are less important than the kind of overall interpretation that Prévost derived from his research. What was novel—in addition to his highly effective visual representation of it—was his consistent attempt to reconstruct the physical and environmental history of the Paris Basin, in a way that made it closely comparable to identifiable conditions in the present world, such as those of northern France and the English Channel. This kind of explanation, as *physique de la terre* or earth physics (*BLT* §2.4), would have been regarded by an earlier generation of naturalists as quite justified

Fig. 10.6 A portrait of Constant Prévost, undated but probably drawn around the time of his public challenge to the interpretation of the Tertiary formations put forward by his mentors Brongniart and Cuvier, and of his unsuccessful candidacy for a place in the Académie des Sciences.

as an end in itself. But for Prévost, as for an increasing number of his contemporaries, causal interpretations were a means to an end, that of reconstructing and then explaining the contingent *geohistory* of a specific region, as part of the still wider project to trace and explain the history of the earth as a whole.

Prévost, then, firmly rejected Cuvier's premature dismissal of actual causes as inadequate for geological explanation. But his criticism was courteously expressed, and does not seem to have damaged his chances of advancement in the world of Parisian science. Cuvier was one of those who reported to the Académie on the first part of Prévost's massive work; it was summarized very fairly and recommended to the Académie for publication, and the report noted Prévost's "remarkable talent" for this novel kind of interpretation.[13]

However, Prévost had certainly ruffled other feathers. While he was expounding his ideas at the Académie, he appealed to Brongniart to support his candidacy for a vacancy in its section for mineralogy and geology. Significantly, however, he told his patron that his credentials were being questioned on the grounds that he was not a true mineralogist but rather a zoologist, that he was not a graduate of the School of Mines, and that he was not sufficiently adroit in politics. "*You have studied only the modern formations*", he reported being told; "*you are not known*", and "*you are a wealthy amateur who does it only for leisure*". Whatever may have been the academic politics involved—which seem not to have changed much in character in nearly two centuries—Prévost failed in this attempt to join the prestigious ranks of the Académie (Fig. 10.6).[14]

12. Fig. 10.5 is reproduced from Prévost, *Candidature de Prévost* (1835) [Cambridge-ES: B.19.281], part of folding pl. ("2me édition") at end of "Formation des terrains" (see Fig. 10.3); he probably used a first version when this paper was read at the Académie in July 1827. Brongniart's earlier ideal sections are reproduced in *BLT*, Figs. 9.3, 9.22.

13. Prévost, "Question géologique" (1827), is in fact a published referees' report signed jointly by Cuvier and Cordier, but probably written by the former.

14. Fig. 10.6 is reproduced from Gosselet, "Prévost" (1896), frontispiece. Prévost to Brongniart, 5 July 1827 [Paris-MHN, MS 1967/601]. It is highly unlikely that *Cuvier* would have criticized Prévost for being too much a zoologist, or not enough of a School of Mines man, or for having focused his attention on the Tertiaries (which Cuvier had publicly recommended as a top priority: *BLT* §9.3). Cordier, a highly "professional" product of the School of Mines, is a more likely culprit (notwithstanding his signing a favorable report on Prévost's work jointly with Cuvier): he might well have regarded even the Secondary formations—on which Prévost had done substantial work in Normandy (§5.5)—as so "modern" as to be unworthy of attention by a true or "hard-rock" geologist. Alternatively or in addition, it may have been Brongniart who blocked him, perhaps out of envy, and, if so, betraying Prévost's trust in him as a patron. Prévost had already tried twice before, in 1822 (when Cordier was elected) and 1824, but he claimed that at that time he had had no serious expectation of winning. On his fourth attempt—the occasion on which many of his earlier articles were published or republished in *Candidature de Prévost* (1835), to support his case—he was beaten by Élie de Beaumont, another but much younger protégé of Brongniart. It was only after Brongniart's death that Prévost was at last elected, succeeding his former patron in 1848.

10.4 CONCLUSION

In the wake of von Hoff's great inventory of the known effects of actual causes within the span of human history (§7.2), the question of their adequacy to explain *all* the events of the deeper past remained a lively issue among geologists. It was clear to many of them that the Tertiaries, as the major group of formations nearest in time to the present world, were the most promising place to start, if the question was ever to be resolved one way or the other. And it was equally clear that the best method would be to subject the Tertiaries to a more detailed scrutiny than had been customary in previous years. They might then serve as a cognitive gateway leading to an adequate understanding of the Secondaries and even more remote eras in geohistory.

In the 1820s, Prévost was the most prominent and active of those who tackled this problem. Building on his earlier studies near Vienna, he focused on the rocks and fossils of the Paris Basin, criticizing the interpretation famously offered by his mentors Brongniart and Cuvier. Prévost denied that there had been any true alternation between marine and freshwater phases in the history of the basin, and therefore rejected the need to postulate repeated natural "revolutions". Instead, he argued that in certain circumstances fossils could be "reshuffled" or derived, and thereby give deceptive evidence of their original habitats. The apparent alternations, he then claimed, were due not to the sea's repeated incursions but to its gradual withdrawal once and for all. He supported this with an extended thought experiment that took the present state of the English Channel into an imagined future, in order to interpret the past geohistory of the Paris Basin; it was an example of actualistic reasoning on an almost unprecedented scale. However, Prévost was not quite alone in his ambition to explain the deep past wholly in terms of the present; Lyell became an ally, after they had met in Paris, and made his debut at the Geological Society with a modest Scottish case that emulated Prévost's approach.

Prévost's research exemplified persuasively how it was now more important than ever to study fossils in terms of their original habitats and environments. The next chapter describes how this approach was also being applied, during these same years, to the interpretation of the rocks and fossils of the Secondary formations, and how it led to the first full reconstructions of entire ecosystems from the deep past.

CHAPTER 11

The geologists' time-machine (1825–31)

11.1 FOSSIL LAND SURFACES AND SOILS

Brongniart and Cuvier had reconstructed the geohistory of the Paris Basin—which remained the reference point for all such interpretations—as a story of repeated invasions of the sea onto the continents. One of Prévost's main arguments for rejecting their inference was his failure to find evidence for true fossil soils overlain by marine sediments. Nowhere in the Parisian sequence could he see unambiguous evidence that a *land* surface covered with vegetation and inhabited by terrestrial animals had been overwhelmed by an incursion of seawater. Cuvier's repeated sudden "révolutions" therefore seemed to Prévost to be illusory. There were indeed sharp boundaries between strata containing marine fossils and underlying sediments with freshwater shells and plant debris, but at such points there were no signs of the erosion surfaces that a sudden and violent incursion might have caused, and no evidence that a land surface had been submerged.[1]

This was a conclusion that Prévost extended beyond the Parisian Tertiaries, for he also questioned whether any of the Secondary formations had been deposited on land. Like many other geologists at this time, he claimed that all the coal seams, in what Conybeare had classed as the "Carboniferous" group of formations, were composed of plant debris that had drifted out to sea from ancient continents that were not themselves preserved. Brongniart, in contrast, had not been surprised to find a spectacular new case of tree trunks being preserved *upright* in strata overlying a coal seam: to him they were a clear proof that an ancient forest had been buried in situ (§4.4). Prévost was hard pressed to offer an

1. Prévost, "Continens actuels" (1828), 268–72.

alternative explanation, though he noted that Brongniart's picture of the locality (Fig. 4.4) did not show the trunks clearly rooted in the underlying beds, and he suggested that they might have been weighted naturally in such a way that they would drift in an upright position. At this point, however, Brongniart and others probably felt that Prévost's explanations were becoming somewhat desperate.

Lyell, primed by his reading of Playfair, did not share Prévost's reluctance to concede any movement in relative sea level during geohistory (other than a gradual fall in the sea itself). In Lyell's view, movements of the earth's crust were not merely likely but positively demonstrated by the observable folding of the Secondary formations. For example, he inferred that the gentle dome-like (anticlinal) folding of the Chalk in the Weald (Mantell's part of southeast England) was of the same date, and probably due to the same forces, as the spectacular case of highly folded strata on the Isle of Wight, which he had seen in Buckland's company just before his visit to Paris. And on the Isle of Wight it could be dated more accurately, to some point *within* the Tertiary era, because—as Webster had shown a decade earlier—it had turned up on end not only the Chalk but even some of the overlying Tertiary formations (*BLT* §9.4). With the earth evidently continuing to be internally dynamic, even in such a relatively recent period of geohistory, Lyell would be as unsurprised as Brongniart to find evidence of former land surfaces and formations of freshwater origin, anywhere in the Secondaries.[2]

Lyell was therefore more than ready to support Mantell's inference that the Weald rocks—including the Tilgate beds with their exciting new fossil reptiles—were of freshwater or estuarine origin, once they had sorted out the relevant formations on simple Smithian principles (§5.3). In effect, this turned one substantial portion of the English Secondary sequence into something like Prévost's interpretation of the Parisian Tertiaries, but on a larger scale and attributed causally to a more dynamic earth. And since these formations were overlain by the ubiquitous Chalk, unquestionably of marine origin, the case for a major alternation between freshwater and marine formations seemed overwhelming. As Lyell put it to Mantell, commenting on his friend's plans for publishing his local research:

> I attribute the [Weald and Purbeck] oysters to the minor oscillations of the land lifting up and depressing the estuary alternately, the grander alterations arose from extensive earthquakes. Do not conceal any evidence of marine [fossils] in the Tilgate, indeed this you will not, I know, but do not throw it into the shade. All will come right & it [the Wealden] is a freshwater formation undoubtedly & the grandest discovery in Geology since Cuv[ier] & Brongn[iar]t came out. If in an estuary it must still have been above the mean level of the sea. How stupendous a conclusion with respect to beds below the *chalk*, so *widely extended* a formation![3]

Right from the start of his hunt for fossil bones in the Tilgate beds, Mantell—or, more often, the quarrymen working for him—had also been finding fragments of large fossil plants (Fig. 5.2): no ordinary plants, but some that were identified for him as similar to palms, treeferns, and other vegetation of an exotic and tropical appearance. Combined with the evidence of the freshwater shells in nearby beds of "Sussex marble", Mantell was therefore following Conybeare's example (§2.3), and indirectly Cuvier's too, in reconstructing the environment in which his giant reptiles had lived. In his book on the *Geology of Sussex* (1827), he sketched in words what his county might have looked like at that time, making this specific episode in geohistory vividly imaginable. But he presented it as a set of conjectural "expectations" derived from the *present* world, which he claimed were confirmed by what he had in fact discovered about the deep past: "What would be the nature of an estuary, formed by a mighty river flowing, in a tropical climate, over sandstone rocks and argillaceous [i.e., clayey] strata, through a country clothed with palms, arborescent ferns, and the usual vegetable productions of equinoctial regions, and inhabited by turtles, crocodiles, and other amphibious reptiles?"[4]

This was just one example of the way in which a few geologists, often taking Prévost's work on the Tertiaries of the Paris Basin as an exemplar, were beginning to turn Smithian stratigraphy as a whole into a truly geohistorical practice. Prévost was not unaware of this, and evidently hoped to encourage the process by publishing an account of his Parisian research in the Geological Society's *Transactions*. Ironically, however, his example generated further evidence that—contrary to his own claims—relative sea level must have fluctuated widely during geohistory.[5]

Lyell's inference that much of southern England had been at or above sea level while the Purbeck and Weald formations were being deposited seemed to be confirmed by a clear case of a former *land* surface being preserved in the Jurassic formations on the Dorset coast. At the Geological Society in 1824, Webster had described the stratigraphy of the Purbeck and the underlying Portland formation, amplifying his earlier and classic work on the area (*BLT* §9.4) in the light of his subsequent fieldwork. The Portland, with its famous giant ammonites (and even more famous building stone) was unambiguously marine in origin; but Webster had interpreted the overlying Purbeck as lacustrine or freshwater. At the base of the Purbeck, he had noted that one specific stratum, which the quarrymen called the "dirt bed", contained what Webster identified as tree stumps in their original position. Lyell had shown them to Prévost during their field trip along the south coast, but the Frenchman had not been convinced that they were truly in situ.

2. Lyell to Mantell, 20 July 1825 [Wellington-ATL].

3. Lyell to Mantell, 16 May 1826 [Wellington-ATL]. Based on their present ecology, oysters were taken to indicate marine conditions. Preparing his Tilgate work for publication, Mantell had found some such hints of marine conditions in the formation that he was keen to establish as freshwater; but Lyell, taking his cue from Prévost, thought such minor alternations were to be expected if the area had in fact been an estuary, and therefore very near sea level.

4. Mantell, *Geology of Sussex* (1827), 51. Converted into a scene from the deep *past*, by the inclusion of his iguanodons in the foreground, this verbal scene was what Mantell later asked John Martin to turn into pictorial form, to illustrate his lectures and form a frontispiece for his *Wonders of geology* (1838): see Fig. 31.8.

5. Lyell to Prévost, 3 March 1826 [copy in Edinburgh-UL: AAF 7], informed the latter that his paper would be welcome for the *Transactions*, but only—with tactful deference to the French—if it had first been read in France; in the event it never appeared in England, even after being read at the Académie in 1827.

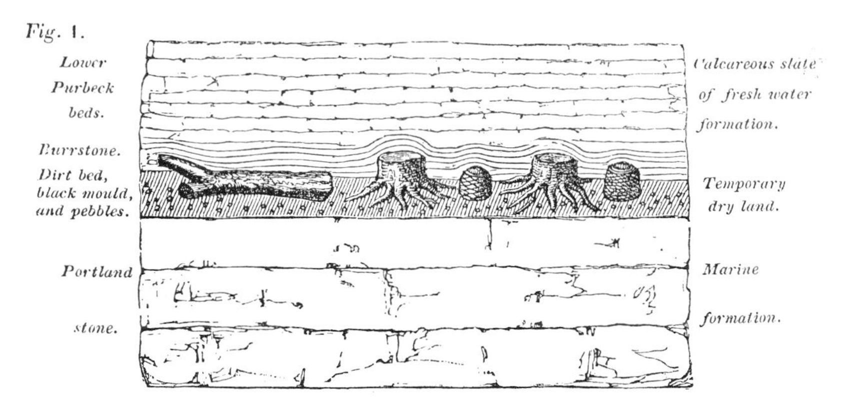

Section of the Dirt-bed in the Isle of Portland, shewing the subterranean remains of an ancient Forest. De la Beche.

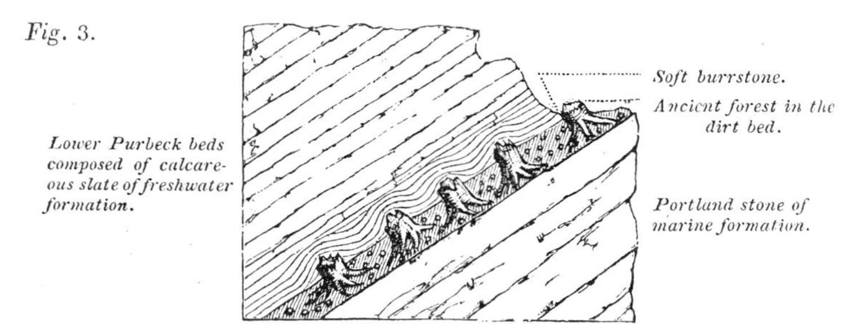

Section of the Cliff east of Lulworth Cove, Dorset. Buckland.

Fig. 11.1 Buckland's small sketch sections of the black "dirt bed" at the base of the Purbeck formation on the south coast of England, interpreted as a fossil soil marking an episode of "temporary dry land" between an earlier period of marine conditions and a subsequent one of fresh water. In the "ancient forest", fossil cycads (see Fig. 11.2) are interspersed among rooted tree stumps and prostrate tree trunks. The section of horizontal strata on the Isle of Portland (fig. 1) was based on fieldwork by De la Beche. The one of tilted strata in the mainland cliffs (fig. 3) was by Buckland himself; he later published both sections in his Bridgewater Treatise (see §29.3).

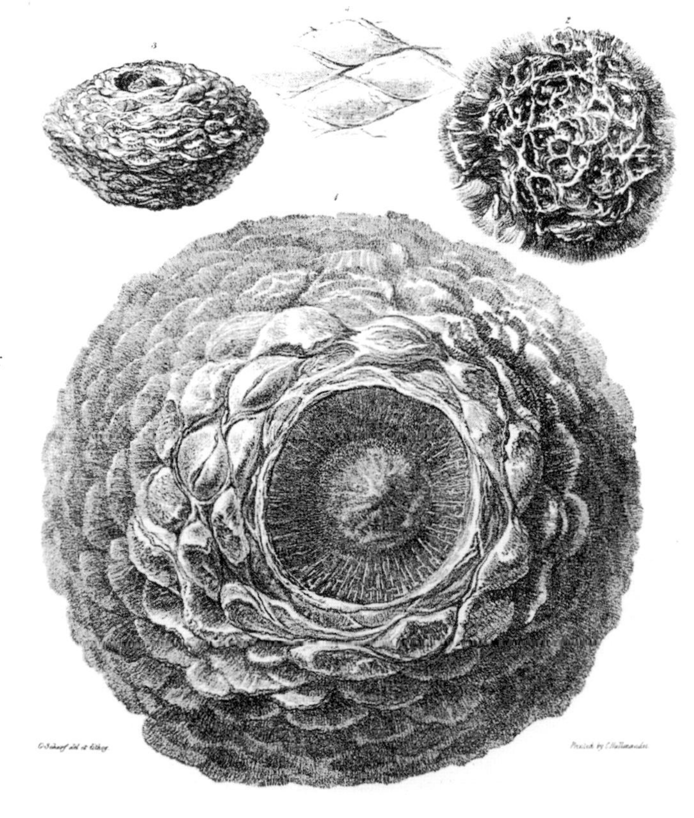

Fig. 11.2 A fossil cycad from the Purbeck formation on the south coast of England: one of the plates, drawn and lithographed by Scharf, illustrating Buckland's paper read at the Geological Society in 1828. The largest image (fig. 1) is of the upper surface, with leaf scars like those of living cycads (fig. 4, top center); the other images (figs. 2, 3) show the same specimen in side view and from below; the original was about 27cm across. Buckland argued that the plants had been preserved in situ on what was then a land surface, and that the climate had been warmer than the present. (By permission of the Syndics of Cambridge University Library)

However, English geologists such as Buckland did not share his skepticism, and took this locality to be clear evidence that here, at least, marine conditions had been followed for a time by a terrestrial environment that supported an "ancient forest", before being submerged again, this time in freshwater (Fig. 11.1).[6]

Four years later, in 1828, in another paper read at the Geological Society, Buckland identified some of the plants rooted in the "dirt bed" as the remains of "petrified" *cycads*. Using the habitats of their living relatives as the key, he inferred that at that time the region had enjoyed a much warmer climate than that of the present day (Fig. 11.2).[7]

11.2 BUCKLAND AND THE FOOTPRINTS OF MONSTERS

Buckland's cycads were an important new indication of what the exotic "former world" might have looked like when the Purbeck beds were being formed. At much the same time, news reached him of an even more exciting find, which suggested that extinct monsters such as his own megalosaur and Mantell's iguanodon (§5.1, §5.4) might be brought back to life, as it were, using more ample evidence than just their fossil bones.

In 1827, Henry Duncan (1776–1846), the Church of Scotland minister at Ruthwell on the coast of Dumfriesshire, was given part of a remarkable slab of sandstone that had been found in a quarry not far away in Annandale. It bore a strange set of large footprints, and he had used it to decorate a summerhouse in the garden of his manse. He reported it to Buckland, the obvious British expert to be informed about such relics of a former world. Buckland was skeptical about it until Duncan sent him a cast of the prints; on seeing this proxy specimen he realized its scientific importance and urged the Scotsman to publish the discovery without delay. From the known geology of the area and the site of the quarry, Buckland would have been confident that the slab came from the New Red Sandstone formation (in modern terms, of Triassic age), which elsewhere in Britain directly underlay the Lias and other "Jurassic" formations. That put the footprints well down in the Secondaries. But the crucial question, of course, was the identity of the quadruped that had made the tracks (Fig. 11.3).[8]

To solve the problem, Buckland applied the strongly actualistic method that he had used so successfully on the bones in the cave at Kirkdale. In that case, he had studied how captive living hyenas gnawed the bones they were given to eat,

6. Fig. 11.1 is reproduced from Buckland, *Geology and mineralogy* (1836), 2: pl. 57, explained on 97. Webster, "Purbeck and Portland beds" (1826), read on 19 November 1824; Prévost's critique is in his "Continens actuels" (1828), 335–36, in a note added after his paper was read.

7. Fig. 11.2 is reproduced from Buckland, "Cycadeoideae" (1828) [Cambridge-UL: Q365.b.12.7], pl. 47, read 6 June; pl. 46 illustrated the structure of living cycads for comparison.

8. Fig. 11.3 is reproduced from Duncan, "Footmarks of animals" (1828) [London-BL: TC.15.a.20], pl. 8, read on 7 January. The find was first published by Duncan's friend James Grierson in an Edinburgh periodical, and a report of Grierson's lecture in Perth appeared in an English-language Paris newspaper, promptly making the discovery known in France: see Pemberton *et al.*, "Footsteps before the Flood" (1996), and Pemberton and Gingras, "Henry Duncan" (2003). Ruthwell is about 15km southeast of Dumfries; the specimen came from a quarry at Corncockle Muir in Annandale. See also Sarjeant, "Fossil vertebrate footprints" (1974), 267–76, and "Footprints before the Flood" (2003), 63–65.

Fig. 11.3 Animal footprints on a slab of sandstone from the New Red Sandstone formation, found near Dumfries in Scotland and described to the Royal Society of Edinburgh in 1828 by Henry Duncan, a local minister and amateur naturalist. The whole slab bore a total of twenty-four footprints; the portion illustrated in this lithograph was about 70cm in length. Duncan reported how Buckland had conducted experiments with living reptiles to try to replicate the tracks, and had concluded that they were not those of a crocodile but might have been made by a large tortoise. These were the first such fossil footprints to be analyzed with some scientific rigor, but many more were found in the following years. (By permission of the British Library)

and he had compared the results with the fossil bones in the cave; it was this experiment that had provided the clinching evidence for his interpretation of the cave as a den of "antediluvial" hyenas (*BLT* §10.6). In the present case, Buckland persuaded a small crocodile and three species of tortoise to walk across a large sheet of "soft pye-crust", and also surfaces of damp sand and soft clay, leaving their tracks to be compared with those on Duncan's sandstone slab. Ever the scientific showman, Buckland made these experiments at Murchison's Saturday *salon*, in the presence of a large party of London geologists and other savants. One of those present recorded how it had been difficult to persuade the animals to move at all, and then how they got stuck in the sticky pastry:

> It was really a glorious sight to behold all the philosophers, flour-besmeared, working away with tucked-up sleeves. Their exertions, I am happy to say, were at length crowned with success; a proper consistency of paste was attained, and the animals walked over the course in a very satisfactory manner; insomuch that many who came to scoff returned rather better disposed towards believing.[9]

On the basis of these experiments, Buckland reported to Duncan—in a slightly more serious vein—that the fossil footprints were definitely not those of any crocodile, but could be those of a tortoise. The match was not perfect; but, as Buckland put it, "I conceive that your wild tortoises of the [New] red sandstone age would move with more activity and speed, and leave more distinct impressions, from a more rapid and more equable style of march, than my dull torpid prisoners of the present earth in this to them unnatural climate". Buckland lost no time in sending his conclusions on the matter to Paris, and they were published in the *Annales des Sciences Naturelles* even before Duncan's own description appeared in Edinburgh. Buckland claimed that the footprints were as clear as those of a hare on snow: while admitting that other savants were still unconvinced, he insisted that "I no longer have the least doubt that these impressions have been made by tortoises". These New Red Sandstone tortoises, like the Purbeck cycads, seemed to bear witness to a climate in the Secondary era much warmer than the present.[10]

11.3 FIRST SCENES FROM DEEP TIME

However, the most complete and reliable reconstruction of a scene from the deep time of the Secondary era was centered on the spectacular fossils from the Lias formation at Lyme Regis in Dorset. Conybeare's earlier verbal reconstruction of the life habits of the bizarre plesiosaur (§2.3) was soon extended by Buckland's research on other specimens that had been found by the indefatigable Mary Anning. One of her more important finds, late in 1828, was of an incomplete but unmistakable "pterodactyle", about the size of a raven. Reporting this promptly at the Geological Society, Buckland maintained that he had long suspected that the fragmentary fossil "bird" bones already known from Lyme, and those from Stonesfield (Fig. 5.4), belonged in fact to Cuvier's "wing-fingered" flying reptile (*BLT* §9.3). The new specimen confirmed this, and some minor differences from Cuvier's original German specimen allowed Buckland to claim a new species. Analyzing its functional anatomy in good Cuvierian style, he inferred that the pterodactyle, like a bird, could have walked with its wings folded, and also perch on the branches of trees. It added variety to his verbal reconstruction of the world at this period in geohistory, which was expressed in far more florid language than was usually permitted in the Society's sober *Transactions*: "With flocks of such-like creatures flying in the air, and shoals of no less monstrous Ichthyosauri and Plesiosauri swarming in the ocean, and gigantic crocodiles and tortoises crawling on the shores of the primaeval lakes and rivers,—air, sea, and land must have been strangely tenanted in these early periods of our infant world."[11]

9. John Murray [to unnamed correspondent], 23 January 1828 [printed in Murray, *John Murray III* (1919), 7–8]. Although sand and clay would have been more authentic, pastry could provide a durable record of the tracks. Buckland was notorious for keeping a wide variety of live animals at his home in Oxford, and the tortoises were probably his; but he may have borrowed the crocodile—presumably a small one—from a London menagerie. The young John Murray "III" [i.e., the third in a long dynasty of publishers called John Murray] (1808–92), returning to London from a year as a student in Edinburgh (attending Jameson's lectures, among others), had just made a detour to see the specimens at Ruthwell. He was the son of John Murray "II", who published the work of many of the leading English "men of science" at this time, including Lyell (see §21.1); later in the century, having taken over the business from his father, he himself published the work of, among others, Charles Darwin.

10. Buckland to Duncan, 12 December 1827 and 17 March 1828 [quoted in Duncan, "Footmarks of animals" (1828), 202–204n]; Buckland, "Traces de tortues" (1828), a letter sent to Thomas Underwood (1772–1835), an English artist and naturalist long resident in Paris.

11. Buckland, "New species of pterodactyle" (1829), 218, read on 6 February. Buckland was, of course, well aware that the Lias was very far from being the lowest or earliest of the Secondary formations (see Fig. 9.1), and that the Transition formations (with at least a few fossils) were older still: "early" and "infant" were characteristically hyperbolic.

At the same meeting of the Geological Society, Buckland also reported on some other specimens from Lyme, which in contrast were relatively common and had long been known to collectors. "By the skill and industry of Miss Mary Anning", however, they could now be interpreted correctly, and indeed sensationally. For Anning—who was now much too well known to be treated anonymously like more ordinary fossilists—had noticed that these phosphatic "bezoar stones", roughly cylindrical and usually with spiral markings, were often preserved in the

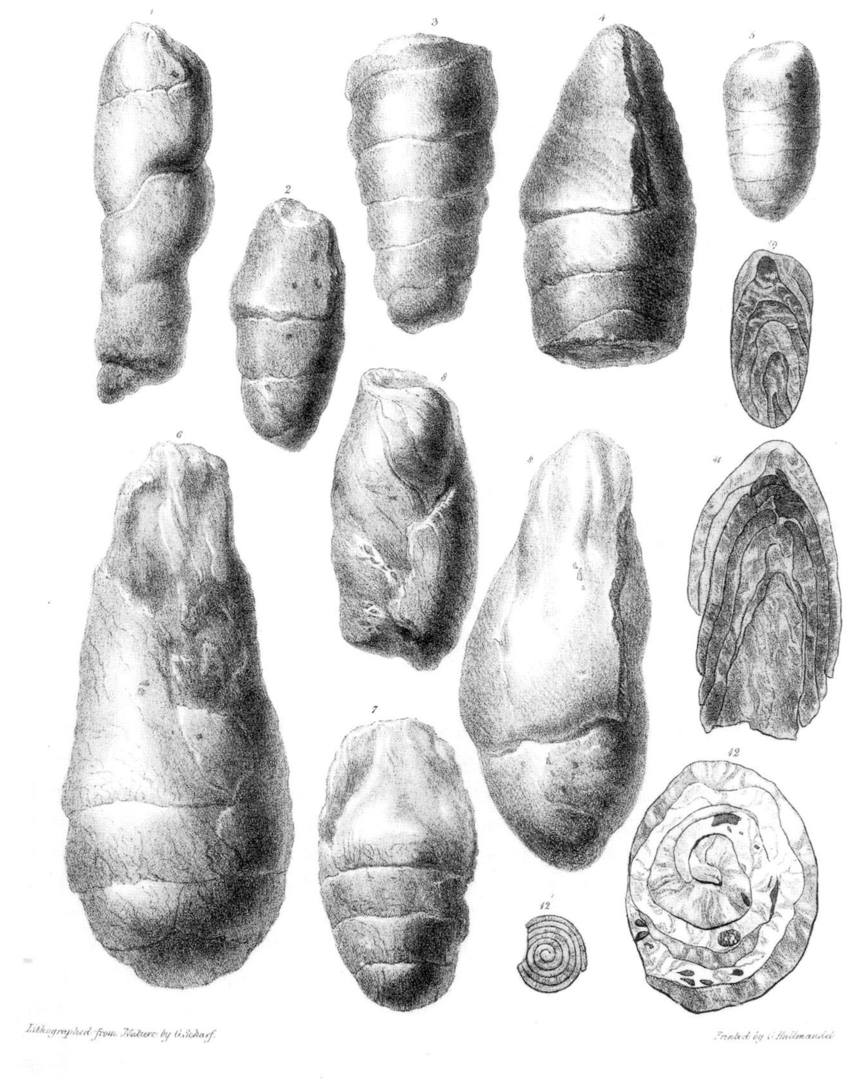

Fig. 11.4 "Coprolites from the Lias at Lyme Regis" in southern England, interpreted by Buckland in 1829 as the phosphatic excrement of ichthyosaurs. The spiral markings on the surface of these specimens (up to 5cm in diameter), and in their internal structure (figs. 10-12 are sections), were attributed to their rotatory movement down the intestine; two specimens (figs. 6, 7) had "minute superficial impressions derived from the vessels of the intestine". Analysis of the coprolites threw light on the feeding habits of the reptiles; they contained the bony scales of fish (fig. 12), and the jet black color of some of them (fig. 4) was attributed to their having also eaten the ink-sacs of squid-like belemnites. (By permission of the Syndics of Cambridge University Library)

abdominal region of ichthyosaur skeletons. And when broken open they were found to contain the distinctive solid bony scales of a fish (*Dapedium*) that was a well-known fossil at Lyme, and the fragmentary bones of that and other fish and even those of small ichthyosaurs. Buckland therefore identified them as what he called "*coprolites*", the petrified excrement of ichthyosaurs, marked by what must have been spiral ridges on the lining of the intestine (like those of living sharks). Furthermore, some coprolites were jet black in color, which Buckland suspected was due to the reptiles' having digested the ink sacs of the squid-like animals to which fossil belemnites (also common in the Lias) were now attributed (§4.3); as an analogy from the present world, he mentioned that octopuses (related to squids) were routinely used as bait on the Newfoundland fishing grounds (Fig. 11.4).[12]

All this threw unexpected light on the food chain of the fauna of the Lias, confirming the existence of predators (and even apparently of cannibalism) at that remote period. And not only at that period: Buckland noted similar objects from various other formations, ranging from the "Mountain" or Carboniferous Limestone up through the younger Secondaries into the Tertiaries and finally to the "diluvium" and his own identification of hyena excrement in Kirkdale Cave (*BLT* §10.6). Collectively they formed the geohistorical "records" of the carnivorous habit. However, far from treating this as a flaw in a prelapsarian Edenic paradise, Buckland turned it into a further sign of the providentially balanced ecosystems that had characterized the organic world throughout geohistory:

> In all these various formations our Coprolites form records of warfare, waged by successive generations of inhabitants of our planet on one another: the imperishable phosphate of lime, derived from their digested skeletons, has become embalmed in the substance and foundations of the everlasting hills; and the general law of Nature which bids all to eat and be eaten in their turn, is shown to have been co-extensive with animal existence upon our globe; the *Carnivora* in each period of the world's history fulfilling their destined office,—to check excess in the progress of life, and maintain the balance of creation.[13]

Buckland's verbal reconstruction of the food chain of the world of the Lias period, based largely on fossils found on the Dorset coast, gave De la Beche—an accomplished artist, unlike Buckland—the inspiration for turning it into a pictorial scene of "a more ancient Dorset". Conybeare's famous earlier scene of Buckland entering Kirkdale Cave (see figure opposite page 1, and *BLT* Fig. 10.24) provided a ready precedent. Whether or not it was his original intention, De la Beche's watercolor drawing was evidently found so striking by those who saw

12. Fig. 11.4 is reproduced from Buckland, "Discovery of coprolites" (1829) [Cambridge-UL: Q365.b.12.8], pl. 28, drawn and lithographed by Scharf; the paper was read on 6 February.

13. Buckland, "Discovery of coprolites" (1829), 235; again, it was hyperbolic (but also traditional and biblical) for Buckland to call the hills "everlasting", although geologists like himself had shown that this was precisely what they were not. He was, of course, attaching his discoveries to an already long established tradition in natural theology, but giving it a novel geohistorical dimension (see §29.3). "Carnivora" here denoted carnivorous animals in general, not a specific group of mammals.

Fig. 11.5 Duria antiquior, or "a more ancient Dorset": De la Beche's imaginative scene of the animals found as fossils in the Lias formation of Lyme Regis, and the plants known as fossils in other Jurassic formations, reconstructed in the modes of life that had been imputed to them through research by Buckland, Conybeare, and others. Here two species of large-jawed ichthyosaurs (1, 2) are feeding on the slender-necked plesiosaurs (3), on the squid-like belemnites and on fish (*Dapedium*, 5); a plesiosaur has also caught one of the pterodactyles (4) on the wing. One ichthyosaur (center) is, conjecturally, spouting like a whale. Coprolites are shown dropping as excrement from one of the plesiosaurs (center). The curious floating animals with sails (right) are an attempted reconstruction of ammonites; their empty shells are lying on the bottom (left foreground). There are also sea-lilies (crinoids) rooted to the sea floor (6, right foreground); on land are a crocodile and a turtle, and a cycad among the palms. De la Beche sold copies of this lithograph to benefit Mary Anning, who had found many of the finest specimens on which the scene was based. The novel format of a view partly in the open air and partly underwater anticipated what later became more familiar in the domestic aquarium. (By permission of the National Museum of Wales)

it that he redrew it and had it turned into a lithograph; he sold copies of the print to other geologists, and gave the proceeds to Anning. It became, in effect, a performance for her benefit; she and her family had again been reduced to near-poverty, since even she could not guarantee a steady supply of specimens of the outstanding quality and novel character that could fetch high prices from collectors. De la Beche's scene, although at first sight little more than a lighthearted joke, was in fact deeply serious in meaning. It assembled and synthesized all the varied research of the previous years, notably but not solely by Buckland, portraying what had been inferred with great reliability about the original life habits and environment of the Jurassic fossils (Fig. 11.5).[14]

Copies of De la Beche's lithograph were certainly purchased by many English geologists, who did not grudge a substantial price in a good cause. Cuvier was kept up to date by being sent one promptly, as usual, and before long copies

Fig. 11.6 Georg Goldfuss's scene of life at the time of the "Jura-Formation" (1831). The pterodactyles (flying overhead, and one clinging to the cliff on the left), on which he had already published a detailed study, are most prominent. The marine reptiles are loosely based on the finds in the English Lias, but many animals are depicted only as shells on the sea floor awaiting fossilization, rather than as living organisms. The scene in the open air blends ingeniously into an underwater view. Allowing for the mirror-image reversal entailed in the process of lithography, this design was clearly inspired by De la Beche's scene of "a more ancient Dorset" (Fig. 11.5). (By permission of the Syndics of Cambridge University Library)

were also in the hands of other foreign savants. Buckland had sent Georg August Goldfuss (1782–1848), the professor of zoology and mineralogy at Bonn, a copy of his pterodactyle paper, and in 1831 Goldfuss published his own verbal reconstruction of the flying reptile, based on a superb new specimen that had been found at Solnhofen. In the margin of the plate depicting the fossil, he included a tiny pictorial vignette of the reconstructed animals, flying and also clinging to a cliff face. He must have seen De la Beche's more comprehensive reconstruction soon afterward; for in the next fascicle (1831) of his long-running monograph on the *Fossils of Germany* he published a closely similar scene of the life of the "Jura-Formation", which was later bound into the completed first volume as its frontispiece. Since this massive illustrated work was aimed at professional geologists

14. Fig. 11.5 is reproduced from De la Beche's print *Duria antiquior* (1830) [Cardiff-NMW], lithographed by Scharf. The original watercolor [Cardiff-NMW, no. 367] has been published at full size by Cardiff-NMW and is also reproduced on the covers of Howe et al., *Ichthyosaurs* (1981), and Rudwick, *Scenes from deep time* (1992). What is probably an earlier version of the lithograph, in which the animals are not numbered or identified, is reproduced in Rudwick, *Scenes*, 45. De la Beche's reconstruction of ammonites was based on an imagined analogy with the living paper nautilus (*Argonauta*) rather than the pearly nautilus (*Nautilus*). The crinoids were based on Miller's work (§4.3) and the cycads on Buckland's (§11.1).

and serious amateur collectors, Goldfuss's Jurassic scene ensured that such reconstructed scenes from deep time soon became familiar to the relevant savants throughout Europe (Fig. 11.6).[15]

When a copy of Goldfuss's print reached the Geological Society and was reported to Buckland, he told De la Beche that this "German parody of your Duria Antiquior" should spur him to produce more of the same kind without delay, to avoid being upstaged any further. He suggested three more scenes, for each of which there was now adequate fossil evidence to make a lively reconstruction. Significantly, Buckland listed them in heuristic rather than geohistorical order, that is, from near the familiar present back into the deepest past. They were (1) "a land piece [of] the Period immediately preceding the formation of Diluvium"; (2) "a lake scene" based mainly on Cuvier's Tertiary animals from the Gypsum formation around Paris; and finally, back beyond the Lias scene that De la Beche had already produced, (3) "a sea scene" from the Carboniferous and even the still more remote Transition period. The first of these—based, of course, on his own "diluvial" research—will give the flavor of Buckland's proposal:

> A land piece—with only rivers plains & mountains—as in Palestrina pavement—exhibiting the gamboled Battles of Elephants & Rhinoceros & Mastodons—Hippopotami jumping into the rivers—Megatherium sitting on his Haunches with one fore Paw against the trunk of a tree and the other reaching down an enormous Branch—Horse Ox and Elk scampering before a Pack of Wolves and falling headlong into fissures—Hyaenas in their Den or dragging into its Mouth their Prey—Tigers crouching to spring on Deer.[16]

In the event, De la Beche never acted on any of Buckland's three suggestions, probably from lack of time. However, taken together with the one scene that he did draw, they signal the invention of a novel pictorial genre that has remained ever since a powerful medium of visual rhetoric, conveying a sense of the reality of geohistory not only to savants and serious amateurs but also in the long run to the general public too.[17]

11.4 CONCLUSION

De la Beche's reconstruction of Dorset in the time of the Lias formation marked the birth of a distinctive new pictorial genre of scenes from deep time. It was also the culmination of years of careful research by Buckland and others, based notably on Anning's spectacular specimens from Lyme Regis, but also on other finds such as Duncan's equally spectacular fossil footprints. However, all this was part of a broader movement within geological practice during the 1820s, by which the straightforward procedures of Smithian stratigraphy were beginning to be transformed into truly *geohistorical* analyses. Although reconstructions of extinct animals were the most striking achievement, as they had been ever since Cuvier's early work on fossil bones, they were being complemented increasingly by reconstructions of the physical environments of the deep past. Prévost analyzed the marine and freshwater sediments of the Paris Basin, and Lyell interpreted the estuarine conditions in which Mantell's Wealden strata had been formed. Lyell also linked his paleoecological reconstructions with equally significant inferences about the tectonic movements of the earth's solid crust that had affected the strata either at the time or subsequently. And Buckland's identification of the Purbeck "dirt bed" as the soil of a former forest with cycads also helped to transform these formations from mere rocks and fossils into the decipherable relics of past environments and ecosystems.

Scenes from deep time soon became a means by which geologists and other savants—and before long the general public too (see §31.3)—could be convinced that it was indeed possible to "burst the limits of time" and to make the deep past vividly knowable in the present, with a reasonable degree of confidence: they were, metaphorically, the products of the geologists' new time-machine. Part of the attraction of such scenes was that they depicted exotic plants and unfamiliar animals, many of them gigantic in size and some of them also monstrous in appearance. This accentuated the alien "otherness" of the deep past: it really had been in sharp contrast to the present world, a foreign country where nature did things differently. Yet, at the same time, these scenes revealed a world that was not in fact totally alien or "other", and in some ways reassuringly familiar. Even in this unimaginably remote past, for example, the world had already been one in which herbivores and carnivores co-existed in bloody conflict but also in harmonious balance. So these careful reconstructions served in a sense to tame the world of the Secondary formations, and to bring it into conceptual and imaginative continuity with the world of the present.

However, in themselves these scenes from deep time merely offered glimpses of selected moments in geohistory: to use metaphors drawn from later technologies, they were mere stills or freeze-frames, abstracted from the continuous flux of geohistory. The next chapter returns to that temporal dimension, to trace how, during these same years, the history of life was coming to be seen ever more clearly as directional in character.

15. Fig. 11.6 is reproduced from Goldfuss, *Petrifacta Germaniae* (1826–44), Theil 1 [Cambridge-UL: MH.3.8], frontispiece, published in Lieferung 3 (1831). The pterodactyle vignette is in Goldfuss, "Reptilien der Vorwelt" (1831), pl. 9, and is reproduced in Langer, "Bilder aus der Vorzeit" (1990), and, with a translation of Goldfuss's verbal reconstruction, in Rudwick, *Scenes from deep time* (1992), 50–51. Nicolas Christian Hohe (1798–1868), the professor of art at Bonn, certainly drew the pterodactyle plate for Goldfuss, and probably the Jurassic scene too. While teaching at Erlangen, years earlier, Goldfuss had written the standard guidebook to the Bavarian bone caves (*BLT* §10.4).

16. Buckland to De la Beche, 14 October 1831 [Cardiff-NMW, no. 183]; the stream-of-consciousness style was characteristic of Buckland's letter writing; the whole relevant text is printed in Rudwick, *Scenes from deep time* (1992), 54. "Palestrina pavement" referred to a well-known Roman mosaic floor in the Italian town (more famously the birthplace of the composer), depicting the fauna of the Nile valley in Roman times: the analogy was yet another sign of the close link in Buckland's mind between the worlds of antiquarianism and geology, the latter now becoming equally *historical* in character, at least in his own scientific practice. Buckland's suggestions show incidentally that a retrospective order, which had been customary in stratigraphy ever since "Conybeare and Phillips" a decade earlier (§3.2), and which Lyell was soon to adopt in his *Principles* (see §26.1), was perfectly compatible with a strongly *geohistorical* approach.

17. Rudwick, *Scenes from deep time* (1992), traces the establishment of the genre as far as the 1860s; it remained prominent in the popular presentation of geology through the rest of the nineteenth century and the whole of the twentieth, and is flourishing more than ever in the age of computer-generated imagery. O'Connor, *Earth on show* (2007), reproduces many more such "scenes" from early nineteenth-century Britain, and analyzes their relation to literary (including poetic) evocations of the deep past.

CHAPTER 12

A directional history of life (1825–31)

12.1 TERTIARY GEOHISTORY

In establishing the directional character of geohistory, or at least the history of life, the Tertiary formations once again played a crucial role. Leaving aside the complications introduced into the records of Tertiary geohistory by shifting environments of seas, lakes, and dry land (§10.3), there were also problems with understanding the more straightforward and unambiguously marine formations and their fossils.

Even before the end of the wars, the classic work of Brongniart and Cuvier on the Paris Basin had been matched by Webster's description of a similar sequence of Tertiary formations in what he called the Isle of Wight Basin (subsequently renamed the Hampshire Basin) in southern England, with similar fossil shells of evidently marine origin. Published coincidentally in the same year had been the magnificent monograph by Brocchi on those of the "Subapennine" formations, in the foothills of the Apennines in northern Italy (*BLT* §9.4). However, Brocchi had pointed out that there were few species in common between his Tertiary fossil shells and those around Paris and in England. He had attributed this to a difference of climate (and, in modern terms, of biogeography) between southern and northern Europe in Tertiary times, just as the living molluskan fauna of the Mediterranean differed in composition from that of the North Sea or the north Atlantic.

This concept of geographically separated Tertiary faunas was reinforced indirectly by the leading botanist Augustin-Pyramus de Candolle [or Decandolle] (1778–1841), Geneva's first professor of natural history. In his major article on "Botanical geography" (1820) for the Parisian *Dictionnaire des Sciences Naturelles*, Candolle had developed Humboldt's idea of "botanical statistics": he had analyzed the world's *present* floras in a quantitative fashion rather like that of German *Statistik* (§7.2), and he had outlined the various "botanical regions" or "provinces" into which he claimed they could be divided. Candolle acknowledged his debt to Humboldt's earlier "botanical arithmetic", but his own work drew the attention of naturalists, more systematically than before, to the spatial distribution of living plants, and it implied that factors other than mere climate or latitude must be involved. His description of floral provinces in the present world could readily be extended to animals as well as plants, and also to the "former world" of fossils; this in effect was what Brocchi had already suggested for his Subapennine fossil shells. It was soon extended further, with explicit acknowledgment to Candolle, on the basis of fossils of another region of Tertiary formations.[1]

In 1825, the young Parisian naturalist Barthélemy de Basterot (1800–1887) reported to the Académie on the fossils of the Tertiary basin of southwestern France, the stratigraphy of which had just been described by Boué (§3.3). Basterot prefaced his fine descriptions and illustrations of the fossil mollusks with a "statistical" analysis of this Tertiary fauna in relation to others. His census revealed what he took to be regional differences between the various Tertiary basins: of his 330 species from the Bordeaux region, 91 were also known from Brocchi's Italian fauna, 66 from Lamarck's Parisian, 24 from Webster's English, and 18 from Prévost's Viennese fauna. Since the degrees of similarity seemed roughly proportional to geographical proximity, Basterot suggested that, as in the present world, so also "in these ancient times horizontal distances had an appreciable influence" on animal distributions.[2]

However, there was another possible explanation for the differences between the various Tertiary faunas. This was that they might be of different *ages* within the Tertiary era. Prévost had suggested this in his first foray into Tertiary stratigraphy, soon after his return to Paris. In his paper on the Vienna Basin, he had claimed that his proposed reconstruction of Tertiary geohistory—in effect, the progressive retreat of the seas off the continents (§10.2)—could be traced by comparing the formations in the different basins: not just identifying a few distinctive and "characteristic" fossils in Smithian manner, but comparing whole *assemblages* of species "statistically", as if they were populations of individuals (*BLT* §9.6). In this quantitative approach he was probably following Brocchi's earlier example, but Prévost had reinterpreted the contrasts as being due to differences of age, not geography. Brocchi's Subapennine fauna and his own from the Vienna Basin both had a much higher proportion of living species than Lamarck's from the Paris Basin; but Prévost argued that this was not because they were spatially far away from Paris, but because they were both nearer the present world *in time* than the Parisian fauna. This seemed to be confirmed by what Brongniart had concluded from his fieldwork in the foothills of the Italian Alps (and more particularly near Turin). There the Tertiary fossil shells had more in common with the Parisian ones, far away on the other side of the Alps, than with Brocchi's on the flanks of the Apennines, just across the plain of the Po: similarities were *not* always proportional to geographical distance (*BLT* §9.6).

So Prévost had concluded that the Tertiary formations represented a much more significant portion of geohistory than had previously been suspected, and that those in the various Tertiary basins were *not* all of about the same age, as had been tacitly assumed. Furthermore, his interpretation had implied that in the course of the Tertiary era the marine molluskan fauna had changed steadily in composition towards that of the present world. It followed that it might be possible to assign relative ages to the Tertiary formations, simply on the basis of a "statistical" analysis of their fossil shells: the greater the proportion of species known alive, the younger the formation. Basterot, for example, provided material for some such interpretation when he reckoned that, of his total of 330 species from the Bordeaux region, 45 were still living in European seas and 21 elsewhere: together, about one in five. This suggested a partial turnover of species—by whatever process—during whatever span of time there might have been since the formations were deposited.

This notion was also supported by Jules Pierre François Stanislas Desnoyers (1800–1887), another young Parisian naturalist concentrating his attention on the exciting new science of geology. Desnoyers, who was Prévost's brother-in-law and a friend and contemporary of Basterot, claimed decisive *stratigraphical* evidence that the Tertiary basins were not all of the same age. In 1825, he reported to the Société d'Histoire Naturelle in Paris (of which he was the secretary) on some fieldwork in Normandy, where he had found patches of a Tertiary formation overlying the Chalk. He concluded that it and several others elsewhere in Europe, among them the Subapennine and the Viennese, all "seem to be the most modern of all the marine formations between the Chalk and the great diluvian debris [*aterrissement*]". This implied that—as Prévost had already inferred—they were *not* of the same age as the Parisian strata. The following year, Desnoyers and Basterot crossed the Channel to see some of the English Tertiaries for themselves, and to meet English geologists and learn what they were doing. Added to what he already knew of the Parisian formations, this gave Desnoyers invaluable first-hand experience of the problems of Tertiary stratigraphy beyond the borders of France.[3]

Three years later, Desnoyers presented his conclusions in a major paper published in the Parisian *Annales des Sciences Naturelles.* Its lengthy title accurately summarized its contents: "Observations on an assemblage of marine deposits more recent than the Tertiary formations [*terrains*] of the Seine [i.e., Paris] Basin, constituting a distinct *geological Formation*; preceded by an outline *of the non-simultaneity of the Tertiary basins*". Desnoyers's conclusion that the basins were of different ages depended—at least in one decisive case—not on inferences drawn from fossils alone but on the traditional and uncontroversial criterion

1. Candolle, "Géographie botanique" (1820), also published as a book; see Browne, *Secular ark* (1983), 58–85.

2. Basterot, "Bassin tertiaire" (1825), 1–11, read at the Académie on 17 January. His father Jacques (1771–1849) knew several of the leading Parisian naturalists, and probably helped the younger Basterot in much the same way in which Brongniart was helping his own son toward a scientific career (§4.4).

3. Desnoyers, "Terrains tertiaires du Cotentin" (1825), 238, with a fine geological map of part of the Cherbourg peninsula, and sections showing the Tertiaries as small local outliers. He and Basterot presented copies of their published work to the Geological Society on 4 November 1825 [London-GS: OM/1]. Desnoyers to Cuvier, 16 July [1826] [Paris-MHN: MS 627, no. 36], later reported on their travels and on the current research of Conybeare and others.

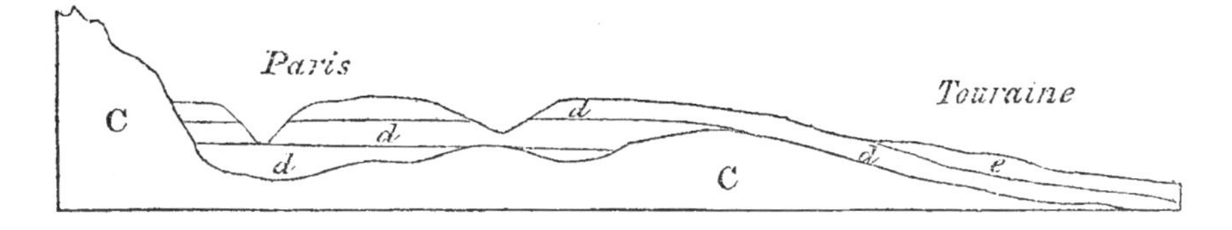

C, Chalk and other secondary formations.
d, Tertiary formation of Paris basin.
e, Superimposed marine tertiary beds of the Loire.

Fig. 12.1 A sketch section showing how the uppermost of the Tertiary formations (*d*) in the Paris Basin (left) extends to the southwest to form the lowest of those in Touraine (right), proving by simple superposition—at least in this case—what Jules Desnoyers in 1829 defined as the "non-simultaneity" of different Tertiary basins. This schematic diagram was published by Lyell a few years later, to explain Desnoyers's unillustrated text.

of superposition. In one direction the outcrop of the Tertiary formations of the Paris Basin was not sharply limited by the emergence of the underlying Chalk. To the southwest, the Parisian formations spilled over, as it were, into another basin of Tertiaries, that of Touraine (the region around the city of Tours in the Loire valley). Or rather, the uppermost and youngest of the Parisian formations spilled over, and then formed the lowest and oldest of those in Touraine. So almost all the sequence in Touraine had to be younger than the sequence around Paris: there was a slight overlap, but as a whole the two basins were manifestly *not* of the same age (Fig. 12.1).[4]

Desnoyers used fossil evidence to generalize from this specific case. He divided the known Tertiary sequences of strata into two distinct "Formations", in the sense defined by Humboldt (§3.1): that is, major *groups* of local formations [*terrains*], recognizable in principle anywhere on earth and formed during distinct and successive periods of geohistory. His older group was represented most clearly by the Paris Basin, but also by the London and Hampshire Basins in southern England. But Desnoyers focused his argument on the younger group: "these deposits belong to a specific period, to a major Formation that is very widespread in Europe and yet scarcely known and not well defined". These "recent Tertiary formations [*terrains*]" included not only those of Touraine, but also those he himself had found in Normandy, the "Crag" formations of eastern England, Basterot's around Bordeaux, Brocchi's "Subapennines", and Prévost's near Vienna. As for the *causes* of the shifting areas of deposition during the Tertiary era, Desnoyers invoked de Luc's earlier concept of occasional "violent commotions", now reinterpreted in terms of crustal upheaval or, in effect, by Élie de Beaumont's newly defined "epochs of elevation" (§9.4). Desnoyers conceded that his concept of the two "periods of Tertiary Formations, and the succession of basins" of different ages, might be difficult to grasp, but it was also fertile in its implications. For it offered the possibility of reconstructing the whole of Tertiary geohistory, by "an insensible and continuous [*progressif*] passage from one to the other, from former nature to present nature", from deep past to the modern world.[5]

Exploring any such ideas required, of course, a careful systematic description of the fossil shells, and rigorous comparisons with living mollusks. For those of the Paris Basin, which remained the agreed point of reference of all studies of the Tertiary formations, this essential database was being provided by the Parisian naturalist Gérard-Paul Deshayes (1796–1875), who was making himself the world's leading fossil conchologist in succession to the now elderly Lamarck. Deshayes's massive *Description of the Fossil Shells of the Paris Region*, which began to appear in 1824, revised and improved on what Lamarck had done some twenty years before (*BLT* §7.4) and it rivaled Brocchi's in the quality of its illustrations. It was the availability of such monographs—Basterot's was another, though on a more modest scale—that might enable Prévost's and Desnoyers's ideas on Tertiary geohistory to be tested.[6]

Deshayes did not restrict his research to the Parisian fauna. To establish his conclusions on the firmest possible foundations, he also amassed a vast collection of fossil shells from the Tertiary formations of other regions, and—for the essential comparison with the modern world—the shells of living mollusks. Emulating Cuvier's famous research strategy with fossil bones (*BLT* §7.3), Deshayes also persuaded other naturalists to lend him their shell collections, repaying them with his own authoritative identifications of their specimens: in total he handled more than 40,000 specimens. Such taxonomic research was slow work (as it still is), and it was several years before he published his results; but it was during the later 1820s that other naturalists learned that Deshayes was convinced that the Tertiary fossil mollusks fell naturally into *three* distinct assemblages (rather than Desnoyers's two), and that he interpreted them as *successive* faunas showing an increasing approximation to that of the present world.

Of Deshayes's fossil faunas, the least similar to the present, and therefore inferentially the earliest, had scarcely any extant species at all (3% of about 1,400); these were the fossils of the Paris, London, and Hampshire Basins. The second fauna was represented by the fossils in the Touraine formations, Basterot's from Bordeaux, and Brongniart's from near Turin; as Basterot had calculated for his own fauna, all these had about one in five species still extant (19% of about 800). The third fauna, inferentially the most recent, had rather more than half its species known alive in the present world (52% of about 700); Brocchi's Subapennine

4. Fig. 12.1 is reproduced from the wood engraving in Lyell, *Principles* 3 (1833), 20, explaining Desnoyers, "Ensemble de dépôts marins" (1829); the crude vertical exaggeration makes the low plateau of Chalk (*C*) to the north of Paris look deceptively like a mountain range. In Desnoyers's title, the word "*Formation*", printed with initial capital, denoted a major group of *terrains* (here translated as "formations") which, it was anticipated, would be identifiable "universally": see below.

5. Desnoyers, "Ensemble des dépôts marins" (1829), 171, 181–83, 188–90. He had earlier proposed calling the younger group "*Quaternaires*", but here he "renounced" that term, having found that it was being wrongly understood to imply a sharp boundary [*limite tranchée*] between the two (193n). Reboul, *Période quaternaire* (1833), later redefined "Quaternary" in what became its modern sense; its approximate equivalent in Desnoyers's time was Buckland's Diluvium and Alluvium, overlying and later than *all* the Tertiaries. The adjective *progressif* did not necessarily imply any sense of improvement, but merely of continuous directional change.

6. Deshayes, *Coquilles fossiles de Paris* (1824–37). On the title pages, the two volumes of text are both dated 1824 and the volume of plates 1837; but text and plates were probably issued together in successive fascicles, as was customary with large monographs on natural history. In the present context, what matters most is that Deshayes's work certainly *began* to be available to other naturalists in the mid-1820s.

fauna and that of the English Crag belonged here. Deshayes also mentioned in passing a fourth and minor group of deposits, for example on the south coast of France, which contained species almost all (96%) of which were known to be alive nearby in the Mediterranean; but he took this to be so recent in geological terms that it hardly counted.[7]

Deshayes recalled that the Tertiary era had long been regarded as a single period with a single fossil fauna, distinct both from the youngest Secondaries before it (in the Chalk and its equivalents) and from the present world. However, his own more thorough research now enabled him to divide it into "three great zoological epochs, perfectly distinct by the assemblage of species in each, and by the constant proportions between the number of living analogue [i.e., extant] species and those that are extinct [*perdues*]". He pointed out that his identification of three distinct faunas, and the corresponding division of the Tertiary into three distinct periods or "epochs", would make it possible to determine the relative ages of further Tertiary deposits, even if they were "never found in superposition". This still novel idea of giving formations relative dates by comparing their fossil contents *en masse*, rather than by observing one lying above or below another or even by finding "characteristic" fossils, embodied a risky prediction: that if and when two of the faunas were found in a single stratigraphical sequence they would in fact be in the predicted order. But Deshayes must have felt confident that his fossil criterion would prove robust, because by the time he presented his results in public his inference had been vindicated, at least in one case, by Desnoyers's field evidence that the Touraine formations were indeed younger than the Parisian: his "second epoch" was clearly later than his "first", by the unproblematic criterion of superposition.[8]

This kind of "statistical" analysis of fossils was not the monopoly of French naturalists. For example, Heinrich Georg Bronn (1800–1862), the young profes-

Gebiete	Ganze bestimmbare Arten-Zahl	Eigene Arten		Verhältnifs-Zahl gemeinsamer Arten mit andern Becken,												oder mit der lebenden Schöpfung	
				Paris	*Bordeaux*	*Montpellier*	*Pohlen*	*Kressenberg*	*Schweitz*	*England* a.	*England* b.	*Italien*	*Wien*	*Maynz*	*Siebenbürgen*	von *Europa*	der *Ferne*
Italien	770	342	0.444	0.104	0.153	0.280	0.961	0.013	0,040	0.032		—	0.126	0.417	0.031	0.240	0.040
Paris	546	388	0.711	—	0.093	0.073	0.013	0.004	0.007	0.070		0.110	0.026	0.411	0.015	0.015	0.018
Bordeaux	296	123	0.416	0.210	—					0.070		0.321	0.128	0.632	0.020	0.200	
Montpellier	529	170	0.321	0.105	0.151	—				0.011		0.528	0.094			0.332	
Pohlen	52	6	0.115	0.135	0.462	0.539	—	0.020	0.096	0.096		0.808	0.308	0.096	0.192	0.346	0.040
Kressenberg	90	48	0.533	0.211	0.033			—	0.011	0.100		0.144	0.			0.033	
Schweitz	46	0	0.	0.109	0.218	5.435	0.109		–	0.174		0.565	0.152	0.022	0.065	0.633	
England a.	134	100	0.747	0.191	0.045	0.037	0.015	0.007	0.022	—		0.097	0.045	0.010	0.015	0.015	0.007
b.	58										—		0.052				
Wien	113	3	0.028	0.124	0.336	0.443	0.142	0.	0.060	0.080		0.858	—	0.018	0.088	0.212	0.
Maynz	70												0.029				
Siebenbürgen	36												0.277				

Fig. 12.2 Heinrich Bronn's "statistical table" (1831) summarizing the Tertiary faunas of twelve different European regions (listed in the first column). The second column shows the total number of species known in each; the third column, the number found only in that region, and the proportion of the fauna expressed in decimal terms. The last two columns (right) give the proportions of each fauna also found in "the living creation", either in Europe or elsewhere in the world. The rest of the table is a matrix showing the proportions found in common between every pair of regions. This kind of tabulation was regarded as an application to the natural sciences (in this case, to geology) of quantitative techniques first developed in the social science of "statistics". (By permission of the Syndics of Cambridge University Library)

sor of natural sciences at Heidelberg, was preparing a massive work on fossils as applicable to stratigraphy. After studying fossil collections around Italy, he summarized his conclusions with an elaborate series of tables [*Tabellen*], analyzing degrees of resemblance at both generic and specific levels. The format of *Tabellenstatistik* was no accident, for Bronn's other teaching responsibility, for business studies [*Gewerbswissenschaften*], brought him into the same world of economic "statistics" as von Hoff a decade earlier (§7.2). His treatment of Tertiary molluskan faunas was just one part of a much broader project, but it was representative of the whole. Bronn distinguished between "older" and "younger" Tertiaries, but he expressed their degrees of resemblance to the fauna of the present world in decimal terms (respectively 0.04 and 0.55) rather than as percentages, which suggests that he reached his conclusions independently of Deshayes (Fig. 12.2).[9]

12.2 ADOLPHE BRONGNIART: PLANT LIFE ON A COOLING EARTH

All this research on the changing molluskan faunas of the Tertiary era contributed to an increasingly persuasive picture of the directional changes that life on earth seemed to have undergone in the course of geohistory. On the larger scale of the whole fossil record, Cuvier had, as usual, led the way, with his claim that there had been a progressive enrichment of the quadruped fauna, first with reptiles and much later with mammals, with the human species as the final addition. In the later 1820s the most striking new example of this kind of large-scale geohistorical interpretation matched Cuvier's vertebrate record with that of plants.

In the years since Adolphe Brongniart made his debut at the Académie with a preliminary account of Tertiary fossil plants and a brief glance at those from the far older Coal formation (§4.4), he had been pursuing this research with a single-minded intensity that recalled that of his hero Cuvier. In 1825 he visited England, like Cuvier a few years earlier (§1.2), to enlarge the range of specimens that he could study at first hand. Despite his youth—like Desnoyers and Basterot he was only as old as the century—he was treated by the English geologists as a respected authority on fossil plants. He dined with leading members of the Geological Society before its first meeting of the new season; Buckland introduced him formally at the meeting itself (Desnoyers and Basterot were also present); and the following morning Lyell gave a breakfast party for him (and for Mantell and others) and got him to confirm that his own *gyrogonites* (the fruits of freshwater *Chara* plants) from the recent Scottish lake deposits (§10.2) were identical to those from the Paris rocks.[10]

7. Deshayes, "Tableau comparatif" (1831), but known to other geologists several years earlier; his second and third epochs together corresponded to Desnoyers's "recent Tertiary".

8. Deshayes, "Tableau comparatif" (1831), 186; *Coquilles caractéristiques* (1831), 18. Lyell later developed this theme much further, with Deshayes's assistance (see §20.2, §26.1).

9. Fig. 12.2 is reproduced from Bronn, *Italiens Tertiär-Gebilde* (1831) [Cambridge-UL: MD.60.27/2], table 11; his distinction of "older" and "younger" Tertiary faunas is on 169–71. Bronn, "Petrefakten-Sammlungen" (1828), described the Italian museums he had visited while compiling his data; *Lethaea geognostica* (1835–38) was the massive handbook of stratigraphical paleontology that emerged later from all this research.

10. Geological Society, 4 November 1825 (London-GS: OM/1); Curwen, *Journal of Mantell* (1940), 56.

Fig. 12.3 Finely preserved fronds of *Nevropteris heterophylla* from the Coal formation of Charleroi, in the great coalfield of the southern Netherlands (now Belgium): a lithographed plate depicting one of the fossil plants named, described, and (in this case) drawn by Adolphe Brongniart, for his monograph on *Fossil Plants* (1828-37). The detailed structure of the leaves (shown in the enlarged drawings), when compared with those of living plants, led him to conclude that these particular fossils belonged to an extinct genus of tree fern. (By permission of the Syndics of Cambridge University Library)

Fig. 12.4 Sketches of living tree ferns in Brazil (left) and on the Indian Ocean island of Bourbon [now Réunion] (right), copied from travel books and used by Adolphe Brongniart in his work on fossil plants (1828-37) as "analogues" for the even larger tree ferns that had apparently characterized the Coal forests in the older Secondary era. (By permission of the Syndics of Cambridge University Library)

In 1828, after much further research back in Paris, the younger Brongniart published a substantial book as a trailer [*prodrome*] for an ambitious illustrated monograph on all fossil plants, the first installments of which had in fact just appeared. And he also summarized his conclusions in a remarkably accomplished paper to the Académie, entitled "General considerations on the nature of the vegetation that covered the earth's surface at the different epochs of formation of its crust".[11]

In their studies of trilobites, belemnites, and other animal fossils, the elder Brongniart, Blainville, and other naturalists had emphasized the importance of combining a zoological with a stratigraphical approach (§4.3). The younger Brongniart extended this by adding a third perspective, the explicitly geohistorical: his use of the word "epochs" was symptomatic. But his first and most basic task was to identify the botanical groups to which fossil plants belonged. This was no simple matter, owing to the usually fragmentary nature of the specimens. He dedicated his large work to Cuvier, not just to curry favor but because he was explicitly adapting his research methods from those that the great zoologist had used for fossil quadrupeds. He identified the affinities of fossil plants on the basis of rigorous comparisons between the best specimens—such as "petrified" stems and roots, and finely preserved fronds—and the corresponding parts of living plants (the latter had been rather neglected by botanists in favor of the more diagnostic reproductive organs, which were very rarely fossilized). Many of Brongniart's illustrations were of living plants, or their constituent parts, juxtaposed with the fossil specimens that he believed had belonged to the same major groups (Figs. 12.3, 12.4).[12]

Brongniart's second perspective was stratigraphical. While certain distinctive plants might be "characteristic" of specific formations, in a simple Smithian sense, Brongniart was more concerned to correlate whole fossil *floras* with the larger *groups* of formations. Leaving aside the second of his floras, which he admitted was poorly defined and limited in stratigraphical range, his major floras were three in number: each had characterized a specific major "epoch" in geohistory. His first flora was dominated by Cryptogams such as ferns, horsetails, and clubmosses, all of a gigantic size: even the tree-ferns had grown to at least fifty feet, and seventy-foot "lycopods" were enormously larger than the little creeping

11. Brongniart (Ad.), "Considérations générales" (1828), read on 8 December, appeared in translation in Jameson's Edinburgh periodical a few weeks later, ensuring that his research was promptly known among anglophone geologists; *Prodrome d'une histoire* (1828), preface dated 10 September. By this time the first two fascicles of *Végétaux fossiles* (1828–37) had already been issued to subscribers; the monograph was never completed, a fate that it shared at this period with many other ambitious and expensive works on natural history.

12. Figs. 12.3 and 12.4 are reproduced from Brongniart (Ad.), *Végetaux fossiles* (1828–37) [Cambridge-UL: MF.50.45], respectively pls. 71 and 39, the former drawn by Meunier "after" Brongniart. The living tree-ferns were copied from images in Spix and Martius, *Reise in Brasilien* (1823–31), and La Thouane, *Voyage de la 'Thetis'*. Brongniart gave all his fossil plants binomial Linnean names, which by this time had become standard practice among botanists dealing with living plants. He made much use of the microscopic cellular structures preserved in "petrified" specimens, and was later (in the 1830s) one of the first paleobotanists to adopt the then new technique of grinding transparent thin sections to study under the microscope. *Neuropteris* (as it is now spelled) was later assigned to the pteridosperms, an extinct group of fern-like but seed-bearing plants.

clubmosses that were clearly their living relatives. This first flora had flourished while the Transition and older Secondary (Carboniferous) formations were being deposited (in modern terms, the later Paleozoic); the debris of these plants had apparently contributed most of the material that had been transformed into coal itself. The second major flora again included many ferns, but it was dominated by Gymnosperms such as cycads and conifers; it characterized the younger Secondary formations, between the German *Muschelkalk* and the Chalk (i.e., roughly, the modern Mesozoic). The third and last major flora was dominated, like the present world, by Angiosperms, particularly the dicotyledonous flowering plants; it characterized the Tertiary formations (the modern Cenozoic).

Adolphe Brongniart's botanical stratigraphy demonstrated, then, that three major floras had flourished successively in the course of geohistory; and new major groups of plants had appeared in succession, bringing the earth's vegetation progressively closer in character to that of the present world. These, he claimed, were "positive results, independent of all hypothesis and all preconceived theory"; that is, the recognition of successive floras was not dependent on any specific theory about what had *caused* the changes. His method for presenting the evidence was unmistakably "statistical" in von Hoff's and Candolle's sense (§7.2, §12.1); he summarized his floras not with Smithian pictures of "characteristic" fossils, but, like Bronn a little later (Fig. 12.2), with a table showing the shifting *numbers* of species in the different botanical groups (Fig. 12.5).[13]

Adolphe Brongniart's grand synthesis underlined the truly *geohistorical* character of plant life: floras had changed in a directional manner, but in a way that could never have been predicted in advance. His research confirmed for plants what other naturalists had already concluded from the fossil record of animals: "we can thus accept that among plants, as among animals, the simplest beings preceded the more complex, and that nature has successively created more and more perfect beings". Life on earth, both animals and plants, had become more

	Première période.	Deuxième période.	Troisième période.	Quatrième période.	Epoque actuelle
I. AGAMES.	4	5	18	13	7,000
II. CRYPTOGAMES CELLULEUSES.	»	»	»	2	1,500
III. CRYPTOGAMES VASCULAIRES.	222	8	31	6	1,700
IV. PHANÉROGAMES GYMNOSPERMES.	»	5	35	20	150
V. PHANÉROGAMES MONOCOTYLÉDONES.	16	5	3	25 (?)	8,000
VI. PHANÉROGAMES DICOTYLÉDONES.	»	»	»	100 (?)	32,000
Végétaux de classe indéterminée.	22	»	»	»	»
Total de chaque Flore.	264	23	87	166	50,350

Fig. 12.5 Adolphe Brongniart's tabular summary of the numbers of fossil plant species that he assigned to six major botanical groups (I–VI) in four successive "periods" of geohistory (*première* to *quatrième*); the quotation-mark symbol denotes zero. This twofold "statistical" distribution defined four successive floras (of which the second was relatively unimportant), which in the course of geohistory had progressively approximated to the flora of the present world. The vastly greater numbers of species known alive (*époque actuelle*, right-hand column) implied that the fossil floras, although probably representative in overall proportions, might be severely impoverished samples from the deep past, owing to the chanciness of preservation and the still underdeveloped state of fossil botany. (By permission of the Syndics of Cambridge University Library)

diverse and more complex in the course of time, probably more abundant and in some sense more "perfect". No naturalist could now claim, with any credibility, that life had maintained an ahistorical stability or steady state, still less a recurring cyclicity, of the kind that Hutton, long before, had conjectured for its physical environment (*BLT* §3.4).[14]

This, however, left the causal questions unanswered: what had generated these great directional changes in the earth's floras? Brongniart's geohistorical perspective included one further element, which he claimed could throw light on this, for he treated his fossils as indicators not only of relative age but also of ancient environments. Like many earlier naturalists, but now with all his new authority as a leading expert on fossil plants, Brongniart argued that those from the Coal formation represented a *tropical* flora. This implied that there had been a major change in climate, at least in the part of the globe that now forms Europe. But Brongniart enlarged that inference into a much broader conclusion. He suggested that the long history of successive floras reflected a long physical history by which the earth had slowly changed from a uniformly tropical state into its more differentiated present climates, most of them much cooler: a gradually falling global temperature must, he argued, have been a major cause of the striking floral changes "since that remote [Coal] epoch to our own day". The huge size of some of the Coal plants even suggested that the climate might have been still hotter than the present tropics (in modern terms, *hypertropical*). Furthermore, the vast stores of carbon now locked up in the coal itself implied that the atmosphere might have been much richer in carbonic acid (i.e., carbon dioxide) at that time than it now is: perhaps up to 8%, he suggested, in place of the present 0.1%. Even in such a fundamental parameter as atmospheric composition, the deep past might *not* have been altogether like the present; and yet of course this did not imply any change in the underlying laws of physics and chemistry.[15]

The physical model of a cooling earth, which the formidable combination of Fourier and Cordier had just rehabilitated and endorsed (§9.2), provided the perfect causal substrate to match the kind of botanical history that Brongniart inferred from his succession of fossil floras. The trend from a tropical-looking flora in the Coal formation, through the intermediate flora of the younger Secondary formations—including the Purbeck cycads (§11.1), as Buckland pointed out explicitly—to the much more temperate and modern-looking flora of the Tertiaries, could be explained as the product of a cooling earth, in just the way that the physicists inferred. At an early stage the earth had still received much of its surface heat by conduction from a hot interior, giving a uniformly tropical or

13. Fig. 12.5 is reproduced from Brongniart (Ad.), *Prodrome d'une histoire* (1828) [Cambridge-UL: MF.54.4], 258. The second flora was confined to the *Grès bigarré* (a sandstone formation of, in modern terms, Permo-Triassic age), and was intermediate in character: see *ibid.*, 218. The floras were also described in his "Considérations générales" (1828).

14. Brongniart (Ad.), *Prodrome d'un histoire* (1828), 221. It was usual among naturalists to refer to the "creation" of new forms of life, a term that was neutral with respect to possible *causes* such as spontaneous generation, transmutation or even, implausibly, unmediated divine action: here, significantly and typically, the creative agency was identified not as God but as *nature*.

15. Brongniart (Ad.), *Prodrome d'une histoire* (1828), 186–87, 221–23, and "Considérations générales" (1828), 249–54; see also Bowler, *Fossils and progress* (1976), 22–26.

perhaps even hypertropical climate: hence the similar Coal plants that had been discovered as far away as India and even Australia. But this source of internal heat had diminished exponentially, until it was insignificant in comparison with the radiant heat received from the sun; the earth had therefore developed much more differentiated climates, far more dependent on latitude and geographical circumstances.

These were no more than conjectures briefly posed; but they were highly suggestive, and symptomatic of a new effort to integrate the geohistorical interpretation of the fossil record with the physical evolution of the earth as a whole, not only its climates but even its atmosphere. In support of his conjectures the younger Brongniart was right to state that very few physicists [*physiciens*] now doubted that the earth had indeed been cooling slowly in the course of time. In correlation with this long-term cooling trend, his geohistory also incorporated the standard model of a globally falling sea level, uncovering progressively larger areas of land (*BLT* §3.5): he assumed that his first flora had been confined to islands or archipelagoes, which had only gradually expanded to form the continents of the present world.

Superimposed on these slow and gradual trends, Brongniart also hinted that the history of plants might have been affected by occasional and perhaps more abrupt events. He claimed that his floras were sharply distinct from each other at the taxonomic level of species, and that they were separated by formations with almost no trace of any plants at all. Most strikingly, the flora of the younger Secondaries had no species in common with that of the Tertiaries; and they were separated stratigraphically by the Chalk, which contained only fossils of marine animals. So Brongniart argued that the periods during which his successive floras had flourished had been separated likewise by great physical "revolutions" linked causally to major geographical changes: in other words, long periods of tranquil conditions had been punctuated by occasional episodes of rapid and perhaps violent change. This was a concept of geohistory obviously derived from what Cuvier and his own father, among others, had formulated earlier (*BLT* §9.1, §9.3). But for the younger Brongniart it was now being reinforced, as it were at just the right moment, by Élie de Beaumont's new synthesis of structural geology, stratigraphy, and paleontology, and his concept of a lengthy geohistory punctuated by sudden "epochs of elevation" (§9.4).

12.3 TROPICS IN THE ARCTIC?

Finally, the physicists' concept of a gradually cooling earth (§9.2), which the younger Brongniart rightly treated as almost consensual among geologists too, was already receiving unexpected support from the efforts of Arctic explorers. Parry's first voyage (1819–20) in search of a Northwest Passage (§7.4) had found the Arctic islands composed largely of Primary rocks. But there were also limestones with shells and corals familiar from the Secondaries of Europe; and most striking of all was the discovery on Melville Island of abundant fossil plants familiar from the Coal formation, and even seams of coal itself. When König reported on the fossils brought back by the expedition, he had claimed that this was "a proof that the inhospitable hyperborean region where they occur, at one time displayed the noble scene of a luxuriant and stately vegetation".[16]

This was confirmed when Parry returned from his third voyage (1824–25). The expedition's surgeon-naturalist—one of Jameson's former students—had found extensive Secondary formations around the spot where the ships had been frozen in for the winter (on Prince Regent Inlet, at the northwest end of Baffin Island, at 73° North). Jameson reviewed their fossils for Parry's published report, and made König's geohistorical inference more explicit. He pointed out that the modern analogues of the fossil plants comprising the "rich and luxuriant vegetation" were all *tropical*, and that the associated limestones rich in fossil corals suggested the same former climate. To find the relics of a tropical flora and fauna at such a high latitude was anomalous in the extreme, unless the earth as a whole had been much warmer at the remote period of the Coal formation. And Parry's discovery was no flash in the pan, for Franklin returned from his second expedition down the Mackenzie with similar specimens collected by his well-qualified physician-naturalist John Richardson (1787–1865). Over wide tracts of their route from Great Bear Lake northwards to the mouth of the Mackenzie on the "Polar Sea", they too had found Secondary limestones with fossil shells, crinoids, and corals that Sowerby identified as similar to those of the Mountain or Carboniferous Limestone in England.[17]

However, the use of this Arctic evidence did not go unchallenged. Fleming, who had already shown himself to be a vigorous critic of Buckland's diluvial theorizing (§6.4), returned to the fray with a hard-hitting article in Jameson's periodical, rejecting the entire analogical argument on which the inference of formerly tropical conditions in the Arctic had been based. Unless the fossils were of the very same *species* as the living organisms, and not merely of the same genus or family, no inference about their original environments could be soundly based: like the mammoth in relation to living elephants, anatomical similarity was no guarantee of ecological equivalence. Conybeare responded, with uncharacteristic asperity, dismissing Fleming as merely a "diligent and meritorious compiler in natural history" rather than a competent theoretician. He emphasized the *cumulative* value of the relevant evidence, with so many different kinds of animals and plants all suggesting the same formerly warmer climate: did Fleming really expect that future Arctic expeditions would discover lions roaming jungles in the high Arctic, or coral reefs in the surrounding seas? Fleming responded in kind, noting that in the years since his earlier critique of Buckland, "with the exception of a very few individuals who may still be found on stilts, amidst the 'retiring waters', the opponents of the [diluvial] hypothesis have become as numerous as were formerly its supporters". That was true, but the reconstruction of former

16. König, "Specimens collected by Captain Parry" (1823).

17. Parry, *Journal of a third voyage* (1826), 91–94; the volume includes Jameson, "Notes on the geology", 145–51. Franklin, *Second expedition to the Polar Sea* (1828), includes Richardson, "Geological notes" (1828) (read at the Geological Society on 21 March 1827). On the coastline west of the Mackenzie delta, the expedition surveyed and named (among others) Mounts Greenough, Conybeare, Sedgwick, and Fitton. Levere, *Science and the Canadian Arctic* (1993), 79–84, 110–25, puts both expeditions into their context. The modern explanation of these tropical fossils at high latitudes, in terms of the subsequent movement of tectonic plates, was not of course available or conceivable at this time.

climates was a different issue; the inference of tropics in the Arctic was unaffected by the fate of the diluvial theory.[18]

12.4 CONCLUSION

Research on the Tertiary formations, since early in the new century, had shown not only that they represented a major era in geohistory, but also that they could act as a cognitive gateway to the still earlier era of the Secondaries (Chap. 10). Variations in the fossil faunas of the different Tertiary basins had seemed to Brocchi, and later to Basterot, to reflect the kind of geographical diversity that Humboldt and Candolle were describing in living plants. But Prévost argued that these variations, when studied "statistically", indicated differences of relative age, not of former geography; and he suggested how the faunas of molluskan species might have changed steadily in composition over time, approximating ever more closely to that of the present world. Following this reasoning, Desnoyers identified two successive molluskan faunas in the various Tertiary basins in Europe, and then he claimed that in at least one case (the Paris Basin and Touraine) the "non-simultaneity" of the basins could be proved by the uncontroversial criterion of superposition. Deshayes, the acknowledged world expert on fossil conchology, refined this by identifying three successive fossil faunas, recognizable by their increasing proportion of extant species as they approached the fauna of the present world. And Bronn independently offered a similar "statistical" analysis of Tertiary mollusks, while also extending it to other kinds of fossils and other periods of geohistory.

This research on Tertiary faunas reinforced and gave greater precision to the increasing sense, among geologists, that the earth's faunas had indeed changed directionally throughout geohistory. The younger Brongniart's ambitious work on fossil plants had the same impact, on the larger scale of the whole fossil record, when he identified at least three major floras that had flourished successively in the course of geohistory, marked by the successive appearance of new major groups of plants and hence an increasing diversity and even, in some sense, progress. And Brongniart linked this with the intrinsic directionality inherent in the theory of a gradually cooling earth, as newly endorsed by the physicists (§9.2): as the earth cooled, climates at its surface would have become progressively more temperate and more differentiated, while even the composition of the atmosphere might have changed in a similar way. And this picture of the changing vegetation on a gradually cooling earth was already getting unexpected support from the discovery in the high Arctic of the characteristic large plants of the Coal formation, and even coal seams, together with equally tropical-looking corals and other marine fossils. This suggested strongly that the whole earth, even at this high latitude, had been tropical in climate at that extremely remote period of geohistory.

However, there was a large fly in this soothing ointment of a smoothly developing geohistory, changing directionally and progressively from unimaginably remote beginnings through ever more familiar phases toward the present world of human life. For the youngest Tertiary formations were separated from the present world by the extremely peculiar and seemingly unparalleled deposits that Buckland had called the "diluvium". Most geologists accepted the geohistorical reality of the "geological deluge" to which he attributed them, even if they rejected his confident identification of this strange event with the biblical Flood, and his consequent dating of it as geologically extremely recent (§6.4). The next chapter picks up this thread of the story: it traces how the "diluvial" enigma was being tackled in these same years, and how the effects of the "geological deluge" were being defined more precisely, or else explained away.

18. Fleming, "Value of the evidence", Conybeare, "Answer to Dr Fleming's view", and Fleming, "Climate of the Arctic regions" (all 1829).

CHAPTER 13

The last revolution (1824–30)

13.1 ALLUVIUM AND DILUVIUM

Cuvier had begun his seminal work on fossil bones, back before the turn of the century, with an analysis of those preserved in the Superficial deposits, which he interpreted as relics of a geologically recent mass extinction (*BLT* §7.1). In his later fieldwork with the elder Brongniart on the Paris Basin, this putative "revolution" had become just the most recent in a series of similar events throughout geohistory (*BLT* §9.1). Nonetheless, in the postwar period the "last revolution" remained for most geologists the most striking event of all, not least because it seemed to mark the boundary or interface between recorded human history and the unimaginably long geohistory that had preceded it. In describing the latter, the earlier language of "the former order of things" had fallen into disuse, because the use of the present as a key to the deep past—by geologists of all stripes—had shown that "the former world" could *not* have been of a different "order" or totally "other" than the present world. Yet the apparent boundary event between them remained an obstacle in the path of those who hoped to demonstrate that the deep past had merged insensibly into the present. Whatever date was assigned to it, and whatever physical character and cause, it remained the most puzzling event in all geohistory, not only for Cuvier but also notably for his prominent English follower Buckland, in his delineation of the last revolution as a "geological deluge".

Buckland's version of the diluvial theory had leapt into prominence beyond Oxford, when his paper on Kirkdale Cave was read at the Royal Society in London in 1822 and published subsequently in its *Philosophical Transactions*, earning him the coveted Copley Medal (*BLT* §10.6). The following year this paper had been reprinted in his book *Reliquiae Diluvianae*, along with other materials that widened the scope of his theory and made it accessible not only to savants but also to serious amateur naturalists (§6.3). Buckland had argued that Europe—and even perhaps the world as a whole—had been ravaged by an aqueous "deluge", probably some kind of mega-tsunami, in the geologically recent past. It had wiped out Cuvier's mammalian megafauna; excavated valleys or at least deepened them; deposited the peculiar Superficial materials that he named the "Diluvium"; and, in some regions, moved huge erratic blocks of rock and smaller distinctive pebbles tens or even hundreds of miles from their obvious places of origin. Much more controversially, Buckland had also dated this catastrophic event only a few millennia in the past; his forceful claim that it was none other than the biblical Flood made his theory a sensitive issue in the peculiar social and political circumstances of Britain, though far less so elsewhere (§6.4).

Buckland's book sold well and he made plans for a sequel; he was evidently confident that he could answer his critics. However, early in 1825 he resigned his college Fellowship and accepted a position in a rural parish, which would have brought his scientific career at Oxford to an end. He took this drastic step in order to marry Mary Morland (1797–1857), a younger naturalist also living in Oxford, and a skillful scientific artist who had long been assisting him in his research. In the event, the Tory prime minister Lord Liverpool, who had earlier helped to get Buckland his readership in geology at Oxford (*BLT* §10.4), saved the day by using once again his powers of patronage (acting this time in the name of King George IV). He awarded the geologist a highly paid position with relatively light duties, as a canon of Christ Church, Oxford's cathedral but also the largest and richest college in the university. This gave Buckland greater financial security and higher social standing, but above all it relieved him of the necessity of exile. For the canonry was an exception to the general rule at Oxford colleges: he was free to marry, and did so on the last day of 1825. He now had the best of both worlds, and he continued to be one of the most prominent geologists in Britain, and one of the best known in the rest of the savant world.[1]

Among Buckland's immediate critics, one of the most trenchant had been his fellow cleric the Scottish naturalist John Fleming, who deplored the Englishman's mixing of geology with religion and was doubtful whether the evidence supported a diluvial interpretation at all (§6.4). Fleming's comments spurred one of Buckland's heavyweight supporters, his counterpart at the other English university, to spring to his defense. Back in 1818, the chair of geology that John Woodward endowed at Cambridge a century earlier had become vacant when its holder had to resign on getting married. Adam Sedgwick (1785–1873) had been elected his successor: he had previously shown no special interest in geology, but as a Fellow of Trinity (Cambridge's largest and richest college) he had garnered more votes than his rival. He was said to have remarked, "Hitherto I have never turned a stone; henceforth I will leave no stone unturned". The story may be apocryphal—he was not in fact ignorant of the science—but he certainly fulfilled it in spirit throughout his long subsequent career as the leading geologist at Cambridge.[2]

Sedgwick had begun an annual course of lectures that matched Buckland's in content and popularity; and he had reinforced his teaching by asking Conybeare to give him an informal crash course in fieldwork. The year after his election he had been one of the founders of the Cambridge Philosophical Society; he and his younger colleague John Stevens Henslow (1796–1861), later the professor of botany, had the phrase "and natural history" added to "[natural] philosophy" in its constitution, thereby ensuring that it would cater to *all* the natural sciences. From the start, Sedgwick was a leading member of what soon became a lively forum for those with scientific interests, in a university that was traditionally more sympathetic to such activities than Buckland's Oxford. His geological interests lay mainly with the hard old Primary rocks—he started on the Dartmoor granite and the slates surrounding it—and his goal was to understand their structural relations in a traditional geognostic sense. But there were no hard old rocks around Cambridge, and anyway he had good reason to participate in the lively debate that Buckland's work on the very youngest deposits had aroused.[3]

In 1825 Sedgwick wrote a lengthy paper in support of one of Buckland's crucial points. It was published in the *Annals of Philosophy*, one of the many scientific periodicals that now competed for the attention of British savants and amateurs; French and German translations of the article soon made it accessible to those in the rest of Europe. Citing his own field observations, which ranged by then over the length and breadth of Britain, Sedgwick argued that the Alluvial deposits *always* overlay the Diluvial ones, rather than alternating with them as some geologists had claimed, and that the two sets of deposits must therefore "belong to two distinct epochs". So Buckland's terms expressed "a natural separation" that was "unconnected with any hypothesis whatever": the recognition of two sets of deposits in recent geohistory—distinct both in physical character and in fossils—could and should be kept apart from the much more contentious issues of the causal origin of the diluvial ones and their relation to human history.[4]

The Alluvial deposits were no problem, in that they were clearly due to the ordinary actual causes that remained observably in action in the present world, such as deposition by rivers and on some coastlines. Sedgwick claimed that the Diluvial ones, in contrast, had been "produced by some extraordinary disturbing forces prior to the existence of any portion of the other class" of deposits. While conceding that there was still no satisfactory causal explanation of the Diluvial deposits, he argued—tacitly against Fleming—that no purely local agencies

1. Boylan, *Buckland* (1984), 147–54. On Morland's earlier work with Buckland, see for example Buckland, "Lickey Hill" (1821), 525 (and *BLT* §10.5). Kölbl-Ebert, "Mary Buckland" (1997), gives useful information about her later, married life.

2. Clark and Hughes, *Sedgwick* (1890) 1: 152–65, describes the election; this joint work by a historian of the university and Sedgwick's own successor as Woodwardian professor is still a valuable source.

3. Sedgwick, "Primitive ridge of Devonshire" (1820) and *Lectures on geology* (1821); Hall (A. R.), *Philosophical Society* (1969), 4–9. Working on the hard old rocks of the Lake District, not far from his family home at Dent (Yorkshire), he met William Wordsworth (whose younger brother was the Master of Sedgwick's college) and transmitted to the poet some of his own enthusiasm for geology: Wyatt, *Wordsworth and the geologists* (1995), 71–84.

4. Sedgwick, "Alluvial and diluvial formations" (1825), 242–49. Buckland's term "diluvial" was of course intrinsically theory-loaded; but Sedgwick's point was vindicated in the event by the continuing use of "*Diluvium*" as a useful descriptive term, even by those who rejected the putative event to which the name referred. Indeed it remained in general use among geologists, notably in the German lands, long after any "diluvial" explanation of the deposits had been abandoned in favor of a glacial or fluvioglacial origin (see §36.1).

would suffice to account for phenomena that were evidently extremely widespread if not universal. Sedgwick again followed his Oxford colleague in claiming that all the Alluvial deposits were very recent, and that the preceding Diluvial event must have been almost equally so. While attributing that important conclusion to de Luc's "labour of many years" in the field (*BLT* §6.2), Sedgwick added acerbically that "had his labours terminated here, he had done great service to geology". By implication, the earlier savant's further attempt to use this scientific evidence to equate the event with the biblical Flood (*BLT* §6.4) was now decidedly questionable.[5]

Like Buckland, Sedgwick was an ordained minister in the established Church of England, working in a university structured around the politics of church and state, though that certainly did not constrain either geologist to be a biblical literalist. However, Sedgwick, unlike Buckland, insisted that geological investigations should be kept totally separate from any appeal to biblical authority. He agreed with Buckland that "the reality of a great diluvian catastrophe during a comparatively recent period" was now well established on purely natural evidence. If such an event was found to be compatible with "the sacred records of the history of mankind", well and good, since "truth must at all times be consistent with itself". But Sedgwick argued that the two lines of investigation—the one physical and geohistorical, the other textual and humanly historical—ought to be kept apart. Only if their respective conclusions were seen to be "derived from sources entirely independent of each other" could a demonstrable "general coincidence" between them carry its full weight and have the impact that it would then deserve.[6]

It cannot be emphasized too strongly that some kind of diluvial theory, if not Buckland's particular version, remained highly plausible throughout the 1820s, in the eyes of almost all competent geologists throughout Europe, and indeed in North America too. The reality of the division of the Superficial deposits into Diluvium and Alluvium seemed to most of them to reflect the reality of the "geological deluge", whether or not its date was recent enough to be equated with the Flood recorded in Genesis (and also, as Cuvier had long argued, in the ancient records of other early human cultures). Far from being undermined by new investigations, the case for some kind of dramatic "last revolution" in recent geohistory was strengthened by much of the newest research, although the character of the putative event or events remained enigmatic.

13.2 ALPINE ERRATIC BLOCKS

The kind of evidence that was most familiar and most widely distributed remained, as it had long been, that of the form of valleys. Some earlier French naturalists had argued that rain, streams, and rivers would be able to erode deep valleys, if only enough time were allowed: the work of Desmarest, Soulavie, Montlosier, and others (*BLT* §4.3, §4.4, §6.1), making this claim in relation to the volcanic landscapes of central France, was well known throughout Europe, as was Playfair's attempt to set the same ideas into a broader theoretical framework. By the 1820s, however, the pendulum had swung the other way. As Fitton noted

Fig. 13.1 A quasi-aerial view of a broad but deep valley of U-shaped profile, drained by a small stream. This illustration, from George Poulett Scrope's album of panoramic views of the volcanic regions of central France (1827), showed in the distance the massif of Mont-Dore in Auvergne, which he interpreted as the highly eroded remnant of a large ancient volcano. Scrope himself was a forceful advocate of the view that all valleys were the products of ordinary erosion by rain and rivers over vast spans of time (see §15.1); but this illustration inadvertently provided others, such as Buckland and Daubeny, with a fine example of those that seemed to them to be inexplicable except by some kind of more violent "diluvial" event.

in his review of Buckland's work (§6.4), almost all well-informed geologists—of course, he included himself—had now concluded that the observable process of fluvial erosion was *not* adequate to account for "valleys of denudation". Certainly the small narrow valleys with V-shaped profiles that many existing streams were observably continuing to excavate bore little resemblance to the most striking kinds of valley topography, particularly the huge deep valleys, common in mountain regions, that had a broad U-shaped profile (Fig. 13.1).[7]

However, the evidence provided by the forms of valleys was clearly ambiguous, since it was recruited by both sides of the argument, both for and against the reality of the putative diluvial event. It seemed to many geologists that the evidence of erratic blocks was far more decisive. Ever since von Buch's great paper

5. Sedgwick, "Alluvial and diluvial formations" (1825), 249, 254n.

6. Sedgwick, "Alluvial and diluvial formations" (1825), [part 2], 33–35.

7. Fig. 13.1 is reproduced from Scrope, *Central France* (1827), part of pl. 10, explained on 178; it shows the headwaters of the Dordogne flowing northwards from the distant peaks of Mont-Dore. Scrope himself suggested that the valley might have been opened up by crustal movement and then enlarged by fluvial erosion. Daubeny, in "Volcanoes of Auvergne" (1820–21), was among those who had doubted whether purely fluvial erosion was adequate to explain such valleys (*BLT* §10.5).

on those in the Alps and the Jura (*BLT* §10.2), published just as the wars ended, erratics had been massive stumbling blocks in the path of any savant who, following Playfair, hoped to account for *everything* geological in terms of actual causes. Now, a decade later, the problem was accentuated by von Buch's recognition that what he had described on the north side of the Alps was more or less mirrored on the south. Revising his earlier paper in the light of further fieldwork (by himself and others) in the north of Italy, von Buch reinforced his inference that there had been massive aqueous events, analogous to the catastrophic flood in the Val de Bagnes in 1818 (*BLT* §10.5) but on a far larger scale.[8]

This was vividly illustrated—in both senses—when in 1830 De la Beche published an album of his own fine lithographs (and a few engravings by others), designed to be "illustrative of geological phaenomena". His book was a visual inventory of sections, maps, and pictures, by which he intended to show concretely what the more theoretically ambitious among his contemporaries would need to explain, if their theories were to be credible. More specifically, some of his images showed how much actual causes were, *and were not*, capable of explaining in the records of geohistory. Many were of British localities; a few came from as far afield as Jamaica, which he had visited a few years earlier. But in the present context the most decisive images derived from his extensive travels on the Continent in 1828–29, during which he seems to have dispelled the sadness of a failed marriage in the exhilaration of geological fieldwork (Fig. 13.2).[9]

De la Beche traveled widely in Italy; like many others making the scientific Grand Tour, the volcanic region of Campania was a major goal. From Naples he visited Herculaneum and Pompeii, seeing for himself the canonical intersec-

Fig. 13.2 De la Beche (brandishing his hammer) and his servant and companion Henri, experiencing the joys of fieldwork: a sketch drawn by the geologist during his tour of Italy in 1828-29. (By permission of the National Museum of Wales)

tion of geology with archaeology, geohistory with human history (*BLT* §4.1); he inspected the Temple of Serapis, now the equally canonical evidence for the dynamic character of the earth's crust (§8.2); and he climbed Vesuvius itself. Standing on the rim of its huge crater, newly enlarged in the great eruption that Scrope had witnessed in 1822, he contemplated the sheer scale and power of the earth's actual causes (Fig. 13.3).[10]

The most significant of De la Beche's subsequently published illustrations, however, were those that hinted that actual causes were *not* adequate to explain all the relics of the deep past. They concerned the erratics that von Buch had recently described, in the region of the Italian lakes on the south side of the main Alpine chain. As on the north side, there were vast numbers of erratics; some were even larger than the one that von Buch had first made famous, the Pierre à

Fig. 13.3 The summit crater of Vesuvius, drawn by De la Beche on 15 February 1829 and published in 1830; it was his first direct experience of any active volcano. He noted that it was still showing residual activity in the aftermath of "the late great eruption" of 1822. A local guide (with stick and cap) is showing a gentleman (in tail coat and top hat)—presumably De la Beche himself—a large block of rock thrown out during that eruption (in modern terms, a volcanic bomb), an indication of the sheer power of this "actual cause". (By permission of the Syndics of Cambridge University Library)

8. Buch, "Alpengeschiebe" (1827).

9. Fig. 13.2 is reproduced from De la Beche, notebook for 2 November 1828 to 1 June 1829 [Cardiff-NMW: no. 363], 116, drawn near La Spezia on 2 May 1829. In De la Beche, *Sections and views* (1830), the Jamaican illustrations came from his time there in 1823–25, which had been primarily to attend to the sugar plantation he had inherited. He had married in 1818 but obtained a divorce—a highly unusual event at this time—in 1826, his wife having left him for another man: see McCartney, *De la Beche* (1977), 12, 26 (this work reproduces many more of his lively sketches).

10. Fig. 13.3 is reproduced from De la Beche, *Sections and views* (1830) [Cambridge-UL: MF.40.7], pl. 22, explained on 34–35, lithographed from his own drawing [Cardiff-NMW: MS 363, 4].

Bot on the flanks of the Jura hills above Neuchâtel (Fig. 6.4; *BLT* §10.2). Despite De la Beche's vaunted theoretical neutrality, his wording shows that in fact he took it for granted that the blocks near Lake Como had somehow been "violently torn" from the bedrock far away, and swept in "so violent a rush" that they had been carried many miles down the valleys, right along the length of the lake, to be deposited finally on lower ground, though in some cases high above the lake. He refrained from offering any further interpretation, but in his *Geological Manual* published the following year he linked it explicitly with Élie de Beaumont's recent conjecture that the Alps had been heaved up at a geologically recent time by a sudden and violent buckling of the earth's crust (§9.4). This might have triggered a huge aqueous event—bearing "masses of floating ice charged with blocks and other detritus"—that could have swept erratics down *both* flanks of the newly elevated mountain chain (Figs. 13.4, 13.5).[11]

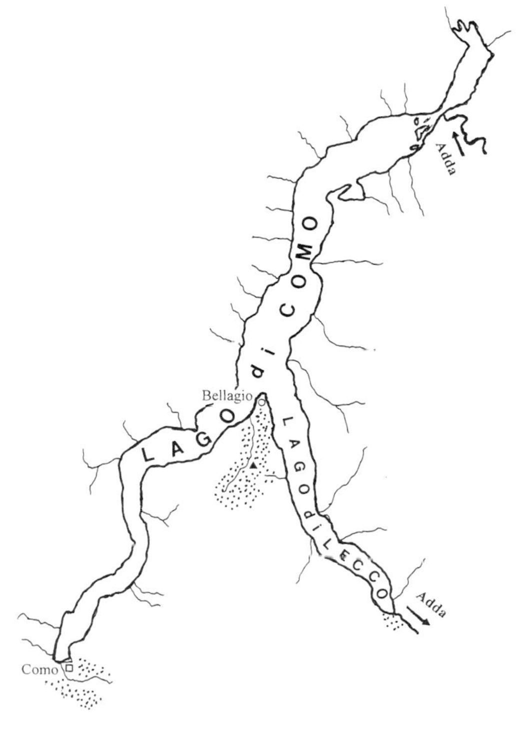

Fig. 13.4 A huge "transported" or erratic block of granite high above Lake Como on the southern flank of the Alps, with De la Beche (and hammer) to show the scale, sketched by him during fieldwork in 1829 and published as a lithograph in 1830: "It is larger than a house, and must weigh much more than 100 tons . . . This block struck me as larger than the celebrated one [the Pierre à Bot: Fig. 6.4] upon the Jura behind Neuchâtel". He inferred that along with innumerable smaller blocks it had been "violently torn" from the solid granite in the central Alpine range; "transported" for many miles on some kind of rapid current, buoyed up on ice and without attrition or rounding; and finally dropped where "this rush from the north" was checked by rising ground (see Fig. 13.5). These Italian erratics matched the better known Swiss ones (including the one cited) that had been transported northward from the other side of the Alps across the Swiss plain and up on to the Jura hills. This suggested that the sudden upheaval of the Alps themselves might have been their common cause. (By permission of the Syndics of Cambridge University Library)

Fig. 13.5 De la Beche's map of Lake Como [Lago di Como] in the Italian Alps, published in 1830, redrawn in modern style to clarify the areas of erratic blocks (stippled) at the outlets of both branches of the lake, but also on ground rising southward from the fork between them (the black triangle marks the position of the erratic shown in Fig. 13.4, well above the lake). He located their source in the central Alpine range far to the north, but their mode of transport remained as puzzling as ever. (By permission of the Syndics of Cambridge University Library)

13.3 ERRATIC BLOCKS IN SCANDINAVIA

However, erratic blocks were not confined to the flanks of the Alps or any other high mountain ranges. As von Buch had noted briefly in his classic paper on the Alpine case, they were also known to be spread out over vast low-lying areas of northern Europe, from eastern England through the Netherlands and Prussia, all the way to northern Russia (see Fig. 13.7). Many of the constituent rocks had been matched with those known to outcrop in situ in Scandinavia, on the far side of the Baltic or even of the North Sea. This made the problem of accounting for erratics far more difficult, and it suggested that the diluvial event might have been far more widespread.

The astonishingly wide distribution of these northern erratics had been vividly described by Gregor Kirilovich, count Razumovsky (d. 1837), a brother of the Russian ambassador in Vienna (and Beethoven's friend and patron). The geological Razumovsky had published a detailed account (in French, of course) of the erratics spread across northern Russia and its Baltic provinces: most could be closely matched with rocks only known in situ far to the north, in Finland and elsewhere in Scandinavia. "These then are the facts, and indisputable facts", he had concluded, "proving in an incontestable manner that the north of our continent, like the south [i.e., the Alpine regions described by von Buch], has been worn down and wreaked by a terrible revolution, [by] aqueous currents of a power that astounds the imagination". Razumovsky conjectured that the erratics had been derived from the catastrophic erosion of mountains far to the north; judging by the magnitude of the effects, he reckoned that the vanished mountains must have been far higher than the present Alps. Any equation with events recorded in human history was ruled out by his inference that the "revolution" had long antedated any human presence, although it was geohistorically so recent.[12]

The title of Razumovsky's little book failed to signal its primary topic, and his description of the erratics seems not to have been noticed widely. But its main conclusions were reported to the Geological Society in London—though without acknowledgment—by William Thomas Horner Fox-Strangways (1795–1865), a young diplomat attached to the British embassy in St. Petersburg. As one of Buckland's former students, he was not surprised to find widespread "diluvium"

11. Fig. 13.4 is reproduced from De la Beche, *Sections and views* (1830) [Cambridge-UL: MF.40.7], pl. 38, fig. 3, explained on 68; Fig. 13.5 is traced from *ibid.*, pl. 31, explained on 52–59; *Geological manual* (1831), 155–80. His field notes on the Como erratics are in his journal for May 1829 [Cardiff-NMW: no. 363], 141–67, the sketch on 158). The coloring of the original map represented the geology of the bedrock, but he noted that this was totally unrelated to the erratics. One of his sections (pl. 8, fig. 3) showed the distribution of the erratics on ground rising irregularly from the point of the fork at Bellagio; the block illustrated here was on the Alpi di Pravolti. Buch, "Alpengeschiebe" (1827), was almost certainly his source of information on the Como erratics.

12. Razumovsky, *Coup d'oeil géognostique* (1819), 67. This was an enlarged edition, published in Berlin, replacing one published in St. Petersburg in 1816 and already out of print; a "table" in two columns (82–96) listed the rock varieties found among the erratics, matched to the localities where they were known in situ. Gould, "Razumovsky duet" (1996), contrasted Prince Andrey Kyrilovich (1752–1836)—Beethoven's friend—with his elder brother, count Alexei Kyrilovich (1742–1822) of Moscow, but sadly missed the *geologist* brother Gregor, who in fact made it a Razumovsky *trio*!

around the city. Nonetheless he, like Razumovsky, was astonished at the "stupendous size and universal distribution" of the erratics. One huge block had been hauled out of a bog to be used as a plinth for the empress Catherine's famous equestrian statue of Peter the Great, the founder of the city; and many were obviously derived from the Primary rocks of Finland to the northwest, where he noted that "the evident traces of diluvian action are on a most astonishing scale". In a later paper, Strangways sketched what was known of the geology of the whole of the Russian empire, noting that "boulders of chert containing shells and corals", clearly derived from the Secondary rocks of the Valdai [Valdayskaya] Hills, were strewn southeastward across another vast tract that included the Moscow region.[13]

In 1824 Brongniart spent the summer doing fieldwork in Scandinavia with his son Adolphe. His main objective was to study the ancient Transition rocks; these yielded some of the finest specimens of trilobites, which were necessary for his continuing research on those fossils (§4.3). But he also gave close attention to the quite separate problem of erratics. While traveling across the north German plain, he noted their sheer abundance, the huge size of some of them, and the astonishing variety of the rocks of which they were composed. Having crossed the Baltic to Sweden, he then identified some of the same rocks in situ; how the erratics had managed to get across a deep arm of the sea remained utterly mysterious.[14]

However, it was not until 1828 that Brongniart published an account of what he was the first to call *erratic blocks*: they were, he wrote, "one of the most widespread, most striking, and most inexplicable of geological phenomena". He declined to speculate on the means by which they had been transported, and simply reported some puzzling aspects of their distribution in Sweden. He described how the Superficial gravelly deposits that contained the erratics were often underlain by smooth and even polished surfaces of bedrock, scratched with parallel grooves like those that James Hall had described around Edinburgh years before (*BLT* §10.2). That the movement in Sweden had been southward was proved, in Brongniart's view, not only by the erratics he had seen far to the south, beyond the Baltic, but also by his observation that the scratched rock pavements were always smooth to the north but often broken off irregularly to the south. He also claimed that the scratches had the same orientation as the *eskers* [*åse*], long narrow ridges of gravelly deposits on which larger blocks were scattered irregularly. The great Swedish savant Jöns Jakob Berzelius (1779–1848), the professor of medicine at Stockholm but also a distinguished chemist and a competent geologist, had been skeptical about this, despite his advantage of local knowledge, until Brongniart showed him instances in the field. Whatever the causative event, it had evidently been strongly directional; but unlike Hall's evidence for an eastward deluge in the Edinburgh region, in this part of Sweden the enigmatic movement had clearly been from north to south (Fig. 13.6).[15]

Brongniart's conclusion was severely negative, and he left the problem open: by whatever agency the blocks had been moved, in his opinion it had *not* been by any "actual aqueous force". It is not clear whether he would have regarded Hall's mega-tsunami sweeping over Edinburgh, or von Buch's mass of turbid

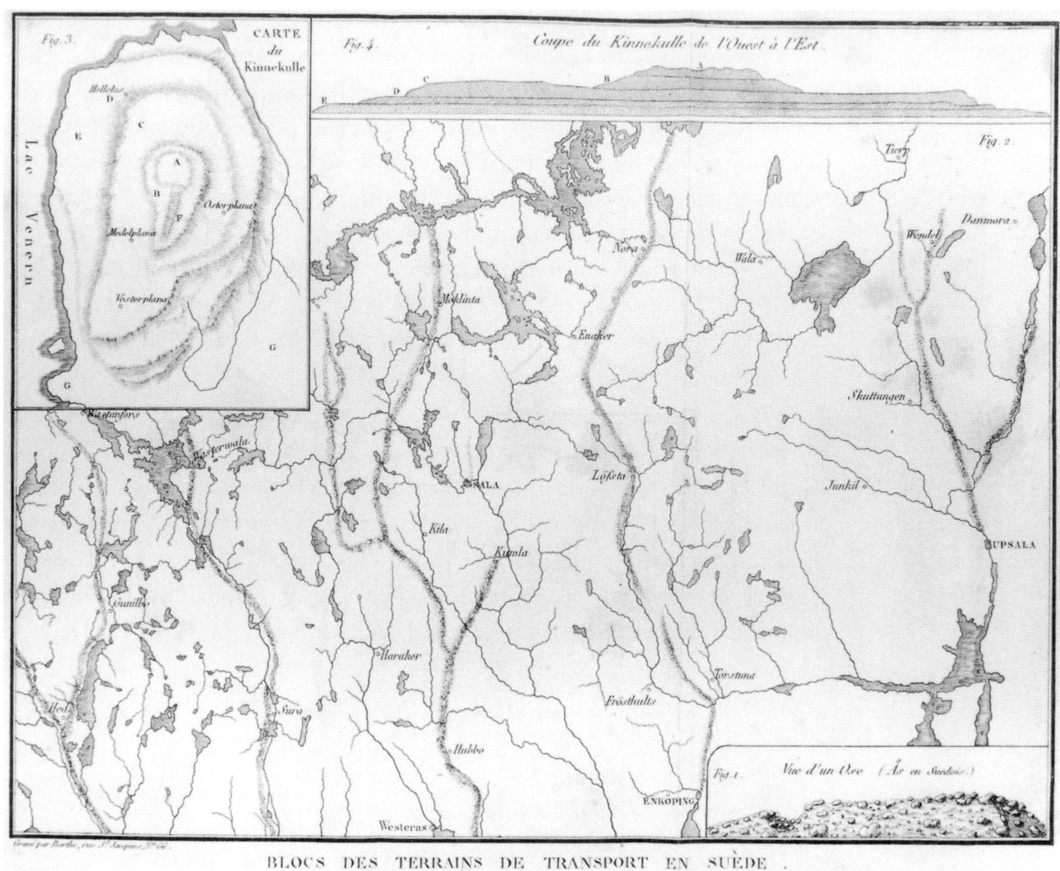

Fig. 13.6 The elder Brongniart's map (1828) of the region to the west of Uppsala in Sweden, showing five roughly parallel ridges (*åse* or *oses*; in modern terms, *eskers*) of Superficial deposits and erratic blocks (see sketch in fig. 1, bottom right), with the same orientation as many scratched surfaces in the underlying bedrock. Whatever the event responsible, it had clearly had a specific southward direction, which was in accord with the well-known occurrence of blocks of Scandinavian rocks on the north German plain, on the far side of the Baltic. The smaller map and section (figs. 3 and 4, top left and right) are of Kinnekulle: a telltale tail of debris (*F*) to the south of the summit (306m) implied that even this substantial hill had been overwhelmed by the puzzling diluvial event. (By permission of the Syndics of Cambridge University Library)

water flushing across the Swiss plain, as plausible agents (*BLT* §10.2): both were scaled up from known actual causes, but both were of a far greater magnitude than had ever been witnessed or recorded in human history, and therefore might have qualified for Brongniart's consideration. Anyway his fieldwork highlighted the mysterious character of the diluvial event, not least because the Scandinavian

13. Strangways, "Environs of Petersburg" (1821), 425–37, read 16 April 1819, and "Geology of Russia" (1822), 18n and pl. 2, read 2 March 1821. It is inconceivable that he did not know of Razumovsky's work, published in 1816 in the same city. He subsequently had a distinguished diplomatic career, and eventually succeeded his father as earl of Ilchester.

14. Brongniart, "Journal de voyage de Suède 1824" [Paris-MHN: MS 2346], e.g. cahier 1, entries for 17, 27, 29 July, the last referring to them as "*blocs erratiques*".

15. Fig. 13.6 is reproduced from Brongniart (Al.), "Blocs des roches" (1828) [Cambridge-UL: CP382.c.44.14], pl. 1, explained on 21–22 (quotation on 7). Kinnekule, which is near the Vänern lake and 18km northeast of Lidköping, is shown as a hill capped with dolerite (*A*) overlying outliers of Transition formations (*B–E*) with fossil trilobites, on a basement of gneiss (*G*).

erratics, unlike the Alpine ones, were on low ground far away from any high mountain chain, and they had even crossed the Baltic into northern Germany.[16]

The erratic blocks on the north German plain itself had in fact just been described by Hausmann in a paper read at the Royal Society at Göttingen; like von Hoff's prize essay for the same body a few years before (§7.2), it was in the now rather archaic medium of academic Latin. Hausmann summarized the evidence that the erratics near Göttingen had indeed come all the way from Sweden: they had not originated in the nearby Harz, as had been supposed, for those hills had merely halted the movement of erratics originating much further north. Hausmann also noted that the trail of erratics extended into the Netherlands, and, as Buckland had suggested, some had apparently crossed the North Sea from Norway to eastern England; in contrast, the Alpine erratics were much more limited in extent and he thought them probably later in date. But Hausmann, like Brongniart, offered no explanation of the phenomenon, concluding that none of those hitherto suggested was adequate. The problem was certainly topical: in the same year the Hollandsche Maatschappij in Haarlem, the main scientific society in the Netherlands, offered a prize for an essay on the origin of the Dutch and German erratics, and its gold medal was later awarded to Hausmann.[17]

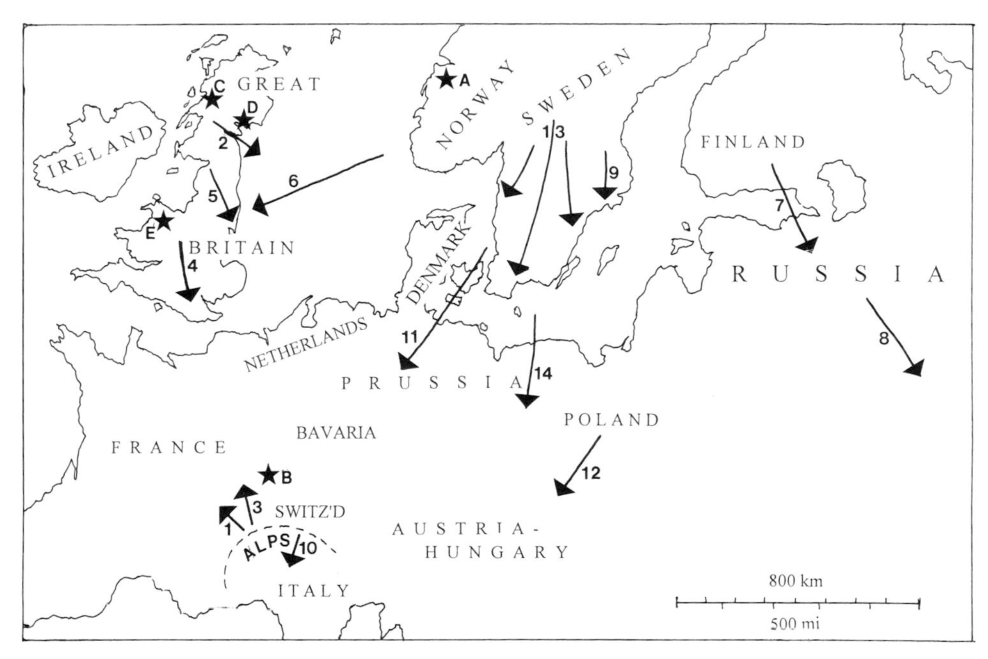

Fig. 13.7 The courses of some of the diluvial currents in Europe (arrows), as they had been inferred by many different geologists by the late 1820s. They are identified as follows, in chronological order: (1) Mont Blanc to Geneva, by Saussure, (2) Edinburgh, by Hall, and (3) Rhône valley to Jura, by von Buch (all *BLT* §10.2); (4) Midlands to Oxford, (5) Shap to Yorkshire, and (6) Norway to Yorkshire, all by Buckland (§6.1, *BLT* §10.5); (7) Finland to St. Petersburg and Baltic provinces, and (8) Valdai Hills to Moscow, both by Razumovsky and Strangways; (9) Sweden, by Brongniart; (10) Alps to Como, by De la Beche; (11) Sweden to north Germany and Netherlands, by Hausmann. Two currents inferred later are also shown: (12) Poland, by Pusch, and (13) Sweden and (14) north Germany, both by Sefström (see §34.1). This map is also used to locate areas (black stars) where former valley glaciers were inferred, mostly later: again in chronological order, (A) Norway, by Esmark (see below); (B) Vosges, by Renoir and others (see §35.2); (C) Lochaber, by Agassiz and Buckland; (D) Forfarshire, by Lyell (see §35.3); and (E) north Wales, by Buckland and Darwin (see §36.1).

Putting together the fieldwork of all these geologists, the evidence of erratic blocks suggested that the diluvial event or events, whatever their exact nature, had been extremely widespread (Fig. 13.7).

Nor was the enigma of northern erratics confined to Europe. John Jeremiah Bigsby (1792–1881), a British physician attached to the commission surveying the frontier between the United States and Canada, had reported in Silliman's *American Journal of Science*, and later to the Geological Society in London, that he had seen vast spreads of erratic blocks on the northern shores of Lake Huron. He attributed them to the usual kind of "debacle": "I am inclined to the opinion that an enormous body of water has rushed over these countries [and] swept from distant lands the colossal fragments of rock so frequent in the Lake". But he also reported that some of these blocks had been seen shifting their position, rafted on ice floes, as the lake's winter ice broke up in spring; so he was evidently thinking also of the explanation put forward long before by the Prussian geognost Wrede (*BLT* §10.2), at least as a supplementary factor. And when De la Beche reported to his local scientific society, and then to the Geological Society, on his trip to Jamaica, he described extensive spreads of gravel that he assigned to the Diluvium. This suggested—at least to him—that the causative event might also have extended into the tropics.[18]

13.4 ESMARK'S GLACIAL CONJECTURE

Around this time, one attempt was made to provide a new causal explanation of the enigmatic diluvial event, using Scandinavian evidence. Its author, Jens Esmark (1763–1839), the professor of mining at Christiania [now Oslo], had been trained at Freiberg before the turn of the century and, while still there, had published a competent account of his "mineralogical" travels in central Europe; he was a Foreign Member of the Geological Society, and no mere novice in geology. Around the time of Brongniart's fieldwork in Sweden, Esmark published a paper in a Norwegian periodical for the natural sciences, suggesting that at some remote time "the earth has been covered in ice". This startling conclusion was supported by local details that he had carefully observed during extensive field

16. Razumovsky promptly criticized Brongniart for failing to mention his own earlier work; he reprinted the relevant passages in the *Annales des Sciences Naturelles* (in which Brongniart's paper had appeared), but conceded that their common conclusion strengthened the case for an event of astonishing magnitude in the geologically recent past: Razumovsky, "Gros blocs de roches" (1829), 142. Brongniart inserted an editorial note (133n) regretting his earlier ignorance of the Russian's work, and denied any intention to claim priority.

17. Hausmann, "De origine saxorum" (1828), read on 25 August 1827; a "free translation" by the younger Jean-André de Luc (nephew of his more famous but deceased namesake), in Geneva's *Bibliothèque universelle*, soon made the work more widely accessible. Hausmann, "Primitieve rotsblokken" (1831), was the Dutch translation of his prizewinning essay in German: see Bruijn, *Prijsvragen* (1977), no. 319. The English erratics of Norwegian origin (on the east coast from Yorkshire to East Anglia) had been noted by Buckland, *Reliquiae* (1823), 191–93.

18. Bigsby, "Lake Huron" (1821), 255–56, and "Lake Huron" (1823), the latter read at the Geological Society on 21 February, 7, 21 March, referring particularly to St. Joseph Island, southeast of Sault Ste. Marie. De la Beche, "Diluvium of Jamaica" (1825), read at the Bristol Philosophical Society on 12 May, and "Geology of Jamaica" (1827), read at the Geological Society on 2 December 1825, 6 January 1826.

trips on the west coast of Norway in 1823. A certain valley near Stavanger, for example, was crossed by a ridge of gravel and boulders, damming a lake and rising 100 feet above the valley floor. Esmark described the ridge as a "glacier-rampart [*Gletscher-Vold*]", which clearly expressed his inference that it was a moraine left by a vanished glacier. He had seen similar moraines close to some of Norway's small existing glaciers at higher altitudes; he was using the standard actualistic method of comparison to infer that such glaciers had once been far more extensive. In another region to the north of Bergen he found strikingly smooth vertical rock surfaces along the shore, which he interpreted as having been polished by a mass of ice right down at sea level.[19]

However, Esmark's paper began with a much longer disquisition on a highly speculative geotheory; his careful local details might therefore have been missed by readers unsympathetic to that outmoded genre. Esmark linked his own ideas to those of William Whiston over a century earlier: he presented his evidence for a formerly greater extension of the Norwegian glaciers as confirmation of Whiston's conjecture that the earth had been alternately very close to the sun and very far away, generating periods of extreme heat and cold. Esmark's claim that "the earth has been covered by ice" was meant literally: it had at one time been a "*snowball earth*" (to borrow a phrase from a modern conjecture of the same kind: see Chap. 35). In fact his concluding summary made no mention of his Norwegian evidence, being confined to a restatement of his hypothetical geotheory: his putative period of extreme cold belonged to the primal part of the earth's history, *before life began*, and even longer before its development was crowned eventually by the human species.[20]

Esmark's paper would probably have gone unnoticed by geologists outside Scandinavia, had it not been translated for Jameson's Edinburgh periodical (and later abstracted in a French one, which Brongniart would certainly have seen). But it is not surprising that Esmark's ideas were not adopted by geologists elsewhere, even when they were made more accessible in this way. Geotheoretical conjectures in the style of Whiston had become unacceptable in the scientific world, and such flights of uninhibited speculation were repudiated; with such an association, even the most careful field observations were likely to be rejected too, or at least overlooked. And finally, to add to its implausibility, Esmark's suggestion that the climate had formerly been much *colder* than at present flew in the face of the cumulative evidence that earlier climates had been hotter, the trend throughout geohistory having apparently been one of progressive cooling, probably reflecting the cooling of the globe itself (§9.2, §12.2).[21]

13.5 CONCLUSION

The sharp distinction between the present human world and the "former world" of the rest of geohistory seemed to most geologists as real in the later 1820s as it had been when Cuvier began his research before the turn of the century. But the "last revolution" or "geological deluge", the interface or boundary event that separated the two "worlds", remained utterly mysterious. Cuvier's and Buckland's dating of the event—equating it with the Flood recorded in Genesis and other ancient records—was widely criticized by other geologists; but almost all of them agreed nonetheless that it had been some kind of natural physical event of exceptional intensity. Most of them were content to describe its observable effects, in the hope that this would help to define its physical character and ultimately reveal its cause.

On this issue Sedgwick diverted his attention temporarily from the geognosy of the old Primary rocks, and sprang to the defense of his colleague Buckland. He affirmed the reality of the distinction between the two classes of Superficial deposits: the more recent "Alluvial" deposits were clearly the products of ordinary actual causes, whereas the older "Diluvial" deposits were surely not. He concluded that there really had been an exceptional diluvial event or period in the geologically recent past. Whether it was recent enough to be equated with the biblical Flood, as Buckland claimed, was something that Sedgwick was content to leave to future research to decide.

The valleys in mountain regions were widely cited as evidence for the reality of the diluvial event, for in many cases their forms seemed incompatible with what was observably produced by ordinary erosion. But they were not decisive, since some geologists continued to claim, as they had since the previous century, that—given enough time—even the deepest valleys could have been excavated by the streams and rivers that still flow in them.

Far more telling evidence for the exceptional nature of the "last revolution" was that of the erratic blocks widely scattered around western Europe, which had unquestionably erred and strayed tens or hundreds of miles from their obvious source areas. De la Beche, following up on von Buch's latest fieldwork, brought to the attention of other geologists the sheer size of some of the Alpine erratics, and the fact that their distribution on the southern flanks of the Alps mirrored what had long been known on the northern side. This suggested, to De la Beche and others, the possibility of a causal link with the recent sudden upheaval of the Alps, as proposed by Élie de Beaumont. But even if such an explanation were valid for the Alpine erratics, it signally failed to cover the far more widespread erratics found on low-lying ground in northern Europe, many of which could be traced back to rocks known in situ in Scandinavia. Razumovsky, Brongniart, and Hausmann were among those who plotted their strange distribution, inferring from their respective fieldwork that the enigmatic causal event had moved the erratics southward from Scandinavia, across Sweden and Finland, over the Baltic onto the plains of northern Russia and Germany, and even into the Netherlands.

19. Esmark, "Jordklodes historie" (1824); his earlier, mineralogical publications were *Reise durch Ungarn* (1798) and "New ore of tellurium" (1816). Worsley, "Jens Esmark" (2006), gives a fine historical account and modern interpretation of his first locality (about 25km east-southeast of Stavanger), where the moraine is now known as the Vassryggen [lake-ridge]; the other is on the island of Gula, south of Sognefjord.

20. Esmark, "Jordklodes historie" (1824), 48–49; Whiston, *New theory of the earth* (1696).

21. Esmark, "Geological history of the earth" (1826); Jameson printed this—as the first in an intended series giving various "opinions on the formation of the earth"—immediately following his own "General observations" (1826) on the recent discovery of tropical-looking fossil corals and coal plants in the high Arctic (§12.3). By about 1827 he was mentioning in his lectures the possibility that there had once been similar glaciers in Scotland: see Davies, *Earth in decay* (1969), 267–70, and this volume, §35.3.

Razumovsky attributed the erratics to an aqueous event of huge power and magnitude; Brongniart and Hausmann declined to speculate on what the event might have been, noting only that no existing explanation was satisfactory. But Bigsby's report of erratics on the shores of Lake Huron showed that the enigma was not confined to Europe, and De la Beche even suggested that the gravels he had seen in Jamaica were also "diluvial" in origin.

The cumulative weight of evidence in favor of the reality of the last revolution in geohistory, and its definition as a widespread "diluvial" event *of some kind*, therefore seemed to most geologists to be overwhelming. One of the few alternative explanations, attributing the effects not to water or mud but to ice, was put forward by Esmark in Norway. But he was peripheral in both geography and language; his local evidence for a formerly greater extension of the Norwegian glaciers was embedded in an unattractively speculative geotheory; and above all such ideas flew in the face of the increasing evidence that in the deep past the earth had been hotter, not colder, than the present. So it is not surprising that Esmark's conjecture had little or no impact on the debate.

The next chapter complements this one by describing what was widely taken to be growing evidence for the reality of a geologically recent "revolution", based this time on how it seemed to have affected the living world.

CHAPTER 14

The last mass extinction (1826–31)

14.1 BONE CAVES FOR BUCKLAND

Cuvier remained the obvious point of reference for any further research on the impact of the putative recent "revolution" on living organisms. He was no biblical literalist; but his identification of the geological event as the "deluge" obscurely recorded by many ancient cultures (the ancient Jewish being just one) did lay him open to having his authority exploited by those with more traditional agendas than his own. In this respect Buckland, one of his most ardent followers, continued to take a characteristically English and Anglican middle way. Like Cuvier, he vigorously repudiated any literalism about the Creation story and the short cosmic timescale allegedly derived from it; yet he was more literal than Cuvier in his interpretation of the Flood story and its putative physical traces. After he published his *Reliquiae* (§6.3), Buckland worked hard to collect further evidence to support his claims that the "geological deluge" had been not only recent enough to be equated with the biblical Flood, but also worldwide in extent and massive enough to have topped the highest mountains. Some geologists, for example Fitton, agreed that the traces of diluvial action were remarkably widespread, but questioned whether they represented a single episode rather than an extended period of an unusually turbulent character (§6.4); and others sided with Cuvier in suspecting that the effects had been confined to relatively low-lying areas, as might be expected even if the mega-tsunami had been very mega indeed.

Anyway, the evidence for the global reach of the diluvial effects became progressively stronger during the 1820s. The evidence within Europe, above all in the form of erratics, was striking enough, and it was vastly extended by reports of diluvial deposits in remote regions, such as Bigsby's from Lake Huron and De la

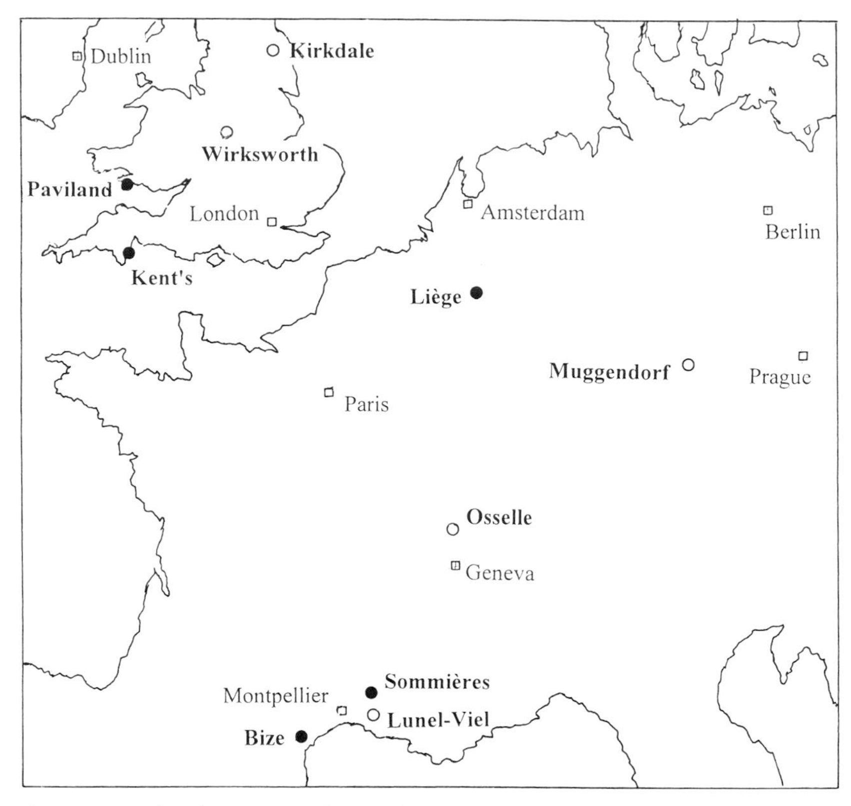

Fig. 14.1 A map of northwest Europe, showing the locations of some of the bone caves excavated in the 1820s and 1830s and cited in discussions about the putative diluvial event. Open circles mark those in which animal bones were found; solid black spots, those in which *human* bones and/or artifacts were also found, and which were cited in discussions of the antiquity of the human species (see Chaps. 16, 29). Like all important debates in geology at this time, this one was highly international, and Buckland—the acknowledged authority on the British caves—made a point of seeing for himself many of the relevant caves on the Continent.

Beche's from Jamaica (§13.3). But even these were outshone by new discoveries of the fossil remains of the fauna that had apparently been wiped out by the diluvial event. Buckland took the lead in searching for further evidence of the immediately "antediluvial" fauna in caves and fissures on the Continent, to match those he had described in Britain, and to add to the already well-known caves around Muggendorf and elsewhere in the German lands (Fig. 14.1).

Early in 1826, not long after Buckland married his research assistant and as soon as he had completed his stint as president of the Geological Society, he—like Scrope before him—took his honeymoon in the form of a lengthy Continental field trip. He began, of course, with a visit to Paris. He had already introduced his future wife to Cuvier as "my friend Miss Morland whose name and Drawings have long been familiar to you", and as Mrs. Buckland she was duly treated with respect as a naturalist in her own right. From Paris, the English couple embarked on a Grand Tour that extended as far south as the classic sights and sites, both cultural and geological, of Campania and Sicily, taking in both Vesuvius and Etna; while in Palermo, Buckland—an Anglican of strongly Protestant sympathies—took evident delight in debunking bogus Catholic pieties, identifying the venerated relics of St. Rosalia as goats' bones, and some periodically miraculous effusions of blood as nothing but bats' urine.[1]

On their return journey, the Bucklands crossed into France to meet Marcel Pierre Toussaint de Serres de Mesplès (1780–1862) in Montpellier. Serres, a lawyer but also the professor of mineralogy and geology at the university, was studying the extensive Tertiary formations and fossils in his native region. However, emulating Buckland's already well-known English research, Serres was also investigating caves and fissures with fossil bones, and encouraging younger local naturalists to look for more. Reporting on his research in the Parisian *Annales des Sciences Naturelles* shortly after the Bucklands' visit, he noted that the caves and fissures of Languedoc contained the bones of a mixture of familiar and exotic animals, together with freshwater shells of still living species. He inferred that they had all been swept in, or overwhelmed on the spot, by some kind of exceptional aqueous current from the north or northeast. This conclusion would have been music in Buckland's ears. At the same time, however, Serres softened any contrast with earlier geohistory by treating these deposits as just the last of the freshwater Tertiary formations that were the main focus of his research.[2]

One young local naturalist, whom Serres noted as having "constantly assisted" him, was Jules de Christol (1802–61), who with Serres's encouragement was excavating a newly discovered bone cave at Lunel-Viel near Montpellier. Buckland visited the cave, watched Christol at work and was impressed by his scrupulously careful methods. After his return to England, Buckland described the Lunel-Viel cave at the Geological Society and recorded his approval of what Christol was doing there. He concluded that, like Kirkdale, it had been an antediluvial hyena den; his brief account appeared in the very first issue of the Society's new *Proceedings* (which henceforth offered rapid publication in summary form within a few weeks, obviating the often long delays before papers were printed in full in the *Transactions*). Christol later published his own account in Paris; like Buckland, he interpreted the Lunel-Viel cave as a den of hyenas, but of a new and even larger species that those at Kirkdale; there were chewed bones, excrement, and all the now well-recognized signs that the animals had lived on the spot until they were overwhelmed by the diluvial event.[3]

Christol's cave therefore took its place as a further addition to Buckland's lengthening tally of antediluvial caves known all across Europe. At a later point

1. Buckland to Cuvier, 5 May 1825 [Paris-IF: MS 3247/4]; Gordon, *Buckland* (1894), 95–96, and Boylan, *Buckland* (1984), 155, 169, 389.

2. Serres, "Cavernes à ossemens" (1826), published in the October issue of *ASN* and probably submitted soon after Buckland's visit. Serres, "Terrains d'eau douce" (1818), had first applied to Languedoc the kind of analysis of Tertiary formations pioneered by Brongniart and Cuvier in the Paris Basin (*BLT* §9.1).

3. Buckland, "Cavern of Lunel" (1827), read on 17 November 1826 and published in *Proceedings* early in 1827. Christol and Bravard, "Nouvelles espèces d'hyènes fossiles" (1828), read at the Société d'Histoire Naturelle on 8 February; his co-author the young Auvergnat geologist Pierre Josephe Auguste Bravard (1803–61) had been trained at the School of Mines in Paris. Serres *et al.*, *Ossemens de Lunel-Viel* (1839), a later full account, made no mention of Christol's part in the research; by that time he and Serres had apparently fallen out, and Christol had left Montpellier (see §28.1). Lunel-Viel is near Lunel (Hérault) and about 20km east of Montpellier. The *Proceedings*, which was instituted during Fitton's presidency and published several times a year, reflected the rapidly accelerating pace of research among English geologists: see Rudwick, "Geological Society's *Journal*" (1995).

on their journey, the Bucklands stopped near Besançon (on the French side of the Jura hills) to see the Osselle cave, in which the spectacular stalactites had long been a tourist attraction. Buckland persuaded those in charge to allow him to break through the stalagmitic layer on the floor of the cave (though not in the largest chamber, where the other visitors who were wining, dining, and dancing might have objected); for he made the risky prediction that deposits with fossil bones would be found beneath it, just as they had been at Kirkdale. The prediction was vindicated spectacularly. The dramatic event matched Cuvier's famous and similarly predictive stagings, many years earlier, of his excavation of a little fossil opossum from a slab of Paris gypsum (*BLT* §7.5) and of the "man a witness of the Deluge", the famous fossil salamander in the Haarlem museum (*BLT* §9.3); those who witnessed the cave excavation were equally impressed. After his return Buckland announced that the Osselle cave had been a den of antediluvian bears, matching those long known around Muggendorf in Bavaria (*BLT* §10.6). And he published in Jameson's periodical an account of a den of *living* hyenas, sent at his request by a British officer serving in India; this supported his interpretation of Kirkdale Cave on the impeccable grounds of a close comparison between past and present, and refuted Fleming's skepticism about it.[4]

14.2 BUCKLAND'S WORLDWIDE ANTEDILUVIAL FOSSILS

It was not only in Continental Europe that Buckland worked hard to collect further evidence in support of his "diluvial" interpretation of recent geohistory. In 1828, only a year after he reported on the French caves, he claimed to have good evidence for the antediluvial fauna in both tropical and arctic regions, which, in combination, strongly suggested—at least to him—that the event had indeed been global in its impact.

Both these new discoveries were incidental products of British imperial strategy. In 1826, after the end of the first Anglo-Burmese war, a diplomatic mission under John Crawfurd (1783–1868), the governor of the East India Company's trading base at Singapore, penetrated some five hundred miles up the Irrawaddy—it was the first time that a steamboat was used for such a purpose—to make a commercial treaty with the King of Burma at his court at Ava (near Mandalay). By the time the protracted negotiations were concluded the level of the river had fallen to its seasonal low, and about halfway back to Rangoon the boat ran aground. While it was being repaired and refloated, Crawfurd, a physician by training but also a member of the Geological Society (and a scholarly orientalist), explored the surrounding area, which was already well known for its petroleum wells. On the river bank, and scattered on the surface of the land nearby, he found abundant petrified fossil bones. Seven large chests full of specimens were shipped back to London.[5]

At the Geological Society the Burmese fossils created a sensation: as Lyell told Mantell, "to say that they surpass in value any collection ever brought to Europe, from any other quarter of the globe, is to say little". Clift—acting in effect as the English Cuvier—identified a rich assemblage of animals such as mastodons, hippopotamus, and rhinoceros (some of them new species), but also turtles and

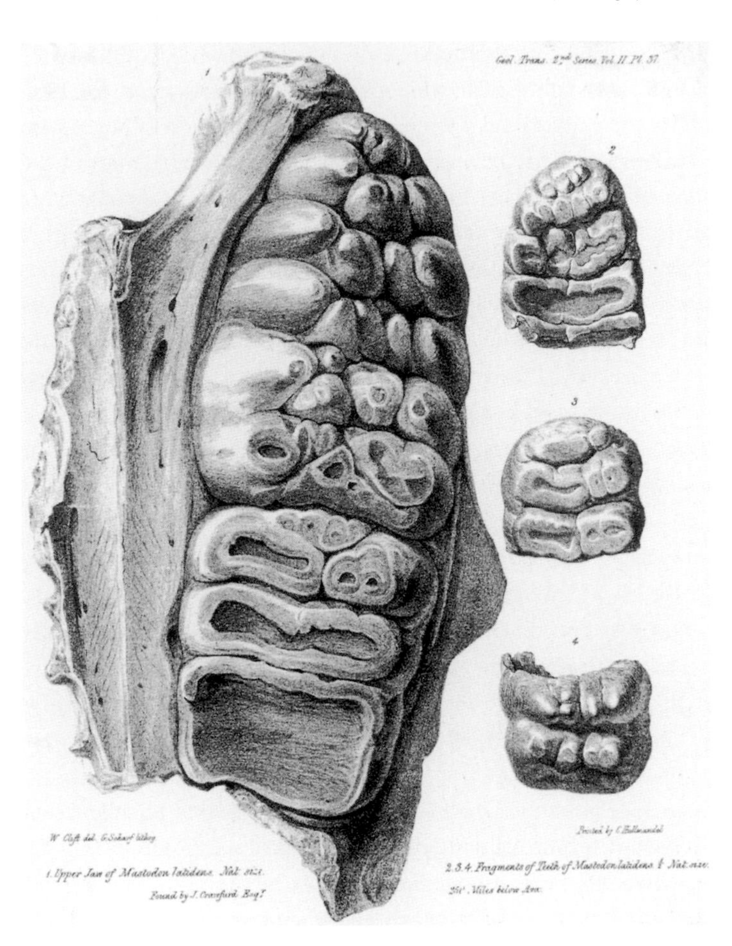

Fig. 14.2 Molar teeth of *Mastodon latidens*, a new species described by Clift in 1828 from the fossil collection shipped back to London by John Crawfurd from the banks of the Irrawaddy in Burma. In the largest specimen (about 26cm long) the teeth are preserved in a fragment of the upper jaw (the other teeth are depicted at reduced size). This and other typically "antediluvial" animals had lived in a tropical latitude, which was taken by Buckland to reinforce his claim that the diluvial effects had been worldwide.

gavials; there was no trace of the region's present mammalian fauna. "Buckland reconciled all to his diluvian hypothesis, as what facts would he not?", Lyell reported sarcastically; but, he added, "be his theory wide of the mark or not, he is always worth hearing". Whether or not Buckland's diluvial theory accounted for them, the Burmese fossils certainly showed that in geologically recent times a fauna of quadrupeds quite distinct from the present one had flourished in tropical Burma, just as it had in temperate Europe (Fig. 14.2).[6]

4. Buckland, "Grotte d'Osselles" (1827), read at the Geological Society on 20 April, published in summary in *Proceedings* but fully in French; Buckland, "Dens of living hyaenas" (1827), dated 5 March. Taquet and Contini, "Buckland et le mégalosaure" (1997), 96, notes the parallel with one of Cuvier's demonstrations. The Osselle cave is in a hillside above the Doubs, about 18km southwest of Besançon; it remains a major tourist attraction, and still gives a sense of the aesthetic and emotional impact of such caves, perhaps better than any of the others mentioned in this narrative.

5. Crawfurd, *Brief narrative* (1827), 15, and *Embassy to Ava* (1829), 325–30. The bones were found on the left (east) bank of the river, north of Magwe; the petroleum wells (some over 200 feet deep) were forerunners of those that later turned Burma (now called Myanmar by its military junta) into a major oil-producing region.

6. Fig. 14.2 is reproduced from Clift, "New species of mastodon" (1828), pl. 37; see also Buckland, "Voyage up the Irawadi" (1828), both read on 18 April, the latter reprinted in Crawfurd, *Embassy to Ava* (1829), appendices, 78–88 and pls. Lyell to Mantell, 17 January, 17 February 1828 [Wellington-ATL], written after the specimens were first exhibited.

This diluvial fauna from the tropics was soon matched by one from the Arctic. Later in 1828 a British naval expedition returned from the region of the Bering Strait, at the Pacific end of what was hoped would be a navigable Northwest Passage around the top of North America. It had been sent there to meet either or both of two other expeditions with the same objective. Parry was trying for the second time to find a passage westwards from the Greenland Sea, through the maze of channels between the Arctic islands; and John Franklin was traveling overland down the Mackenzie River to its mouth, from which he planned to map the northern coastline towards Alaska. In the event, both parties were forced to turn back. But while the relief ship was waiting for them, its scientifically minded commander Beechey—who as a junior officer had been on Parry's earlier expedition (see Fig. 7.6)—explored the northern coast of Alaska, over which the Russian government had been asserting its sovereignty with renewed vigor.

In Eschscholtz Bay at the end of Kotzebue Sound—both named after the Russians who had earlier explored the region—the surgeon from Beechey's ship collected abundant fossil bones from a locality that the Russians had already reported. However, the Englishman found that the bones were not embedded in a grounded iceberg, as they had supposed, but in a deposit of frozen mud: a situation not unlike that of the celebrated frozen mammoth and rhino carcasses in subarctic Siberia (*BLT* §5.3, §10.1). Although there were no carcasses here, Beechey recognized that the bones were similar to those that Buckland had already described before he himself left England. With the geologist in mind—he named the nearby river after him—Beechey brought back a fine collection of the bones, and at his suggestion the Admiralty asked Buckland to report on them. Buckland duly confirmed that they represented much the same "antediluvial" animals that he had described from Kirkdale Cave and elsewhere in England (*BLT* §10.6); but here he attributed the total extinction of the fauna to a sudden cooling of the climate in geologically recent times, "attended by a general inundation" (the bones of musk oxen, a subarctic species still living in the region, were discounted as having been mixed on the beach with the truly fossil bones). Buckland dismissed any human role in the extinction as highly unlikely in such an inhospitable environment.[7]

A third discovery was more ambiguous, but nonetheless potentially important because it came from an even more remote part of the world. Thomas Livingstone Mitchell (1792–1855), the surveyor to the government of New South Wales, collected fossil bones from fissures near Wellington, in a remote inland part of the colony. Specimens were sent to the Geological Society (of which Mitchell was one of the most distant members) and to Jameson, and the latter sent some on to Paris. Cuvier and Clift both identified kangaroo, wombat, koala, and other marsupials similar to the living species of the indigenous fauna. But two large bones were anomalous. Clift attributed one to something like a dugong, the other perhaps to an elephant, or anyway to an animal far larger than any known alive in Australia. As Buckland pointed out, the latter made a strange contrast to the present Australian fauna: it was at least a hint that the "diluvial" phenomena might be found, with further exploration, even at the antipodes.[8]

14.3 FLEMING AND THE COURSE OF EXTINCTION

Following Cuvier's lead, Buckland took it for granted that only a sudden physical change could have wiped out a whole fauna of well-adapted animals: it must have been a *mass* extinction, caused by a violent inundation of some kind, or by a sudden change of climate, or by both. However, like the parallel assumption that the diluvial event had been as recent as the biblical Flood, the inference of a sudden mass extinction was beginning to be questioned by other naturalists. Prominent among them was Fleming, who had already attacked Buckland's diluvial argument, forcefully and in print (§6.4). More importantly, however, Fleming claimed that the remains of the animal victims of the alleged deluge were found just as commonly in "alluvial" materials such as peat, lake marls, and ordinary river deposits as they were in the supposedly "diluvial" gravels. Turned into geohistory, this suggested that at least some of the extinct species had survived into what Fleming assumed was the very recent period of human existence. So he argued that the extinctions had been caused primarily by the hunting activities of early humans, perhaps accelerated by natural changes in the environment, including climatic change. This inference was founded explicitly on the usual actualistic principle, treating the present—or rather, the recent past of human history—as the key to the deeper past. Fleming extrapolated back into deep time what was known from historical records about the *local* extinction, within Britain and in recent centuries, of species such as the bear and the wolf, as a result of *human* activities such as hunting and the clearing of forests. So in Fleming's view no drastic deluge need be invoked: the extinctions had been gradual and piecemeal, and largely if not entirely due to the arrival of the human species on the local scene.

The leading members of the Geological Society in London tended to dismiss Fleming's criticisms as irritating pinpricks (§12.3). But the Scottish naturalist, like some of his counterparts in English rural parishes, was highly competent, and he was making a serious point. He was questioning an important assumption that was currently being made without adequately critical assessment. Cuvier and other naturalists, having fought hard for extinction to be recognized as a real feature of the natural world (or having at least accepted that it was), tacitly assumed that it was an all-or-none effect. Since any given species must be either extinct or extant, it seemed natural to attribute a change in its status to some decisive event.

7. Beechey, *Voyage to the Pacific* (1831), 322–24; Buckland, "Elephants in Eschscholtz Bay" (1831), 605–12; Levere, *Science and the Canadian Arctic* (1993), 63–84, 110–25, describes the expeditions. Fitton, "Address to Geological Society" (1829), 124–25, read 15 February 1829, reported Beechey's gift of specimens to the Geological Society; Buckland first saw them in October 1828, and his opinion on them was widely known, long before his report was published as an appendix to Beechey's official account. Eschscholtz Bay is at the inner end of Kotzebue Sound and about 300km east of Bering Strait.

8. Cuvier's identifications were reported by Pentland to Société Géologique on 4 April 1831 [*BSGF* 1: 144–45]; Mitchell, "Caves at Wellington Valley" (1831), dated 14 October 1830 and read at the Geological Society on 13 April 1831 (his gift of specimens was recorded at the same time). Mitchell was later a distinguished explorer of the interior of Australia (and was knighted for his efforts). Wellington is about 250km northwest of Sydney; the caves (now accessible as a tourist attraction) later yielded many more bones of the anomalous animal, which was then assigned to an extinct giant marsupial, *Diprotodon*.

However, Fleming recognized that there might be many steps on the road to extinction: a gradual reduction in numbers and in geographical range, a dying out in specific areas, and so on. The wolf and the bear, for example, were extinct in Britain, but not yet in the rest of Europe. The final extinction of a species, the demise of its very last individuals, might therefore be an inconspicuous event, local and unnoticed. There was currently some debate, for example, about whether even the dodo—which had long been the very icon of extinction (*BLT* §5.1)—might not still survive in small numbers on Mauritius, unobserved in some remote part of the island. On Fleming's model, extinction might be a gradual and piecemeal process; it would have a natural and almost everyday explanation (counting human activities in this context as natural and everyday), without any need to invoke equally natural but exceptional events such as the putative deluge. If Fleming's argument were accepted, it would clearly undermine the plausibility of any diluvial interpretation of recent geohistory.[9]

This blurring of the boundary between the "diluvial" period and its "alluvial" aftermath—questioning Sedgwick's recent insistence on the sharp distinction between their respective deposits (§13.1)—depended in part on finding, as Flem-

Fig. 14.3 An almost complete skeleton of the "Irish elk" (*Cervus megaceros*), found in 1824 near Limerick in a marl layer underlying peat. It was donated by the archdeacon of Limerick to the Royal Dublin Society; assembled by the Irish surgeon John Hart; and described and illustrated by him in a booklet published in 1826. The antlers measured 9 feet 2 inches [almost 3m] across, far larger than those of a moose or any other living species. The less reduced drawing (fig. 2) showed the structure more clearly: the single antler measured 5 feet 9 inches from skull to tip. The separate bone (fig. 3) was a rib from another individual, with a small hole that Hart, with all his authority as a surgeon, identified as a healed injury: he suggested that it had been caused by a sharp arrowhead, which he took as evidence that this extinct mammal species had been hunted by early humans in Ireland. (By permission of the Syndics of Cambridge University Library)

ing claimed, that the parallel distinction between the "antediluvial" and "postdiluvial" faunas was less clear-cut than Cuvier and Buckland maintained. Cuvier had conceded long before that there were indeed ambiguous cases: notably, the bones of apparently identical ruminants were found in both kinds of deposit (*BLT* §10.1). But he had good reason to argue that, given what was known about living ruminants, the fossils might not be as anomalous as they appeared, because the differences between living and extinct species might not be expressed in the skeletal features that alone were preserved in fossils. Cuvier had also complained, equally reasonably, that many of his specimens had reached him without adequate information about their precise stratigraphical position.

One particular ruminant species, however, was so distinctive that it might well prove decisive. The bones of the gigantic "Irish elk" were known to come from beneath the Irish peat bogs (and from similar situations elsewhere in Europe). It was distinct from any known living species, so Cuvier inferred that it belonged, with the gigantic mammoth and other such animals, to the megafauna that had been wiped out at the last "revolution". Irish naturalists, spurred by pride in the spectacular fossil mammal named after their own country, had intensified their hunt for new specimens. They concluded that the fossils came from deposits that indeed underlay the blanket of peat, but might nonetheless be "alluvial" in character rather than "diluvial". For example, an almost complete set of bones was found near Limerick in 1824 in a marl with freshwater shells, much like the clearly postdiluvial deposit that Lyell described from Scotland around the same time (see Fig. 10.2). The Irish surgeon John Hart (1797–1872), who assembled the bones and published an account of the huge Limerick skeleton, also described another specimen that he took to be strong evidence that the species had been hunted by early humans. This clearly supported Fleming's argument, although Hart himself did not explicitly draw that conclusion. Here was an unquestionably extinct species that had not been the victim of any diluvial event, and which had apparently survived into the human period and might have perished finally at human hands (Fig. 14.3).[10]

14.4 LYELL THE BUDDING SYNTHESIZER

At just this time, Lyell began to emerge, like Prévost and Fleming, as a forceful critic of Cuvier's and Buckland's diluvial theory and a champion of the explana-

9. Fleming, "Geological deluge" (1826) and *British animals* (1828), preface (dated 27 December 1827); Rehbock, "John Fleming" (1985); and Burns, "John Fleming" (2007). On the case of the dodo, see Cuvier, "Note sur quelques ossemens" (1830), and Blainville, "Mémoire sur le dodo" (1835), read at the Académie on 12 July and 30 August 1830 respectively; the latter challenged Cuvier's insistence on the extinct status of the bird (and, perhaps not coincidentally, remained unpublished until after Cuvier's death); Geus, "Animals extinct in historical time" (1997). Grayson, "Pleistocene extinctions" (1984), usefully sets Fleming's ideas in a wider context of later theorizing about "human overkill".

10. Fig. 14.3 is reproduced from Hart [printed in error as "Part"], "Daim fossile d'Irlande" (1826) [Cambridge-UL: T382.b.1.3], pl. 39, a lithograph redrawn by Meunier from the rather poor one in Hart, *Fossil deer of Ireland* (1825), pl. 2. The prompt translation in *Annales des sciences naturelles* made the find widely known throughout Europe, and may have been more easily accessible—even to British geologists—than the original local booklet. The best previous specimen came from the Isle of Man and was on display in Edinburgh; Hart emphasized that his new one was even larger. See Cuvier, *Ossemens fossiles*, 3rd ed. (1825) 4: 77–85.

tory power of actual causes. Yet this did not stop him expressing his admiration for Cuvier's famous "Discourse", in contrast to Jameson's editions of it in English (§1.3), when he sent the French naturalist his own article on Scottish lake marls (§10.2):

> Your octavo 'Discours sur les revo[lution]s' &c is beginning to be known [i.e., in Britain] but I wish an English Translation under your direction could come out. If there was some slight Appendix of your own containing the latest discoveries so as to make it an original edition it would certainly sell well, & do more for the advancement of Geology than any thing which has come out since the Essai Prelim[inair]e [i.e., Essai géognostique, *BLT* §9.2]. That Essay was never known as it should have been as being attatched [*sic*] to so expensive a work [as *Fossil Bones*]. Although French is universally studied here, hundreds of geological and literary readers will lose the information, as long as it is locked up in a foreign tongue; and if Jameson publishes or rather manufactures another of his garbled statements the work will lose half its consequence & real value & popularity, because its genuineness will be suspected.[11]

Although still in his twenties, Lyell was becoming prominent in scientific circles both at home and abroad. His earlier visit to Paris had made him well known to Cuvier and other savants there; he planned to spend time in Dresden, to add German to the languages in which he knew he needed to be fluent; and under the tutelage of his father, a Dante scholar, he was probably already fluent in Italian. In England too he was becoming well known, not least as a result of having joined the Athenaeum soon after that most intellectual of the London gentlemen's clubs was founded. Here the young bachelor not only enjoyed a "genteel elegantly served lunch for s2.6d [12p] with all the newspapers", but also met some of the most intelligent of the London social elite.[12]

In 1824 John Gibson Lockhart (1794–1854), the new editor of the *Quarterly Review*, had renewed his predecessor's invitation to Lyell to contribute to that prestigious periodical, which was read with close attention by leading Tories (Lyell's political sympathies were known to be with the Whigs, but this was no disqualification). In the next three years, Lyell's lengthy essays, on topics of his own choice, gave him an influential platform from which to define his own conception of the sciences in general and of geology in particular. And Lockhart paid him well, which was more than welcome to the fledgling barrister, whose expenses far outran the fees he earned.[13]

In his very first essay, reviewing plans for the new and deliberately secular University of London (now University College London), Lyell strongly supported this attempt to break the Oxbridge and Anglican monopoly on English higher education. His next essay again revealed his reformist political sympathies: he used the recent publications of several provincial institutions as a peg on which to hang a general criticism of the low level of British public interest in the natural sciences, and of state support for them, in comparison with the Continental countries. For example, he asked why it was—given the wealth and worldwide commercial activities of the British—that their museums were "so decidedly inferior not only to those of France, but of several petty states of Italy and

Germany?" Noting the distinguished publications emanating from the Muséum in Paris, he added that "it is humiliating to acknowledge, that no Englishman could even now be the author of similar works, without access to museums such as exist not in his own country". But he gave credit where credit was due, notably to the Geological Society: "the labours of this society, which has ever cultivated geology as an inductive science founded on observation, have tended much to remove the discredit cast upon the study by the wild speculations of earlier authors". In thus rejecting traditional geotheory, Lyell faithfully followed the line laid down by the founders of the Society under Greenough almost twenty years earlier (*BLT* §8.4).[14]

Lyell's third essay focused on his favorite science, using as its peg the latest complete volume of the Geological Society's *Transactions.* His essay was a fine piece of *haute vulgarisation* or high-level popularization. Lyell was well aware that most of his readers, although intelligent and well educated, would not be geologists; and he displayed his credentials as a cultured gentleman by peppering his text with quotations (in the original languages, of course) from literary sources such as Horace, Pliny, Ariosto, Milton, and Shakespeare, as well as those of current political importance such as Talleyrand. "I must write what *will be read*", he told Mantell; and he wanted to be read by just those whom he had defined (when writing to Cuvier) as "[both] geological and literary readers". Emulating his hero's *Discourse*, Lyell's ambition was to write what would not only contribute to debates among savants, and particularly geologists, but also reach the influential social and political elites. His summary of recent research by the Geological Society, like most of his subsequent work, was aimed at both these distinct (though overlapping) audiences. It was in effect his first public attempt to synthesize what the science had so far achieved and where its future efforts should be directed.[15]

Lyell again contrasted geology's earlier reputation for speculative geotheory with the solid results achieved in recent years. Drawing a parallel with human

11. Lyell to Cuvier, [1826] [Paris-MHN: MS 627, no. 68], undated but soon after publication of Cuvier, *Discours sur les révolutions* (1826), and Lyell, "Freshwater limestone" (1826). Lyell's admiration for Cuvier's work was, of course, perfectly compatible with his criticisms of specific aspects of it: only a crude and simplistic historiography depicts them as being diametrically opposed on "uniformity" or anything else.

12. Wilson, *Lyell* (1972), 318; Morrell, "London institutions" (1976).

13. Lyell (K.), *Lyell* (1881), 1: 160–71; Wilson, *Lyell* (1972), 143–44; in the event he did not go to Germany, but he made the effort to learn the language nonetheless (see below). In Italian he was fluent enough, by 1828, to write a review of Gabriele Rossetti's new Italian edition of Dante's *Divina commedia* for the *Quarterly*, although it was never published (Wilson, 187–90). All his essays were anonymous, as was customary in the *Quarterly*, but in practice his authorship was well known in elite circles; his work did not become so well known outside Britain, however, partly because articles in the *Quarterly* were no longer translated for the Genevan *Bibliothèque universelle* as frequently as they had been in earlier years.

14. Lyell, "Scientific institutions" (1826), 155–63. On his first essay, see Wilson, *Lyell* (1972), 144–46.

15. Lyell, "Transactions of the Geological Society" (1826), reviewed the first volume (1822–24) of the new series, in which engravings were for the first time largely replaced by the much less expensive (and often more effective) medium of lithography. The first fascicle (1826) of the second volume had already been published, with Lyell's own "Freshwater limestone" (§10.2) in it, but this volume was not completed until 1828 and was therefore not included in the review. The quoted phrases are in Lyell to Mantell, 22 June 1826 [Wellington-ATL], and to Cuvier, [1826] [Paris-MHN: MS 627, no. 68].

historiography, Lyell emphasized that geology too was based on continual comparison "between what now is and what has been"; this actualistic method was essential to a science which, "although it cannot appeal to demonstrative truth . . . is constantly concerned in weighing a great mass of *probable* evidence". Lyell's perceptive statement of the probabilistic basis of geohistory was strikingly parallel to the legal modes of argument that he was learning to practice in the courtrooms of southwest England. He recognized that reconstructing geohistory, like assigning criminal guilt or innocence where it rightly belonged, depended on building a plausible *case* that would stand up against alternative interpretations of the evidence surviving from the deep past (or from the scene of the crime and its witnesses).[16]

Lyell's claims for the explanatory value of actual causes was based on his conviction that they were still inadequately known. They might *not* be—in the words of Buckland's famous inaugural lecture (*BLT* §10.5)—merely "the last expiring efforts of those mighty disturbing forces that once operated". Lyell respectfully questioned Cuvier's similar assumption, on the grounds that recorded history was too brief to judge the adequacy of actual causes over the far longer periods of geohistory. But even within that short timespan, geological processes had greater explanatory power than had yet been recognized. For example he noted Maria Graham's report that in the great Chilean earthquake of 1822 the land around Valparaiso had risen abruptly by several feet (§8.3); and he suggested that such movements, repeated over the vast spans of deep time, might in fact have elevated whole mountain ranges. Lyell cited evidence, also from the *Transactions*, for similar movements that were recent at least in a geological sense, such as Strangways's description of an apparent former sea link between the Caspian and the Black Sea, which might have been cut by a slight rise in the low-lying region between them. Such examples, and many more, served to show that physical processes observable in the present world might be more potent, and their effects more extensive, than had often been allowed. Lyell's prescription for an actualistic policy for geology was couched in terms that unmistakably echoed those just used by Scrope (§9.3):

> In the present state of our knowledge, it appears premature to assume that existing agents could not, in the lapse of ages, produce such effects as fall principally under the examination of the geologist. It is an assumption, moreover, directly calculated to repress the ardour of inquiry, by destroying all hope of interpreting what is obscure in the past by an accurate investigation of the present phenomena of nature.[17]

Not surprisingly, the main focus of Lyell's essay was on the many "recent and splendid discoveries" of important fossils. He gave a lengthy description of the sensational Secondary reptiles, for which English geologists could take particular credit (Chaps. 2, 5). More generally, he described how the close study of the previously neglected Secondary and Tertiary formations had disclosed the "various and extensive revolutions" that had led towards the present. The language of Cuvierian geohistory was pervasive: the remains of the Irish elk, for example, were "evidently of origin posterior to the last extensive revolution which modified the surface of the land".

Lyell explained that the reality of extinction was now accepted "by all naturalists", because they had decisively rejected the classic alternatives of transmutation and of the survival of "living fossils" (*BLT* §5.1, §7.4, §8.3). His review of the history of life as a whole likewise reflected the current consensus among geologists (Chap. 12). He noted the total absence of human fossils among even the most recent of extinct mammals. It was more surprising, he suggested, that no trace of the *Quadrumana* (in modern terms, non-human primates) had been found, for it had seemed a sound generalization that the fossil record showed "a gradual and progressive scale" from simple to more complex forms of life. "And such is still the general inference to be deduced from observed facts", Lyell added, while hinting that some exceptions—he was probably alluding to the little marsupials from Stonesfield (§5.5)—should caution the geologist not to rely on it uncritically. In short, Lyell, like his colleagues, expressed no serious doubt that the history of life had been broadly *directional* and indeed progressive in character.[18]

As already mentioned, Lyell was inclined to suspect that the earth's "revolutions" had been as powerful in relatively recent geohistory as in the deepest past. However, on the most recent of all such events, the putative "deluge" that might have caused the last mass extinction, he remained prudently silent. But he assigned "great probability" to Cuvier's suggestion that the Tertiary mammals of the Paris gypsum had perished in an earlier marine incursion (*BLT* §9.1); tacitly he rejected Prévost's more gradual kind of explanation of this "revolution" (§10.3). He alluded to Webster's description of a huge fold in the Isle of Wight and to Brongniart's report of even more enormous folding in the Alps (*BLT* §9.4, §9.5); both cases showed that "the disturbing force continued unimpaired even subsequently to the formation of some tertiary deposits, [so that] those geologists who contend it is now in the wane must reason from a very limited number of facts indeed". However, Lyell denied that these powerful events had been worldwide, seeing them instead as products of local or regional movements of the earth's crust, and therefore as vindications of Playfair's "Huttonian" view of crustal mobility in general. So Lyell, like Cuvier and many others, pictured geohistory as dynamic and diverse: "there are proofs of occasional convulsions, but there are also proofs of intervening periods of order and stability".[19]

However, Lyell conceded that, superimposed on this diverse geohistory, there were signs of overall directional trends. For example, he noted that Scrope and Daubeny, in their recent books (§9.3), both agreed that "the effects produced

16. Lyell, "Transactions of the Geological Society" (1826), 507–9.

17. Lyell, "Transactions of the Geological Society" (1826), 518; Buckland, *Vindiciae* (1820), 5; Strangways, "Geology of Russia" (1822), 37–38 and pl. 2.

18. Lyell, "Transactions of the Geological Society" (1826), 510–13. Later in the essay (529–32) he discussed the Stonesfield fossils, dismissing Prévost's doubts about them, as he soon told him: Lyell to Prévost, 23 February 1827 [Edinburgh-UL: AAF 8].

19. Lyell, "Transactions of the Geological Society" (1826), 518. The word "proofs" referred, as usual, to *evidence*, the analogy being legal rather than mathematical.

at present by earthquakes and volcanoes are at least analogous in kind, if inferior in degree, to those that have resulted from similar agents at remote aeras". Less equivocally, he conceded that the many fossils of tropical appearance in the Secondary and Tertiary formations of Europe seemed "to confirm that striking deduction of geology, that the former temperature of the northern hemisphere was much higher than it is at present". He had heeded Fleming's scepticism about this inference (§12.3), recognizing that it depended on there being a valid analogy between, for example, the totally extinct ammonites and their closest living relative the tropical pearly nautilus, or between the extinct plesiosaur and the tropical living crocodiles. Yet after reviewing the evidence, and not least what the younger Brongniart was doing with fossil plants (§12.2), Lyell agreed that it was likely that "when the coal-plants were in existence, the heat was far more intense than is now experienced even between the tropics". As for the physical cause of the cooling trend that had apparently marked the whole of subsequent geohistory, Lyell admitted that he was baffled. He noted that major changes in the distribution of continents and oceans could have affected world climates, but he reckoned that this was quite inadequate as an explanation. He scornfully rejected the newly popular idea of a residual "central heat", which his friend Scrope had adopted from the French physicists (§9.3), and guessed that the true answer would come in due course from the astronomers, or in other words from cosmology and outer space.[20]

In sum, Lyell's review of the *Transactions* showed him to be a consummate synthesizer of what other geologists had published, not just in the volume he was reviewing but also elsewhere and internationally. It showed him to be right in the mainstream of geological opinion. He accepted, as likely to be genuine, both a progressive trend in the history of life and a cooling trend in the world's climates. He portrayed geohistory as having been highly eventful, with long periods of calm punctuated by occasional abrupt "revolutions", although he kept silent on the vexed question of the most recent of these events, the putative "geological deluge". The only points on which his synthesis hinted at a more distinctive viewpoint were his suggestions that actual causes might remain as powerful in the present world as they had ever been, and that in consequence they might have greater explanatory potential than had yet been recognized.

14.5 CONCLUSION

In the middle years of the 1820s, Buckland worked hard to consolidate his claim that the effects of the enigmatic diluvial event had been extremely widespread or even worldwide. On a scientific Grand Tour after his marriage, he and the former Mary Morland inspected several French bone caves, among them those being carefully excavated by Serres and Christol. Buckland interpreted them as the former dens of antediluvial hyenas and bears, closely analogous respectively to his own Kirkdale Cave and the famous caves in Bavaria. Far more exotic evidence for what Buckland claimed as the worldwide reach of the diluvial event came from Craufurd's fossil finds in Burma and Beechey's in Alaska, followed later by the more obscure but still suggestive finds in Australia reported by Mitchell.

However, Buckland's diluvial interpretation of the fossil evidence was continuing to be challenged by Fleming's critique, namely that the "antediluvial" and "postdiluvial" faunas were not as distinct as Buckland (and Cuvier) claimed, and that the alleged mass extinction of the spectacular megafauna might in fact have been gradual, piecemeal, and due to the hunting and other activities of early humans. In this argument the huge and famous "Irish elk" was of potentially decisive importance, for the new finds described by Hart suggested that this species at least had survived into postdiluvial times and had been hunted—perhaps to extinction—by humans.

Finally, Lyell emerged at this time as a shrewd commentator on current research, when he began a series of essays in the influential *Quarterly Review.* In particular, his essay on the latest volume of the Geological Society's *Transactions* presented his own principles and objectives for his science. He was clearly in the mainstream of geological opinion: he expressed confidence in the almost consensual picture of geohistory as *directional,* in terms both of a gradually cooling global climate and of a broadly progressive fossil record. At the same time, he also showed himself to be—like his friends Prévost, Scrope, and Fleming—a strong advocate of the explanatory power of actual causes in geology. Yet he certainly did not rule out the possibility that the vast tracts of deep time might have been punctuated occasionally by relatively violent episodes of sudden change. Only on the most recent of these putative "revolutions"—the enigmatic "diluvial" event—did he remain prudently silent.

Lyell's last and most substantial essay in this series was prompted by a new work by his friend Scrope, which, together with equally significant research by local French geologists, turned the spotlight once more onto the classic ground of central France. This is the topic of the next chapter.

20. Lyell, "Transactions of the Geological Society" (1826), 525–29.

CHAPTER 15

The centrality of central France (1826–28)

15.1 SCROPE'S "TIME! TIME! TIME!"

Scrope's book on volcanoes (§9.3) had not been received with any enthusiasm by other English geologists, who found his ambition to construct a new geotheory far too speculative for their tastes and perhaps over-ambitious in so young a geologist: as Lyell put it later, rather bluntly, "his new facts were either received with scepticism . . . or altogether overlooked amidst the astonishment created at such sweeping generalizations". But Scrope himself treated his next major publication—the one for which he became famous—as the "*pièce justificative*" or prize exhibit in support of the theories in his earlier book. The new one, which he had in fact drafted several years earlier, was his *Memoir on the Geology of Central France* (1827). Its published text was complemented by an album of spectacular colored engravings, mostly wide panoramic views of the volcanic landscapes based on his own drawings in the field. On a small scale these mimicked the panoramas of more general interest that were displayed to a paying public at the eponymous Panorama in London; Scrope himself jokingly recommended one of his own panoramic views (around the city of Le Puy) "when the panoramas of Paris and Petersburg, Spitzbergen and Naples, have ceased to amuse, and novelty is required". With such elaborate illustrations, it was an expensive work, and the number of copies printed may have been small. But as with other scientific publications, impact bore almost no relation to sales, and a little could go a long way. For example, the copy that Scrope presented to the Geological Society could have been studied by plenty of English savants, even if they could not afford copies of their own; and other copies went—by gift or purchase—to foreign institutions such as the Muséum in Paris and Teyler's in Haarlem.[1]

Scrope's illustrations were, at the very least, effective proxies for the experience of seeing the famous volcanic landscapes at first hand, invaluable substitutes for those unable to visit central France in person. More than that, however, they were powerful instruments of visual rhetoric. Their coloring was not naturalistic but conventional in the same style as an ordinary geological map; and they depicted much that in reality could be seen only by going over the ground on foot. Scrope's panoramas therefore enabled his readers (or rather, his viewers) to see the volcanic landscapes *through his eyes*, as he had come to understand the terrain in the light of his fieldwork; the illustrations then became highly persuasive evidence for the geohistorical interpretations that he gave them in his text (Fig. 15.1).[2]

Scrope was economical with his acknowledgments of his predecessors. In line with the Geological Society's ethos, he claimed to have done his fieldwork before reading any earlier work and with nothing but Cassini's topographical map as his guide, being "thoroughly determined to form an opinion exclusively my own". He heightened his own originality by expressing surprise that the French had still done so little research in a region of such geological importance: only Montlosier's little book (*BLT* §6.1) was praised as "profound and attentive", and he used a quotation from it as his epigraph. He mentioned Desmarest's early paper on Auvergne (*BLT* §4.3), now half a century old, but was critical of its incomplete preliminary map of the lava flows. But he failed to cite Desmarest's famous later paper on the "*époques*" of geohistory in Auvergne (*BLT* §8.3), although this

Fig. 15.1 A small part of one of Scrope's panoramic views of the volcanic landscapes of Auvergne, published in 1827 in his *Memoir on the Geology of Central France*. It shows the middle part of the long narrow plateau of La Serre, capped by hard basalt; Scrope, like Desmarest long before him, interpreted this as an ancient lava flow left high and dry by the subsequent erosion of new valleys on both sides (each with a much more recent lava, not visible in this view, on the valley floor). Again like Desmarest, Scrope claimed that these valleys had been excavated over vast spans of time by the slow action of the small streams that still flow in them, and not by any kind of sudden deluge.

had been published in the Institut's own periodical—which even in wartime had been as well known in England as the Royal Society's was in France—and it is inconceivable that he did not know of it. And he did not mention that Desmarest's son had at last published his father's completed map of Auvergne, soon after Scrope visited the region; his own map, being colored, was more striking than the French one, but otherwise it matched Desmarest's very closely. Finally, the local naturalist and prosperous merchant Jacques Mathieu Bertrand-Roux (1776–1862) had been his guide to the area around Le Puy (in Velay, southeast of Auvergne), as he had earlier been Brongniart's (*BLT* §9.6); yet Scrope did not cite Bertrand's book on its geology, with its closely similar if inferior panorama of the volcanic landscape, although like Desmarest's map it was published long before his own.[3]

The bulk of Scrope's text was a detailed description of the extinct volcanoes and their cones and craters, lava flows, and other products. The framework was that of physical geography directed toward questions of earth physics, particularly those concerning the original fluidity of the lavas now preserved as basalts. But his overarching interpretation was geohistorical. The three major areas of volcanic rocks in central France were, he claimed, "the skeletons of three enormous habitual [i.e., long-term] volcanos, the Aetnas of their age": one of these was the massif of Mont-Dore (see Fig. 13.1), which Desmarest had likewise regarded as a highly eroded ancient volcano. But Scrope focused his geohistorical attention on the smaller and more recent volcanoes: those that Desmarest had first mapped in Auvergne, of which the highest was the well-known landmark of the Puy de Dôme (the site of the famous barometric experiment made for Blaise Pascal two centuries earlier); and those much further south in Vivarais, which had been studied closely by Soulavie (*BLT* §4.4). However, Scrope rejected the distinction between "ancient" and "modern" volcanoes, which his predecessors had found useful at least for descriptive purposes, and thereby set himself to disprove decisively, in this classic region, any "diluvial" interpretation of recent geohistory: "The volcanic eruptions appear certainly to have continued from first to last without any very marked interval of quiet, and the excavation of the actual

1. Scrope, *Central France* (1827), Preface [part two] (ix–xiv), dated 1 January 1826; quotation on 180, referring to pl. 12. A manuscript translation at the Muséum [Paris-MHN: MS 176] indicates that someone there, probably Brongniart and perhaps Cuvier too, was seriously interested in Scrope's work. His all-round panoramas (pls. 2, 12) are 193cm (about 6 feet) in breadth. On the huge ones displayed commercially at the Panorama, see Altick, *Shows of London* (1978), 128–40; and Comment, *Panorama* (1999); also, more generally, O'Connor, *Earth on show* (2007), chap. 7. The illustrations in Scrope, *Central France* (1858), uncolored and much reduced in size, are a poor substitute for those in the original (and now very rare) album. The Lyell quotation is from his "*Central France* by Scrope" (1827), 439 (see §18.1).

2. Fig. 15.1 is reproduced from Scrope, *Central France* (1827), part of pl. 2, explained on 174, looking south from Puy Girou (south of Clermont). As in many views published by other geologists at this time (see Buckland's in *BLT* Fig. 10.15), landmarks are identified by conventional "birds", to avoid sullying the landscape with letters or figures. The rhetorical power of the panoramas derived from the interpretative coloring and also, in some cases, from substantial "enhancement" of the view itself, usually involving vertical exaggeration and/or an imagined aerial perspective (see Fig. 13.1).

3. Scrope, *Central France* (1827), vii–viii, 35–39, 2nd map; Desmarest, *Carte du Puy-de-Dôme* (1823), a copy of which was presented to the Geological Society on 5 March 1824; Bertrand, *Environs du Puy* (1823), [2nd pl.]. A charitable explanation of Scrope's omissions would be that he published the text he had completed by 1822, without updating it at all.

Fig. 15.2 Part of another of Scrope's panoramic views of Auvergne (1827), showing (left) some of the many well-preserved cratered cones of loose volcanic ash (the Puy de Dôme, rising above them all, is in contrast a "plug" of solid volcanic rock). From one breached cone, Puy de la Vache, a lava (depicted as rough ground) has flowed mostly to the right, ponding back a small lake, Lac Aidat, before continuing down the valley. Since one late-Roman writer (and bishop of Clermont) was known to have lived by the lakeside, geologists were agreed that even this "modern" eruption dated from before any human textual records. The hill on the far side of the valley, capped with basalt (right), is the upper end of the long narrow plateau of La Serre (seen end-on; in Fig. 15.1 it is seen from the side), which was interpreted as the eroded remnant of a far older lava. At the time of Scrope's fieldwork, the elderly Montlosier was living in his small château ("Maison de M. Montlosier", left) at the foot of one of the cones and at the heart of the volcanic landscape that he had done so much to elucidate.

[i.e., present] valleys, which these [earlier] authors take as an epoch [i.e., a single episode], has *in this country* at least been in continual progress, from the first elevation of its mountains out of the bosom of the sea, to the present moment."[4]

Like the earlier naturalists, Scrope recognized that all the more recent lava flows were on the valley floors, while the basalts capping many of the hilltops above them were the remnants of more ancient flows, left high and dry by subsequent erosion (Fig. 15.1). In fact, both Desmarest and Montlosier had acknowledged that there was no sharp division between "ancient" and "modern", and had treated it as merely convenient; but Scrope ignored all such qualifications and implied that he alone had seen that there were innumerable gradations between two extremes. Nonetheless, it suited his visual rhetoric to distinguish the two classes in his panoramas and on his map, just as Desmarest had on his, for they illustrated vividly his claim that sporadic eruptions had punctuated an uninterrupted process of erosion through vast spans of time (Fig. 15.2).[5]

Ever since the time of Desmarest, naturalists who had seen Auvergne at first hand had agreed that even the "modern" volcanoes, with their well-preserved craters, their cones of loose volcanic ash and their lava flows still as rugged as those on Vesuvius or Etna, were nonetheless prehistoric in date: that is, they antedated the earliest historical records, which in this region went back at least to Roman times (*BLT* §4.3). Both Desmarest and Montlosier had noted that in a few places, where the valley floor was composed of relatively soft sediments rather

than granite, the streams had continued their erosive action *since* the eruptions, carving small ravines between the lava and the side of the valley; and Montlosier had described one spot where a valley had been completely blocked by a lava flow and the stream had carved itself a new channel on another course altogether (*BLT* §6.1). These local details were important, because they demonstrated the continuing efficacy of fluvial erosion, at least on a small scale; and they suggested how, given a far greater lapse of time, the same agency might have excavated the entire valley without any episode of an exceptionally violent character.

Scrope must have been well aware of this evidence: certainly from reading the earlier work, and probably also from being told (and perhaps shown) where to look, by the elderly but active Montlosier, who lived among the volcanoes. But he found more striking examples further south, in the valley of the Ardèche, in the former province of Vivarais that had been Soulavie's territory half a century earlier (*BLT* §4.4). Here Scrope saw for himself how streams and rivers had carved deep gorges along the sides of lava flows. Like Soulavie, he found vivid evidence of the continuing erosion in the beds of the streams themselves: angular blocks of dark basalt eroded from the side of the gorge could be traced downstream, becoming gradually smaller and more rounded by attrition, but always conspicuous against the light-colored boulders and pebbles of granite and other Primary rocks eroded from the bedrock. Again like Soulavie, Scrope was convinced that the erosion was extremely slow by any human standard, and that these lavas must therefore be extremely ancient. Yet several of them were visibly connected to small volcanic cones composed of loose ash, which any more recent

5. Fig. 15.2 is reproduced from Scrope, *Central France* (1827), part of pl. 6, looking north and east from Puy de la Rodde; see Desmarest's map (1806), reproduced in *BLT*, Fig. 4.10. Montlosier probably showed Scrope the volcanoes in the field, as he certainly did with other visiting savants; his château now houses, appropriately, the offices of the Parc Régional Naturel des Volcans d'Auvergne, and a small museum. The rugged surfaces of the "modern" lava flows (known locally as *cheires*) are now less easy to trace from a distance than they once were, and viewpoints such as this are difficult to re-experience, owing—ironically—to Montlosier's efforts as an enlightened landowner to prove that this otherwise infertile land could be used profitably for forestry (he began this afforestation, coincidentally, in the same year that Scrope's book was published).

Fig. 15.3 Scrope's panorama of a lava flow (of basalt with columnar jointing) in one of the tributary valleys of the Ardèche in Vivarais, showing its subsequent erosion into a gorge along the valley side, and its connection to a cratered cone (behind the village) composed of loose volcanic ash. The ensemble gave him an elegant causal argument for the ability of streams and rivers to erode deep valleys, and for the vast spans of time that this must have entailed; and also a geohistorical argument against the reality of any violent deluge in the recent past, at least in this region.

deluge would surely have washed away. So here in Vivarais, Scrope's little "pet volcanoes", as he called them, gave him his most persuasive evidence that the valleys were due solely to fluvial erosion, and that no geologically recent deluge had interrupted the slow working of that actual cause (Fig. 15.3).[6]

The rather deeply eroded lavas in Vivarais gave Scrope one of the intermediates that he needed, in order to erase any sharp distinction between "ancient" and "modern", antediluvial and postdiluvial. They and their cratered cones looked almost as fresh and recent as those in Auvergne, but subsequent erosion had gone much further, so they might be substantially older. But his most effective evidence came from Auvergne, where he plotted the longitudinal profiles of both "ancient" and "modern" flows against an accurate scale of altitude. Most telling in this respect were the many lavas that had flowed eastwards off the plateau of Primary rocks crowned by the Puy de Dôme. All these lavas—whether on a valley floor or up on a hilltop, whether still preserved unbroken or reduced by erosion to isolated remnants—showed profiles that were clearly graded towards baselines at varying heights above the present level of the river Allier, into which all the streams flow. The lava flows did not fall into two distinct groups; they were at many different heights, with no major gaps. Scrope therefore claimed that they "exhibit a natural scale for measuring the duration of the process" of erosion since their respective times of eruption; in effect they had fossilized the past profiles of these lateral valleys, which at each stage had matched what was

then the level of the main valley. Scrope's "natural scale" was—to use de Luc's metaphor from long before (*BLT* §6.2)—a striking "natural chronometer", although it could not be calibrated in millennia or still larger units of time (Fig. 15.4).[7]

In his conclusion, Scrope argued forcefully that the observable actual cause of fluvial erosion was quite adequate to account for even the deepest valleys; and that the occasional eruption of lavas in central France was a happy accident that had preserved many successive phases in an otherwise steady and uninterrupted process. The moral was clear: "surely it is incumbent on us to pause before we attribute similar excavations in other lofty tracts of country, in which, from the absence of recent volcanos, evidence of this nature is wanting, to the occurrence of unexampled and unattested catastrophes, of a purely hypothetical nature!" A diluvial explanation of valleys was, he argued, certainly inapplicable to the Massif Central; and at the very least this undermined claims (such as Buckland's) for the general or universal validity of the theory. Scrope closed his text with a brief summary of the "history" of the region, arranged in "chronological order"; but

6. Fig. 15.3 is reproduced from Scrope, *Central France* (1827), part of pl. 14, showing the village of Jaujac in the valley of the Lignon, which flows into the Ardèche. Scrope's view was much "enhanced" to improve his case: at the village the gorge is not in fact much deeper below the bridge than the houses above it are high, though it does become deeper downstream (left); the cratered cone (Coupe de Jaujac) is much nearer the village than this view suggests; and the view itself has been made quasi-aerial. As Soulavie had discovered, there are several other analogous gorges and cones in the same area, some of them also illustrated by Scrope (pls. 15, 16). The phrase "pet volcanoes" is in Scrope to Murchison, 30 March 1828 [London-GS]. Jaujac is about 20km west of Aubenas (Ardèche).

7. Fig. 15.4 is reproduced from Scrope, *Central France* (1827), part of pl. 18; the altitude scale is in feet above sea level; most of his barometric data came from Ramond, "Nivellement barométrique" (1818). The hill of Gergovia, with the highest basalt, is 7km south-southeast of the city of Clermont [now joined to its neighbor as Clermont-Ferrand] (center foreground).

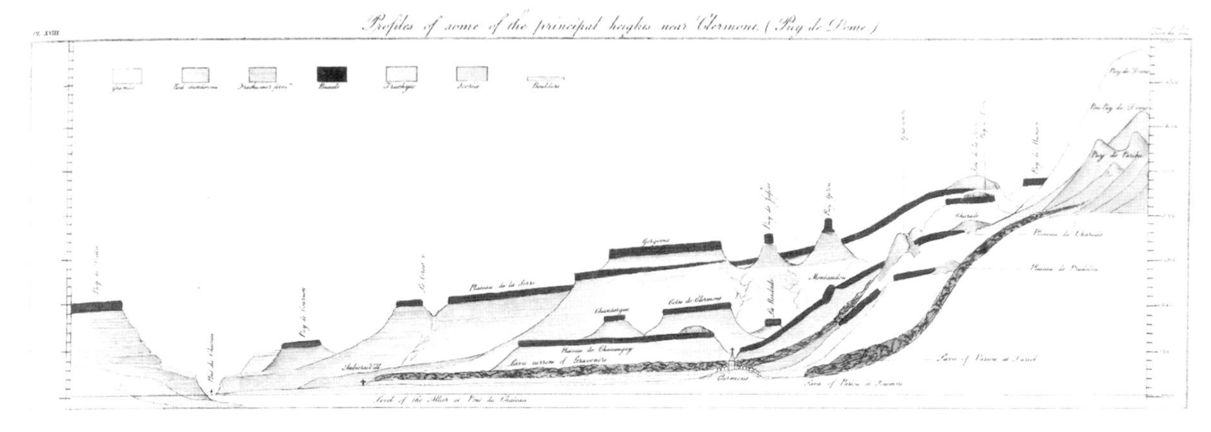

Fig. 15.4 Scrope's diagram of the longitudinal profiles of some of the lavas that have flowed eastward off the plateau topped by the Puy de Dôme (right) in Auvergne, towards the north-south valley of the Allier (left). The long narrow plateau of La Serre (see Fig. 15.1) is in the background, but is exceeded in altitude (and, inferentially, in age) by the more fragmented basalt of Gergovia in front of it. Scrope interpreted these profiles as a "natural scale", preserving many successive phases in the continuous erosion of the valleys, and enabling the successive eruptions to be dated quantitatively (although not to be calibrated in years or millennia). The diagram showed that the lavas were *not* sharply divided into "ancient" eruptions before the erosion and "modern" ones since; this helped to undermine the case for any sudden diluvial event in Auvergne.

this geohistory had no place for any recent interruption of the continuous process of erosion, which he suggested was likely to be punctuated in the future, as it had been in the past, by occasional episodes of volcanic eruption.[8]

Above all, however, Scrope realized perceptively that his inferences would be accepted only if geologists were to learn to comprehend *imaginatively* the vastness of deep time that they already claimed to take for granted. In a purple passage that was to become famous among his fellow geologists, he used current British debates about political economy, and more specifically a metaphor from the world of finance, to bring home the final lesson to be drawn from the geology of Central France:

> The periods which to our narrow apprehension, and compared with our ephemeral existence, appear of incalculable duration, are in all probability but trifles in the calendar of nature. It is Geology that, above all other sciences, makes us acquainted with this important, though humiliating fact. Every step we take in its pursuit forces us to make almost unlimited drafts upon antiquity. The leading idea which is present in all our researches, and which accompanies every fresh observation, the sound of which to the ear of the student of Nature seems continually echoed from every part of her works, is—
>
> Time!—Time!—Time![9]

15.2 FAUNAS AND VOLCANOES IN AUVERGNE

Even before Scrope's book appeared, work of comparable importance had begun to be published on the spot, by local naturalists in Auvergne. It complemented the Englishman's research by focusing on the history of life in this classic region, rather than its volcanic activity.

In 1823, the year in which Scrope returned to England and Buckland published his book on cave bones, two young local huntsmen had noticed fossil bones in a gully on a hillside in Auvergne, near a farm called Boulade. This new and abundant source of fossils soon impinged on the international debate about the most recent periods in geohistory, no less decisively than Buckland's Kirkdale Cave (*BLT* §10.6). Using the terms that Desmarest had introduced when he mapped the area half a century earlier (*BLT* §4.3), the Boulade fossils were found on the flank of a plateau capped by "ancient" basalt, which lay between two valleys in each of which there was a "modern" lava flow that had erupted from a small volcanic cone several miles away. When Buckland revisited Auvergne in 1826 on his way home from his Grand Tour (§14.1), he assumed in advance that the deposit must be alluvial, and the bones therefore "postdiluvial" and of no great interest to him. However, after local naturalists took him to the spot, he changed his mind, agreeing that the Boulade bone deposit, which was well up on the hillside, must be much older than the ordinary alluvial deposits on the floor of the valley. The Académie in Paris had already been informed that the fossils included a hippo tooth as well as abundant bones of deer, which suggested that they might all be "antediluvial" (Fig. 15.5).[10]

A description of the Boulade fossils was soon published in Clermont by these local naturalists. They thanked Cuvier for allowing his assistant Laurillard to identify some of the fossils for them; but in fact the bones were identified only in general terms, sufficient to show a large and varied mammalian fauna but not precise enough to assess its relation to living species. However, the authors themselves put the new fossils in an unmistakably *geohistorical* context. The bones were "veritable monuments of archaeology", yet also the earth's own "documents"; old enough to be of geological significance and therefore able to help "fix the chronology of the globe". A review in the Genevan *Bibliothèque universelle*, which alerted savants throughout Europe to the importance of the new find, enlarged on that theme. Geological work, the reviewer claimed, was usually similar to antiquarian research that had to make do with fragmentary "monu-

8. Scrope, *Central France* (1827), 165–69. Bertrand, *Environs du Puy* (1823), had already argued the same case for fluvial erosion in the area around Le Puy.

9. Scrope, *Central France* (1827), 165. The "drafts" were those of a capitalist banker, not—as some modern geologists have imaginatively but wrongly supposed—those of a heavy drinker. In place of the gold standard of his time, Scrope was a forceful advocate of *paper* currency, which would indeed have functioned in the economy rather like time in geological explanation: see his *Credit currency* (1830) and Rudwick, "Scrope on the volcanos of Auvergne" (1974), 236–42. In a trenchant footnote, Scrope pinpointed the *imaginative* difficulty that his contemporaries experienced in extending to nature what they readily ascribed to God: "There are many minds that would not for an instant doubt the God of Nature to have existed *from all Eternity*, and would yet reject as preposterous the idea of going back a million of years in the History of *His Works*. Yet what is a million, or a million million, of solar revolutions to an Eternity?" (165n).

10. Fig. 15.5 is reproduced from part of Desmarest, *Carte du Puy-de-Dôme* (1823), published by his son Anselme-Gaëtan. Devèze and Bouillet, *Montagne de Boulade* (1827), vii–xii, 1–7. The wooded gully (ravin des Étouaires) northwest of Boulade, together with similar localities nearby, are now collectively known by the name of Perrier, the village at the foot of the hill. Boulade is 2km northwest of Issoire (Puy-de-Dôme), which is 30km south-southeast of Clermont-Ferrand.

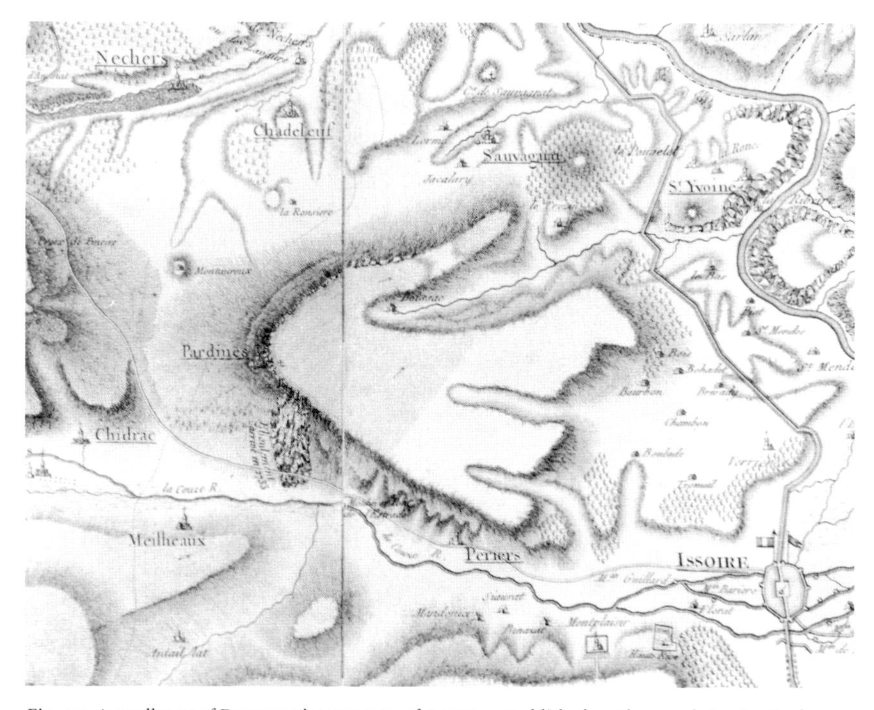

Fig. 15.5 A small part of Desmarest's great map of Auvergne, published posthumously in 1823, to show the position of Boulade, the rich site for fossil bones discovered in the same year. The plateau (center) is capped with "ancient" basalt (shaded). "Modern" lavas have flowed eastward on the floors of the valleys flanking the plateau to both north and south (the end of the northern one, marked with dense stippling, is just visible at Neschers, top left); the present streams in the two valleys likewise flow eastward, into the river Allier (right). Boulade itself is northwest of Issoire (lower right) and on the lower slopes of the plateau, *intermediate* in height between the "ancient" and "modern" volcanic rocks; this position was considered decisive in determining the relative age of the Boulade fauna. This portion of Desmarest's map covers an area of about 7km by 9km.

ments" and often indecipherable inscriptions. By contrast, Auvergne geology was like the celebrated excavations at Pompeii (*BLT* §4.1): "There the material facts are quite evident; they speak quite clearly, the mind has only to assemble them, the imagination has to supply nothing". As the reviewer commented, "It is thus for the geologist who contemplates Auvergne . . . it is thus that geology, piercing the night of time, has been able in its own way to throw light on these remote epochs, well before all historical documents, and on these terrible volcanic cataclysms of which even tradition has not preserved any memory."[11]

However, a second pair of Auvergnat naturalists, apparently working in rivalry with the first, had a greater impact, for they had secured Cuvier's direct patronage and detailed expertise. Jean-Baptiste Croizet (1787–1859), the *abbé* of the nearby village of Champeix, and his friend A. C. G. Jobert, dedicated their work to Cuvier, entitled it *Researches on Fossil Bones* (though limited to those from the département of Puy-de-Dôme), and called their main text a "Preliminary Discourse". But the flattery was sincere, and their work was no amateurish effort: they sent Cuvier their fossil bones (or drawings of them) for identification, and did some highly competent fieldwork to establish their context. Nor was their work merely provincial: unlike their rivals' publisher, theirs (also in Clermont) co-opted three colleagues in Paris with branch offices respectively in Amsterdam, Strasbourg, and London, capable between them of finding subscribers throughout savant Europe. After the first installments had been published, with fine illustrations of the bones, Cuvier summarized their work to the Académie in Paris and gave it his full approval.[12]

Croizet and Jobert duly acknowledged their predecessors, such as Desmarest and Montlosier, and the more recent (and charmingly gallicized) "Ch. d'Aubeny". By the time they completed their text they had also seen Scrope's new book with its "vast and magnificent panoramas". Like their Auvergnat rivals, they gave their work a strongly geohistorical slant, peppering their text with antiquarian metaphors. Cuvier, for example, had shown how to "uncover monuments that time seemed to have wished to conceal and destroy", and by his work "an ancient world has been re-created [*recréé*]". They interpreted the geology of Auvergne in terms of five major "epochs". The Tertiary freshwater limestones that underlay even the oldest volcanic rocks had been deposited during their fourth or penultimate epoch, probably in a large lake, one of a series draining northward into the Paris Basin. However, "in the midst of these antique ruins [are] more recent monuments", namely those of their fifth and final epoch, the period of all the volcanic eruptions in Auvergne. And with no evidence of any human presence at this time, animals instead had acted as "witnesses" of these "ancient catastrophes".[13]

The fossil bones from Boulade, as described by Croizet and Jobert, took the implications of Auvergne geology far beyond what even Scrope had suggested. However, the geohistorical use of the new fossils depended on determining the exact place of the deposit in which they were found: unlike the indoor Cuvierian practice that Croizet and Jobert emulated in other ways, their research also demanded careful outdoor fieldwork. They analyzed the topography around the Boulade locality, plotting accurately the altitudes of the various basalts and lava flows, and the deposits underlying them. They claimed that the volcanic rocks represented a sequence of events that had occasionally interrupted the steady erosion of the valleys; and to explain this erosion they rejected "cataclysms, deluges, or debacles" in favor of "causes analogous to actual causes". They insisted that "*these valleys have not been excavated all at once*", because the occasional eruptions had preserved traces of the form of the valleys at many different stages, thereby providing a natural "chronometer". So Croizet and Jobert aligned themselves with those, such as Desmarest and Montlosier, who had invoked the "constant cause [of] the prolonged action of currents [and] streams, the more or less

11. Anon., "Géologie de l'Auvergne" (1829), 309, 317, reviewing a preprint fascicle of Lecoq and Bouillet, *Vues et coupes* (1830), and another work by Lecoq; also referring to Devèze and Bouillet, *Montagne de Boulade* (1827) but not, surprisingly, to Scrope. The first of these books—otherwise an unremarkable description of Auvergne geology—was innovative in that it was sold with a box of rock specimens: these material proxies complemented its landscape views, though the latter were very crude by comparison with Scrope's. "Piercing the night of time" was a cliché dating back to Buffon (*BLT* §3.2).

12. Croizet and Jobert, *Ossemens fossiles du Puy-de-Dôme* (1826–28). Cuvier, "Rapport sur un ouvrage" (1826), is his report on the first fascicles, read at the Académie on 23 October. Croizet is a reminder that there were at least a few French Catholic counterparts of Fleming the Scottish Presbyterian, and of Conybeare and many other parson-naturalists in the Church of England.

13. Croizet and Jobert, *Ossemens fossiles du Puy-de-Dôme* (1826–28), 10–11, 20–37, 57–58.

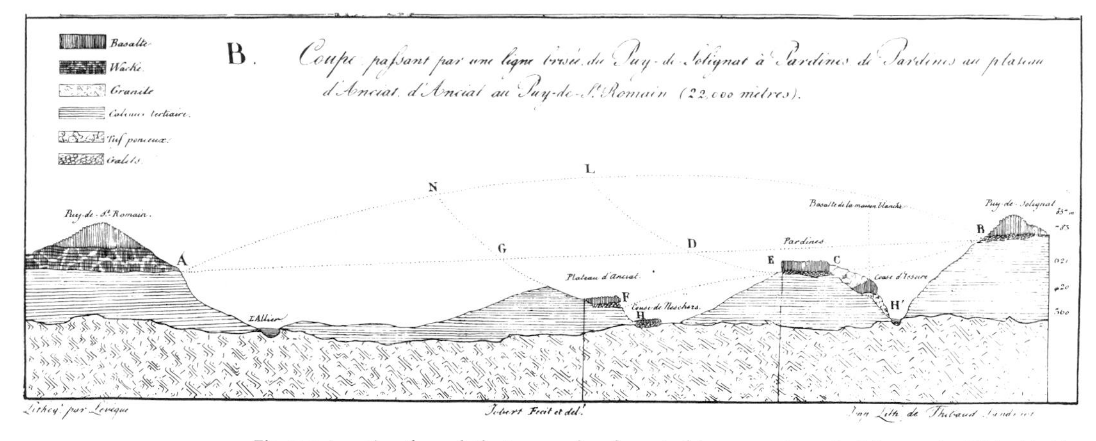

Fig. 15.6 A section through the topography of a part of Auvergne, drawn by Jobert and published in his joint work with Croizet (1826-28), showing their reconstruction of its geohistory in terms of gradual fluvial erosion punctuated by occasional volcanic eruptions. The valleys had been excavated mainly in the soft Tertiary limestone [*Calcaire tertiaire*], but on the line of this section the erosion had just reached the underlying basement of hard granite. Each lava flow, pouring down what was then a valley floor, had covered and preserved the gravelly river deposits [*galets*] of the time, but at successively lower levels. The hilltop basalts at A and B were the highest and oldest, interpreted as relics of an ancient valley floor sloping along the line A-G-D-B, with adjacent hills perhaps rising to A-N-L-B. The next oldest eruption produced the hilltop basalt (E-C) of Pardines, flowing in a valley with the profile L-D-E-C-B. Then came a still lower basalt (F) in a valley with the profile N-G-F-E; and most recently the lava (H) at Neschers in the present valley profiled as F-H-E, which had flowed from a still well-preserved volcanic cone further upstream. The fossil bones came from a deposit preserved on the south (right) flank of the hill of Pardines, which by its height was equated with the third volcanic episode: the Boulade fauna was therefore of an age *intermediate* between the oldest and youngest eruptions. (By permission of the Syndics of Cambridge University Library)

slow but inexorable effect of time". They illustrated their argument in a graphic style quite different from Scrope's (Fig. 15.4), which strongly suggests that they had already developed it before seeing the English work; but it provided eloquent corroboration of Scrope's claim for the adequacy of ordinary fluvial erosion in this region, and hence that no uniquely violent diluvial event needed to be invoked (Fig. 15.6).[14]

Croizet and Jobert claimed that there had been no *uniquely* violent event. Yet several of the beds preserved beneath the basalts were composed of volcanic materials that included huge blocks of volcanic rock (in modern terms, they were agglomerates). These could be traced to their sources in the Mont-Dore massif many miles away to the west, and must have been shifted "by forces that are certainly no longer exemplified in our regions". However, the inferred agents—massive volcanic explosions and mudslides—were not totally unlike anything now known, but simply more intense: "causes analogous to forces now operating, atmospheric events, violent storms, for which those of the present time only provide us with feeble images". Unlike the putatively unique and global "diluvial" event that Daubeny and Buckland had earlier extended to Auvergne (*BLT* §10.5), these events would have been local, and confined to the region around the great volcanic center of Mont-Dore. But not all the deposits were of this kind: some beds looked like ordinary river gravels, and others were of sand and silt, suggesting quiet conditions. So, as Croizet and Jobert put it, "we see behind us [i.e.,

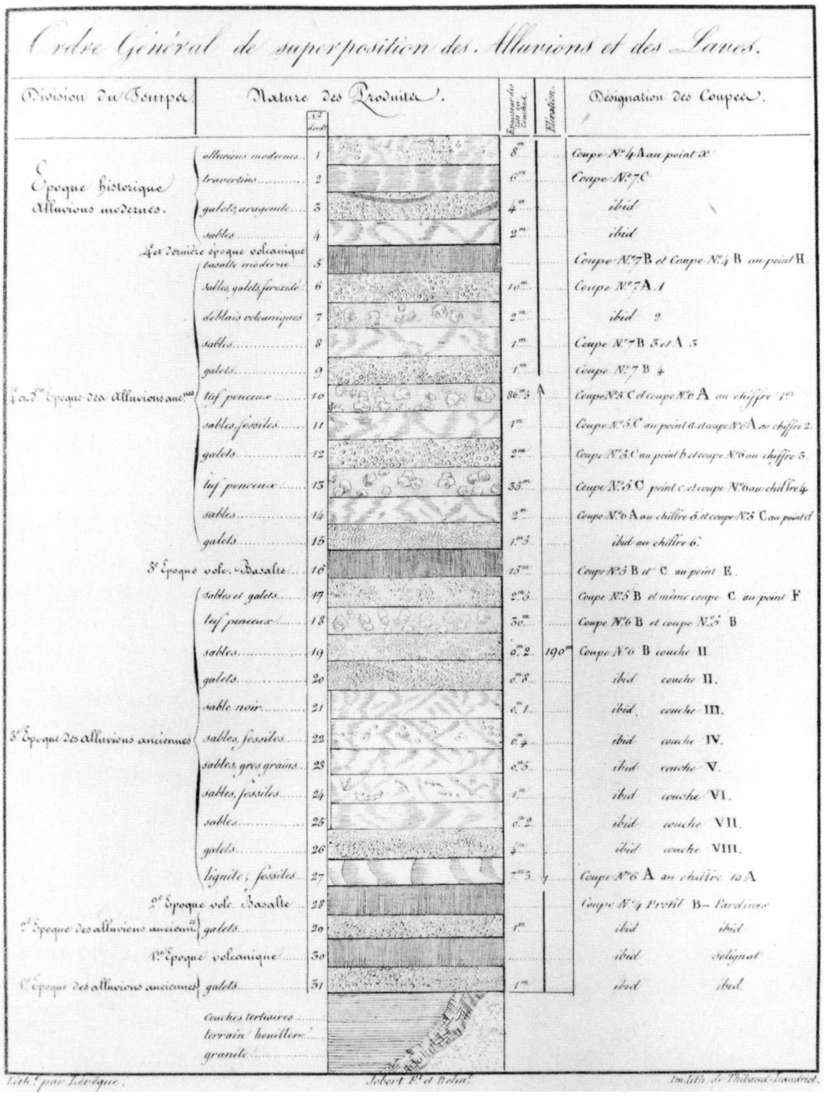

Fig. 15.7 The ideal section constructed by Croizet and Jobert to represent the sequence of volcanic lavas and sedimentary deposits in the part of Auvergne in which their fossil bones had been found. The section was pieced together by noting the deposits underlying each of the successive lavas (see Fig. 15.6); it shows the units as if they were all equal in thickness, being designed above all to display *geohistorical sequence.* There had been four "volcanic epochs" marked by basalts (nos. 5, 16, 28, 30); four episodes that brought large blocks of volcanic rock (in *tuf ponceux* or agglomerate, nos. 7, 10, 13, 18); and many layers of ordinary sand [*sable*] and gravel [*galets*]. The repeated occurrence of the same types of deposit (and of lava) indicated the continuity of overall conditions throughout the sequence. Yet the fossil bones, mostly found in no. 24, were of the "antediluvial" fauna of mainly extinct species. This suggested that the extinction must have been due to causes other than a uniquely violent deluge, for which there was no evidence in this region. The final eruption (no. 5), followed by the "modern alluvia [of the] historical epoch" (nos. 1-4) brought the sequence right up to the present time. (By permission of the Syndics of Cambridge University Library)

14. Fig. 15.6 is reproduced from Croizet and Jobert, *Ossemens fossiles du Puy-de-Dôme* (1826–28) [Cambridge-UL: MF.51.8], pl. 4B, lithographed by Lévêque, explained on 60–62, quoted phrases on 66–71. The figures in the right margin show the altitudes (in meters) of the bases of the four main lavas; the section runs for about 22km, roughly from north (left) to south, with conventional vertical exaggeration.

stretching back in time] a sequence of events, calm and violent in turn, which succeed one another without interruption up to the present time [*l'époque actuelle*]". Or, as they put it elsewhere, they could reconstruct "the history of this period, which is linked [forwards] by an unseen thread to the earliest [human] historical times". They pieced together the scattered evidence from their fieldwork to construct an ideal section, which summarized that long sequence of varied events and located the fossil bones right in its midst (Fig. 15.7).[15]

The fossil bones from Boulade were identified by Cuvier as belonging to individuals of all ages from no fewer than forty species. They included elephant, hippopotamus, rhinoceros, tapir, horse, mastodon, bear, tiger, hyena, dog, cat, otter, beaver, ferret, hare, water rat, ox, and more than ten species of deer: it was a fauna even more diverse than Buckland's famous one from Kirkdale (*BLT* §10.6). Croizet and Jobert noted that the bones were very well preserved apart from breakage, with none of the abrasion that might have shown that they had been swept from elsewhere; and there was no trace of any marine debris. So they rejected the idea of any *geological* deluge and denied that their deposits deserved the name of "diluvium". Since there was also no trace of any humans, they rejected any role for the *biblical* deluge, at least in this region, for "man was necessarily witness to it, and nothing that could have belonged to him has ever been found in the deposits [*alluvions*] of this epoch". If Croizet the Catholic priest was concerned at all about the historicity of the biblical event, he probably adopted the common assumption that it must have been strictly local, and not at all in his part of the world: tacitly, this issue had become marginal and unimportant for geologists (or at least for those outside Britain).

So, finally, Croizet and Jobert concluded that the animals had lived in the area in which their bones were found; the deer, which were most abundant, had probably been preyed upon by various carnivores. The site was in effect equivalent to Buckland's Yorkshire cave, but in the open air; like Buckland, the naturalists had even found what they took to be the excrement [*album vetus*] of the predators. Like Buckland again, they reconstructed the ecosystem of their region, in a verbal scene from deep time: "Hippopotamus and beaver frequented the banks of the ancient Allier; pachyderms and ruminants inhabited our forests and grazed in our valleys; numerous species of carnivores were connected with the latter, each having as it were its own prey".[16]

Going much further back into geohistory, Croizet and Jobert noted that—as Brongniart had reported long before (*BLT* §9.2)—there was also no trace of any marine fossils in the Tertiary limestones that underlay the deposits they had described. So the Massif Central appeared to have been above sea level ever since the start of the Tertiary era (their fourth epoch). This implied that, in whatever way the faunal changes since that time were to be explained, it could *not* be by any major "revolutions" in the distribution of land and sea. Specifically, the most recent set of extinct mammals—on which Cuvier had first built his "re-creation" of the former world—could not have been annihilated suddenly in any recent inundation. Croizet and Jobert did not offer any alternative, except to allude to the possible role of climatic change. The brief essay in geotheory that brought their main text to its close was structured—like its equivalent in Scrope's first book (§9.3)—around their adoption of the newly attractive notion of a gradually cooling earth (§9.2). But if the exotic fossil fauna of Auvergne had disappeared as a result of a cooling of the climate, extinction would have to be radically reconceived as a gradual and piecemeal process.[17]

Although Croizet and Jobert were probably unaware of what Fleming was suggesting around the same time, their research implicitly supported the Scotsman's conception of extinction (§14.3), though with the crucial modification that a gradual process of extinction might not have been dependent on human activity. The Auvergnat hippopotamus, for example, might have declined in numbers quite slowly, struggling against an increasingly unfavorable climate, and finally dying out long before humans first entered the region. So it was now open to other naturalists to apply the same idea to earlier episodes, not just to the most recent times: gradual extinctions might then be accepted as part of the ordinary course of nature throughout geohistory.

However, this kind of alternative to Cuvier's theorizing about extinction remained largely implicit in Croizet and Jobert's volume. Cuvier himself avoided the issue, when he reviewed their completed work. He noted only that he disagreed with the authors' geotheory, and particularly with the notion of global cooling. While conceding that it did not detract from the value of the rest of their work, he judged that—like André's geotheory long before (*BLT* §8.4)—it was too speculative to appeal to his severely scientific audience at the Académie. What he failed to tackle was the implication of the detailed sequence that the Auvergnat naturalists had reconstructed, as summarized in their ideal section. Their fauna of extinct mammals had apparently flourished right in the middle of a lengthy process of slow valley erosion punctuated by sporadic volcanic episodes; much the same kinds of river deposits recurred all the way from before the earliest eruption right up to the present; and there was no trace of any *uniquely* violent event that might have annihilated the animals.[18]

15.3 CONCLUSION

The uplands of central France had been a classic region for the sciences of the earth, ever since Desmarest and Soulavie described and interpreted their extinct volcanoes half a century earlier. Now, with substantial works published almost simultaneously by Scrope in England and by Croizet and Jobert on the spot in

15. Fig. 15.7 is reproduced from Croizet and Jobert, *Ossemens fossiles du Puy-de-Dôme* (1826–28) [Cambridge-UL: MF.51.8], Section no. 8, explained on 76–88; the real thicknesses (in meters) are noted in the column to the right of the section itself; still further to the right are references to the measured sections at particular localities (not reproduced here), which provided the detailed evidence for this deliberately idealized summary and synthesis.

16. Croizet and Jobert, *Ossemens fossiles du Puy-de-Dôme* (1826–28), 88–97, quotation on 95. Some of Croizet's specimens are on display in Paris-MHN.

17. Croizet and Jobert, *Ossemens fossiles du Puy-de-Dôme* (1826–28), 105–22.

18. Cuvier, "Rapport sur un ouvrage" (1828), read on 27 October; the second volume, said to be then in press, never appeared. Subscriptions failed to cover the costs—a common outcome with expensively illustrated works of natural history at this time—and Croizet had already appealed to Cuvier to find him and his co-author funds to recoup their heavy losses: Croizet to Cuvier, 27 August 1828 [Paris-MHN: MS 627, no. 31].

Auvergne, this region was brought back to the attention of geologists throughout Europe. The English and French authors came to similar conclusions, and probably independently. They agreed with their distinguished predecessors that valleys have been excavated very slowly and gradually, by the streams and rivers that still flow in them, acting through vast tracts of deep time. However, this particular region provided exceptionally clear evidence for such a conclusion, because here the slow and steady process had been punctuated by occasional episodes of localized volcanic activity, some of them violent enough to be called catastrophic. These had had the incidental effect of preserving traces of successive phases of the continuous erosive process, thereby providing crucial evidence that there had been no single episode of erosion that could be attributed to an exceptional "geological deluge".

Scrope's work focused on the volcanic activity and its background of slow erosion, emphasizing the need to invoke almost unlimited quantities of "Time!—Time!—Time!" if the observable features were to be interpreted adequately. Croizet and Jobert took the vastness of deep time equally for granted. But they complemented the English work by exploiting a rich new source of fossil bones, reconstructing on Cuvierian lines (and with Cuvier's active assistance) a varied mammalian fauna with many extinct species. Yet in Auvergne they found no sign of the subsequent physical "revolution" or "geological deluge" that Cuvier (and Buckland after him) held responsible for the mass extinction of the "diluvial" fauna worldwide. This suggested that at least in this region the extinctions might have been gradual and piecemeal; but since there was no sign of any human presence in the region at this time, such a process could not be attributed to human activities. All in all, this scrupulously thorough research in the Massif Central, and particularly in the former province of Auvergne, put a large question mark against the dominant interpretation of recent geohistory. Yet at the same time it did nothing to lessen the cogency of other arguments—notably those based on erratic blocks (§13.3, §13.4)—that "actual causes" were *not* wholly adequate to account for all the relevant evidence.

Lying behind all such attempts to understand the most recent phase of geohistory was the sensitive issue of the origin and place of the human species itself. There were in fact two distinct questions, linked only loosely to each other. Had humans co-existed with the extinct "antediluvial" mammals? If so, the origin of the human species would be pushed back far beyond any historical records. And had new species in general been formed by transmutation (in modern terms, evolution) from earlier ones? If so, this would surely apply to the human species too. These are, respectively, the topics of the final two chapters in Part Two of this volume.

CHAPTER 16

Men among the mammoths? (1825–30)

16.1 THE QUESTION OF CONTEMPORANEITY

When in the 1820s the question of "the antiquity of man" became a prominent issue, Cuvier was among those who were deeply skeptical about claims that early humans had been the contemporaries of his megafauna of extinct mammals. But it was not for religious reasons that he was reluctant to concede this.

Being no Frenchman by birth, and a Protestant in a still predominantly Catholic country, Cuvier remained an outsider in French society. Nonetheless he was now at the height of his fame and power (Fig. 1.1). He continued to be at the heart of the Académie's affairs; in 1826–27 he also took his turn for a third stint as the director of the Muséum; and he accepted yet more official positions in the French state. In particular, he was given responsibility for relations between the government and the French Protestant community, a minority in which he retained his cultural roots although on a personal level his religious commitments seem to have been perfunctory. He certainly used his considerable influence to protect the political and civil rights of his fellow Protestants; but when faced with their factional disputes he hardly concealed his impatience with what he regarded as the finer points of theology. Yet, significantly, he promoted reforms in the two Faculties of Protestant theology at French universities, under which their students were required to learn Hebrew, the better to understand the Old Testament and ultimately to help raise intellectual standards among French Protestants. This was not the action of a literalist, but of one who recognized the value of critical biblical scholarship. Amidst all his success and recognition, however, Cuvier was struck by personal tragedy. In 1827 his only surviving child Clémentine died of tuberculosis at the age of twenty-two, only a few weeks before her intended marriage. Her father was devastated by the loss, and ceased for a time to hold his famous weekly *salon*, to which visiting foreign savants had regularly been invited.[1]

When Cuvier revised his "Preliminary Discourse" for the third and definitive edition of *Fossil Bones* (1825), he did not modify his earlier conclusion that no genuine human fossils had yet been found. He had good reason to discount all the more recent reports to the contrary, so the conclusion was more confident than ever. While visiting England in 1818, for example, he had been able to study the celebrated human skeleton from Guadaloupe that König had described; seeing it with his own eyes had confirmed his (and König's) judgment that it was of recent date and, strictly speaking, not a fossil at all (§1.2; *BLT* §10.3). Another skeleton from the Caribbean island, which had later been acquired by the Muséum, reinforced that conclusion. Cuvier also discounted the human bones that had been reported by Schlotheim from Köstritz in Saxony; although they were apparently mixed with the bones of some of the usual extinct mammals, Schlotheim himself had concluded that the case was ambiguous, and that they could have been interred at a much later date.[2]

Other unquestionably human bones, which had been found in a cave at Durfort in southern France, looked "petrified" and therefore "fossil". But one naturalist who studied them had concluded that they probably dated from historic times—he suggested they belonged to Gaulish, Roman, or Saracen victims of war—and were simply encrusted with the stalagmite that was usual in limestone caves: it was a common tourist attraction at "petrifying springs" to leave familiar objects to be "turned into stone" in the same way. And Serres had concurred, putting the case in the context of the long debate about human antiquity for the benefit of readers of the *Bibliothèque Universelle*. The much earlier reports that human bones and artifacts had been found in "regular" Tertiary or Secondary strata (*BLT* §5.4)—clearly much older than *any* Superficial or "diluvial" deposit—had long been dismissed by savants as mistaken or even fraudulent; to judge from the paucity of later references to them, these early claims had almost been forgotten. A more recent claim of the same kind, of alleged human fossils from a Tertiary sandstone near Fontainebleau, was examined by the Parisian naturalist Huot, who dismissed the objects as nothing more than concretions and confirmed the validity of Cuvier's stance.[3]

Yet the question of human antiquity obviously remained one of supreme importance for any understanding of geohistory. Other naturalists agreed with Cuvier that any claim that humans had coexisted with the extinct "antediluvial" fauna needed to be judged by the most rigorous evidential standards. It was all too easy to be misled by assuming that a "petrified" appearance (as at Durfort) or a "stony" matrix (as on Guadaloupe) denoted high antiquity, or by failing to notice signs that bones of recent origin had been mixed with those dating from a far earlier period (as at Köstritz). Cuvier's own firm conviction that the test of genuine contemporaneity had not yet been passed had also been reinforced by the famous case of the human skeleton found by Buckland among "antediluvial" animal bones in Paviland Cave in south Wales (§6.2). Although it might have strengthened their common belief that the "geological deluge" was none other than the biblical Flood, Buckland had rejected it as evidence that there had been antediluvial human life in Europe, on the grounds that it was preserved quite differently from the other bones in the cave deposits; he had concluded that it had been buried in a shallow grave dug into them at a much later time. This case was all the more significant because it had come from a geologist for whose scientific work Cuvier, like many others, had developed great respect.

Having nearly been bitten once, Buckland was justified in being twice shy. The skeleton at Paviland showed how cave deposits could be highly deceptive, and how scrupulously careful their excavation therefore had to be, before they could properly be accepted as yielding positive evidence for the contemporaneity of humans and extinct animals. Caves are quite rare natural features, so it was always likely that they would have been used for shelter, by animals or humans, at many different periods (in Buckland's time, caves in many parts of Europe were still being used as homes by human "troglodytes"). Unless cave deposits were excavated with extreme care, there could be none of the confident relative dating that geologists now took for granted when studying the stratigraphy of "regular" formations. So when Buckland was confronted by stronger evidence for contemporaneity than at Paviland, it is not surprising, and certainly not discreditable, that he remained skeptical about its authenticity.

Among the further British caves excavated in the wake of Buckland's *Reliquiae* was Kent's Hole (now Kent's Cavern) at Torquay on the south coast of Devonshire. Fossil bones were first found there in 1824, and the following year John McEnery (1796–1841), a local Catholic priest and amateur naturalist, told Buckland he had found fossil animal bones and teeth, lying unambiguously beneath the usual hard layer of stalagmite on the floor of the cave. Buckland encouraged him to continue the search, probably just expecting that Kent's Hole could in due course be added to his tally of antediluvial sites. McEnery joined the Geological Society, sponsored by Buckland among others; and several plates of his specimens were drawn (some by Buckland's future wife Mary Morland) in preparation for publication. Then, at some point—the date is unclear—MacEnery made an unexpected discovery: he found stone artifacts mixed with the animal fossils. As he recalled later, "I dug under the regular crust, and flints presented themselves to my hand; this electrified me" (it gave him a severe shock: the metaphor was far more vivid then than it has since become). Working prudently in the presence of witnesses, he excavated flint axes and arrowheads, and had no doubt

1. Negrin, *Cuvier* (1977), 225, 250, 432; Outram, *Cuvier* (1984), 104–9. The students of theology were already expected, of course, to read the New Testament in Greek. Unlike most French Protestants Cuvier's cultural roots were not Reformed (Calvinist) but Lutheran, which put him in a minority within a minority. It is significant that, before her early death, his pious daughter—who must have known him about as well as anyone other than his wife—was recorded as having regularly prayed for his conversion, which hardly suggests that he was the zealously religious person of conventional historical myth.

2. Cuvier, *Ossemens fossiles*, 3rd ed. (1825), 1, "Discours préliminaire", 66–67, n.3, and pl. 1; Schlotheim, *Petrefaktenkunde* (1820), xliii–lxi, 1–4. Grayson, *Human antiquity* (1983), 63–76, 96–113, gives a valuable summary of much of the material reviewed here (and in Chap. 29), though British and Continental research is split rather artificially into separate chapters. Laurent, "Origine de l'homme" (1989), traces the debates in France, particularly in relation to transformist (evolutionary) theories.

3. Hombres-Firmas, "Ossemens humains fossiles" (1821), and Serres, "Observations sur les ossemens humains" (1823), described the bones from Durfort, which is about 20km southwest of Alès (Gard); Huot, "Prétendu fossile humain" (1824).

that the animal bones and artifacts were in the same deposit. But in the absence of human *bones* he doubted their true contemporaneity, and Buckland, perhaps remembering Paviland, also remained skeptical. In any case, McEnery never gained enough financial support to publish his research.[4]

Yet Buckland was not blindly refusing the evidence of his senses; still less was he rejecting the contemporaneity of humans and extinct animals on grounds of biblical literalism. On the contrary, human fossils among the animal ones—if well authenticated—would have given much-needed support to his increasingly embattled identification of the geological deluge with the biblical Flood. It would have required of him no more than a minor modification of his earlier interpretation, conceding that the human species had spread from its putative Asian birthplace as far as Europe *before* it was almost annihilated in the biblical catastrophe. His reaction to another claim, of the same kind as McEnery's, shows that he was ready—even eager—to concede contemporaneity, provided the evidence for it was more satisfactory.

16.2 HUMAN FOSSILS IN LANGUEDOC

The new evidence for human antiquity came from two young naturalists in the south of France. One of them, Jules de Christol, has been mentioned already for his careful excavation of what he (and Buckland) interpreted as the den of an extinct species of hyena (§14.1). But it was the other, Paul Tournal (1805–72), who first reported finding human bones among those of fossil animals. Tournal had been sent to Paris to be apprenticed to a pharmacist, but in his free time had also trained himself as a naturalist, before returning to Narbonne in 1825 to help in his father's pharmacy. It was at the instigation of Serres that he, like Christol, began excavating caves in his own part of Languedoc. Two years later, still only twenty-two, Tournal reported in the Parisian *Annales des Sciences Naturelles* that he had discovered fossil bones in two caves near the town of Bize, not far from his native city. Unsurprisingly, he set his discovery in a Cuvierian framework: he hoped that bone caves such as his might help explain "one of the last catastrophes that has disrupted [*bouleversé*] the globe, and has caused the disappearance of several kinds of animal". The bones were certainly "fossil", or at least ancient, by the rough-and-ready field test that Buckland had recommended: they stuck to the tongue, showing that they had lost most of their organic matter. But they were not of the usual exotic animals such as mammoths, hippos, and rhinos, but of genera still living in Europe, such as bears, wild boar, horses, deer, and cattle. Nonetheless Tournal himself took it for granted that the fauna was "antediluvial", though more experienced naturalists reading his report might have wondered whether in fact it was "postdiluvial". The Bize caves obviously called for comparison with Christol's cave at Lunel-Viel (§14.1), and the two young naturalists were soon visiting each others' sites.[5]

Not long afterwards, Tournal found *human* bones and fragments of pottery among his animal bones. His mentor Serres was the first to mention this in public, but only to discount its significance. In a review of all the recent research in caves in Languedoc, written for the benefit of naturalists at the other end of the country, Serres stated that the "well-qualified geologist" Tournal had found human remains, but only in the stalagmite and uppermost layers of sediment, implying that unlike the animal bones they were not truly "fossil".[6]

However, Tournal himself gave his latest discoveries far greater significance, when a few months later he reported, again in the Parisian *Annales*, that the human remains were unquestionably *mixed* with the animal bones, some of which definitely belonged to extinct species. Christol had by now shown him the Lunel-Viel cave, and Tournal emphasized that his Bize caves were significantly different. Some of the Bize bones looked more "modern" than any previously found as fossils; as he put it, they "seem to link the present [*actuelle*] geological period with the epoch preceding historical times". He was therefore making the important and novel claim that his animal finds were *intermediate* in age between the present world and the world of the antediluvial hyena den that Christol had described. Moreover, his discovery of human bones and shards of pottery threw further doubt on the assumption that all the extinct mammals had died out before the start of the human period in Europe: all the Bize bones, animal and human, were preserved in just the same way, so that unlike Buckland's Paviland case there was no reason to doubt that they were coeval. Like the provincial naturalists in Auvergne (§15.2), Tournal was discreetly challenging the Cuvierian orthodoxy about the sharp break in geohistory at the diluvial moment: "At Bize, in fact, human bones and bones belonging to extinct [*perdues*] species are found in the same beds, both showing the same physical and chemical character . . . The generally accepted proposition that human bones in the fossil state do not exist on our present continents can thus be put in doubt, or at least cannot [yet] be resolved."[7]

Christol had interpreted the Lunel-Viel cave as a den of antediluvial hyenas, but of a new and even larger species than those at Kirkdale (§14.1). He had noted in passing that there were no human bones among the animal ones, and quoted Cuvier in support of his inference that there had been no humans in the region at the time. In this and many other respects, Christol's account of the Lunel-Viel

4. McEnery, *Cavern researches* (1859), 61–62, in a text assembled long after his death from his confusingly jumbled manuscripts: see Clark, "Pioneer of prehistory" (1925). It was eventually published in the *annus mirabilis* of prehistoric archaeology, when his tentative inference of contemporaneity was posthumously vindicated, not least by excavations in the nearby Brixham Cave: see the fine accounts of these later developments in Grayson, *Human antiquity* (1983), 72–76, and Van Riper, *Men among the mammoths* (1993), 74–143; the title of the present chapter is gratefully borrowed from the latter. McEnery was chaplain to a Catholic family living at Torre Abbey in Torquay, a medieval monastery that had been converted to domestic use at the Reformation (Buckland had, of course, no objection to collaborating with a Catholic). McEnery's plans for a volume with thirty plates drawn by Scharf would have needed massive financial support, even with Buckland's backing. Kent's Cavern remains open as a tourist attraction; although greatly altered by later excavations, it still shows clearly the relation of the stalagmite to the underlying deposits.

5. Tournal, "Deux cavernes à ossemens" (1827), 78; the locality was erroneously printed throughout as "Bire" instead of Bize, which is about 20km northwest of Narbonne. Narbonne and Montpellier are only about 100km apart; in an age of efficient mail services, and of public transport by coach, it was not difficult for Tournal and Christol to keep in close contact.

6. Serres, "Observations générales" (1828), 25–43, read at the Linnean Society of Normandy on 7 January.

7. Tournal, "Caverne de Bize" (1828), 348–49, a letter dated 25 October. There is now a commemorative plaque at the entrance of his main cave, 2km north of Bize-Minervois.

cave had simply confirmed Buckland's earlier conclusions. However, in the summer of 1829 Christol reported to the Académie that he, like Tournal the previous year, had found human bones mixed with those of extinct animals: his were in two caves near Sommières, not far from Lunel-Viel. Reflecting their fruitful collaboration, Christol began his report by stressing the importance of Tournal's discoveries. He agreed with his friend that the Bize fauna represented "the passage or link from geological times to historical times"; in fact it was he himself who "in a rigorous manner" had identified among the bones from Bize those of an extinct species of deer, which supported Tournal's inference that the fauna dated from "very remote times". But Christol argued that the Bize fauna was not truly "antediluvial" in character, because there was no trace of the usual extinct species (a single fragment of a bear's humerus could have come from the species still living in the Pyrenees, rather than the extinct cave bear, and a newspaper report of a mastodon bone was simply mistaken). By contrast, the two new bone caves near Sommières had yielded abundant remains of "antediluvial" animals such as rhinoceros, cave bear, and hyena, as well as the more familiar horse, cattle, deer, and badger.[8]

However, by far the most important point in Christol's report, giving it its title, was that in the very same deposits there were *human* bones preserved in ex-

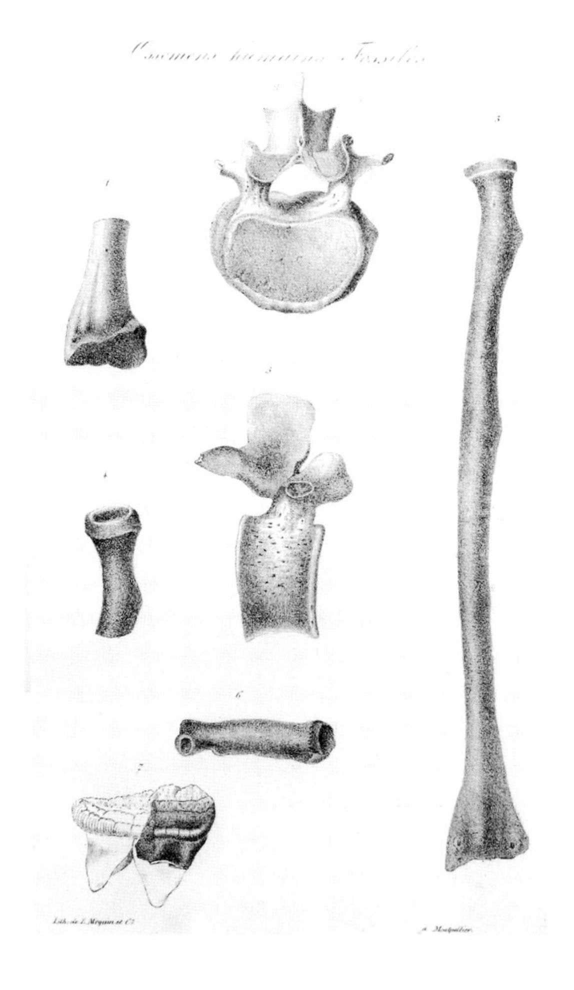

Fig. 16.1 "Fossil human bones" found by Christol in the same deposit as "antediluvial" animal bones, in two caves near Sommières in southern France: lithographed drawings from one of the published versions of his paper read at the Académie des Sciences in Paris in 1829. These specimens—identified for him by Montpellier's professor of anatomy, and found in situ probably by Christol himself—were of the radius (figs. 1, 3, 4), a lumbar vertebra (figs. 2, 5), and a phalange of the left ring finger (fig. 6). These were the first adequate pictures of fossil human bones to be published anywhere, and they supported Christol's claim that humans had lived in Europe at the same period as the "antediluvial" animals. Also shown is a fragmentary molar tooth of a cave bear (fig. 7) from the same deposit, representing that ancient fauna.

actly the same way. He emphasized that if the mammalian cave fauna were to be called "fossil", the same epithet must also be applied to the human bones. Such bones were not common, but shards of crude unfired pottery—representing, he suggested, "the infancy of the art"—were found throughout the thick deposits in the caves, and could be treated as the "true *geognostic equivalents* of human bones", or stratigraphical surrogates for them. But it was the bones that were decisive. Christol himself had found a few of them in situ, near the base of the deposit; and since that material filled the cave almost completely, it was highly unlikely that they had been buried in it at a later time. Furthermore, the issues of trust and gullibility that had made earlier finds so questionable (*BLT* §5.4) were not involved here, unless of course Christol's own honesty were to be impugned. And unlike those reports from the previous century, Christol's booklet gave his readers some fine proxies for the specimens at the heart of his claim (Fig. 16.1).[9]

Christol was cautious, or just modest, about putting forward any grand interpretative conclusions of his own. But in fact he followed Buckland (and Serres) in inferring that the bone deposits were due to a widespread inundation that had overwhelmed the cave-dwelling animals in their dens, and that the bones had not been swept in from far away. As for the co-existence of humans with these animals, he deftly recruited some highly prestigious support by quoting Humboldt's warning that "*disdainful incredulity is as disastrous in the sciences as too readily adopting facts inadequately observed*"; the burden of proof was thus neatly shifted onto those who might still be skeptical about human fossils. In his conclusion, Christol pressed home that powerful but tacit criticism of the great Cuvier:

> I shall abstain at present from all reflection on this discovery of fossil human bones, the only one of its kind that has hitherto been able to satisfy the conditions of emplacement [*gisement*] and [chemical] composition that geologists have specified; I confine myself to reporting the facts, while always recalling that the opinion about the non-existence of man in the fossil state was founded on negative facts, the value of which has sometimes been exaggerated, rather than being deduced legitimately from the certain principles on which the science rests.[10]

8. Christol, "Cavernes à ossements" (1829), summarizes the paper read at the Académie on 29 June, and the subsequent discussion; *Ossemens humains fossiles* (1829) was the full text, published in Montpellier; the text printed in *Annales des mines* would have given it much wider circulation. Christol had been told of the discovery of these bone caves by an even more local naturalist, Émilien Dumas, who lived in Sommières (Gard), which is about 12km north of Lunel-Viel and 25km northeast of Montpellier.

9. Fig. 16.1 is reproduced from Christol, *Ossemens humains fossiles* (1829), pl., explained on last page; all these specimens came from the Souvignargues cave, except the phalange from the nearby Pondres cave. Both caves had been discovered, like Kirkdale, in the course of quarrying. Other human bones found in the two caves included a scapula, humerus, sacrum, metatarsal, and upper molar, but Christol mentioned specifically (23) that he himself had found a radius and a lumbar vertebra at Souvignargues, which may well be the ones illustrated; Joseph-Marie Dubrueil (1790–1852), the professor of anatomy, identified the bones as those of adults of rather small stature. This plate was not published with the text in *Annales des mines*, where it would have looked oddly out of place among illustrations of mining machinery.

10. Christol, *Ossemens humains fossiles* (1829), 25. The claim to uniqueness was probably intended to forestall the objection that Tournal's bones might not be truly antediluvian, rather than to challenge his friend's priority.

Cordier, who was presiding at the Académie and read Christol's paper to the meeting, was immediately impressed. He recognized that Christol's case, if confirmed, would be the strongest yet put forward, since the mammals associated with the human bones were more clearly "antediluvial" than those in Tournal's caves. Buckland was equally impressed when he read Christol's published paper; and his opinion, as the acknowledged leading expert on cave fossils, counted for much more than that of the Parisian mineralogist. Writing to an English geologist long resident in the United States, Buckland was in no doubt about the outstanding significance of Christol's claims. And he was clearly ready and willing to accept them, not least because he judged the young Frenchman to be trustworthy and knew that he understood the pitfalls inherent in cave research. Above all, Buckland emphasized that the new finds, far from being an embarrassment to him, would—if confirmed—strongly support his own theorizing about the putative deluge:

> There is one Discovery of infinite Importance to my Views of the diluvial question just published by Mr. Christol of Montpelier. He has read at the Académie des Sciences a notice on Human Bones in 2 caves near Sommières under circumstances which seem to shew that Men were coeval with the Hyaenas and Elephants. The Cave is described with much modesty and caution and with a full conviction of its importance. I know Christol personally & worked with him 2 or 3 days at Montpelier in the Cave of Lunel and on the surface, and I know him to be an accurate observer and fully up to [i.e., aware of] the nature of Deposits in Caverns. You will see his Report and judge for yourself. It seems to me a Case not to be explained but by admitting what, if it can be established, is the most important Geological Discovery that I can ever hope to witness.[11]

16.3 PROVINCE AND METROPOLIS

Just three months after Christol's paper was read in Paris, one by Tournal followed it at the Académie, and was published soon afterward in the Parisian *Annales*. Returning Christol's compliment, Tournal began his "theoretical considerations" on the caves at Bize by stressing the value of his friend's research on those at Sommières. Both sets of caves showed that humans were the contemporaries of the extinct animals, but Christol's proved that "man lived with much more ancient species of animals, which characterized the antediluvial population". Having seen the site for himself, Tournal insisted that Christol's human bones really did come from the same deposit as those of the extinct mammals, and he praised him for having changed his mind on their contemporaneity: Christol's report had come from one "who saw things as they were and not as he would like them to have been, who visited the Gard [i.e., Sommières] caves without preconceived notions, stripping himself of all idea of system [i.e., geotheory]". Decoding the message, Cuvier's reluctance to concede the reality of the association of human bones with those of extinct animals was no longer justified.[12]

Turning to his own cave fauna at Bize, Tournal claimed in effect that what it lacked in antiquity was compensated by its crucial role as an *intermediate* stage between human history and deep geohistory. Some of the bones from Bize bore cut marks that he interpreted as signs of human butchery; Serres had suggested that some of the animals might already have been domesticated; and shards of pottery indicated a certain level of technical progress. Tournal argued that "wholly simple causes" were adequate to explain the extinctions within the human period and that no special catastrophe need be invoked: as Prévost—and, had he known it, Fleming (§14.3)—had pointed out, the animals could have become extinct quite gradually as a consequence of expanding human populations. Extinctions might have happened throughout: "several species have really been destroyed at different epochs, even during the historical period".[13]

Tournal noted that even those geologists who adopted a diluvial theory now conceded that the Superficial deposits [*terrain diluvien*] had been formed over a long period of time, with much local variety; the study of Auvergne by Croizet and Jobert (§15.2) was a good case in point, though Tournal did not cite it and may not have known of it. He stressed the diversity of bone caves, as illustrated by his own and Christol's examples, and suggested that the causes of the animals' burial might be equally diverse. In various kinds of local inundation, perhaps spread over a long span of time, some animals might have been overwhelmed in their dens, or while just sheltering in caves; others might have been swept from a distance as decaying carcasses, bloated and buoyant: "in fact I do not see why one should want to explain phenomena as varied as those offered by caves and bone breccias [i.e., in fissures] by a unique cause", still less by a cause different from those observable in the present world. These were topics that Tournal promised would be discussed at greater length in a work on the Bize caves that he and Serres now planned to publish jointly. Meanwhile, he concluded that the crucial importance of all such research would be unquestioned, once others came to share his own opinion, "that geology starts where archaeology stops; and that when the latter has exhausted this research and encountered the mysterious and impenetrable veil that covers the origin of nations, geology, supplementing our brief [written] Annals, will revive human pride by showing us the antiquity of our race; for geology alone will be able in the future to give us some idea about the epoch of the first appearance of man on the terrestrial globe."[14]

11. Buckland to Featherstonhaugh, 4 November 1829, [Cambridge-UL: Add.7652.II.LL.23], in reply to Featherstonhaugh to Buckland, 27 June 1829 [London-RS: MS 251, no. 32]. Buckland to Hitchcock, 1 February 1830 [printed in part as "Antediluvian human remains" (1830)], was similar. Cordier's comment is in Christol, "Cavernes à ossements" (1829). George William Featherstonhaugh (1780–1866) of Schenectady in upstate New York was much involved in American canal and railway construction, through which he developed an interest in geology; he visited Europe in 1826–27 and met Buckland and many other leading geologists; from 1834 he was commissioned by the Federal government to make geological surveys: see Eyles (J. M.), "Featherstonhaugh" (1978), and Berkeley and Berkeley, *Featherstonhaugh* (1988).

12. Tournal, "Sur les ossemens humains" (1829), a summary of his letter to Cordier, read at the Académie on 15 February (immediately after one from Serres on the same topic); the full text is in "Cavernes à ossemens de Bize" (1829).

13. Later excavations in the Bize caves showed that their history was much more complex than Tournal imagined, and that their repeated occupation had extended into Neolithic times: hence the pottery.

14. Tournal, "Cavernes à ossemens de Bize" (1829), 258; as the context shows, the word "archaeology" denoted studies of Antiquity *within* the span of literate human history, not before it. His joint publication with Serres never appeared, probably in part because they disagreed on the interpretation of the caves.

This was a bold and novel claim for complete continuity between human history and the nearer reaches of geohistory, inferring a gradual change in the mammalian fauna, perhaps linked causally with the appearance and spread of human populations. But the claim came from a naturalist who was both young and a provincial, and Tournal had yet to convince his seniors in Paris. Still, the combined impact of his papers and Christol's, together with those of the older but perhaps less reliable Serres, built up a certain momentum at the Académie. That august body set up a committee, headed—unsurprisingly—by Cuvier, to examine the whole range of claims for the contemporaneity of humans and extinct animals. Christol and Tournal promptly sent some of their specimens to Paris, to help prove their point. But Tournal's were apparently mislaid or lost; so when Cuvier complained at the Académie that his committee was still waiting to be sent adequate material before it could report back, Tournal sent him another batch, while hinting that he recognized that someone in Cuvier's position was bound to be extremely cautious in judging the issue (Fig. 16.2).[15]

At another meeting several months later, Cuvier still used the same excuse for the delay in producing the committee's report. But he also redefined the problem in such a way as to discount completely what the Languedoc naturalists were claiming. He argued that the wide public interest in the matter was based on a misunderstanding. He asserted that it was not in doubt that human bones were found in "accidental formations, in hardened silts etc.", that is, in the Superficial deposits that Christol and Tournal had been investigating in their respective caves. Cuvier claimed that what mattered to geologists, by contrast, was the alleged existence of "true anthropolites, that is, human bones in the rocks of

Fig. 16.2 An informal portrait of Cuvier, drawn around the time when he was involved in arguments about the authenticity of claims of human antiquity. His main research project at this time, however, was on fossil fish (see §30.1), as indicated by the specimen he chose to be shown holding while being sketched.

regular formation, [of which] there is not yet a single example". But this evaded the point at issue. It was indeed agreed that no genuine human fossils had been found in "regular" formations, even in the Tertiaries, since those claimed half a century earlier (*BLT* §5.4) had long been discounted. But Cuvier knew very well that what was at issue was the contemporaneity of humans with the extinct mammals from the Superficial deposits, the fauna that he himself had been the first to reconstruct. It was a disingenuous, or even devious, manoeuver with which to avoid confronting what Christol and Tournal claimed to have found.[16]

Shortly afterward, and perhaps in consequence of what Cuvier had said, Tournal made his promised trip to Paris, taking with him yet more specimens and confronting the great man in person. On his return home he thanked Cuvier for his kindness to him, but pointedly reminded him that a report was still eagerly awaited. He insisted that it would need to assess, among others, "the fossils of the Bize caves, so remarkable, 1° by the prodigious quantity of bones that they contain, 2° by the difference of population, compared to that of all other known caves, [and] 3° by the mixture of human bones, pottery, and bones of animals belonging to extinct species." The question of the antiquity of the human species remained, for the time being, contentious and unresolved (see Chap. 28).[17]

16.4 CONCLUSION

The sharp boundary that Cuvier, followed by Buckland and many others, had drawn between the "present world" of recorded human history and the "former world" of the extinct mammalian megafauna was beginning to seem questionable in several ways. Scrope claimed that ordinary erosion by rivers had continued smoothly across the alleged boundary, at least in central France, and he therefore rejected the reality of any widespread "diluvial" event (§15.1). Croizet and Jobert argued in addition that in the same region the extinct mammals seemed to have died out without any catastrophic marine incursion, apparently in piecemeal fashion, and perhaps as a result of slow climatic change (§15.2). The sharp division between a human and a prehuman world seemed likewise to be less secure than it had been only a few years earlier. Cuvier and Buckland were justifiably skeptical about claims for the contemporaneity of humans with the extinct megafauna, because they both recognized how deceptive the evidence of bone caves could be, unless extreme care was taken in excavating them. Having almost been deceived by the famous skeleton at Paviland, Buckland was

15. Fig. 16.2 is reproduced from Bultingaire, "Iconographie de Cuvier" (1932), pl. 7; this undated sketch by Lizinka Aimée Zoë de Mirbel (1796–1849) was almost certainly drawn in Cuvier's last years, when his main research was on fossil fish (see §30.1). Meetings of the Académie on 23 November and 28 December 1829 (*RB* 1829: 148–49, 155); Tournal to Cuvier, 3 January 1830 [Paris-IF: MS 3252/80]. Any suspicion that Cuvier conveniently "lost" the first batch of specimens seems groundless: his comprehensive incoming correspondence in Paris-IF and Paris-MHN does not include the covering letter that would surely have accompanied the specimens had they ever reached him.

16. Académie des Sciences, 3 May 1830 (*RB* 1830: 76).

17. Tournal to Cuvier, 13 June 1830 [Paris-MHN: MS 627, no. 104]. Lyon, "Search for fossil man" (1969), is a useful summary of the research described in this section; see also Backenköhler, "Cuviers langer Schatten" (2002).

justifiably cautious about the artifacts that the local naturalist McEnery found with fossil animal bones in Kent's Hole in the south of England.

However, the new caves discovered in the south of France by two other local naturalists were much more difficult to dismiss. Christol reported human bones mixed with those of the unmistakably "antediluvial" species that he had already described. Tournal made a different but equally significant claim, when he found human bones mixed with those of animals that seemed collectively *intermediate* between the "antediluvial" fauna and that of the present day. Taken together, these finds suggested strongly that humans had indeed lived among the extinct mammals, but also that the human presence might have continued uninterrupted into historical times, while the fauna was changing by piecemeal extinction. Buckland, on the other hand, retained confidence in his diluvial theory; but he was explicitly willing to see it modified by accepting that humans might have spread at an early date as far as Europe, *before* the putative diluvial event overwhelmed the megafauna. Cuvier was more resistant to any modification of his long-standing claim that the event was not only real and decisive, but also that it marked a genuine and sharp boundary between the human world and an earlier one that had been wholly prehuman, at least in Europe. He resisted Tournal's contrary claims with delaying tactics that went beyond justifiable scientific caution and bordered on deviousness. The issue of human antiquity was, for the time being, unresolved.

Cuvier's opposition to the new discoveries was not motivated by biblical literalism, or indeed by any other religious concern. But they did threaten his forceful claim that the fossil megafauna on which he had first made his scientific reputation had been wiped out by a purely natural catastrophe, not by any human activity. This alone, in his opinion, could secure his broader claim that extinction was an integral part of the natural world, not just in the recent past but throughout geohistory. At the same time, however, he also needed to defend his interpretation of recent geohistory against another and even more worrying alternative, according to which the fossil animals had not become extinct at all (unless in some cases by human agency), but had *transmuted* (in modern terms, evolved) into the species alive in the present world. The next chapter, the last in Part Two of this narrative, describes this threat from Lamarck's transformist theories, a threat that arose over just the same years.

CHAPTER 17

The specter of transmutation (1825–29)

17.1 GEOFFROY'S NEW TRANSFORMISM

By the 1820s the reality of extinction, as a natural process that had been in operation long before the earth supported any human life, had been accepted by almost all naturalists. Whether it was attributed, with Cuvier, to occasional massive catastrophes, or, following more recent suggestions, to merely local changes in climate and environment, it was agreed that many earlier forms of life preserved as fossils had truly disappeared from the living world. On the other hand, any analogous interpretation of the *origins* of new forms of life remained far more controversial. The debates about the possibility of transmutation (in modern terms, evolution) were only loosely linked to the issues with which the present book is primarily concerned. But they must be mentioned at this point, because in the 1820s one prominent naturalist used *fossil* evidence, more explicitly than ever before, to argue the case for a general theory of transformism.[1]

The large-scale history of life, as it was becoming known in ever more consistent outline (Chaps. 4, 12), raised in a newly acute form the knotty problem of accounting for the origins of species and of larger groups of organisms. No longer could all talk of origins be relegated conveniently to the dream-time of a primal Creation. All the evidence now suggested that origins could be pinned

1. The terminology was fluid, but in the present account the word *transmutation* denotes the process, and *transformism* the theory; the modern term *evolution* cannot be used without confusion, because in the early nineteenth century (and even later, in Darwin's time) it still denoted primarily the *ontogenetic* pathway of development from embryo to adult (an *evolutio* or "unfolding"), not the *phylogenetic* pathway of change from ancestral to descendant forms of life. Laurent, *Paléontologie et évolution* (1987), gives a massively detailed account of the transformist debates in France (where much of the important action took place) in the years up to Darwin's *Origin of species* (1859).

down to specific "epochs" or moments in geohistory: for example, the belemnites at the time of the *Muschelkalk* (§4.3), the mastodon at the time of the Superficial deposits, in either case certainly not later and probably not much earlier. The origin of the mammals had seemed to be timed around the start of the Tertiary era, until the discovery of the little marsupials of Stonesfield pushed it right back into the "Jurassic", among the younger Secondaries (§5.5); but even allowing for the vagaries of preservation it seemed unlikely to have been much earlier than that. And broadly the same history of life appeared to be recorded everywhere that the sequence of formations was explored and their fossils collected. So Cuvier's earlier provisional explanation of apparent origins—in terms of migration from one region or continent to another (*BLT* §10.3)—now looked no better than a stopgap solution, and one that was rapidly losing its plausibility.

In this situation, the reputation of Cuvier's old antagonist might be revived. Lamarck's transformist theories (*BLT* §7.4) had never been eclipsed, let alone suppressed: even in France, despite Cuvier's hostility and powerful institutional position, Lamarck's ideas had a steady following, even a growing one, among the general public and at least some savants. On the other hand, most of the naturalists who made the newly defined science of "paléontologie" (§4.1) their special study continued to regard transformism as highly implausible, for all the reasons that Cuvier had set out in his original "Preliminary Discourse" (*BLT* §10.3) and repeated in its later editions. Transformism was often mentioned only to be condemned at once as a wild and unsupported speculation.

Lamarck himself had described and classified the fossil mollusks of the Paris formations in exemplary fashion at the start of the century, but he had barely hinted at any transformist interpretation of them (*BLT* §7.4). Conversely, he had deployed almost no fossil evidence when in 1809 he set out his general theory of transformism in his *Zoological Philosophy* (*BLT* §10.1). In 1825, however, just as Cuvier published his *Fossil Bones* in its definitive edition, his slightly younger colleague Étienne Geoffroy Saint-Hilaire (1772–1844), the Muséum's professor of vertebrate zoology since its foundation, had the temerity to move the argument unequivocally onto Cuvier's territory. Geoffroy published a major paper in the Muséum's *Mémoires*, raising the question whether living crocodiles—or more precisely, gavials—might not be the direct descendants, "by way of uninterrupted generation", of the fossil crocodiles found in the Secondary formations of Normandy; and he linked this suggestion explicitly with Lamarck's transformist ideas.[2]

Geoffroy's bombshell did not come out of the blue. After his early collaboration with Cuvier lapsed (*BLT* §7.1), Geoffroy had continued his studies of vertebrate anatomy, but they did not compete directly with Cuvier's. Cuvier focused on the relation between structure and function, anatomy and physiology, whereas Geoffroy had traced the more abstract topological connections between corresponding anatomical parts of different kinds of animal (in modern terms, their homologies), regardless of their functions: for example, he had tried to correlate the bones in the lateral fins of fishes with those in the limbs of quadrupeds. A decade later, he had expanded his ideas in two volumes of *Anatomical Philosophy* (1818–22), a title clearly modeled on Lamarck's great work. Geoffroy certainly intended to support the idea of transformism. But it was only when he tackled the crocodiles—on which he regarded himself as no less an expert than Cuvier—that it became clear that he was reformulating Lamarck's theory as a means to explain the *fossil* record.[3]

Geoffroy's treatment of crocodiles, both living and "antediluvial", was primarily anatomical. He used the similarities between living and fossil forms to argue that the fossil species could have been transmuted "by an uninterrupted succession" into the living ones, simply as a result of changes in the shapes and proportions of the bony components. His intervention was triggered by the discovery near Caen in 1823 of a new fossil "crocodile" skull. A local naturalist sent the specimen to Paris, where Cuvier described it briefly (in his new edition of *Fossil Bones*) as an extinct species of gavial. Geoffroy, on the other hand, regarded the "Caen crocodile" as a completely new genus. He named it *Teleosaurus*, because from the deep past of the Jurassic its anatomy pointed towards the goal [*telos*] of the mammals that were yet to come; in certain features it was, he claimed, *intermediate* in structure between reptilian and mammalian forms (Fig. 17.1).[4]

Geoffroy claimed that the anatomical constitution of animals was certainly flexible or labile enough to have undergone such transmutations in the course of geohistory, because analogous structural changes could be directly observed on a small scale even in the present world. "What requires some considerable time in the great operations of nature", he claimed, "is at all times accessible to our senses, and is found in miniature and under our eyes in the phenomenon of monstrosities, whether accidental [i.e., occurring naturally] or induced artificially." For the former he drew on the resources of the great Paris hospitals; for the latter he used the facilities of an egg-hatching factory outside Paris to experiment on chick embryos, inducing monstrosities by altering their environments.[5]

Invoking monstrosities—a phenomenon that in its human form was extremely distressing to all concerned—was hardly likely to commend Geoffroy's version of transformism to naturalists, let alone to any wider public. It was difficult to reconcile with the dominant view (famously expressed by Cuvier but

2. Geoffroy, "Organisation des gavials" (1825); the quoted phrase is in the long subtitle. On Lamarck's followers, see Corsi, *Age of Lamarck* (1988), chaps. 7, 8; those who attended his lectures are listed in its revised edition *Lamarck: genèse et enjeux* (2001), 333–65. On the contrasts between his transformism and Geoffroy's, see Laurent, "Cheminement de Geoffroy" (1977) and *Paléontologie et évolution* (1987), 324–53.

3. Geoffroy, *Philosophie anatomique* (1818–22). Appel, *Cuvier-Geoffroy debate* (1987), focuses primarily on their famous and very public argument in 1830, which concerned alleged homologies between vertebrates and invertebrates; this brought into the open their fundamental disagreement about the nature of living organisms, but did not impinge directly on the issues being traced in the present book; see also Bourdier, "Geoffroy versus Cuvier" (1969). Le Guyader, *Geoffroy* (1998), usefully prints lengthy quotations from Geoffroy's published work but has little on his crocodile papers (chap. 4).

4. Fig. 17.1 is reproduced from Geoffroy, "Organisation des gavials" (1825), pl. 6, drawn by Meunier.

5. Geoffroy, "Organisation des gavials" (1825), 152. He had already dealt with *human* "accidental" monstrosities in *Philosophie anatomique* 2 (1822) and "Sur la monstruosité" (1825); his chick experiments were described in "Déviations organiques" (1826). Rostand, "Geoffroy et la tératogenèse" (1966), aptly terms the theory "transformisme expérimentale". Geoffroy's use of an artificial process as the key to a putative natural one was analogous to Charles Darwin's later use of artificial selection in the practice of animal and plant breeding as a key to "natural selection" in the wild; it met with similar objections that such an analogy between natural and artificial processes could not be valid.

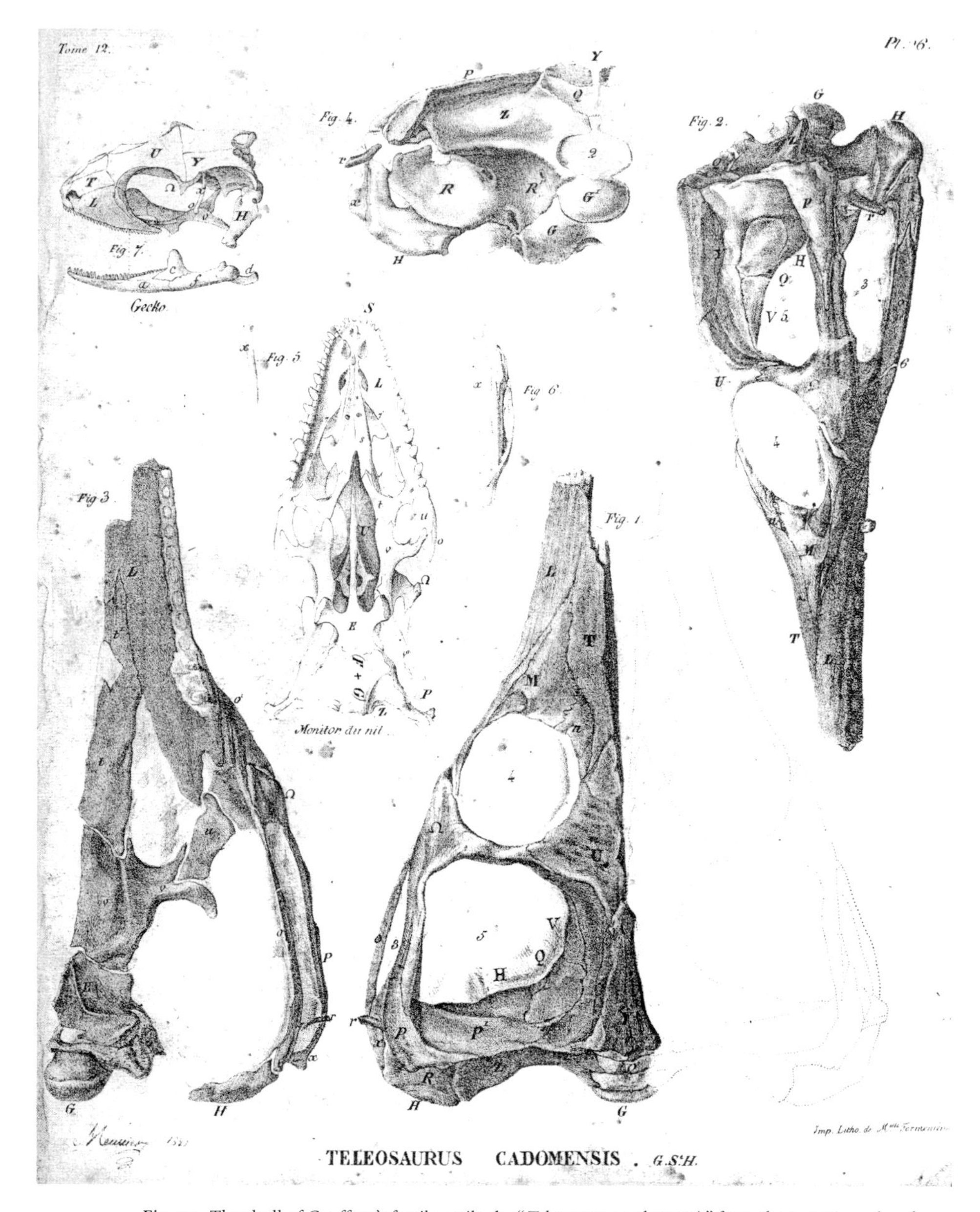

Fig. 17.1 The skull of Geoffroy's fossil reptile the "*Teleosaurus cadomensis*" from the Jurassic rocks of Caen in Normandy (figs. 1-3), with details of the living Nile crocodile and the gecko (figs. 4-7) for comparison, to illustrate his claim that its osteology made it a transformist (or evolutionary) link between true crocodiles *and mammals.* Geoffroy's encroachment onto Cuvier's territory of fossil anatomy, published in the Muséum's *Mémoires* in 1825, aggravated the latter's hostility to this revival of the transformist theorizing of their elderly colleague Lamarck.

held by most other naturalists too) that all organisms were closely adapted to specific modes of life, and were therefore bound to be anatomically stable if they were to survive. And it was utterly repugnant to the even more widely held view (famously expressed, for anglophones, by William Paley: see §29.1) that highly adaptive structures were the marks of ultimately divine design. To make organic change dependent on "accidents"—in the sense of aberrations from the law-governed regularities of nature—was no way to make transformism widely acceptable.

Yet Geoffroy's argument unmistakably followed the actualistic method that was approved, at least in principle, on all sides: the observable present—of monstrosities "accidental" or induced—could act as the key to the unobservable past of deep geohistory. He had earlier conceded that the mummified animals from ancient Egypt, which he himself had carefully collected during Bonaparte's military expedition years before (*BLT* §7.4), were a valid test of the reality of transmutation, although they dated from only a few thousand years ago. But now he claimed that it had been premature for the Muséum's professors to conclude (in his absence) that the mummies were identical to living forms and therefore strong evidence *against* transformism (*BLT* §10.3). He argued that the issue should remain open, until he himself was able to examine the specimens (and particularly the mummified crocodiles) more closely.[6]

Geoffroy was suggesting a *causal* explanation for organic change, and one directly related to the new science of geology. He had long since rejected Lamarck's primary cause of transmutation, namely the alleged inherent tendency of all organic forms to increase inexorably in complexity in the course of time, toward some kind of perfection. Instead, Geoffroy highlighted his senior colleague's secondary factor, the organic response to environmental change; in a footnote he particularly recommended the relevant passage in Lamarck's book "to the meditation of young people". But he went beyond his colleague by explicitly attributing the transmutation of animals such as the gavials to the great physical changes that marked the vast spans of geohistory, particularly "at the epochs signalled as *diluvial*".[7]

In 1828, in a further paper in the Muséum's *Mémoires*, Geoffroy summarized his way of applying Lamarck's transformist theory to the fossil record. The fundamental question *for geology*, he argued, was "whether there has been uninterrupted transmission from antediluvial beings into the animals of modern times". He took Cuvier's *Fossil Bones* as his indispensable baseline, because his colleague, by treating fossils as the "coins" [*médailles*] of geohistory and achieving "a sort of resurrection of the earth's first inhabitants" (§1.3), had brought the question of origins into geohistory and hence into the empirical realm. But Geoffroy argued that the transmutation of "antediluvial" species into modern ones was not as unthinkable as Cuvier and others claimed. What was involved was usually no more than changes in the proportions of anatomical parts, often less radical than those observed or induced as monstrosities in living forms. All the vertebrates (to which his argument was restricted) could be regarded as variations on a single anatomical theme, namely an "abstract being or common type", a ground plan that could be recognized in diverse forms across the whole range of the vertebrates, from fish to mammals. Anatomical transformations right across that range were therefore topologically conceivable, and Geoffroy argued that they

6. Geoffroy, "Organisation des gavials" (1825), 150–55.

7. Geoffroy, "Organisation des gavials" (1825), 150–52. Lamarck had adamantly opposed Cuvier's emphasis on a drastic *révolution* or "diluvial" episode (*BLT* §7.4).

could well have been induced by the physical "revolutions" or major changes in environment that geologists inferred from the record of the rocks.[8]

Geoffroy's intervention in matters of paleontology might have had a greater impact had he shown a better comprehension of the fossil record that he claimed to explain in transformist terms. But he remained a traditional naturalist, working indoors in the Muséum (or indoors watching eggs hatching into chicks). Unlike Cuvier—who at least had the highly competent Prévost acting as his agent outdoors in the field—Geoffroy seems to have had little if any direct experience of finding fossils in situ in their geological formations, even those in Normandy from which his fossil crocodiles had been collected. He suggested that "a sort of chronology" could be derived from a "progressive series" of fossil vertebrates. But the series he proposed, albeit provisionally, started with the ichthyosaur, plesiosaur, and pterodactyl, then passed by way of the Maastricht mosasaur and the Caen crocodile to the American megatherium and megalonix, and ended with the Parisian palaeotherium and anoplotherium. Any competent geologist, familiar with the currently agreed outline of geohistory based on stratigraphy (see Fig. 9.1), would have been baffled or even repelled to see that this supposedly "chronological" series started in what was now being called a Jurassic formation (the Lias), moved forwards to a Cretaceous one (the Chalk), then back to the Jurassic of Normandy, before jumping forwards to the Superficial deposits and finally shifting back to the Tertiaries. Geoffroy's series was not likely to inspire confidence among geologists, nor to persuade them to look to transformism to explain organic change.[9]

Furthermore, there was little in the work of other naturalists to suggest that a better case than Geoffroy's could be made for interpreting sequences of fossils in transformist terms. Cuvier's principal objection—the lack of instances of intermediate forms between living and fossil species—remained conspicuously unmet. However, there was one case that was at least suggestive, although like Geoffroy's crocodilian case it was at a higher taxonomic level than that of species. Among the Tertiary fossils that Deshayes described from the Paris region (§12.1) was a curious mollusk that he had named the *béloptère*. Blainville included it in his monograph on belemnites (§4.3), claiming that it was structurally intermediate between those extinct and enigmatic fossil animals and the living cuttlefish. Blainville himself expressed this in the conventional atemporal language of the animal "series" or "scale of beings": by linking cuttlefish to belemnites, the *béloptère* also showed that the belemnites in turn filled the much larger and more important anatomical gap between the cuttlefish and the pearly nautilus, both of them *living* members of the molluskan group that Cuvier called cephalopods (Fig. 17.2).[10]

In any case, even if Blainville's *béloptère* and Geoffroy's crocodiles were unconvincing, the papers that the latter published in the Muséum's prestigious and widely distributed *Mémoires* ensured that no well-informed naturalist anywhere could remain unaware that transformist theorizing was alive and well, at least in France. Lamarck himself died in 1829—the year after the most important of Geoffroy's papers was published—at the ripe old age of eighty-five, sadly blind

Fig. 17.2 Fossils illustrated in Blainville's monograph on belemnites (1827), showing how the *béloptère* (figs. 2, 3: different views of two species) could be interpreted as anatomically *intermediate* between the "bone" of the living cuttlefish *Sepia* (figs. 1, 1a, 1b: ventral and lateral views, and section) and fossil belemnites (figs. 4-5: longitudinal section, two cross-sections, and the conical series of chambers). Since the *béloptère* was found in a Tertiary formation (the Coarse Limestone of the Paris Basin), it was also *geohistorically* intermediate between the belemnites, found only in Secondary formations, and the living cuttlefish. But Blainville did *not* here interpret it as being transitional in a transformist (evolutionary) sense. (By permission of the Syndics of Cambridge University Library)

8. Geoffroy, "Structure organique et de parenté" (1828); unlike Lamarck, Cuvier was never mentioned by name, though his publications were. In the latter part of this paper, Geoffroy pursued his transformist theme into contemporary controversies about "generation" (very roughly, in modern terms, those of developmental genetics and embryology), which are less relevant in the present context. See the classic analysis in Russell, *Form and function* (1916), 52–78. Geoffroy's transformism was not unlike some twentieth-century ideas of evolution by macro-mutations, which were sometimes disparagingly branded "hopeful monsters".

9. Geoffroy, "Structure organique et de parenté" (1828), 215. It is not clear what exactly he thought his "series" showed, in what sense it was "progressive", and whether he meant it to be (in modern terms) truly phylogenetic or merely morphological. But it was certainly not chronological in any sense that geologists recognized.

10. Fig. 17.2 is reproduced from Blainville, *Bélemnites* (1827) [Cambridge-UL: MF.51.3], part of pl. 1; see 26–31. The two species of *beloptère* shown here had been given trivial (specific) names that further underlined the intermediate character of the genus: Cuvier's *B. sepioidea* (fig. 2) and Deshayes's *B. belemnitoidea* (fig. 3).

but certainly not neglected or scorned by other savants. On the contrary, although Cuvier remained fiercely hostile to transformism and all its works, Lamarck had an enthusiastic following among many younger naturalists on the Continent, and even some in Britain; and his *Zoological Philosophy* was reissued the year after his death, giving it a new lease on life.[11]

17.2 LYELL CONFRONTS LAMARCK

As already emphasized, however, Geoffroy's ideas—and those of Lamarck from which they were derived—seemed to most naturalists to be utterly incompatible with the Cuvierian emphasis on the close integration of structure and function in well-adapted organisms. And to a wider public (and also to many savants) any such transformist theory was utterly repugnant because it reduced the embodiments of divine design to products of mere chance and accident. It was altogether typical that when Conybeare described his new plesiosaur as an "intermediate form [that] adds new links to the connected chain of organized beings" (§2.3), he had immediately dismissed—and banished to a mere footnote—Lamarck's "monstrous" idea that such links were "real [*sic*] transitions from one branch to another of the animal kingdom".[12]

Yet the sheer vehemence of many of these repudiations of transformism points to the underlying anxiety, generally unspoken, that Lamarck's theory had implications reaching into far more sensitive territory than that of extinct Jurassic reptiles. It was perceived as threatening human "dignity" by implying that the human species too might have been derived "by way of uninterrupted generation" from something much less than human. It is beyond the scope of the present book to sketch, even in outline, the complex debates at this time over the possibility of human evolution, and the reasons why it was so often perceived as a threat to the status of humanity itself (and not only by savants such as Conybeare with strong religious commitments). Here it can only be noted that the specter of Lamarck's transformism cast a long shadow over the continuing debates about the interpretation of the history of life, which was being reconstructed with increasing detail and reliability by those who practiced the newly named science of paleontology.

One example of the long shadow of transformism, which turned out to be of great significance for the debates about geohistory, was Lamarck's impact on Lyell. Lyell read Lamarck's *Zoological Philosophy* for the first time while he was working as a barrister in the courts of southwest England, only a few months after he had given the readers of the *Quarterly Review* a positive account of the strongly directional geohistory that had emerged from the latest geological research (§14.4). He must already have been well aware of Lamarck's theory: as just emphasized, it was impossible for any well-informed naturalist not to be. But even the first of Geoffroy's two major articles in the Muséum's *Mémoires* would have alerted him to the *geological* implications of Lamarck's ideas, and he may well have followed Geoffroy's explicit advice to learn about them directly from the horse's mouth.[13]

After reading the book, Lyell told Mantell that Lamarck's theories "delighted me more than any novel I ever read, and much in the same way, for they address themselves to the imagination at least of Geologists". This was a backhanded compliment, as it had been ever since Buffon's geotheory was routinely dismissed in the same way as a mere "novel" [*roman*] (*BLT* §3.2); and at this time the faculty of "imagination" was often contrasted unfavorably, at least in scientific circles, with sober reasoning. But Lyell recognized "the mighty inferences which would be deducible [from Lamarck's theories] were they established by observation". For if transmutation were ever to be *seen in operation*, or at least recorded within the span of human history (as it might have been in the test case of the Egyptian animal mummies), it would be established as a true actual cause. It might then explain the directional and progressive trend in the fossil record that Lyell's review had so clearly endorsed. But Lyell, alluding to his current legal work, now admitted that he read Lamarck's work "rather as I hear an advocate on the wrong side, to know what can be made of the case in good hands". For if the case were valid it would surely have to apply universally, and it might therefore "prove that men may have come from the Orang-outang". Yet he was half persuaded. "But after all what changes species may really undergo!", he exclaimed, in reluctant admiration; "How impossible will it be to distinguish & lay down a line beyond which some of the so called extinct species have never passed into recent [i.e., living] ones".[14]

Having digested Lamarck's arguments, however, Lyell recognized the massive danger they posed. The notes he wrote privately during the months that followed his reading of Lamarck show that he was wrestling with the unnerving *human* implications of transformism. In an important note on "Lamarck's theory", for example, Lyell implicitly dismissed the common objection that transformism diminished or eliminated the divine role in nature, and showed himself to be as much steeped in natural theology as any of his British contemporaries. He pointed out—if only, at this stage, to himself—that "To form a man who shall reproduce himself is a remarkable manifestation of creative Power & to produce animals who may have the faculty of producing [by transmutation] other more

11. On the importation of Lamarck's and Geoffroy's transformism into Britain, and its significance as a Radical alternative to mainstream biology and medicine, see Desmond, *Politics of evolution* (1989), chap. 2. Secord, "Edinburgh Lamarckians" (1991), makes a persuasive case for attributing the anonymous "Importance of geology" (1826), published in Jameson's periodical and quite Lamarckian in content, not to the known Lamarckian Robert Grant but to Jameson himself, making the latter a closet Lamarckian.

12. De la Beche and Conybeare, "New fossil animal" (1821), 561n; Conybeare regarded both the "links" and the "chain" as being, in modern terms, morphological rather than phylogenetic (§2.3). He used the epithet "monstrous" in its traditional emotive sense, but not long before Geoffroy brought it in its technical sense into the debate about transformism.

13. Rather oddly, when Lyell later tackled these issues in *Principles* 2 (1832), he referred to Geoffroy's work (2–3) without citing any of his papers in the usual way in his footnotes, or directly confronting the Frenchman's claims about the transformist interpretation *of fossils* (see §24.2).

14. Lyell to Mantell, 2 March 1827 [Wellington-ATL], printed in *LLJ* 1: 168–69; Bartholomew, "Lyell and evolution" (1973), is still the most valuable analysis of this issue. The orang-utan of Sumatra (now in Indonesia) was at this time the best known of the great apes, and the species usually cited in discussions of human origins and affinities. Lyell read Lamarck in French, of course; there was no English translation until the high tide of neo-Lamarckian biology in the early twentieth century.

perfect [animals], would be as remarkable a display of [divine] power as to create each species of animal [separately], for time is nothing to the Eternal being". In other words, transformism implied a view of the Creator's role that was *more* lofty, not less, than the traditional assumption of divine involvement in the origin of every species. This and many other passages demonstrate that Lyell—notwithstanding modern secularist myths about him—was far from being averse to using religious language.[15]

However, having in his own mind neutralized any theological objection to transformism, Lyell nonetheless set out to protect his favorite science against any Lamarckian reading, in order to defend his own species against the indignity of being assigned a merely animal origin. He evidently recognized that the best way for him, *as a geologist*, to undermine the plausibility of Lamarck's theory was to question the directionality that most other geologists—including, hitherto, himself—attributed to geohistory. He now wondered whether rocks and fossils could be reinterpreted as traces of a geohistory that had *not* been directional, for if so the history of life could not so easily be attributed to any underlying process of transmutation.

Lyell was searching, for example, for an alternative explanation of the oblately spheroidal "figure" or shape of the earth, which had been a key argument in favor of the earth's hot fluid origin ever since Buffon's time (*BLT* §3.2). He reminded himself to ask Herschel, his main informant on matters of physics and astronomy, whether "the *successive* liquefaction by fire & water of every portion of the Earth's surface [could] account for its spheroidal form". If it could, the directional model of a cooling earth could perhaps be replaced by one in which the planet had always been in a steady state, without any hot origin at all. In the privacy of his notebooks he therefore argued in particular with those he called "the Refrigeration men", the physicists who claimed that the earth had been cooling steadily throughout its recorded history (§9.2) and the geologists who interpreted the fossil record in terms of that cooling (§12.2, §12.3). Lyell questioned whether the fossils did in fact imply—as these savants claimed—that "refrigeration has been going on at a regular rate" since the Carboniferous period, and whether the fossils might be made to show instead that the earth "has been alternately warmer & colder & may be so again [in the future]". In place of the cooling curve now adopted (implicitly or explicitly) by most other geologists, Lyell was exploring the possibility of a radically divergent reading of geohistory, "if on the contrary it be contended that all is a fluctuating cycle", or in the long run a steady state like that proposed by James Hutton forty years before (*BLT* §3.4).[16]

To make any such steady-state geohistory convincing, however, Lyell had to explain away the apparently massive evidence for a progressive trend in the fossil record of the history of life. Under the heading of "Human remains"—which, significantly, immediately followed his notes on "Lamarck's theory"—he suggested to himself that since the human species can live in the widest range of climates, "the finding [of] distinct fossil animals of a warmer climate does not prove that man might not then have existed". In other words, Lyell was privately speculating—again in a radically divergent mode—that the vast tracts of deep geohistory, with their apparently tropical faunas and floras, might *not* have been prehuman after all. In the same vein he noted that the Carboniferous "contains in coal field teeth of Shark & high class of fishes, near to reptiles", implying that the quadrupeds might already have been in existence in what he called "the first Zoological era", thereby undermining the reality of any organic progression at all.[17]

The significance of such cases for Lyell's tentative steady-state conception of geohistory depended on an implicit inference that the fossil record was a deeply unreliable guide to the history of life, owing to its intrinsic imperfections. This inference seemed to him to be reinforced—with perfect timing for his private speculations—when his friend and fellow barrister William John Broderip (1789–1859) found the mislaid second specimen of the tiny but outstandingly important Stonesfield marsupials (§5.5). Fitton then revisited the site in the company of the German mining geologists Karl von Oeynhausen (1795–1865) and Ernst Heinrich Carl von Dechen (1800–1889), who were visiting Britain at the time. This international team checked the stratigraphy of the Stonesfield "slate", and agreed that it was definitely an integral part of the Oolite formations, contrary

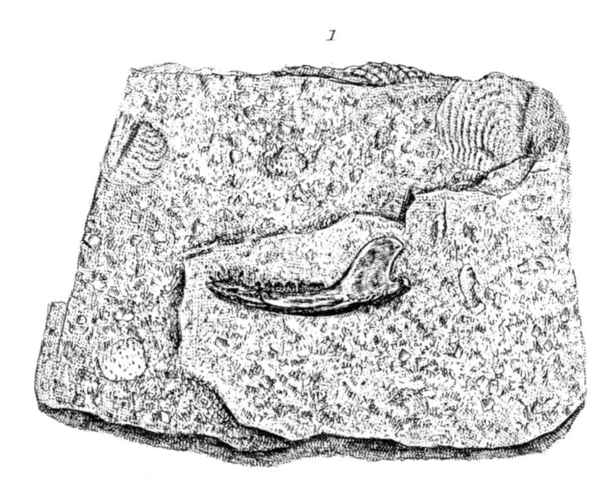

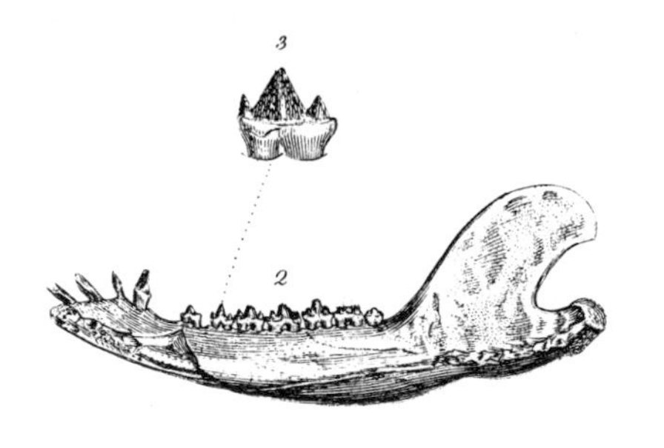

Fig. 17.3 Broderip's specimen of the lower jaw of a small marsupial, found in the Stonesfield "slate" many years earlier, mislaid, but found again in 1827: it is shown (1) at natural size on a piece of limestone with the distinctive marine mollusk *Trigonia*; (2) enlarged as if detached from the rock; and (3) with one of its molar teeth further enlarged. In conjunction with Fitton's stratigraphical fieldwork it confirmed decisively that mammals had already existed not just in the Tertiary era but as far back in geohistory as the time of the great extinct Jurassic reptiles such as Buckland's megalosaur, which came from the very same stratum (§5.1). This supported a slight but significant modification of the generally agreed inference that the history of life had been broadly directional and "progressive", with reptiles preceding mammals, and perhaps marsupials preceding placentals. But Lyell used it privately as evidence that this inference might be radically mistaken, and that a full range of mammals, the "highest" forms of life (apart from humans), might have been present already in the earliest periods of geohistory. (By permission of the Syndics of Cambridge University Library)

15. Lyell, notes for *ca.* July and August 1827 [Kirriemuir-KH: notebook 5: 51–52]. Wilson, *Lyell* (1972), 181, prints an excerpt from Lyell's note on Lamarck but omits the crucial passage in which Lyell revealingly uses religious terminology. Klaver, *Geology and religious sentiment* (1997), 15–84, is the best analysis of Lyell's natural theology and his conception of the human place in nature; it is beyond the scope of the present volume to deal with these topics as fully as they deserve.

16. Lyell, notes for *ca.* September to December 1827 [Kirriemuir-KH: notebook 6: 151; notebook 7: 2].

17. Lyell, notes for July 1827 [Kirriemuir-KH: notebook 5: 52–53, 104.]

to Prévost's earlier doubts about it. So there had already been *some* mammals on earth—however small and apparently insignificant these ones were—well back in the Secondary era. Lyell promptly reported this discovery to his father (who over the years had become more appreciative of his son's scientific interests) in an unmistakable tone of jubilation:

> The lost jaw of the Didelphis of the Stonesfield slate is found again!—a fine thing [Fig. 17.3]. This one seems a true opossum & in a beautiful state. It is another species [different from the first specimen]. So much for the [low] antiquity of terrestrial mammalia, & for the theories of a gradual progress to perfection! There was everything but Man even as far back as the Oolite.[18]

17.3 CONCLUSION

Lamarck's earlier theory of transformism, according to which all organic forms were and always had been in a state of continuous flux in the direction of increasing complexity and perfection, continued to be attractive to some savants, but generally not to those who studied fossils. For the latter, Cuvier's concept of discrete species, each well adapted to a particular mode of life, remained far more persuasive, in part because it was also congruent with the widely held belief that the adaptations of organisms reflected an ultimately divine design. However, this traditional concept had been radically modified by the new dimension of geohistory. Cuvier's insistence on the reality of *extinction* as a perennial feature of the natural world had already won almost universal acceptance. By contrast, the *origins* of species and larger groups remained utterly mysterious; indeed the problem had become even more puzzling, since origins, like extinctions, could now be pinned down (with substantial margins of uncertainty, of course) to specific moments in geohistory.

At this point Lamarck's (and Cuvier's) colleague Geoffroy gave the debate a new turn, by suggesting that fossil species might have been transmuted into living ones under the pressure of severe environmental change, during the occasional drastic "revolutions" that most geologists inferred from the record of the rocks. And he claimed that this kind of transmutation could be seen in operation on a small scale as an actual cause, namely in the monstrosities generated during embryonic development by natural "accidents" (which could be replicated by human intervention). Geoffroy's suggestion was widely regarded as unacceptable, and even treated with repugnance, but his work did ensure that no well-informed naturalist could be unaware of the possibility that a transformist theory *of some kind* might help to solve the mystery of organic origins.

Although Geoffroy did not take his ideas further in this direction, the process he suggested was compatible with inferring that some other fossil species might *not* have undergone any such transmutation and might instead have become truly extinct. So the course of geohistory might have been marked *both* by transmutation *and* by extinction: unlike Lamarck's theory, Geoffroy's did not imply that the two processes were mutually exclusive. In other words, as Brocchi had hinted (*BLT* §9.4), the lifespan of any species might be bounded by its "birth" and its "death", and yet those two natural events might have quite different causal foundations (as they have, obviously, in the case of an individual).

However, those who studied fossils found little to persuade them that Geoffroy's was the correct solution to the puzzle. And there was a further and more prosaic reason for transformist theories to be marginalized among geologists and others who studied fossils. *In practice* it made little difference how individual species or larger groups of organisms had originated; what mattered far more to geologists was that it was now possible to date these events on nature's chronology, to specific points in geohistory. New species might have appeared by transmutation, either continuously and insensibly gradually (as Lamarck supposed), or rapidly at times of environmental upheaval (as Geoffroy suggested); or they might have been formed by some other natural process as yet unknown; or even conceivably by immediate divine action. Old species might have disappeared as a result of environmental change, either sudden and catastrophic (as Cuvier and Buckland maintained), or slow and gradual (as Fleming and the Auvergnat naturalists suggested); or by some internal process analogous to old age and mortality in individuals (as Brocchi had proposed); or in some other way as yet undiscovered. These were unsolved problems, intriguing and important in their own right; but the burgeoning practice of geohistory did not have to wait for their successful resolution. What mattered more immediately to geologists was to determine *when* species had appeared or disappeared, and in what manner: whether suddenly and many at once, or in piecemeal fashion, one by one. In other words, the *history* of life on earth could be reconstructed—"bursting the limits of time"—without first having to solve the *causal* puzzle of how the changes had taken place. The debate about transformism therefore developed in parallel with the reconstruction of geohistory, with only a loose linkage between them. This justifies, or at least explains, the correspondingly marginal position that transformist (or evolutionary) theories occupy in the present narrative.

The possibility of transformism was disturbing to many naturalists—and to many others in the wider society—above all because it would clearly have to apply to the origin of humans as well as animals and plants. The consequent threat to the "dignity" of the human species was recognized with particular clarity by Lyell. In the privacy of his notebooks, he began to consider how the danger might be averted, and transformism rejected, if the apparently directional trends of geohistory were in fact illusory. In place of the almost consensual model of a

18. Lyell to Lyell senior, 14 November 1827 [Kirriemuir-KH], quoted in Wilson, *Lyell* (1972), 182. He was evidently making an exception for the human species, and assuming that its origin was geologically recent, notwithstanding his private speculation (see above) about the fallibility of its fossil record. Fig. 17.3 is reproduced from Broderip, "Observations on the jaw" (1828) [Cambridge-UL: Q384.c.19.3], pl. 11; the specimen had apparently been in his extensive collection all the time, but was somehow mislaid (Prévost's picture of the first specimen is reproduced here in Fig. 5.4). Fitton, "On the strata" (1828), was published in the *Zoological journal* immediately following Broderip's paper; both were promptly translated for the *Annales des sciences naturelles*, making the discovery known throughout Europe. Apart from the Stonesfield marsupials, the oldest known terrestrial mammals were Cuvier's placentals (notably the palaeotherium and anoplotherium) from the Gypsum formation around Paris, some way up into the Tertiaries (§1.3; *BLT* §7.5). Desmond, "Origin of mammals" (1984), interprets the history of the debates about these fossils in the same political terms as his *Politics of evolution* (1989).

gradually cooling earth, on which life had changed "progressively" by the addition (somehow!) of ever "higher" forms, Lyell began to consider a radical alternative, namely a steady-state model similar to the one that Hutton had expounded long before. Lyell's conjecture that even the "higher" forms of life might already have been in existence in the earliest times received a small but—to him—significant boost when Broderip found his mislaid marsupial jaw from Stonesfield, for this confirmed that *some* mammals had definitely been around long before the Tertiary era.

Lyell's private speculations about a possibly steady-state geohistory were powered—at least in part—by his desire to protect human "dignity" from the threat posed by Lamarck's transformist theory. The next chapter traces the further development of this research, in which he explored the implications of his new and radically divergent view of geohistory. This also opens Part Three of this narrative, the whole of which is focused on Lyell, and on his contemporaries and many critics, in the years that saw the gestation, publication, and immediate reception of his great *Principles of Geology*. The focus is justified, because Lyell's work—whether its various component parts were accepted or rejected by other geologists—dominated the 1830s as much as Cuvier's had dominated the years before and since the end of the wars.

Part Three

CHAPTER 18

Lyell and Auvergne geology (1827–28)

18.1 LYELL ON SCROPE'S AUVERGNE

Lyell had first used his platform in the *Quarterly Review* to bring the natural sciences in general to the attention of the influential Tory elites in Britain. He had then focused on his favorite science, summarizing recent research by members of the Geological Society. He had interpreted their work in terms of a picture of geohistory—which he himself clearly accepted—as broadly directional and, in the history of life, broadly "progressive" in character (§14.4). Since reading Lamarck, however, Lyell had begun to explore in private the possibility of a radically different reading of geohistory. This would replace its apparent directionality with a steady state reminiscent of Hutton's geotheory long before; and this in turn might avert the looming threat of a transformist reading of the history of life and of a merely animal origin for humanity itself. But these private speculations were reflected only faintly and indirectly in Lyell's last essay for the *Quarterly Review*.

The subject of this essay must have seemed a surprising choice, for Scrope's new book on the volcanoes of central France (§15.1) was at first glance even more specialized in content than the papers published by the Geological Society. But Lyell worked hard to convince his readers of its wider significance. He scornfully dismissed Scrope's earlier book (§9.3) as having merely "revealed to us a new system of cosmogony", and he contrasted it rather patronizingly with the "maturer judgment" shown in the new one (Lyell was about to celebrate his thirtieth birthday; Scrope had beaten him by a few months). However, Lyell dissociated himself from "that fashion, now too prevalent in this country, of discountenancing almost all geological speculation"; the allusion was to the relentlessly hostile attitude to theorizing adopted by the Geological Society, led by its first president, Greenough (*BLT* §8.4). But Lyell implied that geohistorical reconstructions based on sound fieldwork were in a different category from the fantasies of geotheory, and it was as *geohistory* that Lyell introduced Scrope's new work.[1]

More surprisingly, Lyell's very first point was one that Scrope had hardly touched on at all, namely the implications of his geohistory for the history of *life*. Lyell was largely dependent for this on what Buckland had told him about the recent and ongoing research by local naturalists in Auvergne (§15.2). Coupled with Brongniart's much earlier report that the Tertiary formations of the Massif Central were all of freshwater origin (*BLT* §9.2), the new discoveries of fossil bones in still more recent deposits made this region uniquely important for understanding geohistory, in a way that was unmatched, in Lyell's view, anywhere in Britain:

> The study of this district possesses a peculiar interest, as presenting us with evidence of a series of events of astonishing magnitude and grandeur, by which the original form and features of a country have been greatly changed, yet never so far obliterated but that they may still, in part at least, be restored in imagination. Here, in a word, there has been an entire revolution in the species of plants and animals of a region which has nevertheless preserved its identity from first to last, and survived these extraordinary vicissitudes in organic life.[2]

The region described by Scrope was thus an exception to the general rule, which had suggested that major organic changes were correlated with—and might have been caused by—major changes in physical geography, or by "revolutions" in Cuvier's more catastrophic meaning of the word. In Lyell's estimation, Scrope's analysis of the effects in central France of "rivers and floods, earthquakes and volcanos, during so long a succession of ages" was particularly valuable, "because the various stages in their gradual operation can, in many cases, be pointed out". But still more valuable, he suggested, would be "the light hereafter thrown, by the examination of the organic remains of this district, on a page of the history of animated beings hitherto almost a blank". The missing page that Lyell had in mind was the gap between the living organisms of the present world and those that Cuvier, the younger Brongniart, and others had described from the Tertiary formations: "was this change from the animals and vegetables of the tertiary period slow and gradual", he asked, "or was it effected by a sudden and abrupt transition?". It was this question, and the still more knotty problem of the causation of such organic changes, that Lyell hoped would be clarified by the fossils that were being found in Auvergne. Like Desmarest long before (*BLT* §4.3) and, more recently, the Genevan reviewer of the Auvergnat naturalists' work (§15.2), Lyell expressed it in terms of an antiquarian or *historical* analogy that his cultured readers would have found immediately appealing:

> Already have the showers of ashes that overwhelmed Pompeii, and the congeries of mud and volcanic matter which overflowed Herculaneum, provided us with information concerning the manners and arts of antiquity, which no human efforts could,

> but for them, have rescued from oblivion. And to monuments snatched, by similar catastrophes, from the wasting ravages of time, the geologist may be hereafter indebted for records that may supply many lost links in the great chain that unites the present with the past.[3]

Much of Lyell's long essay faithfully followed Scrope's local descriptions, in a degree of detail that many of his readers may have found exhausting. When he came to summarize the meaning of it all, Lyell's "chronological order" turned out to be a "retrospect", the very opposite of a geohistorical reconstruction. But he could have justified it as following the sound method of Desmarest's *marche analytique* (*BLT* §4.3), and, more recently, of the stratigraphy in "Conybeare and Phillips" (§3.2), for he too moved from the observable present back into the increasing obscurity of the deep past. So the first and most recent of Lyell's "four great periods" spanned the time since the last eruptions in central France; although geologically very brief, this already extended back beyond recorded human history. The second period covered all those geologically recent eruptions, and the third the much earlier ones (Lyell endorsed Scrope's claim that the dividing line between the two categories was arbitrary). The fourth and earliest period was that in which the Tertiary sediments had been deposited in freshwater lakes, comparable in size to the present Great Lakes in North America.

This retrospective geohistory gave Lyell a peg on which to hang a restatement of his own confidence in the total adequacy of actual causes for geological explanation, as already suggested in his paper on the lake marls near his family home in Scotland (§10.2). But his faith in the close analogy between present and past was expressed more forcefully than before, and in a form that hinted at his private doubts about the whole basis for directionality in geohistory. For he now sought to brand his potential critics as breaching all proper norms of inference, claiming that they inferred not a difference in geological conditions between past and present, but a disjunction in the basic laws of physical nature: "The majority of modern geologists are in the habit of classing the tertiary formations as belonging to a state of the earth so distinct from that now prevailing, that they feel themselves released from all necessity of reasoning on strict analogy, and at liberty to give the reins to their imagination, and to refer all inexplicable phenomena, not to their ignorance of the present operations of nature, but to the dissimilarity of her laws in former ages."[4]

1. Lyell, "*Central France* by Scrope" (1827), 438–41.

2. Lyell, "*Central France* by Scrope" (1827), 441; "revolution" was used here in its cyclic sense to denote a complete turnover of species.

3. Lyell, "*Central France* by Scrope" (1827), 442–43; on the excavations that had first suggested the analogy, see *BLT* §4.1. Lyell listed Desmarest among the many French naturalists who had published on Auvergne (439); by the time he wrote *Principles* 1 (1830), 59, he certainly knew of Desmarest's two most important papers.

4. Lyell, "*Central France* by Scrope" (1827), 471. The accusation was of course unfair and unfounded (see below), since other geologists took every opportunity to point to the analogies between conditions in the deep past and in the present, in order to reconstruct the former (Chap. 11). Lyell's real target was the diluvialists, some of whom, notably Buckland, did indeed accentuate the "otherness", not of the Tertiary era itself, but of the subsequent diluvial event.

As an example, Lyell offered a vivid verbal scene from deep time, resurrecting the Tertiary period in central France in fine Cuvierian style. This displayed his geohistorical imagination to good effect, but he also used it as a rhetorical stick with which to beat those geologists who were more impressed by the strange otherness of the deep past than by its modernity:

> What conclusions would they have drawn, had they been admitted to view the tertiary lakes of Central France in all their original beauty and repose? . . . If they had seen myriads of tender insects frequenting the banks, the crocodile and the tortoise emerging from the water, or the lake birds swimming on its surface—if they had marked [i.e., noticed] the herds browsing with security amidst forests of palms,—would they have conceived it possible that all this luxuriance of life, this variety and beauty of design, constituted no more than a transient scene?[5]

Conscious of his intelligent readers, Lyell did not evade the deeper issues that such reconstructions raised about the "place of man in nature". When Scrope had referred to the "humiliating" implications of the immensity of "Time!" (§15.1), he was playing on the common anxiety that the vast scale of geohistory reduced the human species to a cosmically insignificant latecomer. But Lyell argued that this "gloomy" view would be valid only if humans were mere animals (as the unmentionable Lamarck was taken to suppose). If on the contrary they were rational beings, which Lyell equated with a sharp dualism between mind and matter, the conventional anxiety would evaporate. For the rational power to *know* the deep prehuman past—in Cuvier's phrase, "to burst the limits of time"—then became in Lyell's perspective a proof of human cognitive "dominion" over the vast spaces of cosmic history, and therefore proof of ultimate human significance. It was not a new argument, but it was one that appropriately emphasized the cultural meaning of the science that he was bringing to the attention of the British intellectual elite.[6]

However, Lyell maintained that the main issue raised by Scrope's geology was more mundane: the reality or otherwise of the "diluvial" event proposed by Buckland and his allies. This, claimed the budding lawyer, was an issue that "calls for the strict observance of one of our fundamental rules of legal evidence", namely that any case should be founded on the best possible set of relevant facts. The Auvergne volcanoes, erupting at different times in the past, offered just the test case required; for if the valleys had been eroded by any sudden "debacle" it should be easy to determine which volcanic cones and flows dated from before that drastic event and which from after it. Yet Scrope's detailed descriptions (and his highly persuasive panoramic landscapes) showed that the "strong lines of demarcation" implied by Daubeny and endorsed by Buckland (*BLT* §10.5) were not there. So unless and until Scrope's evidence was explained in another way, his inference would have to stand and the putative "geological deluge" be abandoned altogether. Lyell's conclusion made explicit his decisive break not only with Buckland, but also with many others at the Geological Society.[7]

Significantly, however, when Lyell showed Buckland his draft of this essay, his mentor raised no objection to his argument. Instead, Buckland suggested that he add "a hit at the Penn school", alluding to the "scriptural" writers, and in particular the biblical scholar Granville Penn (1761–1844), whose literalistic works were currently seducing the British reading public. Lyell willingly did so, attacking them in withering terms: "they point out the accordance of the Mosaic history with phenomena which they have never studied, and to judge of which every page of their writings proves their consummate incompetence". On this point Buckland and his friends stood on the same side as Lyell and Scrope: they were agreed about the futility of the kind of concordance between geology and Genesis to which the "scriptural" writers aspired, and of the absurdly constricted timescale that they demanded. Whatever else remained controversial in geology, a timescale that was vast beyond human comprehension had long been taken for granted by all who aspired to be known as geologists, and also—at least on the Continent—by all among the educated public who wanted to be regarded as well informed. At Buckland's instigation, Lyell intended to help promote that view more widely in Britain too, by persuading the socially influential readers of the *Quarterly* of its validity.[8]

18.2 LYELL AS GEOLOGICAL REFORMER

At the same time that Lyell was flirting nervously with transformism, he was also planning a book on geology. He had first thought of writing a series of "conversations on geology", modeled on the best-selling *Conversations on Chemistry* by Jane Marcet, the wife of one of his friends at the Geological Society. But, as he told Mantell, he had changed his mind: he now clearly distinguished his own target readership from the burgeoning market for such works of *popular* science. He set his new and more ambitious plans in a context analogous to the movement for political reform that was agitating the British elites at this time: "I felt that in a subject where so much is to be reformed & struck out anew, & where one obtains new ideas & theories in the progress of one's task, where you have to

5. Lyell, "*Central France* by Scrope" (1827), 472; this passage also injected the traditionally static discourse of divine design with the newly dynamic dimension of geohistory (see §29.3). The insects were those (similar to caddis flies) that had constructed tubes of debris that were preserved in the limestones; the herds were of Cuvier's palaeotherium and similar mammals (§1.3), the bones of which Bertrand had found in the Tertiary formations near Le Puy; the palms were attributed explicitly to the younger Brongniart (§12.2).

6. Lyell, "*Central France* by Scrope" (1827), 474–75.

7. Lyell, "*Central France* by Scrope" (1827), 477–81.

8. Lyell, "*Central France* by Scrope" (1827), 474, 482–83; Buckland's suggestion, alluding to Penn, *Mineral and Mosaical geologies* (1822), was reported in Lyell to Murray, 7 August 1827 [quoted in Wilson, *Lyell* (1972), 173]; Lyell to Mantell, 2 March 1827 [Wellington-ATL], printed in *LLJ* 1: 168–69, was much too restrictive in identifying his own view of the timescale only with that of Lamarck. The sharp distinction between scientific and "scriptural" geology, as emphasized by all the leading British savants at this time, justifies the exclusion of the latter from further consideration here. The relation between the two alternative bodies of claimed "knowledge"—like the parallel relation between the modern earth sciences and the views of modern "young-earth" creationists—is a topic of great historical importance in its own right, but it is a different topic, and one that deserves to be treated separately in another work. Lynch, *Creationism and scriptural geology* (2002), usefully reproduces a representative selection of texts published between 1817 and 1857. O'Connor, *Earth on show* (2007), rightly emphasizes the diversity of the publications that the geologists, for their own polemical purposes, treated as homogeneous.

controvert & to invent an argumentation, [original] work is required & one [i.e., a book] like the Conversations on Chemistry would not do."[9]

Lyell explained to Mantell that his book would promote what was beginning to emerge in his mind as a distinctive approach to geology, already foreshadowed in his essays: "I am going to write in confirmation of ancient causes having been the same as modern [causes]". As this narrative has stressed repeatedly, the policy of using the present world as the key to the deep past had long been taken for granted on all sides as the proper method for geology, at least insofar as the science was concerned with reconstructing geohistory. Lyell was in no way an innovator in adopting this method: Prévost had demonstrated its heuristic value (§10.2), and even before Lyell first met the Frenchman his mind had clearly been prepared by his early reading of Playfair. But that "modern causes" were in every respect *identical* to "ancient" ones, and therefore *wholly* adequate to explain the deep past, was much more questionable. However, Lyell proposed to push the actualistic method to its limits. Francophone geologists such as Prévost continued to refer to "*causes actuelles*", but Lyell took to referring to "modern causes", "existing causes", or "causes now in operation", perhaps to dissociate himself from the now outmoded and "scriptural" ideas of de Luc, who had first coined the term "actual causes" (*BLT* §6.2).[10]

Having abandoned his earlier plan for a popular book, Lyell used his essay on Scrope as a dry run for the arguments that he was planning to use in the more ambitious work he now had in mind. In the following months he was hard at work on the much longer text that would justify his claims about the total adequacy of actual causes in geological explanation. Writing to Murray, the publisher of the *Quarterly* and, he hoped, of his book, he assessed the possible competition from other leading geologists, and the likely sales for such a work: his advantage over the likes of Buckland and Sedgwick was his freedom from the constraints of lecturing.[11]

Scrope's book had taken Lyell on what might be termed a virtual field trip of major importance; in imagination he had seen the celebrated extinct volcanoes of the Massif Central through Scrope's eyes, aided by his friend's magnificent interpretative landscapes (§15.1). Scrope's description of the volcanic features, and of the massive Tertiary freshwater limestones that underlay them, gave the region the outstanding strategic importance for Lyell that he hinted at in his review. In the spring of 1828 he had the opportunity he must have hoped for: the chance to make a lengthy fieldwork tour on the Continent, during which the Massif Central could be the first and foremost region on his itinerary. He could hope to improve his book immeasurably by first seeing with his own eyes the classic sites and sights that his potential critics such as Buckland already knew well. It was these places that he would need to see from his own distinctive perspective, if his planned publication were to make its mark.

The geologist Roderick Impey Murchison (1792–1871) was planning to travel with his wife on the Continent, at least as far as northern Italy, and agreed to join forces with Lyell. Murchison had been a soldier until the peace left him on half pay, aimless and idle, but addicted to the expensive sport of foxhunting. However, his highly intelligent wife Charlotte (1788–1869), a good amateur naturalist, had persuaded him to resign his commission, and in 1824 to move to London. Encouraged by Davy at the Royal Institution, he had learned some basic sciences and had then chosen to pursue geology instead of foxes. He had joined the Geological Society and soon replaced Scrope as one of its secretaries; he had learned the field practice of the science by traveling around the Scottish Highlands with Sedgwick. By 1828 he was ambitious for some more challenging fieldwork, which he tackled with military zeal. Charlotte had learned from Mary Anning how to be "a good practical fossilist" or collector, and like Buckland's and Mantell's wives she was also a fine artist; both skills made her a valuable companion in the field. So was Lyell, as a more experienced and knowledgeable geologist and a much better linguist. And the Murchisons' wealth allowed them to travel in their own carriage; like Buckland with the wealthy Greenough years before (*BLT* §10.4), this enabled them to make the best use of time.[12]

Murchison and his wife crossed to the Continent ahead of Lyell, going, of course, straight to Paris; Lyell gave him a letter of introduction to Prévost, his closest ally there. Lyell himself explained to Prévost his current view of the policy that geologists ought to adopt in explaining past geohistory in terms of the observable present world. In language redolent of the courtroom, he argued that the burden of proof [*onus probandi*] should lie with those who did *not* believe in the identity of ancient and modern causes, not with those like Prévost and himself who did: "Surely we are placed in a somewhat unnatural situation, for instead of assuming as we ought the identity of all the causes in nature in the former & present state of the planet, just as we anticipate the correspondence of those causes in all future time, we start by imagining a discrepance, & thus throw the onus probandi on those who assert what all ought to believe without proof until the contrary can be made clearly manifest."[13]

However, there was a profound ambiguity at the heart of Lyell's argument, which had already surfaced in his review of Scrope and which in due course his critics pointed out to him. Other geologists—Buckland can be taken as representative—were as convinced as Lyell was, that the fundamental physico-chemical "laws of nature" had been the same in the deep past as they were in the present and would continue to be in the future. A difference of opinion arose only over the uniformity through time of the far more complex *geological* processes, such as movements of the earth's crust, volcanic eruptions, tsunamis, and so on. All geologists assumed that these were based on unchanging physico-

9. Lyell to Mantell, 5 February 1828 [Wellington-ATL; printed in *LLJ* 1: 177], alluding to Marcet, *Conversations on chemistry* (1st ed. 1805). He urged Mantell to write a similar introductory book and thereby complement his own more ambitious project; Mantell eventually did so in his *Wonders of geology* (1838) and other later works (see §31.3). In the event, the title Lyell first envisaged for his own book was preempted by [Rennie], *Conversations on geology* (1828), an anonymous popular work often erroneously attributed to Penn.

10. Lyell to Mantell, 2 March 1827 [Wellington-ATL], printed in *LLJ* 1: 168–69.

11. Lyell to Murray, 6 June 1827 [quoted in Wilson, *Lyell* (1972), 170–71].

12. Scrope to Murchison, 30 March 1828 [London-GS: S10/2], offered advice on their itinerary. Stafford, *Scientist of the empire* (1989), chap. 1, is a fine summary of Murchison's life and work; see also Kölbl-Ebert, "Charlotte Murchison" (1997).

13. Lyell to Prévost, 20 April 1828 [Edinburgh-UL: AAF 8]; the rhetorical plural referred, of course, not to Lyell and Prévost themselves but to the generality of *other* geologists.

chemical "laws" of a more fundamental kind. But Lyell's critics argued that the surviving evidence from the past should be allowed to show whether the geological processes themselves had been unchanging, or whether for example they had formerly operated at a greater intensity or on a larger scale, or had even been of kinds that had never yet been witnessed in the brief span of human history.[14]

Shortly after writing to Prévost, Lyell too crossed to France and joined the Murchisons in Paris. Before leaving London he sent Murray the manuscript of his book, which was to be centered, like von Hoff's great work a few years earlier (§7.2), on a review of actual causes. But at some point in his subsequent travels Lyell must have asked Murray to put it aside until his return, having realized that what he was seeing would necessitate a fundamental recasting of the work on more ambitious lines. For, as he put it later, his great tour on the Continent, lasting almost a year, "made me what I am in theoretical geology". It was, as it were, Lyell's *Beagle* voyage.[15]

18.3 AUVERGNE THROUGH LYELL'S EYES

Before leaving London, Lyell had made a close study of all the available published work relevant to his tour. He had also spent time "Auvergnising" with Scrope and "pumping" others such as Buckland, Daubeny, and Herschel, who had traveled in the regions he hoped to visit, about where exactly he and Murchison should go and what they should look out for. And they went first to Paris primarily for the same reason, to get further tips from the geologists there (the capital was also the hub of the fine French road system, a legacy of the centralized and militarized Napoleonic regime). So when they reached Auvergne, and further regions beyond it, the way they saw the volcanoes and other features was not a view of raw unmediated nature but a seeing disciplined by the experience of many contemporaries and predecessors. Like other geologists in the field, they were not on their own; even in the most remote places they were surrounded by a cloud of witnesses. Other witnesses were closer at hand, and even accompanied them in person. Lyell was inclined to disparage the contribution of local naturalists, but he and Murchison were in fact heavily dependent, wherever they went, on their detailed guidance and assistance. There was of course much bartering, not only material. Murchison the canny Scot evoked Lyell's admiration for the way he struck hard bargains with innkeepers and others. Lyell too was determined to get as much as he gave; of one local informant, he told his father that "if he pumps unreasonably, I shall find a difficulty in expressing myself in French".[16]

Lyell's highest priority for his geological Grand Tour was also, unsurprisingly, the party's first destination. As he had noted in his essay on Scrope's work, Auvergne was as unknown to English tourists hurrying towards Italy as the interior of Australia was to the colonies on its coast, yet in fact it was "so accessible that [Clermont] may be reached in a journey of less than forty hours by the public conveyance from Paris". After a similar journey, though in the comfort of their own carriage, they based themselves in that city, above which rises the famous Puy de Dôme. That peak and the dozens of other volcanic *puys* had made

Auvergne a focus of intense scientific interest ever since the time of Desmarest (*BLT* §4.3). Lyell and his companions now had that pioneer's posthumously published map (see Fig. 15.5) to guide them in the field, and Scrope's less extensive one; they also met and were assisted by many local naturalists.[17]

Even peasants could give them valuable information. At the start of their fieldwork, deep in rural Auvergne, they were told of a small volcano that was not marked on either map. Although it turned out to be similar to many others in the region, it was an exciting revelation to Lyell, who for the first time saw with his own eyes the kind of evidence that had convinced his predecessors—and contemporaries such as Scrope, and also Croizet and Jobert (§15.2)—that the valleys in Auvergne must have been eroded gradually, not all at once as Buckland and Daubeny had claimed. By the criteria set by the latter, the cone of loose volcanic ash was necessarily "postdiluvian" in date, yet the little river Sioule had already eroded a deep gorge not only through the lava that issued from the cone but also into the underlying gneiss (see Figs. 15.3, 19.1 for similar cases). "This is an astonishing proof of what a river can do in some thousand or 100 thousand years by its continual wearing", he told his father; "No deluge could have descended the valley without carrying away the crater & ashes above". Lyell found a bed of gravel underlying the volcanic rock, which showed that the lava had flowed down a valley that was already there before the eruption; and there was another basalt high on the plateau above, dating from an even older phase in the lengthy process of erosion (see Fig. 15.6 for the similar cases illustrated by Croizet and Jobert). This little volcano was not the new discovery that Lyell imagined, nor were his inferences original. Scrope had described it in detail as "one of the first connecting links between the recent and more ancient eruptions", noting that

14. See the analyses of the manifold historical meanings of "uniformity" in geology in Hooykaas's classic *Natural law* (1959) and "Catastrophism in geology" (1970); Rudwick, "Uniformity and progression" (1971); and, specifically on Lyell, Cannon, "Radical actualism" (1976). Contrary to these and other historical works, Camardi, "Lyell and the uniformity principle" (1999), argues for the indissoluble *unity* of Lyell's position.

15. Lyell, *Principles* 3 (1833), vii, recalled the submission of this early text, which helped establish his priority, but Murray would have known if it had been no more than a post hoc fabrication. Lyell to his fiancée Mary Horner, 2 December 1831 [*LLJ* 1: 354–55], recalled the pivotal importance of the tour. Charles Darwin's later voyage around the world on H. M. S. *Beagle* (1831–36) gave him a similarly decisive store of firsthand experiences relevant to his future theory of transmutation (or evolution); but the parallel is not exact, in that Lyell's ideas were at a much more advanced stage of maturation than Darwin's, before they set out on their respective travels.

16. Lyell to his father, 29 April, 9, 16, 25 May 1828 [*LLJ* 1: 181–88]. This and following sections are based on (1) Lyell's field notebooks [Kirriemuir-KH: notebooks 8–23] and Murchison's [Keyworth-BGS]; (2) Lyell's letters to his family, some of them quoted in *LLJ* 1: 183–251, and in Wilson, *Lyell* (1972), 190–261; and (3) his letters to Murchison after they parted [London-GS], to Mantell [Wellington-ATL], and to other geologists as cited in subsequent footnotes. Dean, "Lyell and Murchison in France" and "Lyell in Italy and Sicily" (both 1999), give useful day-by-day summaries of their itinerary. At several points, however, my interpretation of the impact of what Lyell saw in the field differs from those offered by other modern authors, partly in consequence of my having followed almost the entire itinerary in the field; see also Rudwick, "Travel, travel, travel" (2004). In the following account of the joint part of their tour, Lyell is given more prominence than Murchison, partly because at the time his opinions counted for more, but also because his notebooks and surviving correspondence are much richer sources than Murchison's.

17. Lyell, "*Central France* by Scrope" (1827), 438; Desmarest, *Carte du Puy-de-Dôme* (1823). Scrope, *Central France* (1827), includes a small-scale sketch map covering much of the Massif Central, and a large-scale map closely similar to Desmarest's but covering a much smaller part of Auvergne (§15.1).

the gorge "evinces the immeasurably long continuance of the erosive action, as well as its irresistible power". Its importance for Lyell was that he was seeing it *with his own eyes.*[18]

Lyell and Murchison later spent several days with the elderly but still active Montlosier (§15.1; *BLT* §6.1), at his château among the lava flows (see Fig. 15.2), probably being shown the volcanoes nearby and certainly being told where else to go. In particular, Lyell went to see a locality that Montlosier had made well known, where the Sioule had eroded a completely new course for itself after its old one was totally blocked by a vast lava flow erupted from the Puy de Côme (see *BLT* Fig. 6.2). Like Montlosier, he interpreted this as another fine example of the erosive power of a small river, even in the geologically brief time since this "modern" eruption. Conversely, however, on a point on which Montlosier—with some forty years of local fieldwork behind him—would *not* have agreed with him, and Scrope certainly did not, Lyell had less success. He was already so convinced of the vast antiquity of even Tertiary geohistory that he anticipated finding an *intrusive* contact between the granite underlying the Auvergne volcanoes and the Tertiary limestones to the east. This would imply that the granite was not an ancient "basement" rock at all, but even younger than the limestone; it would prove—sensationally—that granites had continued to be intruded even more recently than the time of Cuvier's early mammals. But although Lyell told his father he had indeed found the kind of junction he hoped for (with, in modern terms, contact metamorphism), he must soon have realized that this had been wishful thinking, for no more was heard of it.[19]

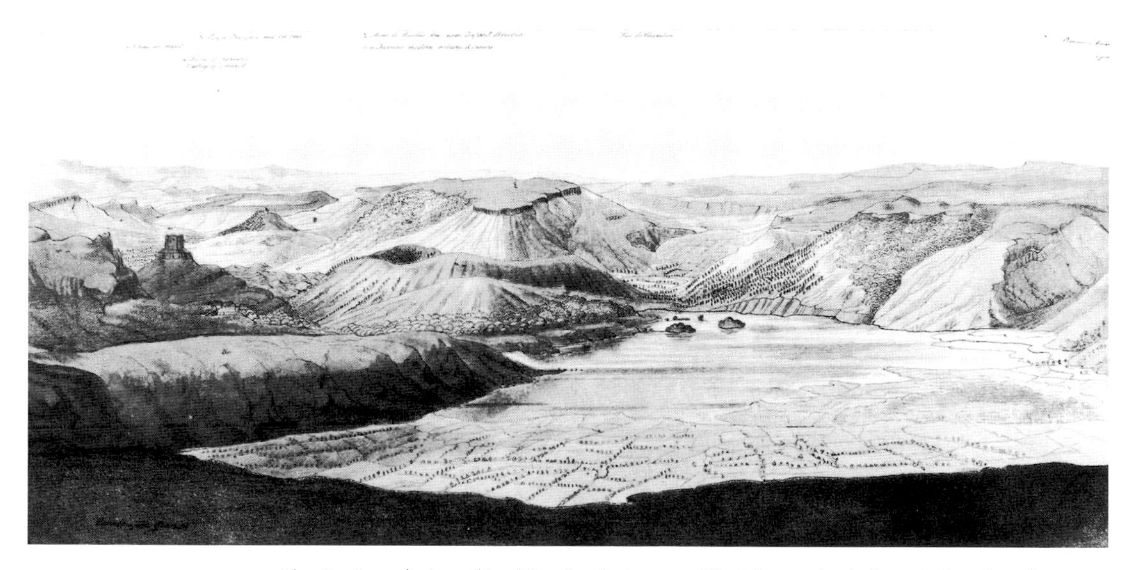

Fig. 18.1 Scrope's view of Lac Chambon in Auvergne. The lake was clearly formed when the valley was blocked by the eruption of the "modern volcano and crater" of Tartaret (just left of the lake, and in front of a plateau capped by "ancient" basalt). A "modern" lava flowed away from the volcano, covering the floor of the valley further downstream and rendering it sterile (it is depicted as rough ground to the left of and beyond the cratered cone). To the left again, on a rocky crag, is the ruined medieval Château de Murol, in the kind of position occupied by several ancient villages in the region. Soon after Lyell had been in this area, he drafted a brief essay on the "analogy of geology and history", using landscapes of this kind to argue that geologists need to reason *historically.*

In Auvergne Lyell saw for himself many of the basalts, lava flows, and volcanic cones that Scrope had depicted so vividly in his proxy panoramas. The experience gave him an occasion for articulating his own equally *geohistorical* way of looking at the landscape. Soon after he and Murchison left Auvergne, he wrote a series of short essays in the back of his current field notebook, in a style that suggests they were intended as drafts of prose that he could use in due course in his book. The longest was entitled, significantly, "Analogy of Geology & History". It started with a recollection of two oddities of the landscapes he had just seen: villages perched on rocky hilltops, and valley floors covered with barren lava instead of fertile soil (Fig. 18.1).[20]

Lyell recognized that these anomalies were similar, for in both cases any causal explanation had to incorporate a *historical* dimension; the present could not be understood except in terms of its own local and contingent past. He expressed this in a way that shows how thoroughly he had absorbed and internalized the geohistorical perspective that had already come to characterize geological argument in general. Not surprisingly, he found an analogy with *legal* history particularly appropriate: if history is not taken into account in geology, he wrote, "we err as much as when we judge of a political constitution without considering the pre-existent state of the laws from which it has grown". What may be more surprising—but only because of the power of later myths about him—is that Lyell, like Buckland and other British geologists at this time, set his argument about geohistory in a context that belongs unmistakably to contemporary debates about theodicy and natural theology:

> If we regard Auvergne & the works of man as merely belonging to one generation, we perceive much that is unintelligible [and] that is not for the best; & the explanation is that the present state of things has grown out of one extremely different, & every thing has not been made merely with a view to the Existing race. The greater number of the cities are built on lofty & often very inaccessible eminences & with narrow streets, as La Serre, Les Roches, Buron &c., instead of [in the] rich valleys which would now be chosen. There is an historical reason for this. They were once fortresses, when to dwell in the plain exposed the inhabitants to plunder & invasion.
>
> So if we turn to Nature we perceive many a valley, where we might expect a fertile alluvial soil & a river, is now filled with a sterile ridge of uncultivable rock. This could not be expected if all the surface were constructed by Nature with a view to the present order of things. But there is a reason for it. This barren rock is a stream

18. Lyell, notebook 9 (18 May to 2 June 1828), 46; and *LLJ* 1: 185–88; Scrope, *Central France* (1827), 85–86. The volcanic cone of Chalusset is 4km northwest of Pontgibaud, which is about 20km west-northwest of Clermont-Ferrand; the volcano was just beyond the western edge of Scrope's large-scale map, but not beyond the limit of his fieldwork.

19. Lyell, notebook 10, 32–35, and *LLJ* 1: 188; Wilson, *Lyell* (1972), 197–98; Lyell to Mantell, 15 June 1828 [Wellington-ATL].

20. Fig. 18.1 is reproduced from Scrope, *Central France* (1827), part of pl. 9. The ensemble of lake, volcano, and lava flow had been described by Desmarest, and was well shown both on his preliminary map of 1774 (see *BLT* Fig. 4.9) and on his complete but posthumous one of 1823; the lava flow reached Neschers (near where Croizet was *abbé*) some 20km east of the volcano (see Fig. 15.5). Lac Chambon is about 25km southwest of Clermont-Ferrand.

of lava which closed the valley & concealed the river. It results from a former system for which the world was made, as for the present, & on which it must depend as will future ones on it. We must take the good & the evil together, of succession to a former system, in human affairs & natural.[21]

In the course of their travels around Auvergne, Lyell and Murchison visited the already famous fossil locality of Boulade, where abundant bones of mainly extinct mammals had been collected and were being published in the "splendid and valuable work" by Croizet and Jobert. The local naturalists claimed that the bone deposit was intermediate—in both position and age—between the so-called "ancient" and "modern" volcanic rocks, providing telling evidence for the gradual character of faunal change in the region (§15.2). There had been some skepticism about this, but Lyell satisfied himself on the spot that the inference was correct (Fig. 18.2).[22]

As Scrope and the Auvergnat naturalists had shown, and Desmarest long before them, even the oldest volcanic rocks were more recent than the underlying Tertiary limestones. Yet the latter contained many freshwater mollusks that were regarded as indistinguishable from living species. This probably triggered the speculations that Lyell formulated later in the essay already mentioned: they were on "the reason of the disappearing of species" and "the laws which regulate the comparative longevity of species". Lyell was clearly aware that these topics, far

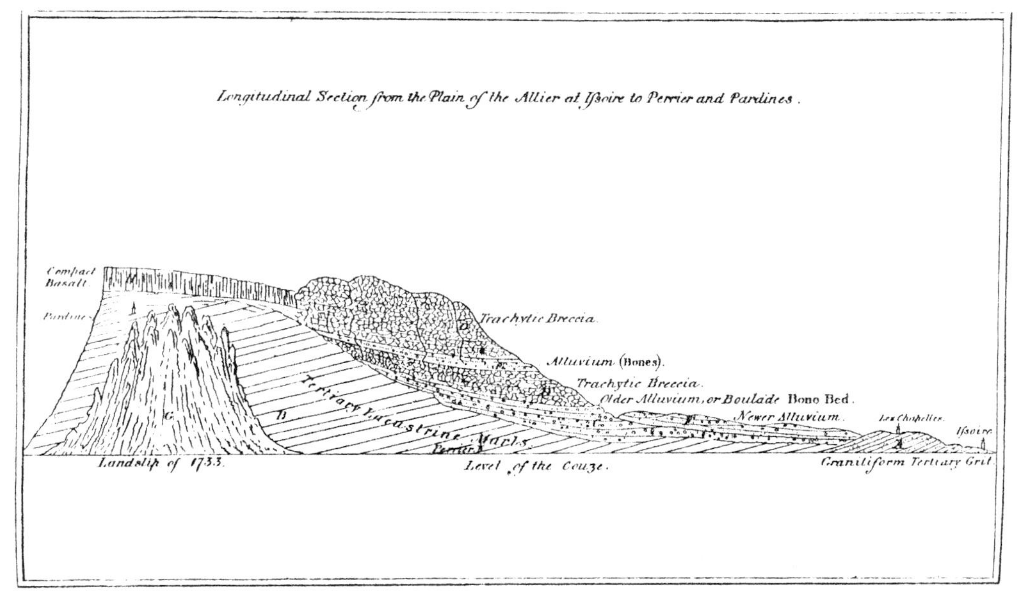

Fig. 18.2 A section to show the geological position of the famous "bone bed" at Boulade in Auvergne, summarizing Lyell's fieldwork in 1828. It expressed his agreement with the local naturalists' conclusion (§15.2) that the "older alluvium" with its rich fauna of fossil mammals was *intermediate* in age between the "compact basalt" capping the plateau (left) and the "newer alluvium" on the floor of the valley (right), both being much younger than the underlying "Tertiary lacustrine marls". The massive "trachytic breccia" (in modern terms, agglomerate) overlying the bone deposits represented the major volcanic activity in the region at the same period. This section was published later by Lyell and Murchison in their nominally joint paper (1829) on the gradual excavation of valleys purely by fluvial erosion and without any "diluvial" action.

from being purely matters of zoology, were also crucial for any reconstruction of geohistory.[23]

Lyell had already recognized from Scrope's work that the Cuvierian explanation of extinction—by a sudden marine incursion of some kind—could not be applied to the changing faunas of the Massif Central, because this region had evidently remained above sea level since early in the Tertiary era (§15.1). Lyell therefore needed to find an alternative explanation of extinction. His essay shows that he was toying with Brocchi's earlier ideas of an analogy between individuals and species: the variable "longevity" of species between origin (or "birth") and extinction (or "death") might be due to factors intrinsic to the organism (*BLT* §9.4). The fossil record in Auvergne, for example, could be explained only if the average longevity of molluskan species was far greater than that of mammals: only such contrasting rates of turnover could account for extinct mammals being preserved in a much younger formation than one containing extant mollusks. Once again, Lyell's argument was framed within a context of "final causes" and natural theology:

> We must bear in mind that the narrower question as to the variable duration of individuals, why so large a proportion perish prematurely & why their natural life is so different in different species in duration, are problems on which as yet no reasonable conjectures can be hazarded as to final causes, much less therefore can we look for an explanation of the laws regulating the term [i.e., span] of existence of a species. Why a raven should outlive a man, or a tortoise a horse, we cannot guess, yet there is no doubt a reason. Certainly it has not been left to accident, as the average term is constant to every species.[24]

18.4 CONCLUSION

In the last of his essays for the *Quarterly Review*, Lyell used Scrope's new book on the extinct volcanoes of central France to introduce the British intellectual elite

21. Lyell, notebook 5, 86–87; the final sentence, like others that express Lyell's engagement with theological issues, is omitted from the long quotation printed in Wilson, *Lyell* (1972), 215–16. The small village of La Serre (a *cité* in the French sense) is shown on one of Scrope's panoramas, perched on an outlier of the plateau of "ancient" basalt of the same name (see Fig. 15.1). This and many similar essays in the back of Lyell's notebooks can be dated and located approximately, by the period and region described in the field notes at the front.

22. Fig. 18.2 is reproduced from Lyell and Murchison, "Excavation of valleys" (1829) [Cambridge-UL: Q340:1.c.6.21], pl. 2; see reference to Croizet and Jobert (45n), and §19.1 below. Desmarest's map of the area (Fig. 15.5), and the similar but more general section by Croizet and Jobert (Fig. 15.6) put this section into its context.

23. Wilson, *Lyell* (1972), 215, mistakes a phrase at the end of a sentence ("laws which regulate the comparative longevity of species") for the title of a separate essay; the passage is in fact an integral part of Lyell's essay on "Analogy of geology & history".

24. Lyell, notebook 5: 83. The itinerary covered by this notebook ("Issoire, by Puy en Velay to Montelimart") indicates that the essay was written after or around the time of their visit to Boulade and before they had seen any Tertiary formations except the freshwater limestones of the Massif Central. There is an unmistakable parallel between Lyell's wording and Brocchi's "l'efemero non campa che poche ore [etc.]": see Brocchi, *Conchiologia fossile subapennina* (1814) 1: 227–28; *BLT*, 527–28; and Rudwick, "Historical analogies" (1977). Lyell had certainly read Brocchi thoroughly in preparation for his planned fieldwork on the Subapennine formations (see §19.3).

to his own similar perspective on geohistory and its wider implications for the "place of man in nature". He emphasized the unique importance of the region that Scrope had described. He endorsed his friend's inference that it showed an unbroken process of slow subaerial erosion of the valleys, recorded more clearly than in other regions because it had been punctuated by occasional volcanic eruptions. There was no trace of any exceptional "diluvial" event, nor indeed of any marine incursion of any kind since before the deposition of the thick Tertiary limestones that underlay and antedated all the volcanic rocks. But Lyell extended Scrope's work by pointing out the parallel inferences to be made about the history of life in the same region: it showed that there had been no drastic episode of mass extinction in the geologically recent past. Lyell also noted the wider implications of the geohistory being pieced together by geologists as a whole: human mental powers gave intellectual "dominion" over the vast tracts of deep time, so that the very recent arrival of their species was no reason for humans to feel insignificant in the cosmos.

Lyell's essay-review of Scrope functioned as a dry run for the book he now planned to write. Rather than the popular introduction to geology that he had considered earlier, it would offer, at a much higher level, a "reformed" perspective on the whole of the science. He would push the use of actual causes—which he preferred to call "modern causes"—to the limit, by arguing that they were *identical* to "ancient causes", and therefore adequate to explain *everything* about the deep past, without invoking anything else. If the strategy succeeded, it would imply that the deep past had always been much like the present, so that geohistory would have to be recast on a steady-state or cyclic model. More immediately, Lyell joined Murchison and his wife in visiting the Massif Central; here he was able to see its extinct volcanoes and other features for himself, though at the same time he was bound to see them also through the eyes of Scrope and his other predecessors.

The volcanoes of Auvergne became Lyell's (and the Murchisons') first destination. Lyell confirmed for himself what Scrope and the local Auvergnat naturalists had already inferred, about the slow erosion of the topography and the similarly slow turnover in the fauna, without any trace of a "diluvial" episode. His vivid firsthand experience of the region also prompted him to write important mini-essays, while still on the spot, as possible drafts for his book. He wrote, for example, on the necessity of *historical* thinking in geological explanation—already a commonplace to himself and other geologists, but not to all those who he hoped would read his work—and on using Brocchi's earlier speculation about the intrinsically limited lifespans of organic species, as a noncatastrophic causal explanation of extinction.

Lyell's tour of Auvergne was just the opening phase of a geological Grand Tour that he recognized as being decisive for the future of his research. It is so important for the theme of the present book that most of the next two chapters are devoted to the further phases of what can be regarded, without exaggeration, as Lyell's equivalent of Darwin's later and more famous voyage on the *Beagle*.

CHAPTER 19

A geological Grand Tour (1828)

19.1 LYELL AND MURCHISON IN SOUTHERN FRANCE

Lyell and Murchison had made Auvergne their first destination, not only because it was Lyell's top priority to see Scrope's ground for himself but also because it was the first region of geological importance to the south of the Paris Basin. But there were almost equally interesting areas in further parts of the Massif Central, which had also been described by Scrope.

What they had seen in Auvergne had convinced Lyell that Scrope and the local naturalists were right to interpret the topography in terms of gradual valley erosion, and that there was no trace of the alleged "geological deluge"; to use a later epithet, he was already a convinced "fluvialist" (see §20.3). But when they traveled south to see the similar features around the city of Le Puy (in the former province of Velay), Murchison proved to be a staunch "diluvialist", even when they were shown the gorges of the upper Loire by the highly competent local geologist Bertrand, who had accompanied Scrope and the Brongniarts in the same way a few years before (§15.1; *BLT* §9.6). "Indeed", Murchison noted at the time, "gradual erosion could never produce such broken irregular effects, & if it is [due] to the existing river, as Monsieur Bertrand attributed it, still less can we fancy so narrow & fantastic a gorge to have been cut". Lyell, on the other hand, conceded that "when in the gorge the height seems so immense & the width of the valley so great that the excavation by modern causes seems immense"; yet he judged that much depended quite literally on one's viewpoint, for he added, "when seen from above it seems more natural". And Bertrand showed them river gravels perched at different heights, like those near Boulade

(Fig. 15.6), which in Lyell's opinion seemed "clearly to indicate the progressive excavation of valleys".[1]

Later, however, Murchison may have been persuaded by his companion—not for nothing was Lyell a trained barrister—or perhaps by the sheer perceptual impact of "the beautiful pet volcanoes" of Vivarais, as Scrope had called them. Murchison's conversion may have been temporary, but he put his name to the nominally joint paper that they later sent back to the Geological Society in London. In this article, Lyell—the style is his—made a strong case for a "fluvialist" interpretation of valley erosion, on the basis of all they had seen in the Massif Central, and hence reinforced the case for rejecting Buckland's "diluvialist" alternative altogether. As he told one of his sisters, the paper was "to reform the Geological Society & afterwards the world" on this crucial issue. However ambiguous some of their examples might seem to those who had not seen the relevant features at first hand, one of the sketches offered by the two travelers, of a gorge on the Ardèche, was a proxy that would be hard for any diluvialist to explain away (Fig. 19.1).[2]

Three related themes were now becoming prominent in Lyell's thinking; none was original to him, but all were to gain in cogency from his persuasive advocacy. One was the continuity and equable normality of the transition from the most recent portion of geohistory into the present, and the absence of any exceptional "geological deluge" at that point; the paper just described showed clearly that Lyell was adopting this view. A second theme emerged from what he saw to the south of Auvergne. In the département of Cantal a deeply eroded ancient volcano, like that of Mont-Dore and as large as Etna, was underlain on one side by the usual freshwater Tertiary limestones; but in Cantal they were at a higher

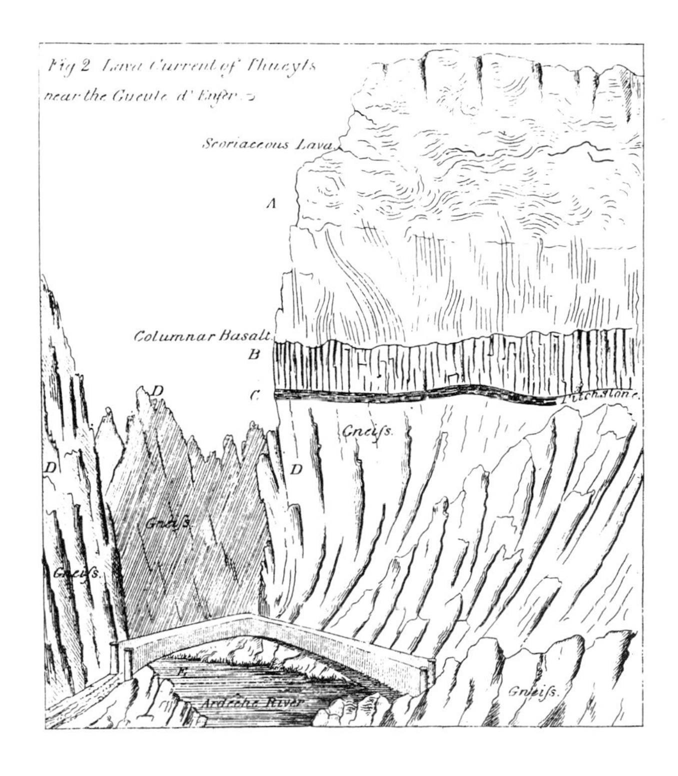

Fig. 19.1 The gorge of the river Ardèche below the village of Thueyts in Vivarais (the little bridge, which carried a mule track, suggests the scale). This sketch was published by Lyell and Murchison in their article (1829) on the power of rivers to erode valleys in hard rocks, even in a geologically brief timespan, without invoking any exceptional "diluvial" agency. Although vertically exaggerated and crude in style, it shows how the gorge was cut through a thick mass of basalt lava but also deep into the underlying hard basement rock ("gneiss"); yet the lava could be traced back to a well-preserved "modern" volcano, a cratered cone of loose volcanic ash, which would surely have been swept away in any "geological deluge". The case was similar to several that Scrope had already described and illustrated (see Fig. 15.3); but this engraving, published with the article in Jameson's Edinburgh periodical, would have been seen by many more geologists than Scrope's expensive album.

altitude than the surrounding basement rocks, and had apparently been elevated since their deposition. To Lyell, this seemed to be evidence for a Huttonian kind of explanation (*BLT* §3.4), which would attribute gradual crustal uplift to equally extended volcanic activity. This was a *causal* theory in earth physics (*BLT* §2.4), but one that might be used to explain events during Tertiary geohistory.[3]

A third theme, which became even more prominent once Lyell and his companions came down off the Massif Central onto the lower land bordering the Mediterranean, was that of the sheer thickness of the Tertiary formations, the signs of their calm deposition, and their apparently vast antiquity. This confirmed what other geologists had already emphasized, namely the crucial importance of the Tertiary era for understanding the deeper past of still earlier geohistory (§10.2, §12.1). Lyell and Murchison made a side trip to Montpellier to meet Serres (§14.1), who knew as much as anyone about the Tertiaries of Languedoc and had come to similar conclusions about them. Everywhere Lyell was looking for analogies with the present, but also for evidence that the Tertiary era had been no different in character from still earlier times. In the lagoons [*étangs*] of the Rhône delta, for example, he saw the kind of environment that in the past could have yielded the mixture of marine and freshwater shells which Prévost had interpreted in similar terms in the Paris Basin (§10.3). And in the thin coal seams that Serres had described (in a work then in press) within the Tertiary formations near Aix-en-Provence, Lyell saw a close analogy with the more important seams in the far older Carboniferous rocks elsewhere, showing that the production of coal had not been confined to that early part of geohistory.[4]

At the same time, Lyell was thinking about the Tertiaries of Provence in relation to those elsewhere in Europe, particularly in terms of their relative ages. Here again he was clearly indebted to Prévost, who had earlier suggested that the Tertiary formations of different regions could in effect be dated by their *assemblages* of fossils (rather than by Smithian "characteristic" ones), even if the direct evidence of superposition was not available (*BLT* §9.6). And Lyell was adopting the idea of a "statistical" comparison of different Tertiary faunas, probably from

1. Murchison, notebook "Puy & Montelimart", 9, and Lyell, notebook 12 (30 June to 21 July 1828), 25–27, both referring to the Loire gorge near Retournac, about 18km northeast of Le Puy. Bertrand, *Environs du Puy* (1823), pls. of "Coupes verticales" and panoramic "Profils". Murchison to Buckland, 4 July 1828 [London-GS: 768/2].

2. Fig. 19.1 is reproduced from Lyell and Murchison, "Excavation of valleys" (1829) [Cambridge-UL: Q340:1.c.6.21], pl. 1, fig. 2, read at the Geological Society on 5, 16 December 1828. This impressive scene, with its elegant ancient bridge, remains almost unchanged; Thueyts, about 14km west-northwest of Aubenas (Ardèche), is built on top of the basalt, which can be traced to the volcanic cone of Gravenne de Thueyts, northwest of the village. Lyell to Eleanor Lyell, 20 August 1828 [*LLJ* 1: 197–98].

3. Lyell and Murchison, "Dépôts lacustres tertiaires" (1829), on Cantal, may have been drafted—probably by Lyell—around the same time as their paper on the excavation of valleys, although it was not read at the Geological Society until 3 April, 1 May 1829, after their return to London (see §20.3). The geological map of the Massif Central in the atlas to Scrope's book would have given Lyell a vivid impression of the broader regional setting of the huge volcanoes of Cantal and Mont-Dore and their associated Tertiary basins. The highest remaining point on the deeply eroded Cantal volcano is the Plomb du Cantal (1855m), 30km northeast of Aurillac.

4. Serres, "Dépôt des terrains tertiaires" and *Géognosie des terrains tertiaires* (both 1829). Murchison and Lyell, "Fresh water formations of Aix" (1829), read at the Geological Society on 19 June after their return; like their Cantal paper it may have been drafted—again, probably by Lyell—soon after their fieldwork around Aix. The coal seams were near Fuveau, about 12km southeast of Aix.

Prévost's work on the Paris Basin and Basterot's study of the Tertiaries in southwest France (§12.1). So when he told Mantell about the formations he was seeing in Provence, Lyell correlated them with others on the basis of a *quantitative* relation between their fossils and living species: "the Aix formation is generally referred to the Paris Upper freshwater [formation] & it is supposed that like the Crag [of eastern England] & Subapennine beds [of northern Italy] a considerable percentage of the animals & plants are still living". His phrasing shows that it was not Lyell himself who had assessed the Provençal fossils in this way; but in any case *percentages* soon became his preferred notation for expressing the relation between any Tertiary fauna and the species living in the present world. Combined with his Brocchian notion that there might be an average "longevity of species" and therefore a steady turnover in the composition of the molluskan fauna, percentages could be used to give quantitative dates (though not in years) to the Tertiary formations.[5]

19.2 LYELL AND MURCHISON IN NORTHERN ITALY

Crossing the then frontier from France into northern Italy (or rather, into the kingdom of Sardinia), Lyell and Murchison reached Nizza [Nice], where illness and the intense heat of high summer forced them to rest for a time. They, or more probably Lyell, used this pause to write their nominally joint paper on the erosion of valleys, which was sent back to London to be read at the Geological Society in their absence; at the same time he may also have drafted the two further joint papers derived from their fieldwork in France, on Cantal and Provence, which were presented at the Society after their return.

While at Nice, Lyell met Giovanni Antonio Risso (1777–1845), a highly competent naturalist who taught at the high school and who had recently published five volumes on the natural history of Nice and its Alpine hinterland. Risso had described a thick Tertiary formation containing 194 species of fossil mollusks, and a "Quaternary or Diluvian" one with no fewer than 257 species. Lyell saw Risso's collections and was shown the formations in the field. Risso had identified the mollusks in his "Quaternary" deposits as being (as Lyell put it) "all modern shells". In contrast, the Tertiary rocks of Nice, which Risso identified as the local equivalent of Brocchi's Subapennine formation, had only (in Lyell's words) "from 15 to 20 per cent in them of modern living *species*". Since he also immediately reminded himself to find out "how many per cent [are] in calcaire grossier", that is, in the Coarse Limestone of the Paris Basin, he was evidently thinking about *all* the Tertiary faunas in the same terms. And he was using such figures not only as possible measures of relative age, but also to estimate what he envisaged as a steady local elevation of the earth's crust: at Nice the "Quaternary" deposit was a mere 25 feet above sea level, whereas the Tertiary fossils were found at altitudes of up to 800 feet. Writing to his father, Lyell was jubilant at what his fieldwork had already achieved:

> The whole tour has been rich, as I had anticipated (and in a manner which Murchison had not), in those analogies between existing Nature and the effects of causes

in remote eras which it will be the great object of my work to point out. I scarcely despair now, so much do these evidences of modern action increase upon us as we go south (towards the more recent volcanic seat of action), of *proving* the positive identity of the causes now operating with those of former times.[6]

Moving eastward through northern Italy, Lyell and Murchison found local naturalists willing to help them in every city of any size. At Genoa, for example, they were shown a fine raised beach 100 feet above sea level, which Lyell took to be further evidence of geologically recent crustal elevation. Having crossed the Apennines to see the Tertiary formations and associated volcanic rocks that the elder Brongniart had made famous (*BLT* §9.6), they followed a route that had since become almost canonical for geologists. At Turin, Franco Andrea Bonelli (1784–1830), the professor of zoology, showed them his fossil shells from the highly tilted Tertiary strata just outside the city (see *BLT* Fig. 9.20), and Lyell carefully made a list of "Subapp. shells identical with recent [i.e., extant species]". At Milan, they were, sadly, too late to meet Brocchi, who had died two years earlier; and they found his Mines Council abolished and much of his collection carried off to Vienna by the Austrians, who had reclaimed Lombardy after the fall of Napoleon. More happily, at Verona they saw collections of the famously well-preserved Tertiary fossil fish from Monte Bolca (see *BLT* Fig. 5.7) and then visited the locality itself. At Padua Lyell made a special effort to see the famous Giotto frescoes, but he also explored the nearby volcanic Euganean Hills.[7]

None of their fieldwork in northern Italy was novel; but, as Lyell told his father, it did enable them to understand and evaluate what other geologists had done before them. And he used a revealing metaphor about what they had just seen in the Euganean Hills: "The volcanic phenomena were just Auvergne over again, and we read them off, as things written in a familiar language, though they would have been Hebrew to us both six months before". The metaphor shows how Lyell, like Soulavie and Cuvier before him (*BLT* §4.4, §9.3), recognized that the work of geohistorical investigation was necessarily interpretative: nature's monuments could yield little insight unless nature's language could be learned and nature's inscriptions deciphered and read. It also shows how he recognized the value of firsthand fieldwork: it was the experience of seeing the Auvergne volcanoes *with his own eyes* that had enabled him to learn nature's language in the first place, and then to read nature's similar inscriptions in the Euganean Hills.[8]

5. Lyell to Mantell, 22 July 1828; the comparison of the faunas was probably due to Serres.

6. Lyell to Lyell senior, 24 August 1828 [*LLJ* 1: 199–200]; Lyell, notebook 13 (24 July to 23 August 1828), 42–44. Risso, *Histoire naturelle* (1826) 1: 119–25, listed the species but did not note the proportion of those extant; his volumes, published in both Paris and the bilingual Strasbourg, were no "merely" provincial work. Nice and its hinterland did not become part of France (as the département of Alpes-Maritimes) until 1860.

7. Lyell to Herschel, 9 September 1828 [London-RS: HS.11.417 and *LLJ* 1: 200–2], and to Lyell senior, 26 September 1828 [*LLJ* 1:202–4]; Lyell, notebook 14 (25 August to 13 September 1828), 68. The isolated *Colli Euganei* stick up prominently out of the alluvial plain of the Po, about 15km southwest of Padua.

8. Lyell to Lyell senior, 26 September 1828 [*LLJ* 1: 202–4]; see Rudwick, "Historical analogies" (1977), 94–97.

At this point the Murchisons cut short their tour, having decided to cross the Alps into Austria and return home by way of Germany, taking the carriage with them. They were feeling the heat, perhaps in more senses than one; and Charlotte, not having heard for some time from her mother, was getting anxious and depressed. For Lyell, however, the tour had become far too important—and too exciting—to abandon it without reaching the active volcanoes further south. He and Murchison may have intended to travel together at least as far as Naples and Vesuvius; but Lyell now wanted to continue still further south, to Calabria—the site of the famous earthquake of 1783 (*BLT* §2.2)—and above all to Sicily and Etna, the largest active volcano in Europe (§7.3). He and Murchison had in fact already decided to go their own ways: Lyell, regretting he had not planned this major extension of his tour before leaving London, had written from Milan to ask Herschel for detailed advice about where, beyond Naples, he might see evidence of recent crustal movement. His description of what he anticipated leaves little doubt that he hoped that Sicily might conclusively close the gap between Tertiary geohistory and the present world: "The disturbance in the freshwater strata of Auvergne and Cantal, due to volcanic action, is so much greater than I had been led to expect, and that of the Subapennine beds from Montpellier to Savona [i.e., the Mediterranean coast, including Nice], containing as they do nearly twenty per cent of decided living species of shells, that I cannot but think that Calabria and Sicily must afford proofs of strata containing still more modern organic remains, raised above the level of the sea".[9]

19.3 LYELL IN SOUTHERN ITALY

Lyell was now on his own. However, from this point onward his tour acquired something of the character of a pilgrimage towards a sacred site, or of a quest for the Holy Grail; with mounting excitement he anticipated that in Sicily he might experience a secular theophany, as it were, and resolve the major problems of his science. But first he continued on the itinerary that he and Murchison had planned together. Indoors at Parma he studied "the finest collection of fossil shells in Italy" with Giambattista Guidotti, the university's professor of chemistry; and outdoors in the field he saw that Brocchi's stratified Subapennine formation (see *BLT* Fig. 9.21), from which the beautifully preserved fossils had come, was at least two thousand feet thick. As with the freshwater limestones in the Massif Central, this impressed on Lyell the sheer magnitude of the Tertiary formations, in comparison with their modest showing in England. As he told Murchison (now by letter instead of face to face), "I begin to look on Subapennine as of immense respectability in point of age"; yet he also called it, significantly, "the most modern formation", thereby implicitly agreeing with Desnoyers (§12.1) that it was more recent than the Tertiaries of the Paris Basin and elsewhere. Lyell then crossed the Apennines again, from Bologna to Florence, sharing the coach with two Frenchmen, an Italian, and a Greek—a typically international party of travelers—and resumed his study on the Subapennines in Tuscany, for example around Siena. This extensive work in the field and in museum collections, always with the help of local naturalists, gave Lyell an enviably thorough knowledge of a classic Tertiary formation, and of the prolific fossil fauna that Brocchi had first made known to geologists everywhere (*BLT* §9.4). This could act as a well-defined benchmark in his emerging sense of Tertiary geohistory.[10]

Lyell continued to fill his notebooks not only with field notes but also with more reflective and theoretical passages of prose, which were clearly intended as drafts for his book. Although they are often rough and incomplete, these essays show how, week after week, he was relating what he saw in the field to the broader uses that he intended to make of these observations. One, for example, clarified the important relation between his geohistorical goals and those of causal explanation or earth-physics. In this draft definition of his science, geohistory was clearly the primary goal, and the discovery of ahistorical physical "laws" strictly derivative:

> Geology has for its object to elucidate the ancient history of the Earth—to enquire into the state of organic & inorganic Nature at successive periods of the past, to trace the changes by which each has been affected, to discern from what causes such changes have resulted, & ultimately to derive from such investigations a more perfect knowledge of the present external structure of our globe, & of the laws now governing its animate & inanimate productions.[11]

From Tuscany, Lyell traveled southward more rapidly. On the stage to Rome he shared the coach with an Italian priest escorting two young Scots novices who as yet spoke no Italian, so he spent the day talking in Latin, their only common language. Continuing to Naples, his companions were four Frenchmen, whom he persuaded to allow the coach to make a short detour to see a well-known flooded volcanic caldera, which they were happy to appreciate as a picturesque lake (Lago di Albano). But he was so keen to get to Sicily without delay, before winter weather made an ascent of Etna impossible, that he planned to cross at once to Palermo. On arrival in Naples, however, he found that the steamship on that route had been commandeered for a government mission, and so he was for several days "condemned [*sic*] to the Phlegrean Fields", the spectacular volcanic region that Hamilton had made famous throughout Europe half a century earlier (*BLT* §2.2).[12]

Lyell made good use of his unplanned time in Campania. He met the local expert on mollusks, Oronzio Gabriele Costa (1787–1867), a naturalist who for

9. Lyell to Herschel, 9 September 1828 [London-RS: HS.11.417; *LLJ* 1: 200–202].

10. Lyell to Caroline and Marianne Lyell, 10, 20 October 1828 [*LLJ* 1: 205–10]; Lyell to Murchison, 14 October 1828. Lyell recorded (notebook 15: 87) that Guidotti gave a figure of only 12% for Subapennine "shells identical with recent" species, but 30% for the *almost*-identical ones that, following Lamarck's practice (*BLT* §7.4), he called "analogues". This would have been a salutary reminder to Lyell that any molluskan "statistics" necessarily depended on the definition of species adopted by taxonomists (see §32.3). Rudwick, "Geological travel" (1996), uses the example of Lyell's tour to explore the analogy with pilgrimage in the light of Victor Turner's extension of Arnold van Gennep's classic anthropological concept of liminality.

11. Lyell, notebook 15: 172. This transcript represents the final text, after many alterations: he evidently regarded it as important to find just the right wording. The published equivalent is in *Principles* 1 (1830), 1, its opening sentence (see §21.1).

12. Lyell to Marianne Lyell, 20–29 October 1828 [*LLJ* 1: 207–10].

political reasons had been ejected from his university position at Otranto. He then crossed to the island of Ischia, where he duly found marine shells high up on its eroded volcanic peak; Costa confirmed that they were all of living species. On Lyell's tacit timescale, this made the deposit geologically very recent, and yet it had been raised some 2600 feet [800m] above sea level; it provided further support for his theory linking crustal elevation with volcanic activity. Around this time, he wrote down a tripartite grouping of the more recent formations. Sandwiched between "Upper Tertiary formations" such as the Subapennines, and "Modern (superior) formations—contemporary with man", such as alluvial river deposits, were "Modern inferior formations of contemporary origin with the species now living, or with a majority of living species", such as the deposits high on Ischia. It was in effect a sketch of how, in stratigraphical terms, the gap between the Tertiaries and the present world might be closed without any "diluvial" disruption. And in one of his frequent letters home, he expressed the same aspiration in geohistorical terms, with Sicily once more as the goal of his scientific pilgrimage: "My wish was to find this peninsula [of Italy] get younger and younger as I travelled towards the active volcanos, and it has hitherto been all I could wish, and I have little fear of bringing the great part of Trinacria [Sicily] into our own times, as it were, in regard to origin."[13]

Lyell took advantage of his enforced stay in Naples to make quick visits to many other well-known geological sites and sights in the surrounding region. He climbed Monte Somma, the huge outer rim half-encircling Vesuvius, and later, Vesuvius itself (see De la Beche's sketch, made only a few weeks later: Fig. 13.3); he saw enough to be convinced that both were the products of gradual accumulation of lava flows, and not—as von Buch claimed (§8.3)—of any sudden crustal upheaval. He visited the famous excavations at Herculaneum and Pompeii (*BLT* §4.1) and saw the museum display of the artifacts that, like fossils from a geological formation, had been retrieved from the volcanic debris (see *BLT* Figs. 4.1, 4.3). In another direction, he saw for himself the famous Temple of Serapis (Fig.

Monte Nuovo, formed in the Bay of Baiæ, September 29th, 1538.

Fig. 19.2 Monte Nuovo ["new mountain"] in the Phlegrean Fields near Naples: a wood engraving published in 1830 in Lyell's *Principles of Geology*, based on a sketch made when he visited this famous site in 1828. The brief volcanic eruption in 1538 that had produced this cratered cone was associated with local crustal elevation that had raised the nearby Roman "Temple of Serapis" (Fig. 8.3) back above sea level. For Lyell, it was a well-known and therefore useful example of the continuing power of the "modern cause" of volcanic action, within historical times.

8.3), checking that it could indeed be explained as the product of recent crustal movements in *both* directions; and also the nearby volcanic cone of Monte Nuovo (Fig. 8.1), the eruption of which in 1538 was suspected of being linked with the Temple's most recent major uplift; and he saw how it was built on a terrace of marine sediments below a line of former sea cliffs, indicating a much earlier episode of local elevation (Fig. 8.4). All this was geological sightseeing with a purpose; Lyell saw nothing novel, but he saw well-known things for himself, and, like other geologists, found them an eloquent display of actual causes in action (Fig. 19.2).[14]

One of Lyell's draft essays, perhaps prompted by what he saw at Pozzuoli, shows how he was now thinking about *all* the geological "causes" that his book would analyze. He conceded that any of them might at times have been violent and sudden in their action; he was *not* assuming that they were always gentle or imperceptibly gradual. The emphasis was rather on their *local* action, not fixed in position but forever shifting to different regions of the globe and affecting each part repeatedly at different times. Lyell contrasted the complexity of geological causes with the relative simplicity of astronomical ones, recognizing that this made the former utterly unpredictable (or unretrodictable) and irreducibly contingent. And the title he gave this essay, "Extinct Causes", shows clearly how his deeply geohistorical perspective entailed a thorough integration of organic and inorganic change:

> In one sense every cause becomes in succession extinct in Nature, viz., locally . . . Every agent in nature becomes locally dormant, but it is not the sleep of death, not even in that locality. Nor are any of the forces above enumerated [volcanoes, earthquakes, fluvial and marine erosion] uniform in their degree of violence in any one spot, not even for any one particular epoch over the whole globe, for they depend, not like the motions of the heavenly bodies on a few simple causes, but on so many, & admit consequently so many modifications, that their possible variety may be conceived infinite, without disparagement to their perfect identity . . . If any one were asked in what state he expects to find the organic world 100,000 years hence he would be rash to answer, & for the same reason we have no right to expect a priori that at an equally distant period of the past we should find it [i.e., retrodictably] in a certain state.[15]

13. Lyell to Murchison, 6 November, and to Marianne Lyell, 9 November 1828 [*LLJ* 1: 210–16]; Lyell, notebook 17: 145, 150. The phrase "into our own times" referred to the emergence of the landmass above sea level, not to the formation of the rocks themselves. Costa, whose *Testacei delle Due Sicilie* (1829) was published shortly afterward, would have made Lyell aware of the crucial importance of an adequate knowledge of *living* molluskan species, as a baseline for any "statistical" comparisons with fossil faunas.

14. Fig. 19.2 is reproduced from Lyell, *Principles of geology* 1 (1830), 335. This is an early use—in a scientific book—of the then new technique of wood engraving; like the long-obsolete technique of woodcuts it enabled illustrations to be printed *on the same page* as text, thus facilitating a close integration of words and images (woodcuts, common in early printed books, are carved along the grain; wood engravings, on the end-grain of hard boxwood, giving a much sharper image). In Lyell's time, wood engravings were often referred to, loosely, as "woodcuts" or just "cuts".

15. Lyell, notebook 17: 183, 177. No brief excerpts can do justice to the conceptual richness of this essay (oddly paginated but in continuous prose).

19.4 LYELL IN SICILY

When Lyell crossed at last to Sicily, he landed not at Palermo but at Messina, and then followed the coastline around the foot of Etna to Catania (see Fig. 19.6). Following suggestions from Buckland and Daubeny, his first major objective, reached with a mule and a local guide, was to explore the Val del Bove, the vast cavity (in modern terms, a collapse structure) in the flank of Etna that faces the sea, its floor covered with lava flows. Lyell studied the rocks exposed in the surrounding cliffs, which offered natural sections through the volcano. Although confused at first by some very confusing appearances, he convinced himself that they did indeed show that Etna, like Vesuvius, had accumulated layer by layer by the addition of successive lava flows running down its flanks. This confirmed that the whole volcano had grown in the same manner as had been documented in the centuries covered by human records, which suggested in turn that its total

Fig. 19.3 "Minor cones on the flanks of Etna": a wood engraving published in Lyell's *Principles of Geology* (1830), based on a sketch made at the time of his ascent of the volcano in 1828. The double cone of Monti Rossi (*1*, right) was formed during the eruption of 1669, when a huge lava flow—the largest in the recorded history of Etna—spread right down to the coast, where it had threatened to engulf the city of Catania. For Lyell, it was a striking example of the enduring magnitude of this "modern cause".

Fig. 19.4 "View from the summit of Etna into the Val del Bove": Lyell's sketch drawn on 1 December 1828 and published later as a wood engraving in his *Principles of Geology* (1833). The vertical "dikes" of volcanic rock jutting out from the cliff on the far side were interpreted as the ducts by which lava had reached the surface in past eruptions on the flanks of the volcano, of which the cratered minor cones in the middle distance were more recent examples.

age must be vast beyond human reckoning. On a later excursion, some peasants pointed him to a place where he found a thick deposit, containing shells of extant species, some 700 feet [210m] above the sea. "This is just what everyone in England, and at Naples and Catania, told me I should *not* find", he told his sister, "but which I came to Sicily to look for—the same which I discovered in Ischia, and what, if my geological views be just, will be found near all recent volcanos." Paying a man and a boy to work for him, he hoped that the further fossils they collected would "fix the zoological date of the oldest part of Etna". He was now using an assemblage of fossil mollusks to calibrate geohistory, as a matter of routine.[16]

Lyell's ascent of Etna followed much the same routine as that of savants ever since the time of Saussure and Hamilton (*BLT* §2.2). Like many more recent geologists, his guides were the Gemmellaro brothers, whose home at Nicolosi on the volcano's southern flank made a convenient base for the climb, and their refuge near the summit a welcome shelter. Near Nicolosi he saw and sketched the double cone of Monti Rossi, from which had issued the famous huge lava flow of 1669. As shown on the Gemmellaros' newly published map of Etna and an earlier quasi-aerial view (Fig. 7.3), this was the largest of all those known from historical records; it vividly exemplified Lyell's belief that geological "causes" had continued into the modern world in undiminished power. Above Nicolosi, Lyell—like Hamilton long before him (*BLT* §2.5)—was impressed by the sheer number of such "minor" cones on the flanks of Etna; unlike Monti Rossi, most of them were so ancient that they could not be dated in human terms, thereby hinting again at the vast antiquity of the volcano as a whole (Fig. 19.3).[17]

The climb was long and tiring, and made more difficult towards the summit by the effects of high altitude. But having passed through the forested zone and the already snow-covered slopes above the tree-line, they reached at last the cratered summit cone, kept free of snow by the warmth of the volcanic action. There, in an icy gale, they got a fine view down into the Val del Bove; as Buckland had requested, Lyell sketched the dramatic scene, his hat held on for him by his guide and his feet almost roasted from the heat below (Fig. 19.4).[18]

Lyell was lucky to be able to make the ascent so late in the year. By the next day the weather had broken, there was heavy snow far below the tree-line, and he was marooned for a time at Nicolosi, though with the Gemmellaros' fine library of local natural history as compensation; on his return to Catania he found he had been elected a member of the local scientific society, the *Accademia Gioenia*.

16. Lyell, notebook 19; Lyell to Buckland, 27 November 1828 [London-KC]; Lyell to Marianne Lyell, 29 November to 3 December 1828 [*LLJ* 1: 216–20]; see Wilson, *Lyell* (1972), 232–37. Lyell published his view of the Val del Bove in *Principles* 2 (1832), frontispiece (see Fig. 25.2); it remains quite difficult to reach and has, of course, changed substantially as a result of many further eruptions since the time of Lyell's visit.

17. Fig. 19.3 is reproduced from Lyell, *Principles* 1 (1830), 364. He was uncertain about the correct name for the cone (*2*) on the left.

18. Fig. 19.4 is reproduced from Lyell, *Principles* 3 (1833), 93; Lyell to Marianne Lyell, 29 November to 3 December 1828 [*LLJ* 1: 216–20]; Gemmellaro (G.), *Quadro istorico dell'Etna* (1828); Rudwick, "Lyell on Etna" (1969).

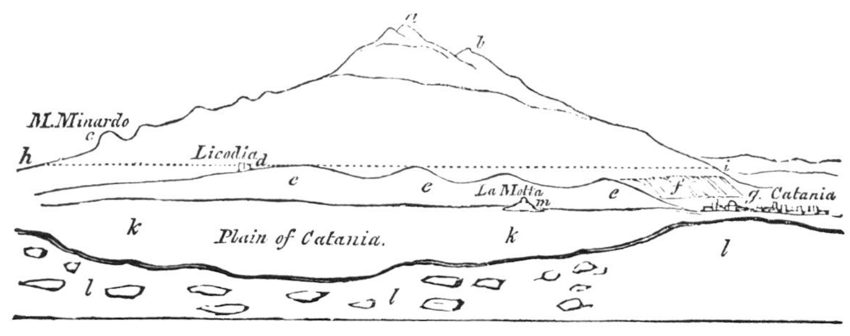

Fig. 19.5 Lyell's view of Etna, sketched from the northern edge of the Val di Noto plateau (*l*), looking across the alluvial plain of Catania (*k*). Above the low hills of marine sediments (*e*) at the base of the volcano, the dotted line (*h-i*) marks "the upper boundary along which the marine strata can be seen". This implied that the whole of the volcano above that line had accumulated *since* the time at which these geologically recent strata had been deposited. Some of the dozens of minor cones are depicted in profile on the western (left) flank. This wood engraving, published later in Lyell's *Principles of Geology* (1833), was based on the sketch that he made on the spot in December 1828.

Lyell's next objective was to place Etna in its regional context, by working out the relation between the whole volcano and the formations that appeared to underlie it. Having crossed the flat alluvial plain of Catania, he climbed on to the hills that formed the northern edge of the Val di Noto (an upland region, in spite of its name). Looking back, he had a fine view of the huge 3300m cone of Etna. Along its base was a line of low hills, composed—as he had already checked—of clayey strata with a few fossil shells of marine species. On his sketch he drew a line on the volcano itself, just above those hills, marking the level up to which he had seen such deposits with shells of extant marine mollusks. This gave visual expression to his conclusion that the whole great volcanic cone (or at least all above the line) had accumulated *on top of* marine sediments that were—by his provisional dating method—geologically very recent (Fig. 19.5).[19]

Like other savants ever since the time of Hamilton (*BLT* §2.5), Lyell was thus convinced that Etna was of literally unimaginable antiquity by any human standards: most of the minor cones on its flanks were earlier than any human records, and all the historically recorded eruptions combined had added very little to its bulk. Yet now he was also convinced that the whole volcano had accumulated on top of marine deposits which, judging by the shells they contained, were extremely recent by any geological standards. Together, these two inferences suggested a striking calibration of geohistory in relation to human history. To establish this point, however, Lyell needed to relate these very recent deposits to the ordinary Tertiary formations, such as the Subapennines on the mainland. As he sketched the distant volcano, he must have judged it likely that the deposits immediately underlying Etna were themselves underlain by rocks such as the limestone of which he stood. So his next task was to check for himself the geology of the Val di Noto and other areas beyond it, in order to integrate Etna into whatever Tertiary geohistory he might be able to reconstruct in Sicily. Once again his fieldwork was not novel: he had been preceded by other traveling geologists, notably his compatriot Daubeny, not to mention competent local naturalists such

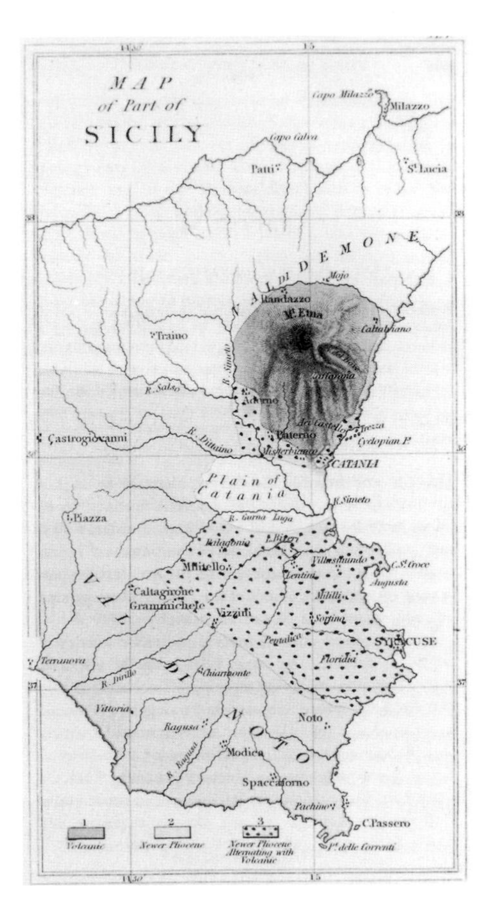

Fig. 19.6 Lyell's map of eastern Sicily, published in his *Principles of Geology* in 1834 but serving here to illustrate his fieldwork in 1828. It shows the great cone of Etna, separated by the alluvial Plain of Catania from the upland region of the Val di Noto, the latter composed of Tertiary formations (by 1834 he called them "Newer Pliocene": §26.1). Agrigento, the furthest point of his main fieldwork, is off the map to the west; Messina, where he had landed from Naples, is off to the northeast.

as the Gemmellaro brothers. But of course he was seeing the region with his own eyes, and with his own profoundly theoretical ideas in mind (Fig. 19.6).[20]

Lyell and his trusty servant Rosario, whom he had hired at Messina, therefore began a long traverse of the east and south coasts, intending thereafter to cross the interior to Palermo. Even as he wrote home describing the execrable conditions he was enduring—much of Sicily was desperately poor—Lyell did not hide his excitement at what he was seeing. His immediate task, however, was quite prosaic, a routine matter of Smithian stratigraphy. The hills of the Val di Noto were largely composed of an almost horizontal limestone formation, deeply eroded by valleys that showed it to be some 800 feet [250m] thick. The rock itself, and the topography it produced, reminded him strongly of the Oolitic (Jurassic) limestones of the Cotswold hills in Smith's native part of England; yet its fossils were clearly Tertiary ones, not Secondary. And on coming down to the coast at Syracuse [Siracusa], Lyell found it underlain by a yellow sandy limestone and

19. Fig. 19.5 is reproduced from the wood engraving in Lyell, *Principles* 3 (1833), 75; quoted phrases from caption of original sketch, which is finely reproduced in Wilson, *Lyell* (1972), fig. 38. The printed caption is not reproduced here, because it represents a distinctly later phase in Lyell's work: for example, it identifies the limestone in the foreground as "newer Pliocene" (see §26.1).

20. Fig. 19.6 is reproduced from Lyell, *Principles*, 4th ed. (1835), 3: pl. 7, first published in 3rd ed. (1834), 3: pl. 5a, explained on 316–24. Daubeny, "Geology of Sicily" (1825); Gemmellaro (C.), "Fisionomia di Sicilia" (1831), read at the Accademia Gioenia on 13 November 1828, i.e., shortly before Lyell's visit.

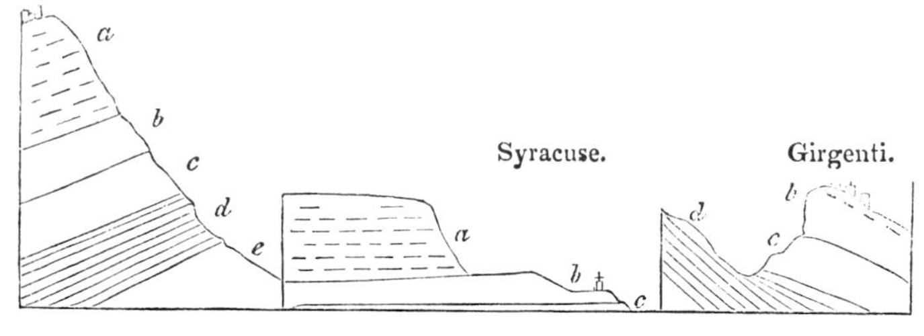

a, Great limestone of Val di Noto.
b, Schistose and arenaceous limestone of Floridia, &c.
c, Blue marl with shells.
d, White laminated marl.
e, Blue clay and gypsum, &c. without shells.

Fig. 19.7 Lyell's diagrammatic sections of the Tertiary formations at three Sicilian localities, based on his fieldwork at the end of 1828 (and published in 1833 in his *Principles of Geology*). The sandy limestone (*b*) and blue marl (*c*), for example at Girgenti [now Agrigento] (section on right), contained abundant fossil shells that were almost all of species still living in the Mediterranean, making them—on Lyell's provisional dating system—geologically very recent. Yet they were overlain by the "Great limestone" (*a*) of the Val di Noto, forming the plateau shown in the Syracuse section (center), which was clearly even younger; and since its deposition this massive formation had been elevated, at Castrogiovanni [now Enna] in the center of the island (section on left), to almost 4000 feet [1450m]. This modest little diagram encapsulated geohistorical reasoning that provided Lyell's Grand Tour with its climax, and that was of decisive importance in his later theoretical development.

then by a distinctive blue marl with abundant and unquestionably Tertiary fossils. This in itself was striking enough, because it showed that the ancient-looking massive limestone must in fact be quite recent in geohistorical terms. Just how striking this bit of stratigraphy was, however, would depend on determining the relative age of the formations, and of their fossils (Fig. 19.7).[21]

Round on the south coast, at Girgenti [now Agrigento], Lyell saw how its famous Greek temples were built on (and indeed of) the same sandy limestone, which overlay the same marl (Fig. 19.7, right section). Daubeny had described the fossils as being "not far, if at all, removed from existing species". If that comment was correct—and Lyell was dependent on expert conchologists such as Costa to confirm it—it implied that the overlying Val di Noto limestone, so ancient in appearance, must in fact be one of the youngest formations in Sicily. "I got so astounded by the results I was coming to", he told Murchison later, "that I began to doubt them; and not without some struggle with my desire to get out of the inns and horse-paths and other evils, I struck back again to Val di Noto right through the centre of the isle". But when he got there he confirmed his inference: "the Val di Noto redeemed all", he told his sister. So he doubled back again across the interior towards Palermo and civilization. On the way, following Daubeny's lead once more, he saw how the town of Castrogiovanni [now Enna] was perched on a high scarp of the Val di Noto limestone, underlain by the usual formations (Fig. 19.7, left section). This was the culmination of Lyell's fieldwork, for it showed that a marine formation of great thickness had been elevated to an altitude of almost 4000 feet [1450m] in geologically very recent times.[22]

"I am come most unwillingly to this conclusion", Lyell told Murchison; but "living shells" were so abundant, even in the sandy limestone and blue marl, that the Sicilian formations must be far more recent in origin than Brocchi's Subapennines, the "numerous extinct species" of which were entirely missing. Although he was dependent on experts for exact identifications, he told Murchison that "I am beginning now to be able, when I see large collections, to distinguish between any marked difference in the proportion of lost [extinct] species and genera". The Subapennines, which he had described only a few weeks earlier as "the most modern formation", now looked very ancient; yet (as Prévost and others had argued) it was in turn much younger than most of the Tertiaries of the Paris Basin and elsewhere (§12.1). In this context, what Lyell had found in Sicily implied a dramatic further expansion of Tertiary geohistory, linking it toward the present world.[23]

Before he left Sicily, or soon afterward, Lyell wrote a draft essay entitled simply "Etna", setting out the evidence he had seen—involving its "minor" cones and lava flows—by which its total age could be estimated as vastly greater than all recorded human history. Yet his stratigraphical fieldwork suggested that the whole volcano was younger than even the youngest of the Sicilian Tertiaries, all of which in turn were younger than the Subappennines, let alone still older Tertiaries such as those of the Paris Basin. If these two chains of geohistorical reasoning were combined, and if his increasingly practiced "eye" for the fossils proved reliable, Sicily would be well on the way to being brought, as he had anticipated, "into our own times". Above all, he had found dramatic evidence that here in Sicily the earth had been just as dynamic in these most recent periods of geohistory as in any earlier times. Sicily had more than fulfilled what Lyell had expected of it in advance. Amid the arduous "evils" of its primitive interior, he had indeed fulfilled his heroic quest and experienced his secular theophany.[24]

19.5 CONCLUSION

Lyell's tour continued to be as productive after he and the Murchisons left Auvergne as it had been in that celebrated region. In Velay and Vivarais, the other regions covered in Scrope's book, they found striking further evidence for the gradual erosion of valleys by purely fluvial action, without the need—or indeed the plausibility—of invoking any diluvial episode. Leaving the Massif Central for Provence and the Mediterranean coast, they also found further evidence for the sheer magnitude of the Tertiary formations and the extent to which they had since been elevated. This supported Lyell's nascent causal theory that linked crustal elevation to volcanic activity, and gave him a further incentive to travel as far as the active volcanoes in southern Italy and Sicily. But he also began to apply the French geologists' method of giving the Tertiary formations relative dates,

21. Fig. 19.7 is reproduced from Lyell, *Principles* 3 (1833), 64.

22. Lyell to Caroline Lyell, 11 December 1828 to 1 January 1829, to Fanny Lyell, 1–10 January 1829, and to Murchison, 12 January 1829 [*LLJ* 1: 221–34]; Wilson, *Lyell* (1972), 240–52, prints excerpts from Lyell's notes (notebooks 20, 21) which convey the confusing nature of this fieldwork, here summarized much too briefly.

23. Lyell to Murchison, 12 January 1829 [*LLJ* 1: 232–34].

24. Lyell, notebook 21: 113–18 [printed in part in Wilson, *Lyell* (1972), 253–54].

by assessing, for each assemblage of fossil mollusks, the proportion—or, as Lyell expressed it, the percentage—of species still extant.

Lyell and the Murchisons continued their tour through northern Italy, visiting many well-known sites, before the Murchisons decided to return to England, leaving Lyell to continue his Grand Tour alone. He first made a close study of Brocchi's celebrated Subapennine formation on both flanks of the Apennines, which gave him a good understanding of what the French geologists (and he himself) regarded as the youngest of the Tertiaries. After seeing further well-known sites around Naples, including the Temple of Serapis, Pompeii, and Vesuvius, he crossed to Sicily, confident that he would find evidence that the island, dominated by the largest volcano in Europe, had been elevated above sea level even more recently.

In Sicily, Etna duly became Lyell's first object of study. He confirmed that its huge mass had indeed been built up by the accumulation of innumerable lava flows, in the same way as had been documented reliably throughout recorded history. Like many earlier naturalists, he was impressed by the sight of the dozens of minor cones on its flanks, most of which antedated all human records. Low on its flanks he found deposits with marine shells of extant species; so he concluded that the whole volcano, although unimaginably ancient by human standards, had accumulated on top of a formation of geologically very recent origin. His stratigraphical fieldwork in the Val di Noto then convinced him that this was one of the youngest of a thick pile of Tertiary formations, which in the center of the island had already been elevated thousands of feet above sea level. Yet he suspected that their fossils were almost all of extant species. If this was confirmed by the experts, the whole of Sicily would have to be attributed to periods of geohistory compared to which even the Subapennines would be very ancient. This chain of reasoning—convincing at least to himself, if not yet to others—seemed decisively to link the Tertiaries to the present world in a correspondingly unbroken chain of geohistory.

CHAPTER 20

Lyell in European context (1829–30)

20.1 LYELL'S HOMEWARD JOURNEY

Lyell's homeward journey began, in effect, with the civilized comforts of Palermo, and not least the hospitality of the marquis of Northampton, its leading English resident and a patron of the sciences (a few years later he became the president of the Royal Society in London). Lyell also met Ferrara, the most important living writer on Sicilian natural history (§7.3), and bought his books. After he sailed back to Naples, Costa duly identified Lyell's Sicilian fossils. The shells he had collected at Castrogiovanni in the center of the island, for example, were of fifty-nine species, all known alive, which confirmed the geologically recent date of the Sicilian Tertiaries and, even more significantly, of their elevation high above sea level.

Lyell wrote to Murchison that "the results of my Sicilian expedition exceed my warmest expectations in the way of modern analogies", that is, in the match between ancient and modern causes that he had hoped to consolidate. But he also related his work to his hopes for his science and his own future plans: "we must preach up travelling, as Demosthenes did [oratorical] delivery, as the first, second and third requisites of the modern geologist, in the present adolescent state of the science". The metaphor was apt, and Lyell intended to practice what he preached: geology was certainly not in its infancy or even its childhood, but he meant to jolt it into maturity by means of his book, which was already "in part written, and all planned". He intended to work hard on it after his return—avoiding if possible the time-consuming tasks of helping to run the Geological Society—in the hope that it would not only reform the science but also earn him enough to recoup the expenses of his Grand Tour. Above all, however, Lyell set out—more uncompromisingly than ever before—the "principles" that would underlie his work and lead to the establishment of the true "system" of the earth. For he was now explicitly wedded to a radical form of explanation by actual causes, which could underpin his belief that geohistory had been steady-state in character and not directional at all; and implicitly his ambition was to rehabilitate the long-despised genre of geotheory:

> It [the book] will not pretend [i.e., claim] to give an abstract of all that is known in Geology, but it will endeavour to establish the *principles of reasoning* in the science, & all my Geology will come in as illustration of my views of those principles, & as evidences strengthening the system necessarily arising out of the admission of such principles, which as you know are neither more nor less than that *no causes whatever* have, from the earliest time to which we can look back, to the present, ever acted, but those *now* acting, & that they never acted with different degrees of energy from that which they now exert.[1]

Lyell traveled fairly rapidly back up the length of Italy, stopping here and there to check the work he had done on his way south, again making full use of local naturalists. Around Siena, for example, he now saw some similarity between the fossils of the youngest beds of the Subapennines and the oldest in Sicily, thus reinforcing his inference about their relative ages. Once again, such apparent details were closely linked to much broader programmatic ambitions: "we want [i.e., are lacking] nothing short of a radical reform in geology", he claimed, alluding once more to the current agitation for radical reform in British politics; and he added, echoing his earlier Demosthenean maxim, "we shall have one soon if honest men will travel and write and travel again". This was a dialectic of outdoor and indoor exertion that he himself, of course, intended to embody and exemplify.[2]

Later, back in Genoa, Lyell met Domenico Viviani (1772–1840), the professor of botany, who told him he was puzzled by the patterns of plant distribution in the Mediterranean region: few species were endemic to Sicily, for example, but there were many on Corsica, an island of similar size and climate. Lyell immediately linked Viviani's problem with his own "new system of geology" for Italy. He now believed on quite independent grounds that Sicily was a very young island, having risen above sea level in geologically recent times, very late in the Tertiary era; Corsica, in contrast, as an island of mainly Primary rocks, might well have been isolated from other land for far longer. The botanical difference between the two islands might therefore be explained in terms of their *geohistory*, provided his Brocchian ideas about a steady turnover in species were valid for plants as well as mollusks.

This new hint of a further integration between the organic and inorganic parts of his nascent "system" was so important to Lyell that he abruptly changed his plans. Instead of following the easy winter route back to Paris, along the Mediterranean coast and then up the Rhône valley, he crossed the Apennines to Turin—where he fitted in another fruitful discussion of Tertiary fossils with Bonelli—in order to cross the Alps by the much higher pass of Mont-Cenis (the coach was equipped with sleds for the deep winter snow). His destination was Geneva, where Saussure's grandson, a link with the heroic days of the science, showed him some of the local geology in the field. But his primary reason for crossing the Alps was to meet Candolle, who had long been the leading exponent of plant "statistics" and plant geography (§12.1). In the light of their discussion, Lyell told Murchison he was "almost certain that my spick-and-span new theory on this subject will hold water". And he told his father—whom as a botanist Candolle knew by repute—how he was elated that his own "new geologico-botanical theory" might explain *geohistorically* the distribution of organisms in general:

> I am now convinced that geology is destined to throw upon this curious branch of inquiry [i.e., biogeography], and to receive from it in return, much light, and by their mutual aid we shall very soon solve the grand problem, whether the various living organic species came into being gradually and singly in insulated spots, or centres of creation, or in various places all at once, and all at the same time. The latter cannot, I am already persuaded, be maintained.[3]

With fresh ideas on the piecemeal "creation" or "birth" of species and on their piecemeal extinction or "death", on the vast timescale of even the Tertiary era, and on the magnitude of the dynamic processes of deposition and elevation during that time, Lyell's "system" for geology was beginning to take shape, at least in his notebooks and in his letters to a favored few. It remained for him to articulate it in a more public form, and to convince his fellow geologists, a wider range of other savants, and if possible the educated public. He left Geneva on the penultimate leg of his journey home determined to achieve these goals as swiftly as possible through the book he had long been planning to write. In the nine months since he left Paris, at least until he met the internationally renowned Candolle, he had been more or less on the periphery of the scientific world: at best in the company of competent but provincial naturalists, and often alone (but for a guide or servant). His return to Paris marked his full re-entry into the savant world, and the start of the process of articulating the "system" that he hoped would bring radical "reform" to his chosen science.

20.2 PARISIAN DEBATES ON THE TERTIARIES

Lyell reached Paris after four days and nights in a coach with only one six-hour rest at Dijon (a journey beside which the longest modern intercontinental flight must be judged quite relaxing). Almost immediately he met Prévost, went to his lecture that evening and to another at the School of Mines the next morning, meeting some of the leading geologists at each; as he told his sister Marianne, "I

1. Lyell to Murchison, 15 January 1829 [*LLJ* 1: 234–35; printed, with corrections, in Wilson, *Lyell* (1972), 256]. Lyell, notebook 21: 104–6.

2. Lyell to Murchison, 22 January 1829 [*LLJ* 1: 239–42; Wilson, *Lyell* (1972), 257].

3. Lyell to Lyell senior, 7 February, and to Murchison, 5 February 1829 [*LLJ* 1: 242–46]; Candolle, "Géographie botanique" (1820), on which see Browne, *Secular ark* (1983), 62–64. His use of the word "gradually" does not imply any hint of Lamarckian transformism: it refers to new faunas and floras appearing by the piecemeal introduction—by whatever means—of discrete new species, *not* (or not necessarily) to the formation of such species by transmutation from preexisting ones.

got thrown in a few hours into the heat of battle". Lyell sensed among the French some apprehension that they might soon be outstripped if the British continued to "travel and write and travel again" as assiduously as he himself had been doing. His manner towards them may have conveyed what he had told Murchison earlier, with staggering self-confidence, or arrogance: that having "fathomed the depth and ascertained the shallowness of the geologists of France and Italy as to their original observations", it remained only to discover "whether Germany is stronger". Combing Paris for geological books to take home, Lyell noted the importance of those published in German and again sensed the urgency of learning that language. He found Cuvier "in great force" at his salon, and was later granted the rare privilege of seeing the naturalist's "sanctum sanctorum" or private workroom, the efficient arrangement of which helped account for a lifetime's prodigious productivity. Lyell also went with other savants to breakfast at the Brongniarts' house, where he talked with both father and son. Back in Paris he was back at the heart of lively geological debate.[4]

Of all the subjects of current discussion, Lyell was most concerned to relate his own recent research to what the French geologists had been doing with the Tertiary formations and their fossils; they too had long recognized the crucial role that the Tertiaries could play in the reconstruction of geohistory as a whole (Chap. 10). So he may have been disconcerted to find that Desnoyers had already reached conclusions similar to his own. He would certainly have known that Desnoyers had earlier described a Tertiary formation in Normandy, relating it to all the others elsewhere in Europe, and claiming that by their fossils they fell into two groups of different relative ages. But the Frenchman had another paper in press for the *Annales des Sciences Naturelles*, taking that inference much further (§12.1). He claimed that the "non-simultaneity of the Tertiary basins" could be confirmed, in one crucial case, on straightforward grounds of Smithian stratigraphy: the formations in the Touraine region were demonstrably younger than those in the Paris Basin to the north (Fig. 12.1). And he backed up that case by a general review of all the Tertiary formations throughout Europe, suggesting the possibility of reconstructing the whole of Tertiary geohistory, by "an insensible and continuous passage", from deep past all the way into the modern world.

Desnoyers expressed in this paper much the same interpretation of the Tertiaries that Lyell had been formulating privately during his tour, except that Desnoyers knew nothing of the formations in Sicily until Lyell arrived. The Frenchman, echoing his fellow Parisian Élie de Beaumont, attributed the shifting succession of basins to a sequence of "epochs of upheaval [*relèvement*]" (§9.4); but although he imagined these events as "violent commotions", rather than the slow and gentle crustal movements envisaged by Lyell, he too treated them as an integral part of a highly dynamic earth. Anyway, there was just time for Desnoyers to add a postscript before his paper was published, noting that Lyell had reached similar conclusions independently. But if priority were to be contested, it lay clearly with the Frenchman, not with Lyell.[5]

However, both Desnoyers and Lyell were dependent on the expert conchologists for any further use of assemblages of fossil shells as criteria of relative age within the Tertiary era. An outstanding expert, the Parisian naturalist Deshayes, was already at work in just this way, focusing his research on the fossils of the Paris Basin but using a vast range of other specimens, living and fossil, for comparison. Deshayes's great monograph on the Parisian fossils had begun to appear several years earlier, and it is likely that by the time of Lyell's visit he had already concluded that there were not two but *three* groups of Tertiaries (Desnoyers's "younger" formations being subdivided), with increasing proportions of extant species as they approached the present world (§12.1).[6]

So the most important event during Lyell's stay in Paris was his meeting with Deshayes. As he told Mantell, Deshayes was "now the strongest conchologist in Europe . . . [and] acknowledged to be the Cuvier of tertiary shells", with a private collection of over three thousand living and fossil species. Deshayes's unfinished monograph on those from the Paris Basin, like many similarly ambitious projects in natural history, was leaving its author impoverished. So he readily agreed to identify the fossils that Lyell had collected during his tour; the two later devised a discreet arrangement by which Lyell could subsidize him without infringing gentlemanly norms. Lyell anticipated, of course, that Deshayes's authoritative identifications would confirm what Costa had already told him, namely that the Sicilian shells were almost all of species known alive, whereas it was already well known that the Subapennines included a substantial proportion of extinct species, and hence were inferentially much older.[7]

Lyell intended to give Desnoyers's notion of "the non-simultaneity of the Tertiary basins" much greater precision through the quantitative analysis of their fossil faunas, which Deshayes's identifications would make possible. Conchology, he was convinced, was the key to Tertiary geohistory. Once again he found an antiquarian or *historical* metaphor vividly appropriate. This time he was among Parisian savants excited by the recent achievement of Jean-François Champollion (1790–1832), who had at last discovered how to decipher the ancient Egyptian hieroglyphs, thereby opening up the oldest known civilization to detailed historical study. Using the trilingual inscription on the Rosetta Stone, which had been found long before during Bonaparte's military expedition in Egypt (*BLT* §7.3), Champollion's crucial key had been the text in the "demotic" or everyday Egyptian script of the Hellenistic period: the text—intermediate in more senses than one—inscribed between the easily legible Greek and the baffling hieroglyphs. Lyell told his sister Marianne that conchology, likewise, was "the ordinary or, as Champollion says, the demotic character in which Nature has been pleased to write all her most curious documents". He suspected that fossil shells, although

4. Lyell to Mantell, 19 February, and to Marianne Lyell, 21 February 1829 [*LLJ* 1: 248–51].

5. Desnoyers, "Ensemble de dépôts marins" (1829), with postscript (214–15) at the end of the first part; in the second, already written but published two months later, he added in a footnote (441n) that Lyell and Deshayes were together developing the theory further (see below). Tasch, "Lyell and Deshayes" (1985), argues for Lyell's priority in Tertiary stratigraphy, above Deshayes, Desnoyers, and Bronn (§12.1).

6. This conclusion was published later in Deshayes, "Tableau comparatif" and *Coquilles caractéristiques* (both 1831). He also distinguished an even more recent set of deposits, with almost *all* species still extant, but he cited only a minor example on the Mediterranean coast—almost certainly Risso's "Quaternary" at Nice (§19.2)—*not* Lyell's far more massive Sicilian formations.

7. Lyell to Mantell, 19 February, and to Marianne Lyell, 21 February 1829 [*LLJ* 1: 246–51]; Deshayes, *Coquilles fossiles de Paris* (1824–37).

less spectacular than the more enigmatic fossil quadrupeds, would yield equally striking results once they were correctly deciphered. Like his hero Cuvier long before (*BLT* §9.3), Lyell was again treating the recovery of geohistory as a matter of hermeneutics, of interpretation, of deciphering nature's own language, a task more subtle than any mere observation or description.[8]

As Lyell's metaphor suggests, his collaboration with Deshayes was intended to go much further than the identification of the fossils he had collected on his Grand Tour. As he told Mantell immediately after his return to London, "we planned together a grand scheme of cataloguing the tertiary shells of the various European basins, that I might draw geological inferences therefrom". The division of labor was bluntly expressed: Deshayes would carry out the painstaking work of exact identification, while Lyell would reveal the deeper meaning of those results. The plan was to include the English Tertiaries (on which he hoped for Mantell's cooperation) as well as the French and Italian formations and others elsewhere. And it would be no mere set of lists of fossil species, but a "statistical" analysis of profound theoretical significance, which Lyell had ambitions to extend to the *whole* of the fossil record. However, the character of Lyell's project was not fully disclosed at this stage, even to his friend, for he had no intention of being further upstaged by Desnoyers or anyone else:

> My results will be an induction from nearly (perhaps more than) 3000 species in the tertiary formations alone, & I hope by other aid than Deshayes to carry it on through older [i.e., Secondary] strata also. No one but yourself & Deshayes is privy to these state secrets as yet, & till I get on further I have no wish to advantage them.[9]

20.3 DILUVIALISTS AND FLUVIALISTS IN LONDON

Back in London, Lyell's primary objective was to get his book as soon as possible into a fit state for publication, by revising what he had submitted to Murray before his Grand Tour, adding new material based on what he had seen in the field and was reading in scientific books and periodicals. While daunted by the magnitude of the task he had set himself, he was not much troubled by the caliber of those whose books might compete with his own. Revealing his own working title, he gave Mantell advice that deftly steered his provincial friend away from any such ambitious plans and towards projects suited to his more modest talents. Lyell discreetly warned Mantell off his own turf:

> Of course [*sic*] you will not attempt to tilt with Fitton and me 'on the general principles of geology', which we mean soon (mine will be soon) to give you. After all my travelling and reading I find it too much to dare, & only excusable when I measure my strength against others & not with the Subject.[10]

Lyell quickly picked up the threads of recent discussions at the Geological Society, which in intensity matched those in Paris. Of course, he had not been totally out of touch with London. On his way back through Italy, for example, he had

Fig. 20.1 "Cause and Effect": De la Beche's caricature portraying the implausibility of Scrope's and Lyell's "*fluvial*" explanation of the origin of valleys. The toddler (identified in a much later hand as Buckland's first child Frank) is urinating into the head of a large valley, while his attendant nursemaid comments, "Bless the baby! what a Walley he have a-made!!!" (spoken in the form that the middle classes often attributed to the lower, and particularly Cockneys). This caricature must have become widely known: twenty years later, critics such as Murchison still referred to the fluvialist position as "the 'piddling' school" of geology. Significantly, De la Beche depicted one of the most difficult cases for the fluvialists to explain, namely a broad valley of U-shaped profile (see Fig. 13.1), for which the existing stream seemed both inappropriate and utterly inadequate as an erosive agent, however much of Scrope's "Time!" (§15.1) was allowed (such valleys were later recognized as characteristic products of *glacial* erosion: see Fig. 35.2). (By permission of the Museum of Natural History, Oxford)

heard from Murchison (who had returned to England in time to be present), and also from Scrope, about the "warm debate" that had followed the reading of their joint paper on valley erosion in central France. As Lyell told his sister, "Buckland and Greenough [were] furious, *contra* Scrope, Sedgwick and Warburton, supporting us", a roster of leading figures that pitted champions of "diluvial" action against those who favored purely "fluvial" erosion. Lyell found on his return that this particular issue in the deployment of actual causes was more contentious in London than any problem about Tertiary geohistory (Fig. 20.1).[11]

8. Lyell to Marianne Lyell, 21 February 1829 [*LLJ* 1: 248–51]; Champollion, *Système hiéroglyphique* (1824); see Parkinson, *Cracking codes* (1999), 31–41.

9. Lyell to Mantell, 24 February 1829 [Wellington-ATL]; Rudwick, "Lyell's dream" (1978).

10. Lyell to Mantell, 23 March 1829 [Wellington-ATL]; he evidently doubted whether Fitton would ever produce a work to rival his own, and he was right.

11. Fig. 20.1 is reproduced from Oxford-MNH, Buckland MSS, Misc.MSS/12. Francis [Frank] Buckland (1826–80) would have been about the right age at the time of the fluvialist-diluvialist debates at the Geological Society; the caricature was presumably given to his father, among whose papers it remained; the note in another hand must have been added after De la Beche was knighted (in 1848). Lyell and Murchison, "Excavation of valleys" (1829), read on 5, 16 December 1828; Scrope to Lyell, 23 December 1829 [Philadelphia-APS]; Lyell to Marianne Lyell, 21 January 1829 [*LLJ* 1: 238–39]. The "piddling" (and identification of the child as Frank Buckland) is in Murchison to De la Beche, 3 April 1851 [Cardiff-NMW: no. 1031]; see Haile, "Piddling school" (1997).

However, the latter remained Lyell's goal: as he told Fleming, he was writing "a general work on the younger epochs of the earth's history". But before he could devote himself fully to this, he had to complete his series of joint papers with Murchison, which could jointly serve as trailers for his larger work. They were also effective progress reports, alerting geologists throughout Europe to what he was doing. Their joint authorship, however nominal in reality, strengthened the authority of the research, by indicating that it represented the agreed conclusions of two competent geologists, not just the possibly idiosyncratic ideas of one. Their remaining two papers, on the geology of Cantal and Provence, were read at the Geological Society early in the summer. They showed, respectively, that substantial crustal movement late in the Tertiary era had been closely related to volcanic activity, and that conditions in that era had not been significantly different from the Secondary era before it. The authors avoided the notorious delays of the *Transactions* by publishing them elsewhere: the first in French, in the Parisian *Annales*, and the second in Jameson's Edinburgh periodical, thus sharing the spoils between the two countries (and the honor of prime authorship between the two geologists).[12]

At the meeting following their paper on Cantal, Conybeare joined in the fray over valley erosion. He dubbed the two parties "*diluvialists*" and "*fluvialists*", as if they were two of the warring sects involved in the notorious religious controversies currently raging in Britain. To make his point, he too chose a specific region that he knew well at first hand, though in his case it was closer to home. He analyzed the puzzling physical geography of the Thames valley, and concluded that diluvial erosion was the most plausible explanation. He found two features particularly telling. As Buckland had shown, distinctive pebbles had apparently been swept into the Thames valley right over the watershed of the Cotswold hills, in a manner that was inexplicable in terms of the present rivers (*BLT* §10.5). Conversely, he claimed that earthworks dating from the "British" (i.e., Iron Age) and Roman periods showed no significant degree of subsequent erosion; and in his opinion it was not good enough just to invoke unlimited "Time!" in Scropean fashion (§15.1), because *some* change should have been perceptible after even two or three millennia, if erosion by rain and rivers were to have the effects being claimed over much longer spans of time. So he supported Buckland in concluding that the evidence still pointed to an exceptional "diluvial current" in the geologically recent past. But he emphasized its normality in the longer run of geohistory, by suggesting that three well-known Secondary and Tertiary formations with massive beds of conglomerate or ancient gravel might be traces of far earlier events of the same kind. Conybeare was staking nothing on the most recent "deluge" having been in any way unique, except in its date and its consequent overlap with human history.[13]

Lyell was therefore scoring much too easy a point when, reporting on the lively discussion of this paper, he told Murchison that "Conybeare admits 3 deluges before the Noachian! and Buckland adds god knows how many *catastrophes* besides, so we have driven them out of the Mosaic record fairly". Conybeare and Buckland probably did still believe at this point that the "geological deluge" had been recent enough to be identified as Noah's Flood as recorded, supposedly by

Moses, in Genesis (*BLT* §10.5); but that identification did not stand or fall by the physical uniqueness of the event. At the next meeting the concluding part of Conybeare's paper was followed immediately by the reading of a letter from a Scottish surveyor, reporting that a stream in the Cheviot hills, in flood after heavy rain, had swept the remains of a bridge—including blocks of stone over half a ton in weight—some two miles downstream. This was a striking example of the sheer power of the actual cause currently in contention; it was a small triumph for the fluvialists, who probably stage-managed the juxtaposition at the meeting. Again there was a lively discussion, and Sedgwick showed that he had indeed deserted the strict diluvialists: there might well have been a diluvial event, but he now considered it had not been worldwide, nor the one to which the Flood story referred.[14]

The following winter, in another paper at the Society, Scrope strengthened the fluvialist case still further by citing two striking cases on the Continent. The Moselle and the Meuse both cut through the massif of hard old rocks that forms the upland of the Eifel and the Ardennes (in the Rhineland and what is now southern Belgium); both rivers were deeply entrenched yet with a meandering course (in modern terms, they showed *incised* meanders). At the most spectacular spot, where the valley of the Moselle was hundreds of feet deep, a loop of the river several miles in length came within a few hundred yards of returning to its starting point. He claimed that such topography was decisive evidence against any diluvial explanation: "Any sudden, violent, and transient rush of water of a diluvial character could only produce straight trough-shaped channels in the direction of the current, but could never wear out a series of tortuous flexures, through which some rivers now twist about, and often flow for a time in an exactly opposite direction to the general straight line of descent, which a deluge or debacle would naturally have taken."[15]

It cannot be emphasized too strongly, however, that Lyell was not being opposed by other geologists for reasons related to the interpretation of Genesis: De la Beche, for example, was a firm diluvialist but also strongly secularist. On the other hand, *all* British geologists, including clerical diluvialists such as Buckland, could indeed expect, and did receive, bitter criticism from what Lyell called "the Granville-Penn school of England", the literalistic purveyors of "scriptural geol-

12. Lyell to Fleming, 10 June 1829 [*LLJ* 1: 253–55]; Lyell and Murchison, "Dépôts lacustres tertiaires" (1829), read on 3 April, 1 May; Murchison and Lyell, "Fresh water formations of Aix" (1829), read on 19 June.

13. Conybeare, "Basin of the Thames" (1829), read on 15 May, 5 June 1829; it was never published in full in the *Transactions*, because Buckland, the referee, lost the manuscript: Boylan, *Buckland* (1984), 175–76.

14. Lyell to Mantell, 7 June, and to Fleming, 10 June 1829 [*LLJ* 1: 253–55], claimed that Conybeare's paper had been written for him by Buckland, but offered no evidence beyond their shared opinions. Culley, "Power which running water exerts" (1829), may have been written soon after the event in August 1827 to which it referred, yet it was not read at a meeting until June 1829. Alternatively, Lyell or another fluvialist may have known of the incident and induced Culley to report it, two years after it took place. In either case there must be a strong suspicion that it was wheeled out specially for the occasion.

15. Scrope, "Excavation of the valleys" (1830), read on 5 February; the cited stretch of the Moselle—familiar to connoisseurs of German wines—is northeast and downstream from Bernkastel, which is about 60km southwest of Koblenz.

ogy" (§6.4, §18.1). Lyell had noted with some bemusement that one new work in this genre was "to prove the Hebrew cosmogony, and that we [geologists] ought all to be burnt in Smithfield", where London heretics had met their fate centuries earlier. The book's Irish author Andrew Ure (1778–1857) was no geologist; he was better known for a volume on chemistry and its industrial application (and he was dismissed privately by Lyell as "an unprincipled hypocrite"). Lyell correctly treated the genre as a peculiarly British aberration, and had taken mischievous pleasure in reporting from Rome how the previous pope had "instituted lectures on the Mosaic cosmogony to set free astronomy and geology" from religious constraints; he had wryly contrasted this surprising liberality at the heart of Catholicism with the situation in benighted Protestant Britain. The relation between geology and Genesis was certainly a problem for some sections of the wider British public, but within the Geological Society and in similar circles of savants it was not, except as an *external* threat (see below).[16]

Both before and after Scrope's intervention, Lyell was continuing to work hard on his book, though still following his own prescription for alternating between writing and fieldwork. Traveling north to spend the summer of 1829 at his family home in Scotland, he first worked his way around the coast of East Anglia studying the "Crag" deposits, the only important English Tertiary formation that he had not previously seen for himself. Much further north he saw for himself the well-known erratic blocks of Shap granite high on Stainmoor; writing to Murchison he explained away these awkwardly diluvial features, in a decidedly ad hoc manner, by inferring that they had been transported at a remote time when that part of England had been far below sea level (he must have had Wrede's ice floes in mind: §13.4; *BLT* §10.2) However, most of the summer he spent indoors, working on his book; and he and his sisters also learned German together, in his case primarily in order to read and exploit von Hoff's great inventory of actual causes recorded in human history (§7.2).[17]

Back in London in the autumn, Lyell was delighted to hear what Sedgwick and Murchison had been doing in eastern Europe, building on the work of Austrian geologists: they had seen the thick formations of rocks of all kinds in the Danube basin and the eastern Alps, looking much like the Secondaries in Britain but by their fossils all clearly Tertiary in age. This reinforced Lyell's own claim that the Tertiaries represented a major portion of geohistory, not significantly different in character from the Secondaries before them. He told Fleming that Sedgwick had returned "full of magnificent views; [he] throws overboard all the diluvian hypothesis; is vexed he ever lost time about such a complete humbug". In fact, Sedgwick—by now as much a heavyweight as anyone at the Geological Society—had a mind of his own, and believed there was good evidence for one or more "diluvial" events of some kind, even if they were all too ancient to be recorded in Genesis or any other human record. But Lyell was correct in judging that the Cambridge professor's defection from the position of Buckland, his Oxford counterpart, might lessen the impression that the argument had become a personal vendetta on the part of the London geologists against those at the two English universities (towards which Lyell's anticlerical antipathy was now notorious).[18]

20.4 SEDGWICK'S ANNIVERSARY ADDRESS

Early in 1830 the Society's "Anniversary" meeting gave Sedgwick, the president for the previous twelve months, the duty—and opportunity—to give the annual address. Fitton, his immediate predecessor, had set a precedent by summarizing all the papers read during the year, but Sedgwick set another by using the papers as the basis for evaluating the current state of the science as a whole, far beyond the Society itself and also beyond Britain. Two of the debates that he singled out for particular comment were relevant to what Lyell was compiling for his book.

On the origin of valleys, Sedgwick accepted that Lyell and Murchison had presented "evidence which seems to me not short of [conclusive] demonstration", that *no* "great denuding wave" had been involved in central France (thus Scrope's work was also implicitly endorsed). But he also accepted that Conybeare had presented a strong case that some such event, or more probably several such events spread over "successive epochs", *had* been responsible for valleys such as the Thames. And other valleys appeared to have been caused, at least in part, by folds or faults in the earth's crust, subsequently enlarged by other means such as fluvial erosion. In short, Sedgwick pointed to the sheer variety of the forms of valleys already recognized (*BLT* §2.4), and suggested that their causal origins might be equally diverse: in effect, he argued that a *multicausal* explanation of valleys was more likely to be correct than any monocausal one. This conclusion matched Sedgwick's irenic character, in that it helped to make peace between fluvialists and diluvialists by pleasing both parties; but it was no merely facile compromise, for it recognized the complexity and multiplicity of the likely causes and of their observable outcomes.[19]

Another issue on which Sedgwick commented at some length was the relation between the Tertiary formations and the Secondaries. He noted that Fitton had found that the Chalk at Maastricht, long celebrated for its fossils (*BLT* §2.1), graded down into other "Cretaceous" formations (as they were now being called) but not up into the overlying Tertiaries. However, he explicitly hoped and expected that the remaining "great chasm" between Secondaries and Tertiaries

16. Lyell to Marianne Lyell, 21 January [*LLJ* 1: 238–39], and to Mantell, 19 March 1829 [Wellington-ATL (not in *LLJ*)]. At Murray's request, Lyell drafted a review of Ure, *New system of geology* (1829), for the *Quarterly*, but it was never finished or published: Wilson, *Lyell* (1972), 265. To repeat a point made earlier: on the issue of "geology and Genesis", the perceived distinction between savant circles and the wider public in Britain at this period justifies my decision to focus here on the former. Of course, some of the geologists felt obliged to respond to the external threat posed by the scriptural literalists (see below), but only in the same way that their modern American counterparts feel obliged to respond to homegrown creationists.

17. Lyell to Murchison, 11 August 1829 [London-GS: L17/3].

18. Lyell to Fleming, 10 June 1829 [*LLJ* 1: 253–55], and to Marianne Lyell, 2 November 1829 [Kirriemuir-KH: quoted in Wilson, *Lyell* (1972), 269]; in the former, Lyell added that Sedgwick "says he lost two years by having also started a Wernerian", but this simply meant that he, like most British geologists a decade or two earlier, had taken specific types of rock to be "characteristic" of particular periods (*BLT* §8.1), an assumption long since generally abandoned. Sedgwick and Murchison, "Vale of Gosau" (1829), read on 6, 20 November, 4 December.

19. Sedgwick, "Address to Geological Society" (1830), 189–92. At this time, and particularly for the mathematically trained Sedgwick, the word "demonstration" meant a proof in a mathematical sense, i.e., one that in effect was irrefutable. For more probabilistic forms of evidence, "proofs" was the usual term.

would be filled elsewhere in due course, and he offered some of his own and Murchison's recent work in the eastern Alps as a possible candidate. He recalled the pioneer study of the Paris Basin by Brongniart and Cuvier some twenty years earlier (*BLT* §9.1), and he argued that subsequent research had reinforced Cuvier's prediction that the Tertiaries could act as the key for understanding still earlier periods of geohistory and for discerning the earth's general "laws". All this would have been music to Lyell's ears.[20]

In fact Sedgwick knew very well what Lyell was doing, and wanted to endorse it. For when he sketched provisionally the Tertiary geohistory that was now emerging from research all over Europe, he listed first the classic Parisian formations and those in the Isle of Wight, followed by Desnoyers's ones in Touraine and the Crag of eastern England; then came the Subapennines in Italy; and finally the deposits with fossils so much like the living fauna that they hardly deserved to be called Tertiary, among which were the Sicilian formations that Lyell was about to describe. Sedgwick mentioned the importance of studying whole assemblages of fossils, as advocated by Prévost and Desnoyers among others, and claimed that transitions from one fauna to the next were now well established empirically. All these Tertiary formations underlined the sheer magnitude of the time involved. But Sedgwick also argued more generally that any break in the sequence of formations should no longer be taken to imply a "short period of confusion". Even substantial breaks—in the form of major unconformities—were being progressively filled up by research in other regions; so the breaks were merely *local* or minor, where "a leaf seems to be torn out of the volume of her [nature's] history". Lyell could hardly have had the ground better prepared for him than by Sedgwick's combination of sympathy and authority. Lyell was certainly not facing a phalanx of opponents, even if some of the geologists at the Society were critical of certain of his ideas; in some ways he was even pushing at an open door, at least in the company of those who mattered most to him.[21]

Those who also mattered at the Geological Society, but in a profoundly different way, were the "scriptural geologists". They were an *external* threat that could not be ignored, because they offered the wider public a radically alternative version of the science, and their published works were enjoying great popularity. Like Lyell, Sedgwick correctly treated these "monuments of folly" as a novel and specifically British development, with "no parallel in the recent literature of continental Europe". He ended his address with a long tirade of startling vehemence, not only against the genre in general, but specifically against Ure's recent book, for the very good reason that Ure had been elected a member of the Society, and had now revealed himself to be a Trojan horse. Sedgwick criticized the book for being full of gross and elementary errors, and, where it was not, for blatant plagiarism. Fundamentally, however, he attacked the whole genre for its "unnatural union" of geology and scripture: "to seek for an exposition of the phaenomena of the natural world among the records of the moral destinies of mankind, would be as unwise, as to look for rules of moral government among the laws of chemical combination". Like Fleming (§6.4), Sedgwick invoked Francis Bacon—the patron saint of the natural sciences—to censure this compounding of "fantastical [natural] philosophy" with "heretical [i.e., literalistic] religion". Sedgwick the pious evangelical had no truck with the creationists of his time. The ground was thus well prepared for Lyell to defend his science against the same threat, as he sought to produce a book which—like Cuvier's famous Discourse (*BLT* §9.3) and his own earlier essays in the *Quarterly* (§14.4)—would appeal not only to knowledgeable savants but also to the educated public.[22]

20.5 CONCLUSION

Lyell's return journey, in the later part of his Grand Tour, was marked by a dramatic broadening of his theoretical ambitions as a geologist, though for the time being they were expressed only in private. He told Murchison of his confident belief that actual causes were totally adequate to account for everything about the deep past, not only in kind but also *in degree*; geological processes had *never* been more powerful than they were observed to be in the present world. And he saw this as "necessarily" leading to a global "system" or geotheory according to which the earth would be seen as being in a steady state: this in striking contrast to the directional model of geohistory favored by other geologists. His encounter with the botanist Viviani then suggested to him how his own concept of ceaseless slow changes in physical geography might be combined with similar changes in faunas and floras. This could lead to a general theory linking the distribution of organisms to the history of their physical environments, a theory that was strengthened by his meeting with Candolle in Geneva.

Back in Paris, Lyell found that Desnoyers had largely upstaged him in his interpretation of the Tertiary formations. However, the French geologist's concept of the "non-simultaneity" of the various Tertiary basins, and their geohistorical sequence in relation to their changing molluskan faunas, was significantly amplified by Lyell's new fieldwork. For the conchologist Deshayes confirmed that, on the agreed faunal criterion, Lyell's Sicilian formations must be even more recent than any of those considered by Desnoyers, thus improving the gradual sequence from deep past into the present world. Lyell now intended to use Deshayes's expertise to consolidate this interpretation of the Tertiary era, and then to extend it back into the era of the Secondaries in order to prove the gradual character of organic change throughout geohistory.

20. Fitton, "Observations on Maestricht" (1830), read on 18 December 1829; Sedgwick and Murchison, "Vale of Gosau" (1829).

21. Sedgwick, "Address to Geological Society" (1830), 192–206; Lyell was duly credited for his unpublished work in Sicily (199n); transformism was dismissed in the usual manner (§17.1) as "opposed to all the facts of any value in determining such a question" (205).

22. Sedgwick, "Address to Geological Society" (1830), 206–12. Klaver, *Geology and religious sentiment* (1997), 102–31, gives a fine analysis of Sedgwick's religious position, and relates it to those of Lyell, Buckland, and Whewell. John Bird Sumner (1780–1862), the bishop of Chester (and later archbishop of Canterbury), was the probable author of a closely similar critical review, "*New system* by Ure" (1829), in the Anglican *British critic*: this was another sign that scriptural geology was *not* supported by those who mattered in the Church of England. Ure, a competent chemist, was already a Fellow of the Royal Society, so his admission to the Geological must have seemed unproblematic; however, there was growing concern in the latter body that new members were being admitted without adequate scrutiny, since in practice all that was required was a couple of existing members willing to vouch for the candidate's interest in geology and his social status as a gentleman.

Back in London, geologists were more concerned with the origin of valley topography than with the Tertiary formations or their fossils. The lively arguments between "diluvialists" and "fluvialists", as Conybeare dubbed the two sect-like parties, were aggravated in England, in that the diluvialist interpretation was still linked to the historicity of the biblical story of Noah's Flood, and the fluvialist to secularizing efforts to sever any such link. But this was no simplistic conflict between geology and Genesis. Conybeare, who like Buckland equated the "geological deluge" with the biblical Flood, also claimed that there had been other events of the same kind in prehuman geohistory, so that the last one was not unique except in its date; and Sedgwick was also convinced of the reality of natural "deluges" in the deep past, although he had come to doubt whether any of them had been recent enough to be identified with the Flood.

Sedgwick took advantage of his position as president of the Geological Society to review all these contentious issues. On the question of erosion, he was convinced by Lyell's (and Scrope's) fluvialist interpretation of central France, but equally by Conybeare's diluvialist claims in the case of the Thames. The sheer diversity of valley forms suggested to him that a multicausal explanation was more credible than any monocausal one; this was an irenic solution that pleased both parties, but it was no merely facile compromise. On the geohistorical interpretation of stratigraphy, Sedgwick argued that further fieldwork would in due course fill many of the apparent gaps, and that even major unconformities might reflect local, not global disturbances. On the Tertiary formations in particular, he endorsed Lyell's still unpublished view that they represented a vast span of geohistory, but one that was yielding to careful research. All in all, Lyell could hardly have asked for a more sympathetic trailer for the major work that he was now completing. The next chapter describes how that work first appeared in public.

CHAPTER 21

Geology's guiding principles (1830)

21.1 INTRODUCING LYELL'S *PRINCIPLES*

By the time that Sedgwick's first presidential address to the Geological Society gave Lyell's forthcoming work a warm endorsement, its author expected it to fill two volumes. He evidently intended the first to be on actual causes, and the second on the Tertiary geohistory to which they would be the key. The preparation of the volumes dragged on into the spring of 1830, because Lyell kept adding new material and lengthening what he had already written; but his publisher Murray gave him a generous contract and was tolerant of the delays. Lyell reluctantly abandoned a plan to visit Iceland in the summer, to complete his firsthand study of the active volcanoes of Europe, because he realized he could not finish his volumes beforehand. He then formed a plan to travel through Russia and down to the Crimea, mainly to see the steppes and their vast alluvial deposits. But in the end he followed Scrope's suggestion to go instead to Spain, to see some extinct volcanoes that might be as instructive as those in Auvergne. This meant, however, that only a part of his work could be completed and in press before he left. The first volume of his *Principles of Geology* was duly published by Murray in the summer of 1830, while he himself was in the wilds of Catalonia.[1]

Any potential purchaser of Lyell's volume would have glanced first at the title page and the frontispiece—in effect, the equivalents of a modern dustjacket and its publisher's blurb—and would have found both somewhat unexpected. Even those who knew that the thirty-three-year-old had a high opinion of his own abilities might have been surprised that Lyell had entitled his work "Principles

1. Lyell to Egerton, 26 April, and to Eleanor Lyell, 11 May 1830 [*LLJ* 1: 265–68]; the complex history of the volume is described in Wilson, *Lyell* (1972), 265–73, on the basis of Lyell's many letters on the subject to Murchison, Mantell, Murray, and others.

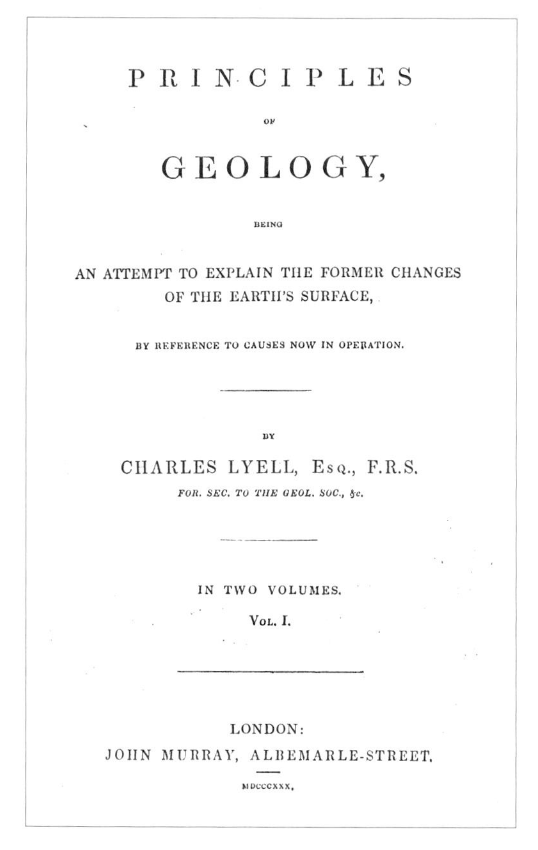
PRINCIPLES
OF
GEOLOGY,
BEING
AN ATTEMPT TO EXPLAIN THE FORMER CHANGES
OF THE EARTH'S SURFACE,
BY REFERENCE TO CAUSES NOW IN OPERATION.

BY
CHARLES LYELL, ESQ., F.R.S.
FOR. SEC. TO THE GEOL. SOC., &c.

IN TWO VOLUMES.
VOL. I.

LONDON:
JOHN MURRAY, ALBEMARLE-STREET.
MDCCCXXX.

Fig. 21.1 The title page of the first volume of Lyell's *Principles of Geology* (1830), with its ambitious title, and its subtitle summarizing the aim of the whole work. The volume was priced at one guinea [£1.05], a substantial sum that showed it was aimed at the scientific and social elites who subscribed to Murray's *Quarterly Review* and similar publications, not those whom Lyell and his friends often called "the vulgar".

of Geology"; for "principles" still reverberated with echoes of the great Newton's great *Principia*, revered as the very foundation of natural philosophy. Lyell's title suggested a work with comparable ambitions to remodel his own science on new foundations, as indeed he had privately admitted, or boasted, to Murchison and others. The subtitle was more modest, in that it conceded that the author was only making "an attempt"; but the attempt was none other than "to explain the former changes of the earth's surface, by reference to causes now in operation", which boldly implied a combination of the whole of earth-physics (*BLT* §2.4) with the more recent genre of geohistorical reconstruction. However, his explanations would be confined to the earth's "surface", which hinted that he would exclude speculations about its deep interior on the grounds that this was beyond direct observation. His description of himself, as "Foreign Secretary to the Geological Society", advertised both his international standing and his respectable credentials (in contrast, for example, to any "scriptural geologist"), while suggesting that the scope of the work would also be worldwide. And John Murray's name would have assured purchasers that the book came from one of the foremost scholarly publishers, a highly respectable Tory with premises in the same street as the Royal Institution (Fig. 21.1).

Any frontispiece was expected to act as a visual summary or epitome of a book's contents or argument. What was evidently a serious work on geology would have been expected to depict a dramatic volcanic eruption, for example, or a section through the earth's crust to show its successive rock formations,

Fig. 21.2 The "Present state of the Temple of Serapis at Puzzuoli", the frontispiece of the first volume of Lyell's *Principles of Geology* (1830), copied from Jorio's *Tempio di Serapide* (1820). Part of the way up each column the limestone was bored by marine mollusks, which proved that the Roman building was for a long period submerged, before re-emerging almost, but not quite, to its original level (the marble floor was still covered by shallow seawater). By Lyell's time, geologists attributed both changes to local movements of the earth's crust, rather than to global (eustatic) changes in the level of the sea itself; Lyell intended the ruin to be seen as a small-scale example of the effects of the actual cause of crustal movement, operating *in both directions* within the span of recorded human history, and equably enough for these columns to have remained upright. Jorio's picture also neatly encapsulated Lyell's ambition to integrate geohistory with human history.

or a geological map to show their distribution (see Fig. 3.2). Instead, potential purchasers of Lyell's book found an engraving of a Classical ruin in Italy, which might have seemed more appropriate in a work on Roman architecture or some such archaeological topic. It might indeed have reassured Lyell's general readers that the book would not be wholly alien to the world of their Classical education. Far more important, however, at least to the author, was that this was no ordinary Roman ruin, but one that epitomized the geological "system" that he intended to advocate. He had chosen to have a copy made of Jorio's picture of the so-called Temple of Serapis near Naples (Fig. 8.2). Around Pozzuoli the sea had both risen and fallen again in relation to the land, within the mere two millennia since Roman times (§8.2). Lyell intended to use the case of Serapis, interpreted as the result of movements of the land in relation to a stationary sea level, as an epitome of his concept of a ceaselessly dynamic earth. It displayed the traces of a crucially important actual cause, the most recent movement having been recorded historically, and plausibly attributed to volcanic activity nearby. Above all, the picture symbolized Lyell's intention to use human history as the key to geohistory, and to show that the latter was not marked by any overall directional trend (Fig. 21.2).[2]

2. Fig. 21.2 is reproduced from Lyell, *Principles of geology* 1 (1830), re-engraved from Jorio, *Tempio di Serapide* (1820), pl. 7A (reproduced here as Fig. 8.3). The copying is quite exact, except that the engraving is much reduced in size. This image (or, later, a new updated engraving) continued to be used in the successive editions of *Principles* through the rest of Lyell's long life, being also embossed in gold on

Beyond the title and frontispiece, Lyell's aims were set out in a very brief introductory chapter, which potential purchasers might well have read before deciding whether to buy the volume. He began with a carefully phrased definition of the science, based on what he had drafted while in the field (§19.3). Lyell put the reconstruction of geohistory into first place; it was a clear bid to assert the priority of geohistory, as an interpretative practice, over the strictly descriptive goals that the founders of the Geological Society had prescribed for the science almost a quarter-century earlier (*BLT* §8.4). But he set a second goal for geology, which transcended not only any merely descriptive practice but even the geohistorical aims of those who had been inspired by Cuvier "to burst the limits of time" (*BLT* §9.3). This further goal was to incorporate into geohistory the previously rather separate practice of earth-physics (*BLT* §2.4). He proposed making a much more consistent effort to find adequate *causal* explanations, not just for what could be observed in the present world, but for all the contingent events that marked the course of geohistory in the deep past: "Geology is the science which investigates the successive changes that have taken place in the organic and inorganic kingdoms of nature; it enquires into the causes of these changes, and the influence which they have exerted in modifying the surface and external structure of our planet".

At the same time, however, Lyell made geohistory subservient to earth-physics, in the sense that a full understanding of the workings of the present world depended on a knowledge of the deep past: "By these researches into the state of the earth and its inhabitants at former periods, we acquire a more perfect knowledge of its *present* condition, and more comprehensive views concerning the laws *now* governing its animate and inanimate productions". In other words, the past was the key to the present just as much as the present was the key to the past. The point was clarified by means of an extended analogy with human history, developed from what he had drafted in the wilds of Auvergne (§18.3): his synopsis of the chapter began with the words "Geology defined—Compared to History". In both the human world and the natural, the present carried within it the traces of the past, and could not be understood without taking that past into account; the present itself was historicized. An analogy with human history also served to explain the wide range of other sciences on which the geologist might need to call, just as the historian needed a comparable array of human and social sciences. In both cases Lyell's imagined ideal was not a polymath, but rather a savant who—like himself—worked within a network of others who could be "pumped" for diverse forms of specialized knowledge.

Lyell ended his introduction by claiming that geology had been retarded most of all by being confused with theories about the earth's primal *origin*; using a historical analogy yet again, he promised to show that "geology differs as widely from cosmogony, as speculations concerning the creation of man differ from history". He credited Hutton with having been the first to renounce any such ambitions for geology, and predicted that this view "will ultimately prevail". So at the very start of his volume Lyell gave notice that Hutton would be his model, at least in some respects, while also hinting that his own ambition was to offer an improved version of his hero's geotheory.[3]

21.2 THE LESSONS OF HISTORY

Lyell first set out his principles, and prepared the ground for the "system" to which he had claimed they "necessarily" led, in the introductory chapters that he had written since his return from the Continent. He used almost half his volume to go on the attack against three conceptions of geology that he opposed, before even starting on the inventory of actual causes that was to substantiate his own position. The first part of his attack was skillfully disguised as history. Lyell offered a lengthy survey of the history of ideas about the earth; its purpose, however, was not to provide historical background but to argue a current case as forcefully as possible. The primary target of Lyell's history was not the diluvialism of his former teacher, and Buckland's allies such as Conybeare, but rather the scriptural or "Mosaic" geology of writers such as Ure, and their fundamentalist allies among the less enlightened English clergy. Coaching Scrope on how he should review his volume for the *Quarterly*, Lyell noted that one of the bishops had criticized Ure in the conservative *British Critic*, as vehemently as Sedgwick at the Geological Society, and he commented that "they see at last the mischief and scandal brought upon them by Mosaic systems". He explained that he had decided long ago that "if ever the Mosaic geology could be set down [i.e., repudiated] without giving offence, it would be in an historical sketch". So he made a clear distinction between the "bishops and enlightened saints" on the one hand, and the "modern physico-theologians" on the other. If Scrope, like Lyell himself, were to handle the issue with tact and courtesy, Lyell was sure that the former would join them in repudiating the latter. His and Scrope's battle was not with the religious, but with the "modern offenders" who had revived a literalism that had long been abandoned on the Continent.[4]

Lyell began his history—in Enlightened multicultural fashion—with ancient Sanskrit cosmogony, before passing to the Classical authors; the Creation story in Genesis was not mentioned at all. But Lyell concluded that "the ancient [i.e., prehuman] history of the globe was a sealed book" to *all* the writers of Antiquity; and he gave that traditional metaphor his usual antiquarian twist by adding that "although written in characters of the most striking and imposing kind, they were unconscious even of its existence". He then passed with conventional rapidity over the medieval period, "wherein darkness enveloped almost every department of science", and the story only quickened with the Renaissance. Here his survey was heavily weighted towards the Italian naturalists, who were compared

---

the cloth cover of some of the later ones as a kind of secular icon or emblem of his geology: see Rudwick, "Lyell and the *Principles*" (1998). After Lyell's death it was chosen as the design for the reverse of the Geological Society's Lyell Medal (1875), funded by his endowment (and bearing his own image on the obverse): Woodward, *Geological Society* (1907), 250. See Ciancio, *Teatro del mutamento* (2005), and Gould, "Lyell's pillars of wisdom" (2000).

3. Lyell, *Principles* 1 (1830), 1–4. The chapters have no titles, but each has a synopsis, usually running to several lines.

4. Lyell to Scrope, 14 June 1830 [*LLJ* 1: 268–71], referring to Sumner, "*New system* by Ure" (1829). The term "physico-theology" had first been used to denote the genre that had flourished a century earlier, in which the still older tradition of natural theology (*BLT* §1.4) was applied to the then new sciences of nature ("physics"), to demonstrate the providential goodness or at least the existence of God.

favorably with the "physico-theologians" in England; most of his detail on the former was taken, often almost verbatim and with scant acknowledgment, from the long historical introduction to Brocchi's great work on the Subapennine fossils (*BLT* §9.4). For more recent times his praise was more thinly spread, and the great naturalists of the generation of Pallas and Saussure were given little credit. In sum, Lyell's history of the sciences of the earth—he failed to recognize that "geology" itself was a historical construct—was squarely in the Enlightenment tradition; it was a story of the invincible progress of rational thought, retarded only by the forces of superstition and obscurantism. Lyell's history was a polemical tract for the times: it was designed to show how geology needed the radical reform that he promised to provide, because it had been pursued hitherto on the wrong principles, or only fitfully on the right ones.[5]

Lyell ended his history, tactfully and tactically, just short of his own times; Smith was one of the few living figures to be mentioned by name. In anticipation of the next phase of his argument, his target now shifted from the scriptural geologists to those of his colleagues at the Geological Society who continued to stress the disjunction between present and past and the "otherness" of the latter. Here he deftly recruited the great Cuvier onto his own side (identifying him unmistakably by a famous phrase): "if many [naturalists] continued to maintain, that 'the thread of induction is broken' yet in reasoning by the strict rules of induction from recent to fossil species, they virtually disclaimed the dogma which in theory they professed". Lyell concluded that it was only since the turn of the century that geology had made real progress, comparable to that of astronomy at an earlier period; like Cuvier he stressed the parallel between the discoveries of deep time and deep space. So, in the end, the role of the history of geology for Lyell was not so much to anchor his own work in that of his predecessors—even Hutton—as to emphasize the break between his own times and all that had gone before. Predictably, he expressed that disjunction in terms of an analogy with human history: "the charm of first discovery is our own, and as we explore this magnificent field of inquiry, the sentiment of a great historian of our time may continually be present to our minds, that 'he who calls what has vanished back again into being, enjoys a bliss like that of creating'".[6]

Lyell's quotation was from *Römische Geschichte* [*Roman History*] (1811–12) by the professor of history at Berlin, Barthold Georg Niebuhr (1776–1831). This great work had had an impact on the writing of human history as profound as that of Cuvier's *Fossil Bones* (1812) on geohistory, and for a similar reason, in that it was based on a scrupulously careful evaluation of original materials. Oxford's professor of history had reviewed Niebuhr enthusiastically in the *Quarterly*—the year before Lyell's first essay there—as part of a belated English discovery of German critical historiography; and evidently Lyell read the work itself, at least in part, after it was translated by two of Sedgwick's colleagues at Trinity College, Cambridge. The sentence that inspired Lyell expressed perfectly, in analogical form, the *geohistorical* goals that he clearly hoped to achieve:

> He who calls what has vanished back again into being, enjoys a bliss like that of creating; it were a great thing, if I might be able to scatter, for those who read me, the

cloud that lies on this most excellent portion of [hi]story, and to spread a clear light over it; so that the Romans shall stand before their eyes, distinct, intelligible, familiar as contemporaries, with all their institutions and the vicissitudes of their destiny, living and moving.[7]

21.3 THE IDENTITY OF PAST AND PRESENT

Having made his bid "to free the science from Moses"—the patriarch acted in Lyell's private writing as code for the scriptural geologists—he turned next to those of his fellow geologists who had given aid and comfort to the enemy, however unintentionally, by equating one particular event in the biblical narrative with what Lyell regarded as an illusory event in the geological record. In other words, he now attacked the diluvialists, but also more generally all those who inferred that geohistory had been marked by occasional catastrophic events of kinds unmatched in the modern world, or at least of greater magnitude or violence than any known from human records. He argued that the greatest single factor that had lent plausibility to such ideas had been an inadequate sense of the earth's timescale. Using a vivid analogy with human history, he imagined Champollion deciphering the monuments of ancient Egypt, or other antiquarians studying the written documents of other nations, while under the illusion that these records spanned only a single century rather than several millennia:

> A crowd of incidents would follow each other in thick succession. Armies and fleets would appear to be assembled only to be destroyed, and cities built merely to fall into ruins. . . He who should study the monuments of the natural world under the influence of a similar infatuation, must draw a no less exaggerated picture of the energy and violence of [past] causes, and must experience the same insurmountable difficulty in reconciling the former and present state of nature.[8]

It was a fair point, up to a point. The geologists whom he was criticizing, including diluvialists such as Buckland, often spoke casually of millions of years; but Lyell claimed in effect that they had not yet fully grasped the implications of what they professed, or the extent to which the sheer application of "Time!—

5. Lyell, *Principles* 1 (1830), 5–74; McCartney, "Lyell and Brocchi" (1976). Porter, "Lyell and the history of geology" (1976), is still the best analysis of Lyell's historiography. Lyell's history was taken at face value by some anglophone historians later in the nineteenth century; in consequence of the enduring popularity of Geikie, *Founders of geology* (1897), its argument was copied by countless historians and geologists in the twentieth, fitting readily into the heroic mold of "Science versus Religion".

6. Lyell, *Principles* 1 (1830), 71–74, quoting a phrase from Cuvier's "Discourse" (*BLT* §9.3) and a sentence from Niebuhr (see below). Throughout the *Principles*, Lyell used "our" and "we" not as a quasi-royal plural, but as a rhetorical device to recruit all reasonable geologists onto the same side as himself. The analogy between geology and astronomy had been a favorite trope since the time of Hutton's appeal to ceaseless "revolutions" in time and space (*BLT* §3.4) and Cuvier's reference to "bursting the limits" of both (*BLT* §9.3).

7. Niebuhr, *History of Rome* (1828) [translated by Julius Charles Hare and Connop Thirlwall], 5; his original work was reviewed anonymously by Thomas Arnold, "*Römische Geschichte* von Niebuhr" (1825). See Rudwick, "Historical analogies" (1977) and "Transposed concepts" (1979).

8. Lyell, *Principles* 1 (1830), 76–79.

Time!—Time!"—as Scrope had memorably put it (§15.1)—might iron out the irregularities and dissolve the appearance of sudden and violent events. Like Sedgwick in his anniversary address, Lyell pointed to the progressive enlargement of the sequence of formations that intensive fieldwork throughout Europe was currently disclosing, and hence the extension of the timescale needed for their accumulation: "No sooner does the calendar appear to be completed, and the signs of a succession of physical events arranged in chronological order, than we are called upon to intercalate, as it were, some new periods of vast duration". Lyell's own recent Continental fieldwork was a striking case in point: a geologist such as himself, as he recalled with modest anonymity, "returns with more exalted conceptions of the antiquity of some of those modern deposits [in Sicily], than he before entertained of the oldest of the British series" (§19.4). It was above all the human imagination that needed to be stretched, as Lyell felt that his own had been, to comprehend the vast timescale of geohistory. He was confident that the future would prove to be on his side: with further research, objections to what he called "the doctrine of absolute uniformity"—using, significantly, a religious term to describe his own position—would in due course melt away.

Lyell argued that a further effort of the imagination was also needed in order to recognize a systematic bias in our perspective on the causes at work on and in the earth. As terrestrial beings we are bound to underestimate the submarine and subterranean processes that are beyond our direct observation; an intelligent aquatic being (perhaps a dolphin?), or a subterranean "Gnome", would each construct quite different but equally distorted "systems" of geology. So "an effort both of the reason and the imagination" was needed, to transcend these limitations. Specifically, Lyell claimed that the human viewpoint habitually led geologists to overestimate the destructive or "wasting" aspects of geological processes, such as erosion, and to underestimate the constructive or renovating aspects such as deposition. By compensating for this bias, Lyell intended to demonstrate that such processes are in reality *balanced*, and that the earth is therefore in a state of dynamic equilibrium. Only by discovering the true scale of the less accessible actual causes, by penetrating deep time with something analogous to the astronomer's telescope, would the real character of the natural world be disclosed. Not for the first time, Lyell claimed the moral high ground by implying that his was the way of virtuous effort, in tacit contrast to the mental laziness of his opponents:

> It is only by becoming sensible of our natural disadvantages that we shall be roused to exertion, and prompted to seek out opportunities of discovering the operations now in progress, such as do not present themselves readily to view. We are called upon, in our researches into the state of the earth, as in our endeavours to comprehend the mechanism of the heavens, to invent means for overcoming the limited range of our vision.[9]

Lyell's "doctrine of absolute uniformity" asserted that events in the deep past had *never* been of greater extent, suddenness, intensity, or violence—let alone of different kinds—than actual causes. The latter were processes that could be directly observed in action or that had been reliably recorded in human history; in either case, only those that were demonstrably adequate to have produced the observed effects could count as "true causes" [*verae causae*] on the Newtonian model. However, Lyell's definition of actual causes was in practice significantly enlarged, for he was prepared to include events that might conceivably happen *in the future*. Nothing could show more clearly that his concept of geological "uniformity" went far beyond any mere rule of right reasoning from the present to the past. His example was that of the Great Lakes, which had the potential to "lay waste a considerable part of the [North] American continent" in the geologically near future, once they were breached by the steady erosion of the gorge and falls at Niagara; no "hypothetical agency" would be needed to produce what Lyell did not shrink from describing as a future catastrophic "deluge". To be consistent, he might therefore have been expected to concede the possible historicity of the alleged "deluge" in the geologically recent past, provided that it too was attributed—following von Buch, Hall, and others (*BLT* §10.2)—to a straightforward physical cause such as a mega-tsunami. But this option was excluded by Lyell's prescriptive rejection of any putative past events of a greater magnitude than those that had been observed in the human present or that could be confidently predicted for the future. In practice, therefore, his "absolute uniformity" implied a constancy that could only come from an earth in a steady state.[10]

21.4 REFUTING A DIRECTIONAL GEOHISTORY

Lyell briefly dismissed the "doctrine of universal formations"—the earlier geognosts' expectation that the same sequence would be recognizable everywhere (*BLT* §8.1)—before turning to what he conceded were much "weightier objections" to his own "doctrine of absolute uniformity". Any interpretation of the earth as being in a state of dynamic equilibrium had to face what most geologists regarded as strong evidence that its history was, on the contrary, one of more or less steady *directional* change, in both the inorganic realm and the organic. Lyell now tackled those two aspects in turn.[11]

Some of the strongest evidence for directional change in the inorganic world was, paradoxically, the *fossil* evidence for major changes of climate. Specifically, the fossil record suggested a dramatic cooling of the climate, at least in the regions that had been explored geologically: many Tertiary mollusks were of kinds now living only in tropical seas; there were abundant corals in some Secondary

9. Lyell, *Principles* 1 (1830), 76–85; he had in mind not only conceptual instruments but also material ones: he alluded to the glass-bottomed boat with which Donati, *Storia naturale marina* (1750), had studied the shallow floor of the Adriatic and discovered its close similarity to Tertiary sediments and fossils now high above sea level.

10. Lyell, *Principles* 1 (1830), 87–88. Laudan, "Methodology in Lyell's science" (1982), makes a persuasive case for the importance of Herschel's advocacy of *verae causae* on Lyell; the two savants were in contact informally while Lyell was writing the *Principles*, even before Herschel's influential *Preliminary discourse* (1830) was published. Years later, when Lyell was able to visit Niagara, he made a special study of the gorge and falls, and used a Scropean quasi-aerial panorama of them as the frontispiece of his *Travels in North America* (1845); it is finely reproduced in Wilson, *Lyell in America* (1998), pl. opp. 22.

11. Lyell, *Principles* 1 (1830), 90–91. On the directionalist synthesis as his major target, see Rudwick, "Uniformity and progression" (1971), and Lawrence, "Lyell versus central heat" (1978).

formations, and huge tree-ferns in the Coal strata, in what looked like the remains of swampy tropical jungles, not only at temperate latitudes but even in the Arctic (Chap. 12). All this was generally taken to support the physicists' inference that the globe had cooled slowly from an initially hot fluid state (Chap. 9). Fleming, a naturalist as much in favor of "uniformity" as Lyell, had denied the validity of this fossil evidence, on the grounds that the fossil species were not exactly the same as the living ones with which they were being compared, and might have had quite different ecological preferences; but Conybeare and others had effectively demolished that argument (§12.3). So Lyell had to admit, grudgingly, that here the balance of plausibility lay with his opponents: he told Fleming that "as a staunch advocate for absolute uniformity in the order of Nature, I have tried in all my travels to persuade myself that the evidence was inconclusive, but in vain". A cooling trend in geohistory, at least in the parts of the globe that had been adequately studied, therefore had to be accepted as real: "I am more confirmed than ever, and shall labour to account for vicissitudes of climate, not to dispute them". The unwelcome facts would have to be explained away; but first they had to be set out, which Lyell did fully and fairly.[12]

Lyell next suggested "how such vicissitudes can be reconciled [*sic*] with the existing order of nature"; or, to put it another way, how an apparent cooling trend could be reconciled with a geohistory of "absolute uniformity". For this purpose he deployed what he described to Mantell as "my grand new theory of climate". This was based on Humboldt's earlier and famous synthesis of the climatic observations made on his own and others' extensive travels. In an innovative kind of thematic cartography, the great Prussian geographer had drawn what he called "isothermal lines" across the globe, which showed graphically (in both senses) how climates were far from being dependent simply on latitude. For example, Britain was on the same latitude as Labrador, but their climates were in sharp contrast, for climate was also dependent on the distribution of land and sea, the direction of winds and currents, and many other factors. This Humboldtian insight had been amply confirmed by more recent work in physical geography; like other London savants, Lyell had access, for example, to the unpublished reports of British naval expeditions, thanks to the scientific interests of Francis Beaufort (1774–1857), the new official "Hydrographer" at the Admiralty. So he speculated—the verb was his—that the climates of the deep past had been quite different from the present, simply as a result of perhaps radically different distributions of the continental landmasses, and consequently of winds and ocean currents. The explanatory potential of the idea was almost unlimited, as he hinted, tantalizingly, to Mantell:

> I will not tell you how, till the book is out—but without any help from a comet, or any astronomical change, or any cooling down of the original red-hot nucleus, or any change of inclination of axis, or volcanic hot vapours and waters and other nostrums, but all easily and naturally. I will give you a receipt [recipe] for growing tree ferns at the pole, or if it suits me, pines at the equator; walruses under the [tropic] line, and crocodiles in the arctic circle.[13]

In his book, Lyell developed these speculations in a more concrete form, by working out the kinds of distribution of land and sea that could have produced extremes of global climate, both hot and cold. That any such extremes really had existed in the deep past—or might, with equal likelihood, exist in the remote future—depended on what modern geologists would call his *tectonic* theory: namely that landmasses (of any size from oceanic islands to whole continents) were always in process of rising from beneath the sea by crustal elevation, and at other times and places of being eliminated by crustal depression or erosion and replaced by sea. The extent of Lyell's commitment to "absolute uniformity" became apparent in two otherwise surprising assumptions in his hypothetical model: that the proportions of land and sea would have remained roughly constant throughout; and that the changes would not have been totally random, but arranged in the regular rhythm of what he called "the winter of the 'great year,' or geological cycle" and an equivalent "summer" (Fig. 21.3).[14]

Although Lyell did not make his strategy explicit until later in the work, his "new theory of climate" was designed to subsume the whole of the fossil record within just one phase of a single round of his "geological cycle". On Lyell's geotheory, the apparent cooling of global climate—from the older Secondaries (the earliest with fauna and flora that were adequately known) to the present—was due not to a cooling of the deep interior but to changes in the surface distribution of land and sea: these had caused global climates to swing slowly from the "summer" to the "winter" of the earth's unimaginably lengthy "great year". If all this were true, it followed that a comparably lengthy future might see the climates swing back eventually to another "summer". The implications were startling: as he had put it to Mantell, "All these changes are to happen in future again, & iguanodons & their congeners must as assuredly live again in the latitude of Cuckfield [Fig. 5.2] as they have done so [in the past]." This was no rash exaggeration or casual joke, safely confined to a private letter, for Lyell repeated and amplified his claim in public, embedded in the otherwise sober prose of his book:

> Then might those genera of animals return, of which the memorials are preserved in the ancient rocks of our continents. The huge iguanodon might reappear in the woods, and the ichthyosaur in the sea, while the pterodactyle might flit again through umbrageous groves of tree-ferns. Coral reefs might be prolonged beyond the arctic

12. Lyell to Fleming, 3 February 1830 [*LLJ* 1: 259–61]; *Principles* 1 (1830), 92–103.

13. Lyell to Mantell, 15 February 1830 [Wellington-ATL; printed in part in *LLJ* 1: 261–62]; *Principles* 1 (1830), 104–13. All the alternative explanations mentioned had been mooted at one time or another, but the steadily cooling earth was currently by far the most widely supported; to lump them all together as mere "nostrums" or quack remedies was deliberately dismissive. The "map" in Humboldt, "Lignes isothermes" (1817), is highly diagrammatic and shows no landmasses; the superb isotherm maps in Berghaus, *Physikalischer Atlas* (1845–48), pls. 1–3, for example, supported Lyell's argument far more clearly, but were not printed until 1837–38, too late for his use at this point: see Robinson, *Early thematic mapping* (1982), 71–73. Beaufort devised the standard scale for strengths of wind, from calm to hurricane, now known by his name.

14. Fig. 21.3 is reproduced from Lyell, *Principles*, 4th ed. (1834), 1 [Cambridge-UL: Ant.c.48.152], pl. 1 (at 80); see text at 172–84, and 1st ed., *Principles* 1 (1830), 116–23.

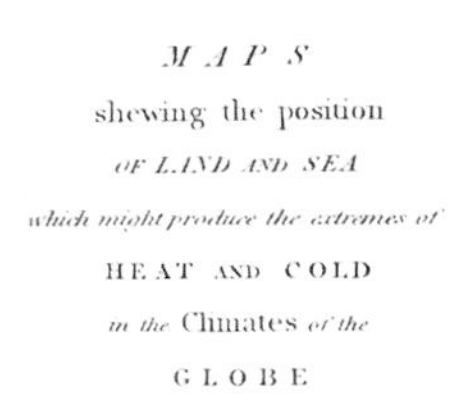

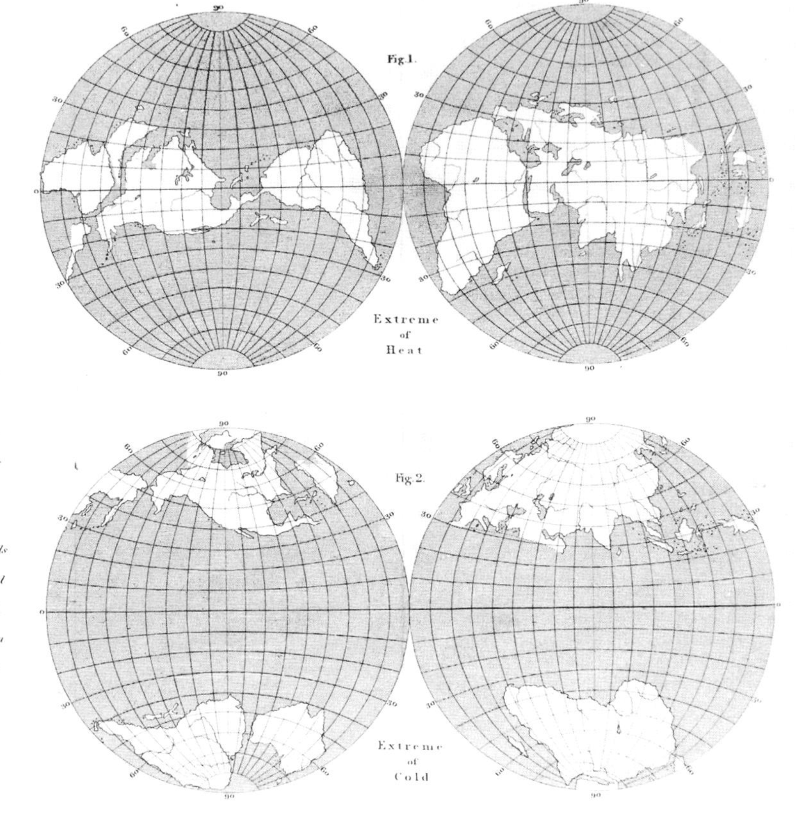

Fig. 21.3 Lyell's illustrations of his openly speculative climatic theory. His maps of the "summer" (fig. 1) and "winter" (fig. 2) of the "great year" or "geological cycle"—applicable to both past and future—showed the *present* continents in different positions. This was simply to illustrate how climates might be affected by contrasting geographies, in both of which the ratio between land and sea is roughly the same as it is at present (any similarity to modern maps showing reconstructions of continental displacements due to the movement of tectonic plates is fortuitous). This illustration, published in the third edition of Lyell's *Principles* (1834–35), explicated visually what he had already described verbally in the first edition (1830).

> circle, where the whale and the narwhal now abound. Turtles might deposit their eggs in the sand of the sea beach, where now the walrus sleeps, and where the seal is drifted on the ice-flow.[15]

No claim could have been more startling—or more implausible—to the geologists among Lyell's readers, for whom it was self evident that the history of life was essentially linear, and did not show any signs of cyclicity. On a more mundane level, even Smithian stratigraphy would have been impossible, if the fossils supposedly "characteristic" of one formation were found to reappear in another of quite different age. But Lyell was properly careful in his wording: it was *genera* that might reappear, not species. His own intended reconstruction of Tertiary geohistory would have collapsed if, for example, the mollusks of the Parisian formations could not be distinguished *on the level of species* from those of the Subapennines. Lyell's verbal *jeu d'esprit*, by contrast, simply embodied his belief that the ecological conditions of the world that had long ago sustained the iguanodon might in the distant future be repeated with sufficient precision to sustain animals of the same general kind; Mantell's particular species had become truly extinct once and for all, but a roughly similar animal might appear at some future time, if conditions again became suitable for that form of life.[16]

Although he risked anticipating the later parts of his work, Lyell sketched briefly the ways in which the known fossil record might be explained in terms of his highly speculative theory of climate. As a first example he took Conybeare's "Carboniferous" period. Rather surprisingly, he interpreted its environment not as a continent with low-lying swampy forests bordered by shallow seas, but as one of archipelagoes of oceanic islands. He noted that botanists such as the younger Brongniart (§12.2) regarded its fossil plants as having "the characters of an insular, not a continental flora"; and that the apparent absence of large land quadrupeds was analogous to what was known about island faunas in the present world. But this was the Carboniferous in those regions that had been explored geologically; to infer that the whole globe had been covered at that time with an ocean dotted with islands was, he suggested, as unsound as it would be for a New Zealander (i.e., a Maori, before Europeans arrived) to imagine that the northern latitudes of the present world had the same ratio of land to sea as the southern. By contrast, Lyell interpreted the world of the younger Secondaries as one of much larger islands with large rivers; for instance, the "Wealden" formation in which Mantell's iguanodon was preserved had evidently accumulated in a large estuary or delta (§11.1). A third example was the apparently abrupt transition between the youngest Secondaries (the Chalk at Maastricht) and the oldest Tertiaries, which Sedgwick had mentioned in his address; here Lyell conceded that the sharp faunal contrast might not be entirely due to a break in the fossil record, but also in part to a relatively rapid pace of environmental (and faunal) change, which he correlated with the upheaval of the Alps around that time.[17]

These examples (here summarized too briefly) were sufficient to show in outline how the whole fossil record could be explained in terms of an ever-changing physical geography and its environmental consequences, rather than as the result of any overall directional trend. So, finally, Lyell considered "the gradual diminution of the central heat of the globe", as recently proposed by the physicist Fourier and the geologist Cordier (§9.2). He accepted the evidence that temperature in mines increased with depth, and the inference that the earth had a hot interior, but he denied that this need be attributed to a *residual* heat. Lyell noted that Cordier had found much local variation in the temperature gradient, and attributed this to similarly local differences at depth, probably related to local igneous activity. Little was yet known about these unobservable processes,

15. Lyell, *Principles* 1 (1830), 123; Lyell to Mantell, 15 February 1830 [Wellington-ATL; *LLJ* 1: 261–62]. Ospovat, "Lyell's theory of climate" (1977), is an important analysis of its relation to his cyclic geohistory.

16. Lyell's formulation offered greater flexibility than may appear to modern biologists, because what the taxonomists of his generation defined as "genera" have often become "families" (or even broader taxa) in the work of their modern successors. Lyell's reasoning, though clear enough to the savants of his time (see §22.3), may have been lost on his less knowledgeable readers, as it has been on some modern historians.

17. Lyell, *Principles* 1 (1830), 125–41.

so Lyell claimed that it was "more consistent with philosophical [i.e., scientific] caution, to assume that there is no instability in this part of the terrestrial system", but rather the stability of a steady state. He rejected any geohistorical inference from central heat: "let us not hastily assume that it has reference to the original formation of the planet, with which it might be as unconnected as with its final dissolution". In effect, to infer a directional geohistory was to be rash or intellectually lazy; to infer a dynamic equilibrium—to be Lyellian—was to be rigorous and scientific.[18]

21.5 REFUTING A PROGRESSIVE HISTORY OF LIFE

This completed Lyell's critique of interpretations of the inorganic realm that pictured geohistory as being in any way directional. However, it remained to demolish, or explain away, his opponents' strongest argument of all, which was based on the organic realm: the apparently directional character of the fossil record itself. Cuvier's early hunch that an age of mammals had been preceded by one dominated by reptiles (*BLT* §9.3) had been vindicated by the spectacular crop of strange reptiles from the younger Secondary formations (Chaps. 2, 5), while still older ones contained only fish. This insight had been enlarged by bringing plants and invertebrates into the picture, as the plausibility of the living fossil argument (*BLT* §5.1) ebbed away. An age of ammonites and belemnites, an even earlier age of trilobites, and most spectacularly the younger Brongniart's age of gigantic Carboniferous tree-ferns, all took their place in the broader panorama of a directional history of life (Chap. 4; §12.2). Above all, the much earlier rumors of a human presence in deep geohistory (*BLT* §5.4) had long been discounted; the authenticity of human fossils in the Superficial deposits was currently a matter of lively controversy (Chap. 16), but even if they were genuine the human species would still be the most recent major new arrival on the scene. This was a formidable array of evidence in favor of a directional—and indeed "progressive"—history of life, which Lyell would somehow have to explain away, if his radically alternative vision of a world in steady state were to gain any plausibility at all.

Lyell defined his own "doctrine" as the belief that "all former changes of the inorganic and organic creation are referrible to one uninterrupted succession of physical events, governed by the laws now in operation". But of course his opponents among the geologists did not deny the constancy of the basic laws of nature. Everything therefore hinged on the meaning to be assigned to what Lyell called "the assumption of uniformity in the order of nature". This was far from straightforward. In summarizing the progressive picture that he intended to refute, Lyell did not refer to the detailed research of his fellow geologists, as they might have expected, but cited instead a very general scientific essay by Davy, the recently deceased president of the Royal Society. However, Lyell split the picture in two. He claimed that "the progressive development of organic life, from the simplest to the most complicated forms . . . though very generally received, has no foundation in fact". But he accepted that the human species was genuinely a recent arrival, while denying that this was "inconsistent with the assumption that the system of the natural world has been uniform from the beginning", or at least as far back as there was any evidence to go on.[19]

Lyell sought to undermine the credibility of a progressive history of life as a whole, by appealing to the chanciness of preservation and the consequent imperfection of the fossil record. Conversely, he seized on even the most isolated and dubious reports to argue that "advanced" organisms had already been in existence in early times: years earlier, for example, Fleming had found fragments of what he claimed to be dicotyledonous wood, from coal strata in Scotland and from Irish greywacke of Transition age, both far too early to fit into the younger Brongniart's synthesis of plant history (§12.2). The tiny marsupial mammals from the Secondary (Jurassic) strata at Stonesfield were much better authenticated (§5.5); but rather than treating them, as other geologists did, as a modest anticipation of the more varied and abundant mammalian faunas of the Tertiary era, Lyell regarded them as vindicating his inference that mammals of all kinds might have been living elsewhere, even during the Secondary era (§17.2). "It is, therefore, clear", he concluded with characteristic confidence, "that there is no foundation in fact for the popular theory of the successive development of the animal and vegetable world, from the simplest to the most perfect forms".[20]

Although neither Hutton nor Lamarck was mentioned here by name, the ghosts of both lurked between the lines of Lyell's argument. Davy's book, which Murray showed Lyell in proof while Lyell was completing his own, revived much earlier fears that Hutton's kind of earth—for ever in a steady state—implied a denial of the createdness of the world and was likely to lead to materialism or even atheism (*BLT* §6.4). Conversely, the progressive trend in the history of life revealed by the fossil record was being canvassed, however surreptitiously, as evidence for its "successive development" (in modern terms, evolution) by Lamarck's kind of transmutation, which was widely regarded as subversive for the same reason (Chap. 17). Lyell knew that any acceptance of his ideas in geology, by the British social elites whom he wanted to reach, was contingent on distancing himself from such dangers. His denial of the reality of "successive development" in the fossil record implied a steady state in the organic world to match what he had claimed for the inorganic, thus keeping Lamarck at bay. And he had other stratagems in mind for denying that a steady state in the whole "system", inorganic and organic, implied the kind of eternalism for which Hutton had been criticized.[21]

So, finally, Lyell had to deal with the thorny issue of the place of Man in the kind of nature he was outlining. In this case, as already mentioned, he accepted

18. Lyell, *Principles* 1 (1830), 141–43.

19. Lyell, *Principles* 1 (1830), 144–45; Davy, *Consolations in travel* (1830), 133–51, part of an imaginary conversation set among the ruined temples at Paestum.

20. Lyell, *Principles* 1 (1830), 145–53. Fleming, "Neighbourhood of Cork" (1821).

21. It is beyond the scope of the present book to deal adequately with these wider implications of Lyell's position; but see Secord's important "Introduction" to his abridged edition of *Principles* (1997), xxix–xxxii; and Klaver, *Geology and religious sentiment* (1997), 61–74.

that the fossil record reflected the true course of geohistory, rather than being an artifact of imperfect preservation: there were no genuine human fossils, and the human species was truly a geologically recent arrival. Yet he argued that this event had made no difference on the purely physical level, because "the superiority of man depends not on those faculties and attributes which he shares in common with the inferior animals, but on his reason by which he is distinguished from them". The sharp dualism which Lyell had already expressed in his earlier essays (§17.2) served to detach humanity from the rest of nature: "the animal creation may . . . be supposed to have made no progress by the addition to it of the human species, regarded merely as part of the organic world", because "the union, for the first time, of moral and intellectual faculties capable of indefinite improvement, with the animal nature", was so unprecedented an event that it could *not* be treated "as one step in a progressive system". To descend from what Lyell conceded were the heights of metaphysics to the mundane level of geology and its associated sciences, it followed that the arrival of the human species was in a material sense a quite ordinary event: "We have no reason to suppose, that when man first became master of a small part of the globe, a greater change took place in its physical condition than is now experienced when districts, never before inhabited, become successively occupied by new settlers".[22]

21.6 LYELL'S REVIVAL OF GEOTHEORY

Lyell brought his weighty introductory essays to a close by reiterating his "doctrine of absolute uniformity", expressed now in prescriptive terms as the most reliable *method of reasoning* for geologists to follow. He argued that in this respect the recent arrival of the human species was almost irrelevant: "In reasoning on the state of the globe immediately before our species was called into existence, we may [i.e., *should*] assume that all the present causes were in operation, with the exception of man, until some geological argument can be adduced to the contrary". The present deserved to be treated as the key to the past; only in that way could the adequacy of actual causes for geological explanation be properly judged. It followed that—as Desmarest had prescribed long before (*BLT* §4.3) and Cuvier after him (*BLT* §9.3), though Lyell did not cite either—the "natural order of inquiry" was the *reverse* of the chronological or geohistorical: "if . . . we cautiously proceed in our investigations, from the known to the unknown, and begin by studying the most recent periods of the earth's history, attempting afterwards to decipher the monuments of more ancient changes, we can never so far lose sight of analogy, as to suspect that we have arrived at a new system, governed by different physical laws".[23]

Lyell argued that the same kind of reasoning, carefully applied, would eventually be found to eliminate all the apparent discrepancies between past and present, not only in terms of the kinds of causes involved but even in terms of their intensity of action. Since his goal was ultimately to integrate causal explanation with geohistorical reconstruction, it is not surprising that antiquarian analogies were as pervasive as ever; it was nature's monuments that had to be explained, nature's archives that had to be interpreted:

> When we are unable to explain the monuments of past changes, it is always more probable that the difficulty arises from our ignorance of all the existing agents, or all their possible effects in an indefinite lapse of time, than that some cause was formerly in operation which has ceased to act; and if in any part of the globe the energy of a cause appears to have decreased, it is always probable, that the diminution of intensity in its action is merely local, and that its force is unimpaired, when the whole globe is considered. But should we ever establish by unequivocal proofs, that certain agents have, at particular periods of past time, been more potent instruments of change over the entire surface of the earth than they now are, it will be more consistent with philosophical caution to presume, that after an interval of quiescence they will recover their pristine vigour, than to regard them as worn out.[24]

Once again, prescriptive rules for reasoning about the deep past were intricately interwoven with Lyell's convictions about what that past had in fact been like. To reason in a Lyellian manner was to exhibit proper "philosophical caution", but Lyell was convinced that this method would in fact disclose a world that had always been more or less the same kind of place, at least as far back as there was any evidence to go on, and as far forward into the future. So Lyell ended with a piece of purple prose that unmistakably echoed Cuvier's similar and famous passage (*BLT* §9.3), celebrating the ability of the human spirit "to burst the limits of time" through the practice of the science of geology, just as the limits of space could be transcended through another science:

> Thus, although we are mere sojourners on the surface of the planet, chained to a mere point in space, enduring but for a moment in time, the human mind is not only enabled to number worlds beyond the unassisted ken of mortal eye, but to trace the events of indefinite ages before the creation of our race, and is not even withheld from penetrating the dark secrets of the ocean, or the interior of the solid globe; free, like the spirit which the poet described as animating the universe, "pervades all things, earth and sea's expanse and heaven's depth".[25]

21.7 CONCLUSION

Lyell's long introductory chapters—almost half the volume—gave his readers a foretaste of the massively elaborate "system" that the rest of his work would try to substantiate. Despite his renunciation of any speculation about the primal origin of the earth, or "cosmogony", the ambitious scope of Lyell's work left his contemporaries in no doubt about the genre to which the *Principles* belonged.

22. Lyell, *Principles* 1 (1830), 153–57; one of Lyell's principal models was the history of Australia, considered (as usual at this time) as having been, in effect, uninhabited before Europeans arrived.

23. Lyell, *Principles* 1 (1830), 158–60.

24. Lyell, *Principles* 1 (1830), 164–65.

25. Lyell, *Principles* 1 (1830), 166, quoting (here given in Loeb translation) Virgil, *Georgics*, bk 4, 221–22: "[Deus namque] ire per omnes / Terrasque tractusque maris, coelumque profundum". Surely not by accident, Lyell replaced the theistic reference ("Deus namque . . .") with mere "spirit". As in his earlier essays (§14.4, §18.1), he used such Classical tags to associate himself with the cultured readers he hoped to reach.

What he himself described repeatedly as a "system" was indeed what that word by long tradition denoted: a "theory of the earth" or *geotheory* (*BLT* §3.1). The genre had been banished to the margins of savant discourse, ever since Cuvier's devastating critique of its pretensions, two decades earlier (*BLT* §8.4). Geotheories had continued to be produced and published since that time, though less often than before; but they were no longer written by leading savants and had little impact on the savant world. As the preceding narrative has shown, geologists had become more concerned to discuss the increasing flood of technically competent papers that were presented at the Geological Society and in other savant settings, and published in the *Annales des Sciences Naturelles* and many similar periodicals. But now Lyell was making a bid to revive the genre of geotheory, and doing so with impeccable scientific and social credentials. The next chapter summarizes the next phase of Lyell's exposition of his geotheory, and describes how his first volume was received by his fellow geologists and other savants.

CHAPTER 22

"The Huttonian theory rediviva" (1830–31)

22.1 LYELL'S SURVEY OF ACTUAL CAUSES

Toward the end of the essays that introduced his *Principles of Geology*, Lyell stressed once more that the key to deciphering the "archives" of geohistory was "to examine with minute attention all the changes now in progress on the earth". To these he now turned. However, his introductory essays had grown to such a length that his exhaustive survey of the earth's actual causes had to be truncated, deferring those involving the organic world to his second volume. The purely inorganic causes surveyed in the remainder of the first included both the aqueous and the igneous. This reflected the general recognition that the old argument between Neptunists and Vulcanists (*BLT* §2.3) had been misconceived, since *both* classes of process were now acknowledged to be important. Unlike Hutton, however, Lyell emphasized that both were "instruments of decay as well as reproduction", so that each contributed to both sides of the balance of forces that maintained the earth in a steady state. Thus his set of chapters on the effects of running water on land, and then on the effects of marine currents and tides, each dealt with the action of those processes first in erosion and then in deposition. These and his later chapters on volcanoes and earthquakes all took the form of detailed inventories, not only of what had been observed by geologists and travelers in his own time but also what had been reliably recorded in earlier centuries: causes counted as actual—or, in Lyell's terms, as "modern", "existing", or "now in operation"—provided they had been reliably witnessed within the two or three millennia of recorded human history (Fig. 22.1).[1]

1. Fig. 22.1 is reproduced from Auldjo, *Sketches of Vesuvius* (1833) [Cambridge-UL: MF.38.34], pl. opp. 27 (in the original the flows are distinguished by a striking range of colors). John Auldjo (1805–86), a wealthy young man who traveled widely and had climbed Mont Blanc in 1827, may have used as his

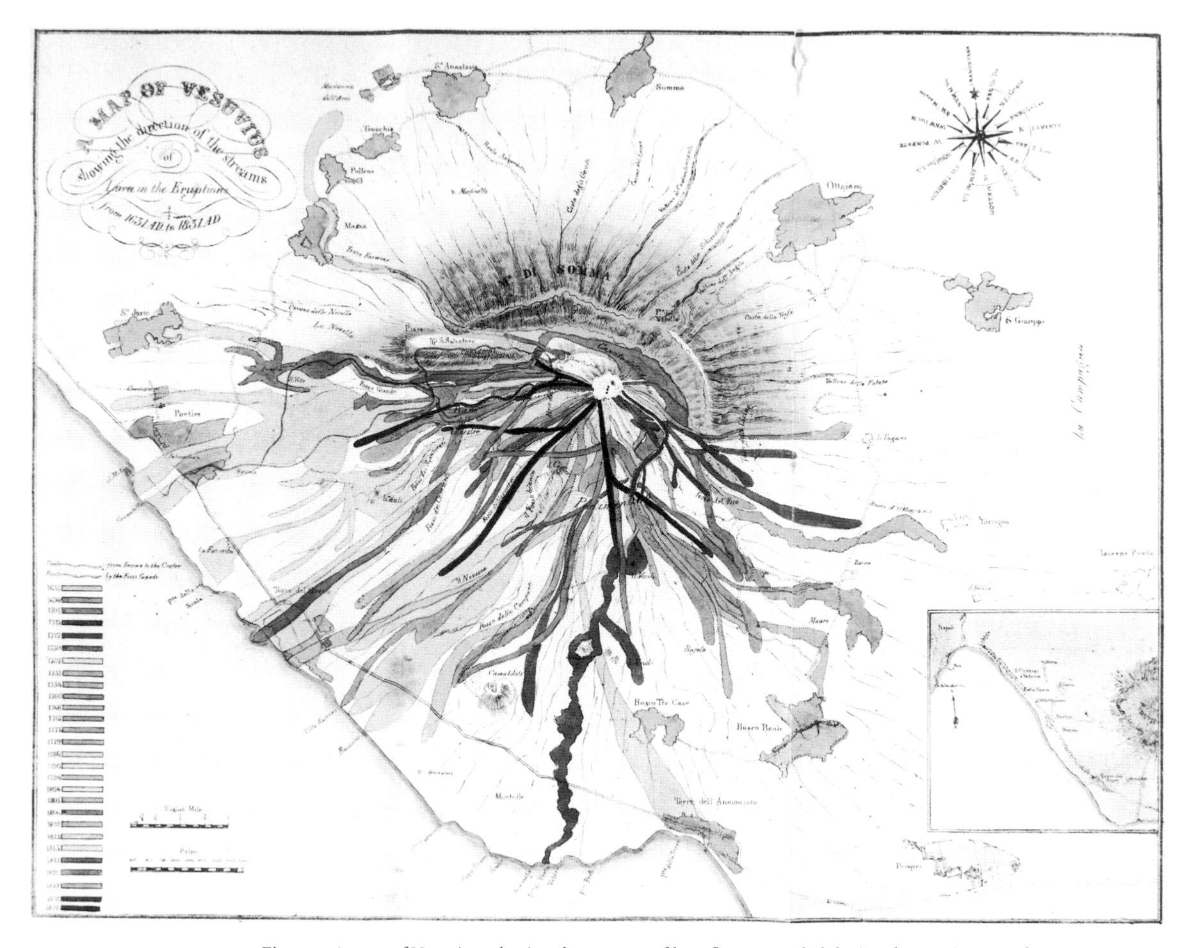

Fig. 22.1 A map of Vesuvius, plotting the courses of lava flows recorded during the previous two hundred years. Although it was published in 1833, too late for the survey of actual causes in the first edition of Lyell's *Principles*, he was familiar with the accurate records from which it was compiled, and it epitomizes the kind of documentation with which much of his work was concerned. In this case he argued that the whole volcanic cone had been built up, layer by layer, by an even longer succession of similar eruptions; and that the arcuate ridge of Monte Somma, which was generally interpreted as the remnant of a still older and larger cone, had been formed in just the same way and not by any sudden upheaval (see below). Pompeii is near the southeast corner of the main map, and the small inset shows the position of Vesuvius in relation to the city of Naples. (By permission of the Syndics of Cambridge University Library)

Lyell's sources for all this factual material were remarkably varied. They ranged from his own observations in the field, through oral information from his acquaintances and the published work of geologists and travelers of many nations and in several languages, to the works of Classical antiquity and other ancient records. More specifically, he made extensive use of reports of the overland travels and voyages of exploration of the previous hundred years or more, which provided textual and pictorial proxies that greatly enlarged his secondhand experience of actual causes: for example he had never seen an iceberg, but he was familiar with them from published reports (see Fig. 7.4). He also reviewed an impressive range of the ordinary geological publications of his own century and the previous one. Preeminent among such printed sources was von Hoff's famous inventory of actual causes (§7.2), to which Lyell's recent efforts with

German had at last given him belated access. However, when Lyell coached Scrope on how he should review the *Principles* for the *Quarterly* (see below), he pointed out that von Hoff's compilation was purely descriptive and geographical, whereas his own was interpretative and theoretical, in that it was arranged according to the physical causes that he claimed had been responsible for the varied changes. His private assessment of his debt to von Hoff, although strictly accurate, was hardly generous: "Von Hoff has assisted me most, and you should compliment him for the German plodding perseverance with which he filled two volumes with facts like tables of statistics; but he helped me not to my scientific view of causes, nor to my arrangement . . . My division into destroying and reproductive effects of rivers, tides, currents, &c. is, as far as I know, new."[2]

Throughout these lengthy chapters, Lyell was trying to demonstrate the sheer *power* of actual causes, and the magnitude of the effects they have produced even within the geologically brief period of recorded human history (see Fig. 7.1). His intention was clear: to persuade his readers that these directly observable agencies of change could have effected far more than had generally been assumed, in the course of what all geologists agreed was the inconceivably longer span of geohistory. He argued that Cuvier, for example, had concluded *prematurely* that actual causes were inadequate to account in full for the deeper past (*BLT* §9.3); Lyell claimed that if their real scope were appreciated their total adequacy would also be conceded. As in previous decades, the argument was not about the validity of actual causes in the interpretation of geohistory, but about their adequacy or inadequacy to explain *all* the preserved traces of real events in the deep past.

Lyell showed no marked preference for slow and gradual processes over sudden and violent ones. On the contrary, the more he could demonstrate the power of actual causes within the span of human history, even in events of catastrophic intensity, the more he would make it plausible to suppose that the deep past had not been significantly different. In his inventory of the effects of earthquakes, for example, he devoted a whole chapter to the catastrophic impact of the great Calabrian quake of 1783, the effects of which had been described at the time by Hamilton and the Neapolitan savants (*BLT* §2.2) and which—even half a century after the event—he had hoped to see for himself. Conversely, in the classic case of the falling sea-level in the Baltic, already well documented in the previous century (*BLT* §3.5), Lyell rejected what he called von Buch's "extraordinary notion" that the land was "slowly and insensibly rising" (§8.1), because there were no historical records of earthquakes or volcanic activity in Scandinavia. By the rigorous criteria that Lyell had set himself, he could not attribute the alleged

---

model the similar but more diagrammatic map by Gemmellaro (G.), *Quadro istorico dell'Etna* (1828). "Aqueous" and "igneous causes" are in Lyell, *Principles* 1 (1830), 167–311, 312–479, respectively; the corresponding terms "Neptunist" and "Vulcanist" continued to be used loosely to denote differing emphases about geological processes, although the specific problem at the heart of the original controversy—the origin of the puzzling rock basalt (*BLT* §2.4)—had long ago been settled in favor of the Vulcanists.

2. Lyell to Scrope, 14 June 1830 [*LLJ* 1: 268–71]; in the printed volume, too, his acknowledgments of von Hoff were decidedly economical. The variety of his sources can best be appreciated from the full bibliography listed in modern format in Rudwick, "Lyell's sources" (1991). Probably not by coincidence, von Hoff belatedly donated his volumes to the Geological Society around the same time (the gift was recorded on 4 October 1830, after the summer recess).

effect to the one and only cause that he accepted as competent to produce it, and so he denied its reality altogether.[3]

In his inventories of both volcanic eruptions and earthquakes, Lyell followed his own prescription by moving from the well known to the more obscure; or, in practice, from the recent past back into earlier centuries where the records were generally sparser and less reliable. For example, he described the histories of Vesuvius and Etna in retrospective detail, because without reliable human records of their frequent eruptions geologists might have wrongly inferred that these great volcanoes had been formed by processes different from anything observed in the modern world. Vesuvius was particularly important in this context, because von Buch had used it as an example in his theory of "craters of elevation" (§8.3), treating the older incomplete cone of Monte Somma as proof that there had been a sudden episode of crustal upheaval, before the present cone was formed by successive eruptions (see Fig. 22.1). Lyell illustrated his text at this point with a sketch section designed to show that the structures of the more recent and the older cones were identical, and therefore that the causes responsible for their formation were also unchanged (Fig. 22.2).[4]

In his detailed inventory of actual causes, Lyell was also concerned to counter any impression that they always acted in one direction. Each class of causes included some processes that tended to build up the land areas at the expense of the sea, while others had the opposite effect. Deposition in deltas, for example, was balanced at other times and in other places by coastal erosion; elevation of

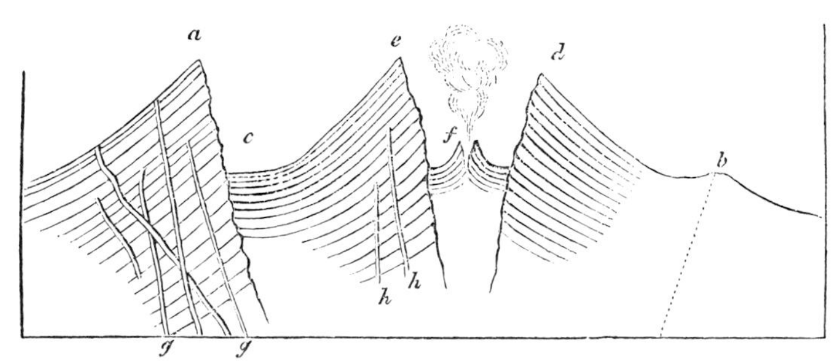

Supposed section of Vesuvius and Somma.

a. **Monte Somma, or the remains of the ancient cone of Vesuvius.**
b. **The Pedamentina, a terrace-like projection, encircling the base of the recent cone of Vesuvius, on the south side.**
c. **Atrio del Cavallo*.**
d.e. **Crater left by eruption of 1822.**
f. **Small cone thrown up in 1828, at the bottom of the great crater.**
g.g. **Dikes intersecting Somma.**
h.h. **Dikes intersecting the recent cone of Vesuvius.**

Fig. 22.2 Lyell's section through Vesuvius, showing the ancient (and now incomplete) cone of Monte Somma (*a*) as being identical in structure to the more recent cone (*e*, *d*) of Vesuvius itself (see Fig. 22.1). Lyell claimed that both had been formed simply by the addition of successive layers of lava and volcanic ash (the internal structure of the very recent little cone within the crater of Vesuvius could not be seen but was assumed to be the same). The vertical "dikes" of volcanic rock (*g*, *h*), which Lyell had seen for himself on the face of Monte Somma but not on Vesuvius (so they were shown failing to reach the surface of the latter), were taken to have been the conduits of lava in earlier eruptions. The section was designed to counter von Buch's claim that Monte Somma had been formed as a "crater of elevation" [*Erhebungscrater*] in a sudden episode of crustal upheaval, quite different from the later formation of Vesuvius: such alleged causal contrasts between past and present were anathema to Lyell.

the land was balanced by its subsidence (as demonstrated eloquently on a small scale by his frontispiece of the Temple of Serapis: Fig. 21.2), both kinds of crustal movement being attributed to igneous processes at depth. It was essential for Lyell to persuade his readers of the reality of "modern" changes in *both* directions, if his concept of a ceaselessly dynamic earth were to seem plausible. But the rhetorical effort was worthwhile. For if local crustal movement could be shown to be a Newtonian "true cause" [*vera causa*], in the sense that it was observably effective within the span of human history, then it could also be deployed across the vastly greater spans of geohistory, to explain the elevation of whole mountain ranges, or conversely the subsidence of continents to form the seas within which the familiar thick piles of rock formations could have accumulated. So he had to establish the observed reality of recent crustal *elevation*, rather than just a subsiding sea level: for example, he cited Maria Graham's report to the Geological Society on the great 1822 earthquake in Chile (§8.3). But crustal *subsidence* likewise required verification by reliable witnesses: in the instructive case of the 1819 earthquake in the Indus delta, Lyell was able to cite reports by British officers serving in India, who as gentlemen were of course treated as impeccably trustworthy (Figs. 22.3, 22.4).[5]

Lyell ended his exhaustive—and, to some of his readers, surely exhausting—survey of inorganic actual causes by claiming to have shown that "the renovating as well as the destroying causes are unceasingly at work, the repair of the land being as constant as its decay": the earth was in a state of dynamic equilibrium, so there was no reason to suppose that it had been fundamentally different at any earlier period of geohistory. His concluding passage recalled Hutton even more clearly: not only the steady state that the Scotsman had attributed to the earth, but also his deistic natural theology, according to which, in Panglossian fashion, all was for the best in the best of all possible worlds (*BLT* §3.4). Lyell claimed that earthquakes, although humanly destructive, were in fact necessary for the maintenance of a constant balance between land and sea, which in turn was essential for "the subserviency of our planet to the support of terrestrial as well as aquatic species", and implicitly for the human species above all others: "This cause, so often the source of death and terror to the inhabitants of the globe, which visits, in succession, every zone, and fills the earth with monuments of ruin and disor-

3. Lyell, *Principles* 1 (1830), 227–32, 412–35. Having, in the event, crossed to Sicily direct from Naples, he had never set foot in Calabria (§19.3). Criticism of his dismissal of the Baltic shoreline changes later put that issue high on his fieldwork agenda (see §33.1).

4. Fig. 22.2 is reproduced from the wood engraving in Lyell, *Principles* 1 (1830), 344, explained on 340–60, the vertical scale being, as usual, greatly exaggerated; it was one of a small number of "woodcuts" embedded in the text of Lyell's first volume.

5. Fig. 22.3 is reproduced from Lyell, *Principles*, 4th ed. (1835), 2 [Cambridge-UL: Ant.c.48.153], pl. 6, explained on 237–42; it was given the privileged (because expensive) treatment of being engraved not on wood but on copper, from a drawing by the topographical artist William Purser, based on "a sketch taken on the spot by Capt. Grindley in 1808", and given to Lyell by Capt. Alexander Burnes, one of his many military and naval informants. Fig. 22.4 is reproduced from Lyell, *Principles*, 3rd ed. (1834), 2, pl. 5, drawn by Burnes. Lyell's original brief description of the case is in *Principles* 1 (1830), 406. The Rann of Kutch is now within India, though most of the Indus delta is in Pakistan; Juna Sindri, the site of the fort, is about 40km northeast of Lakhpat. On the centrality of the *vera causa* principle in Lyell's work, see Laudan, "Methodology in Lyell's science" (1982) and *Mineralogy to geology* (1987), chap. 9.

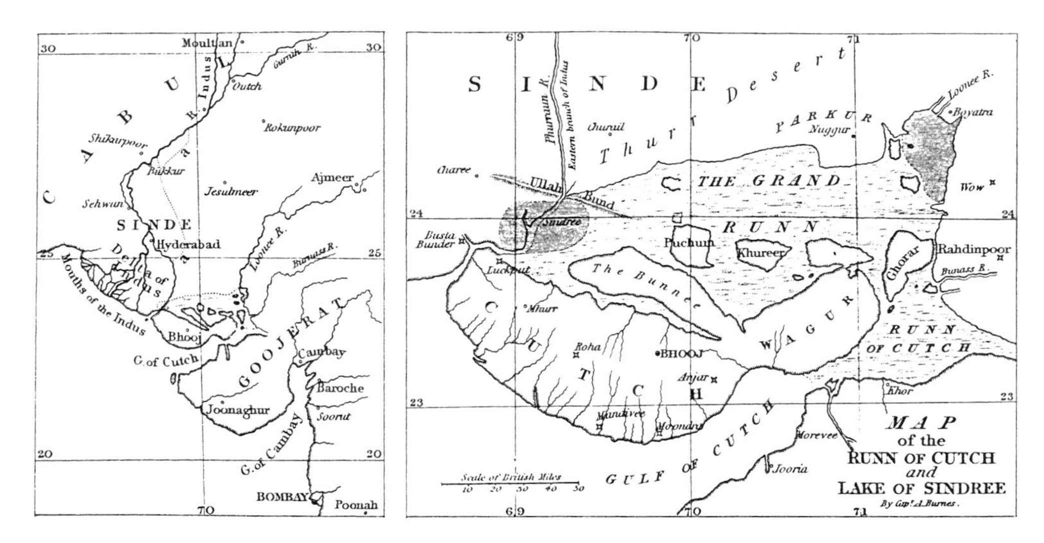

Fig. 22.3 The fort of Sindree on the Indus delta, drawn by a British officer in 1808, *before* the earthquake of 1819 abruptly submerged all but the highest parts of the building and caused the strongpoint to be abandoned. Lyell cited this in his *Principles* (1830) as a well-documented case of a local subsidence of the earth's crust; he was later given this drawing by another officer, and published it in his fourth edition (1835). Such illustrations also served to exemplify the worldwide scope of his work. (By permission of the Syndics of Cambridge University Library)

Fig. 22.4 A map of the Indus delta (left), to show the regional setting of the vast salt flats of the Runn of Cutch (right); the latter map shows the area (shaded) around Sindree, which sank in the earthquake of 1819, while the long low ridge of Ullah Bund to the north was raised at the same time (the scale is of 50 miles). The maps were drawn by a British officer serving in the region; Lyell used them in the third edition (1834) of his *Principles* to amplify his verbal account of the earthquake in the first (1830). He pointed out that no movement of the earth's crust would ever have been perceived, had the affected area not been at sea level; this reinforced his argument that most earthquakes were probably accompanied by crustal elevation or depression or both, even if no such movement was recorded.

der, is, nevertheless, a conservative principle in the highest degree, and, above all others, essential to the stability of the system".[6]

22.2 SCROPE ON LYELL

Like any book, then as now, the *Principles* was read in many different ways: the general educated public often understood Lyell's work in quite different ways from his savant readers, above all the knowledgeable geologists. The published reviews reflected the same wide range of reactions. Those that catered to the reading public tended to focus on Lyell's attack on the scriptural geologists, and to deal mainly with the implications of Lyell's geology for the interpretation and authority of the Bible and other such issues. In contrast, what was far more interesting to the leading members of the Geological Society and to other savants—including those who regarded themselves as religious believers—was Lyell's evident intention to rehabilitate and improve Hutton's geotheory. To them it was obvious that this was the goal of Lyell's work.[7]

There was no disjunction or discrepancy between, on the one hand, the private reactions of geologists and other savants, as expressed in their letters (and doubtless in their unrecorded conversations), and, on the other, the responses to Lyell that they expressed in public, for example in reviews. Throughout the gradient between private and public, their reception of Lyell's volume was in many ways appreciative. Even those who were critical of his refusal to accept any causal contrast between past and present acknowledged that his thorough survey of actual causes established a valuable baseline for that comparison. But almost all Lyell's savant critics were dubious about his attempt to explain away the evidence for a broadly directional or even progressive trend in geohistory. Above all, they were highly critical of the "system" or geotheory to which Lyell claimed that his method of interpretation "necessarily" led. It was one thing to test for the possibly total adequacy of actual causes in the explanation of the deep past; it was quite another to claim "absolute uniformity" in the overall state of the earth, apparently from and to all eternity. The remainder of this chapter describes some of the very first of these highly knowledgeable reactions to Lyell's first volume.[8]

Of all the English geologists, Scrope had shown himself to be one of Lyell's strongest supporters, so it is significant that even he was critical of Lyell's geothe-

6. Lyell, *Principles* 1 (1830), 473, 479. It is of course possible to dismiss this providentialist natural theology as mere window-dressing, designed by Lyell to make his geology more palatable to some of the readers he hoped to attract; but it is more realistic, and certainly more consistent with his private notes, to regard the sentiment as sincere, and Lyell's own.

7. A valuable list of all known British reviews is in Secord's abridged edition of *Principles* (1997), 459–61; his important "Introduction" evaluates the reception of the book in relation to the broad range of publics in Britain to which these reviews were addressed. In contrast, the focus here—in line with the rest of the present book—is on Lyell's interactions with his scientific and intellectual equals. The finest and most thorough analysis (at least for the sciences of the nineteenth century) of the diverse "readings" that a specific book could receive, is in Secord, *Victorian sensation* (2000), which deals with *Vestiges* (1844), Chambers's famously anonymous evolutionary work (see §36.3).

8. Secord's "Introduction [to Lyell's *Principles*]" (1997) argues for an alternative interpretation, which gives priority to Lyell's actualistic method and—relative to the present narrative—downplays the significance of his neo-Huttonian geotheory.

orizing. Murray, as the publisher of both the *Quarterly* and the *Principles*, had wanted the former to carry a review of the latter as promptly as possible. Lockhart, the editor of the periodical, had consulted Lyell beforehand about whom he should ask to write it. The two English professors of geology, Buckland and Sedgwick, had both professed to be too busy, and Conybeare had the reputation of failing to deliver reviews on time, so Lockhart suggested Scrope. Lyell had told him that Scrope was indeed competent, and admitted that he himself was pleased with the choice, since Scrope was likely to be more positive about the book than Conybeare might have been. Scrope accepted, although he had spent little time on geology since his own books were published. He had become much involved with national politics and with local affairs in rural Wiltshire; he had just completed his book *On Credit Currency*, which developed the ideas on political economy that had provided him incidentally with a vivid metaphor for geological time (§15.1). "I have neglected geology of late," he admitted to Lyell, "and am not quite *au courant* of the progress made in it latterly on the Continent"; he was well aware that French geologists such as Prévost had been "fighting away for 'existing causes' for a long time past", and was therefore uncertain just how original Lyell was being.[9]

However, Scrope was impressed by what he read, as proofs of Lyell's text arrived: "What I admire in you is the assurance with which you speak of doctrines still supported by the Bucklands, Conybeares &c. as exploded errors, past praying for". Scrope approved of Lyell's attack not only on these diluvialists, but also on the "scriptural" geologists beyond the pale of the Geological Society: "If between us we can succeed in freeing Geology once and forever from the clutches of Moses, we shall have deserved well of the science". He was delighted with Lyell's assessment of the power of actual causes, but he was not convinced by Lyell's dismissal of the distinctive character of the Primary rocks, and indeed by his total rejection of all directional interpretations of geohistory, such as Scrope himself had adopted in his first book (§9.3). In short, while he expressed his criticism delicately, Scrope was highly skeptical about Lyell's attempt to revive a Huttonian "system" that denied any sign of the earth's origin:

> You are aware that I think you carry your principle rather too far when you argue for the endless or rather *beginning-less* succession of past changes. I shall be obliged to fight this point a little, particularly as you are rather unjust to those who *look for a beginning*, classing them with the miracle mongers, as if the actual [i.e., present] order of Nature could not admit of the creation of new worlds as well as new genera or new continents.[10]

Lyell replied at once, explaining to Scrope how he regarded his own work as original in relation to Continental geologists (his assessment of von Hoff has been mentioned already); he also encouraged his friend to help him to attack both the diluvialists and the "scriptural" writers. Most important, however, was Lyell's claim that his apparent eternalism was just a question of sound method, not—as he put it dismissively—one of theology:

> Probably there was a beginning—it is a metaphysical question, worthy a theologian—probably there will be an end. Species, as you say, have begun and ended, but the analogy is faint and distant. Perhaps it is an analogy, but all I say is, there are, as Hutton said, 'no signs of a beginning, no prospect of an end' . . . All I ask is, that at any given period of the past, don't stop inquiry, when puzzled, by refuge to a 'beginning', which is all one with 'another state of nature', as it appears to me. But there is no harm in your attacking me, provided you point out that it is the proof I deny, not the probability of a beginning.[11]

Scrope's long essay on Lyell's work appeared in the next issue of the *Quarterly*. Reassuring its conservative readers, he began by claiming that geology was leading to enhanced "proofs of a Designing Intelligence", quite contrary to ignorant accusations that it led instead to impiety and atheism. Scrope then promoted Lyell's work as marking "the beginning of a new era in geology", though he chided his friend for having raised the question of uniformity "rather prematurely", before completing the survey of actual causes that might settle the issue. Scrope devoted most of his review to a long and appreciative synopsis of Lyell's still incomplete inventory, before returning to that crucial question. He distinguished two separate meanings of uniformity. The first was that rocks and fossils, and the past events that they represent, "are to be attributed to the operation of existing causes". To this claim, he wrote, "we give, with Mr. Lyell, and, we believe, the great body of European geologists, our unqualified concurrence"; and he expected it to be confirmed by reference to Lyell's enlarged inventory of actual causes. The second claim, on the other hand, was that "the existing causes of change have apparently operated with absolute uniformity from all eternity", and that "there are no traces of any beginning to this series of changes and productions, or of any variation in the ratio of its progress as regards the whole". This aspect—of what he identified unequivocally as "Huttonian theory"—failed to convince Scrope, pending "much more decisive proofs" of its validity.[12]

Scrope argued that Lyell had confused these two distinct meanings of uniformity, and had unfairly tarred all his critics with the same brush. Uniformity of the basic physical laws of nature was not denied by anyone, or at least not by any savant; indeed such an assumption was the necessary precondition for any scientific work. He pointed out that it was in no way violated by the inference that the earth had gradually changed in character, for example in temperature, in the course of its long geohistory; to accept the strong evidence for overall directional change was, in Scrope's view, much more "philosophical" than Lyell's contrary

9. Lyell to Scrope, 6 May 1830 [quoted in Wilson, *Lyell*, 273–74]; Scrope to Lyell, 11 June 1830 (see below); the preface to his *Credit currency* (1830) is dated 30 May.

10. Scrope to Lyell, 11 June 1830 [Philadelphia-APS: B.D25.L, printed in part in Wilson, *Lyell* (1972), 274–75]. The quoted passage illustrates again how "creation", in ordinary scientific discourse at this time, did not necessarily imply any non-natural event: no one treated continents as supernatural in origin.

11. Lyell to Scrope, 14 June 1830 [*LLJ* 1: 268–71]; he misquoted Hutton, possibly because he had never read the original and only recalled the famous aphorism from his reading of Playfair.

12. [Scrope], "*Principles* by Lyell" (1830), 411–13, 426.

assumption, and ought not to be ridiculed as mere "cosmological reveries". The astronomical perspective highlighted the dubious grounds for Lyell's insistence on "uniformity" in the sense of the earth's alleged steady state: "the eternal stability of this speck of matter in its present condition, appears to us [i.e., to Scrope himself] as unreasonable an assumption as the eternal duration of its actual [i.e., present] divisions of land and water, justly stigmatised by Mr. Lyell".[13]

Scrope thought Lyell's climatic model ingenious, but doubted if it could sustain his steady-state theory. He argued that Lyell had failed to explain away the strong evidence for "a progressive state and a limited existence" for the earth, and particularly for its hot origin and subsequent slow cooling (§9.2, §9.3). Finally, Lyell's dualistic treatment of the human species seemed "somewhat too wire-drawn" to carry conviction: "We should say it is exactly the moral character of man which presents the greatest anomaly and novelty, and tends most strongly to exhibit the progressive march of creation". All in all, Scrope gave Lyell's book a generous welcome, but he left the readers of the *Quarterly* in no doubt that he was deeply skeptical about Lyell's ultimate goal. In Scrope's judgment, the book was designed not just to demonstrate the explanatory value of actual causes as the best *method* for geology, but more fundamentally to promote a version of Hutton's *geotheory*, which the progress of the science had now made questionable if not untenable.[14]

22.3 DE LA BECHE AND CONYBEARE JOIN IN

Even before Scrope's review was published, the same kind of criticism began to circulate more informally among English geologists, when De la Beche drew a telling caricature and distributed it as a lithographed print. De la Beche had joined the Geological Society, and risen to become a leading member (§2.2), more or less in tandem with Lyell, his almost exact contemporary. More recently, he had been in France and Italy on a geological Grand Tour at almost the same time as Lyell, though their paths had not crossed (§13.3). He had just published a fine album of *Sections and Views*, most of them his own lithographs, designed to illustrate the kinds of "geological phaenomena" that any adequate geotheory would have to explain, for example the notorious problem of erratic blocks (Figs. 13.4, 13.5). And he was working on an introductory *Geological Manual*, in which he would outline the now almost consensual conception of the science, treating geohistory as unimaginably lengthy, broadly directional or even progressive, and probably marked by occasional catastrophic episodes.[15]

When De la Beche read the first volume of *Principles*, he, like Scrope, was highly critical of Lyell's attempt to explain away all the evidence for a directional geohistory and to revive Hutton's steady-state geotheory in its place. He had recently drawn a scene of "A more ancient Dorset" (*Duria antiquior*, Fig. 11.5), which admiringly caricatured the achievement of Buckland and others in reconstructing the vanished world of the ichthyosaurs and other bizarre fossil reptiles. So he set out to compose a similar—but less flattering—scene in criticism of Lyell's work. It is worth describing in some detail, because De la Beche's final design, and the draft sketches that led up to it, form a series that displays, in

Fig. 22.5 De la Beche's first draft sketch for a caricature in criticism of Lyell's Huttonian geotheory. The author of a "Theory of the earth" (left), in barrister's wig and elegant tailcoat, is contrasted with the geologist (right) in field clothes with hammer and collecting bag. The tinted spectacles (of a design that was common at the time) being offered by the theorist implied that the landscape—representing the observable "facts" of geology—would be seen by other geologists in a false light. De la Beche never completed the sketch, because he abandoned this design in order to try out others. (By permission of the British Geological Survey. © NERC. All rights reserved. IPR/83/23C)

vividly free-associative style, the themes in Lyell's book that he thought it most important to criticize.[16]

The first of De la Beche's draft sketches showed Lyell as an indoor gentleman, indeed a barrister, standing on ground marked "Theory", confronting a plain outdoor field geologist (such as De la Beche considered himself to be), and offering him colored spectacles *through* which to view the plain facts of geology; and the theorist concealed behind his back a volume which (like Hutton's work) was entitled "Theory of the earth". De la Beche was clearly intending to caricature the *Principles* as a revival of Hutton's geotheory, and as entailing a distortion of what impartial observation would suggest (Fig. 22.5).[17]

De la Beche's next two sketches (not reproduced here) made further play with the themes of tinted spectacles, impaired perception, and even the "cookery"—in a pejorative sense—suggested by Lyell's own "recipe" for climate change (§21.4). But after them came a sketch that took a different tack, showing a chemical

13. The use of the quasi-royal "us" or "we" for an author's own opinions was standard style in the *Quarterly* and similar publications.

14. [Scrope], "*Principles* by Lyell" (1830), 464–69. In effect he was distinguishing between the actualistic *method* in geology and what Whewell was soon to dub a "uniformitarian" *theory* about geohistory (see §24.4).

15. De la Beche, *Sections and views* (1830) and *Geological manual* (1831); the preface of the former, which is critical of geotheorizing, was written not earlier than April 1830 and perhaps after Lyell's volume appeared. See also McCartney, *De la Beche* (1977).

16. Rudwick, "Caricature as a source" (1975), reproduces and analyzes the complete series of sketches and sets out the evidence for inferring that the target of the sketches was indeed Lyell and his book; but the date of the drawings now seems most likely to be 1830, soon after Lyell, *Principles* 1, was published (see below).

17. Fig. 22.5 (and Fig. 22.6 below) are reproduced from De la Beche, notebook of 1830–31 (Keyworth-BGS: GSM 1/123); the whole series was sketched in the back of this field notebook. De la Beche's depiction of Lyell in indoor clothes was of course unfair to a seasoned field geologist, but it made an effective polemical point. The caption reads, "Take a view, my dear Sir, through these glasses, and you will see that the whole face of nature is as blue as indigo"; this may refer to Lyell's claim that even high mountains, such as those shown, are the products of crustal elevation from beneath the deep blue sea.

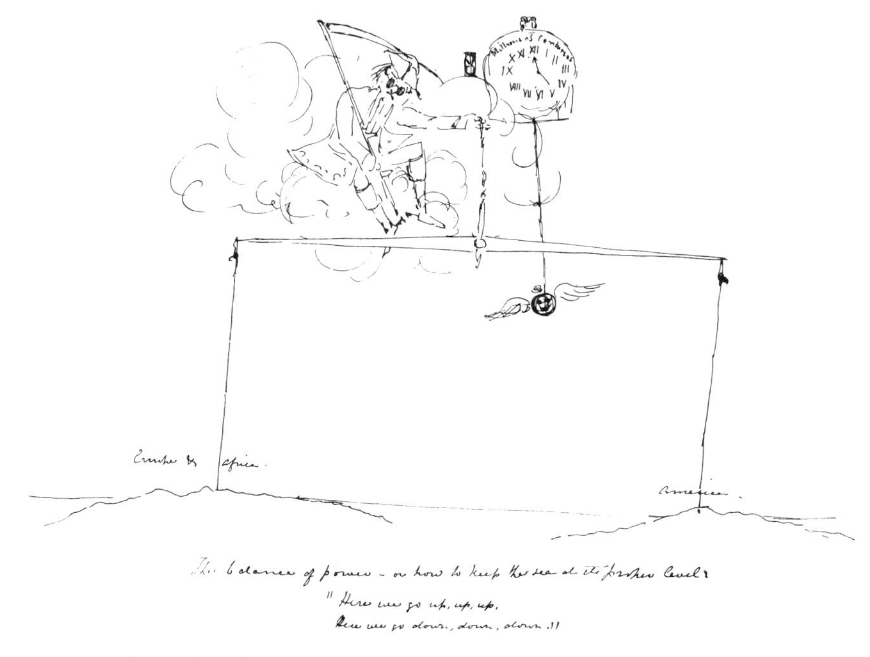

Fig. 22.6 De la Beche's draft sketch of Lyell's geotheory as a chemical balance held by Father Time (seated on clouds, with traditional sickle and hourglass, but also colored spectacles), coupled with an hourglass and a clock marked in "Millions of Centuries", with its pendulum driven fancifully by angelic or fairy wings; Europe and Africa are on one side of the balance and the Americas on the other, rising or falling in endless oscillatory motion. The caption includes a further political pun, and derides Lyell's tectonic theory as only fit for a rhyme to go with a child's seesaw: "The balance of power—or how to keep the sea at its proper level: 'Here we go up, up, up, / Here we go down, down, down.'" (By permission of the British Geological Survey. IPR/83/23C)

balance with Europe and Africa in one pan and America in the other. This combined Hutton's (and Lyell's) conception of the earth as a balanced system in dynamic equilibrium with the latter's tectonic theory of continents moving ceaselessly up and down throughout geohistory. De la Beche then developed this theme by combining the balance with an evocation of Lyell's prescription of virtually unlimited time as a panacea for all geological problems (Fig. 22.6).

The following sketch showed Father Time again, accompanied now by a ghostly figure representing Space, both being accosted by the theorist in barrister's wig and gown, presenting them with a volume: "Behold my book, Sirs, Time & Space". This clearly alluded to Lyell's frequent appeals to the parallel between geology and astronomy, and to his ambitions to make his theorizing universal. In the next sketch De la Beche parodied Lyell's claim that even the geologically recent "Alluvium" represented an immensely lengthy period, and his transformation of the still older period of the "Diluvium" into just the more recent part of a vastly enlarged Tertiary era: those two sets of deposits were personified respectively as Father Time (now with angelic wings to suggest his fanciful status), and as a huge bearded figure—possibly modeled on the "Ancient of Days" in William Blake's *Creation*—with puny geologists clambering irreverently over him. Lastly, changing tack yet again, De la Beche sketched a parodied meeting of the Geological Society, with its members transformed into ichthyosaurs and other extinct monsters. But he hardly began this sketch, probably because it suggested the design that finally satisfied him, and that he lithographed for distribution among his friends.

De la Beche's final design was inspired by Lyell's fanciful forecast of what the world might be like in the unimaginably distant future, when the "great year" of the "geological cycle" of global climate might return to the "summer" it had enjoyed at the time of the ichthyosaurs in the distant past (§21.4). Lyell, who had in fact considered a possible appointment at the University of London (now University College), was depicted as "Professor Ichthyosaurus", lecturing to an audience of other Jurassic reptiles on the subject of a human skull. This was a *fossil* specimen, for the scene was set in a post-human *future*: as the subtitle put it, "Man found only in a fossil state,—Reappearance of Ichthyosauri". And the professor was explaining, in a parody of Cuvierian functional anatomy, that the fossil must have belonged to an animal of "a lower order" than his audience, thus ridiculing Lyell's denial of the progressiveness of the real fossil record. Finally, De la Beche quoted—from Byron's famous poem *The Dream* (1816)—the line that introduced each successive scene as the dreamer moved forward in time; he implied that Lyell's "theory of the earth" was equally fanciful. On receiving a copy of the print, Buckland wrote: "Many thanks and much praise for your Caricature of Actual Causes & the Huttonian Theory rediviva; the Book is written in a very seductive Style & will no doubt make many Converts". He and other recipients were in no doubt that De la Beche had hit his target (Fig. 22.7).[18]

Another recipient of De la Beche's print would have been Conybeare, whose caricature of Buckland in Kirkdale Cave (*BLT* Fig. 10.24) had first begun the fashion for serious visual jokes among the English geologists. Even before Scrope's review appeared in the *Quarterly*, Conybeare launched a trenchant critique of Lyell's work in the long-established *Philosophical Magazine*; his essay appeared in installments spread over nine months, rather like contemporary novels (such as those by Charles Dickens a few years later) in another kind of monthly. Conybeare praised the *Principles* as a work that would turn geological research in valuable directions, although he regarded Lyell's "theoretical system" as premature. He hoped the new work would promote serious discussion, which alone would enable the merits of rival theories to be evaluated. However, he was "altogether unconvinced" by Lyell's rigorous concept of "absolute uniformity", and criticized him for dismissing any alternative as unscientific. Above all, he identified the *Principles* as "an expanded commentary on the celebrated Huttonian axiom" that there was no sign of a beginning or end to the terrestrial "system". Treating Hutton's theory "philosophically"—he dismissed the earlier "moral" or religious objections to it as unfounded—Conybeare stated that "I have ever

18. Fig. 22.7 is reproduced from De la Beche's print *Awful changes* [Cardiff-NMW] (1830). Buckland to De la Beche, 15 September 1830 [Cardiff-NMW: MS 179], can hardly refer to any caricature but this, and therefore dates it (and the draft sketches that led up to it) within a few weeks of the publication of Lyell's volume. The quotation from Byron's poem, "A change came o'er the spirit of my dream", to which the title of the print also alludes, is the refrain with which "The Dreamer" moves from one fantastic vision to another. See Rudwick, *Scenes from deep time* (1992), 48–49, and "Caricature as a source" (1975); on Lyell's professorial career, see §25.1.

Fig. 22.7 "Awful changes", De la Beche's lithographed caricature (1830) of Lyell's work. It shows Lyell as "Professor Ichthyosaurus", lecturing in a tropical setting to an audience of other Jurassic reptiles, and analyzing a fossil human skull as belonging to a lower class of animal than themselves. Unlike De la Beche's pictorial reconstruction of the reptiles' real environment in the deep past (Fig. 11.4), this was a scene set in the remote post-human *future*, when—so Lyell conjectured—animals similar to these extinct reptiles might *return*. The caricature, which was distributed widely among geologists, ridiculed those aspects of Lyell's geotheory that his contemporaries found most implausible: his denial of any progressiveness in the fossil record, and his claim that geohistory was cyclic rather than directional, and the earth in an overall steady state. (By permission of the National Museum of Wales)

regarded it, and continue to regard it, as one of the most gratuitous and unsupported assertions ever hazarded". Like Scrope and De la Beche, and indeed Buckland, Conybeare located the core of the new work unambiguously in Lyell's Huttonian geotheory.[19]

Conybeare conceded that Lyell's retrospective method had its advantages, but so did the alternative strategy of surveying the evidence for geohistory in true chronological order, to determine whether there had in fact been "uniformity" in Lyell's radical sense. Conybeare protested that those who reasoned as he did were *not* guilty of proposing "different laws of nature" in earlier times, but only that the same laws might have produced agents of widely varying intensities in different circumstances. Contrary to Lyell's rhetoric, it seemed to Conybeare to be more scientific to let the evidence itself decide the issue: "whether they all indicate a uniform and constant operation of the same causes, *acting with the same intensity*, and *under the same circumstances*; or rather evince that there has been a gradual change in these respects, and that the successive periods have often given rise to such new circumstances, as must have in very great degree modified the original forces."[20]

Over the following months Conybeare set out his alternative interpretations, first of igneous causes and then of aqueous ones. He reviewed the evidence, both old and new, for the earth's hot fluid origin, ranging from its oblate spheroid shape to its apparent long-term cooling trend, but he was skeptical about Lyell's "very ingenious explanation" of climatic change. Likewise he continued to doubt the adequacy of Lyell's "fluvial theory" for the erosion of valleys, even allowing "a sufficient number of millions (I should rather say *infinit-illions*) of years"; he claimed that the observable effects showed instead the involvement of "violent currents and of vastly extended sheets of water" (he conceded that the valleys in Auvergne were exceptions). Yet he emphasized that this "diluvial theory" was strictly scientific. For among such physical events, which appeared to have occurred at intervals throughout geohistory, it was a quite separate question whether or not the very last could be equated with the Flood recorded in Genesis. Conybeare was not going to let Lyell's rhetoric couple him with the literalistic "scriptural" writers.[21]

22.4 CONCLUSION

After Lyell's lengthy introductory essays, the first volume of his *Principles* only had space left for the first part of his planned inventory of actual causes; but it was an impressively thorough survey of those involving inorganic agents. Unlike von Hoff's inventory (from which much of Lyell's information was derived) it was so arranged as to support his theoretical claim that all classes of geological processes contributed to *both* sides of a dynamic equilibrium. Aqueous and igneous processes, for example, contributed both to the formation of new land areas and to their destruction. Far from describing these processes as invariably gentle or their action as gradual, Lyell's strategy was to emphasize their power, even their violence, provided this had been witnessed by reliable observers within the span of human history. The greater the power that could be ascribed to them, the more they could be regarded as adequate to account for everything in the unobserved prehuman past of geohistory. The latter would then be seen to have been

19. Conybeare, "Lyell's 'Principles'" (September 1830), continued as "Phaenomena of geology" in issues for November 1830 through April 1831. He repeated his critique a decade later, more informally and with relatively minor changes in the light of Lyell's later editions, in Conybeare to Lyell, February 1841 [Philadelphia-APS, printed in full in Rudwick, "Critique of uniformitarian geology" (1967)]; this is one of the most eloquent statements of the position of Lyell's geological critics.

20. Conybeare, "Phaenomena of geology" (1830–31), November 1830 issue, 359–62.

21. Conybeare, "Phaenomena of geology" (1830–31), March 1831 issue, 111–17. In his "Inaugural address" (1831) at the opening of the theological college in Bristol (published by Murray), he mentioned in passing that in fact he did regard the most recent diluvial episode as likely to have been the biblical Flood, but he treated this, correctly, as separate from the geological issue.

fundamentally similar to the present world, and the earth would be established as a steady-state system recalling that of Hutton.

Among the earliest reactions to Lyell's volume, Scrope's is particularly revealing, because in many ways he was closest to Lyell in his emphasis on the explanatory power of actual causes and in his ardent desire to free their science from the ignorant "scriptural geology" that hampered its wider acceptance in Britain. But although Scrope duly welcomed Lyell's treatment of actual causes, he also criticized his friend's rejection of the cumulative evidence that the earth had *not* been in a Huttonian steady state throughout geohistory, but on the contrary had undergone broadly directional change from a probably hot origin. De la Beche, who was about to summarize the same almost consensual view of geohistory in his own book on geology, was equally critical of Lyell's geotheoretical ambitions. He lampooned them in a celebrated caricature, criticizing Lyell's claim that the earth was in a Huttonian steady state, from and to eternity. So did Conybeare, who like Scrope was a friendly critic, but was unconvinced by Lyell's ingenious attempt to explain away what most other geologists regarded as clear evidence for directionality. The next chapter continues this account of early criticisms of Lyell's *Principles* and takes it beyond England.

CHAPTER 23

Promoting Lyell's Principles (1830–31)

23.1 TWO CRITICS FROM CAMBRIDGE

If Conybeare was the theological heavyweight at the Geological Society, Sedgwick's colleague William Whewell (1794–1866) was its polymath. Whewell, a northerner of lowly social origins, had had a meteoric career at Trinity College, becoming well known beyond Cambridge for his part in promoting the new Continental physics and mathematics. Elected to the Royal Society in 1820, at the age of twenty-six, he had not joined the Geological until 1827, but the following year had been appointed to the chair of mineralogy at Cambridge. Like Sedgwick and most other Cambridge dons, he had also been ordained in the Church of England. Early in 1831 he reviewed Lyell's *Principles* for the *British Critic*, the Anglican monthly that long ago had published some of de Luc's work in English (*BLT* §6.4). Significantly, Whewell wasted no space on defending Lyell (or himself) against the "Mosaic" or "scriptural" writers of recent years; he noted that all "intelligent persons" were now agreed that the science properly belonged "in the hands of the natural philosopher", at least insofar as geological speculation was "independent of the condition and history of man". With that important proviso, Whewell judged that geology had come of age as an autonomous natural science, no less than, for example, astronomy.[1]

Whewell welcomed Lyell's work as making geology newly accessible and attractive to a more general public, because it offered a broad synthesis in place of

1. [Whewell], "*Principles* by Lyell" (1831), 206; as in the *Quarterly* and other such periodicals, review essays in the *British critic* were anonymous, but their authors were usually known to those who mattered. Yeo, *Defining science* (1993), sets Whewell effectively in context. See also, more generally, the seminal essay on the informal "Cambridge network"—in which both Whewell and Sedgwick were prominent—in Cannon, *Science in culture* (1978), chap. 2.

"many a long and dreary communication" that he had sat through at the Geological Society's meetings. After the Society's prolonged and rigorous focus on boring "facts", Whewell saw heuristic value in the "fresh outbreak of the spirit of theorizing" signaled by Lyell's work, if indeed it led to a more probing discussion of the "principles" of the science. Whewell gave unstinting praise to Lyell's "masterly" synthesis. Yet he, like Conybeare, was profoundly skeptical about the particular geotheory that underlay it, namely "the well-known Huttonian doctrine" of the *total* adequacy of actual causes, and the geohistory of "indefinite and boundless cycles" that was alleged to follow from their application. And while ready to agree that the timescale of geohistory was inconceivably vast, Whewell was skeptical about Lyell's use of virtually infinite time in order to solve all the awkward problems of the science. However, he anticipated that Lyell would prove to be right to expect that some of the apparent major breaks in the sequence of formations—and hence in the geohistory that they represented—would eventually be bridged. Summarizing that sequence, he neatly shared the credit among the major European nations: the Germans (such as Werner) for the Primaries and Transition rocks, the English (such as Smith) for the Secondaries, the French (such as Brongniart) for the Tertiaries. And now—newly made known in Britain by Lyell's work—the Italians deserved credit for what he proposed to call the "*penultimate*" formations, those closest to the present (he was evidently unaware that some naturalists, such as Risso, were using "Quaternary" for these deposits: §19.2).

Whewell highlighted three points on which Lyell would need to satisfy other geologists, if his geotheory were to carry conviction. The first was to show that actual causes were indeed capable of producing the observable effects from the deep past, "in kind and magnitude". He praised Lyell's inventory of actual causes as constituting almost a new science, which he proposed to call "*Geological Dynamics*". Yet he doubted if even this impressive "machinery" was adequate to effect all that it was being called upon to explain: for example, whether the small vertical movements of the earth's crust witnessed in human history (such as those affecting the Temple of Serapis), multiplied by even the vastest spans of deep time, could have produced such spectacular effects as Webster's up-ended Tertiary strata on the Isle of Wight (*BLT* §9.4). Lyell's second task would be to show that the same array of actual causes could account for the increasingly strong evidence for a long-term cooling trend in the world's climate, to which Lyell himself had added when he reported the "tropical" size of many of his Sicilian fossil shells. Here again, Whewell, like most of Lyell's critics, doubted whether Lyell's "ingenious but somewhat venturous theory" of climate change would suffice. Lyell's third and hardest task, which was needed "to give even a theoretical consistency to his system", would be to "supply us with some mode by which we may pass from a world filled with one kind of animal forms, to another, in which they are equally abundant, without perhaps one species in common". Whewell anticipated that this might be the one point in geology at which it would be necessary to invoke "a manifestation of creative power, transcending the operation of known laws of nature"; in effect, the problem of sudden faunal change (and of the origin of species) might be the Achilles' heel of Lyell's naturalistic synthesis.[2]

Hard on the heels of Whewell's penetrating evaluation of Lyell's work came Sedgwick's second "anniversary address" to the Geological Society, marking the end of his two-year presidency. He followed his own precedent by again reviewing the whole range of current research, but he paid particularly close attention to Lyell's new work. However, he introduced it in the context of Herschel's recent astronomical theorizing about the possible terrestrial effects of long-term cycles of planetary eccentricity. Sedgwick noted that such speculations, however inconclusive, did show how geology was bursting the limits of time: "man seems to be no longer a worshiper at the portal of Nature's temple, but is allowed to pass within, and to be so far a partaker of her mysteries, as to see with his intellectual eye both the past and the future". Dropping such Classical imagery, Sedgwick emphasized that he believed that "all the primary modes of material action" owed their unquestioned uniformity as laws of nature to the constancy of their divine origin. But the elegant simplicity of astronomy—for example in Newton's laws of gravitation—was in stark contrast to the irreducibly complex messiness of geology, because on earth the basic laws of nature combined to form processes of "indefinite complexity". Without yet mentioning Lyell by name, Sedgwick criticized—in magnificent purple prose—the assumption that volcanic and other *geological* agents have always acted with uniform intensity:

> This theory confounds the immutable and primary laws of matter with the mutable results arising from their irregular combination. It assumes, that in the laboratory of nature, no elements have ever been brought together which we ourselves have not seen combined; that no forces have been developed by their combination, of which we have not witnessed the effects. And what is this but to limit the riches of the kingdoms of nature by the poverty of our own knowledge; and to surrender ourselves to a mischievous, but not uncommon philosophical scepticism, which makes us deny the reality of what we have not seen, and doubt the truth of what we do not perfectly comprehend?[3]

The target of Sedgwick's critique, already obvious to his audience, was now revealed: these fundamental issues had been confused, and "their spirit sometimes violated", in Lyell's recent volume. Sedgwick first acknowledged the "delight" with which he had read it; the work merited "a distinguished place in the philosophic literature of this country; higher praise than this I know not how to offer". Specifically, he followed Whewell in giving Lyell credit for the "general system of geological dynamics" set out in his inventory of actual causes. Beyond that primarily descriptive task, however, Lyell had revealed himself as "the champion of a great leading doctrine of the Huttonian hypothesis", and Sedgwick found it regrettable that Lyell wrote as "the defender of a theory", using throughout "the language of an advocate" (which, of course, he was). In other

2. On Whewell's "geological dynamics" in relation to causal "physical geology", and on Sedgwick's role (see below), see Smith (C.), "Geologists as mathematicians" (1985).

3. Sedgwick, "Address to the Geological Society" (1831), 301, read on 18 February; Herschel, "Astronomical causes" (1832), read on 15 December 1830.

words, Sedgwick highlighted the *rhetorical* character of Lyell's work: the trained barrister had tried to present the most persuasive *case* for Huttonian uniformity. But Sedgwick was not persuaded: "to assume that the secondary combinations arising out of the primary laws of matter, have been the same in all periods of the earth, is, I repeat, an unwarranted hypothesis with no *a priori* probability, and only to be maintained by an appeal to geological phaenomena".

In Sedgwick's judgment, Lyell's theory failed utterly in that crucial test against the records of geohistory: the "indefinite succession of similar phaenomena" that it predicted was simply not there. Most obviously, the known fossil record indicated ever more clearly a broadly progressive history of life, of which "the recent appearance of man on the surface of the earth (now universally admitted)" was the culmination. "Were there no other zoological fact in secondary geology", Sedgwick added, "I should consider this, by itself, as absolutely subversive of the first principles of the Huttonian hypothesis". Adopting Lyell's own retrospective method, Sedgwick then saw a first "violation of continuity" in the spreads of

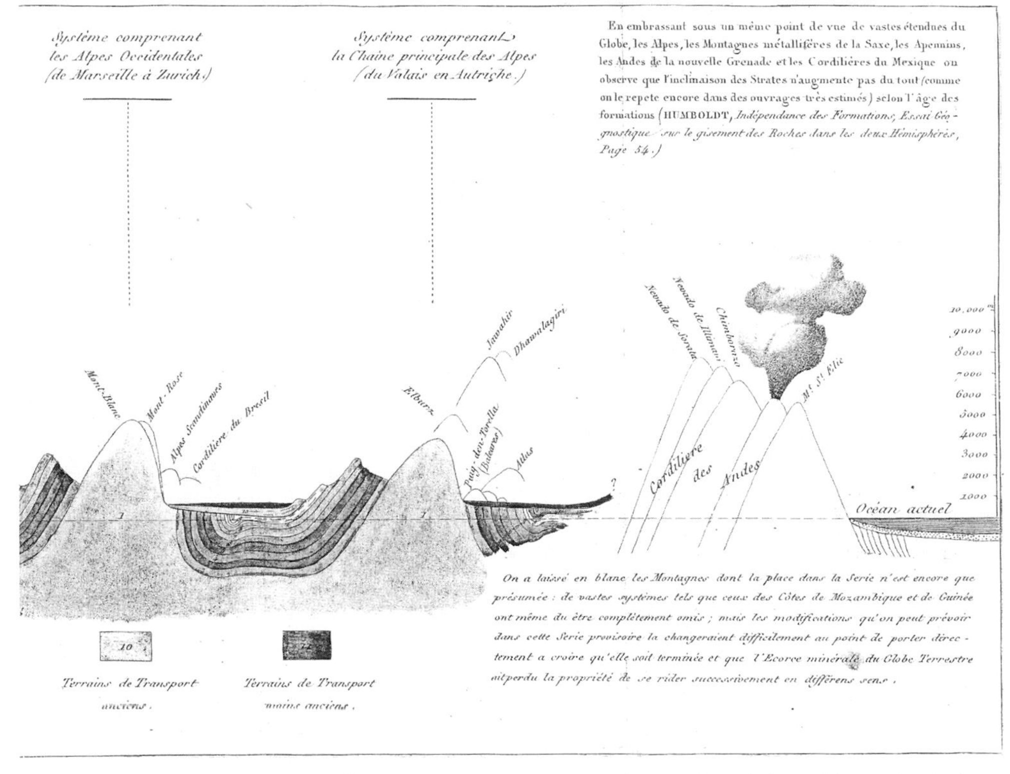

Fig. 23.1 One end of Élie de Beaumont's great theoretical section through the earth's crust (1830), showing his interpretation of the most recent phases of geohistory (see Fig. 9.3 for the preceding, Secondary and Tertiary, phases). After the elevation of the western Alps, including Mont Blanc (left), the older Superficial deposits (*Terrains de Transport anciens*, layer 10), were deposited, and then buckled by the recent elevation of the main Alpine chain (center, with the Himalaya, and other mountain ranges beyond Europe, possibly of the same date, added in outline). Their erosion in turn produced the younger Superficial deposits (*Terrains de Transport moins anciens*, layer 11). Élie de Beaumont suspected that the Andes (right, one of them shown as an active volcano) might have been elevated still more recently, perhaps even folding layer 11 (note the question mark). In any case he saw no reason to regard the earth's occasional "revolutions" as having ceased; the sediments now accumulating on the floor of existing oceans (far right) might be elevated in the future.

"diluvial gravel" and erratic blocks scattered so widely across northern Europe: "to talk of river action [to account for them] appears to me, in a case like this, little better than a mockery of my senses".[4]

In pursuing this theme, Sedgwick switched his attention from Lyell to Élie de Beaumont, whose recent synthesis of stratigraphy and tectonic geology had made a deep impression on geologists everywhere. The Frenchman had claimed that each of the successive periods of folding and mountain elevation had a distinctive orientation; and that they coincided with the major breaks and unconformities between successive sets of formations (§9.4). "The steps by which he reaches this noble generalization", Sedgwick maintained, "are so clear and convincing, as to be little short of physical demonstration"; he regretted that he had not yet been in possession of "this grand key to the mysteries of nature" during his recent fieldwork in the Alps. Reviewing Élie de Beaumont's results, Sedgwick noted specifically that he had distinguished two separate deposits of "diluvial gravel" and had attributed them to *two* distinct periods of mountain elevation and consequent diluvial action. This was of crucial significance, for it showed that no single episode of this kind could be regarded as unique (Fig. 23.1).[5]

Sedgwick pointed out that Élie de Beaumont's new research indicated not only a sequence of "comparatively short periods of violence and revolution", during which the earth's crust had buckled in successively different directions, but also intervening "long periods of comparative repose", during which the thick formations of sediment had accumulated slowly (see Fig. 9.3). Comparing this with Lyell's "directly opposed" views, Sedgwick had no hesitation in preferring Élie de Beaumont's, "because his conclusions are not based upon any *a priori* reasoning, but on the evidence of facts", that is, of observable features, some of which Sedgwick had seen for himself. However, he was not uncritical of the Frenchman's work; he suspected, for example, that the idea of precise parallelism in the trends followed by contemporaneous folding in different regions had been "pushed too far". Conversely, he conceded that "nineteen twentieths" of Lyell's work was unaffected by his criticism, and he hoped only that its sequel would be stripped of "that controversial character, by which, in my opinion, some pages of his present volume are disfigured".[6]

Toward the end of his long address, Sedgwick made his own contribution to damping the fires of controversy, by stating more firmly—backed now by Élie de Beaumont's authority—what he had hinted in his address the previous year, namely his conclusion that the diluvial deposits dated from at least two distinct epochs, not one. He felt no shame in admitting such a change of opinion, since the "ardent generalizations of an advancing science" should always be subject to revision and correction. In this case there had been an apparent match or "double testimony" between the diluvial gravels and "the record of a general deluge"

4. Sedgwick, "Address to the Geological Society" (1831), 302–6.

5. Fig. 23.1 is reproduced from Élie de Beaumont, "Révolutions du globe" (1829–30), part of pl. 3.

6. Sedgwick, "Address to the Geological Society" (1831), 307–12. There is no good reason to suppose that he dragged Élie de Beaumont into his argument simply as a stick with which to beat Lyell; on the contrary, the Frenchman's synthesis made much better sense of what Sedgwick had seen for himself in the field, and it is not surprising that he was impressed by its explanatory power.

in Genesis; so it had been reasonable to equate the two, and thereby forge a valuable link between geohistory and human history. But in Sedgwick's opinion the failure to find any human fossils in the so-called "diluvial" deposits ought to have made geologists—including himself—more cautious; and the discovery that the deposits were not all of the same age had now finally undermined the case for the equation. Yet conversely the recognition that geohistory had been punctuated by many *local* episodes of violence had "taken away all anterior incredibility from the fact of a local deluge". So the biblical story might still contain a kernel of historicity, even if the event had been far from universal and the evidence for it was not to be found in Europe.

In *scientific* terms, to abandon the alleged identity of "geological deluge" and biblical Flood therefore entailed no great change of opinion, however significant it might be in the peculiarly English public debate provoked by "scriptural" geology. In the somewhat jocular ethos of the Geological Society—as expressed, for example, in De la Beche's caricatures—Sedgwick therefore adopted the same wry metaphor that Lyell had used when he imagined how the "scriptural" writers wanted to see all the real geologists burnt at the stake. "Having been myself a believer," said Sedgwick, "and, to the best of my power, a propagator of what I now regard as a philosophical [i.e., scientific] heresy, I now think it right, as one of my last acts before I quit this Chair, thus publicly to read my recantation." His declaration made plain his defection from one distinctive element of the "diluvial" theory still maintained by Buckland and Conybeare; but of course he remained firmly on their side—and that of all the leading British geologists other than Lyell—in regarding geohistory as having been a long story of broadly directional change punctuated by occasional episodes of violent upheaval. As Sedgwick handed over the presidency to Murchison, his mock-heroic "recantation" on the specific point of the identity of deluge and Flood, not greatly emphasized in his lengthy address, hardly altered the balance of opinion at the Geological Society.[7]

23.2 LYELL'S CONTINENTAL RECEPTION

The leading English geologists whose views have now been summarized showed remarkable unanimity about Lyell's work. Although they ranged from his strong supporter Scrope to his dogged but perceptive critic Conybeare, with De la Beche and Sedgwick positioned in between and Whewell representing the wider community of savants, they all welcomed Lyell's inventory of actual causes, while reserving judgment about the adequacy of these agents to explain *all* the traces of events in the deep past. But they were highly critical of Lyell's attempt to explain away what they regarded as the clear evidence of a *directional* geohistory. And they all rejected his proposal to replace it with a steady-state or cyclic conception, which they regarded as being at the heart of his project and identified without hesitation as being inspired by Hutton's earlier geotheory.

Their counterparts on the Continent would soon have joined in this consensus, had it not been for an unfortunate coincidence in the wider world. Just when

Lyell's volume was published in London, Paris erupted in the brief violence of the "July Revolution", which toppled the unpopular and authoritarian Bourbon monarch Charles X and brought in a new and constitutional monarchy under the "citizen-king" Louis-Philippe. Prévost, who as Lyell's closest French ally had been sent proofs of *Principles* and had promised to translate it, was diverted into political activism. As Lyell found to his dismay on his way back from Spain less than three months later, the savant world in Paris was almost at a standstill and Prévost had achieved nothing. In the event, a French translation of *Principles* was not published for another decade (and then not by Prévost), though a German one began to appear in 1832, about as promptly as could be expected for such a major undertaking. Consequently the impact of Lyell's ideas was more gradual and more muted on the Continent than it was in the anglophone world.[8]

A happier outcome of the July Revolution was that the new monarch's patronage helped consolidate the infant Société Geologique de France. This had been founded in Paris by some of the younger French geologists—Boué and Prévost were the main driving force—some three months before the revolution. Their intention was certainly to emulate the Geological Society in London, but also to break what they perceived as a tight grip on the science by some of the older and more established Parisian savants; however, they did persuade Cordier and Brongniart to join them, the former as their first president. From the start, the Société adopted its sister body's recent innovation of allowing free discussion of the papers that had been read (§2.2). But in contrast, and as a mark of French *égalité*, the new Société had no distinct classes of membership, and foreigners were welcome to join on the same terms as Frenchmen; within the first year all the leading British geologists did so.[9]

The Société Géologique was a forum in which at least some notice was taken of Lyell's work. A copy was displayed at a meeting late in 1830, after the Parisians' summer break and not long after the book was published in London; but Sedgwick—the first of the English geologists to join and get its monthly *Bulletin*—must have been as disconcerted as Lyell to see that the very next gift recorded was Ure's notorious "scriptural" book (§20.3). But when Boué—the Société's most cosmopolitan and multilingual member (§3.3) and one of its secretaries—read

7. Sedgwick, "Address to the Geological Society" (1831), 312–13. Daubeny, "Diluvial theory" (1831), dated 15 January (a month before Sedgwick's address), was likewise prompted by Lyell's volume, and restated a moderate case for diluvialism with reference to the valleys of Auvergne. Some modern historians, fixated on the idea of "conflict" between geology and Genesis and unfamiliar with the ethos of the Society at this period, have taken Sedgwick's "recantation" much too literally.

8. Lyell to Marianne Lyell, 2 October 1830 [*LLJ* 1: 301–3]; Wilson, *Lyell* (1972), 301. Lyell, *Lehrbuch der Geologie* (1832–34), translated by Carl Friedrich Alexander Hartmann (1796–1863) from the second (1832–33) edition, but misleadingly termed a "textbook". This was corrected to "principles" in his translation of the sixth edition (1840): Lyell, *Grundsätze der Geologie* (1841–42). Lyell, *Principes de géologie* (1843–48), translated by Tullia Meulien, was also based on the sixth edition. See Vaccari, "Lyell's reception" (1998).

9. Lapparent, "Société Géologique" (1880), summarized its early history on the occasion of its semi-centenary; see also Laurent, "Société Géologique" (2007). Fox, "Scientific societies in France" (1980), sets it in its national context; see also Rudwick, *Devonian controversy* (1985), 27–30, and "Year in the life of Sedgwick" (1988).

a comprehensive "Report on the progress of geology" for 1830, emulating what Sedgwick had done with his "anniversary" addresses in London, he did not mention Lyell's volume at all.[10]

A year later, however, in the second of his annual surveys of new publications worldwide, Boué did briefly mention the *Principles* along with other "general treatises on geology". He suggested that time alone would tell whether Lyell would be able to meet the obvious objections to his theories; but like Lyell's other critics Boué welcomed the controversy it was sure to stir up, as likely to lead to progress in the science. Specifically, and again like other critics, Boué praised Lyell for "applying the philosophical movement [*marche*] from known to unknown, long since followed by many geologists on the Continent", which put Lyell's actualistic method firmly in its proper place. On the other hand, Lyell would have been pleased to read Boué's later paper to the Société on "The deluge, diluvium, and the older alluvial epoch"; for here he attacked—even more vehemently than Lyell—what he saw as the peculiarly English attempt to identify the biblical Flood with *any* of the Superficial deposits. Later still, Boué published what purported to be a review of the *Principles*, though most of it was in fact a translation of Conybeare's critique (§22.3). He noted Lyell's Huttonian claim that "nature would always have remained the same", and characterized his work as "a philosophical treatise on geogeny". Strictly speaking, "geogeny" was inappropriate, since Lyell explicitly denied that there was any evidence of the earth's *origin*; but Boué did place the *Principles* firmly where he thought it belonged, in the genre of geotheory. His brief comments were not out of line with the consistent criticisms made by the wide range of knowledgeable English geologists whose views have already been summarized. However, with only the first volume published, Lyell's "system" or geotheory was not yet fully revealed.[11]

23.3 THE GOAL OF TERTIARY GEOHISTORY

The first volume of Lyell's *Principles* was published while he himself was in Catalonia, chasing some more extinct volcanoes, though in the event his Spanish fieldwork had no impact on his later work comparable to his earlier tour through France and Italy. On his way back to London he was in Paris, as just mentioned, less than three months after the July Revolution. His disappointment at finding that Prévost had got nowhere with his promised translation was compensated by the satisfaction of spending much time with Deshayes and his huge shell collection. For Lyell's eyes were now firmly fixed on what he planned as the culmination of his second volume, namely his reconstruction of Tertiary geohistory by means of its changing molluskan faunas. Here he was dependent on Deshayes: Lyell's own qualitative impressions of the different faunas, however confident (§19.4), were no substitute for the quantitative "statistics" that an expert conchologist could provide. While working with him, Lyell realized just how impoverished Deshayes had become. So he arranged to pay him the substantial sum of £100 for three months' work on his behalf, to teach him conchology while he was still in Paris and to identify further specimens after he returned to London; he saved the Frenchman's gentlemanly face by implying that the payment came from his publisher Murray rather than himself. He was soon able to report to Scrope on what the arrangement was yielding: "The results are already wonderful as confirming the successive [i.e., non-simultaneous] formation of different [Tertiary] basins, and the gradual approximation of [i.e., to] the present order of things, and will settle, I hope for ever, the question whether species come in all at a batch, or are always going out and coming in".[12]

Lyell's comment to Scrope epitomized his intention to integrate geohistory with earth-physics, reconstruction with causal explanation. The sequential origin of the Tertiary basins was a matter of geohistory, of specific contingent events in the deep past. But the means by which he intended to plot these events was the ever-changing character of the fauna; and this in turn depended on a process that Lyell assumed was operating alike in past, present, and future (and which he therefore expressed in the present tense), namely the timing of the appearances and disappearances of individual species. To discover the unchanging ahistorical "law of nature" underlying this process of faunal change required—by his own criteria—a close study of the actual causes observably governing the origin, survival, and extinction of organisms in the present world. So here was material both for the deferred part of his inventory of actual causes, and for the geohistory that would follow from it and bring the whole work to its climax.

Back in London in time for the start of the Geological Society's new season, Lyell took advantage of his standing as an unmarried barrister—although he was no longer practicing—to move into an apartment in Gray's Inn, on the same staircase as his fellow lawyer Broderip. Since Broderip was also one of the best amateur conchologists in England, Lyell had a competent substitute for Deshayes on his doorstep. He then settled down to work on his second volume, while his first was receiving the shrewd but friendly criticisms that have already been summarized.

One topic on which Lyell sought expert advice of another kind concerned the terms in which he could describe Tertiary geohistory. He had of course read Whewell's review of his first volume with close attention, and noted the Cambridge savant's suggestion that the most recent Tertiary deposits might be called "Penultimate" or "Liminal" (on the *threshold* of the present). This prompted him to ask Whewell's advice on naming *all* the Tertiary formations around Europe in a way that would reflect his own conclusions about their relative ages. Adopting the usual retrospective order from the present back into the past, what came first—even before the youngest Tertiaries—were the alluvial deposits formed

10. Société Géologique, 22 November 1830 [*BSGF* 1: 45–46]; Boué, "Progrès de la géologie" (1831). The other secretary was Élie de Beaumont. The *Bulletin* was sent to the Geological Society from the start, ensuring that British geologists were kept aware of the French society's work (the donation of the first issue was recorded on 1 December 1830).

11. Boué, "Progrès en 1831" (1832), 182–84, read on 10 January, 7, 21 February 1831; "Le déluge, le diluvium" (1832), read on 23 December 1831; "Principes par M. Ch. Lyell" (1832), 317–19, written after the publication of Lyell's second volume (§24.1) but scarcely referring to it.

12. Lyell to Scrope, 8 October 1830; see also Lyell to Marianne Lyell, 2 October, and to Mantell, 10 October [*LLJ* 1: 301–7]; Wilson, *Lyell* (1972), 301–3. Lyell's payment to Deshayes for three months' work was equivalent to one-quarter of his father's annual allowance to him, or in other words to the whole of his own reliable basic income for the same three months: a striking indication of the importance he attached to Deshayes's work.

during the times of human history. He proposed to call these "Contemporary"; being consistent, he extended this geological period *into the future*, "to take in all since Adam & all that shall be till his posterity die out, whether roasted by the summer heat or frozen out by the winter of the annus magnus [§21.4]". Turning then to the true Tertiaries—dating from before the brief human period—he summarized the provisional results of his joint study with Deshayes by dividing them into four groups that he inferred to be of increasing age, with about 95% (or at least 65%), 30%, 1%, and 0% of living ("recent") species. He proposed using four Greek prefixes that would reflect these percentages: for example the second group, with a minority, might be called "Meiosynchronous" [*meion*, lesser]. Whewell, however, thought this "long, harsh and inappropriate"; he suggested in its place either "Meioneous" or "Meiotautic", before adding in a postscript that "Miocene" [*kainos*, recent] would be "shortest and best". Fortunately for geologists ever since, Lyell adopted Whewell's afterthought, together with "Pliocene" [*pleios*, full] for the group with a large majority of extant species and "Eocene" [*eos*, dawn] for those with only a tiny minority. Still older formations, with no extant species at all, would then be "Acene" (in place of Lyell's "asynchronous").[13]

This nomenclature would give clarity and consistency to Lyell's proposed outline of Tertiary geohistory; it would also give his geology a respectably Classical aura that would appeal both to academics and to the educated public, while the names themselves would reflect its endorsement by a highly respected and versatile savant. However, the "statistical" basis for the successive periods might be difficult for his readers to grasp, without some more concrete illustration of the corresponding molluskan faunas. So Lyell asked Deshayes to provide him with pictures of a few fossil shells "characteristic" of the successive Tertiary periods—selected from the many hundreds of species on which in fact they were based—which he could use in his forthcoming volume (see Fig. 25.4).[14]

23.4 AN ACTUAL CAUSE IN ACTION

However, neither the rest of geology, nor indeed the earth itself, was standing still and holding its breath while Lyell completed his work. Since the present narrative might suggest otherwise, an example is worth mentioning at this point. During the summer of 1831, one of the inorganic actual causes that Lyell had reviewed in his first volume sprang spectacularly into action, catching the attention of the educated public throughout Europe. Although too late to be included in Lyell's inventory, it was for him a welcome public demonstration of the continuing dynamism of the terrestrial world.

In July 1831, at a point some 40km off the south coast of Sicily and far from any active volcano, boiling seas were reported and then a series of violent explosions. Soon a cone of volcanic ash emerged above sea level, growing to about 100 feet (30m) in height as more material was spewed thousands of feet into the air. British naval ships operating out of Malta soon fixed the position of this new navigational hazard, and it was duly named Graham Island after the First Lord of the Admiralty in London; others, however, named it in honor of Ferdinand, the King of the Two Sicilies (Fig. 23.2).[15]

Fig. 23.2 The new submarine volcano off the coast of Sicily, erupting violently on 6 August 1831, with a column of steam rising more than 1500 feet; by this time it had already ejected enough loose material to form a cratered cone rising about 100 feet above sea level. This lithograph, based on a drawing made on the spot, was published by a British officer who, being in the service of the Bourbon king of the Two Sicilies, called it "Ferdinandea", though he tactfully dedicated his account to William IV of Britain. (By permission of the British Library)

The new volcano was studied more scientifically by Carlo Gemmellaro, one of the brothers whose hospitality Lyell had enjoyed on Etna (§19.4), and by the Prussian geologist Friedrich Hoffmann (1797–1836) of Halle, who happened to be working in Sicily at the time. One of the British naval parties sent specimens to London by one of the fast new steam-packets; John Barrow (1764–1848), the Arctic explorer who was now secretary to the Admiralty, forwarded them at once to Lyell, who thus saw them only "eight or nine days after they had been thrown up hissing hot". The geologist immediately brought the new volcano into the service of his wider objectives. He displayed astounding self-confidence by assum-

13. Lyell to Whewell, 21 January, 17, 20 February 1831 [Cambridge-TC: a.208/108–10]; Whewell to Lyell, 31 January 1831 [Philadelphia-APS]; extracts are printed in Wilson, *Lyell* (1972), 305–7. Deshayes had later reported revised percentages (see below).

14. It is significant that Deshayes—left to himself, as it were—published at the same time a general work on "Shells characteristic of formations" which treated them in a simple Smithian manner with no trace of Lyell's "statistical" approach: *Coquilles caractéristiques* (1831).

15. Fig. 23.2 is reproduced from Smythe, *Late volcanic island* (1832) [London-BL: 10163.d.10], pl. opp. 6; in a similar scene (pl. opp. 2) the column of steam was said to be 1500–1800 feet (about 500m) high; other plates show quiescent moments, revealing a cratered cone just like those on the flanks of Etna and in Auvergne.

ing quasi-Messianic status—however jokingly—and recruiting Mantell under his own crusading banner:

> I congratulate you, one of the first of my twelve apostles, at Nature having in so come-at-able [i.e., accessible] a part of the Mediterranean thus testified her approbation of the advocates of modern causes. Was the cross which Constantine saw in the heavens a more clear indication of the approaching conversion of a wavering world?[16]

The Académie in Paris sent Prévost, on board a French naval vessel, to report on the phenomenon. But he was too late to see the short-lived volcano in action: the flag that his compatriots planted on the summit, claiming "Île Julia" for the French (their name referred to the month of its eruption), was somewhat misplaced, for by then the sea was rapidly washing away the loose volcanic ash (there having been no eruption of lava, which would have been much more resistant). By the end of the year the volcanic island, under whatever name and flag, was no more. Still, its brief existence provided geologists such as Lyell with further evidence that volcanoes are formed, from the start, solely by the accumulation of newly ejected material, and not—as von Buch claimed—first by a local upheaval of the earth's crust (§8.3). On this issue Prévost switched from von Buch's side to Lyell's, after studying Etna and Vesuvius on his way back to Paris. By that time the new volcano had also been discussed in this context at the Geological Society, after Horner paraphrased Hoffmann's German report. In sum, the three months' wonder of a totally new volcano reminded the whole scientific world, and indeed the reading public, that the earth could still spring a spectacular surprise, however temporarily (Fig. 23.3).[17]

Fig. 23.3 "The Île Julia on 29 September 1831 at 2 p.m.": a lithograph after a drawing made on the spot, to show the new volcano at the moment when a party from the naval brig *La Flèche* planted the French tricolor on its summit and the geologist Prévost landed to make scientific observations and collect specimens (small figures can just be seen on the island). By this time the volcano was no longer active, and the cone of loose volcanic ash was rapidly being washed away. This print was dedicated to the Académie des Sciences in Paris, which had sent Prévost there.

23.5 "BISHOPS AND ENLIGHTENED SAINTS"

Meanwhile, Lyell had been hoping to increase the impact of his work, not only among his fellow geologists but also on the general educated public, by seeking a position that would yield some income and raise his profile in scientific London. Whewell's favorable review in the *British Critic* had helped Lyell to gain the approval of other intellectual Anglicans—the "bishops and enlightened saints", as he had called them—from whom, as he jokingly told Whewell, he might have become estranged over the past years "by falling into the society of Lawyers, Geologists and other sinners". So he asked Sedgwick to sound out Charles Blomfield (1786–1857), the scholarly bishop of London, about proposing himself for a chair in geology at the new King's College in London: this was being planned as a moderate Tory and liberal Anglican counterweight to the predominantly Whig and ostentatiously non-denominational London University (now University College London). Blomfield and the rest of the governing Council of King's were, with one exception, enthusiastic; as Lyell told Whewell, "they were disposed to allow the utmost latitude to a geologist provided he came by his theories as straightforward deductions from facts & not warped expressly to upset scripture". The proviso was not unreasonable, since geology had notoriously been exploited in that way within living memory (*BLT* §6.4) and threatened to be so again; and the Council felt responsible for ensuring that their new institution was not misused as a base for anti-religious propaganda. Only Buckland's former colleague Copleston (§6.4), who was now the dean of St. Paul's cathedral in London (and also bishop of Llandaff in Wales), was suspicious about the kind of geology that Lyell represented, but after Conybeare reassured him he too fell into line. Like many of the educated lay people who were reading the *Principles*, these churchmen found some of Lyell's conclusions novel and even "startling", but they were satisfied that he had reached them "in a straightforward manner" and not from any anti-religious motive. So Lyell joined several other London savants who, as part-time professors, were to give courses on the various natural sciences when the College opened later in the year.[18]

16. Lyell to Mantell, 30 August 1831 [Wellington-ATL; *LLJ* 1: 329–30]. Gemmellaro (C.), "Fenomeni del nuovo vulcano" (1834), read at Catania on 18 August 1831; Hoffmann, "Vulcanische Eiland" (1832), called it "Isola Ferdinandea" but listed in his title no fewer than five other names that it had been given. Lyell first analyzed it substantially in *Principles*, 3rd ed. 2 (1834) 145–50. Dean, "Graham Island" (1980), describes its history in detail but reproduces almost none of its rich iconography.

17. Fig. 23.3 is reproduced from Prévost's print *L'Île Julia* (1831), based on a drawing by Edmond Joinville; similar but smaller (and less accomplished) scenes are in Prévost's preliminary report, "Nouvel islot volcanique" (1831), pl. 4, and his full paper, "Notes sur l'île Julia" (1835), pl. 5 (or "A"); Gohau, "Constant Prévost" (1995). See also Horner, "New volcanic island" (1831), read 30 November; the minutes [London-GS: OM/1] confirm that it summarized Hoffmann, not von Buch; Lyell to Mary Horner, 2 December 1831 [*LLJ* 1: 354–55] describes the meeting. There has since been some volcanic activity on the same spot, in 1863 and 1999, but no new island.

18. The eternalistic geotheory of the notoriously anti-religious George Hoggart Toulmin (1754–1817), first published around the same time as Hutton's (*BLT* §3.4), had been republished (in 1824–25) by the equally notorious radical republican and atheist Richard Carlile (1790–1843), so the churchmen's anxiety about the possible misuse of geology was far from paranoid: see Porter, "Politics of a geologist" (1978).

However, despite Conybeare's efforts on Lyell's behalf, Copleston then mounted a rearguard action to try to block the appointment, unless Lyell could convince him that his lectures would not undermine traditional doctrine on two specific points. The first was that of "a positive act of creation, particularly of the creation of man, [at] about the period usually assigned to it". Lyell replied that he knew of no "physical" (i.e., geological) evidence to doubt the traditional dating of human origins derived from textual evidence (*BLT* §5.4); the answer was probably honest, for he either discounted or was still unaware of current French claims for a possibly deeper human antiquity (Chaps. 16, 29). On the mode of origin of *any* species (including the human), Lyell told Copleston that in his second volume he would attack the "leading hypothesis which has been started [i.e., proposed] to dispense with the direct intervention of the First Cause"; the volume would soon show that he meant Lamarck's transformism. This reassured Copleston that Lyell was not a closet Lamarckian, and implied that he was following Whewell in anticipating that the origin of species might turn out be an exception to the purely natural or "secondary" causes invoked for the rest of the terrestrial world. On this point Copleston then expressed himself satisfied.

Copleston's second worry concerned Lyell's attitude to the traditional idea of "a universal deluge" subsequent to the creation of the human species. He was of course familiar with Buckland's "diluvial" research, having reviewed it for the *Quarterly* while he was still at Oxford (§6.4), and he must have known it was now being strongly criticized by other geologists. Lyell replied, again with honesty, that he, along with other "eminent foreign and English geologists", did now have strong evidence for rejecting Buckland's idea that "a deluge has passed over *the whole earth* within the last 4000 years", but that he knew of "no physical evidence" that such an event might not have affected "the whole inhabited [*sic*] earth". On that crucial verbal difference he played Copleston at his own game by repeating the latter phrase in Greek [πασῃ ἡ γῃ οικουμενῃ], thereby hinting that Copleston was being ignorantly literalistic in his construal of "universal". Here Lyell was following what had long been suggested by biblical scholars and other savants, and most recently by Sedgwick at the Geological Society: if, as was widely assumed, early humans had been confined to a limited geographical region (probably Mesopotamia), they might indeed have been victims of some kind of *local* watery catastrophe, of which the story of Noah's Flood might be a faint but genuine record. On this point Copleston still expressed reservations about Lyell's response; but he had to fall back on a traditional appeal to "the language of Scripture as understood and interpreted by the Church", which was in tacit contrast to the interpretation indicated by contemporary scholarship. He saved face by expressing his pastoral concern at doctrinal changes that "perhaps would put in jeopardy the faith of many"—the "weaker brethren" argument often resorted to, then as now, by religious conservatives in a tight corner—but in the end he withdrew his objections and Lyell's appointment was confirmed.[19]

This episode has been worth recalling because it indicates the sensitive boundaries that had to be negotiated in England—in contrast to Continental Europe, and even Scotland—as Lyell sought to address his *Principles* to the educated public as well as to his fellow geologists. But the outcome reflected the fact that the "enlightened saints" were a powerful party within the Church of England, and Copleston an exception among its bishops and cathedral deans (some of whom were members of the Geological Society). Lyell evidently had no need to expect problems at King's, provided he showed gentlemanly tact and refrained from being triumphalistic about the adjustments to traditional biblical interpretation that the progress of geology might require.

23.6 CONCLUSION

Among Lyell's fellow geologists—who were more familiar with his evidence than most of the "enlightened saints"—perceptive critics such as Scrope and Conybeare were soon followed by others, such as the heavyweight polymath Whewell and, for the second time, Sedgwick. They all continued to praise Lyell for his analysis of actual causes—a topic that Whewell elevated in scientific status by calling it "geological dynamics"—while criticizing him for claiming that the application of such known processes to the deep past necessarily led to a Huttonian steady-state geohistory. They claimed that Lyell, in arguing for the "uniformity" of nature, had confused the highly complex processes of geological agency with the basic physico-chemical "laws of nature" on which they were founded. The latter, they agreed, must indeed be assumed to be stable from and to eternity; but the former might have varied greatly in power and intensity in the course of geohistory, and only a close empirical study of the surviving traces of their action could or should settle the question one way or the other. Only the thorny problem of the origin of species, and particularly of the human species, might turn out to lie beyond the bounds of natural agency.

All Lyell's critics agreed in rejecting the claims of "scriptural geology", and often did so with great vehemence; Boué in Paris, like Lyell himself, correctly identified it as a peculiarly English aberration. Sedgwick deserted the other diluvialists when he explicitly abandoned the claimed identity of "geological deluge" and biblical Flood, but in savant circles his jokey "recantation" caused only a minor ripple. He adopted Élie de Beaumont's distinction of at least two separate "diluvial" episodes in the geologically recent past, and inferred that neither was recent enough to have overlapped with recorded human history. Whether Lyell would persuade other geologists to abandon diluvial episodes altogether was, however, quite another matter.

Lyell's *Principles* would probably have been evaluated in much the same way by geologists beyond the anglophone world, had its publication not coincided with the July Revolution in France, which aborted Prévost's plans to translate it. But Boué did later bring it to the attention of the fledgling Société Géologique, a Parisian body that was modeled on the Geological Society and soon counted most of the leading British geologists among its members. Meanwhile Lyell

19. Lyell to Mantell, 16 March 1831 [Wellington-ATL; *LLJ* 1: 316–17]; and other letters [Kirriemuir-KH] quoted in Wilson, *Lyell* (1972), 308–13. Wilson's account of this episode, shaped perhaps more by encounters with American fundamentalism than with the Church of England, should be compared with Rudwick, "Lyell and his lectures" (1975), 231–38, which exonerates Lyell from the charge of dishonesty and Copleston from that of hypocrisy.

himself was working hard on his second volume. He intended it to culminate in his reconstruction of Tertiary geohistory, calibrated by means of its steadily changing molluskan fauna; for this in turn he was increasingly dependent on the expertise of Deshayes in Paris, while adopting Whewell's suggested names (e.g., "Pliocene") for its successive periods. And Lyell's confidence in the undiminished power of actual causes was neatly reinforced, at just the right moment, by the sudden and sensational eruption of a totally new volcano off the coast of Sicily, which caught the attention of the whole scientific world and indeed the general public.

Around this time, King's College in London, a newly founded Anglican institution, was recruiting savants to teach the natural sciences. Lyell hoped to have himself appointed to a part-time chair in geology, which might raise his scientific profile in London and even generate some income. After suitable lobbying by Conybeare and other geologists, most of the "bishops and enlightened saints" (as Lyell dubbed them) on the college's Council were keen that it should have the services of such a prominent up-and-coming geologist. Only Copleston, Buckland's former colleague, demurred. He asked Lyell to confirm that the geology he would expound in his lectures would not contradict traditional scriptural interpretation on two points, the creation of Man and the universality of the Flood. Lyell tackled a tricky situation with punctilious tact—even trumping the bishop on a point of scholarly exegesis—and eventually convinced him that he would not misuse his position for anti-religious propaganda. The episode illustrates the still delicate nature of the relation between geology and the wider English public; within the scientific world, on the other hand, the status of geology had long been secure, even in England. However, Lyell's ambitious plan to "reform" his science with a comprehensive new geotheory was still only partly in the public realm; the next chapter describes the second volume of his *Principles* and its reception.

CHAPTER 24

The uniformity of life (1831–32)

24.1 THE SECOND VOLUME OF LYELL'S *PRINCIPLES*

Lyell's role as a reviewer for the *Quarterly*, had given him prestige in the eyes of the English social elites. The status of King's College, in contrast, was as yet uncertain; but since his lectures there would in practice be open to all, they could at least help give his ideas valuable publicity among the same elites. He cannot have expected the fees to yield much income, but a short annual course would not cost him much time to prepare; his friends persuaded him that it would help him organize his ideas, and perhaps act as a dry run for the short introductory book on geology that he hoped to write later. Anyway he could postpone writing these lectures, since the College was not due to open until the autumn and the course could be given still later. Much more urgent was the speedy completion and publication of the second and final volume of his *Principles of Geology*, while the first was still in the public eye and still being talked about in British households of the higher social classes, and of course at the Geological Society.

Lyell made steady progress with his work, despite two contrasting distractions. He was becoming closely attached to Mary, one of the well-educated daughters of Leonard Horner (1785–1864), the Warden (i.e., head) of London University and previously a prominent Whig in Edinburgh; Horner was also a competent geologist who, years earlier, had made Brocchi's important work on the Italian Tertiaries known in Britain through an essay in the *Edinburgh Review* (*BLT* §9.4). But he had recently resigned, his health undermined by acrimonious politicking at his infant institution, and he had moved with his family to Bonn to recuperate. Mary's unexpected departure from London caused Lyell to make an uncharacteristically impulsive decision: he postponed his plan to spend the summer in Scotland, and crossed to Germany, ostensibly to add the extinct volcanoes of the

Eifel region to the tally of those he had seen for himself. But he also visited the Horners not far away in Bonn, where Mary accepted his proposal of marriage. (The new paddle-steamers made travel across the North Sea and right up the Rhine much easier than it had been just a few years earlier.)

The other and less welcome distraction was political. These were turbulent times in Britain, and therefore in the political circles in which all the leading geologists moved. Serious social unrest, and even riots, had raised realistic fears of a revolution in England to match the recent one in France and others elsewhere on the Continent (the one in the Netherlands had just split that country in two and produced the new kingdom of Belgium in the southern half). The Tories had been replaced in government by Whigs committed to a moderate reform of the electoral system. Just as Lyell's storm with Copleston was brewing in a teacup at King's, the Whigs' Reform bill was raising a much larger storm in the House of Commons, and the bill was only passed after another general election had increased the Whig majority. Since it had yet to pass the House of Lords before the reforms could come into effect, any by-election was seen as an important indicator of opinion among the limited electorate. Unhappily for Lyell, while he was at his family home in Scotland and busy writing, a by-election arose in the local constituency, in which his staunchly Tory father was one of the very few electors under the unreformed franchise. Lyell himself, a discreet Whig despite his links with Tories such as Whewell and his publisher Murray, was therefore put in an awkward position, though in the event he managed to avoid overt engagement with either side (the Tory candidate won). Trying to create for himself a career as a "*professional*" savant—who might be recognized as having serious commitments analogous to those of the established professions of medicine, the law, and the Church—Lyell's priority was to keep himself disengaged from the political world in order to give undivided attention to his scientific work.[1]

Despite these distractions, Lyell made steady progress with his second volume. Towards the end of 1831 all his chapters on actual causes in the organic world were complete and set in type, whereas those on Tertiary geohistory still needed more work. Lyell was anxious that a review of his new volume should appear promptly in the *Quarterly*, like Scrope's the previous year; so after consulting Lockhart, the editor, he asked Whewell to write a review in advance of publication. He knew that the Cambridge savant would give the new volume a fair but critical evaluation; like his review of the first, it would carry great weight among both the geologists and the wider reading public. Whewell agreed, liked what he read in the proofs he was sent, and began writing. Lyell then made another impulsive decision: he asked Murray to publish these three hundred pages as they stood, deferring his geohistory for what would have to be a third volume. At the cost of failing to offer the complete "system" he had promised, this would at least complete his published inventory of actual causes without further delay, and also keep his project in the public eye.

Lyell's truncated second volume was published in the first days of 1832. He proudly mentioned his new appointment at King's on the title page, and he found another quotation from Playfair to act as a suitable epigraph, though it was a platitude that did not need any Huttonian endorsement: "a change in the animal kingdom seems to be part of the order of nature, and is visible in instances to which human power cannot have extended". The volume's planned dedication to Murchison was deferred, along with the Tertiary research to which it would have alluded; he was replaced by Broderip, in recognition of assistance with the natural history that was now the main subject matter of the new volume. On the other hand, the planned frontispiece view of Etna could not be replaced at such short notice; it was left high and dry with no relevance to what the volume now contained, though it might have reminded readers to buy the sequel (see Fig. 25.2). At least it made the book look more attractive, since otherwise there were very few illustrations to relieve its dense text.

24.2 THE BIRTHS AND DEATHS OF SPECIES

Lyell's survey of those actual causes that involved living organisms followed the same pattern as his earlier survey of aqueous and igneous causes; it was an inventory designed to display the full range and power of the processes "now in operation", in order to deploy them later as resources for reconstructing and explaining the deep past, the final objective that had now been postponed. Lyell's second volume was not a digression into zoology and botany for their own sakes, let alone into "biology" (Lamarck's neologism was as yet rarely used). For Lyell, the sciences of living nature were strictly subservient to his *geological* objectives: in his preface he stated clearly that his second volume would "be found absolutely essential to the understanding of the theories hereafter to be proposed" in the third. Although he did not set out his plan at this point, he intended to chart the course of geohistory—or at least the Tertiary era as the nearest part to the present—by dating the various formations by means of their molluskan faunas, plotting their gradual approach to the faunas of present seas (§20.2). This dating method, combined with a Scropean emphasis on the vastness of "Time!" (§15.1), would, he hoped, enable him to show that there was no need to invoke any causes more sudden or violent than those known in human history, let alone other *kinds* of cause, in order to account for everything that had left any trace from the deep past. Only after showing that the earth had remained much the same kind of place as it now is, throughout the Tertiary era, could this "absolute uniformity" be extrapolated plausibly to still earlier parts of geohistory—or, conversely, be projected into the distant future—and thereby demonstrate that the "system" of the earth showed no sign of either beginning or end.

Lyell's dating method depended for its efficacy on his conviction that faunas had changed *gradually*—literally so, *per gradum* or step by step—in the course of time, by the piecemeal appearances and disappearances of individual species, rather than suddenly and all at once at particular moments, for example by mass extinctions followed by mass replacements of whole faunas. The Smithian method of using "characteristic" fossils in stratigraphy (§3.1; *BLT* §8.2) highlighted the

1. Wilson, *Lyell* (1972), 314–26, gives an account of both distractions; Morrell, "London institutions" (1976), is an insightful analysis of Lyell's attempts to become a "professional", and Porter, "Charles Lyell" (1982), of his management of his "public and private faces".

distinctiveness of successive Secondary formations, which in practice tended to divert attention from the many species that straddled the boundaries between them; so it had often been taken to support the idea of occasional episodes of sudden change. In contrast, the more recent research on the Tertiary formations and their fossils had focused on whole assemblages of fossils; it had highlighted the presence of the same species in many formations and particularly the gradual increase in the proportion of species still known alive (§12.1). Brocchi had offered an effective way of thinking about this kind of gradual faunal change when he suggested that species, like individuals, had "births" and "deaths", and determinate lifespans in between (*BLT* §9.4). Lyell had privately adopted Brocchi's model while he was on his Grand Tour (§19.1), but to make it serve his public purposes he now had to establish three points about the present world. These were, first, that species themselves were truly natural units, no less distinct and real than individuals; second, that actual causes could account for the piecemeal "deaths" of species, without the need to invoke any Cuvierian mass extinctions; and third, that the "births" of species could at least be shown to have occurred—by whatever means—in the same piecemeal fashion.

Lyell began by arguing that the big questions about the history of living things as part of the "system" of the earth—for example, whether it showed a progressive trend or "absolute uniformity"—needed to be tackled at the level of species. Just as the big questions about the origins of mountains or valleys were, in Lyell's view, best resolved by the summation of the small-scale observable effects of crustal movement or fluvial erosion, so likewise the big questions about the history of life had to be approached by the summation of events involving individual species. Lyell's first task was therefore to establish the reality of species as discrete and stable units in the living world. However, an awkward obstacle lay in his way, namely the covert presence of a theory that rejected that fundamental point. The reality and stability of individual species were taken for granted by most naturalists, not out of respect for the biblical story of Creation but simply because the ordinary practice of natural history—identifying, naming, and classifying—seemed to depend on it (*BLT* §2.1). This was why Cuvier, for example, had regarded Lamarck's transformist theory as incompatible with good science (*BLT* §7.4). Lyell too had to confront this threat: he had scientific reasons for rejecting Lamarck, in addition to his worries that the Frenchman's "theory of the transformation of the Orang Outang into the human species" (as Lyell put it in his synopsis) might threaten the "dignity of man" (§17.2). Lyell had to take the Lamarckian bull by the horns.[2]

Lyell recognized in fact that transformist theory, if true, would be even more unsettling for geologists than for zoologists and botanists. The latter treated organisms on the single time-plane of the present world, in which species could at least be treated operationally *as if* they were truly distinct. But geologists had incorporated the dimension of time and geohistory into their practice, and for them the problem of the long-term stability or otherwise of species could not be evaded. Lyell conceded that Lamarck's theory had the advantage, at least in the eyes of some, of avoiding the need for "the repeated intervention of a First Cause": with transmutation, divine Creation could be conceived as a primal act, rather than as punctuating geohistory at intervals. Lamarck's invocation of virtually infinite time (*BLT* §7.4) was also attractive, but Lyell was compelled by his own actualistic principles to oppose him here. Like Cuvier (*BLT* §9.3), Lyell concluded that since no transmutation could be seen to have occurred during the period of human records, there was no warrant for inferring that it would have been effective over the far longer periods of geohistory. Transmutation did not qualify as an actual or "true" cause [*vera causa*]. Lyell drew on other naturalists for his claim that variability within a species (in modern terms, intraspecific variation) never went beyond quite narrow limits, at least under natural conditions; so here too the evidence told against transformist theory. He therefore concluded that "species have a real existence in nature", each being adapted unchangeably to a particular mode of life.[3]

Lyell next sought to make sense of the patterns of distribution of organisms (in modern terms, their biogeography) and to explore the actual causes that might be responsible for them. Here again he drew on the work of other naturalists, notably Candolle; but Lyell gave biogeography an added temporal or *geohistorical* dimension, following the way he had been speculating privately even before he visited the Genevan (§20.1). For example he reviewed the means by which species can be dispersed to new regions or, conversely, prevented from spreading, pointing out that even rare or improbable kinds of events might become likely in the long spans of geohistory.[4]

This led to the wider problem of changes in the specific composition of faunas and floras, or the processes responsible for the introduction and elimination of species. Lyell reviewed the various theories about the circumstances in which new species seemed to "come into existence" (this and similarly vague phrases served to shelve, for the time being, the knotty problem of just how they had originated); he sided with those who inferred that each had had its origin at a single point in space and time, where conditions were—providentially—suitable for it to exist and survive. He implied that these points were scattered "uniformly" in space and time, rejecting the idea that there were localized "*centres* or *foci* of creation [at which] the creative energy has been in greater action than in others". Like all his contemporaries, Lyell used the language of "creation" routinely and even casually; it left conveniently ambiguous the question whether he believed it denoted an immediate "intervention of a First Cause" or the operation of an ordinary "secondary cause" of some kind. Writing to Whewell before the volume was published, he referred to his own ideas as "the successive creation theory".[5]

2. Lyell, *Principles* 2 (1832), 1–17. This and the following paragraphs summarize the analysis in Rudwick, "Introduction" [to facsimile reprint of *Principles*] (1990), xxix–xxxv; the interpretation tries to do justice to Lyell's own strategy of argument, rather than the uses to which it was put by later naturalists, notably Charles Darwin.

3. Lyell, *Principles* 2 (1832), 18–65; this and later references are to whole chapters; as in Lyell's first volume, the detailed synopses given under "Contents" (vii–xii) make it easy to locate specific topics and quoted phrases. On variation and other topics in natural history, one of Lyell's most frequently cited sources was Prichard, *Physical history of mankind* (1826). See Coleman's classic article, "Lyell and the 'reality' of species" (1962); also Corsi, "French transformist ideas" (1978) and "Before Darwin" (2005).

4. Lyell, *Principles* 2 (1832), 66–122.

5. Lyell, *Principles* 2 (1832), 123–75; Lyell to Whewell, 26 November, 5 December 1831 [Cambridge-TC: a.208/113, 115].

Lyell's next step was to apply to living organisms the picture of ever-changing environments that he had presented in his first volume. The ceaseless flux of physical geography and climate would always tend to cause gradual changes in ecosystems (to use the modern term), for example by eliminating the conditions on which a particular species depended for its survival, or creating new conditions favorable to others. Small and gradual physical changes might even sometimes precipitate sudden and drastic effects, as when a new link between separate land areas allowed a predator to spread to regions where other species had no defence against it. Lyell described many cases recorded in human history, to support the plausibility of such scenarios. He therefore concluded that extinction, generally piecemeal but occasionally on a larger scale, must be as pervasive as the processes of geographical and environmental change by which it was caused: "amidst the vicissitudes of the earth's surface, species cannot be immortal, but must perish one after the other, like the individuals which compose them".

While such arguments retained much of Brocchi's fruitful analogy between species and individuals, Lyell now abandoned the Italian naturalist's speculation that species might have intrinsically limited lifespans and become extinct by something analogous to old age. Lyell's *ecological* concept of extinction owed more to Fleming's analysis of the impact of human activities on animal distributions (§14.3), and indeed to Cuvier's notion that only a sudden change of environment could wipe out a well-adapted species. Lyell pointed out that Lamarck's theory offered no escape from the reality and omnipresence of extinction, because in ever-changing physical circumstances a previously well-adapted species would be eliminated by competition from others better adapted to the new conditions, long before it could be transmuted by the imperceptibly slow process that Lamarck had postulated.[6]

If the successive extinction of species was thus "part of the constant and regular course of nature"—even more so than in Cuvier's concept of occasional "catastrophes" of mass extinction—then it was necessary for Lyell to consider next "whether there are any means provided for the repair of those losses". As with the maintenance of a uniform ratio of land and sea by means of antagonistic physical processes (§21.3), so too for the living world Lyell assumed that there must be a similar balance of forces. If he was to demonstrate that the "system" as a whole was in a steady state, he had to show that the origin of new species was as constant and regular as their extinction. Yet he was on difficult ground at this point, because he could not produce any positive evidence for the origin of species as an observable actual cause, and he had to explain why "so astonishing a phenomenon can escape the attention of naturalists". This lack of observed cases of the formation of new species had to be explained *away*, by inferring that such events, like extinction, must be widely scattered in space and time—and in most cases hidden in the depths of the sea—and that it was intrinsically unlikely that a single instance would have happened under the scrutiny of naturalists within the span of modern human history. So the principle of actualism on which Lyell's whole work was based had to be inverted at this point: the present had to be interpreted in the light of the past, namely the geological evidence that new species *had* somehow appeared in the course of geohistory.[7]

In any case, Lyell's text implied that the origin of species, like their extinction, was a relatively sudden kind of event, at least on a geological timescale. In contrast to Lamarck's notion of continuous transmutation, in which the boundaries between species were in the long run illusory and none became truly extinct (unless by human agency), Lyell conceived species as real natural units. He assumed that each remained stable in form and habits throughout its lifespan, between an enigmatic "birth" at a providentially favorable point in time and space, and a more comprehensible "death" as a victim of local environmental change. While allowing his readers to assume—if they so wished—that he attributed the origin of species to direct divine action, he could equally well be read as believing that some quite ordinary actual cause might be responsible, although the culprit had yet to be identified. Significantly, however, he omitted to mention—even to refute it—the one culprit currently under discussion among naturalists, namely Geoffroy's speculations about an analogy with the production of monstrosities, which did suggest a possible natural mechanism for the origin of species, and one that might operate quite swiftly (§17.1). Lyell certainly knew of Geoffroy's work, but he referred to it only in conjunction with Lamarck, as if the two had proposed identical theories. Geoffroy's version of transformism, however attractive it might have been to Lyell in other circumstances, was too heavily tarred with a Lamarckian brush to be considered at all.[8]

In effect, therefore, Lyell completed the transformation of a traditional *mystery* (in the original and proper sense of the word) into a mere *problem*. The "creation" of species was no longer an utterly transcendent primal act at the origin of the cosmos, or even a rare and special intervention in the wake of an episode of mass extinction, but an event as ordinary and common as the decline and final dying out of species under the pressures of environmental change. Lyell's ambiguity about the status of the events by which new species "come into existence" was certainly convenient for his broader strategy for making his concept of geology attractive to the wider public, for it deflected attention away from Lamarckian transformism and dissociated himself from that socially and politically suspect theory. But the origin of species, however fascinating a problem in its own right, was ultimately irrelevant to Lyell's project. It was of course frustrating for him not to be able to include it in his otherwise comprehensive inventory of actual causes that were well attested and more or less fully understood. But what mattered to him much more was that he had established—at least to his own satisfaction—that it was a category of event that mirrored extinction in being scattered in piecemeal fashion across space and time. With species as real units that came and went at frequent intervals, each remaining stable in form between the

6. Lyell, *Principles* 2 (1832), 141–75.

7. Lyell, *Principles* 2 (1832), 176–84.

8. Lyell, *Principles* 2 (1832), 2, mentions Geoffroy by name and quotes his notion of "uninterrupted succession" (§17.1), but—uncharacteristically—fails to give the customary footnote reference to any of his papers; Corsi, "French transformist ideas" (1978), 231, suggests that he was paraphrasing a French review of one of them, but Lyell must surely have known of them more directly from his thorough reading of the *Mémoires du Muséum*. Lyell's concept of species has some similarity to the theory of "punctuated equilibrium" within a modern Darwinian framework, which makes as good sense to many modern paleontologists as Lyell's did to some of their intellectual forebears.

limits of "birth" and "death", the hundreds of molluskan species that Deshayes was busy identifying for him could be relied upon to give Lyell a "statistical" basis for Tertiary geohistory.

24.3 ORGANIC PROGRESS AS AN ILLUSION

However, Lyell's model of faunal change (and, by extension, floral change) did not yet address the most important objection raised by his critics, namely that the fossil record showed an overall directionality, and indeed progress, with "lower" organisms appearing long before "higher" kinds, and with the human species appearing last of all. Although Lyell himself had earlier adopted that almost consensual viewpoint (§14.4), he was now claiming instead that the organic world, like the inorganic on which it depended, was subject to perpetual flux but without any directional change in its overall character. Lyell defended his new position obliquely in the last part of his survey of actual causes, which dealt with those that were neither wholly inorganic nor wholly organic in character, but involved interactions between the two realms. Most of this second half of the volume was in fact an analysis of the circumstances in which organisms were likely—or unlikely—to be preserved as fossils (in modern terms, the topic was taphonomy). The effect of Lyell's lengthy review was to highlight the sheer chanciness of the process and hence the intrinsic imperfection of the fossil record. But the chanciness was not spread evenly, for the biases were systematic: terrestrial organisms were far *less* likely to be preserved than aquatic ones, and particularly marine organisms; and those with shells or bones were far *more* likely to become fossilized than those without any such "hard" or mineralized parts.

Lyell applied this analysis to the rest of his "system" in two major ways. First, his insistence on the very *low* chances of preservation of terrestrial animals and plants helped undermine his critics' inference that organisms were formerly of "lower" kinds and less diverse overall: Lyell suggested that "higher" organisms such as mammals and dicotyledonous plants could well have existed in earlier periods without happening to be preserved as fossils. Second, his emphasis on the relatively *high* chances of preservation of marine mollusks—which combined two favorable factors, having easily fossilized shells and living in the marine environments in which most sediments were formed—suggested that these, of all organisms, could provide the most fully representative (or, least unrepresentative) samples of the world of life at earlier periods of geohistory. This foreshadowed the crucial use that Lyell planned to make of marine mollusks in his reconstruction of Tertiary geohistory.

Finally, tackling the special case of the human species, he claimed that its chances of preservation were, if anything, *higher* than those of most other mammals, because it could adapt itself to such a wide range of climates; and that the continuing failure to find any authentic human fossils therefore reflected its genuinely recent origin. He now knew that Christol and Tournal had recently claimed to have found human fossils in the south of France (§16.2); but he followed Buckland in discounting their alleged antiquity, since cave deposits were always likely to have been disturbed. However, for Lyell the geologically very

The accompanying section will enable the reader to comprehend the usual form of such islands. (No. 6.)

No. 6.

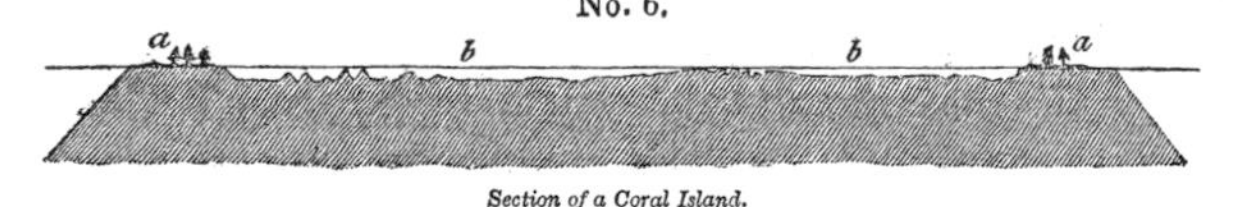

Section of a Coral Island.

(*a a*) Habitable part of the island, consisting of a strip of coral, inclosing the lagoon. (*b b*) The lagoon.

The subjoined cut (No. 7) exhibits a small part of the section of a coral island on a larger scale.

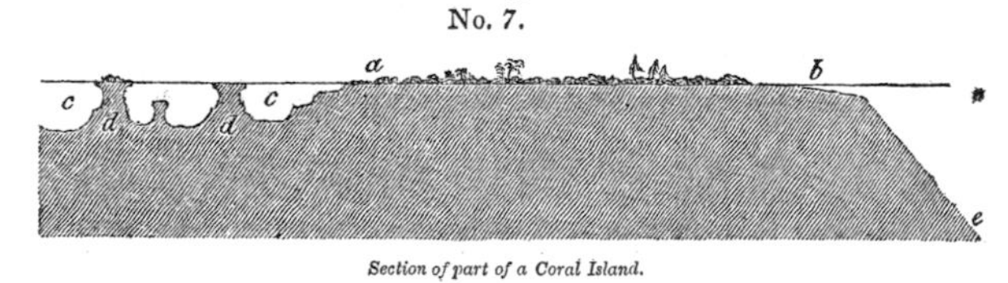

(*a b*) Habitable part of the island.
(*b e*) Slope of the side of the island, plunging at an angle of forty-five to the depth of fifteen hundred feet.
(*c c*) Part of the lagoon.
(*d d*) Knolls of coral in the lagoon, with over-hanging masses of coral, resembling the capitals of columns.

Fig. 24.1 Lyell's sections through a coral atoll, published in his *Principles* (1832) and based on illustrations of Whitsunday Island in Beechey's account of his Pacific voyage (1831). Lyell interpreted atolls as being built up by the coral organisms on the rims of the craters of extinct volcanoes.

recent origin of humans was no sign of an overall directionality in the history of life, and therefore no concession to his critics; for, as he had argued in his first volume, the outstanding novelty of the human species lay purely in its unprecedented mental powers, not in its physical features as just another mammal (§21.6).[9]

The last chapter of Lyell's volume rounded off his survey of actual causes with a brief discussion of coral reefs and limestones, features of organic origin that contributed to the inorganic processes of physical geography and stratigraphy respectively. Once again, this was no bald inventory in the manner of von Hoff, but a review geared to Lyell's overall theoretical "system". He analyzed the organic debris (broken shells, etc.) from which future limestones might now be forming, and of which those of earlier periods are composed, in such a way as to undermine the idea that there had been any progressive *increase* in the abundance of organisms secreting limey shells in the course of geohistory. And he interpreted coral atolls as "nothing more than the crests of submarine volcanos, having the rims and bottoms of their craters overgrown by corals", thereby suggesting once more how volcanic action might have affected every part of the earth at one time or another, without any directional trend (Fig. 24.1).[10]

Such topics, like the more important ones that had occupied the bulk of Lyell's volume, all went to show that the organic world, like the inorganic, was a theater of ceaseless change, yet without any overall directionality. With this vast and varied repertoire of actual causes fully surveyed, in both inorganic and

9. Lyell, *Principles* 2 (1832), 185–282 (see esp. 224–27 on human antiquity).

10. Fig. 24.1 is reproduced from Lyell, *Principles* 2 (1832), 290, which was based on Beechey, *Voyage to the Pacific* (1831), pl. opp. 188; these and a couple of others in the same chapter (283–301) are almost the only wood engravings ("cuts") in the entire volume.

organic realms, the way was clear at last for Lyell to demonstrate in his final volume that the terrestrial "system" was in a steady state of balanced equilibrium throughout.

24.4 CATASTROPHISTS AND ONE UNIFORMITARIAN

Murchison was the first British geologist to have an opportunity to comment in public on Lyell's new volume, when, a few weeks after it appeared, he delivered his first presidential address to the Geological Society. While giving the now customary review of all the papers that had been read during the previous twelve months, Murchison followed Sedgwick's example and also noted many other publications worldwide. But he gave special attention to Lyell's work. He recalled their joint tour on the Continent, "when the first idea of this arduous task began to germinate in his [Lyell's] mind", and he emphasized the importance of Lyell's focus on the Tertiary molluskan faunas and his subsequent discovery in Sicily of deposits that had "continued . . . uninterruptedly to the historic aera" (§19.4). It had then become clear to Murchison that "Mr. Lyell was beginning to unfold the true papyri of geological history", a historical metaphor based appropriately on the most famous of all the earlier discoveries at Pompeii (*BLT* §4.1).

Murchison suggested indeed that the real geohistorical significance of Lyell's work was that his friend had "completely effaced from his mind as arbitrary and untrue, those lines of demarcation between, what had been termed, the ancient and existing orders of nature". The binary structure of de Luc's geohistory (*BLT* §3.3), still echoed in Buckland's diluvial theory (Chap. 6; *BLT* §10.5), was to be eliminated altogether. But the completion of Lyell's important Tertiary research had been delayed—to Murchison's barely concealed regret—by his exhaustive study of actual causes, and in the event the president said little about the new volume beyond, predictably, praising "the clear and impartial manner in which the untenable parts of the dogmas, concerning the alteration and transmutation of species and genera, are refuted".[11]

Whewell's review, published in the first issue of the *Quarterly Review* to appear after the book itself, took the same anti-transformist line. He began, however, by noting that any work as comprehensive as Lyell's had become conceivable only in the wake of the astonishing progress of geology in the previous two decades. Much of this he attributed to the profound impact of Smith's pioneer work in stratigraphy, an English achievement with an international impact: "the study of organic fossils as the right-hand of our [i.e., geologists'] philosophy . . . has of late been the general admission throughout geological Europe". Whewell then had no difficulty teasing apart the different senses of "uniformity" that Lyell, for his own purposes, had creatively confused. The simplest and least controversial was that of the actualistic method, which, Whewell noted, all geologists (not just Lyell) had long adopted "upon the most pregnant evidence" of its heuristic value. The contrasts between present and deep past might in some cases appear formidable, but "the theorist is not so easily daunted . . . the adventure is, at least, worth a trial". Whewell was thus in no doubt about the virtue of *trying* to explain the past by means of actual causes, as expressed in Lyell's subtitle;

he clearly recognized that the real question at issue was whether they would be found wholly adequate. "The advocate [*sic*] of the geological adequacy of the existing *dynamical* laws of the world"—that is, Lyell—might account for a multitude of physical "revolutions" in past geohistory, but there were also changes in the organic world to be explained.[12]

So Whewell turned to the crucial case of the origin of species, which in his view was the greatest single problem that Lyell's new volume had raised. He devoted most of his essay to a forceful rejection of the claims of the "*Transmutationists*", among whom, he noted, were "no small number of continental geologists". Clearly he knew of Geoffroy's recent work (§17.1) as well as Lamarck's classic case for transformism, for he asked rhetorically—expecting a negative answer—"may we not thus, through natural causes, obtain a transition from the plesiosaur of the lias to the crocodile of the Nile?" Much of Whewell's review followed the lines of Lyell's own argument. He too insisted that all the evidence suggested that "the mutability of the species is finite", and that "*no one instance* can be produced, or pointed out with probability, of such an establishment of a new species" as would demonstrate that transmutation was a true actual cause. As with Lyell's inorganic changes, here too Whewell had no problem with allowing a Scropean generosity with "Time!": "we have no occasion to embarrass ourselves for want of thousands, or if necessary, of millions of years" to produce a new species. Far from trying to defend a traditional reading of the Creation story, he concluded that "we [i.e., he himself] do not conceive that those who endeavour to fasten their physical theories on the words of scripture are likely to serve the cause either of religion or of science". In the end, Whewell dismissed transformism primarily on the grounds that it was no better than "a visionary and unauthorised speculation": unauthorized, that is, by any solid empirical support.[13]

Whewell treated the origin of species as the "one remarkable exception, indeed, to the illustration of the past by means of the present". He was quite prepared to be convinced by Lyell that actual causes were adequate to explain many if not all the major changes in the inorganic world, and also many if not all cases of extinction; but he insisted that "the creation of new species, fitted to new conditions of the elements, in successive periods of the earth's history, [is] utterly out of the reach of any known laws of physiological action". He therefore concluded with purple prose that had both Cuvierian and scriptural overtones: "we find that even if our [natural] philosophy is allowed to burst the barriers of time, and to summon to its aid the energies of the elemental world, it is still unable to touch even the skirts of the garment of creative power which envelopes the Supreme Being". Descending from such heights to the more mundane business of tracing the lives of species through geohistory, Whewell made explicit what he believed Lyell had implied: "so far as we *can* trace the history of the new species

11. Murchison, "Address to the Geological Society" (1832), 373–76.

12. [Whewell], "*Principles* by Lyell, vol. II" (1832), 103–7.

13. [Whewell], "*Principles* by Lyell, vol. II" (1832), 117–18. Of course, Whewell, like Lyell, also had further reasons for rejecting Lamarck, relating to the "place of man in nature".

and families which have inhabited the earth, they have made their appearance exactly *as if* they had been placed there, each by an express act of the Creator—each provided by its Author with such powers and habits, with such organs and constitutions as adapted it precisely to the condition of things in which it was to live".[14]

Yet it cannot be emphasized too strongly that—apart from this one "striking exception"—Whewell aligned himself with Lyell "and other geologists" in asserting that "all the facts of geological observation are *of the same kind* as those which occur in the common history of the world". Like the actualistic method itself, "uniformity" in the *kinds* of geological processes was uncontroversial, at least among the savants in whose name Whewell was addressing the educated public that read the *Quarterly*. But having disposed of "uniformity" in that sense, he had next to confront Lyell's much more contentious use of the word:

> Are the extent and the circumstances of the geological phenomena *of the same order* as those of which the evidence has been collected? Have the changes which lead us from one geological [i.e., geohistorical] state to another been, on a long average, uniform in their intensity, or have they consisted of epochs of paroxysmal and catastrophic action, interposed between periods of comparative tranquillity?[15]

On this specific point, Whewell saw geologists divided into what he called "two sects", arguing with each other with all the fervent intensity that marked contemporary religious controversy in Britain. Ever the master of neologisms, he suggested calling these scientific sects "the *Uniformitarians* and the *Catastrophists*". In effect, these labels were derived from Conybeare's "fluvialists" and "diluvialists" (§20.3), but enlarged in scope from the specific problem of the origin of valleys to the pattern of geohistory as a whole. Whewell had no hesitation in identifying the catastrophists as those with "prevalent doctrine" on their side; and he added that "Mr. Lyell will find it a harder task than he appears to contemplate to overturn this established belief". In fact, Whewell and others clearly considered Lyell to be in a minority of one: at least at this moment in history, there was only one uniformitarian. This deserves emphasis, because Whewell's memorable terms later passed into general use among geologists, often without regard to the specific meaning that Whewell attached to them. Whewell implied that he himself was on the side of the catastrophists, because he found it plausible to suppose that natural agents might, in the right circumstances, be capable of producing the "great and distant catastrophes" that the evidence seemed to demand.[16]

As in his review of Lyell's first volume (§23.1), Whewell was not averse to what he saw as a revival of geotheorizing among geologists; on the contrary, he welcomed it, anticipating that unlike the much earlier rash of controversy between "Neptunians and Plutonians" his contemporaries would "probably now be content to work their way back, step by step". He meant this quite literally, or rather, geohistorically: he recommended starting with "the history of those strata which are uppermost [i.e., the Tertiaries] and come nearest to our own time, [because] by this path of investigation alone can they hope to ascend to the higher and more remote ages of geological antiquity". Whewell was well aware that this was just what Lyell planned to do in his final volume, but he rightly set Lyell's research in the broader context of what Brongniart and Cuvier had begun (*BLT* §9.1) and many others had since extended. And so, after reviewing "the complexity of causes" that Lyell's great inventory had set out, Whewell anticipated "no difficulty in conceiving how all the vast variety of the tertiary formations may have been produced". Only by starting with the Tertiaries could the theorist hope to resolve the even larger issues of the causal foundations of geohistory as a whole: "To retrace the history of this [Tertiary] period, thus depending on all these causes, is the first question which the theoretical geologist has to solve; and his problem is here reduced to its most simple form—inasmuch as, in this case, the events approach nearest to those of our own time, both in date and in kind, and are least perplexed with succeeding mutations".[17]

Thus the projected climax of Lyell's *Principles* received a ringing endorsement, well in advance of its publication, from perhaps the most powerful intellect to grace the Geological Society. But Whewell was nobody's poodle, and Lyell recognized that he and other critics were also raising formidable objections to his own cherished notion of "absolute uniformity". After reading Whewell's essay, Scrope told Lyell that "between you as a *uniformitarian* & the *Catastrophists* I do not see any but an imaginary line of separation"; it seemed "only a dispute about degree"—that is, about the degree of past intensity that could properly count as "uniform" with actual causes—and Scrope thought that "a little concession on either side will unite you in perfect cordiality".[18]

Lyell, however, rejected any such irenic move. Quoting Scrope's letter, he told Whewell that "on this point I cannot budge an inch". For Lyell it was "a most important point of principle" that events witnessed in some three millennia of recorded human history be treated rigorously as the defining standard for the rest of geohistory; he insisted that more catastrophic causes were "uncalled for", and inconsistent with the evidence, "at least if deluges must follow paroxysmal eruptions". Here he revealed the deepest dynamic underlying his insistence on "uniformity": the most recent of all putative deluges, as championed by his former mentor Buckland, was still his bête noire, because it could be, and had been, used to retain a link, however tenuous, with "Moses" or biblical literalism. Telling his fiancée how he was drafting a text that would respond to his most formidable critics, he used the vivid Pauline imagery of a spiritual struggle: "I am grappling

14. [Whewell], "*Principles* by Lyell, vol. II" (1832), 117, 125.

15. [Whewell], "*Principles* by Lyell, vol. II" (1832), 126.

16. [Whewell], "*Principles* by Lyell, vol. II" (1832), 126. Whewell evidently assumed that at least some of the causes responsible for catastrophic events would prove to be similar in *kind* to those known in the present world, although quite unknown in *degree*: von Buch's and Hall's earlier conjectures about mega-tsunamis (*BLT* §10.2) would have been likely models for this. Bartholomew, "Singularity of Lyell" (1979), traces his increasing intellectual isolation.

17. [Whewell], "*Principles* by Lyell, vol. II" (1832), 132.

18. Scrope to Lyell, 20 March 1832 [Philadelphia-APS].

not with the ordinary arm of flesh but with principalities and powers, with Sedgwick, Whewell & others for my rules of philosophizing as contradistinguished from theirs & I must put on all my armour".[19]

24.5 CONCLUSION

Lyell made good progress with what he intended to be the second and final volume of his *Principles*, despite the happy distraction of becoming engaged to Leonard Horner's daughter Mary and the unhappy distraction of the political turbulence surrounding the Reform Bill. However, his exhaustive inventory of those actual causes that involve living organisms became so bulky that Murray agreed to publish it separately, postponing to a third volume his reconstruction of Tertiary geohistory, the planned climax of the work.

In the second volume of the *Principles*, Lamarckian transformism was forcefully rejected. Lyell argued instead that organic species were real natural units, stable in form and habit throughout their span of existence; and that both their "births" or points of origin and their "deaths" or points of extinction were events that were scattered piecemeal across space and time, rather than being concentrated in sudden episodes of mass origins and mass extinctions. But Lyell's adoption of Brocchi's analogy between species and individuals was modified, in that he now attributed extinctions not to anything analogous to old age but to purely environmental factors consequent upon the ceaseless changes in climate and physical geography that he had outlined in his first volume. The mode of origin of new species was a more knotty problem, because Lyell could not point to any case in which such an event had ever been observed. So it could only count as an actual cause by inverting the logic of the actualistic method and inferring the present from the past: the fossil record showed that new species *had* somehow come into existence at determinate points in past geohistory, and presumably continued to do so in the present, but the process itself was unobserved and utterly obscure.

Lyell countered the dominant view that the history of life had been strongly directional and even progressive, by arguing that the fossil record was extremely incomplete. An analysis of the conditions necessary for fossilization proved that there was a radically systematic bias in the record, such that "higher" forms of life (e.g., mammals and dicotyledonous plants) might well have existed, even at the earliest known periods of geohistory, without leaving any trace. Conversely, however, marine organisms with "hard" parts, such as the mollusks that Lyell intended to use in his analysis of Tertiary geohistory, could give a far more reliable record of life. Finally, Lyell completed his inventory with some actual causes that were both organic and inorganic in character, such as coral reefs and limestones; they too were used to support his view of the history of life as non-directional.

Murchison was the first to welcome Lyell's second volume in public, but Whewell's was a more penetrating review. Like Lyell himself, Whewell emphatically rejected transformism. But he also argued more explicitly that the origin of new species—each accurately adapted to environmental conditions at that point in space and time—must involve some exceptional transcendence of purely natural processes. Whewell also defined what he saw as two "sects" emerging among geologists, the "*catastrophists*" and the "*uniformitarians*", though he treated the former as clearly dominant and implied that Lyell himself was the sole example of the latter. As always, the main point at issue was the explanatory *adequacy* of the actual causes that both parties agreed should be invoked as far as possible. Scrope considered that the distinction between the two "sects" was ultimately arbitrary—a question of just how much the intensity of causes might have varied in the deep past—and that the parties were therefore open to conciliation. But Lyell insisted that his own position was radically opposed to that of his critics, and he intended to demonstrate the all-explanatory role of actual causes, at no more than their *present* intensity, in the reconstruction of a steady-state geohistory. This was to be embodied in the third and climactic volume of his *Principles*, which is the subject of the next two chapters; they will complete Part Three of the present book.

19. Lyell to Whewell, 24 March 1832 [Cambridge-TC: a.208/120]; Lyell to Mary Horner, 26 March 1832 [Kirriemuir-KH]; both quoted in part in Wilson, *Lyell* (1972), 350–51. The allusion was to Ephesians 6:12–13, though Lyell probably stopped short, even in private, of equating his critics with "spiritual wickedness in high places".

Fig. 25.1 "Lecture room at the King's College": George Scharf's drawing of preparations for Lyell's London lectures, probably those given in 1832. Ammonites and other specimens are on the table; a map of Europe is on the easel (left); sections are being pinned up below the windows, and others are being drawn—probably Lyell himself is depicted—on the triptych blackboard (right). The lecture theater, with arcuate tiered benches, was similar to the one at the Royal Institution (and to traditional anatomy theaters). (By permission of the Department of Prints and Drawings, British Museum)

CHAPTER 25

Completing Lyell's Principles (1832–33)

25.1 LYELL'S LECTURES

The publication of the second volume of the *Principles*, and Whewell's favorable review of it, kept Lyell's work in the eye of the savants at the Geological Society and of the educated British public who read periodicals such as the *Quarterly Review*. In the spring of 1832, people drawn from both these overlapping groups met at King's College to hear Lyell expound what was still to come (Fig. 25.1).[1]

The general interest in geology was such that many of these people wanted to come with their wives or daughters. This put Lyell in a dilemma. Like Buckland a decade earlier, faced with an invitation to lecture to a mixed audience at the Royal Institution (§6.3), Lyell thought the presence of women would make his course "unacademical" and detract from its prestige, yet he wanted as many fee-paying subscribers as possible. The College too felt the dilemma: opinion on its Council was divided, and it later ruled against women, though without retrospective effect; so in the event there were many in Lyell's audience. His first lecture, which by convention was a free sample, was given in foul weather and at the height of the political crisis over the Reform Bill; there was a disappointingly small audience. So he repeated it, this time to a crowd of about three hundred, but fewer than seventy then paid to attend the rest of the course (and only two

1. Fig. 25.1 is reproduced from Scharf's undated pen and wash drawing [London-BM: Royal Book 3, no. 182]. The props are clearly geological, and Lyell's were the first and only such courses given there in the early 1830s. Jackson, *Scharf's London* (1987), 89, reproduces this drawing in the context of the wide range of Scharf's other work at this period. The following account of Lyell's lectures is based on Rudwick, "Lyell and his lectures" (1975), 240–51, which in turn is based on his lecture notes [Edinburgh-UL: Lyell MSS, 8]; most quotations are from the parts that Lyell judged important enough to write out in full. Wilson, *Lyell* (1972), 353–60, prints important extracts from letters to his fiancée Mary Horner about the lectures.

were students at the College).

Lyell began with a quite conventional summary of the structure of the earth's crust and its constituent rocks; but he also made the unconventional claim that the Primaries did not deserve their name, since he maintained that they were of many different ages and gave no evidence whatever of the earth's ultimate origin. However, he followed Playfair in using the analogy between space and time, astronomy and geology, to reduce this apparent adoption of cosmic eternalism to a question of the limitations of human knowledge; and he made it even more innocuous by casting it in the language of natural theology:

> It may be perfectly true that there may be a boundary to the material Universe, and yet the most powerful telescope that man may ever be able to invent may only serve to disclose to us the myriads of new worlds . . . There is no termination to the [human] view of that space which is filled with manifestations of Creative power; why then, after tracing back the earth's history to the remotest epochs, should we anticipate with confidence that we shall ever discover signs of the beginning of the time that has been filled with acts of the same creative power?

Following Whewell's lead, Lyell argued that the creation of each organic species, at just the right time and place to allow it to survive, required "a higher exertion of creative power than the mere formation of an uninhabited planet"; so geology was superior even to astronomy in the wealth of its evidence for divine activity in nature. Conversely, following Sedgwick's lead, Lyell also argued that in the rapidly changing state of the science it was foolish to try to correlate geological details with biblical texts, as the "scriptural" writers notoriously did. So geology was vindicated on both counts as a suitable science for a college of Anglican foundation, and acquitted of any suspicion of undermining religious faith. Lyell tactfully ended his lecture by quoting what he described to his fiancée as "a truly noble & eloquent passage" from the sermon that had been given at the opening of the College by Blomfield, the bishop of London, to the effect that truth from any source could only enhance an appreciation of "the glory of the Creator". All this was, of course, adapted to the occasion, but there is no good reason to doubt Lyell's sincerity. Much as he despised the "scriptural" writers whom he personified as "Moses", and much as he criticized the diluvialist geologists, Lyell's private writings confirm what he put into the public realm, and make it clear that he shared the undemanding natural theology of most of his fellow savants in Britain.[7]

After this introduction, Lyell used the rest of his course to outline some of what he planned to publish in his final volume. He first explained the actualistic principles on which he would extract geohistory from stratigraphy, and then used the Tertiaries of Europe as his main example. Between the well-established formations (e.g., the Subapennines) and the present, his own recent research in Sicily (§19.4) had revealed "monuments of an intervening period", and further "intercalations of new periods" were to be expected as fieldwork was extended. This implied that any apparent discontinuities in the series of formations, or in their fossils, were due simply to imperfect knowledge about an imperfect record. This led in turn to Lyell's sharp criticism of the catastrophists. Their alleged sudden events were rejected on grounds of scientific method, as "inventions not simply without value but as extremely mischievous in the present state of the science". Into this methodological wastebasket Lyell discarded indiscriminately all "revolutions, catastrophes, deluges, periods of repose, refrigeration, annihilations [and] paroxysmal elevations"; in other words, most of the explanatory categories invoked by most of his fellow geologists, not only sudden events but also the long-term directional trend of a cooling earth (§9.2). Such "irregular causes", he said, "we [i.e., he himself] avoid because we are assured that we have not yet exhausted the resources which the study of present causes [offers]"; to infer such events was premature, because the vast repertoire of actual causes that he had described in his first two volumes had not yet been fully exploited. Whether or not Lyell's sweeping criticism was justified, this was certainly not an argument about natural causation, still less about "Science versus Religion".

Lyell claimed that his own "method of philosophizing" had already proved its worth, for example when applied to the origin of basalt or of fossils: "this plan has put geologists into the right road, the other has led to contradictory systems". He now showed its efficacy in solving the more complex problems of understanding geohistory. He argued that, just as in the present world, the Tertiary formations had always accumulated locally (i.e., in the various basins: see Fig. 3.3), whereas the process of organic change (by the piecemeal creation and extinction of individual species: §24.2) was continuous and global. Hence the fossil record was bound to appear discontinuous. To illustrate this, Lyell combined the Brocchian analogy between species and individuals with the system of decennial censuses that had operated in Britain since the turn of the century. Like the periodic visits of census "commissioners", the natural recording of continuous change in the population of species (i.e., by the preservation of fossils) was bound to look discontinuous, because of the "commemorating process visiting and revisiting different tracts in succession". Conversely, every time a region was exposed to erosion and its sediments destroyed, it was "like the burning of documents" in human history (the destruction of the great manuscript library in Alexandria in late Antiquity was a notorious example). The analogy with history—the deployment of the metaphors of *nature's* documents, monuments, and archives (*BLT* §4.1)—was now so much taken for granted among savants that it rarely surfaced except in their relatively popular scientific writing; but in a case like this it could still have a useful role in explicating a quite difficult concept for the savants themselves.

Lyell next considered what fossils were most effective for the purpose of plotting the course of geohistory. But mention of fossils provided the occasion for a lengthy eulogy on Cuvier, the news of whose death (see §28.1) had just reached London. Lyell thought that "many of his geological speculations may [in future] require to be modified or rejected", because they had been based on little first-

2. The pervasive natural theology expressed in Lyell's unpublished manuscripts is rarely apparent from the quotations in Wilson, *Lyell* (1972), a biography much influenced by the posthumous use of Lyell as an icon of anti-religious secularism.

hand fieldwork. But he had high praise for the Frenchman's work on paleontology: "when the facts came under his own observation, nothing could exceed the caution with which he drew his inferences, and I may venture to say that there are few authors who will have less to retract [posthumously!]". This was not only generous but also judicious, and the rest of Lyell's remarks confirmed the depth of his admiration for the great naturalist, and not least for his professionalism. In fact Lyell used the occasion to make his own trenchant contribution to the current controversy on the alleged "decline" of the sciences in England compared to the Continental countries. As his notes show, he attributed it to the lack of governmental support for the sciences, and to the more professional approach of the French:

> In France though the numbers [of savants] are few they are systematically organized. It is not left to chance—sciences subdivided—each [savant] given professionally to [i.e., specializing in] some branch—not wealth but honour—and a competency. Compare French to small standing army—English to desultory multitudinous host of irregulars.[3]

After this digression Lyell returned to Tertiary geohistory, arguing that marine mollusks were the most suitable fossils for dating, particularly if they were all identified by a single competent naturalist, who could apply uniform standards of taxonomy throughout. Of course, this was just what Deshayes—a fine example of a French scientific specialist—was currently doing for him in Paris (§23.3). So here in his lectures, for the first time in public, Lyell revealed an outline of the geohistory that Deshayes's work had made possible. It was expressed in the nomenclature that he owed to Whewell (§23.3); it was novel above all in being based explicitly on *geohistory*, not on distinctive types of rock (like chalk for "Cretaceous", coal for "Carboniferous"), nor on specific geographical regions (like "Jurassic" after the hills of the Jura).

Lyell outlined Tertiary geohistory in the usual retrospective manner, in terms of five periods of increasing disparity from present faunas and inferred increasing age. These were the "*Recent*" period of human history (here he adopted other geologists' term in place of his own "Contemporary"); the "*Newer Pliocene*" for the formations in Sicily, Nice, and elsewhere (Deshayes now defined this as having 96% extant species); the "*Older Pliocene*" for the Subapennines, English Crag, and others (52%); the "*Miocene*" for the formations near Vienna, Turin, Bordeaux, and elsewhere (19%); and the "*Eocene*" for those in the Paris, London, and Hampshire Basins (3%). The most important effect of this scheme was that it implied continuity through the alleged break between present and deep past, and left no room for any uniquely disruptive "diluvial" episode. Lyell summarized the evidence that similar marine and freshwater sediments, and volcanic rocks, dated from each of these periods, indicating that conditions had remained broadly similar throughout the Tertiary era. He made this geohistory more comprehensible by taking Auvergne and its adjacent regions as an example, showing how they could be used to provide a "perfect restoration of [the] Eocene period, [its] lakes, seas, volcanos, land quadrupeds, reptiles, testacea [mollusks], plants, insects [and] Geography". This clearly showed how Lyell intended to reconstruct geohistory in a series of such vignettes: "to burst the limits of time" (in Cuvier's famous phrase) by moving backwards from the known present into an ever deeper unknown past.[4]

Lyell then described this Tertiary geohistory in greater detail. He devoted one whole lecture to the Newer Pliocene formations of Sicily, and the next to Etna, because they jointly straddled the evanescent dividing line between present and deep past. He estimated that the huge cone of Etna might represent some sixty or seventy thousand years, though even this would only be the tail end of the Newer Pliocene period, itself only the tail end of the Tertiary era, which was the tail end of the known record of geohistory. He described the dozens of minor cones of loose volcanic ash on the flanks of Etna (see Fig. 7.3), and concluded—with all the authority conferred by his own fieldwork (§19.4)—that "no [diluvial] wave has passed over the forest zone of Etna". Significantly, he told his fiancée afterwards that Copleston, who was attending the lectures and was neither a bigot nor a fool, had been convinced by his reasoning: this scholarly prelate, although theologically conservative, evidently had no problem with a vast timescale for the earth as a whole, or even, now, with Lyell's inference that the biblical Flood could not after all have been worldwide. In his final lecture, Lyell emphasized how radically the geography of the region around Sicily had changed, even within the lifetimes of organic species still in existence; but the course ended without any grand conclusions.

This lecture course was repeated at King's in 1833, just after Lyell's third volume was published; but by then the College's general ban on women had come into effect, the content of the lectures was no longer so novel, and the course attracted few subscribers. On the other hand Lyell—unlike Buckland a decade earlier (§6.3)—accepted an invitation to lecture at the Royal Institution, which with Davy as director had long been famous for its fashionable mixed audiences. This course, given in parallel with his second one at King's but pitched at a more elementary level, attracted much praise and a large number of subscribers (including more women than men). But he later decided not to repeat it, and he also resigned his position at King's. In both cases he withdrew on the strictly practical grounds that the net fees he received did not justify the time that the courses took to prepare; at neither institution did his lectures provoke any criticism on religious grounds.[5]

3. The "declinist" controversy had been started by Babbage, *Decline of science* (1830). The versatile savant Charles Babbage (1792–1871), who was currently constructing his famous mechanical computer, was an enthusiastic member of Lyell's audience; Dolan, "Representing novelty" (1998), describes his studies of Serapis (not fully published until 1847) in relation to Lyell's views.

4. Rudwick, "Lyell and his lectures" (1975), 249–50; the percentages noted here are those reported in Deshayes, "Tableau comparatif" (1831), read at the Société Géologique on 2 May. The "Acene" had dropped out of sight in Lyell's public scheme, though not out of his mind (see §26.1). He later renamed his Newer Pliocene the "Pleistocene" (see §32.3), so that the Older then became just "Pliocene"; the sequence familiar to modern geologists was completed still later by the intercalation (by others) of "Paleocene" and "Oligocene", respectively before and after the Eocene.

5. Lyell's two 1833 courses are described in Rudwick, "Lyell and his lectures" (1975), 251–59; two extended extracts that he wrote out in full for the Royal Institution, which give a good impression of his lecturing style, are transcribed in Rudwick, "Charles Lyell speaks" (1976).

Lyell's London lectures spanned only two years. But they were nonetheless almost as important in spreading his ideas, among those who mattered most to him, as Cuvier's famous Parisian course on geology had been nearly three decades earlier (*BLT* §8.3). Just as Cuvier's lectures had acted as a trailer for his later *Discourse* (*BLT* §9.3), so Lyell's played the same role in relation to the culminating volume of his *Principles*. But from this point onwards, Lyell chose in effect to rely on his books alone to propagate his geology and to provide him with income. However, he had good reason to be confident that he could indeed become a "professional" savant, in a financial as well as a social sense. His first two volumes had sold out and were reprinted in a second edition (with minimal changes) even before the third volume was ready; indeed they were selling so well—by the standards of fairly expensive non-fiction—that Murray was already planning to reissue the whole work in a much cheaper format, which would greatly increase its range of readers.

25.2 A CONTINENTAL INTERLUDE

After his first lecture course was completed in the spring of 1832, Lyell took a break before resuming work on his third volume: he was now regularly following his own prescription to "travel and write and travel again" (§20.1). Putting first things first, he crossed to Germany and married Mary Horner in Bonn. Following the example of Buckland and Scrope, he then took her on a geological honeymoon through Germany and Switzerland into northern Italy and back through France. This itinerary gave him an opportunity, as usual, to make or renew contacts with many Continental geologists. For example, on his way up the Rhine from Bonn he stopped in Heidelberg, and met for the first time the joint editors of a *Jahrbuch* that was rapidly becoming the premier German periodical for the earth sciences. In 1829 Karl Cäsar von Leonhard (1779–1862), the professor of mineralogy and geognosy (and a Foreign Member of the Geological Society), had started a periodical "for mineralogy". But he had soon recruited his younger colleague Bronn, who was making a name for himself with his studies of fossils (§12.1), and the scope of the periodical was then expanded to cover "mineralogy, geognosy, geology, and paleontology [*Petrefaktenkunde*]", or, in other words, the whole range of the earth sciences. Both men were important for Lyell, and Bronn had just published his tabular "statistical" summaries of all fossil faunas, but especially of those of the Tertiaries (Fig. 12.2). Renewing old contacts was equally important to Lyell: for example, at the end of his tour he stopped to see Prévost and Deshayes in a Paris now deprived—or, for some savants, relieved—of the dominating or domineering presence of Cuvier.[6]

Lyell's itinerary was also designed to fit in some useful fieldwork, and not least to see more of the Alps, Europe's greatest mountains. More specifically, he wanted to see the field evidence on which the Swiss geologist Bernhard Studer (1794–1887), the highly respected professor of geology at Bern, had inferred that some "Primary" schists and gneisses in the Alps were really Secondary in age, altered in composition and appearance by the intrusion of granites. This was vital evidence for the idea to which Lyell had alluded in his lectures, that the Primaries in general might not be truly "primary" at all, so that geohistory might indeed be left in Huttonian fashion "with no vestige of a beginning". He found Studer doing fieldwork in the Alps behind Bern, and was gratified to find him well aware of the *Principles* and therefore pleased to meet its author. Studer directed him to Saussure's classic locality of Valorsine near Chamonix (*BLT* §2.4). Lyell was duly convinced by what he saw there: as he told his father-in-law Horner, "I believe the whole series, gneiss & all, to be altered secondary sandstone & conglomerate, of age perhaps of lias. So much for *primary* rocks". It was a striking instance—and not the last—of how he could benefit from the outstanding fieldwork of his Swiss contemporaries.[7]

25.3 THE FINAL VOLUME OF LYELL'S *PRINCIPLES*

On their return to London, Lyell and his wife set up home so as to be (as he told Murchison) "as near to Somerset House [i.e., to the Geological and Royal Societies] and Athenaeum [the intellectual gentlemen's club] . . . as health & gentility of situation & my purse would permit". Even in his domestic arrangements, Lyell was becoming a dedicated professional to the core. With an increased allowance from his father, and funds from capital settled on Mary by her father, the couple had enough steady income to sustain a modest degree of "gentility", but of course Lyell hoped to supplement it significantly from royalties on the *Principles* and other future books. He naturally remained concerned with sales figures, and therefore continued to be careful to make his work attractive to both his overlapping audiences: to make it intelligible to the general book-buying public without sacrificing its scientific standards in the eyes of other geologists. As a telling sign of this dual purpose, his third volume included a glossary of the technical terms that he had used throughout the work, inserted at the request of some of his less expert readers. But it is significant that in introducing this glossary Lyell clearly distinguished such "general readers" from the "men of science" who would not need its aid: the latter were those whom Whewell shortly afterwards suggested should be called "*scientists*".[8]

6. Wilson, *Lyell* (1972), 361–72, describes his tour. Leonhard and Bronn's *Jahrbuch* soon became the *Neues Jahrbuch*, which has remained ever since one of the most distinguished in its field (now under the name of *Zentralblatt*). Bronn had been in Italy at the same time as Lyell, studying museum collections, but their paths had not crossed.

7. Lyell to Horner, 25 August 1832 [Kirriemuir-KH: quoted in Wilson, *Lyell* (1972), 370–71, with a reproduction of his sketch section (fig. 47)]. Studer, "Nördlichen Alpenkette" (1827), was prompted in part by Necker, "Filons de Valorsine" (1826); the latter, however, was mainly a confirmation of Brongniart's famous discovery of Chalk fossils high on the Alps above Chamonix (*BLT* §9.5). Later on this trip, Lyell met Charpentier at Bex and heard about Venetz's theory of former glacial extension (see §34.3).

8. Wilson, *Lyell*, 372–76; Lyell, *Principles* 3 (1833), Appendices, 61–83. [Whewell], "*Connexion of the sciences* by Mrs Somerville" (1834)—a review of a book by the distinguished female savant Mary Somerville (1780–1872)—recorded (59) that "*scientist*", a word devised by analogy with "artist", had been proposed (in fact by himself) after the moral philosopher and poet Samuel Taylor Coleridge (1772–1834) had criticized the leading members of the infant British Association for the Advancement of Science for calling themselves "philosophers": see Morrell and Thackray, *Gentlemen of science* (1981), 20; Yeo, "Whewell's philosophy of knowledge" (1991); and Ross, "Scientist" (1962). However, Whewell's "inclusive" term was rarely used, and the gendered (but, at the time, generally accurate) phrase "man of science" remained the dominant anglophone term until the twentieth century: see Barton, "Men of science" (2003).

Lyell's third volume was published at last in the spring of 1833, just before he gave his two parallel lecture courses at King's and the Royal Institution. It was dedicated, as planned long before, to Murchison, who had already seen it in proof and welcomed it in his presidential address to the Geological Society a few weeks earlier. The frontispiece was a pretty landscape, geologically colored in Scropean style, showing the extinct volcanoes that Lyell had studied at Olot in Catalonia in 1830; but it had no particular importance in the text. The landscape that should have occupied that strategic position had been rather wasted, as already mentioned, on a second volume that in the event had had no relevance to it. In what had become a third volume, on the other hand, its importance was decisive. It was a fine picture of Etna and the Val del Bove, based on a sketch that Lyell had made on the spot before he climbed the volcano late in 1828 (§19.4). It was intended to illustrate his argument—formulated privately while he was in the field, sketched in his lectures, but only now to be set out fully in public—in which Etna would form the crucial link between Tertiary geohistory and the present world of human history (Fig. 25.2).[9]

Fig. 25.2 "View of the Valle del Bove, Etna": the frontispiece of the second volume of Lyell's *Principles* (1832) but relevant only to the then still unpublished third volume (1833). This engraving, based on a sketch that Lyell made on the spot in 1828, was in the style made popular among geologists by Scrope's panoramas (§15.1). The colors distinguished the relatively recent minor cones (middle distance, 5, 7) and lava flows (foreground), which Lyell claimed had been erupted since this side of Etna collapsed, from the great bulk of the 3000m volcano (background, and "inliers" at 4, 6, 9), which was of far more ancient origin though still extremely young by geological standards. The upper slopes (2) were covered with snow, except on the summit cone (3) with its plume of vapor. The smoking minor cones (7, and another to the left) were said to date from eruptions in 1811 and 1819 respectively. A full Scropean panorama would have been needed to portray the whole breadth of the vast arena; here only the northern part was shown, and Lyell feared that it would "give no idea of the extraordinary geological interest, still less of the picturesque grandeur of this magnificent scene of desolation".

The volume itself began with a lengthy preface in which Lyell summarized the course of his work, ever since he had sent Murray an early version on the eve of his Grand Tour (§18.2). This was ostensibly to explain his long delay in completing the work, but between the lines Lyell's account also staked his claims to priority, particularly on the Tertiary formations and their fossils. In his opening chapter, however, he turned first to his quarrel with his catastrophist critics, claiming a sharp dichotomy between "two distinct methods of theorizing" in geology; this was the contrast that he had expressed privately in Pauline terms of cosmic struggle (§24.4). As in his first volume, the position of his critics was thinly disguised as history: a "false method of philosophizing" had arisen from "the entire unconsciousness of the first geologists of the extent of their own ignorance respecting the operations of the existing agents of change". And as usual, Lyell's critique put himself on the moral high ground, contrasting dogmatism with patience:

> Never was there a dogma more calculated to foster indolence, and to blunt the keen edge of curiosity, than this assumption of the discordance between the former and the existing causes of change . . . The course directly opposed to these theoretical views consists of an earnest and patient endeavour to reconcile [*sic*] the former indications of change with the evidence of gradual mutations now in progress.[10]

Lyell's strategy was, of course, that the explanatory adequacy of actual causes would be demonstrated by deploying the full repertoire that he had set out in his first two volumes. He had noted earlier that these volumes would "be found absolutely essential to the theories hereafter to be proposed"; now, similarly, he described their inventories of actual causes as "preliminary treatises", subservient to what was to come. Adopting his familiar analogy with the antiquarian task of deciphering an unknown ancient language, he regarded them "as constituting the alphabet and grammar of geology". Actual causes were indispensable, because nature's documents could be deciphered, and nature's own history reconstructed, only by tracing their action "in an indefinite lapse of ages". The goal of the culminating volume of the *Principles* was defined unmistakably as the reconstruction of geohistory.[11]

25.4 LYELL'S METHODS FOR GEOHISTORY

Extending Lyell's linguistic metaphor, the first major section of his volume was devoted to the *syntax* of geology, or to the ways of reconstructing the sequence of geological events. After an introductory summary of the main categories of

9. Fig. 25.2 is reproduced from Lyell, *Principles* 2 (1832), frontispiece (explained on 303–4); the Val del Bove is described more fully in *Principles* 3 (1833), 83–97. See also Murchison, "Address to the Geological Society" (1833), 443–44.

10. Lyell, *Principles* 3 (1833), 2–3.

11. Lyell, *Principles* 3 (1833), 7; see Rudwick, "Historical analogies" (1977), 94–97. Secord, "Introduction [to Lyell's *Principles*]" (1997), argues, contrary to the interpretation offered here, that a geohistorical narrative was *not* Lyell's objective.

rock formations, Lyell focused on the Tertiaries. He began with the Paris Basin, and with the work of Brongniart and Cuvier a quarter-century earlier (*BLT* §9.1); for this, he noted, had marked an "era" or decisive milestone in the history of geology. In fact, in Lyell's opinion, it had been so successful that for a time it had, paradoxically, retarded the science: geologists had tried to correlate Tertiary formations elsewhere with one or another part of the Parisian sequence, before eventually realizing that the Tertiaries might be of quite different ages in different basins. So Lyell summarized and synthesized some twenty years of highly international research on these formations, starting with Webster on the Isle of Wight and Brocchi on the Subapennines (*BLT* §9.4), fitting in Prévost on the Vienna Basin (*BLT* §9.6), and coming up to the recent work by Basterot and Desnoyers (§12.1). But since even the Subapennines had a fauna distinctly different from the

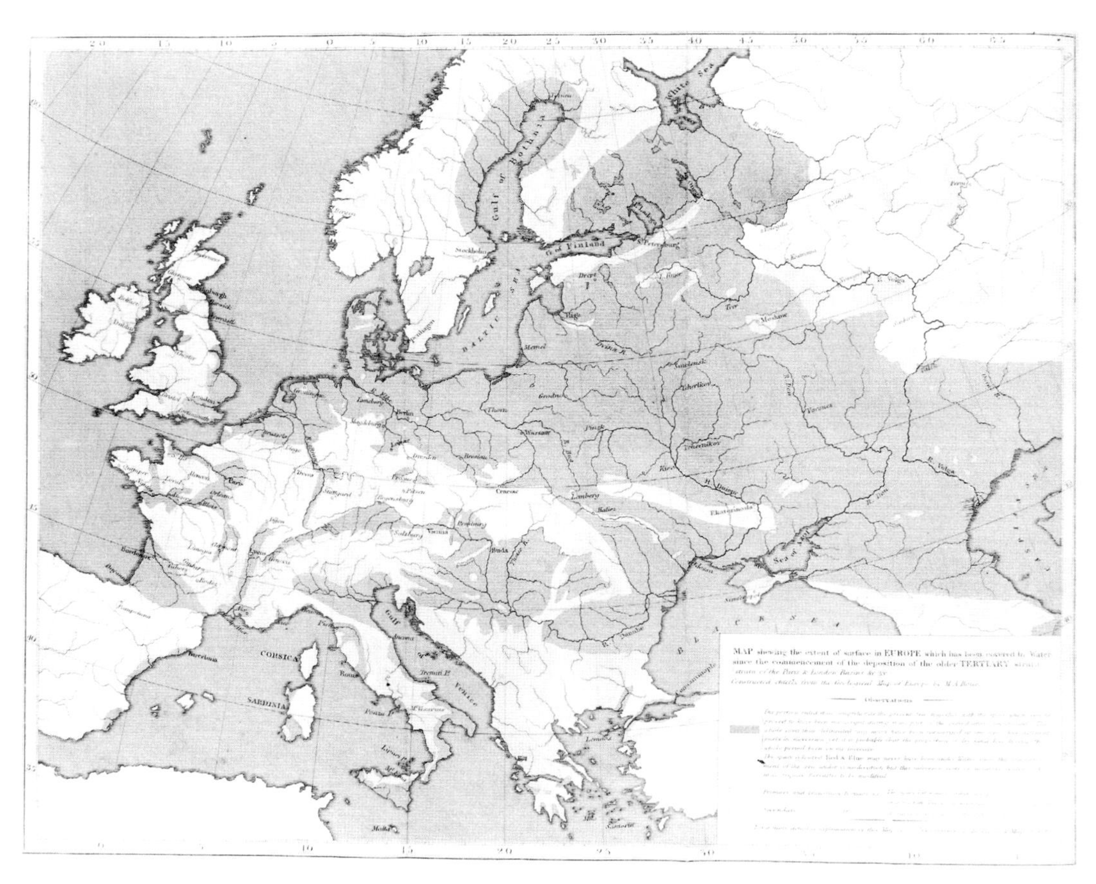

Fig. 25.3 Lyell's map of Europe, showing (by the darkest tone) how much of this part of the globe was known to have been submerged below sea level *at some time* during the Tertiary era (or at present), as a result of perpetually shifting areas of deposition, now preserved in part as the various Tertiary basins. The map was published in the second volume of Lyell's *Principles*, but was relevant only to the third (1833); it was based on the latest and best geological map of Europe, by Boué (1831); areas not yet surveyed geologically (North Africa and much of Spain and the further parts of Russia) were left blank. Lyell noted that the areas of outcrop of Primary and Secondary rocks (depicted with very pale color washes) "may never have been under Water since the commencement of the aera under consideration [i.e., the Tertiary], but this inference rests on negative evidence & may require hereafter to be modified": sediments might have been deposited there but later eroded away. He also emphasized that, despite appearances, this was *not* (in modern terms) a paleogeographical map, since it did not depict the putative geography at any specific moment or period in geohistory.

present, Lyell promised to describe "monuments of an intervening period"—he was alluding of course to those of Sicily—that would give "evidences of a gradual passage from one condition of the animate creation to that which now prevails"; and he concluded that "the line of demarcation between the actual [i.e., present] period and that immediately antecedent, is quite evanescent".[12]

However, Lyell's use of the Tertiaries as the main testbed for his "system" or geotheory depended on showing that their era was not untypical of geohistory as a whole. So he had first to account for the striking contrast between the quite localized Tertiary basins and the far more uniform and widespread Secondary formations (notably the Chalk, the youngest of the latter, which outcropped with little variation over vast areas of northwestern Europe). Lyell's explanation was quite simple: the part of the globe that is now Europe had been predominantly covered by sea during much of the Secondary era, but predominantly continental—with only occasional and local incursions by the sea—during the Tertiary. By implication, there was nothing unique about either situation, and in another part of the globe the two might have been reversed, with a mainly continental regime followed later by a mainly marine one. Apparently abrupt junctions between successive formations—"violations of continuity", in Lyell's emotive phrasing—were likewise due not to any sudden event of the kind that catastrophists invoked, but simply to shifting areas of deposition. A region once covered by sea, accumulating marine sediments on its floor, might have been elevated later above sea level; but still later it might have been submerged again, whereupon new marine sediments would accumulate on top of the older ones, but would be more or less distinct from them. Lyell illustrated this argument with a novel kind of map—like his frontispiece of Etna, it had been published prematurely in his second volume—showing that much of the European landmass had been covered by sea during the Tertiary era: not all at any one time, but *at one time or another* (Fig. 25.3).[13]

This model of continually shifting areas of deposition was next combined with Lyell's "hypothesis of the gradual extinction of certain animals and plants, and the successive introduction of new species", which he had set out in the second volume (§24.2). If the marine areas—in which species were most likely to be preserved—were continually shifting in position, "the fossilizing process . . . may be said to move about, visiting and revisiting different tracts in succession". As in his lectures, Lyell explained "the working of this machinery" by deploying the Brocchian analogy: "Let the mortality of the population of a large country represent the successive extinction of species, and the births of new individuals

12. Lyell, *Principles* 3 (1833), 8–22. Lyell converted to graphical form what Desnoyers had described verbally, the stratigraphical evidence for the "non-simultaneity" of the Tertiaries of the Paris Basin and Touraine, which served as a control on the validity of the purely fossil criterion for the relative ages of Tertiary basins. His section (here reproduced as Fig. 12.1) was among the first of the large number of wood engravings that illustrated his third volume, in contrast to the few in the first two: the new technique, with its great advantages of cheapness and close integration with the text, was only just beginning to be used widely in books published in Britain.

13. Fig. 25.3 is reproduced from Lyell, *Principles* 2 (1832), map opp. 1 (explained on 304–10), based mainly on Boué, *Carte géologique de l'Europe* (1831). The word "violation" was (and is) far more emotive in British than in modern American usage: appropriate, say, to rape but not to speeding.

the introduction of new species". Lyell imagined "commissioners" visiting different provinces in turn, not in fact in a decennial census but in a somewhat random but never-ending itinerant task of compiling and updating "statistical documents" for each province. Then, for any specific province, the composition of the population recorded on one visit would differ from the next, roughly in proportion to the lapse of time between the two: the longer the interval, the more individuals would have died, while others would have been born. In the same way, two successive marine formations in the same Tertiary basin (for example, those in the Paris Basin) would have fossil faunas differing in species roughly in proportion to the lapse of time between the two successive marine incursions. Lyell conceded that "some other causes besides the mere lapse of time" would affect the rate of change; but he claimed that the analogy, although not perfect, was nonetheless valid.

Significantly, Lyell reinforced this point with a historical or antiquarian analogy. He imagined ancient Greek inscriptions being found beneath a modern Italian town, rather as the Roman site of Herculaneum had been excavated beneath Portici at the foot of Vesuvius. Antiquarians would be grossly mistaken if they then inferred that the language of the region had changed abruptly from Greek to Italian. If on another site "*three* buried cities" were found superimposed, with inscriptions successively in Greek, Latin, and Italian, historians might begin to realize that the "catastrophes" that had annihilated each settlement might have been totally unrelated to the change of language. The ruins might simply preserve three phases in a process of gradual linguistic change: "the passage from the Greek to the Italian may have been very gradual, some terms growing obsolete, while others were introduced from time to time". Lyell's readers, with their Classical education and, in many cases, experience as tourists in Italy, hardly needed him to interpret this little parable. His catastrophist critics were as mistaken as the imagined antiquarians, if they assumed "that it is part of the plan of nature to preserve, in every region of the globe, an unbroken series of monuments to commemorate the vicissitudes of the organic creation"; and therefore equally mistaken to infer mass extinctions or other sudden events between successive but distinct formations. Lyell concluded that the fossil record must be inherently and intrinsically incomplete; his critics supposed that it was more or less perfect, but "we must shut our eyes to the whole economy of existing causes, aqueous, igneous, and organic, if we fail to perceive *that such is not the plan of nature*".[14]

Two further introductory chapters dealt with the criteria used by geologists to determine the relative ages of formations, and hence the relative dates of events in geohistory. Superposition was fundamental, but not always applicable; distinctive rock types were helpful on a local level but often deceptive, not least because they might be repeated in formations of quite different ages; fossils were most reliable but not infallible. This much was familiar to all but the least well-informed of Lyell's readers. But he refined the criteria by explaining why the species of marine mollusks were likely to be the most reliable of all, because their shells were abundant and easily preserved, and—in the present world—generally had much wider geographical ranges than terrestrial species. In addition, however, Lyell claimed that they seemed on average to have much longer lifespans than,

Fig. 25.4 Fossil mollusk shells selected by Deshayes as characteristic of the "Pliocene Tertiary Period", a period named by Lyell (following Whewell's suggestion) to denote its high percentage of species known alive in present seas: eight of the twelve shown here were in that category. But these were no more than a tiny sample of the total of 777 species that Deshayes had identified from Pliocene formations. This set of drawings, together with similar ones for the still older Miocene and Eocene periods, were supplied by Deshayes in 1831 for Lyell to publish in his *Principles*; in the event, in its third and final volume (1833).

14. Lyell, *Principles* 3 (1833), 23–34. It is significant that Lyell's conception of linguistic change involved the piecemeal introduction and dying out of discrete linguistic units (words, etc.)— ananlogous, of course, to his concept of organic change—rather than their continuous transformation in the course of use, which would have been analogous to Lamarckian transformism. Yet the new German philological research (by Grimm, Bopp, and others), which suggested just such a process of imperceptibly gradual change, was a topic of great interest at the time in the British intellectual circles in which Lyell moved: see Rudwick, "Historical analogies" (1977), and more generally, Amsterdamska, *Schools of thought* (1987), chaps. 2, 3.

for example, the species of mammals. He had thought hard about this Brocchian point during his earlier fieldwork (§18.3), but its significance now became clear.

In general, geologists used the Smithian method of finding the most reliable "characteristic" fossils to identify specific formations (Chap. 3). This was not Lyell's method, but he made a minor concession to his readers at this point, by publishing Deshayes's pictures of a small selection of the species "characteristic" of the Pliocene, Miocene, and Eocene. Significantly, however, he recruited them not for stratigraphy but for *geohistory*, labeling them as characteristic not of (say) the Pliocene formations, but of the Pliocene *period* (Fig. 25.4).[15]

These illustrations represented a regression to the conventional Smithian concept of "characteristic fossils", but evidently Lyell thought the concession worth making. Like Prévost before him (*BLT* §9.6), and Desnoyers and Deshayes more recently (§20.2), Lyell was concerned to determine the relation between an entire population of fossil species and the corresponding population of species known to be alive in present seas, in order to use that relation for dating the formations. But although the constituent species changed from Eocene through Miocene and Pliocene to Recent, there was nothing here to suggest that the world of life—or at least of molluskan life—had changed significantly in character since the earliest part of the Tertiary era. This part of geohistory looked decidedly steady-state, or so Lyell intended to claim; at least this far back, the deep past did *not* look like a foreign country.[16]

25.5 CONCLUSION

Lyell's lectures at King's College gave him a valuable opportunity to reach an influential audience drawn from the scientific and social elites in the capital. He used the occasion to outline some of the contents of the still unpublished final volume of his *Principles*, making geology palatable in the College's Anglican environment by stressing its contribution to traditional natural theology. Reacting to news of Cuvier's death, Lyell praised the work of the great naturalist, and took the opportunity to compare the professionalism of French savants—which he himself hoped to emulate—with the less well-organized and allegedly "declining" state of the sciences in Britain. In the substance of his lectures, he outlined his reconstruction of the Tertiary era, now divided into a sequence of periods marked by their increasing approximation to the present world. This naturally led to a focus on his own fieldwork in Sicily, which provided the crucial link between past and present.

As an interlude before completing his work, Lyell married Mary Horner in Bonn and then took her on a geological honeymoon around western Europe. He made several important new contacts with Continental geologists and renewed old ones; improved his firsthand knowledge of Europe's highest mountain range; and approved Studer's inferences about the Secondary age of some altered and deceptively Primary-looking rocks. Back in England, Lyell at last completed and published his third and final volume, hoping the whole massive work would establish him as a truly "professional" savant.

As foreshadowed in his lectures, the climactic volume of the *Principles* would show that a full appreciation of the power of observable actual causes would demonstrate their total adequacy—at no more than their present intensity—to account for everything in the deep past. Lyell criticized the catastrophists more trenchantly than ever; he deployed the Brocchian analogy between species and individuals, and the antiquarian analogy between geology and archaeology, to argue that the appearance of sudden changes in the deep past was an illusion founded on a deeply imperfect fossil record. The Tertiary era—and more particularly its division into the successive periods that Deshayes's research on fossils had made it possible to distinguish—would be the testbed for Lyell's concept of geohistory as essentially uniform and steady-state. Now, with all his geohistorical procedures set out in full, he could bring his argument to its climax with a detailed reconstruction of geohistory, focused on the Tertiary era. This is the subject of the next chapter, which brings Part Three of this narrative to its conclusion.

15. Fig. 25.4 is reproduced from Lyell, *Principles* 3 (1833), pl. 1, dated December 1831 (explained on xxvii–xxviii); the shells were drawn in Paris by Paul Louis Oudart but engraved in London. Only four species (8, 10–12) were *not* known alive; four species had been named by Brocchi ("Broc"). The Continental origin of this plate, and of the similar ones of Miocene and Eocene shells (pls. 2–3), was immediately apparent to knowledgeable readers, because the gastropod shells were depicted with their spires pointing downward (the British convention was to orient them the other way up). Deshayes also supplied novel pictures of "Microscopic fossil shells" (in modern terms, foraminifera) from the Eocene of the Paris Basin, drawn under a low-power microscope (pl. 4).

16. Lyell, *Principles* 3 (1833), 35–52.

CHAPTER 26

Geohistory in retrospect (1833)

26.1 LYELL RECONSTRUCTS THE TERTIARY ERA

Lyell began the final and climactic part of his *Principles* by explaining the Eocene, Miocene, and Pliocene, Whewell's novel terms for the Tertiary portion of geohistory. With the Pliocene now divided into Older and Newer, they denoted four distinct and successive phases of the Tertiary era. But Lyell emphasized that they were most unlikely to represent *contiguous* periods (as many later geologists were, erroneously, to take them as being). Rather, they were scattered samples preserved from a far longer and largely unrecorded past: "like chasms in the history of nations", the gaps between them arose simply from the incompleteness of the preserved record. "We have little doubt", Lyell wrote, "that it will be necessary hereafter to intercalate other periods, and that many of the deposits, now referred to a single era, will be found to have been formed at very distinct periods of time." The four named periods recorded four discrete moments, as it were, in the process by which the specific composition of the molluskan fauna had changed smoothly and continuously, in a manner "strictly analogous, as we before observed, to the fluctuations of a population such as might be recorded at successive periods, from the time when the oldest of the individuals now living was born, to the present moment". And since "the Recent strata form a common point of departure in all countries", it followed that the same method could in principle be applied globally. Tertiary formations in India or South America, for example, could be confidently dated as Eocene, if they too were found to contain a similarly tiny percentage of extant species. Even if their fossils included *no* species in common with the Eocene formations of the Paris Basin (which was, in modern terms, the "type" Eocene), "yet we might infer their synchronous origin from the common relation which they bear to the existing state of the animate

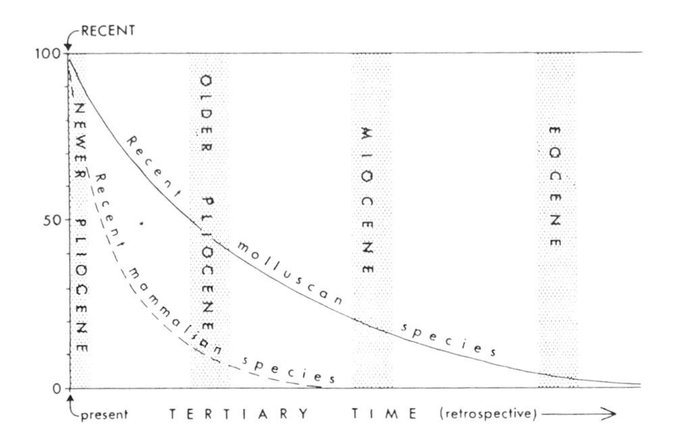

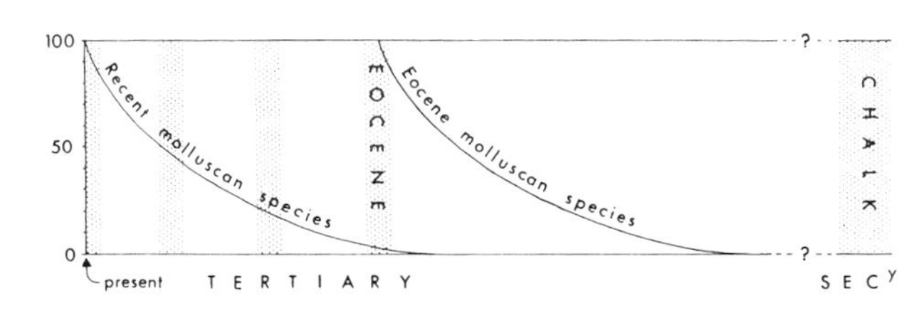

Fig. 26.1 An interpretation of Lyell's natural chronometer (in modern graphical format), to show his conception of his four Tertiary periods as mere fragments, preserved by chance from a continuous geohistory, and dated quantitatively (though on a scale not able to be calibrated in millennia or millions of years) by their percentages of extant molluskan species (vertical axis). Lyell's retrospective reconstruction—starting at the present and penetrating in Cuvierian manner back into the past—goes from left to right; so real time flows, unconventionally, *from right to left*. The inferred gaps left plenty of room for further periods to be intercalated, if faunas with intermediate percentages were to be discovered in future. The shorter average lifespan of mammalian species (shown by the more rapidly decaying survivorship curve of living or "Recent" species) made them less suitable than mollusks for dating purposes: even in the Miocene they were all of extinct species. The lower diagram shows (on a more condensed timescale) how Lyell interpreted the *total* disjunction between the oldest Tertiary (Eocene) and the youngest Secondary (Chalk) faunas as the product of an extremely long unrecorded gap, longer than the *whole* of Tertiary geohistory.

creation". Nothing could show more clearly the radical contrast between Lyell's geohistorical project and the Smithian method of using "characteristic fossils" in stratigraphy.[1]

Deshayes's lists of Tertiary molluskan species, printed as a massive appendix to the final volume of the *Principles*, set out the empirical basis for Lyell's "statistical" analysis of the gradually changing molluskan faunas and their occasional preservation in particular formations. Lyell was proposing a *quantitative* dating method that in principle would allow *absolute* ages to be attributed to formations, even if that absolute scale of time could not easily be calibrated in millennia or millions of years; Scrope's similar "natural scale" of erosion in Auvergne (Fig. 15.4) provided him with a ready precedent. But Lyell (like Scrope) never used the classic metaphor of a "natural chronometer", probably for the same reason that he never referred to "actual causes": both concepts were central to what he was doing, but both phrases would have been fatally flawed in his eyes by their long association with de Luc, whose much earlier approach to geology (*BLT* §3.4, §6.3, §6.4) was an influential example of the kind of theorizing that Lyell was determined to refute. In contrast to de Luc, Lyell intended to show that there was no radical disjunction between the "present world" and the "former world"; that actual causes were wholly adequate to explain all the apparent contrasts between them; and that a natural chronometer could be extended back from the present, not merely to some alleged boundary event near the dawn of human history, but into the furthest depths of prehuman geohistory.[2]

However, in Lyell's opinion his chronometer revealed a record of geohistory even less complete than that of human history; hence his anticipation that further research would require the intercalation of further periods between those he defined. Other geologists such as Sedgwick often conceded that, metaphorically, a page had been torn here and there from the volume of nature's history (§23.1). But Lyell argued in effect that only a few pages here and there had been preserved at all, or at least discovered hitherto (Fig. 26.1).[3]

After all his introductory chapters, setting out his general methods for reconstructing geohistory, Lyell devoted the bulk of his volume to a detailed analysis of the Tertiary era. He moved backwards from the present world of directly observable causes, for the same heuristic reason that Cuvier (*BLT* §9.3), and before him Desmarest (*BLT* §4.3), had long since established: as Lyell put it, "this retrospective order of inquiry is the only one which can conduct us gradually from the known to the unknown". For each period, Lyell described in turn its marine formations, its freshwater deposits, and the contemporary igneous activity (either volcanoes, or the crustal movements that he attributed to igneous processes at great depth). This was not just for the sake of giving a comprehensive account, but primarily to demonstrate the uniform action of the same processes throughout the Tertiary era. Lyell devoted by far the greatest space to the Newer Pliocene period, the most recent part of the Tertiary era. This reflected his own intensive and largely original fieldwork on them, but of course he had given them that attention on account of their theoretical significance in his broader interpretation of geohistory. The Newer Pliocene period and its record of rocks and fossils had a crucial role in breaking down the Delucian contrast between present and deep past, turning disjunction into continuity, and proving that no exceptional or sudden change—specifically, no "geological deluge"—had marked the recent past.

This retrospective survey of Tertiary geohistory began with the Newer Pliocene marine strata that Lyell had studied in Sicily (§19.4). He stressed three points about these strata: that their fossil mollusks were almost all of species known alive, many in the nearby Mediterranean; that they evidently accumulated slowly and therefore represented an extremely long period of time; and that although

1. Lyell, *Principles* 3 (1833), 52–60. His geochronology (to use the modern term) would have been unworkable if its statistical units (i.e., molluskan species) could not be taken to be morphologically stable and therefore taxonomically recognizable throughout their lifespan. As already suggested (§24.2), this was an extremely weighty reason for Lyell to insist on the stability of species and to reject Lamarckian transformism, although arguably not the most profound of his reasons.

2. Lyell, *Principles* 3 (1833), "Instructions for using M. Deshayes's tables of shells" (395–98), followed by the tables themselves and a "statistical" analysis of them (Appendices, 1–52). Rudwick, "Lyell's dream" (1978), reconstructs his argument, and "De Luc and nature's chronology" (2001) suggests his predecessor's unacknowledged role in it.

3. Fig. 26.1 is reproduced from Rudwick, *Meaning of fossils* (1972), fig. 4.5. Lyell's provisional "Acene" (not marked on this diagram) would be available as an appropriate name for any pre-Eocene Tertiary formations with no extant species at all. He did not refer to anything like the survivorship curves used in these diagrams, but he would have been familiar with the concepts underlying them: demographic "statistics" based on the first four British decennial censuses (1801–31) underlay the Malthusian arguments about population dynamics, which were evoking intense controversy in the political and intellectual circles in which Lyell moved. An exponential curve also has the effect of spreading Lyell's four preserved periods fairly evenly, which he seems to have thought they were. Rudwick, "Lyell's dream" (1978), text-fig. 1, uses instead a graphical format analogous to the "revolutions" of the hands of a clock; this too captures Lyell's imagery of a steady turnover of species. His interpretation of the relation between Chalk and Eocene (and hence between Secondary and Tertiary) is in striking contrast to the modern concept of a relatively sudden mass extinction (supposedly wiping out the last dinosaurs) at the Cretaceous-Tertiary ("K/T") boundary, *without* any major gap in the record.

geologically so recent, they had already been elevated high above sea level in the center of the island (Fig. 19.7). Lyell then argued that these massive formations must underlie, and therefore be older than, the whole of the huge cone of Etna (Fig. 19.5). He described that great volcano with all the authority of one who had studied it at first hand and climbed to its summit (Fig. 19.4). He argued that its internal structure—exposed in the walls of the Val del Bove (Fig. 25.2)—showed that it had grown gradually in size, by the accumulation of successive lava flows just like those witnessed in human history, and not by the kind of sudden crustal upheaval invoked by von Buch and Élie de Beaumont (§8.3, §9.4). Lyell tried to estimate the age of Etna, as a very rough guide to the timescale of geohistory: he concluded that it "must have required an immense series of ages anterior to our [human] historic periods, for its growth; yet the whole must be regarded as the product of a modern portion of the newer Pliocene epoch". In his lectures he had estimated Etna's age in tens of millennia, but in print he cautiously avoided putting any figure on it. Yet the implication was still clear: the timescale of geohistory was vast beyond human imagination. Finally, in this lengthy analysis of Sicilian geology, Lyell recalled the Huttonian theory he had expounded in his first volume (§22.1), to the effect that any volcano is merely the surface manifestation of massive igneous processes at great depth. Just as Etna had grown slowly by the accumulation of lava flows erupted from below, so in another part of the island the underlying crust could have been elevated at the same time and equally gradually, by intermittent earthquakes; and Lyell used the topography of the Val di Noto and the stratigraphy of Sicily as a whole to claim that this was just what had happened (Figs. 19.6, 19.7)[4]

Lyell continued his lengthy analysis of the Newer Pliocene by reviewing its volcanic rocks and marine strata elsewhere, noting that the latter were visible above sea level only in areas currently subject to earthquakes, as expected on his theory of crustal elevation. Turning to non-marine deposits of the same age, he conspicuously avoided using Buckland's theoretically loaded term "diluvium", which he had been content to adopt in his earlier work (Fig. 10.2). Instead, he criticized the idea that there had been a single "alluvial epoch": he concluded, with Sedgwick and others (§23.1), that the coarse Superficial gravels had been formed in several different episodes, not all at once; but unlike those other geologists he insisted that ordinary actual causes were adequate to account for all such deposits, so that no diluvial event of any kind need be invoked.

The huge erratic blocks found in northern Europe and around the Alps (Fig. 13.4) were more difficult for Lyell to explain away. The usual explanation in terms of some kind of catastrophic mega-tsunami, as first mooted some twenty years earlier by von Buch and Hall (*BLT* §10.2), was firmly rejected. Instead, Lyell adopted the idea suggested at that time by Wrede, that even large erratics could have floated great distances if they were buoyed up on ice floes. This kind of explanation fitted neatly into Lyell's inference that large areas of Europe had been below sea level at one time or another in the quite recent past (Fig. 25.2), and with his view that as a consequence local climates might have been different from those of the present: for example, as Wrede himself had suggested, erratics might have been rafted on ice floes from Scandinavia right across the Baltic, while the future north German plain was submerged by a shallow sea. However, the otherwise similar Alpine erratics were far more of a problem: they were often found high up, and to attribute them to floating icebergs would entail a deep submergence of the whole region at a very recent time. Very awkwardly, Lyell was therefore forced to account for them in a quite different way. He suggested that they had been swept from the mountains rafted on ice released by the sudden collapse of barriers damming temporary Alpine lakes, as had happened on a small scale in the notorious Val de Bagnes disaster of 1818 (*BLT* §10.5). It was a tacit concession that explanations of a decidedly catastrophist character might sometimes be unavoidable.[5]

After this lengthy analysis of the Newer Pliocene, Lyell dealt more briefly with the records of the next two preserved samples from the Tertiary era. Brocchi's Subapennines and the Crag deposits of eastern England were among the marine formations that he dated as Older Pliocene; and he assigned to the same period the extinct volcanoes he had studied in previous summers in the Eifel and Catalonia (the latter, as already mentioned, were pictured in his frontispiece of this volume, but received no special emphasis in the text). For the Miocene he was able to cite examples of marine formations from widely scattered regions such as Desnoyers's Touraine and Basterot's Bordeaux, Bonelli's Turin and Prévost's Vienna. Freshwater deposits of the same age, and contemporary volcanic rocks, came for example from the work of the local naturalists in Auvergne (§15.2). Lyell noted that while a substantial minority of the marine molluskan species of this period were still known alive—indeed the period was defined by that proportion—its terrestrial mammals were all extinct, which he interpreted as the result of a higher rate of turnover among the latter (Fig. 26.1). Significantly, he suggested that one set of marine formations (near Montpellier) might belong in the otherwise unrecorded gap between Miocene and Older Pliocene; he commented that "we [i.e., he himself] are fully prepared for the discovery of such intermediate links". The geohistorical record was intrinsically imperfect, but Lyell expected it to become less imperfect in the course of further research.[6]

Shifting further back in his retrospective geohistory, this time to the Eocene period, Lyell reversed his usual order of topics by dealing first with the freshwater deposits underlying all the famous volcanic features of central France. This allowed him to argue—as he had in his review of Scrope's work (§18.1)—that here there had been no marine incursions at all, at any time in the Tertiary era, so that sudden Cuvierian "*révolutions*" could not be invoked to account for the manifest faunal changes that had affected the region. As for the alternations of marine and freshwater formations in the Paris Basin—the classic stronghold, as it were, of catastrophist interpretations of Tertiary stratigraphy—Lyell respectfully rejected the repeated sudden changes of environment famously proposed by Brongniart

4. Lyell, *Principles* 3 (1833), 62–117; Rudwick, "Lyell on Etna" (1969), reproduces most of the wood engravings that illustrated this reasoning.

5. Lyell, *Principles* 3 (1833), 118–54.

6. Lyell, *Principles* 3 (1833), 155–224. As already mentioned, later geologists duly inserted the "Paleocene" before the Eocene, and the "Oligocene" after it.

and Cuvier (*BLT* §9.1) in favor of Prévost's picture of sediments accumulating steadily in an ever-changing geography of slowly shifting environments (§10.3). The latter, as he put it, made it "more easy to explain the manner of their origin and to reconcile [*sic*] their relations to the agency of known causes". Echoing his review of Scrope (§18.1), and his own lectures (§25.1), Lyell emphasized the sheer normality and tranquility of the world revealed by the Eocene formations: contrary to what he claimed his critics believed, "we are naturally led to conclude, that the earth was at that time in a perfectly settled state, and already fitted for the habitation of man". The Eocene period was *not* a foreign country where nature did things differently: apart from minor contrasts—and, of course, the absence of the human species—the world had been much as it is at present.[7]

Turning to the record of Eocene volcanic activity, Lyell attributed to that period all the earlier eruptions that Scrope had described in central France, such as the Etna-sized but deeply eroded ancient volcanoes of Mont-Dore and Cantal (§15.1, §19.1), which were built on top of freshwater limestones that he also dated as Eocene. On the other hand, the much later eruptions, including the famously recent-looking cones of loose volcanic ash and their still fresh lava flows (Fig. 15.2), were attributed not to the Newer Pliocene, as might have been expected in view of their similarity to the recent cones on the flanks of Etna, but to the Miocene. Other geologists were bound to find this claim surprising and highly implausible, but Lyell's reason for proposing it soon emerged. He was determined to eliminate any grounds for claiming, as Daubeny and Buckland had done (*BLT* §10.5), that the volcanic features of the region were sharply divided into those formed before and after the putative diluvial event. By re-dating the allegedly "postdiluvial" eruptions as not merely Newer Pliocene but as ancient as Miocene, their silent witness to the absence of any trace of a subsequent "diluvial wave" became more eloquent than ever.

After this rather blatant piece of special pleading, Lyell ended his review of Eocene geohistory with a description of the formations of that age in the London and Hampshire Basins (*BLT* §9.4), and a long analysis of the great eroded dome of Chalk—Mantell's home region of the Weald, between the North and South Downs—that now separates them. Here again he was concerned to refute any ideas of sudden upheaval or violent diluvial erosion, and to argue instead that the elevation had taken place concurrently with slow marine erosion, some of the eroded material having gone to form the Tertiary sediments in the basins to the north and south.[8]

26.2 GEOHISTORY WITH "NO VESTIGE OF A BEGINNING"

Having thus concluded his retrospective reconstruction of Tertiary geohistory, Lyell could afford to treat the Secondary era far more briefly, in a mere dozen pages, just "to show that the rules of interpretation adopted by us for the tertiary formations, are equally applicable to the phenomena of the secondary series". The total discontinuity between the Eocene molluskan fauna and that of even the youngest Cretaceous formation (the Chalk of Maastricht) was explained away as the natural result of a gap in the fossil record even longer than the timespan between the Eocene and the present (Fig. 26.1). However startling this conclusion may seem to modern geologists, it followed directly from Lyell's theory that the rate of turnover of species had always been steady and uniform. The apparent contrast between the widespread Secondary formations and the localized Tertiary basins was again explained as a natural consequence of contrasting geographies—in the part of the globe that is now Europe—during those two successive eras; and the often greater degree of dislocation of the rocks (folding, faulting, etc.) was attributed simply to the greater span of time that had elapsed since the Secondaries were formed. Lyell concluded that there was nothing to suggest that conditions or processes were significantly different in the Secondary era; the stability and uniformity of the "system" was thus upheld for that portion of geohistory too.[9]

Before pressing his retrospective geohistory back to the Primary rocks, Lyell digressed to deal with the latest and perhaps greatest threat to his geotheory, namely the ambitious synthesis of paleontology, stratigraphy, and tectonic geology recently proposed by Élie de Beaumont (§9.4). Like all serious geologists of any nationality, the Frenchman took it for granted that geohistory had been unimaginably lengthy, much of it being recorded in the sequence of sedimentary formations. But he claimed that these long and tranquil periods had been punctuated by a sequence of occasional sudden episodes of mountain-building upheaval, when the earth's crust had buckled along certain lines, leaving each time as its trace a widespread unconformity between the formations dating from before and after that time (Fig. 9.3). Élie de Beaumont's catastrophist theory was particularly worrying to Lyell because it was based on fieldwork as extensive and thorough as his own, if not more so. Specifically, Lyell was alarmed that the most recent of all the Frenchman's putative "*révolutions*" was tentatively equated with a sudden elevation of the Andes (Fig. 23.1), which in turn suggested a possible natural explanation—by means of a mega-tsunami of some kind—for the diluvial deposits around the world. This threatened to allow diluvial theory to be rehabilitated on an impeccably naturalistic basis. Lyell deflected the threat, rather unconvincingly, with a renewed insistence on the magnitude of geological time and the fragmentary character of the record offered by the sequence of formations. The putative suddenness of each episode of upheaval did indeed depend on proving the brevity of the time between the youngest formation affected and the oldest one unaffected; but Lyell concluded that "even if all the facts appealed to by [Élie] de Beaumont are correct, his intervals are of indefinite extension". The appearances could thus be saved—at a pinch—for a non-catastrophic explanation.[10]

7. Lyell, *Principles* 3 (1833), 225–56.

8. Lyell, *Principles* 3 (1833), 257–323. A conventional geological map of southeast England (pl. 5), based on Greenough's and "exhibiting the denudation of the Weald", is the only copper engraving in the volume, apart from the frontispiece of Etna and Deshayes's Tertiary fossils. Lyell's interpretation of the Weald as the product of slow *marine* erosion became, much later, the basis for Darwin's famous estimate of some 300Ma for the Tertiary era alone (a figure several times *in excess of* modern radiometric dating): see Burchfield, "Darwin and geological time" (1974).

9. Lyell, *Principles* 3 (1833), 324–36.

10. Lyell, *Principles* 3 (1833), 337–51.

Having disposed of this latest challenge to his concept of "absolute uniformity", at least to his own satisfaction, Lyell turned finally to the Primary rocks, a term that he criticized immediately as being without sound basis. He proposed that "unstratified" Primary rocks such as granite should be called "*Plutonic*" (after the god of the underworld), to denote their deep-seated igneous origin. Like Hutton, he denied that granite was "the oldest of rocks"; on the contrary, "it is now ascertained that this rock has been produced again and again, at successive eras". It often penetrated Secondary strata (as he had seen in the Alps), and Lyell inferred that it was still being formed beneath the main centers of volcanic activity and crustal elevation.

The "stratified" Primary rocks such as schist and gneiss were more difficult for Lyell to explain: although they looked rather like ordinary layered sediments, in places they could be seen apparently merging imperceptibly into granites. This puzzle could be solved, Lyell suggested, only by adopting "the Huttonian hypothesis" that they were sediments that had been more or less radically altered at great depth (and hence under great pressure) by the intense heat coming from molten granite. This kind of process had long been recognized on a smaller scale, for example near the edges of small intrusive "dykes" composed of other igneous rocks. But Lyell, following earlier suggestions by Boué (§3.3) and Studer (§25.2), proposed that the stratified Primary rocks as a whole should now be renamed "*Metamorphic*" [i.e., transformed]. He then proposed the term "*Hypogene*" [i.e., formed at depth] for the Plutonic and Metamorphic categories combined. All these neologisms—as respectably Classical in derivation as his Tertiary periods—had a clear purpose. They embodied Lyell's claim that *all* the kinds of rock commonly regarded as characteristic of the earliest part of geohistory might in fact have been formed at *any* period, and were likely to be forming still in the present world. The so-called Primaries were therefore not primary at all.[11]

Lyell summarized his geohistory in a diagram and a couple of tables appended to his text. One table was simply a summary of the stratigraphical sequence of formations, listed in the customary retrospective order (as in "Conybeare and Phillips": §3.2) from Recent down to Carboniferous, together with lists of localities culled from all over Europe and around the world. There was nothing remarkable or controversial about this, except at the very base. Here Lyell included, in Conybeare's "Carboniferous Group", all the formations that other geologists classed as *Transition*, including even the ancient-looking slates of the "Grauwacke". The effect of this deft sleight-of-hand was to eliminate the Transition category altogether. With the transformation of the Primaries into Hypogenes (Plutonics and Metamorphics), geohistory was therefore left with *no* clear records at all, older than those that Lyell defined as Secondaries.[12]

The other table was more deeply interpretative. Here the same major stratigraphical "groups" were listed again; but their constituent rocks were classified under six headings according to their inferred origin: "alluvial" (i.e., subaerial); aqueous, either marine or freshwater; volcanic; and hypogene, either plutonic or metamorphic. This created a kind of matrix plotting times against origins, though many of the spaces in the matrix had to be left blank. For example Lyell could not cite any "alluvial" deposits older than the fossil soil of the famous "dirt

DIAGRAM

Shewing the relative position which the Plutonic and Sedimentary Formations of different ages may occupy ; (in illustration of TABLE I.)

No. 91.

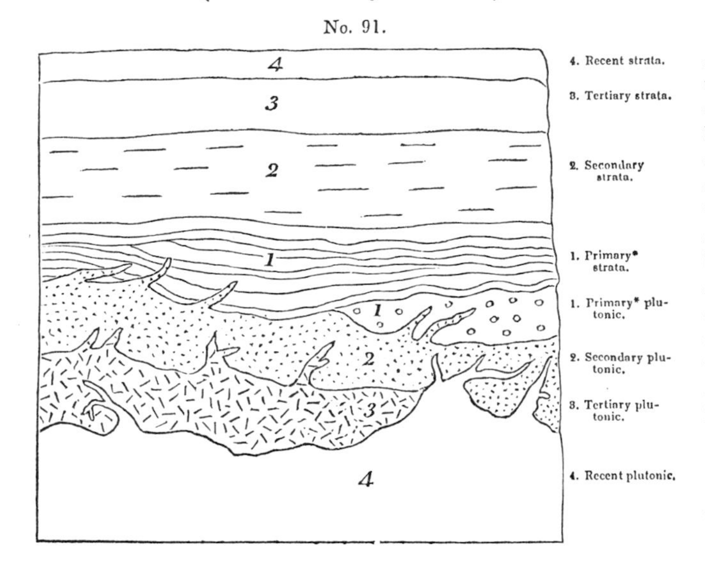

Fig. 26.2 Lyell's diagrammatic ideal section through the earth's crust, to illustrate his claim that there is no preserved record of any beginning to geohistory. In the upper part of the section is the ordinary stratigraphical succession of sedimentary strata, from the Primaries (1) up to the Recent (4). But in the lower part these are shown underlain by "plutonic" igneous rocks, in an inverted sequence with the most recent intrusions (4) in the deepest position, and the oldest (1) nearest the surface (and therefore most likely to have been exposed by subsequent erosion and to be observable). The *plutonic* intrusions had transformed all the oldest sediments into *metamorphic* rocks, thereby destroying or making unintelligible any records of that or still earlier periods of geohistory, let alone of the earth's ultimate origin.

bed" at Portland (§11.1), well up in the Secondary series. Nor could he list any visible "hypogene" rocks younger than the Alpine granites that intruded the Lias (§25.2), well down in the Secondaries; for all the later periods he had to be content with suggesting localities where plutonic and metamorphic rocks might be "*Concealed*". Nonetheless, Lyell's intention was clear: the matrix summarized his claim that every category of rock had been formed at every period, thus demonstrating again the steady-state uniformity of the "system" throughout geohistory. Finally, Lyell included a diagram that showed graphically—in both senses—how, by classing the Primaries as Hypogene and absorbing the Transition into the Secondaries, he had in effect knocked the bottom out of geohistory (Fig. 26.2).[13]

In his brief "concluding remarks", Lyell therefore had to defend himself against the criticism that his steady-state "system" was as eternalistic as Hutton's had been long before (*BLT* §3.4). He readily admitted the Huttonian affinities of his geotheory; but he protested that the charge of eternalism—he cited Scrope's review (§22.2), but De la Beche and many others had made the same point (§22.3)—was as unfair to him as it had been to Hutton. Playfair long before had insisted, albeit somewhat disingenuously, that Hutton had merely claimed that there was no *evidence* of a "beginning" to the world, not that there had never been one. Likewise, Lyell insisted that he too was only claiming that no evidence

11. Lyell, *Principles* 3 (1833), 352–82.

12. Lyell, *Principles* 3 (1833), "Table II" (389–93).

13. Fig. 26.2 is reproduced from Lyell, *Principles* 3 (1833), 388; the interpretative "Table I" (386–87) that immediately precedes it is not printed as a matrix but it could have been (as Boué's had been: Fig. 3.5).

of the earth's origin could be found in the rocks now preserved at its surface. As in his lectures, he deployed the classic analogy with astronomy: "in vain do we aspire to assign limits to the works of creation in *space* . . . we are prepared, therefore, to find that in *time* also, the confines of the universe lie beyond the reach of mortal ken". Any apparent eternalism was nothing more than an expression of the limitations of human knowledge. Echoing Hutton once more, Lyell claimed that the divine design of the earth was demonstrated by its very stability: to assume that humans would ever find evidence of its beginning or end was "inconsistent with a just estimate of the relations between the finite powers of man and the attributes of an Infinite and Eternal Being". It was appropriate that Lyell's vast extended argument should conclude on that unmistakably Huttonian note of deistic natural theology.[14]

26.3 CONCLUSION

The final part of the climactic volume of Lyell's *Principles* presented his reconstruction of the Tertiary era, offered as the most reliable sample of what could be done with the whole of geohistory. His concept of the ever-changing composition of the Tertiary molluskan fauna, combined with his claim that the stratigraphical record was an intrinsically imperfect record of geohistory, led him to argue that the successive periods of the Tertiary era (Eocene, Miocene, Older Pliocene, and Newer Pliocene) were no more than temporally scattered samples of geohistory, snatched from the ravages of time and preserved by chance in spatially scattered "basins". For each period, reviewed retrospectively from the present back into the deeper past, he reconstructed in turn the same array of physical processes and environments, demonstrating that the earth had been essentially in a steady state throughout the Tertiary era.

The most recent period, the Newer Pliocene (which he was later to rename the Pleistocene), received the most detailed analysis, because it served to efface the alleged contrast and discontinuity between the present world and the rest of geohistory. The Sicilian formations and the huge cone of Etna that overlay them—which together had been the culmination of his own fieldwork (§19.4)—provided the crucial evidence that tied the present back into the past in an unbroken sequence of ordinary events and processes. However, Lyell's rejection of any diluvial episode at this point in geohistory left him with some awkward problems. He explained the erratic blocks of northern Europe as having been drifted on ice floes or icebergs at a time of higher relative sea level; but the otherwise identical Alpine erratics, found at much higher altitudes, were attributed to the quite different cause of natural dam bursts on a necessarily vast scale. This represented a tacit concession to the catastrophists; erratic blocks might turn out to be the Achilles' heel of Lyell's ambitious geotheory of "absolute uniformity".

Having demonstrated his uniformitarian case for the Tertiary era—at least to his own satisfaction—Lyell moved back briefly into the Secondary era, just enough to claim that the earth had then been in much the same state. But the still earlier Transition era, which other geologists increasingly recognized as distinct (§4.3), was deftly eliminated altogether by being silently assimilated into the Secondary. And the Primary rocks, which other geologists regarded as the traces of the very earliest part of geohistory, were eliminated in a more radical way. Lyell interpreted them all as either "plutonic" rocks intruded in Huttonian fashion from the depths of the earth, or else "metamorphic" rocks formed by the drastic transformation of other rocks in the same depths of the earth, which had destroyed whatever record of earlier geohistory they might otherwise have borne. So Lyell argued that geohistory could not be reconstructed any further back than what he redefined as an enlarged Secondary era (extended backward by the inclusion of the Transition rocks). Back beyond that point, there was, in Hutton's famous phrase, "no vestige of a beginning"; while for the rest of the record of geohistory, from the earliest Secondaries to the present world, everything pointed to an earth in an essentially steady state.

However, Lyell denied that his steady-state geotheory was eternalistic: he claimed that his rejection of any beginning was a matter of the limitations of human knowledge, not an assertion that the earth (or the universe) was in fact uncreated and eternal. Nonetheless, it was clearly Hutton's vision (as mediated in sanitized form by Playfair) that inspired Lyell to construct an even grander geotheory on the firmer foundations of the geology of his own time. It was a "system" of almost infinite complexity, but one in which both inorganic and organic "causes" balanced and interacted to produce an enduring dynamic stability.

Yet, in contrast to Hutton's geotheory, Lyell's represented the total integration of that genre with the highly contingent *geohistory* that had been constructed during the intervening half-century. Lyell's "system" was fully geohistorical, and no less so for presenting its geohistory for heuristic reasons in a retrospective order, as a series of vignettes penetrating ever further back in deep time. Unlike Hutton's "succession of former worlds", in which the whole point was that nothing significant had distinguished any one "world" from any other, either earlier or later, past or future, the successive periods of Lyell's geohistory were each distinctive, individually knowable, and, in principle, even dateable. They might be quite similar—rhetorically, even the iguanodons of the deep past might perhaps return in the remote future (§21.4)—but in detail each was nonetheless distinguishable and contingently unique.

Lyell's *Principles* therefore represented not just the revival of the genre of geotheory, but its transformation; and it provoked other geologists to articulate their own attitudes to geological method and to geohistory more clearly than ever before. The vast scope of Lyell's work, the fertility of its constituent theories, and its substantial originality on several levels, have justified devoting the whole of Part Three of the present book to an account of its slow gestation and immediate impact, and to a detailed analysis of its extended and discursive argument. Part Four, which now follows, traces how Lyell's geotheory—now fully in

14. Lyell, *Principles* 3 (1833), 382–85; the last quotation is the concluding phrase of the entire work (excluding appendices, etc.).

the public realm—fared at the hands of his fellow geologists; how he was forced by their criticisms to modify it in important ways; and how, in consequence, a synthesis that combined both "uniformitarian" and "catastrophist" approaches was eventually forged. It is this that has endured in its essentials into the world of modern geology.

Part Four

CHAPTER 27

Challenges to Lyell's geotheory (1832–35)

27.1 CONTESTED MEANINGS OF "UNIFORMITY"

The publication of Lyell's *Principles of Geology*, in three volumes spread over four years, was certainly an important event for the science, not least on account of its massive scale and ambitious scope. But in the eyes of the knowledgeable geologists who read it, both at the time and more reflectively in later years, it did not represent a "revolution" in any sense analogous to the political upheavals across Europe with which it coincided. On the other hand, it certainly focused debates that were already under way among geologists, and provided them with new materials for argument; as several of them commented, it marked an "epoch" or milestone in the development of their science. It synthesized empirical observations and theoretical inferences drawn not only from Lyell's own work but also from a wide range of his predecessors and contemporaries. And it did so in a highly consistent manner, subjecting them all to the rigorous control of an interpretative rule that allowed no deviation from a strict concept of "absolute uniformity". In Whewell's terms, it defined Lyell (and only Lyell) as a "uniformitarian" (§24.4). Its exhaustive inventories of geological processes "now in operation" forced other geologists to consider, more carefully than before, *how far* such "actual causes" might indeed be adequate to explain the preserved traces of geohistory, and therefore how far the deep past might have been similar to the present. However, Lyell's contemporaries—like modern scientists faced with a comparably bold theoretical synthesis—picked and chose from his massive work, adopting his viewpoint on some issues and rejecting it on others.[1]

1. Wilson, in *Lyell* (1972)—subtitled "The revolution in geology"—presents Lyell's work as "revolutionary". It was mentioned in passing as the founding "paradigm" of geology in Kuhn's classic *Scientific revolutions* (1970), 10; see also Cohen, *Revolution in science* (1985), 313–15, and Elena, "Imaginary Lyellian

One point on which they were agreed was the genre to which the *Principles* belonged and the tradition that it followed. As a reminder that these debates were not confined to Britain—a reminder much needed, in view of the massively anglocentric and anglophone bias of much modern historical writing about them—a first example can be taken from one of the summaries of new geological work worldwide, which Boué was writing annually for the Société Géologique in Paris. By the time he presented his report for 1833, it had grown from some thirty pages to just over five hundred, forming a substantial supplementary volume to the Société's monthly *Bulletin*. Nothing could indicate more clearly the rapid rate of growth of the science by the early 1830s (and, in consequence, the necessarily selective character of the present narrative).

In his earlier reports Boué had briefly noted the first volumes of the *Principles* (§23.2). In his report for 1833, compiled after the final volume was received in Paris, his review of the whole work was immediately preceded by a tirade worthy of Lyell himself, against the inanities of "*géologie Mosaïque*". Not for the first time, Boué treated the "scriptural" genre as a peculiarly English aberration. He ridiculed its ignorant authors for presuming to challenge such heavyweight savants as Conybeare, Buckland, Sedgwick, and Whewell: these were his own examples, all of them Anglican clergymen. He then placed Lyell's work on its own under the heading of "*Géogénie*", claiming that Lyell had devoted "the finesse of his mind and the elegance of his pen" not to geology proper but to the speculative genre of geotheory: "in a word, it is a theory of the earth in the style of Buffon's, although adapted to the modern manner of treating the sciences". In a literal sense "geogeny" was an inappropriate label, since Lyell—in contrast to Buffon—denied that there was any evidence of the earth's origin; but Boué hit the nail on the head in assigning the work to the same *genre* as Buffon's more than half a century earlier (*BLT* §3.2), albeit in an updated form.[2]

Boué's summary of Lyell's *Principles* was fair, and he correctly pointed out its heavy dependence on the work of other savants, not least those beyond Britain. He noted that the second volume was concerned with the question, "Is the *species* a reality in nature, or should Lamarck's ideas on the transmutation of species be adopted?" And Lyell's third volume used Deshayes's expertise to synthesize all the research—by geologists of several nations—that had been modeled on "the greatest star in the scientific halo [*sic*] of Mr Al[exandre] Brongniart", namely that senior saint's pioneer work on the Tertiaries of the Paris Basin (§12.1; *BLT* §9.1). Like Sedgwick, Boué juxtaposed Lyell's work with that of Élie de Beaumont: whatever the wider reading public in Britain may have thought, for *geologists* everywhere the important debate was not about natural causation, let alone "geology and Genesis", but about the pattern of geohistory. In addition to the fundamental question whether the overall pattern had been directional or cyclic, the other principal question was: had geohistory always been smooth and equable, or had its long tracts of deep time been punctuated by occasional sudden events of greater intensity than anything yet witnessed in the extremely brief span of human history? Boué noted that Lyell rejected not only "the system of great perturbations, of deluges, of cataclysms, and of *diluvium*, the great warhorse of many of his compatriots", but also Élie de Beaumont's far more weighty theory of occasional "epochs of elevation" (§9.4). While the former had indeed often been tinged with "Mosaic" or scriptural concerns, towards the latter Boué was as positive as Lyell was negative.

"Everyone [*sic*] is agreed and has long recognized", Boué claimed, "that most disruptions [*redressemens*] are due to violent and sudden movements, or to a sequence of such movements". In contrast, Lyell—implicitly in a minority of one—attributed *all* crustal disruptions to "the indefinitely prolonged repetition of local, slow and continuous effects". Boué regarded Lyell's idea as quite plausible, at least for the elevation of whole continents (as opposed to linear mountain ranges). But again he hit the nail on the head when he implied—as Scrope had suggested directly to Lyell himself (§24.4)—that the rival interpretations really differed only in degree. For example, how many separate movements had been involved in the elevation of a mountain range? A single huge buckling, or a "sequence" of several, or hundreds of very small ones? And just how sudden had these movements been? Had each lasted months, or years, or centuries? These were matters that it was simply not yet possible to determine. In sum, Boué's brief but balanced review was typical of the reception of Lyell's work among geologists, not only in Britain but internationally. Whatever Lyell himself might wish, the new geotheory was *not* to be treated as a package deal; on the contrary, its component parts were to be accepted, modified, or rejected on their separate merits.[3]

When the British Association for the Advancement of Science was founded with an inaugural meeting in York in 1831, the assembled "men of science" had asked relevant British savants to review the most important current issues in each of the sciences that were to be admitted to the magic circle of "Science" (the singular term, and its restricted scope, were soon contentious). Geology was of course prominent among these favored sciences; at the second meeting, in Oxford in 1832, it had been assigned to one of the four specialist "Sections" which met in parallel sessions.[4]

Sedgwick and Conybeare were asked to report on Élie de Beaumont's theory; in the event it was Conybeare who produced a review, assessing it in relation to

---

revolution" (1988). Wilson's Manichaean view of the argument between Lyell and his critics is summarized in his "Geology on the eve" (1980), which includes an intemperate attack on the less revolutionary interpretations offered (successively) by Hooykaas, myself, and Porter. Bartholomew, "Singularity of Lyell" (1979), is still one of the most insightful essays on this point.

2. Boué, "Progrès de la géologie" (1833), lxxiv, read 4, 18 February; "Sciences géologiques" (1834), 166–75, read 17, 24 February, 7 April; in the latter, he noted the urgent need for a French translation. He must have known that Omalius, in *Éléments de géologie* (1831), 1–2, had just proposed redefining "*géogénie*" to cover *all* kinds of causal analysis in geology (i.e., what had earlier been called *physique de la terre* or earth-physics: *BLT* §2.4); but this was not the meaning that he chose to apply to Lyell's work. Klaver, *Geology and religious sentiment* (1997), rightly puts Lyell in the same category as Buckland, Sedgwick, and Whewell (Conybeare could have been included too, but is less generally known to historians).

3. Boué, "Sciences géologiques" (1834), 216–43, referring to the revised version of Élie de Beaumont's "Révolutions du globe" (1829–30) inserted in De la Beche, *Manuel géologique* (1833) (see below).

4. Morrell and Thackray, *Gentlemen of science* (1981), 58–94, 451–60; this is the superb and definitive account of the early years of the BAAS; see also Cannon, *Science in culture* (1978), chap. 7, an insightful essay in an important and under-valued volume. Geology's Section, which it shared with geography, was initially given the label "III" (from 1835 it became "Section C", and has remained so ever since), being ranked after mathematics and the "exact" physical sciences but ahead of the biological ones; the number of Sections increased in later years.

the British evidence for folding and unconformities, which he knew at first hand as well as anyone. Having emphasized his great respect for the Frenchman, whom he had earlier met and seen at work in the field, Conybeare concluded that the theory seemed to be generally valid, although the successive phases of crustal movement had not been oriented as precisely and invariably as Élie de Beaumont claimed. In the classic case of the Isle of Wight, Webster's careful fieldwork (*BLT* §9.4), and that of others since, even allowed the spectacular "derangement" of the rocks to be dated, by Élie de Beaumont's criteria, to a point *within* the Tertiary era, or, still more precisely, within what Lyell was soon to define as the Eocene period (the London Clay had been turned up on end, but not the overlying formations that were correlated with others in the Paris Basin). Conybeare therefore felt justified in describing this movement as "a single and most violent convulsion". On the other hand, he also concluded that in the equally classic case of the Weald in southeast England (Mantell's home region) the broad dome of Chalk had probably risen slowly and gently, being gradually eroded at the same time, just as Lyell claimed (§26.2). The implicit moral of the story was that each case deserved to be evaluated on its own merits.[5]

The infant British Association had also commissioned about a dozen of its leading lights to write some much more substantial reports, which would review how "the Advancement of Science" could best be effected. Conybeare, who in 1830 had been honored with election as a "corresponding" (i.e., foreign) member of the Académie des Sciences in Paris, and who was widely regarded as one of geology's finest intellects, was an obvious choice. His "Report on the progress, actual [i.e., present] state, and ulterior prospects of geological science" was presented at the Oxford meeting; other savants surveyed other sciences in the same way.[6]

Conybeare's report covered the whole field of geology, but toward the end of it he reviewed Lyell's work. While criticizing his concept of "absolute uniformity" as over-rigorous, Conybeare praised the work as a whole for marking "almost a new aera in the progress of our science". More specifically he was impressed by Lyell's reconstruction of Tertiary geohistory, recognizing its theoretical importance as the era nearest to the known present; he approved Lyell's use of fossil mollusks as a kind of natural chronometer, and was convinced by his argument that there had been a gradual transition from the Tertiary into the present world. Applied to the Tertiary era, he found Lyell's idea of uniformity "felicitous and satisfactory".

However, Conybeare had reservations about the other meanings that Lyell gave to "uniformity". As in his more specific report, he praised von Buch and Élie de Beaumont for their theory of the sudden elevation of mountain chains, though he considered it had been rash or "unguarded" to have initially attributed the rise of the Alps, for example, to a *single* movement rather than to "the repeated disturbances of a long succession of geological epochs". Such natural events might have been of greater magnitude than anything witnessed in human history, but their frequency and intensity were matters for investigation, and their possible occurrence was certainly not to be ruled out in advance. Likewise he praised Fourier and Cordier for having rehabilitated the idea of a gradually cooling earth (§9.2), so he was critical of Lyell's treatment of the older rocks, and particularly of his proposed dissolution of the Primaries (§26.2). This revealed their other fundamental disagreement. Unlike Lyell, Conybeare was convinced that most of the evidence pointed towards a geohistory that had been directional, and that the earth had had a hot fluid origin: this was reflected in the peculiar character of the Primary rocks, and in the tropical character of many of the Secondary fossils, even those found in the Arctic (§12.3). Like Lyell's other critics, Conybeare recognized that the whole argument of the *Principles* was designed to reject any such directionality, and to claim instead that the "system" of the earth had been one of dynamic stability, from as far back in the deep past as there was any evidence to go on. In contrast to that Huttonian view, Conybeare was convinced that there *was* some "vestige of a beginning".[7]

27.2 DE LA BECHE AND "THEORETICAL GEOLOGY"

Another critic whom Lyell could not lightly dismiss was De la Beche, who, unlike Conybeare, was a man of almost his own age and, furthermore, equally anticlerical in his opinions. De la Beche had produced his album of *Sections and Views* (1830) in full knowledge of what Lyell was then about to publish (§13.2). While acknowledging the heuristic value of theorizing, he had offered his maps, sections, and landscape views explicitly as a kind of neutral inventory for aspiring theorists: these physical features, he said in effect, are some of those that must be taken into account in any satisfactory theoretical explanation. What he thought of Lyell's geotheory had become apparent in the caricatures that he sketched, and particularly the one he circulated, lampooning its Huttonian cyclicity (§22.3). Soon afterwards he had published a *Geological Manual* (1831), a very dry and densely printed volume, which Lyell—quite unrealistically—feared would compete with his own larger and more attractive work. However, one advantage that De la Beche's volume did have was that its publisher had offices in Paris and Strasbourg as well as London, so that the book became known at once on the Continent. It was successful enough to reach a third edition in as many years, with much enlargement and updating, and it was soon translated into both French and German.[8]

However, De la Beche's *Manual* was, as the title made plain, a compendium or work of reference; the author's own theoretical opinions were kept in the background. So he then wrote a volume of *Researches in Theoretical Geology* (1834),

5. Conybeare, "Parallelism of lines of elevation" (1832–34), esp. 122–23, in August 1832 issue of *Philosophical magazine* (the article was concluded two years later).

6. Conybeare, "Report on geological science" (1833). On the BAAS reports, see Morrell and Thackray, *Gentlemen of science* (1981), 474–79, 491–96.

7. Conybeare, "Report on geological science" (1833), 398–407; the report was accompanied by a very long colored geological section from the north of Scotland across the Alps to Venice, synthesizing international research in much the same way as his earlier map (Fig. 3.3). By the time he revised his report for publication he was able to take account of all three volumes of Lyell's work.

8. De la Beche, *Sections and views* (1830) and *Geological manual* (1831), both published by Treuttel & Würtz; the second edition of the latter (1832) was translated by von Dechen, as *Handbuch der Geognosie* (1832), and by Brochant, as *Manuel géologique* (1833). Brochant also inserted (616–68) Élie de Beaumont's revised version of his theory of occasional "époques de soulèvement" (§9.4).

Fig. 27.1 De la Beche's drawing of the earth, "Supposed to be seen from Space"; it depicted the very slight polar flattening (into an oblate spheroid) that was generally attributed to the planet's hot fluid origin. This engraving was published as the frontispiece—and therefore implicitly the epitome—of his *Theoretical Geology* (1834), a volume that presented his version of the directional kind of geohistory that was adopted at this time by almost all leading geologists other than Lyell.

tacitly in response to Lyell's work. At the very outset he declared that "the theory of central heat and the former igneous fluidity of our planet" was more plausible than any alternative. In fact, the book was structured around that physical theory and the directional geohistory that it entailed. Although De la Beche may not have seen Lyell's work in full before his own volume went to press, the latter was a highly competent summary of current opinion among geologists, and not least of the almost consensual geohistory that Lyell's neo-Huttonian cyclic model opposed. Like his rival, though much more briefly, De la Beche reviewed the repertoire of actual causes and the stratigraphical sequence of rock formations that might be explained by them. But unlike Lyell he dealt with the strata in true chronological order, all the way from the "Grauwacke group" (the Transition rocks of other geologists) through to the Tertiaries (for which he preferred his own term "Supracretaceous group").

In keeping with the same geohistorical order, all this was preceded by De la Beche's review of the "Inferior stratified rocks" (the Primaries of other geologists); he interpreted them in the usual way, explaining their often distinctive composition and structure as the result of their having been deposited while the world's oceans were still hot (and perhaps of a quite unmodern composition), and inferring that they dated from before the first appearance of life. And he began the book at the very start of geohistory, with an astronomical assessment of the earth, considered as one of the planets in the solar system. The frontispiece showed it in that light, and with characteristic precision De la Beche drew the planet with the very slight polar flattening that had long been treated as one of the most telling signs of its original fluidity (Fig. 27.1).[9]

27.3 SCROPE AND THE REVISED *PRINCIPLES*

Lyell's dialogue with his knowledgeable critics had of course already begun (Chaps. 22–24). In the face of further shrewd criticism from geologists such as Boué, Conybeare, and De la Beche, his next move had to be made without delay. As soon as his third and last volume was off his hands he started revising the whole work for the new edition that Murray had promised to publish in a cheaper format (since the first two volumes had already been reprinted in an almost unchanged second edition, the new one was to be called the third). Murray delayed bringing it out until most of his stock of the earlier and more expensive volumes had been sold, so Lyell had time to revise the whole work. It was duly published in 1834–35, in four smaller volumes and at less than half the price; it was almost as cheap per volume as the good non-fiction books in the same format in Murray's "Family Library" series, and could be expected to sell as well and as widely. He added much new information about actual causes, describing for example the sudden eruption and rapid demise of Graham Island or Île Julia (§23.4); and he provided new pictorial illustrations of some of the examples he had already described in words (Fig. 22.4).[10]

Other changes in Lyell's text took account of the informed criticism that the work had already received. For example he included hypothetical world maps (Fig. 21.3) to illustrate and thereby strengthen his controversial model for explaining climatic change without conceding directionality. But most significant in this respect were two largely rewritten chapters on volcanic and seismic processes, which were likewise designed to bolster his steady-state interpretation against the directionality that almost all other geologists saw in the traces of geohistory. While conceding the reality of the earth's internal heat (as evidenced by the temperature gradient in mines), Lyell rejected Cordier's claim that it pointed to a "central heat" that in turn could be regarded as a residue of the earth's hot origin (§9.2). But here in the depths of the earth Lyell was almost out of his depth scientifically, because the issues required knowledge of physics and chemistry that he did not possess; and these "causes", while certainly "actual", were not directly observable at all, and could hardly be discussed except in the speculative style that he lost no opportunity to censure in others.[11]

9. Fig. 27.1 is reproduced from De la Beche, *Theoretical geology* (1834), frontispiece; on the original engraving the diameter measures 70mm on the equator, 69mm from pole to pole. Since he made no reference to Lyell's "metamorphic" interpretation of the Primaries (§26.2) it is likely that when his book went to press he had not yet seen the third volume of *Principles*.

10. On its price, see Secord, "Introduction [to Lyell's *Principles*]" (1997), xiv.

11. Lyell, *Principles*, 3rd ed., 2 (1834): 273–322; it is beyond the scope of the present work to describe and analyze all his revisions (listed in 1: xvi–xix) in the detail they deserve, or those of his subsequent editions.

However, perhaps the most significant change in Lyell's volumes was one of the smallest: the subtitle defined it no longer as "an attempt to explain the former changes of the earth's surface by reference to causes now in operation", but more modestly as "an inquiry how far" they might be so explained. Those two little words *how far* represented a tactical retreat on Lyell's part, a tacit admission that the crucial issue was the collective *adequacy* of actual causes, and a recognition that the onus was on him to demonstrate that adequacy in each specific case.

Lyell's new edition provided an occasion for Scrope, his closest ally in England, to return to geology by reviewing his friend's completed work. Since reviewing the first volume (§22.2), Scrope had had little time for anything scientific, and had excused himself from reviewing the third volume. He had unsuccessfully contested two Parliamentary seats near his Wiltshire home against those he called "the rotten borough mongers" (one was so "rotten" that there were only thirteen electors). More actively than Lyell, he was strongly on the Whig side in the campaign for electoral reform, though he was too independent ever to be a party man. He was more than content to write for the Tory *Quarterly Review*, evaluating many new publications on political economy, because for him (as for Lyell) that periodical reached the elites he most wanted to influence. After the Reform Bill was passed and the electorate enlarged, Scrope contested a more promising seat (for the industrial town of Stroud) but was narrowly defeated; that result was overturned, however, on grounds of electoral malpractice, and in 1833 he was elected unopposed, becoming at last a Member of Parliament. He was then more easily persuaded to take time off to review Lyell's third edition, again for the *Quarterly*. Like Conybeare but, as it were, from the opposite direction, Scrope showed himself to be appreciative of Lyell's work, yet far from uncritical. While Conybeare was on Whewell's definition a leading "catastrophist" (§24.4), Scrope was far from being a "uniformitarian" in the same sense as Lyell.[12]

Scrope echoed Conybeare when he judged that the publication of Lyell's *Principles* "will always form an epoch in the history of geology", because it challenged the previously dominant Cuvierian assumption that actual causes were *not* wholly adequate to explain the deep past. In fact, Scrope noted that earlier savants, "in their ardour for explaining every thing"—that is, for geotheory—had caused the Geological Society to go to the opposite extreme, so that "for a time theory was *tabooed* by common consent" and the science had "shrunk into little else than a barren descriptive arrangement of the rocks which coat our planet". But although Scrope himself found such sober stratigraphical work tedious, it had in fact laid the foundations "for the building which Mr. Lyell, in a happy moment, undertook to raise". The metaphor expressed a flattering interpretation of the recent history of Lyell's science; once again the *Principles* was treated as a revival of the genre of geotheory, albeit on firmer ground than before. Echoing his own earlier purple prose (§15.1), Scrope agreed with Lyell that "TIME is, in truth, the master-key to the problems of geology" and that "the concession of an unlimited period for the working of the existing powers of nature has permitted us to dispense with the comets, deluges, and other prodigies which were once brought forward *ad libitum*". The familiar slur on catastrophist theorizing, while more justly applicable to much earlier generations than that of Buckland and Conybeare, certainly showed where Scrope's loyalties lay.[13]

However, while most of Scrope's long essay was devoted to an appreciative summary of Lyell's massive work, he was skeptical about those aspects that were alleged to undermine a directional interpretation of geohistory. Referring for example to that part of "the theory of Hutton" represented by Lyell's "*metamorphic* theory", Scrope stated that "we [i.e., he himself] are certainly no converts to it yet". While disclaiming any ambition to offer an alternative, he did suggest a "hint" for Lyell to consider for his next edition. He sketched a possible scenario for the formation of a stratified Primary rock such as gneiss, not by Lyell's metamorphic alteration from something quite different, but by the accumulation of material derived from the erosion of newly erupted granite on the floor of a very hot ocean very early in geohistory; "the resulting rocks must partake very much of the character of his metamorphic class". Likewise, Scrope balked at his friend's dismissal of Cordier's "theory of central heat". Noting the diversity of explanations of heat itself—a topic that was currently contentious among physicists—he declared himself in favor of "the notion that the globe is gradually cooling down, and still retains an intense temperature below its surface", though he suggested that the equally intense pressure at great depth might mean that the interior was no longer in a fluid state. All this was unavoidably speculative; but once the reality of a central heat was conceded, it was available as a possible explanation not only for crustal movement and igneous activity, but also for the gradual cooling suggested by the earth's changing faunas and floras (§12.2) and other apparently directional features of geohistory.[14]

Scrope argued that the "practical difference" between Lyell and himself was a strictly empirical question: "whether or not there *are* traces on the earth's surface of former changes of a more violent and tumultuary character than such as habitually occur at present"; or, putting it another way, "whether the present order of change is *cyclical*, and uniform in amount through equal periods, or progressive [i.e., directional] and, on the whole, diminishing in violence". Scrope emphasized that the latter possibility—the option preferred, of course, by himself and almost all other geologists—did not "involve any doubt (as Mr. Lyell seems to imagine) of the permanency of the existing laws of nature": ordinary physical laws were not violated in any way by the hypothesis of a gradually cooling earth. The real point at issue was not "the constancy of the laws of nature"—which, Scrope pointed out, "no one disputes"—but the pattern of geohistory as either directional or cyclic; "and on this point Mr. Lyell must be content to join

12. Scrope to Lyell, 18 May 1833 [Philadelphia-APS]. His articles on political topics in the *Quarterly* are listed in Sturges, *Bibliography of Scrope* (1984), which also summarizes his political career; he retained his seat for Stroud (Gloucestershire) until he resigned in 1867 at the age of seventy. His *Principles of political economy* (1833)—the title paralleled Lyell's—was his most substantial publication, but Poulett Scrope earned the nickname of "Pamphlet Scrope" for his prolific output of shorter (but more influential) works. In contrast, his spoken contributions to Parliamentary debate were rare: he jokingly excused himself by asserting that "a Parliamentary reputation is like a woman's; it must be exposed as little as possible" (Sturges, 25n).

13. [Scrope], "*Principles* by Lyell" (1835), 407–10.

14. [Scrope], "*Principles* by Lyell" (1835), 413–15, 443–46.

issue with other geologists, under the disadvantage of all analogy being against him".[15]

Scrope therefore went on to tackle the thorny issue of eternalism. In his earlier review he had criticized Lyell for asserting that "the existing causes of change have operated with absolute uniformity from all eternity" (§22.2). Lyell had denied this, insisting that "no vestige of a beginning" referred simply to the limits of human knowledge, and pointing out that Playfair had long ago defended Hutton in the same way (§26.2). Scrope now replied that if Lyell had in fact confined himself to that modest claim, he himself would have been content; "but he [Lyell] went farther, and declared it to be unphilosophical [i.e., unscientific] to look for traces of a beginning, or to imagine it possible that we should discover such". This, Scrope argued, was certainly misguided. All the component items of geohistory—"the different states of the earth's surface, and the different species by which it has been inhabited"—had evidently had beginnings and endings at determinate points in time, so the same might well be true of the earth itself. The usual analogy between space and time made the point: "as astronomy has proved this planet to be a mere speck in the immensity of space, so geology *may* prove that . . . it has had a beginning, and will probably therefore have an end". In short, the issue was an empirical one, open to investigation, and not to be foreclosed by dismissing even the search for traces of a beginning (and thereafter of directionality) as "unphilosophical". In *this* sense of "uniformity", Scrope argued, it was Lyell's position, not his own, that was truly unscientific.

In conclusion, Scrope turned Lyell's arguments against him by borrowing his clothing of natural theology. To search for "the first-formed strata"—though without assuming that they had yet been found or perhaps ever would be—was no more irreverent or unwarranted than "Mr Lyell's subterranean cookery of sedimentary strata into granite" or any of his other legitimate speculations. The divine attributes were "not degraded, but rather exalted" by a vision of countless worlds being successively "called into existence" while others "decay and become extinct". So Scrope neatly turned Lyell's grandiloquent ending (§26.2) on its head, by inverting one crucial word. He claimed that what was really "inconsistent with a just estimate of our own [cognitive] powers and of the attributes of the eternal and infinite Creator" was to assume that evidence of a beginning or ending of the earth must lie "*without* [i.e., beyond] the reach of our philosophical speculations" (Lyell had written "within"). In short, Lyell had no right to call on natural theology to bolster his claims for "uniformity" or to undermine those of his opponents.[16]

Scrope's review was a devastating critique of Lyell's claims for "absolute uniformity", all the more so for coming from his closest ally among the English geologists and for being expressed, no doubt sincerely, "in the same friendly spirit" as the earlier exchange between the two. Geologists who adopted the usual interpretation of geohistory—certainly as directional, and probably as both hotter and more turbulent in its earlier phases—were not to be dismissed as indulging in miracle-mongering, or as believing that the laws of nature had formerly been different, or as having an inflated view of human cognitive capacities. According to Scrope, they were no less "philosophical" than Lyell, and their modes of reasoning no less rigorous. Above all, the issues dividing them from Lyell were strictly empirical, and therefore capable of being resolved one way or the other in the course of further research. On the other hand, Scrope had nothing but praise for his friend's inventory of actual causes. He agreed that they would prove adequate to explain far more than had hitherto been thought possible, provided they were combined with the vast spans of time that all geologists professed to deploy but that many still failed in practice to appreciate adequately. In effect, Lyell's "uniformity" confused several distinct meanings, and they were not all equally plausible. In distinguishing between the fully acceptable components of Lyell's geotheory and those that were far more questionable, Scrope's review epitomized once again the response of many other geologists to what was on offer in the *Principles*.

27.4 SEDGWICK AND "SUBTERRANEAN COOKERY"

As already emphasized, the most serious obstacle in the path of Lyell's geotheory was the strong opinion of almost all other geologists, not least his friend and ally Scrope, that geohistory had been essentially directional in character, not cyclic or steady-state. The evidence that they found most persuasive was of two kinds. As organic evidence, there was the directional—and perhaps also, in the modern sense, progressive—fossil record in the Secondary and Tertiary formations (see Chap. 30). The inorganic evidence focused on the distinctive character of the Primary rocks. Here, near the apparent start of geohistory, Lyell's metamorphic theory was generally given a rough ride by other geologists. It had long ago been agreed that *some* granites were far from Primary in age, since they were clearly intruded into Transition rocks, and even into Secondaries; and their putatively intense heat appeared to have "cooked" or altered those surrounding rocks (in modern terms, this was *contact* metamorphism). On that specific point it could be said that all geologists were now Huttonians. But it was quite another matter to agree with Lyell that *all* the apparently ancient-looking schists and gneisses were nothing more than altered Secondary sediments (in modern terms, the products of *regional* metamorphism), or that in consequence the bottom deserved to be knocked out of conventional geohistory (§26.2).

A few years earlier, for example, while Brongniart was exploring what he took to be diluvial features in Sweden (§13.4), he had also followed local geologists in studying the Transition formations, particularly in pursuit of their fossil trilobites (§4.3). In Sweden, unlike most other regions, these strata were still in their original horizontal position (see inset map and section in Fig. 13.8); they were almost as unaltered as, say, Smith's best English formations (those that by now were being called "Jurassic"), and their fossils were almost equally well preserved. But the most striking Swedish fossils were not ammonites but trilobites, and the

15. [Scrope], "*Principles* by Lyell" (1835), 447–48; the word "progressive" was used at this time, in widely diverse contexts, to denote changes consistently in a certain direction, without any necessary overtones of improvement; "analogy" here meant convergent lines of evidence.

16. [Scrope], "*Principles* by Lyell" (1835), 446–47.

dating of the strata as Transition—by definition the oldest with any fossils—seemed beyond question. Yet they overlay gneiss and similar rocks, exposed over vast areas in Sweden, which had obviously reached their present state even further back in the depths of geohistory. Here, at least, the Transition formations did *not* shade gradually into the Primary rocks; on the contrary, there was a complete discontinuity between them, in the shape of an unmistakable major unconformity. To most geologists, Scrope included, Lyell's metamorphic explanation of rocks such as gneiss therefore seemed forced and unnatural. It was much simpler to treat gneiss at face value, as a truly Primary rock that had been formed *somehow* at that very early period in geohistory, long before even the archaic trilobites had appeared on earth.[17]

On the other hand, Lyell found at least a crumb of comfort in the fieldwork that Sedgwick was doing, trying to unravel the structure of the Transition rocks of north Wales. These were more typical of that category, in being much disturbed and folded, and with fossils that were rare, poorly preserved, and often difficult to find. Yet the fossils were there, mostly trilobites and some of them recognizably similar to the far better specimens from Sweden; so there was little doubt that the rocks were roughly of the same age. Sedgwick had begun to work in north Wales in 1831, hoping to find a clear stratigraphical sequence downwards from the Secondaries. Instead he found that the usual Carboniferous formations were underlain, below a major unconformity, by an apparently chaotic mass of slates and greywacke. In successive summers away from his teaching work at Cambridge, Sedgwick made the best of a bad job: he described it later as "like rubbing yourself against a grindstone" (and, like a knife on a wet grindstone, he also had to work in frequently atrocious Welsh weather). His reward for all this hard work was not only a sequence of Transition formations—though they remained difficult to unravel (see §30.3)—but even more an understanding of why they were so difficult. Sedgwick, like other experienced field geologists,

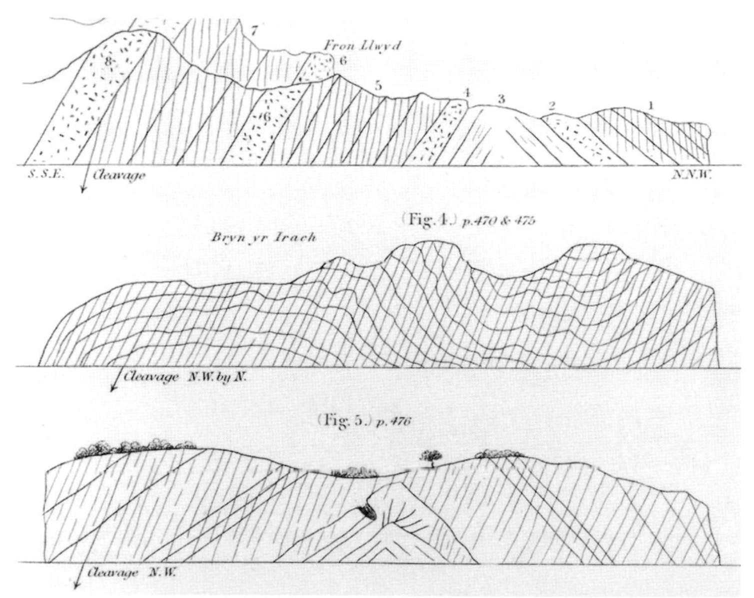

Fig. 27.2 Some of Sedgwick's sections of Transition (or "Grauwacke") strata in Wales, published by the Geological Society in 1835. He argued that the original bedding of these highly folded rocks had later been obscured by the imposition of "slaty cleavage" at a uniformly high angle. If slaty cleavage were mistaken for bedding—as it often had been previously—serious errors would be made about the structure and sequence of the rocks. Although the slates split most easily along the cleavage planes (making them useful for roofing), the true bedding could be detected on closer scrutiny by slight differences in color and texture, or by beds of coarser material (top section, stippled) relatively unaffected by the cleavage. Slaty cleavage was widely interpreted as a mild form of what Lyell called "metamorphic" change.

recognized that the planes on which the slates split most easily (often making them valuable as a roofing material) were unrelated to the bedding or layering of the original muddy sediment, and rarely coincident. His major paper explaining structural features such as this "slaty cleavage" was read at the Geological Society early in 1835. It was considered so important that it was published with almost unprecedented speed later the same year, taking priority over other papers already in the pipeline (Fig. 27.2).[18]

This apparently arcane point had profound implications. The geohistorical kind of explanation now routinely applied to strata and formations was here extended to matters of rock structure and composition: the rocks had obviously been deposited first as sediments (and the trilobites entombed in them), then folded or buckled, and still later affected by the slaty cleavage. Although the physical and chemical processes involved in that final stage remained obscure, it could be regarded as a mild form of the "metamorphic" effect that Lyell was claiming as a much more general phenomenon. If the same kind of processes had acted more intensively, or repeatedly, it was conceivable that *all* trace of the original bedding might be lost, along with all trace of fossils, and that the end product might then be an apparently "Primary" rock. However, as on other issues, most geologists only followed Lyell some of the way down that road, not the whole way. They conceded that Sedgwick's old Transition formations had been altered by slaty cleavage, but not that the Swedish gneiss and other basement rocks were nothing more than overcooked slates, nor that the final product of Lyell's "subterranean cookery" might be nothing less than granite.

27.5 CONCLUSION

The full publication of Lyell's *Principles*, while causing no scientific revolution, certainly provoked intense argument among geologists. They declined to regard his geotheory as a unified package, and instead they continued to accept some components while criticizing or rejecting others. In general, they welcomed his enlarged repertoire of actual causes as capable of explaining more than had hitherto been supposed; but they rejected his claim that there had never been any events more sudden or violent than those witnessed in human history, and even more his claim that there had been no overall directionality in geohistory. Above all, therefore, they unpacked his unitary concept of "absolute uniformity" into its component meanings, some of them more plausible than others.

17. In modern terms, the ancient Precambrian metamorphic rocks of Sweden were planed off by erosion before the deposition of the Lower Paleozoic formations; and the crust in the Baltic region has been so stable ever since that these ancient fossiliferous rocks are still preserved with their original horizontality.

18. Fig. 27.2 is reproduced from Sedgwick, "Large mineral masses" (1835), pl. 47, figs. 2, 4, 5, (read 25 March); see Rudwick, *Devonian controversy* (1985), 51–52, 69–73, and, for Sedgwick's broader project, Secord, *Victorian geology* (1986), esp. 46–68. Bedding, jointing, and cleavage were routinely distinguished by other geologists but had not previously been analyzed so clearly; "slaty" cleavage was so called to distinguish it from the quite different phenomenon of cleavage within individual crystals of minerals such as mica.

Boué was typical of Lyell's critics—and a salutary reminder that the debate was not confined to Britain—in that he identified the *Principles* as a revival of the old genre of geotheory, albeit updated to take account of the most recent research. Boué also juxtaposed Lyell's concept of uniformity with Élie de Beaumont's influential theory that the long tracts of geohistory had been punctuated by occasional sudden episodes of crustal upheaval. When the British Association for the Advancement of Science was founded, one of the first actions of the geologists among the assembled "men of science" was to ask Sedgwick and Conybeare to evaluate Élie de Beaumont's ideas in relation to British geology; Conybeare did so, confirming its plausibility in many cases while also endorsing Lyell's more gradual concept of crustal movement in others.

Conybeare was also asked by the British Association to report on the current state of geology in general, in the course of which he again gave Lyell's work high praise but discriminate criticism. So did Lyell's friend Scrope, when he turned briefly from politics to geology and reviewed the third (and inexpensive) edition of the *Principles*. Although Scrope and Conybeare approached Lyell's work from opposite directions in the debate about "uniformity", they reached convergent conclusions. They welcomed his emphasis on the explanatory power of actual causes, but criticized his rejection of Élie de Beaumont's occasional "revolutions", and even more his rejection of any directionality in geohistory. So did De la Beche, in his own restatement of the evidence for a hot fluid origin for the earth and its consequently directional geohistory. And Lyell's metamorphic theory, as a way of undermining that evidence, was generally given a rough ride by other geologists. Following Sedgwick's analysis of the old Transition rocks of Wales, slaty cleavage was soon accepted as a mild form of metamorphic alteration; but Lyell's much more radical idea that *all* the so-called Primary rocks were the products of even more intense "subterranean cookery" was found deeply implausible.

However, the most striking evidence that geohistory had *not* been cyclical or steady-state in character as Lyell claimed, but unmistakably directional, continued to come not from rocks and mountains but from fossils, from the records of the history of life. And this evidence, which is the subject of the following chapters, became ever stronger and clearer during just the years when Lyell was expounding his contrary argument. The next chapter begins this review by describing the further accumulation of evidence that seemed to tie the human species, and indeed the primates as a whole, back into geohistory.

CHAPTER 28

The human species in geohistory (1830–37)

28.1 TOURNAL CONFRONTS THE SAVANT WORLD

In the early 1830s, the question of human antiquity remained one of the most refractory problems in geohistory. The much earlier claims that traces of humans could be found in the "regular" strata (*BLT* §5.4) had long since been dismissed by savants. By the 1830s, and in terms of the stratigraphy that had been constructed during the preceding half-century, geologists were agreed that no authentic human fossils or artifacts had been found even in the Tertiary formations; in terms of the corresponding geohistory, even the most recent part of the Tertiary era—Lyell's Newer Pliocene period—was taken to belong to *prehuman* geohistory. Human history was tacitly equated with the history of human records, or with the history of *literate* humankind; even in the wake of Champollion's successful deciphering of the hieroglyphs (§20.2) it could not be traced back further than the early dynasties of ancient Egypt, only a few millennia in the past. Between that point and the most modern Tertiary formations lay the disputed territory of those deposits—neither clearly "modern" nor clearly Tertiary—to which Buckland had given the theoretically loaded name "diluvium". It was a name that had stuck because it was convenient, although many of those who used it were becoming skeptical about the character and even the reality of the "geological deluge" that it denoted, and still more about the equation of any such event with the biblical Flood (§23.1). Whether the geohistorical period represented by the diluvium had been human or prehuman remained highly contentious. In practice the question of human antiquity devolved into that of the contemporaneity—or otherwise—of humans and the extinct "antediluvial" mammals.

While Lyell was planning and writing his *Principles*, this problem of the antiquity of the human species—and hence the relation between humankind and geohistory—refused to go away. The young French naturalists Christol and Tournal were insisting, with increasing confidence based on their excavations of caves in southern France, that the cumulative evidence for the co-existence of early humans with the extinct mammals of the Superficial deposits was becoming almost irresistible (§16.2). But Cuvier did resist it, sticking firmly to his long-standing belief that the evidence was insecure; Tournal visited him in person, armed with yet more specimens, but still failed to convince him. The issue was important enough, however, for the Académie des Sciences in Paris to appoint a committee to assess it, though with Cuvier as its chairman its impartiality was uncertain (§16.3).

Cuvier's committee, perhaps over-conscious of the Académie's responsibility for giving sound judgments on controversial scientific issues, certainly dragged its feet, with the convenient excuse that it needed still more evidence; Cuvier himself was certainly extremely reluctant to concede that his entrenched position on human antiquity might be mistaken. But he and his colleagues were not the only ones to drag their feet. Serres, who had been holding many of Tournal's best specimens in preparation for their proposed joint publication, did not send any fossils to Paris until after his young protégé had visited Cuvier. By that time the capital was in the turmoil that was about to erupt into the July Revolution of 1830 (§23.2). It was not the most propitious time for the calm consideration of scientific issues, however important they might be. Still less was it a time at which it was prudent for metropolitan savants—least of all, for one as deeply involved in national politics as Cuvier—to leave Paris and travel to remote parts of Languedoc to see the field evidence for themselves.[1]

Tournal did not give up his efforts to convince the savant world. He sent a paper to the provincial Académie in Toulouse, in which he summarized his own and Christol's finds. He reported a horse-sized fossil bear in the Bize cave, defended his major claim that "*man also is found in the fossil state*", and argued that "man already lived in society . . . [and] even in a quite advanced state of civilization". The permanent secretary of the Toulouse Académie, Jean-François d'Aubuisson de Voisins (1769–1841)—Cuvier's contemporary and a respected geologist in the Mines Council—supported Tournal's conclusion that "when extinct races of bears, rhinoceros and hyenas existed, man lived in our regions [i.e., Languedoc]" at a level of civilization similar to that of existing Native Americans; and he added that "it appears that the honor of having been the first to have shown this belongs to Mr Tournal". Privately, however, Aubuisson criticized Tournal for continuing to use the theoretically loaded term *terrain diluvien*, and urged that a clear "wall of separation must be built" between scientific and religious claims. It was advice that the younger man took to heart.[2]

Perhaps emboldened by Aubuisson's positive response, Tournal sent another paper to Paris, this time to the newly founded Société Géologique (§23.2). He cut through the convoluted arguments over the correct meaning of "fossil", with which Serres—not for nothing a lawyer by profession—had confused the question of human antiquity. What mattered most, Tournal argued, was not the state of preservation of the specimens but the contemporaneity of human bones and those of extinct animals, for which he claimed that the *field* evidence had become overwhelming. More explicitly than before, and following Aubuisson's advice, he abandoned any literally "diluvial" explanation of the Superficial deposits; like his compatriots Croizet and Jobert (§15.2), he claimed that they represented a long period of time, not a single sudden event, and that there was complete continuity between them and the deposits still being formed in the present world: "there is an insensible passage between the present epoch (historical) and the former [*ancienne*] epoch (geological)". Above all, his own and Christol's caves proved the case for contemporaneity, and showed not only that the human species already existed but also that "at that epoch man already lived in society".[3]

However, Tournal's forceful restatement of the case for human antiquity did not go unchallenged at the Société Géologique. Desnoyers, one of its secretaries, who had recently made his mark with his research on the Tertiary formations (§12.1), argued that the crude pottery found by Tournal in the Bize caves was no sign of high antiquity: it was "not at all an antediluvial industry", since similar wares were found in burial mounds and other sites from the "Gallo-Roman" [Iron Age] period, and the Roman literary sources proved that the indigenous cultures of that time had been technically quite primitive. In effect, Desnoyers suggested that Tournal's caves contained a mixture of materials from two quite different periods: genuinely fossil animal bones, mixed with pottery and other human remains from the periods of recorded history. A few days later he repeated this rejection of what the young provincial was claiming, and gave his own argument the widest publicity, when he included a review of the question of human antiquity in his comprehensive survey of the society's recent research. This was printed promptly in its *Bulletin* and was therefore widely noticed by geologists everywhere.[4]

Tournal responded without delay. He criticized Desnoyers's claim that the deposits in the Bize caves must be mixed in origin, pointing out that if this inference continued to be made—on no better basis than the a priori assumption that humans could not have lived among the extinct mammals—it would prevent the issue from ever being resolved. In contrast, his own interpretation was based not only on the identical preservation of the animal and human bones, implying

1. Académie des Sciences, 28 June 1830 [*RB* 1830: 87–88]. Cuvier would have been unlikely to visit the caves, even if the times had been less turbulent: he had spent his whole career primarily as a museum naturalist, and had rarely ventured into the field since his work with Brongniart around Paris in the first years of the century (*BLT* §9.1). In fact, with either prudent foresight or just good luck, he traveled to England with his stepdaughter Sophie Duvaucel just before the revolution erupted, and returned after calm had been restored.

2. Tournal, "Cavernes à ossemens" (1834), is in fact mainly Aubuisson's report (dated 1831) on Tournal's papers. Aubuisson to Tournal, 28 April 1831 [Narbonne-AM], quoted (in translation used here) in Grayson, "Time depth for paleoanthropology" (1990), 7; this summary of his *Human antiquity* (1983) adds valuable material on Tournal.

3. Tournal, "Observations sur les ossemens humains" (1831), read on 16 May. Serres's confusing semantic arguments about the term "fossil" are set out in, for example, his "Lettre adressée au Président" (1830) and "Sur les ossemens humains" (1830).

4. Desnoyers, "Ossemens humains des cavernes" (1832), read on 30 January; "Travaux de la Société Géologique" (1832), 250–55, read on 6 February, complemented the more wide-ranging international survey in Boué, "Progrès de la géologie" (1832), read on 30 January.

their contemporaneity, but also on his observation in the caves themselves that the sediments had clearly accumulated gradually and without any mixing. But whereas the former claim might be evaluated by a museum study of specimens, the latter could be judged only *in the field*, which is just where Tournal's critics could not or would not go.[5]

Less than a month later, however, the most prominent and persistent of Tournal's critics was removed unexpectedly from the debate. Cuvier died suddenly on 13 May 1832, after a brief illness, at the age of only sixty-three. Whatever his human failings may have been, his positive impact on the savant world over a span of almost four decades had been immeasurable. His death was a dramatic loss for all the sciences in Paris, in France, and around the world, and it marked the end of an era (Lyell's sincere tribute, inserted into his London lectures on hearing the news, has already been mentioned: §25.1).

A few months later, the distinguished astronomer and mathematician Dominique François Jean Arago (1786–1853)—the other permanent secretary of the Académie in Paris, but no friend of the deceased Cuvier—invited Tournal to send him a paper on the Bize caves for the *Annales de Chimie et de Physique*, the prestigious periodical that he co-edited with the chemist Louis Joseph Gay-Lussac (1778–1850). In effect, this would enable Tournal to bypass the geologists, appealing over their heads to a wider scientific audience. He duly sent Arago a major paper entitled "General considerations on the phenomena of bone caves". He noted that the question of the antiquity of man, and of human contemporaneity with the extinct mammals, had regrettably turned savants into "partisans [of] opposed camps". He was careful to praise Buckland as a pioneer of cave research, while deploring the fact that his title *Reliquiae Diluvianae* had prejudged the "diluvial" issue (and also regretting that this valuable book, a decade after its publication, had still not been translated into the international scientific language). For Tournal, the question of the biblical Flood had now become an incubus on the study of the most recent portion of geohistory; copying some of Aubuisson's very words, he wanted to avoid further confusion by keeping science and religion apart. Once again he argued that the misnamed *terrains diluviens* or Superficial deposits had been formed over a very long period, by local agencies that had produced equally local and varied deposits; no single or unique deluge had been involved. But the heart of his argument, like Christol's, was the evidence for contemporaneity in the caves in Languedoc: "we did not hesitate, I say, to proclaim, and we still maintain today, that *man exists in the fossil state*".[6]

In the long run, however, what was even more important was Tournal's proposed periodization of recent geohistory, which he had previously hinted at only in passing. A "former [*ancienne*] geological period", prehuman and inconceivably vast in duration, was contrasted with the subsequent and relatively brief "modern geological period" of human existence. This had long been conventional and uncontroversial. But Tournal now proposed that the "modern" period should be redefined and divided into a "*période antehistorique*", beginning at the first appearance of the human species and probably very long in duration, followed by the "historic period" of recorded human history, beginning—as suggested by Champollion's decipherment of the Egyptian hieroglyphs—no more than about seven millennia in the past. This was a supremely important proposal, because it explicitly defined a conceptual space that had been opening up implicitly for some time, for what would later be called *prehistory*. It ascribed to human history a very lengthy *preliterate* period, for which—as Tournal had already claimed—the evidence would have to be exclusively *geological* in character, not literary and not antiquarian (or, in the then current sense, archaeological). And by contrasting his own caves with the inferentially much older ones described by Christol, he offered at least a hint of how the "antehistoric period" itself might need to be differentiated into a sequence of distinctive phases, marked for example by the piecemeal extinction of the more exotic mammals and by the technical progress of the human population.[7]

Notwithstanding what had been Cuvier's dogged objections to the last, the evidence for a great antiquity for the human species was becoming ever stronger. But the case was retarded by contingent circumstances, though whether as cause or effect is difficult to judge. Christol and Tournal never published full accounts of their finds; indeed both dropped out of the debate. Christol, having published his preliminary account of the Sommières caves in 1829 (§16.2), produced nothing further on human antiquity, though he continued to work on bone caves until he moved from Montpellier on being appointed professor of geology at Dijon; he tried to get Buckland's *Reliquiae* translated, but seems never to have received its author's promised updating of its argument. Tournal's final broadside was the paper just described, published in 1833 in a prestigious periodical that enabled him to bypass the geologists who had summarily dismissed his report. In 1834 he was appointed curator of a new museum in his home city of Narbonne (where he remained for the rest of his life), but he became increasingly involved in local politics. However, it is not clear whether these two provincials withdrew from the fray because they were disheartened by the opposition they had encountered in the capital, or whether their case faltered because they were not there to press it further.[8]

Serres, on the other hand, did not give up. That the problems posed by bone caves were prominent in the minds of geologists throughout Europe is evident from van Marum's proposal, in 1831, that the Hollandsche Maatschappij in Haarlem should offer a gold medal and cash prize for an essay on this topic. Four

5. Tournal, "Cavernes à ossemens de Bize" (1832), read on 16 April.

6. Tournal, "Phénomènes des cavernes à ossemens" (1833); Arago to Tournal, 15 September 1832 [Narbonne-AM], quoted (in this translation) in Grayson, "Time depth for paleoanthropology" (1990), 7.

7. Tournal, "Phénomènes des cavernes à ossemens" (1833). Reboul, *Période quaternaire* (1833), 1–5, divided a post-Tertiary "Quaternaire" into "antehistorique" and "historique" periods at almost the same time as Tournal; but "Quaternary" was slow to enter the vocabulary of geologists. Stoczkowski, "La préhistoire" (1993), rightly emphasizes Tournal's conceptual innovation in relation to the later, and better known, work of Boucher de Perthes. It was only then, later in the century, with the belated acceptance of the coexistence of early humans with the Pleistocene megafauna, that the word "archaeology" was extended in meaning to cover the vast spans of preliterate "prehistory": see Van Riper, *Men among the mammoths* (1993), chap. 7.

8. Christol to Buckland, 12 August 1836, 20 February 1837, and other letters [Oxford-MNH]. Serres *et al.*, *Ossemens de Lunel-Viel* (1839), 48, mentioned Christol only in a brief note on his Sommières work (§16.2).

years later Serres was announced as the winner, for his book-length survey of the evidence from caves all around Europe. He concluded that there were several kinds of "diluvial" deposits, representing not one but several "grandes cataclysmes"; but that some of the animals found in caves had become extinct since the appearance of the human species, some of them after humans had developed technical skills, and some even within the times of recorded human history. In this way Serres further blurred the formerly sharp distinction between the human past and deeper geohistory, while retaining a loosely "diluvial" kind of causal explanation. But he remained a provincial, and his attempts to correlate his geohistory with Genesis put him in an increasingly marginal position.[9]

28.2 SCHMERLING'S HUMAN FOSSILS IN BELGIUM

Only a year after Cuvier's death, and before Serres won his prize, geologists were confronted with new field evidence for human antiquity, surpassing in quality even what Christol and Tournal had produced. The new finds were made by Philippe-Charles Schmerling (1790–1836), the son of an Austrian merchant who lived in Delft. Schmerling had studied medicine at Leiden, the premier Dutch university, and had settled in Liège while it was still in the unified post-Napoleonic kingdom of the Netherlands. He stayed there, practicing as a physician, when in a brief revolution in 1830 the southern Netherlands split from the north and declared independence as the new kingdom of Belgium. The previous year a quarry manager had sent Schmerling some fossil bones—he must have been known to be interested in such things—which had been found in a limestone cave near the village of Chokier, in the valley of the Meuse upstream from Liège. These led Schmerling to many more caves in the same area, and to a major research project on their deposits. He was well aware that fossil bones in caves were the focus of much scientific argument right across Europe, and in 1831 he published a brief description of the caves, mentioning that he had found two human skulls, and flint and bone tools, among the animal bones. But this report was tucked away in small print at the back of a general book about the province of Liège, and was apparently not noticed outside Belgium.[10]

Two years later, however, Schmerling sent a brief paper on his fossils to be read at the Société Géologique, and another to be published in Leonhard and Bronn's *Neues Jahrbuch*, which had already become the German geologists' leading periodical (§25.2). These made his claims widely known among geologists across Europe, and acted as effective trailers for the two impressive volumes that he published soon afterward in Liège. Like Croizet and Jobert (§15.2), he paid tacit homage to the deceased Cuvier by calling his own work *Researches on Fossil Bones*, although he too qualified that title as referring just to his local region. Most important, of course, was his claim to have found human bones and also flint and bone tools, all in close association with the bones of extinct mammals. Christol and Tournal had not published anything but preliminary reports, illustrated only by the few specimens on Christol's single small plate (Fig. 16.1); in contrast, Schmerling's new Belgian evidence was fully described, and illustrated by large and superbly lithographed plates depicting his best specimens. These made his claims much more difficult to ignore or dismiss. He was well aware of the potential pitfalls involved in cave research, and stressed that he had personally excavated many caves that had not previously been disturbed in any way. He emphasized that the bones were scattered at random in the sediment, and he inferred that they had been swept by some kind of "diluvial" current into caves that had not previously been occupied by the living animals. But there was no reason to suppose that the bones and artifacts were not all of the same age.[11]

Schmerling's volumes were largely devoted to detailed descriptions of his fossils, analyzed bone by bone in what had long been the standard Cuvierian fashion; with Cuvier's *Fossil Bones* as his guide, and with his own medical training, Schmerling was well qualified to make the necessary identifications. He reported that rhinoceros, horse, and many ruminants were abundant, bear and hyena were rare, but there were many other mammals ranging in size from rabbit to mammoth. Not surprisingly, however, it was the human bones that were given pride of place. Schmerling criticized "the gratuitous hypothesis" that there had been no humans on earth until the modern period, arguing that until the Superficial deposits had been more thoroughly explored that assumption was premature and unjustified. He insisted that his specimens had unquestionably come from the same deposits as the bones of the extinct mammals, since he himself had excavated them with the greatest care, and all the bones were preserved in exactly the same way; there was no trace of any later disturbance of the sediment, let alone any later burial.

Above all, Schmerling had had the good fortune to find a couple of human skulls in a cave at Engis, the next village upstream from Chokier. The larger and better specimen, which he identified as that of an elderly adult, was too incomplete in the facial area to be sure of its racial type (as defined by Blumenbach's physical anthropology); but he thought it likely to be "Ethiopian" and relatively primitive, as might be expected from its apparent antiquity. What was much more important was that he had found it *under* a "bone breccia" one meter thick containing the bones of rhinoceros, horse, and ruminants; so it was unambiguously at least as old as those mammals, some of which were of extinct species.

9. Serres, "Cavernes à ossemens" (1835); also in book form, which reached a third edition in 1838; his was the only essay submitted: Bruijn, "Prijsvragen" (1977), 229–31. Another Haarlem prize, offered the same year for an essay on *human* fossils, attracted, rather surprisingly, no entries at all; Serres may have been judged to have answered both. Serres, "Contemporanéité de l'homme" (1832–33) and "Mémoire sur la question" (1833–34), had already argued for extinctions within recorded human history, by analyzing verbal descriptions and pictorial representations of animals in texts and artifacts from Antiquity (among them, for example, the famous Palestrina mosaic that had inspired Buckland: §11.3); this was a topic that Cuvier must have thought he had settled decisively the other way (*BLT* §10.3). Serres, *Cosmogonie de Moïse* (1838), subtitled "compared with geological facts", reached a third edition in 1859, but this kind of writing remained far less common on the Continent than in Britain.

10. Schmerling, "Cavernes à ossemens fossiles" (1831); see also Henderickx, "Schmerling" (1994). Chokier is on the Meuse, about 12km west-southwest of Liège (in the other direction, downstream, the river enters the [modern] Netherlands, changes its name to the Maas, and flows past the city of Maastricht).

11. Schmerling, "Cavernes à ossemens" (1833), read on 18 March; "Knochenhöhlen bei Lüttich" (1833); *Ossemens fossiles* 1 (1833), 1–23. He correctly contrasted his own inference with what Buckland had concluded for caves such as Kirkdale and Muggendorf (i.e., that they had been animal dens), but his was in fact identical to the Englishman's interpretation of other caves such as Wirksworth and Paviland (Figs. 6.2, 6.3).

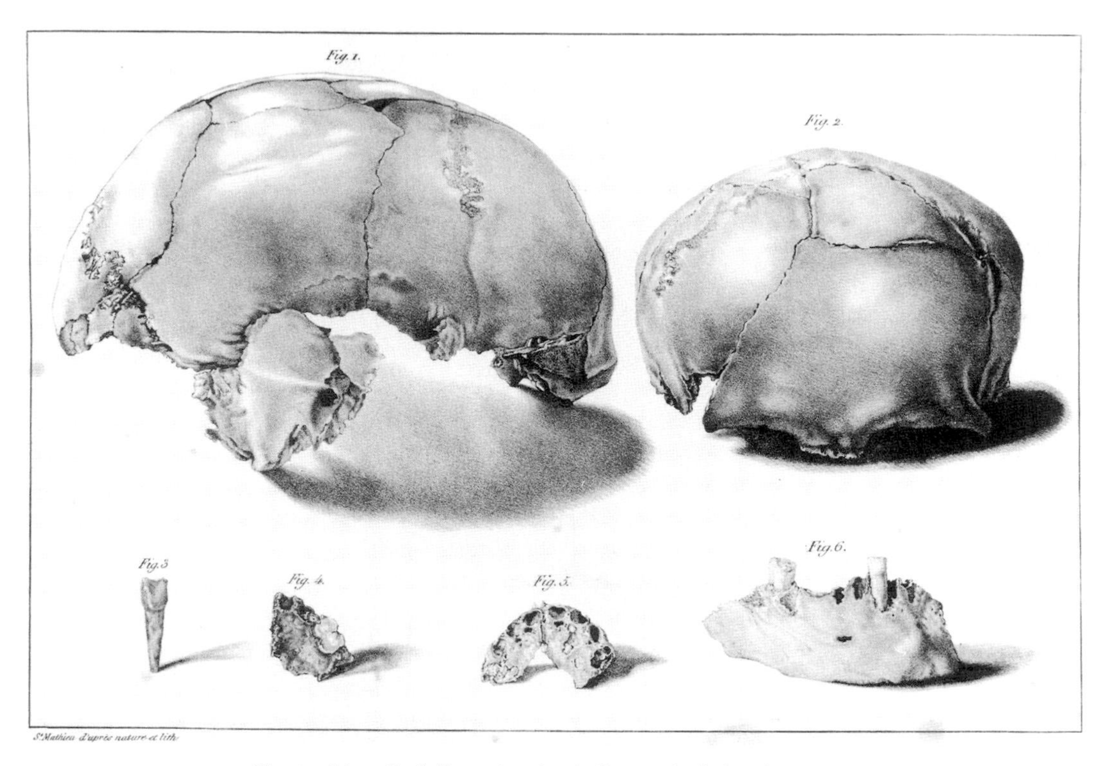

Fig. 28.1 Schmerling's illustrations (1833) of human fossils from bone caves near Liège. This lithographed plate showed—at full size in the original—an adult cranium from the Engis cave, in lateral and frontal view (*figs. 1, 2*); the maxillaries from a smaller and juvenile skull (*fig. 5*); a large incisor (*fig. 3*); and fragments of upper and lower jaws (*figs. 4, 6*). These drawings, on the first of the seventy-six superb folio plates in Schmerling's work, had a huge impact on the debate about human antiquity. (By permission of the Syndics of Cambridge University Library)

The smaller human skull, from a young individual, had crumbled away while being extracted, though fragments had been retrieved; as if to compensate for this, he had found a molar tooth of a mammoth right beside it "in the depths of the bone-bearing silt". Schmerling's human fossil specimens were by far the most impressive yet seen, and his superb pictures of them were effective proxies for those savants who could not see the originals (Fig. 28.1).[12]

Schmerling's other and almost equally impressive kind of evidence for human antiquity, namely "débris worked by the hand of man", was mentioned briefly at the end of the work, after his thorough description of the highly diverse animal bones from the caves. The artifacts were of two kinds. First, there were chipped flint tools, which "could have served to make arrows or knives". Significantly, they had a uniform bleached crust, which Schmerling noted was quite different from that of flint nodules in the Chalk or flint pebbles in Tertiary gravels. As his readers would have understood at once, this proved that the tools were far more recent in origin, and likely to be contemporary with the fossil bones. Yet the fact that there was any patina at all proved that the tools could not be the work of the craftsmen who at this time still practiced the same skills, producing chipped flints for flintlock firearms; so fraud was ruled out. Second,

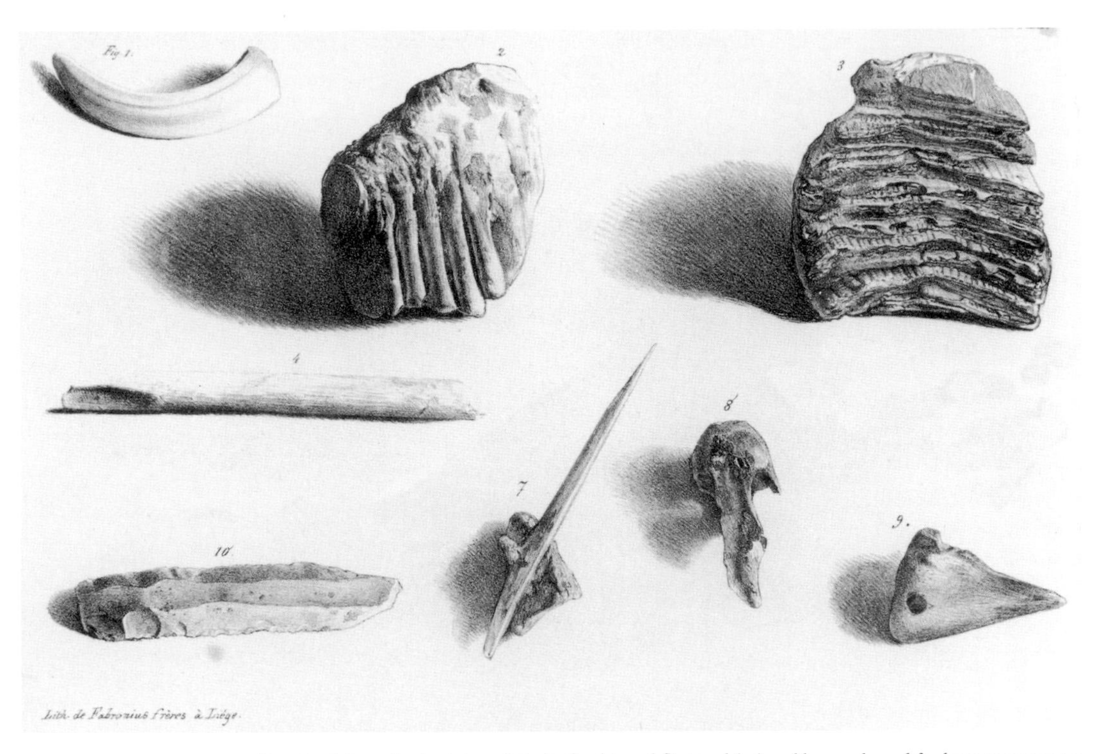

Fig. 28.2 Schmerling's pictures (1834) of a chipped flint tool (*10*) and bones shaped for human purposes (*4, 7, 9*), found with the bones of extinct animals in caves near Liège. He suggested that the flints were for arrows or knives, and that among the worked bones were a coarse needle (*7*) and perhaps an ornament pierced for hanging on a necklace (*9*): the putative needle was still stuck to a fragment of stalagmite, which proved that it was not a recent introduction. Also shown here are two molar teeth of Blumenbach's *Elephas primigenius* or mammoth (*2, 3*), one of which (*3*) Schmerling had found close beside the smaller of the two human skulls (see Fig. 28.1), which made it decisive evidence for contemporaneity and hence "a great prize". (By permission of the Syndics of Cambridge University Library)

there were also various animal bones that Schmerling took to be products of "the industry of the antediluvian race", for they showed unmistakable signs of having been shaped for human use; at least one specimen, he suggested, might have been purely for ornament, which tacitly underlined the full humanity of those who had made it. Taken together, the two kinds of artifact were conclusive: "even if we had not found human bones in situations completely favorable for regarding them as belonging to the antediluvian epoch, worked bones and fashioned flints would have been given us this evidence [*preuves*]" (Fig. 28.2).[13]

In Schmerling's view the conclusion was unambiguous, despite what Cuvier had continued to insist to the last. True human fossils did exist, and an "antediluvian race" of humans, skillfully fashioning sophisticated tools, had lived in a

12. Fig. 28.1 is reproduced from Schmerling, *Ossemens fossiles* 1 (1833) [Cambridge-UL: MH.5.3], pl. 1, drawn "from nature" and lithographed by Saint-Mathieu, described on 60–62 and in the list of pls.; the human fossils are described more generally on 53–66.

13. Fig. 28.2 is reproduced from Schmerling, *Ossemens fossiles* 2 (1834) [Cambridge-UL: MH.5.3], combined from parts of pls. 22 (figs. 2–4) and 36 (figs. 7–10), explained on 124, 177–78. Among these specimens were some from the caves at Engis (figs. 2, 3, 7), Chokier (fig. 9), and Fond-le-Forêt (fig. 4).

world with equally "antediluvian" and now extinct mammals. Noting the similar reports from the south of France, he considered that the case for human antiquity was approaching moral certainty.

In the event, however, even Schmerling's outstanding evidence failed to overcome the dogged resistance to his conclusion, among geologists of all persuasions. Lyell, for example, visited him in Liège in 1833, on his way home from Bonn at the end of a trip up the Rhine, after completing the first edition of his *Principles.* He was impressed by the size and quality of Schmerling's collection, and sympathized with his provincial lack of intellectual company. As he told Mantell, there were "none who take any interest in his discoveries save the priests—and what kind *they* take you may guess, especially as he has found human remains in breccia, embedded with the extinct species, under circumstances far more difficult to get over than any I have previously heard of". Yet Lyell did "get over" this awkward evidence: to concede that there were genuine human fossils in these deposits would have given aid and comfort to the priestly enemy, by supporting a theory that had often (though not always) equated the diluvial event with the biblical Flood, and human fossils with the Flood's victims. Not for the first or last time, Lyell's anticlericalism blinded him to the significance of what he was seeing with his own eyes.[14]

Schmerling sent the Société Géologique a copy of his work, with its superb proxy specimens, as soon as it was published, followed in 1835 by a report on further specimens from the Liège caves. He sharply criticized the dogmatism of the unnamed "theory men" [*hommes à système*] or "museum men" [*hommes de cabinet*] who rejected his claims, and predicted that in spite of their opposition the case would one day be won by scrupulously careful observation *in the field.* That was the crucial point, as it was for Tournal: even the finest specimens, real or proxy, would never convince the skeptics if they doubted that the human bones had truly been found in the very same deposits as those of extinct mammals, and that these deposits had not mixed materials of different ages.[15]

Later the same year Schmerling took his argument to Bonn, to the large annual meeting of the German Association of Researchers and Physicians [Gesellschaft Deutsche Naturforscher und Ärzte], the peripatetic body on which the British Association had been modeled. But although he and his specimens were given pride of place in the geologists' opening session, his reception there was the same as in Paris. Buckland was among those who dismissed the case, on the grounds that the human bones could have come from a later burial, as he believed they had at Paviland (§6.2). Later, having visited the Liège caves on his way back to England, Buckland added a forceful postscript to his Bridgewater Treatise (see Chap. 29), rejecting Schmerling's conclusions. This was in effect a slap in the eye for the provincial naturalist, for it implied either that he was incompetent to excavate the deposits carefully enough to rule out the possibility of a later burial, or, worse still, that he could not be trusted to have found what he claimed. Sadly, in the same year that Buckland's book appeared, Schmerling died at the age of only forty-five, and therefore never witnessed the later fulfillment of his prediction about the crucial role of *fieldwork* in the vindication of his case for human antiquity.[16]

28.3 THE FIRST FOSSIL PRIMATES

The evidence that humans had co-existed with the extinct "antediluvial" mammals was growing steadily, even if it still failed to convince those whose opinions mattered most. If it was eventually confirmed (as in the event it was, a quarter-century later), it would pull the duration of the human species back into geohistory proper, long before the brief era of recorded human history. Even this, however, would still leave a puzzling major gap in what could otherwise be interpreted as a "progressive" history of life. The discovery of new fossil localities in Tertiary formations was rapidly enlarging the range of mammals known from that era, beyond all those that Cuvier had described; yet, surprisingly, no fossil "*quadrumana*" or non-human *primates* had ever been found. This left a glaring gap in the fossil record at the very point at which it was closest to the human species. But the gap was convincingly plugged on cue, as it were, by almost simultaneous discoveries on three continents, thousands of miles apart. However, the timing was not altogether a coincidence, because by this time many geologists were actively on the lookout for any trace of fossil primates.

In 1834 Édouard Amant Isidore Hyppolyte Lartet (1801–71), a French provincial landowner, lawyer, and amateur naturalist, alerted Geoffroy in Paris to the discovery of a rich new Tertiary deposit of fossil bones at Sansan, near his home in southwest France. Three years later, in January 1837, Lartet reported to the Académie in Paris that among these bones he had found the fossil jaw of a gibbon; further research, he suggested, might therefore show that "this former [*ancienne*] nature, still so little known, was no less complete, no less advanced in the organic scale, than that in which we live". The implication was that if monkeys had been around, humans might have been too. This made the specimen so important—if its identification was correct—that the Académie appointed the customary high-powered committee to report on it; one of its members was Blainville, who had, in effect, replaced Cuvier as the world's foremost expert on comparative anatomy. In June 1837 the committee confirmed Lartet's conclusion, calling it "one of the happiest and least anticipated discoveries that have been made in paleontology in recent times" (Fig. 28.3).[17]

14. Lyell to Mantell, 16 September 1833 [Wellington-ATL; printed in *LLJ* 1: 401–3]. Lyell, *Principles*, 3rd ed. (1834) 3: 159–61, noted Schmerling's work but without drawing any conclusion from it; thirty years later he belatedly conceded its significance, and excused his earlier skepticism—rather lamely—on the grounds that at that time it had "seem[ed] to contradict the general tenor of previous investigations": *Antiquity of man* (1863), 67–69.

15. Société Géologique, 16 March 1835 [*BSGF* 6: 170–73]. Schmerling may have weakened his argument in the eyes of his critics by expressing surprise that the human bones found in Bavarian caves, reported by Esper over half a century earlier (*BLT* §5.4), had almost been forgotten; his own evidence was much stronger and less ambiguous than Esper's.

16. GDNA, 19 September 1835 [*Isis*, 1836, col. 708]; Buckland, *Geology and mineralogy* (1836) 1, 103–6, 598, claimed inter alia that the human bones were less decayed than the animal ones. The decisive evidence for contemporaneity was found a quarter-century later, from 1859 onward, not in Belgium or Languedoc but in the Somme valley in northern France and at Brixham Cave (not far from Kent's Hole) in southern England: see Grayson, *Human antiquity* (1983), chap. 9, and Van Riper, *Men among the mammoths* (1993), chaps. 4–6.

17. Fig. 28.3 is reproduced from Duméril *et al.*, "Ossements fossiles de quadrumanes" (1837) [Cambridge-UL: CP340:2.b.41.4], pl. at 996, read at the Académie on 26 June, reporting on Lartet,

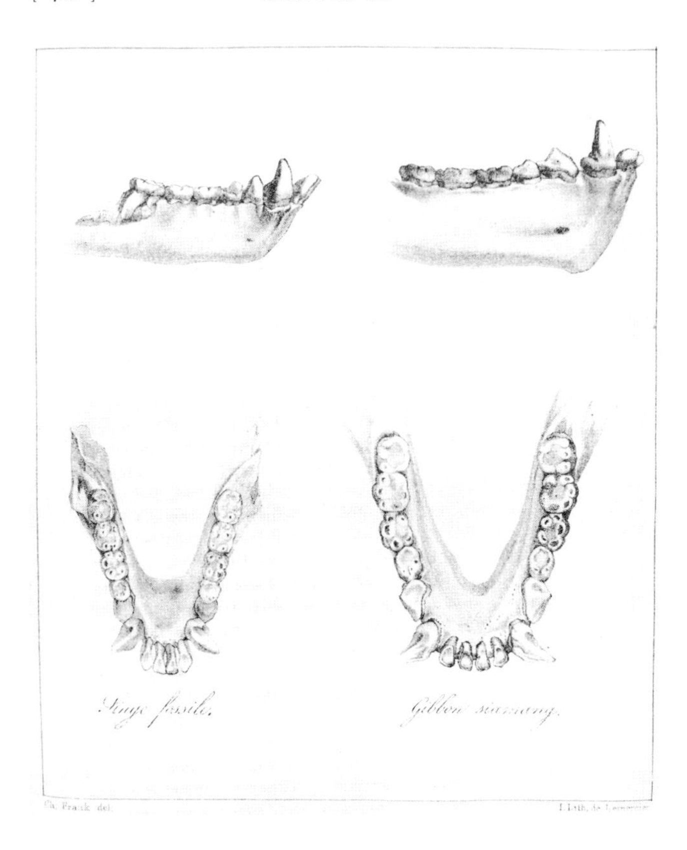

Fig. 28.3 The lower jaw of a fossil primate, found in 1837 by Edouard Lartet in a Tertiary deposit in southwestern France (left), compared with that of a living gibbon (the Sumatran siamang, right). It was one of the first primate fossils to be discovered; others were found at almost the same time in India and Brazil. The mammalian group to which the human species belongs thus acquired—for the first time—a fossil record stretching back into the Tertiary era. (By permission of the Syndics of Cambridge University Library)

A few months after Lartet's discovery was announced in Paris, and shortly before its validity was authoritatively confirmed, a similar find was reported at the Geological Society in London. It was in a letter sent from India by Proby Thomas Cautley (1802–71), a military engineer serving in the Bengal Artillery and one of the Society's most distant Fellows, and his friend Hugh Falconer (1808–65), a physician in the Bengal medical service and the director of the botanic garden at Saharanpur. They had been collecting fossils from a Tertiary formation, rich in the bones of mammals and reptiles, which outcropped in the Siwalik Hills along the edge of the Gangetic plain. Their specimens, many of them sent back to London, had already so impressed the leading figures at the Geological Society that earlier in 1837 the expatriates had each been awarded a Wollaston medal (like Mantell two years earlier).[18]

What Cautley and Falconer now reported was even more sensational: among all their Siwalik bones they had found an astragalus of a fossil monkey. They treated even this single specimen as enough for good Cuvierian reasoning: "although only a solitary bone of the foot, the relations of structure are so fixed that the identity of the fossil is as certain as if the entire skeleton had been found". And they used good Lyellian reasoning to suggest why primate fossils were so rare: they noted that tropical scavengers (such as bone-crunching hyenas) were so effective that the remains of dead monkeys were rarely seen in India even in places where they were living in abundance (Fig. 28.4).[19]

In fact, Cautley and Falconer noted that their find had already been confirmed by two other expatriate naturalists, civil engineers working in Bengal: the

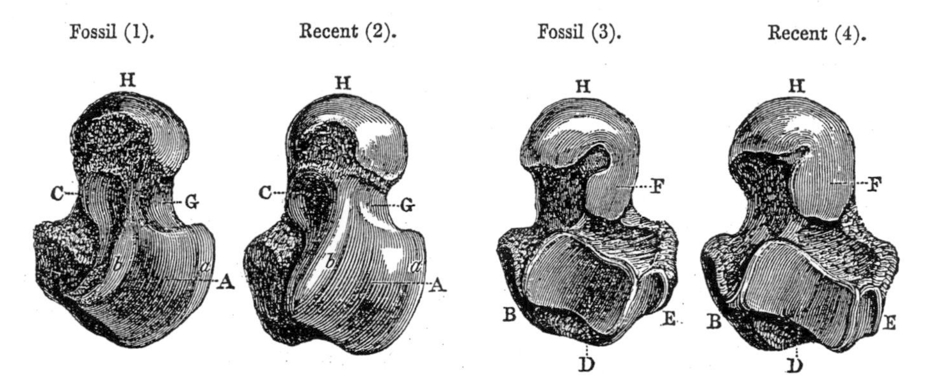

Fig. 28.4 Two views of an astragalus of a fossil monkey found in a Tertiary formation in India (1, 3), compared with the same ankle bone in a species living in India (2, 4). These specimens were sent to London by the expatriate English naturalists Proby Cautley and Hugh Falconer, and were shown at the Geological Society in June 1837. Their find, matching Lartet's in France at almost the same time, confirmed the existence of primates well back in the Tertiary era. (By permission of the Syndics of Cambridge University Library)

latter had already reported to the Asiatic Society in Calcutta—the premier savant institution in Asia—that they had found a fragmentary skull and other primate bones in the same formation. The secretary of that Society noted with triumph that an "immense step" had been made with this "discovery of the remains of a quadrumanous animal, the nearest approach to the human being that has yet been found in the fossil state in company with the extinct monsters of primeval antiquity". But although this find was published before Cautley and Falconer's (or Lartet's), and although the Indian periodical regularly reached Europe without much delay, it had less impact because the specimens themselves remained in India and could not be examined at first hand by the European experts.[20]

"Ossements fossiles de Sansan" and "Machoire inférieure fossile" (both 1837), read on 16 January and 17 April (he found a second specimen the following year); the rich mammalian fauna was first reported in Lartet to Geoffroy, 13 February 1834 [printed in *BSGF* 4: 342–44]. The primate fossil was later named *Pliopithecus* but it is now assigned to an extinct group not closely related to gibbons, and the deposit is dated middle Miocene: see Ginsburg, *Faune Miocène de Sansan* (2000). Sansan is about 12km south of Auch (Gers); many other specimens from Lartet's collection are on display at Paris-MHN, including a complete (though composite) skeleton of a huge mastodon. Many years later, Lartet became famous as the co-discoverer of Paleolithic art, in the caves of Périgord; Fischer, "Édouard Lartet" (1872), is still a valuable source.

18. Falconer to Asiatic Society, 3 January 1835 [printed in *JASB* 4: 57–59] first reported the fossil deposit (now dated as Miocene); Falconer and Cautley, "Sivatherium giganteum" (1836), described and named a huge new fossil ruminant from it; *Fauna antiqua Sivalensis* (1845–49), their later full account with superb lithographed plates by Scharf, remained incomplete and never described the primates. Nair, "Siwalik fossil collecting" (2005), describes the social context of their research and analyzes the interactions between India and England, natives and expatriates. Cautley was later knighted for his work on irrigation projects.

19. Fig. 28.4 is reproduced from Cautley and Falconer, "A fossil monkey" (1840) [Cambridge-UL: Q365.b.12.11], read at the Geological Society on 14 June 1837, wood engraving at 501; the living species was "*Semnopithecus Entellus*", the temple langur.

20. Editorial by James Prinsep [*JASB* 5 (1836): x]. The primate jaw was described and illustrated in Baker and Durand, "Sub-Himálayan fossil remains" (1836), and was later reproduced alongside Lartet's Sansan specimen in Blainville, *Ostéographie des mammifères* (1839–64) 1: "Primates: Pithecus" (dated 14 June 1839), pl. 11.

Fig. 28.5 "Fossil man": an imagined portrayal of an ancestral human being—inspired by Schmerling's fossils from near Liège—combining a bipedal gait, a stone axe, and a cloak of animal fur with strongly *simian* (and negroid?) bodily features. This was the first and most striking of a series of reconstructions of fossil animals, illustrating a *popular* account of the history of life written by Pierre Boitard (1789-1859) and published in the *Magasin Universel* in 1838. As geologists realized—with either alarm or delight—any such imaginative reconstruction made vividly apparent what the wider public might conclude from the new evidence of fossil primates, if it were combined with the still ambiguous evidence of human fossils.

The third discovery of fossil primates was made in Brazil, at almost the same time as those in France and India. The Danish naturalist Peter Wilhelm Lund (1801–80) had already found abundant fossil bones of a wide variety of extinct mammals in caves near Lagoa Santa (Minas Gerais). In July 1836 he found the femur and humerus of a primate. But his preliminary report on all his fossils, including what he named *Protopithecus*, was not written until late in 1837, and not published until 1838, and then in Danish. Like Esmark's paper in Danish on former glaciers (§13.4), it might then have gone unnoticed in the wider savant world, if Lund had not sent translated excerpts to the editor of the *Annales* in Paris, where they were published in 1839. By that time, however, the other primate discoveries were well known, and Lund's was chiefly significant for showing that the fossil record of primates extended, like their living relatives, to the New World.[21]

All these primate finds, from three continents, brought the fossil record unnervingly close to the orang-utan and hence also to *Homo sapiens* (§17.2). They therefore reinforced the claims that continued to be made by French naturalists led by Geoffroy (§17.1), that the fossil record could not be explained adequately without invoking transmutation. The ghost of Lamarck was still looking over geologists' shoulders. More precisely, and more alarmingly, fossil monkeys could readily be combined with what Tournal, Christol, and Schmerling had claimed for the authenticity of fossil humans, to argue—at least in popular scientific journalism—that human origins had been positively bestial (Fig. 28.5).[22]

28.4 CONCLUSION

During the years when Lyell was writing and publishing his *Principles*, the argument about human antiquity rumbled on. The earlier reports by the provincial French naturalists Christol and Tournal—that they had found genuine human bones and pottery along with the bones of extinct mammals in undisturbed deposits in caves in Languedoc—had been rejected by Cuvier, though others such as Buckland had been inclined to accept their authenticity (§16.2, §16.3). The Académie's committee on the problem, chaired by Cuvier, continued to sit on the fence, despite receiving further specimens. Undeterred, Tournal restated his case at the provincial Académie in Toulouse, where Aubuisson lent him his support, advising him to drop all "diluvial" language and thereby sever the residual link with the question of the historicity of the biblical Flood. Tournal took this advice and sent a further report to the Société Géologique in Paris, but its secretary Desnoyers dismissed the case. Around the same time, Cuvier was removed from the debate, when he died suddenly without having softened his long-standing skepticism about all such claims for the contemporaneity of early humans with the extinct megafauna that he had been the first to describe.

At Arago's invitation, Tournal then sent a final statement of his case to a scientific periodical that enabled him to bypass the skeptical geologists. He insisted again that there was now very strong *field* evidence for contemporaneity, and every indication that these early humans had been as technically advanced as, for example, existing Native Americans. Above all, Tournal proposed that the human portion of geohistory—coming at the tail end of an inconceivably vast span of prehuman geohistory—now needed to be redefined to include a lengthy "*antehistoric period*", preceding the few millennia of recorded human history. This explicitly opened up a conceptual space for a *preliterate* history of the human species, or what would later be termed *prehistory*. In other words, Tournal's proposal dissolved the equivalence of human history with the history of *literate* human civilizations, which Cuvier and many others had almost taken for granted; it blurred the previously sharp boundary between the human and prehuman

21. Lund, "Brasiliens dyreverden" (1838) and "Espèces éteintes de mammifères" (1839). Hartwig, "*Protopithecus*" (1995), claims Lund's as the "first" fossil primate find, but in historical terms the date of the discovery itself is less significant than the date at which it became known to geologists internationally. In modern terms, Lund's late Pleistocene platyrhine is also much less "deep" in geohistory than the other early finds (now dated as Miocene-Pliocene).

22. Fig. 28.5 is reproduced from Boitard, "L'homme fossile" (1838), 209, drawn and engraved by Susemihl. Its reappearance in 1841 in a notorious English atheist publication would have confirmed the worst fears of the anti-traditionalists: see Secord, *Victorian sensation* (2000), 311–312. Rudwick, *Scenes from deep time* (1992), 166–69, and Moser, *Ancestral images* (1998), 135–39, both reproduce the similar engraving from Boitard, *Paris avant les hommes* (1861), a much later (and posthumous) popular work that enlarged on this theme; this in turn was re-used (in 1887) to illustrate Haeckel's hypothetical "ape-man" *Pithecanthropus* (Moser, 139). See the important analysis of Boitard's work in Blanckaert, "Avant Adam" (2000); also, more generally, his "Fossiles d'imaginaire" (1999), and Delisle, "Origines de la paléontologie humaine" (1998). On Geoffroy's transformism in the 1830s, see his *Recherches sur de grands sauriens* (1831), *Études progressives* (1835), "Études sur l'orang-outang" (1836), and, specifically on Lartet's finds, "Singe trouvée à l'état fossile" (1837); however, his contemporaries were increasingly aware of what Bourdier, "Geoffroy versus Cuvier" (1969), tactfully calls Geoffroy's "tired and seething brain".

parts of geohistory. Yet in the event both Tournal and Christol faded from the savant scene, either because they grew disheartened at their failure to convince the metropolitan naturalists, or perhaps just because their attention and interests were diverted in other directions. Serres remained active in pursuing the problem, and his general review of cave bones won him a prize at Haarlem; but his concern with relating his geology to Genesis marginalized his voice.

However, the issue was kept alive, and indeed sharpened, by striking new claims. Schmerling reported that he, like Christol and Tournal, had found human fossils mixed with the bones of extinct mammals, in caves near his home city of Liège in the new kingdom of Belgium. Schmerling's human bones were far better specimens than those from Languedoc; they were accompanied by unquestionable flint tools and plausible ones of bone; and Schmerling described them all in authoritative detail and depicted them on superb lithographic plates. Yet even this fine publication failed to convince the geologists whose opinions counted most. As before, their skepticism was based on the allegedly ambiguous character of the field evidence, and the likelihood that materials of quite different ages had become mixed up during the long history of cave occupation. There was thus continuing resistance to the possibility that the history of the human species might have overlapped substantially with that of the "antediluvial" mammals; or, in other words, that human history might extend back into the part of geohistory represented by the Superficial (or "diluvial") deposits. The case for the contemporaneity of the human species with the extinct mammals remained unresolved (and continued to be for another quarter-century).

On the other hand, what had long been a puzzling gap in the near end of the fossil record of vertebrate life was decisively closed. Lartet's find of fossil primate bones in southwestern France was matched by similar and almost simultaneous finds by Cautley, Falconer, and other British expatriates in India, and by Lund in Brazil. These were promptly accepted as good evidence that non-human primates had already existed well back in the Tertiary era, further reinforcing the view—held by almost all geologists other than Lyell—that the history of life as a whole had been broadly directional or "progressive". As Geoffroy's example showed, however, this could all too readily be turned into evidence for a transformist reading of the fossil record; a popular article by Boitard confirmed the worst fears of those who opposed any such interpretation, by depicting a bestial evolutionary ancestry for the human species itself. Conversely, however, the same evidence for directionality could be harnessed for less disturbing ends; the next chapter describes how Buckland made one such interpretation widely accessible.

CHAPTER 29

Buckland's designful geohistory (1832–36)

29.1 NATURAL THEOLOGY AND "SCRIPTURAL" GEOLOGY

While Lyell was defending and reinforcing the steady-state geotheory set out in his *Principles*, Buckland was compiling a major work which, while not designed as a rejoinder, did in fact synthesize the evidence for a *directional* geohistory and make it more accessible. Since his book was presented within an explicit context of natural theology, it has often been misinterpreted as a work on the "Religion" side of a supposed conflict with "Science". This view is so profoundly mistaken that it must first be cleared out of the way.

The root of the misunderstanding lies in a frequent failure by modern readers to distinguish between natural and revealed theology. In the early nineteenth century many "men of science" were at least content, and some were enthusiastic, to see the traditional "argument from design"—as embodied, for example, in William Paley's already classic *Natural Theology* (1802)—supported with evidence drawn from their own scientific research. That features of the natural world bore signs of ultimately divine design seemed to many of them almost self-evident. Treating providential purpose as the "final cause" of such features was, of course, perfectly compatible with searching for—and in many cases finding—their "efficient causes" in terms of purely natural processes. Savants such as Lyell and Scrope were not dissembling, or merely pandering to their intended audiences, when they used the language of natural theology. The discourse of design was not only useful in practice for analyzing living organisms (as it still is for biologists today). It also served at the time as a metaphysical glue that helped to bind savants of all religious persuasions into a scientific *community*: the British Association for the Advancement of Science (§27.1), for example, soon recognized this when it brought Anglicans, Unitarians, and Quakers together in amicable collaboration.[1]

On the other hand, matters of revealed theology—doctrine based on what believers interpreted as divine self-disclosure in and through human history, above all as recorded in scripture—had a quite different status. For the most part (an exception will be mentioned shortly), such properly *religious* issues had little significant contact with the natural sciences, unless scripture itself was taken to be an authoritative source of *scientific* information. In the science of geology, that *misuse* of scripture—as its critics considered it to be—was based on the literalism adopted by the "Mosaic" writers or self-styled "scriptural geologists". As already emphasized (§18.1, §21.2), the scientific geologists, whether pious clergymen or anticlerical skeptics, drew a sharp distinction between themselves and these literalistic usurpers of the fair name of "geologist". Boué, as an independently minded foreigner, made just this distinction when he castigated the "scriptural" genre as a specifically English phenomenon and mentioned four Anglican clergymen, in contrast, as distinguished examples of those he regarded as real geologists (§27.1).[2]

At just this time, Conybeare vigorously defended himself and the other members of this exemplary quartet (Buckland, Sedgwick, and Whewell) against an anonymous attack in print from the "scriptural" quarter, which alleged, as usual, that the science of geology had an inbuilt "infidel tendency". Going on the offensive, Conybeare protested that it was a gross misuse of scripture to quarry it for scientific ideas; "in a word", he insisted, "the Bible is exclusively *the history of the dealings of God towards men*". For Conybeare this was not a retreat by "Religion" in the face of advancing "Science", but a restatement of what should have been the supreme consideration all along. For this country parson (and renowned geologist) was also a forceful advocate of the kind of biblical scholarship that had long ago liberated Continental savants from the incubus of literalism. Lecturing at the theological college that he had recently helped to set up in Bristol, Conybeare introduced its students to the biblical hermeneutics that German scholars had long taken for granted, which made nonsense of the unhistorical literalism adopted by the "Mosaic" writers (and, of course, by modern fundamentalists). "The true science of Comparative Philology", Conybeare pointed out, "has been warmly and successfully cultivated by the Germans, undoubtedly the most learned of nations, but has been unaccountably neglected in this island . . . and the consequence is, that our literati are disgracefully behind their Continental brethren in all philological discussions".[3]

The attitude of Conybeare and other "enlightened saints"—as Lyell had dubbed them (§23.5)—to the "scriptural" or "Mosaic" literalists was also well expressed by Mary Buckland, when she told Whewell in the customary jocular manner how one of the latter was using a prestigious lecture series in Oxford to attack her husband and other geologists. It was not only Lyell and Scrope whom the literalists wanted, at least metaphorically, to see burned at the stake:

> By way of encouragement to my husband's labours, we have had the Bampton Lecturer holding forth in St Mary's against all modern science (of which it need scarcely be

said he is profoundly ignorant), but more particularly enlarging on the heresies and infidelities of geologists, denouncing all who assert that the world was not made in 6 days as obstinate unbelievers, &c. &c. . . . Alas! my poor husband—Could he be carried back a century, fire & faggot would have been his fate, and I dare say our Bampton Lecturer would have thought it his duty to assist at such an 'Auto da Fé'. Perhaps I too might have come in for a broil as an agent in the propagation of heresies.[4]

Buckland was furious at these "absurd and mischievous Bampton lectures", and considered it essential that "men of science" should now publicly repudiate not just the offending lecturer Frederick Nolan (1784–1864) but the whole genre of "scriptural" geology. Writing to William Vernon Harcourt (1789–1871), the son of the archbishop of York and a prominent co-founder of the British Association, Buckland insisted that "the time is now arrived when this school must be put down—singly they are unworthy of the notice of any scientific man, but it is not unworthy of any one to take up the question on its general bearings". He urged Harcourt to write an essay on some of these "anti-philosophical volumes" for the *Edinburgh Review*; an equally angry Murchison was urging Sedgwick to do the same in the *Quarterly*. In the event, nothing came of all this, but Buckland's Oxford colleague Baden Powell (1796–1860) did later answer Nolan with a powerful defense of the natural sciences as compatible with non-literalistic religious belief.[5]

1. See, for example, Brooke, "Natural theology of the geologists" (1979), and, more generally, *Science and religion* (1991), 192–225; Morrell and Thackray, *Gentlemen of science* (1981), 224–45. The idea of "intelligent design", favored by some modern American fundamentalists as a covert substitute for cruder varieties of creationism, is a rewarming of Paleyan natural theology, merely extending it from the organismic to the molecular level.

2. Paley himself had exemplified the distinction being made here: his *Natural theology* (1802) complemented his *Evidences of Christianity* (1794), an equally popular and influential work dealing with the doctrinal interpretation of the New Testament; both books were reprinted many times during the early nineteenth century. Paley's episcopal patron had been Barrington, the dedicatee of Buckland's *Reliquiae* (§6.3).

3. Conybeare, *Lectures on the Bible* (1834), xv; the lectures had been given in 1832–33 at Bristol College; see also Conybeare, "On geology" (1834), 309, responding to Anonymous ("A layman"), "Infidel tendency" (1834). Moore, "Interpreters of Genesis" (1986), is a valuable analysis of the relation between geologists and biblical critics at this period (Conybeare being, of course, both).

4. Mary Buckland to Whewell, 12 May 1833 [Cambridge-TC: a.66/31]; she exaggerated for effect, of course, by placing the era of heretic-burning only one century in the past rather than three. The lectures, delivered in the university church and published as Nolan, *Analogy of revelation* (1833), were an Evangelical counterblast to the liberal or Broad Church "invasion" of Oxford during the meeting of the British Association in 1832, at which Buckland had acted as president: see Morrell and Thackray, *Gentlemen of science* (1981), 234–35.

5. Buckland to Harcourt, 12 May 1833 [printed in Harcourt, *Harcourt papers* (1880–1905) 13: 321–23]. Powell, *Revelation and science* (1833), *Natural and divine truth* (1838), and later works: see Corsi, *Powell and the Anglican debate* (1988), chap. 10. Conybeare, who was engaged in serious theological work in parallel with (but separate from) his geology, gave the same endowed lecture series a few years later, which was published as his substantial *Writings of the Christian Fathers* (1839). Here he defended the traditional Anglican view of the Patristic writings as important but subordinate to scripture, against the new Tractarian (or "Romanizing") claims for their authority on a par with the Bible; but this important theological controversy had no significant link to contemporary scientific debates. In these same years, Sedgwick's immensely influential *Studies of the university* (1831; 4th ed., 1835) represented his defense of the natural sciences, and particularly geology, against the charge that they fostered impiety; his argument was somewhat similar to Buckland's.

Buckland's main "labours" at this time (ably assisted by his wife) were on a book that was to have a format as handsome as that of Lyell's first edition, and to be aimed at a similar range of readers, spanning both geologists and the upper-class British public. Buckland was one of eight British savants who were invited to write for a series that was exceptionally well funded under the will of Francis Henry Egerton (1756–1829), the eighth earl of Bridgewater. This recently deceased aristocrat had hugely enlarged his inherited wealth by enterprising investment in coal mines and the new railways. But he had also been a Fellow of the Royal Society and a clergyman, and he had wanted to see the traditional "argument from design" brought up to date by taking into account the recent rapid progress in all the natural sciences. The contributors to his series were to be highly paid, and the project attracted a distinguished set of savants. Whewell, for example, was invited to write on astronomy and physics; Buckland's volume, like his lectures at Oxford, was to cover geology and mineralogy. When asked by his wife what he would write for his Bridgewater Treatise (and thereby earn his huge fee of £1000) Buckland was said to have replied, "Why, my dear, if I print my lectures with a sermon at the end, it will be quite the thing". But in fact he took his task far more seriously than that characteristically flippant remark might suggest; so seriously, in fact, that his book was not completed until after all the others were published. It was then the subject of intense discussion in savant circles in Britain; Lyell, for example, praised and welcomed it in his next presidential address at the Geological Society. It was soon translated into French and German, and became a major contribution to the international debate about geohistory (Fig. 29.1).[6]

Before describing the main themes of Buckland's book, one minor topic must be mentioned briefly. In fact it was relegated to a footnote, which reflected its

Fig. 29.1 William Buckland in 1838, around the time that French and German translations of his Bridgewater Treatise on geology (1836) made him more widely known than ever among the educated public, not only in Britain but throughout Europe, as well as among geologists everywhere. The artist captured his deeply serious character, which belies his later and unwarranted dismissal as a mere buffoon. (By permission of the National Portrait Gallery, London)

increasingly marginal place in the lives of geologists; yet it remained important in the eyes of the wider literate public in Britain (though not elsewhere in Europe). The topic was the relation between the "geological deluge"—whatever form it might have taken—and the biblical Flood. Buckland's famous claim that the historical reality of the biblical event could be supported by geological evidence (Chap. 6; *BLT* §10.5, §10.6) had met with trenchant criticism from other geologists; from Fleming, for example, on the grounds that it was inconsistent with *both* "the testimony of Moses" *and* "the phenomena of nature" (§6.4). And subsequent fieldwork on the allegedly diluvial deposits had led diluvialists such as Sedgwick to conclude that there was evidence of *several* successive "diluvial" events, of which only the last—if any—could be the trace of the biblical Flood (§23.1).

This gradual withdrawal—by geologists—from any simple correlation between geological deluge and biblical Flood culminated, in effect, in Buckland's concession that his own earlier claim that they were identical now needed drastic revision in the light of more recent research. A footnote in his Bridgewater Treatise showed the wider public that he, like Sedgwick before him, had now abandoned the equation. He described the event that had entombed the diluvial fossils as "the last of many geological revolutions that have been produced by violent irruptions of water, rather than the comparatively tranquil inundation described in the Inspired Narrative [in Genesis]". Turning to the most recent set of extinct animals, he also claimed that there were "strong reasons for referring these species to a period anterior to the creation of man". And the strongest of those reasons was in fact none other than "the non-discovery of human bones" with those of the extinct animals. In other words, the absence of human fossils—if the earlier alleged cases (Chap. 16) and the latest (Chap. 28) were discounted—had become proof that the "diluvium" was not strictly "diluvial" after all: Buckland's revision severed the last link between his own "diluvium" and the *biblical* "deluge" or Flood. The historical reality of the latter could still be maintained, however, simply by adopting the interpretation that biblical scholars in any case preferred—as Lyell had implied (§25.1)—restricting the "universality" of the Flood to whatever limited area the human species had occupied at that time.[7]

6. Fig. 29.1 is reproduced from a sketch by William Brockedon, dated 7 February 1838 [London-NPG: 2515/87]. Lyell to Mantell, 18 January 1832 [Wellington-ATL; printed in *LLJ* 1: 367–68]; Buckland, *Geology and mineralogy* (1836), of which translations were published in 1838; Lyell, "Address to the Geological Society" (1837), 517. Topham, "Science and popular education" (1992) and "Bridgewater Treatises" (1998), are focused on the production of the books and their reception by diverse classes of readers; Topham argues that in the event the Paleyan theme was *not* a central topic for the series, though it did contribute to its perceived status as authoritative "safe" science. See also Brock, "Authors of the Bridgewater Treatises" (1966).

7. Buckland, *Geology and mineralogy* (1836) 1, 94–95n; in a longer footnote earlier in the book (22–25n) he recruited the formidable learning and unquestioned piety of his colleague Edward Bouverie Pusey (1800–1882), Oxford's professor of Hebrew and one of its leading Tractarians, to support a non-literal interpretation of the *Creation* narrative at the start of Genesis, thereby making the geologists' vast timescale more acceptable to the biblically conservative segment of the educated public in Britain. Dating perhaps from around this time were his manuscript amendments to his own copy of *Reliquiae* (§6.3)—probably intended for its revised edition—in which "diluvial" terms were replaced by the neutral word "Inundation": two such passages are reproduced in Dean, "Rise and fall of the deluge" (1985), 90. Buckland's long path from Flood to Ice Age (see §36.3) is traced briefly in Gould, "Freezing of Noah" (1985).

29.2 STRATIGRAPHICAL FOUNDATIONS

Buckland's main argument, no less than Lyell's, depended necessarily on a factual foundation of stratigraphy. Unless the sequence of formations was reliably established, no inferences about time and geohistory could be regarded as secure. Scrope might grumble that geology had "shrunk into little else than a barren description of the rocks which coat our planet", to the neglect of what he found to be more interesting features, such as volcanoes (§27.3). But most geologists were impressed by, and many contributed to, the rapid expansion of straightforward stratigraphical research during these years, based on fieldwork all over Europe (and to a much lesser extent in other parts of the world). In effect, this vindicated the Wernerian geognosy from which stratigraphy had emerged, after Smith's work had demonstrated the outstanding added value of the fossil criterion (Chap. 3; *BLT* §8.2).

Buckland was well aware that stratigraphy was the indispensable foundation for any inferences about geohistory. He was, of course, convinced that the world had *not* always been much the same kind of place, as Lyell claimed, and specifically that the forms of life had changed irreversibly in the course of geohistory: symbolically, the iguanodons had *not* returned—and would not in the future (§12.4)—but had characterized just one period and no other. Although this was taken for granted by all geologists other than Lyell, Buckland regarded it as one of the most important concepts to get across to the educated public that was expected to buy the Bridgewater books. So he asked Webster to adapt for him the large diagrammatic section through the earth's crust that Webster used in his lectures in London, to show the sequence of formations. The resulting long foldout colored engraving was the epitome of Buckland's whole work, and was given pride of place in the volume of plates that accompanied his text (Fig. 29.2).[8]

Webster's visual summary for Buckland's Bridgewater Treatise embodied the kind of collaboration that underpinned stratigraphy. During the 1830s the correlation of specific formations from one region to another became increasingly

Fig. 29.2 "Ideal section of a portion of the earth's crust", drawn by Webster in 1833 for Buckland's Bridgewater Treatise on geology (1836). This reproduction, although greatly reduced in size, shows the general design, with a "basement" of Primary granite (left and right) penetrated by many igneous rocks that emerge at the surface as extinct or active volcanoes (see the portion reproduced on a larger scale in Fig. 29.3 below). As in Alexandre Brongniart's analogous section (Fig. 9.1), the older rocks are portrayed as immensely thick in comparison to the younger ones, reflecting the general conception of the relative spans of time that the formations represented. Groups of animals and plants characteristic of successive periods of geohistory are depicted above the section itself (see Fig. 29.4). This design—combining stratigraphy, structure, igneous rocks, and fossils—epitomized the directional geohistory that Buckland intended to convey to the educated public; this panoramic engraving was the very first illustration in his highly successful work. (By permission of the Wellcome Library, London)

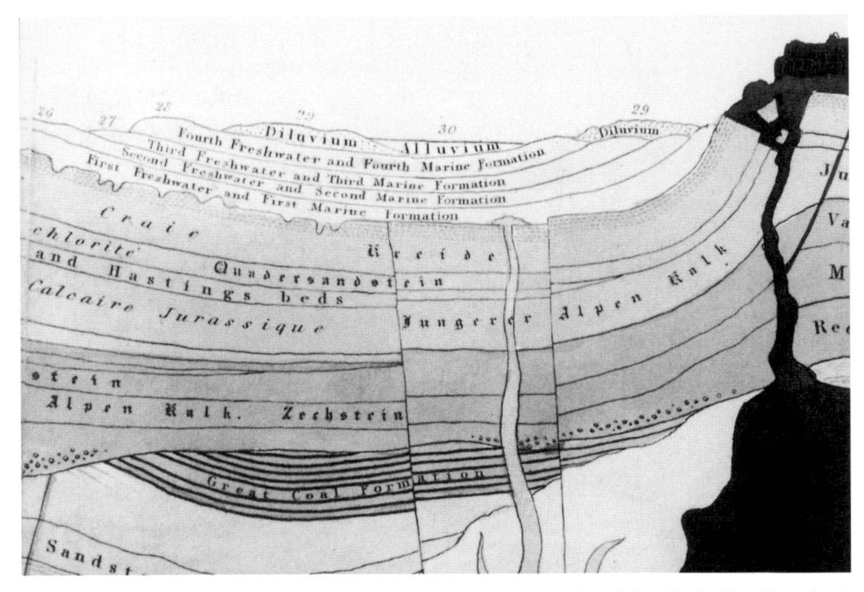

Fig. 29.3 A small portion of Webster's great ideal section (see Fig. 29.2) published in Buckland's geological Bridgewater Treatise (1836), to show the current state of stratigraphy. The Secondary formations are named trilingually (English names in Roman, French in italics, German in Gothic), in recognition of the internationally agreed status of many of the larger units. This portion of the section also shows the widespread major unconformity above the "Great Coal Formation" (which was responsible for the isolated coalfield "basins" in which this economically invaluable deposit was generally found), and the less obvious one at the base of the Tertiary, above the Chalk/*Craie*/*Kreide*. The section was drawn by Webster in 1833, too soon to incorporate Lyell's Tertiary research: the only Tertiary formations shown are those originally described by Brongniart in the Paris Basin (which Lyell dated as "Eocene"). The "Diluvium" and "Alluvium" at the top of the pile are as thin as the design allowed; the whole of human history was tacitly represented as an extremely brief final phase of geohistory.

confident and precise, and still larger units began to be defined as "Groups" of possibly global validity, such as "Carboniferous", "Jurassic", and "Cretaceous" (the "systems" of modern geologists). Greenough, who was working hard on the revision of his great geological map of England (which had long ago superseded Smith's for most purposes), was active in trying to get international agreement on stratigraphy and mapping. In 1835, for example, he raised such matters during the annual field meeting of the Société Géologique, held that year at Mézières in eastern France. Excursions across the frontier into Belgium enabled the party to see the spectacular folding of the older Secondary formations in the Ardennes, which led to discussion of their correlation with English and German sequences. Greenough and Buckland then traveled on to Bonn for the much larger annual meeting of the German Association, where Lyell was also present, together with Frenchmen such as Élie de Beaumont, Prévost, and both Brongniarts (and Schmerling exhibited his human fossils: §28.2). At Greenough's instigation, von Buch, the doyen of the German geologists, convened an ad hoc international committee to try to standardize stratigraphical nomenclature and the conven-

8. Fig. 29.2 is reproduced from Buckland, *Geology and mineralogy* (1836) 2: pl. 1 [London-WL: V.25106], 116cm in width; Buckland to Webster, 14 August 1833 [printed in Challinor, "Correspondence of Webster", 1961–64, letter 88], comments on the design, then in draft but already engraved.

tions used in mapping. Mundane negotiations such as these underlay all the stratigraphy that Webster summarized graphically—in both senses—for Buckland (Fig. 29.3).[9]

A few years earlier, Brongniart's rather similar ideal section had shown fossils as tiny icons within the formations in which they were found (Fig. 9.2). But Buckland asked an Oxford artist to draw the fossils separately, above Webster's section, in groups representing the faunas and floras of successive eras of geohistory. The difference of design reflected the far more deeply *geohistorical* meaning that Buckland attached to his stratigraphy. His fossils were not just objects char-

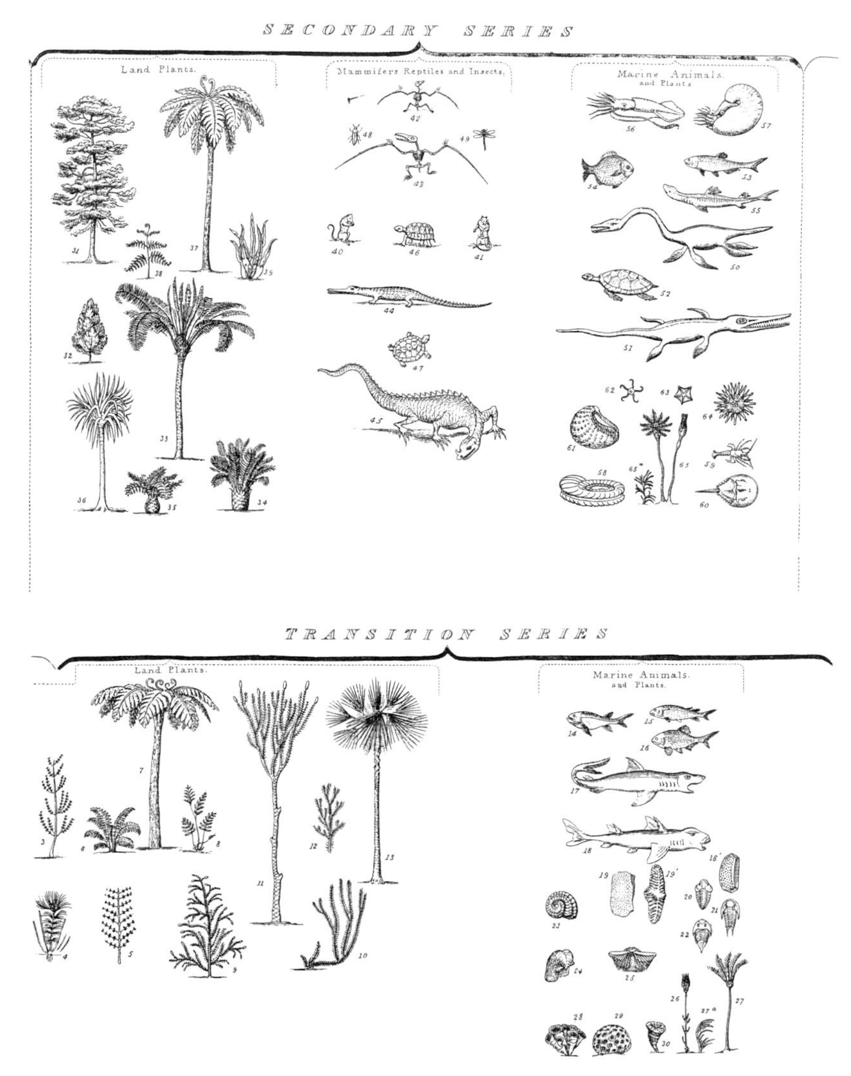

Fig. 29.4 The earlier parts of Buckland's great sequence of life on earth, as summarized visually in his Bridgewater Treatise (1836): assemblages of the fossil animals and plants—most of them shown reconstructed as if alive, though not drawn to scale—that had characterized the periods of the younger (above) and older (below) Secondary formations (Buckland, idiosyncratically, called the latter "Transition series"); other pictures, not reproduced here, continued the sequence (see Fig. 29.2) to cover the "Tertiary series" and even the very recently extinct dodo. Buckland argued that *all* the organisms portrayed—the earliest no less than the most recent—were demonstrably well designed for their diverse ways of life: the "progress" shown was in increasing diversity, *not* in degrees of perfection (as Lamarck's transformist theory was taken to imply).

acterizing certain rocks, but remains of organisms that had lived at certain times; on his engraving they were reconstructed, wherever possible, to show what they had been like when alive. The successive groups of organisms summarized the directional, and indeed apparently progressive, character of the history of life. Among the animals, an assemblage that included fish was succeeded by one with the now well-known Secondary reptiles, and they were succeeded by the Tertiary mammals and birds. Among the plants, the archaic tree-ferns of the coal forests were followed by successively more modern floras, as the younger Brongniart had outlined a few years earlier (§12.2). The progressiveness lay not in any greater "perfection" in the later forms, but in an increasing diversity and the successive addition of "higher" kinds of life. And he claimed that this was not the product of any gradual transmutation; somehow or other, new forms of life had been fashioned providentially to suit new environmental conditions as they developed on the ever-changing surface of a slowly cooling planet (Fig. 29.4).[10]

Buckland took the opportunity to assimilate into this progressive picture of life's history the famous little fossil marsupials from Stonesfield, which Lyell had claimed as evidence that mammals as a whole might already have existed even in the very earliest times in geohistory (§17.2). Buckland cited a study of living marsupials by Clift's assistant at the Royal College of Surgeons, the anatomist Richard Owen (1804–92), as evidence that marsupials were physiologically "inferior" to placental mammals, and in "an intermediate place between viviparous and oviparous animals" (in effect, between placental mammals and reptiles). So it was to be expected that they would have appeared at an earlier point in geohistory, before the environments on the slowly cooling planet were suitable for more advanced (placental) forms of mammalian life. In short, Buckland regarded the Stonesfield marsupials as Secondary forerunners of the Tertiary "age of mammals". Far from undermining the generally progressive trend of geohistory, as Lyell had claimed, they positively reinforced it.[11]

9. Fig. 29.3 is reproduced from Buckland, *Geology and mineralogy* (1836) 2: part of pl. 1. On the meetings at Mézières and Bonn, see Rudwick, *Devonian controversy* (1985), 135–41, and "International arenas" (1987); French was used as the working language for such sessions, overriding—for the sake of international communication—the pan-German nationalistic sentiments with which the GDNA had been founded in 1822. When the early members of the BAAS were wondering what to call themselves, the GDNA's word "*Naturforscher*" was considered as a possible model, but its derisively over-literal translation as "nature-poker" was enough to rule it out; Whewell's suggested alternative, "*scientist*", failed to be adopted (§25.3).

10. Fig. 29.4 is reproduced from Buckland, *Geology and mineralogy* (1836) 2: part of pl. 1; the fossils, drawn by Joseph Fisher, are identified on 12–17. Unlike most other geologists at this time, Buckland included the "Carboniferous group" in his "Transition series", so the early plants shown (lower left) are those of the Coal formation. All these fossils were drawn just too soon to distinguish the "Silurian" forms (see §30.3), which—so Murchison claimed—antedated even the first fish and first land plants; a Silurian assemblage would have emphasized still further the directional and progressive character of the whole of geohistory. In fact, the "Marine animals and plants" (lower right) shown under the heading of "Transition" include three trilobites (20–22) from Murchison's Silurian rocks, as well as fish from much later Carboniferous formations.

11. Buckland, *Geology and mineralogy* (1836), 1: 72–75; the Stonesfield "*Didelphis*" specimens were given space on the very first plate (2: pl. 2A, B) following his fold-out ideal section, on which they were reconstructed as marsupial mice (see Fig. 29.4, numbers 40–41); see also Owen, "Marsupial animals" (1834). The monotremes (the Australian egg-laying platypus and echidna) were already well known, and recognized as mammals even "lower" in status than marsupials, but they had not been found as fossils.

29.3 PALEY GEOHISTORICIZED

Buckland had been commissioned to write a Bridgewater Treatise on geology and mineralogy; in the event he wrote mostly on the fossils that were his first love in his science. Paley had famously begun his *Natural Theology* by imagining an intelligent person unfamiliar with clockwork, who might pick up a watch left lying on stony ground, take it carefully to pieces, and correctly conclude that unlike a mere pebble such an intricate mechanism must have had a designer. The argument was that, by analogy, the equally well-adapted structures found in living organisms must have had a divine Designer. In effect, Buckland set himself the task of giving Paley's argument a newly geohistorical dimension. In one of his lectures he said, "If Paley had been a geologist he would not have commenced his [Natural] Theology as he does, for he would have known that a stone might not have been there for ever: it might have been a coral once formed by an animal [and] since rolled [smooth] by a deluge". In other words, stones now had *geohistorical* stories to tell, far beyond what Paley's generation could have imagined.[12]

Buckland intended to use the recent progress of paleontology to show that organisms had *always* had the same well-adapted and apparently designful character, as far back in geohistory as fossils had been found. This would not only extend the scope and persuasiveness of Paley's traditional argument, but also build another bulwark against the threat of Lamarck's transformism (Chap. 17): Buckland would show that earlier organisms had *not* been crude imperfect forms, but creatures as well adapted as any in the modern world. Much of his book was therefore devoted to the kind of functional analysis of fossils embodied in De la Beche's famous reconstructed scene of life in "a more ancient Dorset" (Fig. 11.4), which in turn stemmed from Cuvier's pioneer reconstructions of the same kind (§1.3; *BLT* §7.5); the analyses were supported by a profusion of plates in a separate volume of illustrations. Two of Buckland's many examples will serve here to illustrate his argument; they came from opposite ends of geohistory.

The megatherium, the bones of which had been found in Superficial deposits in South America and reconstructed into a skeleton conserved in Madrid, was one of the very first of the fossils that Cuvier had famously used, forty years earlier, to argue for the reality of extinction (*BLT* §7.1). In 1832 Woodbine Parish (1796–1882), the British consul in Buenos Aires (and a Fellow of the Geological and Royal Societies), sent to London some of the first new megatherium bones to reach Europe since the time of Cuvier's early work; Clift exhibited and described them at the Geological Society soon after the great naturalist's death. A few months later, when the British Association met in Oxford, Buckland explained them to the assembled savants on the last evening of the meeting; the monstrous size of the animal was demonstrated in characteristically lighthearted ways when Buckland's five-year-old son Frank sat in the pelvis and Clift crawled through the birth-canal. Then, and subsequently in his book, Buckland used the megatherium to out-Cuvier Cuvier, extending and improving the Frenchman's functional analysis. However ungainly and clumsy the animal might have looked at first glance, Buckland argued that in fact it had been very well-adapted for its inferred mode of life of digging up plant roots, while its bony armor—

discovered since Cuvier's early work—showed that its ponderous build would have been no disadvantage. In short, it offered "fresh proofs of the infinitely varied, and inexhaustible contrivances of Creative Wisdom" (Fig. 29.5).[13]

Fig. 29.5 The first megatherium skeleton from South America, as reconstructed by a Spanish naturalist and displayed since the 1790s in the royal museum in Madrid: this German drawing (1821) was reproduced in Buckland's Bridgewater Treatise (1836) to help explain new but less complete specimens that had recently been sent to England. Buckland argued that the huge ponderous animal had been well adapted to a life of digging up plant roots, and was further proof of divine design throughout geohistory.

12. Paley, *Natural theology* (1802), 1–8; J. E. Jackson, notes on Buckland's lecture, 22 May 1832 [Keyworth-BGS: 1/365; quoted in Boylan, *Buckland* (1984), 597]. Buckland may well have shown his students pebbles of the kind he mentioned, eroded out of the "Corallian" limestone that outcrops near Oxford, rounded by attrition, and deposited in the nearby Thames gravels that he regarded as "diluvial" in origin (*BLT* §10.5).

13. Fig. 29.5 is reproduced from Buckland, *Geology and mineralogy* (1836), 2: pl. 5, fig. 1; the megatherium is described and analyzed in 1: 139–64; 2: 19–20 and pls. 5–6; the frontal view was more dramatic, though less informative anatomically, than a conventional lateral view. Buckland's picture was copied by Scharf from Pander and d'Alton, *Riesen-Faulthier* (1821), pl. [2] (where the animal had been assigned to *Bradypus*, the same genus as the living sloths). This famous skeleton had been reconstructed in the 1790s by Juan-Bautista Bru and is now on display, still—admirably—in the quadrupedal pose that Bru

At the other end of geohistory, Buckland's analysis of trilobites from the Transition formations, which the elder Brongniart had been one of the first to study seriously (§4.3), was even more significant, because it demonstrated superbly designful adaptations in some of the earliest animals known, which had become extinct "perhaps millions of years ago" (Fig. 29.6).[14]

Buckland, profoundly moved "with feelings of no ordinary kind", described the compound eyes of trilobites as the "instruments of vision, through which the light of heaven was admitted to the sensorium of some of the first created inhabitants of our planet". He argued that they proved the presence of providential design even in the depths of geohistory (and also showed that the water had been as normal and transparent as it is in present seas). These extremely ancient animals had not been the crude imperfect beings that Lamarck's transformist theory was assumed to entail: "We do not find this instrument [of the compound eye] passing onwards, as it were, through a series of experimental changes, from more simple into more complex forms; it was created at the very

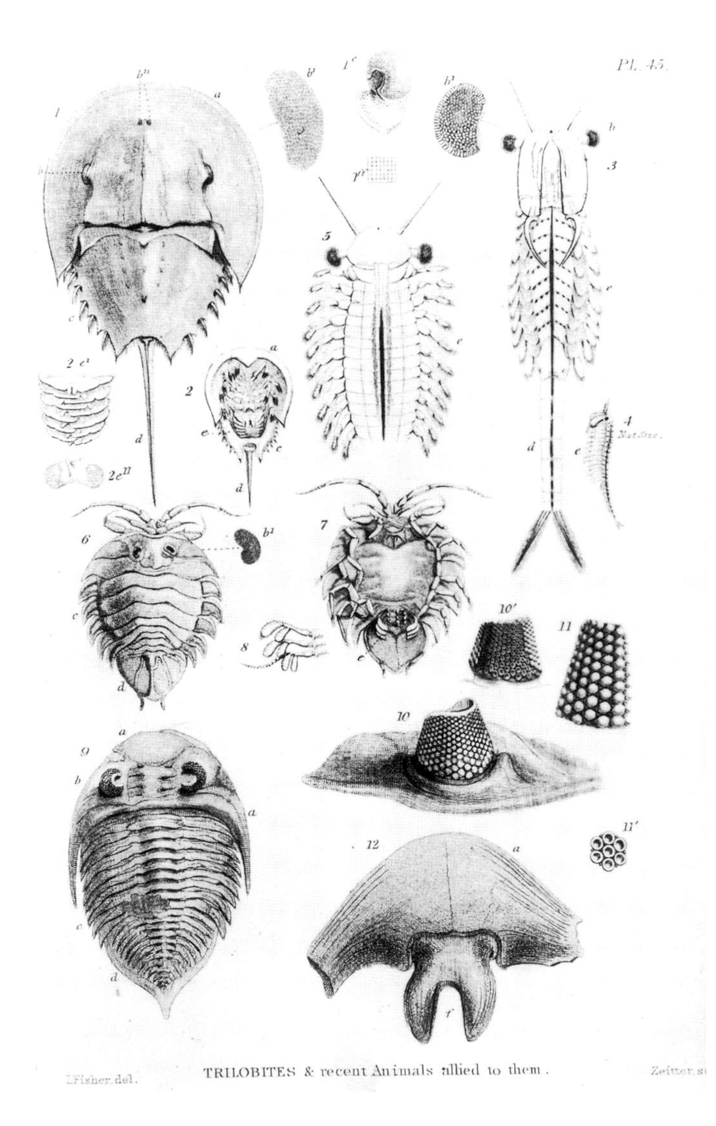

Fig. 29.6 A trilobite from the ancient Transition strata (in fact from Murchison's "Silurian": see §30.3) compared with living arthropods: engravings from Buckland's Bridgewater Treatise (1836). The trilobite *Asaphus* (*9*, bottom left) had large conical compound eyes (*10*) composed of rows of lenses (*11*); Buckland pointed out that they would have given the crawling animal highly effective all-round vision of the sea-floor, where both its food and its enemies would have been. The living king-crab *Limulus* (*1*, *2*, top left) and the crustaceans *Branchipus* (*3-5*, top right) and *Serolis* (*6*, *7*, center) were somewhat similar in general form, though recognized as not being closely related to trilobites; they too had compound eyes adapted to their respective habitats.

first, in the fulness of perfect adaptation to the uses and condition of the class of creatures [i.e., arthropods], to which this kind of eye has ever been and still is appropriate."[15]

29.4 CONCLUSION

The Huttonian geotheory put forward by Lyell was matched in part by Buckland's Bridgewater Treatise, which presented some of the evidence that geohistory had been strongly directional, and not steady-state at all. His book was one of a series commissioned to bring up to date, from the latest scientific research, the traditional claim that divine design could be discerned in all parts of the natural world. This kind of argument was widely accepted by savants, notably but not only in Britain. They all understood the distinction between this kind of natural theology and the more contentious issues of revealed theology that often divided people of different religious persuasions or none. More specifically, geologists who were clergymen, such as Buckland and Conybeare, and those who were inclined to be anticlerical (though not necessarily anti-religious), such as Lyell and Scrope, were united in drawing a sharp distinction between themselves and the "Mosaic" or self-styled "scriptural" writers, whose literalistic claims the geologists forcefully repudiated. In this context, one point had become so marginal in the eyes of geologists that it was relegated in Buckland's book to a mere footnote. Here he publicly abandoned his famous earlier claim that the "geological deluge" had been the very same event as the biblical Flood. In his opinion the "deluge" had been real enough, and just the most recent of its kind, but it had long antedated the local event dimly recorded in Genesis.

Buckland was well aware that any interpretation—including his own—of still earlier geohistory depended necessarily on a foundation of well-established stratigraphy, and on the study of fossils reliably located within that framework. The naming of formations, and of larger "groups" such as Cretaceous and Carboniferous, was a matter of intense international discussion, in which Greenough was prominent. The results of this kind of negotiation formed the basis for Buckland's summary of current stratigraphy, and hence for his interpretation of the fossil record—and the history of life—as unmistakably directional. In this context, he used Owen's studies of living marsupials to argue that the famous little fossils from Stonesfield were just where they might have been anticipated in the fossil record: they were Secondary forerunners of the "higher" (placental) mammals of the Tertiary era, and not—as Lyell had claimed—evidence that mammals of all kinds might have existed from the earliest times.

~ gave it (and Buckland reproduced), in Madrid-MCN (*BLT* §7.1). Later research, in contrast, interpreted it as an almost bipedal animal (and the bony armor was found to belong to another animal): a reconstruction of ca. 1850, incorporating casts of Parish's bones, is now on display in London-NHM. See Clift, "Remains of the megatherium" (1835), read 13 June 1832; Buckland, "Megatherium recently imported" (1833); and Boylan, *Buckland* (1984), 197–98.

14. Fig. 29.6 is reproduced from Buckland, *Geology and mineralogy* (1836), 2: pl. 45, drawn by Joseph Fisher and explained in 1: 396–404 and 2: 71–73.

15. Buckland, *Geology and mineralogy* (1836) 1, 402–3.

However, Buckland also set out to demonstrate that throughout geohistory all living organisms had been well-adapted to their specific modes of life. So Paley's classic evidence for the providential designfulness of the living world was no longer confined to the present but could be extended back in deep time to the earliest known periods of geohistory. Buckland's Cuvierian analyses of form and function—interpreted in the service of this renewed natural theology—showed, for example, that even the earliest trilobites had been as well designed for their allotted mode of life as any megatherium from the geologically recent past. Organisms had not progressed from crude beginnings to ever more "perfect" forms, in the way that Lamarck's transformist theory was assumed to entail. But the fossil record did show directionality of another kind, namely in the increasing diversity of the fauna and flora by the successive addition of new and "higher" forms of life.

During just the years when Buckland's Bridgewater Treatise was being written and then making its impact on geologists and the wider reading public, throughout Europe and even beyond it, further evidence for this kind of directionality was accumulating in a striking manner. The discovery of the first fossil primates in the Tertiary has been mentioned already (§28.3); the next chapter describes research that helped to clarify the history of life much further back in time, by analyzing the very earliest known vertebrates and then penetrating back towards the origin of life itself.

CHAPTER 30

The progression of life (1833–39)

30.1 AGASSIZ AND THE AGE OF FISH

Buckland and almost all other geologists were convinced that the apparent directionality of the fossil record was *not*—as Lyell claimed (§24.3)—a mere artifact of systematically defective preservation. Their confidence was strengthened at just this time by new research that closely paralleled the younger Brongniart's earlier analysis of fossil plants (§12.2).

In his last years, the indefatigable Cuvier had been hard at work on a massive research project on fossil fish, which was intended to parallel and complement his earlier work on fossil quadrupeds (see Fig. 16.2). The year before he died, however, perhaps sensing the limitations of mortality, Cuvier recognized the outstanding talents of a young Swiss zoologist, Louis Jean Rodolphe Agassiz (1807–73) of Neuchâtel, and, in effect, handed this research project over to him. Agassiz had been well trained in zoology and other natural sciences at Heidelberg and Munich; like all educated Swiss he was, of course, bilingual. By the time he visited Cuvier in Paris, the ambitious twenty-five-year-old had already published competent work on some fish brought back from Brazil, and had also started to study fossil fish. In 1832, just after Cuvier died, Humboldt used his Prussian influence to get Agassiz appointed professor of natural history at the newly opened college in Neuchâtel. Agassiz returned to his native city and soon made his mark among naturalists; the importance of his research was quickly recognized, as he traveled around Europe studying all the main collections of fossil fish, comparing them with museum specimens of living forms (Fig. 30.1).[1]

1. Fig. 30.1 is reproduced from Rivier, "Société Neuchâteloise" (1932), frontispiece; the original painting [in Neuchâtel-BV] is attributed to Henri Beltz and dated 1835. See Gaudant, "Agassiz fondateur de la paléoichthyologie" (1980), and Coleman, "Cuvier and Agassiz" (1963). Agassiz (E. C.), *Louis*

Fig. 30.1 Louis Agassiz at the age of about twenty-eight: a portrait painted at the time that his travels to study fossil collections, and the early installments of his published *Researches on Fossil Fish* (1833-43), were making him a familiar figure in scientific circles around Europe.

In 1833 Agassiz staked his claim to his research field by appealing for subscribers to what would clearly be a major publication, and he issued the first part of it to tempt them. Its title, *Researches on Fossil Fish*, was obviously modeled on Cuvier's great work, which Agassiz took explicitly as his guide. This was no mere flattery of his recently deceased patron. Agassiz followed Cuvier's methods, basing his work on detailed anatomical comparisons between fossil and living forms. He also adopted Cuvier's strategy of moving back into successively older and therefore stranger faunas. So he started with the famously well-preserved Tertiary fish from Monte Bolca near Verona (*BLT* §5.2); as a result of Napoleon's cultural and scientific plundering of Italy, some of the best specimens were now more conveniently available for study at the Muséum in Paris.[2]

In his preface, Agassiz emphasized the fundamental importance of fossil fish, in terms that not only boosted his project but also showed a shrewd assessment of its scientific importance. Fish, being vertebrates, were animals of "high" organization, yet they were known as fossils "without interruption in all the sedimentary formations from the most ancient to the most recent"; and at certain points in the sequence—the Monte Bolca formation was a good example—they were abundant and well preserved. In contrast, fossil mammals were usually fragmentary, reptiles were rare, and with mollusks and other invertebrates the shells and similar "hard parts" were usually the only parts preserved. So Agassiz claimed that fossil fish offered an outstanding opportunity to make reliable inferences about the animals' functional anatomy and about the environments in which they had lived. Beyond the usual basic aims of classifying and naming his fossils, his goals were ambitious and clearly geohistorical: to find "the laws of succession and organic development of fish during all the geological epochs", and hence ultimately "to know what connection there is between the organic development of the earth and that of the different classes of animal".[3]

When Agassiz first visited Britain in 1834, he anticipated that his study of further collections of fossil fish would simply amplify the broader conclusions he had already reached. But in the event the sixty-three public and private collections that he studied in Britain were so rich and some of the specimens so fine—many were from famous Continental localities such as Bolca, Oeningen, and Solnhofen—that he felt he almost had to make a fresh start; it was a tribute to the flourishing culture of fossil collecting among wealthy amateur naturalists in Britain. Buckland ensured that funds from the British Association enabled Agassiz to bring his Swiss artists to England to draw the best specimens on the spot (the owners would have been unlikely to accept the risks involved in shipping them to Neuchâtel).[4]

The leaders of the Geological Society likewise considered Agassiz's provisional results so important that his major paper on fossil fish was the first to be read when the Society reconvened after the 1834 summer recess. Building on Cuvier's earlier classification, Agassiz defined four main groups of fish, based primarily on the kinds of scales on the skin: he called them placoids, ganoids, ctenoids, and cycloids. Of these, the first two groups were known right from the start of the fossil record, and were still represented in the present world though not in great variety (apart from the shark family). The last two groups were known from the start of the Cretaceous, but were abundant only in the Tertiary and comprised three-quarters of all living forms. So the fossil record of fish showed the usual directional increase in diversity (see his later graphical representation, Fig. 36.7). Above all, as Agassiz emphasized, fish provided "the only example of a great division of vertebrated animals in which we may follow all the changes experienced in their organization during the greatest lapse of time of which we possess any relative measure".[5]

In effect, Agassiz's history of fishes closely paralleled what Adolphe Brongniart had outlined for the history of plants a few years earlier (§12.2): new and "higher" groups had made their appearance successively in the course of time, increasing the diversity of the organisms and bringing them ever closer to those of the present. There was no good reason to suppose that the conditions under which fish were likely to be preserved as fossils had changed significantly in the course

---

Agassiz (1885), and Marcou, *Agassiz* (1896), print valuable extracts from his correspondence. Lurie, *Agassiz* (1960), is a useful biography; on his local context, see Schaer, *Géologues de Neuchâtel* (1998). Neuchâtel had long been Prussian territory, and did not join the Swiss Confederation until 1857.

2. Frigo and Sorbini, *600 fossili per Napoleone* (1997). Agassiz, *Poissons fossiles* (1833–43), was his major publication, supplemented later by *Poissons fossiles du Vieux Grès Rouge* (1844–45), on those from the Old Red Sandstone, the oldest of all fish then known.

3. Agassiz, *Poissons fossiles* (1833–43) 1, preface dated 12 July 1833; Jeannet, "Ouvrages d'Agassiz" (1928–29), tabulates the extremely complex publishing history of the work.

4. Agassiz was the *only* foreigner to benefit from the BAAS's substantial research grants in the years to 1843, receiving a total of £520, in return for which he attended and performed at several annual meetings: Morrell and Thackray, *Gentlemen of science* (1981), 320, 465–66, 551. He also repaid his obligations by translating, editing, and publishing Buckland's Bridgewater Treatise in German as *Geologie und Mineralogie* (1838–39).

5. Agassiz, "New classification of fishes" (1834), read on 5 November, and *Poissons fossiles* (1833–43), 1: 10–23; Gaudant, "Agassiz fondateur de la paléoichthyologie" (1980). In the long run, Agassiz's classification was not adopted by other naturalists, who preferred that of the anatomist Johannes Müller (1801–58).

of geohistory. So the evident directionality—and even progressiveness—of their fossil record could not be explained away in Lyellian fashion as a mere artifact of imperfect preservation. In short, Agassiz's massive work added powerful support to the opinion of almost all geologists other than Lyell, that fossils provided a broadly reliable record of a clearly directional history of life. In 1836 the Geological Society awarded him its Wollaston medal for this research; Broderip expressed the general opinion when, accepting the medal in Agassiz's absence, he referred to the work as "the powerful zoological lever which he has placed in the hands of Geologists". As Agassiz himself put it emphatically, his latest detailed research was confirming—in the case of fish—the general principle "of the *progressive and organic development of life on earth*".[6]

Agassiz had good reason to be confident that what he was finding in the fossil record of fish was also valid more generally. Cuvier's early and prescient hunch that there had been an age of reptiles before the first appearance of mammals (§1.2; *BLT* §8.1) had been amply confirmed by the sensational identification and reconstruction of the reptilian "monsters" of the Secondary era: the ichthyosaur and plesiosaur in the sea (Chap. 2), the megalosaur and iguanodon on land (Chap. 5), the pterodactyl in the air (§11.3). Mantell, the discoverer of the iguanodon,

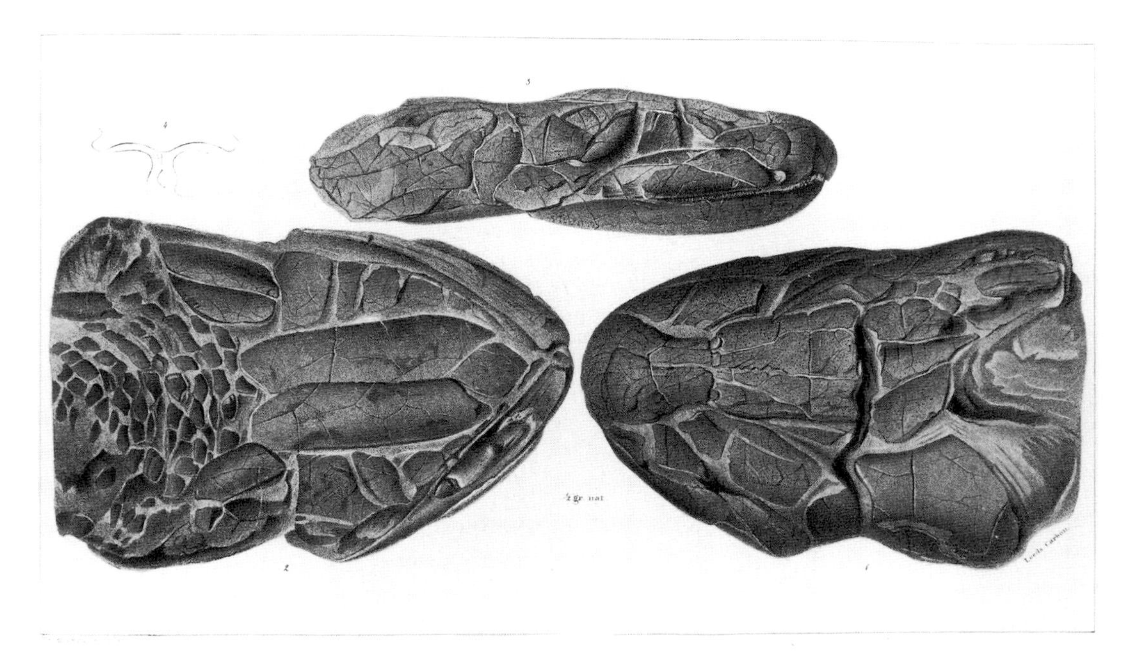

Fig. 30.2 A skull of the *Megalichthys* ["huge fish"] from the Coal formation of Yorkshire, found by Agassiz and Buckland in the museum in Leeds, soon after Agassiz had seen Hibbert's more fragmentary specimens in Edinburgh and judged them to belong not to a reptile but to a very large "ganoid" fish. Agassiz honored the Scottish naturalist by naming the fish *M. Hibberti*; he regarded it as related to the living freshwater gar fish, the "ganoid" *Lepidosteus*. This lithograph was published in 1839 in one of the installments of his superbly illustrated *Researches on Fossil Fish*, which was modeled explicitly on Cuvier's similarly massive work on fossil quadrupeds. The skull, about 35cm long, is shown in dorsal (*1*), ventral (*2*), and lateral (*3*) views. Agassiz's reinterpretation pushed the apparent start of the "age of reptiles" forwards, leaving the period of the Carboniferous formations as an "age of fish". The sheer size and anatomical complexity of this and many other Carboniferous fish showed that these earliest of all known vertebrates had not been the crude "imperfect" animals that Lamarckian transformism was generally taken to entail. (By permission of the Syndics of Cambridge University Library)

had more recently defined the Secondary era as "the geological age of reptiles", describing it as a period when they had been "the Lords of the Creation", long before the earth was supposedly fit for human habitation. However, one outcome of Agassiz's research trip to Britain was that it confirmed what Buckland's artist had already depicted on his great panorama of geohistory (Fig. 29.4), namely that before the start of the age of reptiles there had been a distinct "*age of fish*".[7]

This inference was soon reinforced when Agassiz removed an apparent exception to it. The Scottish naturalist and antiquarian Samuel Hibbert-Ware (1782–1848) had identified some fossil teeth from Carboniferous rocks near Edinburgh as those of a large unknown "saurian"; it was the first reptile to have been reported so early in the fossil record. But when Agassiz saw the specimens, while attending the British Association meeting in Edinburgh, he stated—with all his new authority as the acknowledged expert—that they belonged in fact to a huge ganoid fish. The earliest reliable record of reptiles was thereby pushed forwards out of the Carboniferous, leaving that period as a time when the vertebrates had apparently been represented *only* by fish. Since Agassiz's research was showing that the Carboniferous had been a time when ganoid (and placoid) fish had already been abundant and diverse, it might indeed count as an age of fish, and big and complex ones at that (Fig. 30.2).[8]

30.2 PHILLIPS'S CARBONIFEROUS BENCHMARK

Still further back in geohistory, what appeared to be the earliest well-defined period in the history of life, and even a hint of its very start, also emerged during the 1830s. These were unforeseen outcomes of research that was purely stratigraphical in its initial objectives. Murchison and Sedgwick, at first independently and then in a rather tense collaboration, tried to unravel the Transition formations or "Grauwacke group", which they and other geologists (except Lyell, as usual) regarded as containing the traces of the earliest forms of life.

However, any such clarification of the Transition formations was dependent on a firm reference point in the overlying Secondaries, of which the lowest (and oldest) major "group" was the Carboniferous. This necessary benchmark was provided on cue by the up-and-coming geologist John Phillips (1800–1874), the nephew of William Smith and like him a man of humble social origins. Phillips had initially been Smith's assistant and informally his apprentice. But he had

6. Geological Society, 19 February 1836 [*PGS* 2: 354–55]; Agassiz, *Poissons fossiles* (1833–43), fascicle 6, "feuilleton additionel, mars 1836". [Broderip], "*Poissons fossiles* par Agassiz" (1836), an enthusiastic review in the *Quarterly*, promptly made Agassiz's interpretation widely known in Britain, beyond the Geological Society.

7. Mantell, "Age of reptiles" (1831), although published in Jameson's periodical, was only a short semi-popular article.

8. Fig. 30.2 is reproduced from Agassiz, *Poissons fossiles* (1833–43), Atlas 2 [Cambridge-UL: OA.5.12]: pl. 63, published in livraison 10/12 (1839), discussed in [text] 2: 89–96; the matching pl. 63a has line drawings of the same specimen, keyed to identify the various bones in the skull. The specimen was drawn and lithographed by J. C. Weber, one of the Swiss artists employed by Agassiz: see Purcell and Gould, *Finders, keepers* (1992), 110–23. Hibbert, "Limestone of Burdiehouse" (1836), read at the Royal Society of Edinburgh on 2 December 1833, 17 February, 21 April, 1 December 1834; in the published text he recorded his acceptance of Agassiz's interpretation.

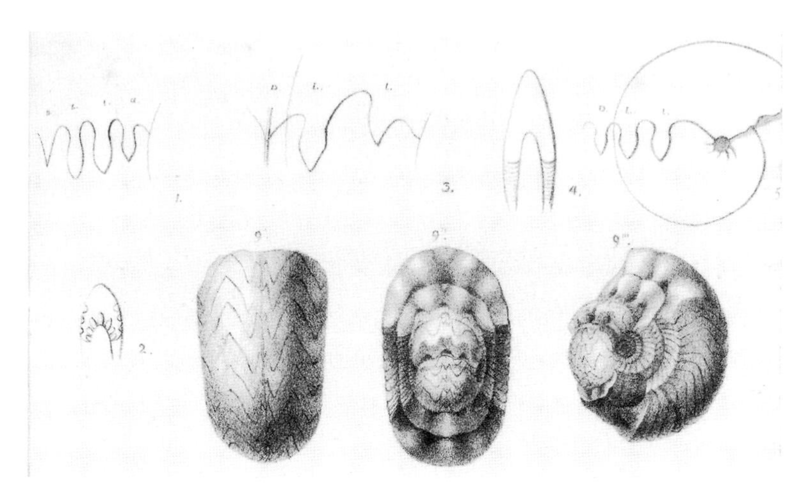

Fig. 30.3 Leopold von Buch's illustrations of several distinct species of the ammonite shells that he termed "*goniatites*", so named on account of the angular "suture" lines marking the edges of the partitions that divided the original shell (like those of the living pearly nautilus). The specimen shown "solid" belongs to the species that Goldfuss had named *carbonarius* in allusion to its occurrence in the Carboniferous formations of Germany. The other species are shown just in outline, or just by the form of their characteristic suture line, "unwrapped" from around each whorl of the spiral shell. In fact all these goniatites came from rocks of about the same age, so they had flourished much earlier in geohistory than the better-known Jurassic ammonites (with much more elaborate suture lines: see Fig. 4.1). (By permission of the Syndics of Cambridge University Library)

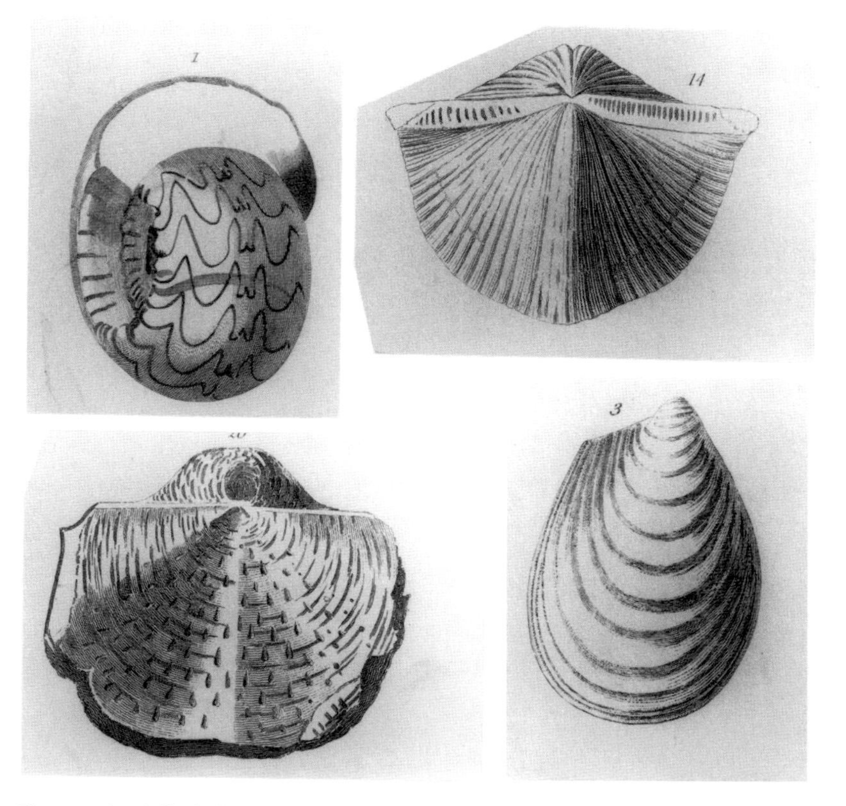

Fig. 30.4 John Phillips's illustrations (1836) of some of the fossil shells characteristic of the "Mountain" or Carboniferous Limestone of northern England: the mollusks *Goniatites* (upper left) and *Posidonia* (lower right), and the brachiopods *Spirifer* (upper right) and *Productus* (lower left). These were characteristic components of what Phillips described as the rich and diverse marine fauna of the Carboniferous period. (By permission of the Syndics of Cambridge University Library)

begun to carve out an independent career when in 1825 he was appointed curator of the museum in York, which had been set up in the wake of the discovery of fossil bones at Kirkdale (*BLT* §10.6); he then augmented his small salary by giving public lectures on geology around Yorkshire. In 1831 the British Association held its inaugural meeting in his home city, and a year later he was appointed its salaried administrator, serving under its "general secretary" Murchison. This work brought Phillips into contact with the leading British "gentlemen of science"; although he was their social inferior, his evident talent soon led to his acceptance in these savant circles. In 1834, while continuing to live in York, he added yet another part-time position when he was appointed Lyell's successor as professor of geology at King's College in London; and two years later he began working for De la Beche's governmental Survey, unpaid at first but later as an employee. In short, Phillips was a fully "professional" scientific savant, a species already well established on the Continent but not yet common in Britain.[9]

Phillips's rapidly growing reputation as a geologist was based in part on several useful introductory books and encyclopedia articles, but even more on some fine original research. The first volume of his *Geology of Yorkshire* (1829) had described the stratigraphy and fossils of the younger Secondary formations in his county (in area, by far the largest in England). Its epigraph from Cuvier, combined with praise for Smith on its opening pages, suggested how he intended to develop his uncle's empirical methods in a direction that would treat fossils, as his French hero had done, as the traces of living organisms and the markers of geohistory. The second volume of this work, published in 1836, extended his stratigraphy downwards into Yorkshire's thick and widespread Carboniferous formations. The spectacular tree-ferns and other large land plants of the Coal formation itself were already being described by Sternberg in Germany and the younger Brongniart in France (§4.4, §12.2); and their work was being supplemented in Britain by John Lindley (1799–1865), who taught botany at the London University (the secular rival of King's College), in collaboration with William Hutton (1798–1860), a resident of Newcastle and therefore at the heart of the world's most productive coalfield. Phillips left the fossils of the Coal formation in these competent hands, and focused his attention instead on the Carboniferous formations underlying it, particularly the thick "Mountain" or Carboniferous Limestone of the Pennine hills that form the central spine of northern England; he made a point of studying the same formation elsewhere in England and on the Continent. This widespread marine limestone of the Carboniferous period contained a distinctive fossil fauna, characterized not only by mollusks, among them the ammonite-like shells that von Buch had recently defined as "*goniatites*", but also by abundant and diverse "brachiopods" (Figs. 30.3, 30.4).[10]

9. Morrell, "Phillips in the 1820s" (1989), is a valuable account of his earlier work; *Phillips* (2005), describes and analyzes his whole career in detail, and is the indispensable biographical source.

10. Fig. 30.3 is reproduced from Buch, *Über ammoniten* (1832) [Cambridge-UL: MB.9.70/4], part of pl. 2; Fig. 30.4, from Phillips, *Geology of Yorkshire 2* (1836) [Cambridge-UL: MF.40.6], pl. 6, fig. 3; pl. 8, fig. 20; pl.9, fig. 14; pl. 20, fig. 1; see also Rudwick, *Devonian controversy* (1985), 84–86, 146–48. Sternberg, *Flora der Vorwelt* (1820–38), Brongniart (Ad.), *Végétaux fossiles* (1828–37), and Lindley and Hutton, *Fossil flora of Great Britain* (1831–37), were all completed (or left uncompleted) in the later 1830s; Chaloner

30.3 MURCHISON'S SILURIAN AND SEDGWICK'S CAMBRIAN

Phillips's distinctive and well-described Carboniferous fauna served Murchison as his benchmark for comparison with even older fossils; it was a crucial component in his and Sedgwick's joint attempt to unravel the stratigraphy of the Transition formations in which the fossil record seemed to begin (Fig. 30.5).[11]

When Murchison returned from his Continental fieldwork with Lyell (Chaps. 18, 19) he had cast around for a topic of research in which he could shine in savant circles in London. After meeting Smith, who was living in retirement in Yorkshire, the ambitious Scotsman became convinced that Smithian stratigraphy could be taken much further, to prove that fossils were invariably more reliable than rock types as guides to stratigraphy. In 1831 he had persuaded the Geological Society to make Smith the first recipient of its then newly endowed Wollaston medal, thus healing the long-standing rift between them (*BLT* §9.5); and at the inaugural meeting of the British Association later that year Murchison had publicly dubbed Smith the "father" of English geology. He hoped to vindicate that iconic attribution, but how he would do so was initially quite uncertain. In the event, it was almost by accident that he focused on using Smithian methods to bring better order (in all senses) to the obscure Transition rocks.[12]

Following a lead suggested by Buckland, Murchison found in the Welsh Marches (along the border between England and Wales) a sequence of formations that he could trace downwards below the Old Red Sandstone, without any unconformity to break the continuity; in fact he had the sequence demonstrated to him in the key area by the local parson and amateur geologist Thomas Taylor Lewis (1801–58), but later took all the credit for himself. The "Old Red" was generally classified as the lowest formation in the Carboniferous "group" and hence the lowest and oldest of all the Secondaries in Britain. It had been the last

Fig. 30.5 Roderick Murchison (left) and Adam Sedgwick (right): portraits drawn in the 1830s, when they jointly clarified the stratigraphy of the Transition or "Grauwacke" rocks; defined the upper formations as "*Silurian*" and the lower as "*Cambrian*"; and claimed that their fossils represented the earliest known phases in the history of life. (By permission of, respectively, the British Museum and the National Portrait Gallery, London)

formation to be described in the retrospective stratigraphy of "Conybeare and Phillips" (§3.2; *BLT* §9.5), a reference work of which, after a decade, the planned second volume had failed to appear, partly because the underlying Transition formations were still so poorly known. However, in the Welsh Marches there was a clear sequence of formations below the Old Red, composed of shales alternating with limestones. Far from being highly folded or confused by slaty cleavage, these formations were only gently tilted; and their often abundant fossils were well preserved and had long been prized by fossil collectors. In the key area of Shropshire the formations even had a topography of parallel scarps (one of them Wenlock Edge, much later made famous by Housman's poem) not unlike those of the far younger (Jurassic) formations in Smith's home area. In short, Smith's mapping some thirty years earlier was a perfect model for sorting out these ancient formations (Fig. 30.6).[13]

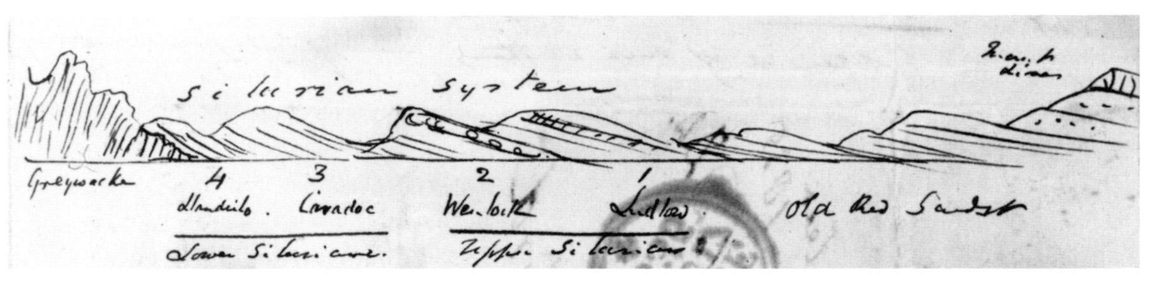

Fig. 30.6 Murchison's sketch section of the Transition formations that he was mapping in the Welsh Marches, drawn in 1835 to show Buckland (and, subsequently, other geologists) how they lay without unconformity directly below the lowest Secondary formation, the Old Red Sandstone (right). In the letter to Buckland that included this section, he called them for the first time the "*Silurian System*"; its four divisions (named after places in the region) were numbered from top down, his method being the usual one of penetrating from the known towards the more obscure, not upwards in geohistorical order. The lowest and oldest formation (4, Llandeilo) is shown resting with a major unconformity on Sedgwick's still older "Greywacke" (left), the Lower Transition or "Cambrian" rocks. The unconformity later proved to be illusory, which led to a prolonged and famously acrimonious boundary dispute between Murchison and Sedgwick. (By permission of the Museum of Natural History, Oxford)

and Pearson, "John Lindley" (2005). The Carboniferous Limestone in Belgium was described by Dumont, "Province de Liège" (1832); and its fossils, by Koninck, *Terrain carbonifère de Belgique* (1842–44), the latter based on research already well known in the 1830s; see also Rudwick, *Devonian controversy* (1985), 135–41. Cuvier, in his work for *Règne animal*, had shown that brachiopods ("lamp-shells") were quite distinct from bivalve mollusks (clams, mussels, etc.); their bilaterally symmetrical shells are dorsal and ventral, not left and right; they are very abundant as fossils but only a few survive in present seas.

11. Fig. 30.5 is reproduced from a lithograph after a portrait of Murchison by William Drummond, published in June 1836 [London-BM: 1865.6.10.1219], and from a portrait of Sedgwick by Thomas Phillips dated 1833 [London-NPG: 33648].

12. Secord, *Victorian geology* (1986), 47–51; Rudwick, *Devonian controversy* (1985), 63–69. Mantell was the second recipient of the Wollaston medal, and Agassiz in 1836 the third.

13. Fig. 30.6 is reproduced from Murchison to Buckland, 17 June 1835 [Oxford-MNH]; the postmark has leached through from the outer side (envelopes were not yet in general use). The highest formation of all (far right) is "Mount[ain] Limes[tone]". Murchison, "Silurian system of rocks" (1835), included a wood engraving similar to this sketch; he had already summarized the stratigraphy in "Transition rocks of Shropshire" (1834), read at the Geological Society on 22 January, and in still earlier progress reports. Secord, *Victorian geology* (1986), is the definitive account of all this research, and of the later disputes between Murchison and Sedgwick; see also Thackray, "Lewis and Murchison's *Silurian system*" (1979), Torrens, "Historiography of *The Silurian system*" (1990), and Rudwick, *Devonian controversy* (1985), 82–87, 127–30.

At first, Murchison referred to his formations as "Upper Transition" or "Upper Grauwacke", to distinguish them from Sedgwick's older and more confusing rocks. Later, however, he tried to rid geology of these traditional terms, because in his opinion both were loaded with dubious connotations. In 1835 he proposed calling his own formations "*Silurian*", after an ancient British tribe that had famously resisted the Roman occupiers in the Welsh Marches (significantly, such historical or antiquarian associations remained characteristic of a science that was becoming ever more *geo*-historical in its practice). Under whatever name, they represented one of the first sequences of pre-Carboniferous formations to be described in the kind of detail that was now taken for granted in Secondary stratigraphy, and with a similarly close attention to their fossils. Élie de Beaumont invented an appropriate French verb ("*secondariser*") when, writing to Sedgwick, he praised Murchison's work for helping "to *secondarise* the barbarous group of the Transition formations".[14]

Murchison's "Silurian" was originally intended to describe the "system" of rocks found in the specific region after which it was named. When Murray published Murchison's research in full, it was as a handsome volume entitled *The Silurian System* (1839), with a prolix subtitle that listed all the English and Welsh counties covered by his survey (and by the estates of the aristocrats and landed gentry who, along with geologists, were his principal subscribers). Although the "Silurian" formations were given the most detailed attention in the text, Murchison's descriptions covered everything in the region from Superficial deposits to Primary rocks; a superb but quite conventional geological map also showed clearly that this was a work in the established genre of regional geology. It was

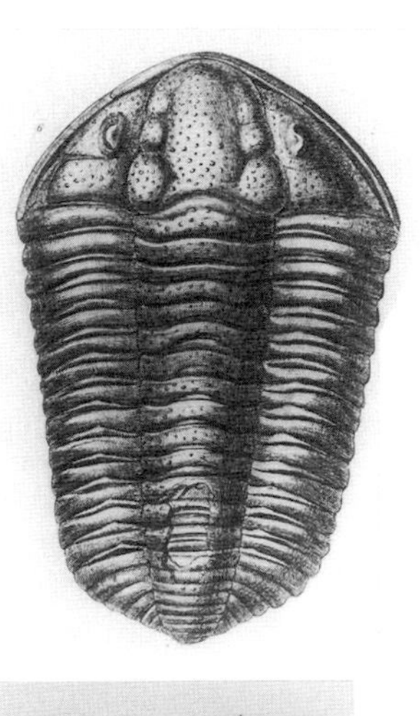

Fig. 30.7 Murchison's illustrations (1839) of some fossils characteristic of what he defined in 1835 as the "*Silurian*" formations underlying the Carboniferous in the Welsh Marches, and subsequently recognized elsewhere: the trilobite *Calymene* (upper left), the cephalopod mollusk *Orthoceratites* (upper right), the coral *Favosites* (lower left), and three different brachiopods (lower right). The corals in particular suggested deposition in warm or even tropical seas.

only gradually, as the widening scope of Murchison's research took him beyond his original area of fieldwork, that he shifted the primary meaning of "Silurian" from local towards potentially global stratigraphy. This brought his word into line with terms such as "Jurassic", which, notwithstanding their local origin, were being used increasingly to denote *all* the rocks, of whatever kind and wherever found, that occupied specific places in the stratigraphical pile (see §36.2).

In the present context, the importance of the Silurian lies less in the stratigraphy itself (which had a complex and contentious history in subsequent years) than in the character of the fossils that Murchison described from his formations, first from the Welsh Marches and then in other regions further afield. The constituent Silurian formations each had their own "characteristic" fossils, which enabled him to map their outcrops in the now conventional Smithian fashion. More importantly, however, he also claimed that the Silurian fauna *as a whole* was itself distinctive, and distinctly different from that of Phillips's overlying Carboniferous formations. The Silurians had abundant and varied fossil shells, both mollusks and brachiopods, and also diverse trilobites, corals, and many other forms of invertebrate life; these fossils were already well known to collectors and highly valued by them, but their stratigraphical position had previously been unclear (for example, most of Brongniart's trilobites reproduced here in Fig. 4.2 were identified retrospectively as Silurian in age). The corals, found sometimes in reef-like masses, suggested a tropical environment, confirming what was already suspected on other grounds, and fitting neatly into the overarching theory of a gradually cooling earth (Fig. 30.7).[15]

Much more surprising, however, was the total absence of another kind of evidence for that climatic reconstruction. The Silurian formations seemed to be totally lacking in plant fossils of any kind, let alone the huge tree-ferns and other exotic plants of tropical appearance that characterized the Coal formation further up the sequence, or later in geohistory (see Fig. 29.4). Murchison therefore claimed that the Silurian "system" represented a distinctive major period in geohistory, dating from *before* the first appearance of any large land plants. Even in the early years of his research this inference seemed quite plausible to most geologists, although, of course, it was based on negative evidence. As Murchison pursued his Silurian imperiously into other regions, by his own fieldwork or in the proxy form of collections sent to him from around Britain and eventually from around the world, it gained cumulative weight.[16]

Meanwhile, Sedgwick had been tackling the still older and more difficult rocks of the "Lower Grauwacke" or earlier Transition formations. Once he gained in the field a practical understanding of how to detect the true structure of the rocks when it was obscured by slaty cleavage (§27.4), he was able to unravel a

14. Élie de Beaumont to Sedgwick, 26 July 1835 [Cambridge-UL: Add. 7652.IF.36], written in the context of his own current fieldwork for the geological survey of France; see Rudwick, *Devonian controversy* (1985), 127–30.

15. Fig. 30.7 is reproduced from Murchison, *Silurian system* (1839), pl. 7, fig. 6; pl. 9, figs. 2, 5; pl. 15bis, figs. 3, 4; pl. 12, figs. 1, 2, 5.

16. The imperial style of Murchison's geological practice is analyzed in Secord, "King of Siluria" (1982), and, in a broader biographical context, in Stafford, *Scientist of the empire* (1989).

sequence of thick and highly folded Transition formations in north Wales, and another in northwest England. Rather confusingly, he called them respectively "Cambrian", after the Roman name for Wales, and "Cumbrian", after the name of the region also known—not least, to the growing crowds of tourists hoping to share Wordsworth's experience of the sublime—as the "Lake District". However, as Sedgwick, like Murchison, began to broaden the scope of his stratigraphy, he took to using "*Cambrian*" to refer to the rocks of both regions. The upper parts of these sequences contained a few poorly preserved trilobites and other fossils, but in the otherwise similar rocks in the lower parts he found no fossils at all. Yet this absence of fossils could hardly be attributed to their destruction by Lyell's putative metamorphic effects (§26.2), because the lower parts appeared to be no more affected by "subterranean cookery" than the upper. So Sedgwick inferred that these sequences might bear genuine witness to the very start of the fossil record, though not perhaps to the beginning of life itself.[17]

Murchison explained his Silurian, and Sedgwick his Cambrian, when they both attended the British Association's meeting in Dublin in the summer of 1835. Immediately afterwards, Buckland and Greenough took their ideas to the Continent. Buckland suggested at the field meeting of the Société Géologique how the new terms might be applied to similar rocks in the Belgian Ardennes, and then explained them to an even wider range of geologists at the German savants' meeting in Bonn (§29.2). Buckland, for one, was clearly treating "Silurian" and "Cambrian" as stratigraphical terms of potentially global validity, and not—despite their etymology—in a purely local or regional sense. And "Silurian", at least, was soon adopted by Continental geologists, for they recognized that the fossils they already knew well in some of their own formations were those that Murchison described and illustrated in his *Silurian System*.[18]

Since Sedgwick's Cambrian fossils, in contrast, were poorly preserved and not clearly distinct from those in the Silurian formations, Murchison proposed that the two sets of rocks should jointly be termed "*Protozoic*", that is, containing the traces of "original life". Sedgwick was more cautious, and preferred "*Palaeozoic*" or "ancient life"; he want to reserve "Protozoic" for the underlying Primaries (by definition the oldest rocks of all), which might in future turn out to have some traces of the very earliest organisms. But whatever the name, the Silurian and Cambrian formations did seem jointly to represent at least an approach to a true "vestige of a beginning" for life on earth, notwithstanding Lyell's rejection of any such possibility.[19]

30.4 CONCLUSION

The directional and even "progressive" character of the history of life—as it was perceived by almost all geologists other than Lyell—was powerfully reinforced during the 1830s. Buckland argued in his Bridgewater Treatise that the oldest known forms had already been complex in anatomy, well adapted to their appropriate modes of life, and certainly not the crude imperfect beings that Lamarckian transformism was taken to entail (§29.3). At much the same time, Agassiz, having inherited Cuvier's massive research project on fossil fish, classified them both zoologically and stratigraphically, and argued that their diversity had increased progressively in the course of geohistory, in just the way that the younger Brongniart had already outlined for fossil plants. Agassiz also argued that the Carboniferous period had been a true "age of fish", apparently antedating even the earliest quadrupeds (in the "age of reptiles"). This made the fossil record of the vertebrates as a whole more clearly progressive than ever.

Any attempt to understand the still earlier phases in the history of life, as recorded in the Transition formations underlying the Secondaries, depended on a reliable point of reference in the Carboniferous formations, the lowest and oldest group of Secondaries. Complementing Agassiz's work on their fossil fish, and that of the younger Brongniart and others on their plants, Phillips—Smith's nephew and a rising star of English geology—described in exemplary fashion the abundant invertebrate fossils of the Carboniferous. Murchison profited from this benchmark when he sought to extend Smithian or fossil-based stratigraphy downwards into the poorly known Transition rocks. The sequence of undisturbed formations (with abundant fossils) that he found in the Welsh Marches and named the "*Silurian* system" turned out to have wider validity, first in other parts of Britain and then across the rest of Europe and even further afield. Murchison took it to be the record of a Silurian *period* of geohistory, characterized by a distinctive fauna of marine invertebrates but with no trace of the large terrestrial plants that had flourished during the subsequent Carboniferous period.

At much the same time, Sedgwick tackled the more disturbed and difficult formations below the Silurians, both in Wales and in England's Lake District. What he defined as the "*Cambrian* system" of rocks contained few fossils, and those were poorly preserved and not clearly distinct from Murchison's Silurian ones. Nonetheless Sedgwick interpreted his formations as representing an even earlier Cambrian period of geohistory. However, he prudently declined Murchison's proposal that their two periods should jointly be called "*Protozoic*", since even the Cambrian fossils might turn out not to be the very first. He preferred the more cautious label of "*Palaeozoic*": the Cambrian and Silurian were certainly a realm of "ancient life" and—notwithstanding Lyell's skepticism—seemed to represent at least "a vestige of a beginning".

In the very same years, all this work was complemented by research that led—whether unexpectedly or by intention—to still fuller ways of imagining what the earth and its life had been like in the depths of geohistory. This is the subject of the next chapter.

17. Sedgwick, "Cumbrian mountains" (1835), 66 (read at the Geological Society on 5 January 1831 but later revised); Secord, *Victorian geology* (1986), 99–101.

18. Sedgwick and Murchison, "*Silurian* and *Cambrian* systems" (1836), read at the British Association on 14 August 1835; on the French and German meetings, see Rudwick, *Devonian controversy* (1985), 135–41, and "International arenas" (1986).

19. Murchison to De la Beche, 6 April 1838 [Cardiff-NMW: no. 1014], and Murchison, *Silurian system* (1839), 11; Sedgwick, "English series of stratified rocks" (1838), read at the Geological Society on 23 May: see Secord, *Victorian geology* (1986), 125–32, and Rudwick, *Devonian controversy* (1985), 242–46. The later dispute between Murchison and Sedgwick stemmed in part from Sedgwick's inability to define a Cambrian fauna distinct from Murchison's Silurian one (see §36.2).

CHAPTER 31

Imagining geohistory (1831–40)

31.1 THE "GREAT DEVONIAN CONTROVERSY"

Murchison's claims for his "Silurian system" of rocks carried enormous economic implications. For if the Silurian truly represented a period of geohistory before there were any of the large plants from which coal seams had apparently been formed (Fig. 4.4), it followed that Silurian strata and fossils, anywhere in the world, could confidently be used to mark the point beyond which it would be futile to bore expensively in search of coal, the essential energy source for the newly recognized "Industrial Revolution". When a major anomaly on this point emerged at the heart of Murchison's Silurian research, it was therefore a serious threat to his ambitions to become a public figure and a statesman of science. But it also threatened to undermine his concept of geohistory and the stratigraphical methods on which it was based. This many-sided threat came from De la Beche.

With the recent abolition of slavery throughout British territories, De la Beche had lost much of the inherited wealth that he had derived from his plantation in Jamaica. But he had persuaded the British government to pay him in an ad hoc way to make a geological survey of southwest England; it was a one-man band that was a very pale imitation of the well-funded governmental project in which Élie de Beaumont and other French geologists were currently surveying the whole of their country. Late in 1834, only months after Murchison had first reported the results of his Silurian research, De la Beche sent the Geological Society a collection of fossil plants of typical Coal species (see Fig. 12.3) from Devonshire. What made them immediately contentious was that he reported finding them in "Lower Grauwacke" rocks, which made them even older than Murchison's "Silurian" and probably equivalent to Sedgwick's "Cambrian" in Wales. De la Beche could no longer afford to travel back to London frequently, and he was

Fig. 31.1 "Preconceived Opinions *v[ersus]* Facts": De la Beche's caricature of the scene at the Geological Society when in 1834 his report on the fossil plants of Coal species that he had found in ancient-looking "Lower Grauwacke" in Devonshire was criticized (in reality, in his absence) by Murchison, Lyell, and others. "This, Gentlemen, is my *Nose*", announces De la Beche; to which his assembled critics reply, "My dear Fellow!—your account of yourself generally may be very well, but as we have classed you, *before we saw you*, among men *without noses*, you *cannot possibly have a nose*". This caricature, which was sent to Sedgwick (and another version to Greenough), employed the same visual rhetoric—of distorting spectacles and contrasted clothing—that De la Beche had used in those he had sketched earlier in criticism of Lyell's Huttonian geotheory (Figs. 22.5, 22.6). His claim that he had found Coal plants in strata far more ancient than any previously known was a serious threat, for different reasons, to the theorizing of both Murchison and Lyell. (By permission of the Syndics of Cambridge University Library)

not present when his specimens were discussed at the Society. So he was furious to hear from Greenough—who was never slow to stir up trouble—that Murchison and Lyell, neither of whom had yet seen the relevant part of Devonshire, had flatly denied that the fossils could have come from the rocks in which De la Beche claimed to have found them. To De la Beche this was an outrageous case of plain factual observation being willfully distorted or even denied because it was inconveniently incompatible with his critics' theories. He promptly sent both Greenough and Sedgwick a caricature ridiculing the situation, anticipating correctly that they would show it around at the Society and that it would help recruit support for his own position (Fig. 31.1).[1]

Murchison's skepticism is easy to understand. He was convinced that his failure to find large plant fossils in his Silurian formations reflected their genuine absence from the world at that "Protozoic" period of geohistory, not just a failure of preservation. He had already dismissed some other reports of coal seams or coal plants in ancient Grauwacke or Transition strata elsewhere, on the grounds that the relevant rocks were not as ancient as had been supposed, and in fact belonged to true Coal strata, unrecognized as such because they were untypical in appearance. So he assumed that De la Beche had made a similar mistake in his fieldwork in Devonshire. This was a serious charge to bring against a fellow geologist who was more experienced in the field than himself; and it was all the more alarming for De la Beche, in that his still precarious employment depended on the government being assured that he was fully competent.

Lyell's skepticism is at first glance more surprising, since he might have been expected to welcome a new find of fossil plants in rocks that on his reckoning would be almost inconceivably *older* than the Coal formation: in effect, the tree-ferns might then have returned on a later (Carboniferous) round of the earth's stately Huttonian cycle, just as he thought it conceivable that the iguanodons would return at some time in the distant future (§21.4). But for Lyell the sticking point was that Lindley, the botanist who was England's answer to the younger Brongniart, had identified De la Beche's specimens as *identical* to species well known from the Coal strata. Lyell was prepared to speculate that reptiles *similar* to iguanodons might return, to fill the same ecological niche in some future world, but not the very same species: to concede otherwise, here or in the Devonshire case, would imperil the kind of natural chronometer on which he had founded his reconstruction of the Tertiary era and, he hoped, the rest of geohistory (§26.1).

De la Beche's report from Devonshire therefore threatened to upset some heavily laden apple-carts. The "great Devonian controversy", as it came to be called, rumbled on for several years, frequently flaring up in heated arguments at the Geological Society, regularly at the British Association's annual meetings, occasionally also at the Société Géologique in Paris, but above all in intensive correspondence between all the major figures and many minor ones too. In the event, the anomaly was resolved in three distinct and contrasting phases. In 1836, Murchison at last went to Devonshire to study the relevant area in the field, taking Sedgwick as his companion and somewhat reluctant ally. Immediately afterward, in the full glare of publicity at the British Association's meeting in Bristol, and in the presence of politicians, Murchison claimed triumphantly that De la Beche had made a huge mistake, that he had radically misinterpreted the field evidence, and that there was a large and previously unrecognized area of Coal strata in the middle of Devonshire (no immediate practical consequences were at stake, because the coal seams were thin and sparse, and uneconomic except for local use). De la Beche later conceded, grudgingly, that he had indeed misread the structure of the rocks, and that the strata in which he had found the fossil plants were not among the lowest but some of the uppermost in the Devonshire sequence (in the event, his governmental support survived this admission). Murchison had therefore saved the phenomena, or at least some of them, and also his concept of the Silurian as a period that had antedated the first terrestrial flora.[2]

1. Fig. 31.1 is reproduced from De la Beche to Sedgwick, 11 December 1834 [Cambridge-UL: Add. 7652.IA.125]; see Rudwick, *Devonian controversy* (1985), 95–107.

2. De la Beche had misinterpreted a major syncline for an anticlinal structure, having mistaken the "way up" of strata that were highly ambiguous because they were tightly folded and often in a vertical position: see Rudwick, *Devonian controversy* (1985), 164 (fig. 7.6). Cases such as this induced later geologists to search for, and find, "way-up criteria" that can show unambiguously in which direction a highly folded sequence is "younging", and hence help unravel the large-scale structure.

However, this explanation of the previously anomalous fossil plants—as coming from true Coal strata of untypical appearance—still left the age of the rest of the Devonshire "Grauwacke" unresolved. Murchison, citing the evidence of its fossils, annexed most of it to his growing Silurian empire, leaving Sedgwick the residue (with few or no fossils) for his Cambrian. But this solution created a new anomaly. As De la Beche doggedly continued to point out, there was no trace in the field of the major unconformity that was to be expected, if there was such a huge gap in the middle of the Devonshire sequence (between the Silurian and the Coal). It was here that his colleague Phillips came to his rescue, by identifying typical Carboniferous Limestone fossils in the strata immediately underlying those with the Coal plants, preserved again in rather unfamiliar types of rock but in the right position in the sequence. This at least justified De la Beche's insistence that the main anomaly was still unresolved, because there was no unconformity at the point at which Murchison insisted that there must be one. So the two major protagonists were quite evenly matched.

Indeed, far from being resolved, the problems with Murchison's explanation grew, when his allegedly Silurian fossil shells, corals, and trilobites—made known largely through the collecting efforts of local amateurs—were studied more closely by Phillips. For they turned out not to be Silurian species after all. There were then two possibilities: that they were pre-Silurian or post-Silurian, Cambrian or Carboniferous. Lyell for a time adopted the first, which for him had the welcome implication that the Cambrian fauna had been far from primitive in character (see §32.2). But the second option was suggested by puzzling signs that the fossils in question were rather like those of the Carboniferous formations that Phillips was currently describing from Yorkshire (§30.2). Eventually, after further intense debate and tortuous argument, Murchison again saved his own phenomena (with Sedgwick as his nominal collaborator once more), by proposing a solution that was far more radical than the face-saving compromise it first appeared to be.

In 1839, not long after his big *Silurian System* was published, Murchison claimed that the older Devonshire rocks (below all the Carboniferous ones) represented a major period of geohistory, previously unrecognized, *intermediate* between the Silurian and the Carboniferous. He proposed that they be called "*Devonian*" (an adjective previously used, as it still is, in a purely geographical sense, to denote the inhabitants or other features of Devonshire). At the earliest opportunity Murchison extended his Devonian "campaign"—the belligerent metaphor was characteristic—to the Continent, where similar fossils were already known in rocks that were less disturbed and easier to unravel. With the help of many other geologists in the field and in museums, he identified Devonian formations in Belgium and the Rhineland, and later as far away as Russia; further collections, and fieldwork by others, eventually extended the concept to North America. With steadily growing evidence from these less ambiguous regions, Murchison's new solution for the anomaly ultimately gained almost universal support among geologists.[3]

The Devonian controversy (here summarized very briefly and all too inadequately) raised issues of great importance for the practice of fossil-based stra-

Fig. 31.2 De la Beche at the height of his career as director of the state-supported Geological Survey in Britain: a portrait drawn around the time that the "great Devonian controversy" finally subsided in a near-consensus, after about eight years of often vehement argument. Alone among leading British geologists, De la Beche had for many years worn spectacles, a characteristic that he exploited by metaphorically transferring them to his critics in the serious caricatures for which he was well known (Figs. 22.5, 22.6, 31.1). (By permission of the National Portrait Gallery, London)

tigraphy and its refinement from its relatively crude Smithian beginnings: at one point Lyell—no mean judge in such matters—told De la Beche that in his opinion it was the most important theoretical issue ever to have been debated at the Geological Society. However, it is significant that in the event it did *not* undermine the growing consensus about geohistory in the way that Murchison had feared. The Devonshire case remained for several years an extremely puzzling anomaly, first because the Coal fossils appeared to be seriously out of their expected place in the sequence of formations, and then because the other fossils were enigmatic and difficult to place at all. But most of the leading geologists tacitly anticipated that it would indeed prove to be an *anomaly*, which would eventually be resolved. While it was being studied intensively, and often vehemently debated, most geologists—with Lyell as the usual prominent exception—continued to act on the assumption that the ancient Cambrian and Silurian formations described by Sedgwick and Murchison did contain a record of the very beginnings of life itself, or at least of its first "Protozoic" or "Palaeozoic" manifestation; and that the puzzling formations in Devonshire would somehow fall into line in due course. Even De la Beche eventually accepted, however grudgingly, his critics' interpretation of Devonshire itself, if not Murchison's broader theorizing. And his governmental career not only survived but flourished; his Geological Survey was continued on an ever more permanent basis, eventually extending its work to the rest of Britain (Fig. 31.2).[4]

3. Rudwick, *Devonian controversy* (1985), traces the course of these long and complex arguments in the great detail made possible by an exceptionally well-preserved documentary record, including field notebooks and prolific correspondence. This controversy, and the Cambrian-Silurian controversy described in detail in Secord, *Victorian geology* (1986), which was loosely linked with it, were both concerned primarily with questions of stratigraphical method and classification, not with geohistory, and are therefore only tangentially relevant to the present narrative.

4. Fig. 31.2 is reproduced from a lithograph after a pencil sketch, dated 22 April 1842, by William Brockedon [London-NPG: 2515/94].

The recognition of a "Devonian" system and period, where none such had been suspected or anticipated, had two important theoretical consequences. First, it supported a conception of relatively gradual change in the long-term history of life: the newly recognized "Devonian" fauna of invertebrates had some elements in common with both the Silurian fauna and the Carboniferous. Phillips, who was recruited by De la Beche to describe the Devonshire fossils in detail for the Survey, concluded that the faunas had not changed suddenly, but by piecemeal replacement of species and genera. His "statistical" view of faunal change—which clearly distanced him from his uncle and early mentor Smith—was of course compatible with Lyell's views, as the latter had applied them to the test case of the Tertiary faunas (§26.1); but it clashed with Murchison's proprietorial claim that his Silurian was sharply distinct from anything else.[5]

Second, the "Devonian" solution to the anomaly reinforced an ecological interpretation of geohistory, in a way that geologists did not at first fully appreciate. For the recognition of a Devonian "system", sandwiched between the Silurian and the Carboniferous, required that the contentious Devonshire rocks be correlated with the very thick Old Red Sandstone formation, which occupied the same position in the stratigraphical sequence elsewhere in Britain and in mainland Europe. Yet there was an almost total contrast in rock types between the slates and thin limestones in Devonshire (and the Ardennes and the Rhineland), and the distinctive red sandstones and conglomerates of the Old Red in, say, the Welsh Marches (the nearest relevant region) and large tracts of Scotland. Even more worryingly, there was also a total contrast between the fossil shells, corals, and trilobites that Phillips was describing from the one, and the strange fossil fish that Agassiz was busy working on from the other. The difference was so complete that Murchison's solution gained credibility in the eyes of other geologists only when, in 1840, he finally found *both* sets of fossils within the *same* formation (though in separate strata). This discovery in the Baltic provinces of Russia (now the Baltic states), was confirmed authoritatively by Agassiz at the British Association meeting in Glasgow on his return; Murchison triumphantly welcomed it as the "Q.E.D." or final and indisputable proof of his "Devonian" stratigraphical interpretation.[6]

31.2 GRESSLY'S CONCEPT OF "FACIES"

What was equally important about the "Devonian" solution to the controversy was that Murchison, like other geologists, attributed the contrast in both rock types and fossils to an original difference in physical environment, though they remained uncertain what the difference had been. It was the latest example, and by far the largest in scale, of a principle that had slowly been gaining ground in stratigraphical practice: the recognition that formations occupying the same place in a continuous sequence might be of quite sharply contrasting appearance in different areas, because they had been deposited simultaneously under different conditions.[7]

Two of the earliest cases had involved the classic description of the Tertiaries of the Paris Basin by Brongniart and Cuvier. They themselves had treated two specific formations as lateral equivalents—although they were totally distinct in appearance and had no fossils in common—because they were sandwiched between the same formations above and below (*BLT* §9.1). And Webster had then correlated the Coarse Limestone of the Paris Basin with the London Clay in the London and Isle of Wight (Hampshire) Basins, on the grounds that the fossils of these formations were closely similar, although the sediments were quite different (*BLT* §9.4). But in neither of these early cases was the correlation given a clearly environmental explanation: the first was just passed over as an anomaly (and later rescinded altogether: *BLT* §9.6), and the second was simply treated as one in which fossils took proper priority over rock types as criteria of relative age.

Brongniart and Cuvier had treated the Tertiary formations around Paris as the products of an alternation between marine and freshwater conditions, with putative "révolutions" causing sudden changes from one state to the other (*BLT* §9.1). Later, and in contrast, Prévost had envisaged both kinds of sediment accumulating *simultaneously* in different parts of the Paris Basin, according to local environmental conditions; and he explained this by referring to analogous situations in the present world and in the conceivable future (§10.3). In 1835, as a candidate for a coveted place in the Académie, he reprinted his articles on these topics, which gave his ideas further publicity among geologists and other savants around Europe. Although he was unsuccessful in the competition, Prévost's example was important, because his work showed that detailed stratigraphy could be turned into equally detailed reconstructions of ancient geographies and ancient environments. It represented an extension of a geohistorical approach in which actual causes were explicitly taken as the key to the deep past; it is no wonder that Lyell treated Prévost as his closest ally in France.[8]

This far more geohistorical approach to stratigraphy was generally "in the air" in the 1830s; or rather, it was gradually emerging from the changing practice of the science. Certainly it was developed still further, and in effect codified, by the young Swiss geologist Amanz Gressly (1814–65). In 1837, having undertaken intensive and detailed fieldwork in his native Jura hills, Gressly presented the scientific society in Neuchâtel with a dramatically novel account of the "Jurassique" formations, on the strength of which Agassiz later hired him as the first

5. Phillips, *Palaeozoic fossils* (1841), the Survey's first paleontological publication, contained his illustrations of key specimens, his "statistical" analyses, and his broader inferences: see §36.2; also Morrell, *Phillips* (2005), chap. 6.

6. Rudwick, *Devonian controversy* (1985), 358–66; Agassiz's identification of Murchison's specimens was later confirmed definitively in his monograph on the fish of the Old Red, *Poissons du Vieux Grès Rouge* (1844–45).

7. The nature of the difference remained unclear in the Devonian case. It was only much later that detailed studies of Devonian sediments led to the modern inference that the Old Red was a formation of "*continental*" (i.e., terrestrial or freshwater) origin, in contrast to the evidently *marine* formations of the Devonshire rocks and most of those in the rest of Europe and North America.

8. Prévost, *Candidature de Prévost* (1835), included, in addition to reprinted articles (and one not previously published), an autobiographical "Notice préliminaire" (5–12) commenting on the difficulty of opposing the dominant (i.e., Cuvierian) viewpoint, and insisting that only the "géologue-observateur" or *field* geologist could judge the issues involved. His *Histoire des terrains tertiaires* [1835] is another collection of the same reprints, probably issued at the same time.

of his many assistants. Gressly described the rock types and structural relations [*géognosie*] of the formations, their fossils, and their correlations [*synonymie*] with other regions. This was now standard and conventional in any serious work on regional stratigraphy. And the practical internationalism that was equally taken for granted in the science was exemplified by Gressly's use of "*Portlandien*" for a part of the "Jurassic" sequence in the Jura: this well-established name was derived from the Isle of Portland (in fact, a peninsula) on the south coast of England, where the famously fine Portland stone was quarried and shipped all over Europe for grand building projects, and its almost equally famous giant ammonites and other fossils were shipped to museums all over the world.[9]

However, Gressly's work was much more than conventional stratigraphy. He also discussed what he called the different "*facies*" (a Latin word meaning faces, countenances, or expressions) of each formation, and he argued that this new concept explained why Jurassic stratigraphy and paleontology needed "a complete reform". For the rocks comprising one and the same formation, and the fossils found in them, could show quite different facies in different areas. Gressly interpreted facies as the products of contrasting physical and organic environments: in the Portlandian part of the Jurassic period, for example, coral reefs had been growing in some areas, at the same time that fine muddy sediments were being deposited in others. In favorable cases—Gressly regarded his own as one—

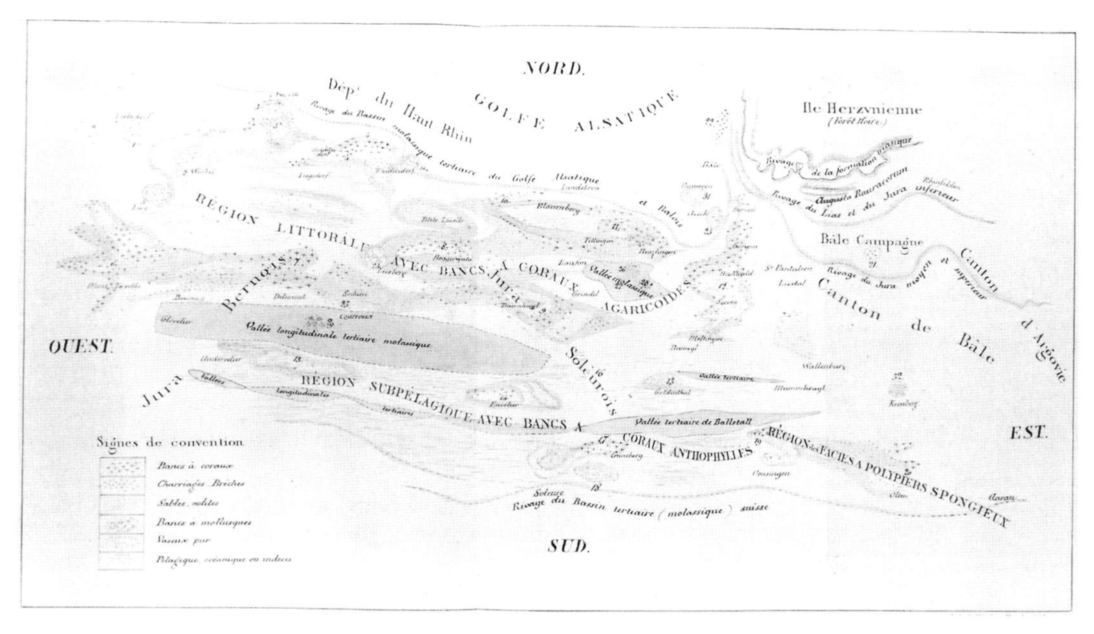

Fig. 31.3 Amanz Gressly's lithographed map of the eastern Jura (1840), plotting the original distribution (after "ironing out" the subsequent folding of the rocks) of the varied "*facies*" of the late Jurassic [*Portlandien*] formations: coral reefs [*Bancs à coraux*], beds with mollusks, sands, shales, etc. (see key). It also marks two parallel zones of coral reefs with corals of different kinds, interpreted as shallow-water [*littorale*] and deeper-water [*subpélagique*] environments. In the northeast is the massif of older rocks in the Black Forest [*Forêt Noire*], interpreted as part of an island [*Ile Herzynienne*] and fringed to the south by three successive positions of the retreating Jurassic shoreline [*rivage*]. The map represents an area about 80km across; in modern terms, it reconstructed both the paleogeography and the paleoecology of this region, turning stratigraphy into detailed geohistory, and it became a model and precedent for many later studies of the same kind. (By permission of the Syndics of Cambridge University Library)

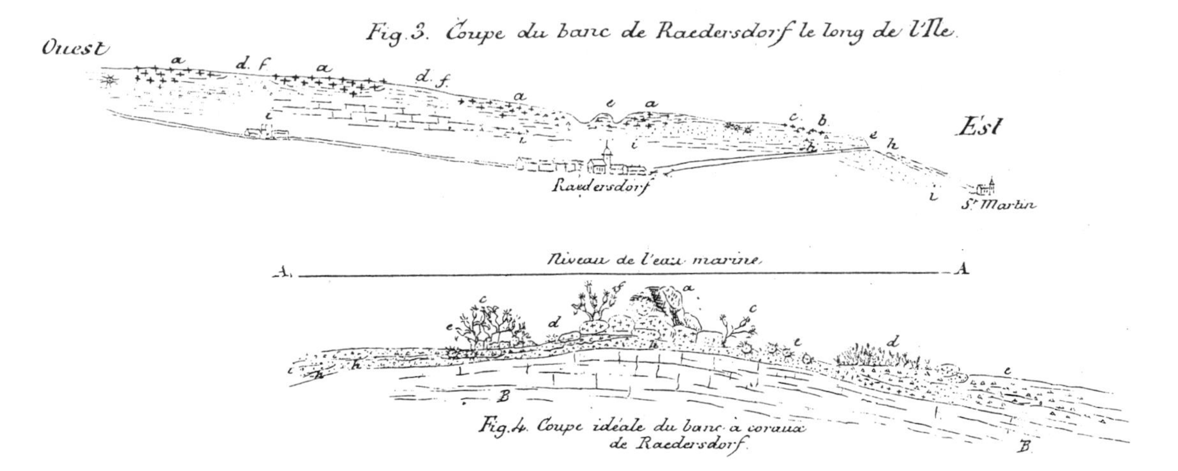

Fig. 31.4 Gressly's reconstruction of a late Jurassic (Portlandian) coral reef in the Jura hills (published in 1840). *Fig. 3* (above) is his sketch of a long ridge behind the village of Raedersdorf, showing the outcrops of rocks of different "*facies*"; the patches of massive coral are marked by the pattern of small crosses (*a*). *Fig. 4* (below) is an "ideal section" through his reconstruction of the same reef, based on his detailed fieldwork, and showing many distinct zones characterized by different organisms. The highest (and shallowest) part was built of massive corals (*a*); in deeper water were delicate crinoids (*c*, *d*); there were also areas with sea-urchins and oysters (*e*), and sands with other molluscs (*h*). By analogy with similar organisms in modern environments, Gressly reckoned that the Jurassic sea level (*A-A*) at this point was "only a few fathoms or even no more than a few feet" above the highest part of the reef. Semi-diagrammatic reconstructions of this kind—this was one of the first, if not the very first—were much more useful to geologists than the striking pictorial scenes that they used increasingly in their popular publications (see below). (By permission of the Syndics of Cambridge University Library)

it was therefore possible to plot not only the ancient geography of a region, but also the distribution of distinctive habitats and different kinds of sea floor, each with its own characteristic marine organisms (Fig. 31.3).[10]

Gressly's mapping in the Jura hills, plotting diverse ecological zones in the deep past of Jurassic time, was based, of course, on his fieldwork. In favorable situations he felt able to reconstruct the environments in even greater local detail. Using coral reefs in the present world as his actualistic point of reference, he reconstructed some of those in the Jurassic world, suggesting how the diverse communities of organisms had been grouped in comparable micro-environments, although the species themselves were different from living ones (Fig. 31.4).[11]

9. Gressly, "Jura Soleurois" (1838–41), read (in outline) to Société des Sciences Naturelles de Neuchâtel on 15 April 1837 [*MSSNN* 2: 14]. See Wegmann, "Notion de facies" (1963); and Cross and Homewood, "Gressly's role" (1997).

10. Fig. 31.3 is reproduced from Gressly, "Jura Soleurois" (1838–41), pl. 6 (1840) [Cambridge-UL: Q382.b.10.3], explained on 235–36; the concept of facies is explained on 8–26 (1838). An earlier stage in Gressly's reconstruction is represented by his section (pl. 7) through the Jurassic formations, as they would have been *before* their strong folding in the Jura; in modern terms, the section was palinspastic (it was one of the first of its kind, if not the very first).

11. Fig. 31.4 is reproduced from Gressly, "Jura Soleurois" (1838–41) [Cambridge-UL: Q382.b.10.3], pl. 10, figs. 3, 4, explained on 239; figs. 1, 2 (not reproduced here) are a north-south section through the ridge and a map of the areal distribution of the facies. Raedersdorf is marked on the smaller-scale map (Fig. 31.3), towards the north edge and below "Haut Rhin"; it is just on the French side of the frontier, about 25km south of Mulhouse (Haut-Rhin). In fact, Gressly in landlocked Switzerland had no *first-hand* experience of contrasting marine environments in the present world until he visited Sète on the Mediterranean coast many years later: Wegmann, "Notion de facies" (1963), 93.

Gressly's concept of "facies"—which soon entered both the vocabulary and the practice of geologists everywhere—gave formal and explicit shape to the profound way in which stratigraphy was being transformed by an increasingly geohistorical perspective. This was reshaping geology in conformity with Lyell's actualistic prescription, although often independently from Lyell and in ignorance of his work. The traces of the deep past were being interpreted increasingly in causal terms, and those causes were referring to processes observably active in the present world. In the hands of a small but growing band of geologists, the stratigraphical descriptions that at this time formed the bread-and-butter of geological practice were being turned into geohistorical reconstruction and geohistorical narrative.

31.3 MORE SCENES FROM DEEP TIME

However, the most striking sign that geologists in the 1830s were indeed developing a deeply geohistorical outlook came from a direction rather different from the primarily stratigraphical work that has just been described. Its point of origin was the kind of reconstruction that Conybeare had pioneered with his famous scene of Buckland crawling into an antediluvial hyena den (see figure facing page 1, and *BLT* §10.6), and which De la Beche had developed in his equally famous scene of the bizarre reptiles living in "a more ancient Dorset" than Mary Anning's ("*Duria antiquior*", Fig. 11.5). The latter print, although distributed widely for Anning's benefit, had apparently been thought too speculative, or too frivolous, to be published formally. Others modeled on it began to appear soon afterwards, but they too were published on the margins, as it were, of sober geological discourse. The first such borrowing, and the first to be aimed at geologists and serious fossil collectors, was the scene that Goldfuss inserted into his large and otherwise decidedly sober work on German fossils (Fig. 11.6). But it was marginal in the sense that what it showed was not related to the fossil specimens described and illustrated in the rest of his volume, and it was apparently just a marketing gimmick. Goldfuss's print prompted Buckland to urge De la Beche to produce some more scenes, depicting other moments in geohistory (§11.3). Sadly, however, De la Beche never acted on these imaginative suggestions. He did insert three little vignettes into his *Geological Manual*, but they were hardly integrated into this dry stratigraphical text.[12]

Two further adaptations of De la Beche's Dorset scene, one of them by Phillips, soon appeared in publications aimed at the general reading public, but for just that reason they were less significant for the debates among savants. A third was like Goldfuss's print in being intended for geologists and serious fossil collectors, though it too was in all senses derivative. It formed the frontispiece of a lavishly illustrated description of the superb ichthyosaur and plesiosaur specimens which one of Anning's best customers, Thomas Hawkins (1810–89), had amassed in his private collection. Within the next few years several more scenes were published, likewise centered on the Jurassic reptiles. But they were becoming a pictorial cliché, and in the present context are important only as evidence that geologists were now well aware that fossils could indeed be used to help the public imagine an otherwise unimaginably deep past.[13]

Much more significant was a scene from a quite different part of geohistory; unknowingly, it followed to some extent one of the suggestions that Buckland had made privately to De la Beche. Johann Jakob Kaup (1803–73), a curator at the grand-ducal museum in Darmstadt, had for several years been studying the Tertiary deposits in that part of the Rhine valley. Among many other mammalian bones, he had found a huge lower jaw with downward-curving tusks; he had exhibited it at a meeting of the German Association in Berlin in 1828 and had interpreted it as belonging to an elephant-like animal, the largest land mammal yet known. Buckland, in his Bridgewater Treatise, chose this *Dinotherium* ["terrible beast"] as his very first example of designful adaptation (§29.3). He inferred that the strange tusks were well adapted for "raking and grubbing up by the roots large aquatic vegetables" from the bottom of lakes or rivers, and suggested imaginatively that this action would "combine the mechanical powers of the pick-axe with those of the horse-harrow of modern husbandry". In the same year, a complete skull and other bones of the dinotherium were found by August Wilhelm von Klipstein (1801–94), the professor of mineralogy at Giessen (where he had the famous chemist Justus von Liebig as a colleague). Klipstein and Kaup judged the new finds important enough to merit a separate monograph under their joint authorship.[14]

The report by Klipstein and Kaup, in which they approved of Buckland's Cuvierian functional inferences, was in the customary form of superb lithographed illustrations of the bones, accompanied by a brief explanatory text; as usual, this made the specimens available in proxy form to savants everywhere. However, literally outside this conventional format was a pair of lithographed scenes that encapsulated what was involved in this deeply geohistorical style of research. On the front cover was a lively scene from Tertiary time, showing the dinotherium reconstructed in standard Cuvierian fashion but set—as in De la Beche's scene of ancient Dorset (Fig. 11.5)—in a landscape with details indicated by the other fossils and the associated rocks. Matching it on the back cover was an equally lively scene from the present world, showing the recent recovery of the spectacular

12. De la Beche, *Geological manual*, 2nd ed. (1832), wood engravings on 231, 383, 385, based on Cuvier's Parisian mammals (Fig. 1.2) and the ichthyosaur and plesiosaur from the Lias (Fig. 2.6); also published in the French edition, though not in the German. Buckland borrowed Goldfuss's cliff-hanging pterodactyl for the sole (and very small) reconstructed scene in his Bridgewater Treatise, *Geology and mineralogy* (1836) 2, pl. 22P. Goldfuss, in a later part (1844) of his *Petrefacta Germaniae* (1826–44), published a scene showing a Coal forest and the Carboniferous marine fauna in great detail. All these images are reproduced in Rudwick, *Scenes from deep time* (1992), 55, 57, 69, 89.

13. The frontispiece of Hawkins, *Ichthyosauri and Plesiosauri* (1834), and the other scenes alluded to here, are all reproduced and analyzed in Rudwick, *Scenes from deep time* (1992), 59–85. O'Connor, *Earth on show* (2007), a fine study of "popular" science that in many ways complements the present account of "elite" research, puts these and many other "scenes" into a much broader social and aesthetic context, at least for the case of Britain.

14. Klipstein and Kaup, *Schädel des* Dinotherii gigantei (1836); a French edition (1837) ensured that it was noticed internationally. Buckland, *Geology and mineralogy* (1836), 1:135–39.

Fig. 31.5 "*Dinotherii gigantei*" in their reconstructed Tertiary environment: the front cover of the volume of plates accompanying the description of the huge fossil mammal by Klipstein and Kaup (1836). The other animals, and the plants, were all based on fossils found in the same deposit, and the active volcano in the background was indicated by volcanic deposits in the same sequence. The word "Atlas" is composed jokingly of letters constructed from bones, picks, and shovels. This scene exemplified the deeply geohistorical style that now characterized much research on strata and fossils, though such images were still tacitly treated as too speculative to be published in the main body of a scientific publication. (By permission of the Syndics of Cambridge University Library)

new specimens. Together, these two scenes neatly symbolized the task of trying "to burst the limits of time" and reconstruct geohistory. The "before" of the reality of the deep past was matched by the "after" of its fragmentary survival into the present; or, alternatively, the "before" of careful excavation was matched by the "after" of an imaginative but disciplined reconstruction (Figs. 31.5, 31.6).[15]

One further example of the burgeoning genre of reconstructed scenes from deep time must be mentioned here, because it indicates the more popular direction that the genre took, once its place in geological practice was established. In 1838 Mantell moved his collection and his family to the fashionable seaside resort of Brighton. There he supplemented his income from a less than flourishing medical practice by offering scientific lectures; he also exhibited his fossils in

Fig. 31.6 The recovery of "the colossal skull of the *Dinotherium giganteum*" from Tertiary deposits at Eppelsheim in Hesse: the scene on the back cover of the report by Klipstein and Kaup (1836). The specimen being raised is the skull (upside down); a huge femur (right) has already been extracted. The operation is being watched by one of the gentlemanly authors (below the skull), while laborers do the hard work, though one of them (left) is taking time off to celebrate the find. The clouds in the sky show the extinct mammal as a proboscidean Cheshire Cat observing its own resurrection; the visual joke reflected the still marginal status of such reconstructions. (By permission of the Syndics of Cambridge University Library)

15. Figs. 31.5 and 31.6 are reproduced from the covers of Klipstein and Kaup, *Schädel des Dinotherii gigantei* (1836) [Cambridge-UL: S365.bb.83.1], drawn and lithographed by Rudolf Hofmann and Ludwig Becker: see Langer, "Bilder aus der Vorzeit" (1990), and Rudwick, *Scenes from deep time* (1992), 68–72. Eppelsheim is about 40km southwest of Darmstadt. The new specimens were so important for Buckland's argument that he promptly reproduced the skull, and the Germans' reconstruction of the whole animal (though not, rather surprisingly, its scenic setting), in the *only* new plate added in the new edition of his Bridgewater Treatise, *Geology and mineralogy*, 2nd ed. (1837) 2: 603 and pl. 2<prime>.

16. Fig. 31.7 is reproduced from Mantell, *Wonders of geology* (1838), frontispiece; this is, unusually, a *steel* engraving, which allowed the long print run required for a popular book to be economic. Martin's own designs, including his famous *Deluge* (1828) and his work for Mantell, are described and further illustrated in Rudwick, *Scenes from deep time* (1992), 21–25, 74–83; see also Feaver, *Art of John Martin* (1975). O'Connor, *Earth on show* (2007), chap. 7, and Freeman, *Victorians and the prehistoric* (2004), trace the cultural history of the genre, though only for Britain.

Fig. 31.7 John Martin's imaginative rendering of "the country of the iguanodon", which Mantell used for the frontispiece of his *Wonders of Geology* (1838). The prehistoric monsters are depicted with scant regard for anatomical accuracy, but their huge size is suggested by the pterodactyl, cycad, and small tortoises (right foreground), and their tropical habitat by the palm trees. The estuary in the background indicates the environment in which (according to Mantell and Lyell) their fossil bones had been preserved in the "Wealden" formation; the ammonites on the ground are already fossilized, and refer to the previously marine conditions in the area. The unearthly strangeness of this dark and fearsome scene set the emotional tone for many popular pictorial representations of the deep past in the rest of the nineteenth century (and beyond it).

a private museum that generated further funds from admission fees. One of the early visitors to his museum was the painter John Martin (1789–1854), who was famous for vast *historical* canvases depicting in melodramatic style such events as the fall of Babylon and the destruction of Pompeii. After he had exhibited any of these huge oil paintings to a fee-paying public, Martin would convert it into a mezzotint of a more suitable size for domestic decoration, and the sale of these prints earned him more than the original paintings. When he visited Mantell's museum, its imaginative proprietor seized the chance to commission the artist to design a *geohistorical* scene depicting "the country of the iguanodon". Turned into a small engraving, this acted as an alluring frontispiece for his book on the *Wonders of geology* (1838), which was designed, as the title suggests, to catch the rapidly growing middle-class market for popular science (Fig. 31.7).[16]

The pastoral tone of earlier scenes—even when De la Beche showed his animals busy eating each other (Fig. 11.5)—was here replaced by the nightmarish Gothick melodrama of what was known as the "Martinesque" (or *martinien*) style. Martin's rendering of the iguanodon owed less to the sober Cuvierian practice of careful functional anatomy than to the long pictorial tradition of depicting the dragon confronted by St. George. Mantell himself still had very few bones or teeth on which to base his reconstruction of the animal, which he had simply scaled up to a monstrous size from the living iguana (§5.4). Even Mantell's liberties with the hard evidence paled beside Martin's. But it was Mantell the geologist who chose to use Martin's scene to give potential purchasers a taste of what his book had to offer. For the first time in the history of this genre, but certainly not the last, scientific precision was sacrificed to serve commercial interests of popularization. Martin's scene may have boosted Mantell's sales, his income, and his popular fame, but it did almost nothing for his science. And that charitable "almost" is justified only in that the scene did perhaps help to establish, in the mind of the reading public that bought Mantell's and similar books, the idea that geology was now a reliably *geohistorical* science: the deep past of "the country of the iguanodon" had been as real as biblical Babylon or Roman Pompeii.[17]

31.4 CONCLUSION

The work of Murchison and Sedgwick on the Transition or "Grauwacke" formations, defining the Silurian and Cambrian systems respectively (§30.3), seemed at first to represent an unproblematic extension of Smithian stratigraphy into the very oldest rocks with any clear record of life. In particular, Murchison claimed that his Silurian formations dated from a period of geohistory preceding the first appearance of the large land plants that had formed the coal seams of the economically vital Coal formation. De la Beche's report, that during his official survey of Devonshire he had found Coal plant fossils in "Lower Grauwacke" rocks, was therefore an unwelcome and obtrusive fly in Murchison's Silurian ointment, and its validity was attacked vehemently by the latter and, for different reasons, by Lyell. The "great Devonian controversy" that raged almost continuously for the next eight years forced geologists in Britain, and increasingly those elsewhere, to examine as never before the foundations of their use of fossils in stratigraphy. The huge anomaly of the Devonshire fossils was eventually resolved to the satisfaction of all those well qualified to judge the issue, but only by acknowledging a totally unanticipated and previously unrecognized "Devonian" fossil fauna, and a corresponding Devonian period of geohistory, *intermediate*—in both character and age—between the Silurian and the Carboniferous.

16. Fig. 31.7 is reproduced from Mantell, *Wonders of geology* (1838), frontispiece; this is, unusually, a *steel* engraving, which allowed the long print run required for a popular book to be economic. Martin's own designs, including his famous *Deluge* (1828) and his work for Mantell, are described and further illustrated in Rudwick, *Scenes from deep time* (1992), 21–25, 74–83; see also Feaver, *Art of John Martin* (1975). O'Connor, *Earth on show* (2007), chap. 7, and Freeman, *Victorians and the prehistoric* (2004), trace the cultural history of the genre, though only for Britain.

17. In 1834 a newly found iguanodon specimen, with some of the limb bones, had somewhat improved Mantell's evidence: Norman, "Mantell's 'Mantel-piece'" (1993). On the transformations of the iguanodon in the service of arguments about evolution, see Desmond, "Designing the dinosaur" (1979), and Torrens, "Politics and paleontology" (1997). Knowledge of the osteology of the iguanodon remained deeply imperfect until almost complete skeletons were found in Belgium much later in the century. Martin was commissioned by Hawkins to make a similarly *martinien* frontispiece for the latter's *Great sea-dragons* (1840), showing the same reptiles as in De la Beche's original scene but utterly different in style and feeling: see Rudwick, *Scenes from deep time* (1992), 81–83, which also traces the later history of the genre as far as the 1860s. By that time all the familiar visual tropes of modern scenes of the same kind—including the latest computer-generated animations—were well established.

The Devonian controversy had two important consequences for the developing picture of geohistory as a whole. First, in the hands of Phillips its successful resolution supported the Lyellian claim that faunas had changed in a piecemeal fashion in the course of time, by the appearance of new forms of life and the disappearance of old ones, generally without any sharp break in continuity. And second, it presented geologists with a new case—of unprecedented magnitude—of a phenomenon that Prévost and other geologists had already described on a smaller scale. This was the apparently contemporaneous origin of strongly contrasted kinds of deposits and fossils in different geographical areas: the Devonian limestones and shales, and their marine fossils, occupied the same place in the sequence in some regions as the Old Red Sandstone and its fossil fish in others. That these contrasted formations and fossils were indeed of the same age was then treated as a large-scale example of what the Swiss geologist Gressly defined as the phenomenon of "*facies*". In the case of the Jurassic formations in the Jura hills on the Franco-Swiss border, Gressly interpreted the contrasting facies, each with its own distinctive rock types and fossils, as the result of simultaneous deposition under contrasting physical and environmental conditions, for example as coral reefs growing close to sea level, muds accumulating in deeper water, and so on.

This signaled a new level of *geohistorical* insight in the interpretation of stratigraphy. In effect, Gressly plotted in space and time what naturalists might have seen, had they been able to explore the Jurassic seas in a suitable underwater time-machine. His work was therefore not unrelated to further attempts to reconstruct specific environments from the deep past in the form of pictorial scenes. Such scenes were certainly helpful in aiding the geohistorical imagination of geologists, but they continued to be a genre on the margins of scientific practice, because they were widely regarded as too speculative (or even too frivolous) to be incorporated fully into respectable scientific work. At the same time, however, the genre took off, and began a spectacular separate trajectory, as an effective vehicle for conveying the reality of deep geohistory to a wider public, in works of popular science. But this public favor and popularity came at a high price, namely that of dumbing-down: Martin's darkly Gothick interpretation of Mantell's extinct monsters was an early example of how scientific accuracy came to be sacrificed to the demands of commercially driven showbiz and popular hype.

The preceding chapters have described how the 1830s saw a vast expansion and deepening of the geohistory that could be reconstructed from the stratigraphical record of sedimentary rocks and their fossils. Most of this work seemed to support and consolidate the dominant view of geohistory as directional in character, and even in some sense "progressive". The next chapter describes how Lyell—the lonely proponent of an ambitious geotheory centered instead on a cyclic or steady-state geohistory—responded to all this cumulative evidence for directionality.

CHAPTER 32

Lyell's geotheory dismembered (1834–40)

32.1 THE TRANSFORMATION OF THE *PRINCIPLES*

There was a sharp contrast between the perceptions of Lyell's work by other savants and by the educated public, the two distinct audiences that he wanted his work to reach (§25.3). The public were impressed above all by his eloquent verbal portrait of an earth operating under purely natural processes through inconceivably vast spans of time. This was a concept that was still relatively novel outside savant circles in Britain (though not on the Continent). It was also in striking contrast to what was currently being purveyed by the popular "Mosaic" writers (again, almost exclusively in Britain). This "scriptural" standpoint had long been rejected by geologists and other savants—whether or not they were religious—and such matters were of importance to them only for managing the public relations of their science. For geologists, the chief value of the *Principles* was that it contributed to "geology proper" (as both Scrope and Lyell himself called it) by clarifying the most recent part of geohistory and by showing that much of it—according to Lyell, all of it—was explicable in terms of "actual causes" or processes observable in the present world.

Among geologists, the selective acceptance of some of Lyell's ideas, and the rejection of others, was reflected in the Royal Society's decision, late in 1834, to acknowledge the importance of Lyell's work by awarding him one of its first two Royal Medals (the other was for physics): not as high an honor as the Copley Medal that Buckland had won for his cave research over a decade earlier (*BLT* §10.6), but a signal honor all the same. Greenough, the Geological Society's current president, also presided over the committee at the Royal that made the recommendation; he was still very much a scientific *eminence grise* (except that he was bald). Among those whose opinions he sounded out was Conybeare, whose judgment—as always, both perceptive and generous—was that Lyell's work, being the best in Britain since Playfair's, was the obvious candidate; but Sedgwick argued that Murchison deserved the award for his work on the Transition rocks (§30.3). Greenough later told De la Beche that "the award of the Royal Medal occasioned us a great deal of difficulty".[1]

The committee recommended that the medal be awarded to Lyell, and specifically for his *Principles*. Four points about the work were cited: (1) its wide scope and "philosophical spirit and dignity"; (2) the value of his having "directed the attention of geologists to effects produced by existing causes"; (3) his "admirable description of many tertiary deposits", partly based on his own fieldwork; and (4) his new method of "discovering the relative ages" of Tertiary formations by "determining the relative proportion of recent [i.e., extant] and extinct species". In contrast to these positive grounds for the award, "the Committee declin[ed] to express any opinion on the controverted positions contained in that book". The latter were not specified, but judging by the reactions of the leading geologists on other occasions at least three would have been prominent: Lyell's assertion that actual causes were *wholly* adequate for causal explanation; his claim that there had *never* been any event more intense or "catastrophic" than those recorded in human history; and above all his insistence that there was *no* overall directionality in geohistory.[2]

As his earliest critics had pointed out, Lyell's concept of "uniformity" bundled together several distinct meanings of the word, which he had creatively confused for his own purposes. Leaving aside the constancy of the basic physical "laws of nature"—which no savant disputed (§27.3)—the least controversial meaning was related to Lyell's advocacy of actual causes. As suggested earlier, Lyell was here pushing at an open door, in the sense that their explanatory value had long been accepted in principle by all geologists. The only questions at issue were whether actual causes were *wholly* adequate, and whether they had indeed been "uniform" in their action, in intensity or at least in kind. Lyell himself had tacitly conceded this, when in his third edition he altered the subtitle of the *Principles* from being an "attempt to explain" the past—implicitly, *everything* in the past—in terms of actual causes, into a more modest enquiry *how far* they were adequate for explanation (§27.3).

Lyell therefore had to show by further examples that actual causes were more fully adequate than his knowledgeable critics acknowledged. When Murray published fourth and fifth editions of the *Principles*, in response to the rapid sale of the third, Lyell incorporated new material drawn from his own work and the most recent publications of other geologists and travelers, citing new cases and enlarging on examples he had already used (Fig. 22.3). At the same time, these new editions satisfied the growing interest of the reading public, and earned Lyell a welcome flow of income. In contrast, in the absence of international copyright the author received no royalties for an American edition (copied from Lyell's third), which satisfied the appetite of a different public; American interest in Lyell's approach to geology had been boosted by the lectures that Benjamin Silliman of Yale had given (as he told Lyell) both in the "fine intellectual city" of Boston and in the "money-loving, money-seeking & money-getting" city of New York.[3]

Fig. 32.1 A portrait of Charles Lyell, drawn in 1836 around the time that he was revising his *Principles* for successive editions and planning his introductory *Elements*. (By permission of the National Portrait Gallery, London)

At much the same time, with similar British readers in mind, Lyell revived his original plan to write an elementary account of his science (§18.2), which could now serve as an introduction to his larger work. After several changes of plan, the smaller one was published with the title *Elements of Geology* (1838). It only covered "geology proper", in that it dealt with the origins and classification of the different kinds of rock, and the sequence of stratigraphical formations and their geohistorical interpretation (a French edition soon made the new work widely accessible on the Continent, in advance of the delayed translation of the earlier and larger work). In effect, the *Elements* covered much the same ground as Lyell's original third volume (Chap. 26); but he condensed his detailed description of the Tertiaries and expanded his analysis of older rocks, making it a more balanced survey of stratigraphy and geohistory. In view of this overlap between his two main publications, Lyell dropped all the "geology proper" from his next (sixth) edition of the *Principles* (1840), which thereby became a three-volume treatise on actual causes, with no directly geohistorical component at all. To complement this at the same relatively high level, he planned a separate work

1. Conybeare to Greenough, 22 November 1834, and Sedgwick to Greenough, 25 November 1834 [London-UCL]; Greenough to De la Beche, [4 December 1834] [Cardiff-NMW, no. 621]; Lyell to Mantell, 10 December 1834 [Wellington-ATL; *LLJ* 1: 443]. The other members of the committee were De la Beche, Sedgwick, König, and Gilbert, though in the event De la Beche was unable to take part. The medals had been endowed by George IV in 1825, but were first awarded under his successor William IV: see MacLeod, "Medals and men" (1971), and Hall (M. B.), *All scientists now* (1984), 28–29.

2. Council of Royal Society, 28 March, 20 November, 1 December 1834 [London-RS: Council minutes, 12: 40, 137, 147–50].

3. Lyell, *Principles*, 4th ed. (1835) and 5th ed. (1837); Silliman to Lyell, 15 June 1836 [Kirriemuir-KH: quoted in Wilson, *Lyell* (1972), 426–28]. Lyell himself lectured to similarly large and appreciative audiences when he visited the United States for the first time in 1841: Wilson, *Lyell in America* (1998), 38–46.

on "the study of fossil shells and their application to Geology". Had such a book ever appeared, it would probably have extended his "statistical" method of geohistorical reconstruction—as he had originally planned (§20.2)—back in deep time from the Tertiary into the Secondary era (Fig. 32.1).[4]

This complex publishing history is important, because the split between Lyell's *Principles* and *Elements*, and the refashioning of both, reflected the selective uptake of his ideas by other geologists, and the partial failure of his grand geotheoretical vision. In its original form, his *Principles* had set out a geotheory based on a wide repertoire of actual causes, applied to the reinterpretation of geohistory in terms of a Huttonian cyclicity or steady state. In his successive revisions, Lyell was able to enlarge and reinforce the scope of actual causes, making their applicability to the deep past more persuasive. In many ways, other geologists were duly persuaded; over the years they applied Lyell's methods of reasoning to an increasingly wide range of features, accepting that actual causes were indeed adequate to explain many of them. Yet some of the most important of these features were recalcitrant, and Lyell's contemporaries remained convinced that in *some* cases it was necessary to invoke causal events more sudden and intense than any yet witnessed in the very short span of human history.

32.2 CATASTROPHES AND DIRECTIONALITY

Lyell received unstinting praise from other geologists for the way that he had, in effect, updated von Hoff's inventory of actual causes (§7.2), while making it more effective by linking it closely to his own causal analysis (§22.1). In fact von Hoff completed his own great work at this time—a full decade after the earlier volumes that Lyell had quarried so fruitfully—by publishing a final volume dealing with features he had not covered previously, such as coral reefs and glaciers. Von Hoff told von Buch that this volume was his "geological swansong" (he died four years later at the age of sixty-eight), but he was still fully up to date with the latest research. He praised Lyell's *Principles* for showing how, with a proper appreciation of the vast timescale of geohistory, actual causes were indeed adequate to explain much about the deep past. Yet he also noted how von Buch and Élie de Beaumont had jointly developed a coherent and persuasive theory of crustal elevation during the years since his earlier volumes were compiled; and he agreed with them that "general catastrophes" of the kind they proposed (§8.3, §9.4) were the best explanation for the structures seen in mountain ranges.[5]

Von Buch had famously argued—on the basis of his fieldwork in the Canary Islands and other volcanic regions—that at least some volcanoes had been formed by localized upheavals of the earth's crust, later augmented by the lava and volcanic ash ejected from the aperture formed by the upheaval itself (§8.3). To many geologists this seemed a better explanation of the vast craters (in modern terms, calderas) seen on some volcanoes, than to suppose that the whole structure was formed entirely from materials ejected from below. They recognized, however, that in some cases a smaller volcano might indeed have been formed by the latter process, within the earlier crater: the conical Vesuvius within the arcuate Monte Somma was a well known and often cited example (Fig.

22.1). Von Buch had regarded such "elevation craters" [*Erhebungscratere*] as, in effect, a special case of the process by which mountain ranges had been formed: in the latter case the crustal buckling would have been linear, rather than being focused around a specific point (§8.3). Above all, he and other geologists almost took it for granted that in either case the crustal movement must have been sudden and violent. This was an inference that seemed to be powerfully supported by the astonishing structures that they saw with their own eyes in mountains such as the Alps: the huge overfold that Saussure had famously depicted half a century earlier remained an impressive example for all who traveled to see Mont Blanc (*BLT* Fig. 2.25), and the even larger structure that the young Élie de Beaumont had discovered on the Diablerets must have remained a vivid memory for him and for other geologists who climbed high enough to see it (*BLT* Fig. 9.17).

This theory of the origin of volcanoes and mountain ranges belonged to the long-established science of earth-physics (*BLT* §2.4): it was a causal explanation that was claimed to be valid wherever and whenever volcanoes and mountains might be formed, even if in practice the causal processes remained obscure because they had taken place at great depth. However, von Buch had also made the theory geohistorical, by demonstrating that rocks had been disturbed and folded at several different times in different regions: mountains did not all belong to the Primary era, or, as it were, to the earth's primal dream-time. Élie de Beaumont had then given much greater precision to the theory's geohistorical dimension, when he used the stratigraphical evidence of unconformities to distinguish buckling movements that had acted along different lines in the earth's crust at distinct and determinable points in time (§9.4). He claimed that these "epochs of elevation" [*époques de soulèvement*] had punctuated geohistory from the earliest times to the geologically recent past; but the thick formations of ordinary sediments that had been deposited in the long intervals between such movements proved that they had been very rare events. It was therefore not surprising that no such violent episode had been recorded within the short span of human history. The theory counted as "catastrophist" on Whewell's definition (§24.4), but the explanation was impeccably natural; even if it could not appeal directly to actual causes, it certainly did not deserve to be censured as "unphilosophical" or unscientific.

Nonetheless, Lyell did reject the theory of epochs of elevation (§26.2); but since it was founded on careful fieldwork more extensive than his own, he could not just dismiss it. He took the opportunity to study the structure of mountain

4. Fig. 32.1 is reproduced from a sketch by John Massey Wright (1777–1866) [London-NPG: no. 1865.6.10.1215]. Lyell, *Elements of geology* (1838); *Élémens de géologie* (1839); *Principles*, 6th ed. (1840); see Wilson, *Lyell* (1972), 409, 503–15. As a rather poor compensation for the non-appearance of his work on fossil shells, he later enlarged the *Elements* into two volumes (2nd ed., 1841), putting back much of his detailed Tertiary stratigraphy, which, of course, *was* based on fossil conchology. The history of the successive editions (to the end of Lyell's life) is summarized in Rudwick, "Lyell and the *Principles*" (1998); and their impact on the Continent, in Vaccari, "Lyell's reception" (1998).

5. Hoff, *Veränderungen der Erdoberfläche* 3 (1834), 1–3, 238–52 ("Über allgemeine Katastrophen der Erde"); Hamm, "Hoff's *History*" (1993). Two posthumous supplements to Hoff's work (1840–41) provided a chronological inventory of all recorded volcanic eruptions and earthquakes from the 4th millennium B.C. to 1832.

ranges for himself, whenever he could fit such fieldwork into his itineraries. But he focused his attack on the theory's vulnerable point, its interpretation of volcanoes as point-sources of crustal upheaval. In 1828 he had searched for evidence that Vesuvius and Etna had been formed wholly by the accumulation of lava flows and layers of volcanic ash, and not at all by upheaval from below (§19.4). Later, he seized on the reports of the new volcano off the coast of Sicily as further evidence that no crustal movement was involved, and he welcomed Prévost's conversion to that view (§23.4). In 1835 he took advantage of his time in Paris, and then at the international gathering of savants in Bonn, to debate the idea of "elevation craters" with its proponents. In Paris, as he told Sedgwick, he and Prévost had "warm discussions" with von Buch and Élie de Beaumont; in Bonn, the same teams had a "famous fight" (in French, of course) before a large audience in the meeting's geological section. Mary Lyell reported that her husband and Prévost had beaten their opponents, and Lyell himself told Mantell, "I am as convinced as ever that their views are quite erroneous". But from the other side the score looked far less decisive. Élie de Beaumont had done extensive fieldwork on Etna the previous year, and had assembled weighty evidence—convincing at least to him—that such volcanoes had indeed originated in major crustal upheavals. So that issue was as yet unresolved: "each party persevered in its opinion, without convincing the other".[6]

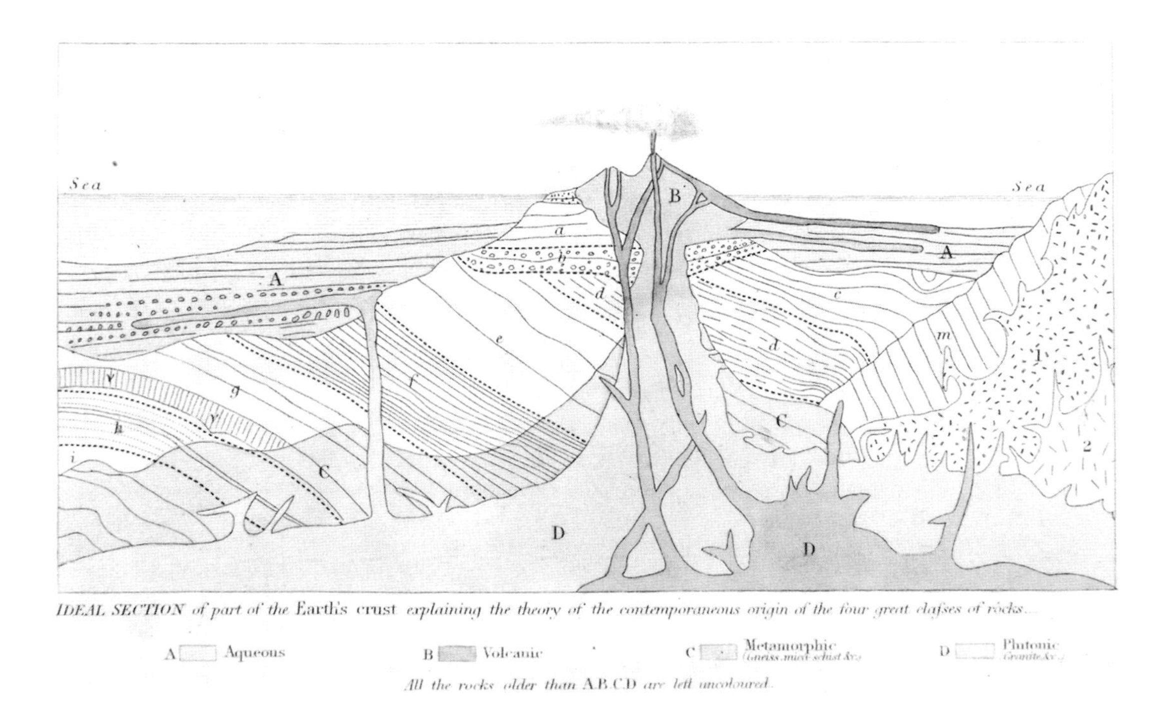

Fig. 32.2 Lyell's "Ideal section of part of the earth's crust", engraved as the frontispiece—and therefore the epitome—of his *Elements of Geology* (1838). It illustrated his continuing claim that *all* the main classes of rock (*A*, aqueous or sedimentary; *B*, volcanic; *C*, metamorphic; *D*, plutonic) have been formed in the geologically recent past and are still being formed at present; much earlier products of the same processes are also shown (*a-i*, aqueous [in retrospective order]; *v*, volcanic; *m*; metamorphic; *1*, *2*, plutonic). Even this introductory book embodied Lyell's highly controversial Huttonian geotheory of an earth in steady state, which entailed a non-directional geohistory.

Even more seriously, Lyell had to defend his geotheory against a mounting tide of criticism directed against what he regarded as its most significant feature, namely his rejection of any directionality in geohistory and his claims for a Huttonian earth in steady state. Yet he defiantly reiterated that position in the *Elements*. He took the little theoretical sketch with which he had concluded the first edition of the *Principles* (Fig. 26.2), and epitomized his new work by designing a more detailed and elegant version and placing it in the most prominent position (Fig. 32.2).[7]

Between the lines, as it were, the whole of Lyell's *Elements* was designed to reinforce his advocacy of a steady-state geotheory. For example, the text was peppered with small pictures of representative fossils, which effectively conveyed the message that more or less comparable organisms had flourished at every period of geohistory (the illustrations had maximal impact by being embedded at the right points in the text, thanks to the still relatively new medium of wood engraving). Since Lyell's review of the stratigraphical evidence was arranged in the usual retrospective order—from the youngest formations to the oldest—he had to sustain this impression of a stable world of life right to the end of the text (and the start of the known fossil record). Somewhat scraping the barrel, he achieved this by illustrating his brief account of Sedgwick's recently defined Cambrian rocks (§30.3) with one fossil that was clearly distinct from any of the subsequent Silurian forms. Unfortunately for his grand strategy, within a couple of years the formation in which his chosen Cambrian fossil was found was reassigned to the newly defined Devonian system (§31.1). This eliminated Lyell's slender evidence that the earliest known fauna was not significantly different in character from a much later one (Fig. 32.3).[8]

This was just one small example of the way that Lyell's Huttonian geotheory was given a rough ride in the decade that followed its first publication. For in those same years the directional interpretation of geohistory, to which Lyell was deeply opposed, was receiving cumulative and almost overwhelming support, as the previous chapters have suggested. Above all, it seemed to other geologists to be beyond question that the history of *life*, at least, had been directional and even, in some sense, progressive. But the same directionality also seemed to them to be detectible in the inorganic history of the earth: in the steadily cooling climates suggested by its faunas and floras, and in the evidence that mountain-building movements and volcanic activity had been on a much larger scale in the earlier eras of geohistory than in more recent times.

6. GDNA, 22 September 1835 [*Isis* 1836: cols. 711–13]; Lyell to Sedgwick, 23 July 1835 [*LLJ* 1: 450–53]; Mary Lyell to Frances Lyell, 21 September 1835 [Kirriemuir-KH: quoted in Wilson, *Lyell* (1972), 420]; Lyell to Mantell, 14 October 1835 [Wellington-ATL; *LLJ* 1: 455–57]. Élie de Beaumont, "Origine du Mont Etna" (1836), based on fieldwork there in 1834, was his massive rejoinder.

7. Fig. 32.2 is reproduced from Lyell, *Elements of geology* (1838), frontispiece, explained on 266–70, drawn and engraved by James Lee.

8. Fig. 32.3 is reproduced from Lyell, *Elements of geology* (1838), 465. The fossil had just been named *Endosiphonites* by Sedgwick's young Cambridge colleague David Ansted (1814–80), but Phillips then pointed out that Münster had already named it *Clymenia*: Ansted, "Fossil multilocular shells" (1838); Münster, "Mémoire sur les clymènes" (1834); Rudwick, *Devonian controversy* (1985), 245–46. Secord, *Victorian geology* (1986), 108, reproduces Ansted's pictures of the fossil. The work of Continental geologists such as Münster soon confirmed that *Clymenia* was indeed characteristic of some Devonian formations.

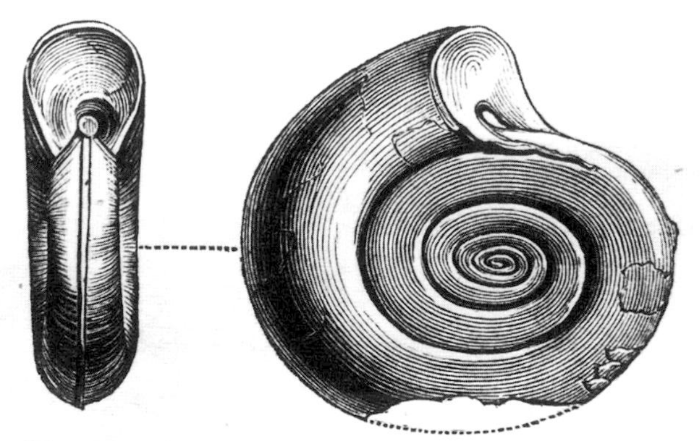

Endosiphonites carinatus, **Ansted.†** **Cambrian strata, Cornwall.**

Fig. 32.3 A distinctive ammonite-like fossil shell (more generally known as *Clymenia*) from what Sedgwick provisionally identified as Cambrian strata in Devon and Cornwall; it was the last (and therefore the oldest) of the fossils used by Lyell to illustrate the retrospective survey of stratigraphy and geohistory in his *Elements of Geology* (1838). This cephalopod mollusk, which was almost as complex as von Buch's "goniatites" in the much younger Carboniferous formations (Fig. 30.3), seemed to support Lyell's claim that the history of life had not been directional or "progressive" at all. Soon afterward, however, with the resolution of the Devonian controversy (§31.1), the relevant strata were sensationally redated as Devonian, so that *Clymenia* could no longer be claimed as part of the earliest known fauna.

Scrope had from the start been strongly critical of Lyell's Huttonian reading of geohistory (§22.2, §27.3), even though he had preceded Lyell in advocating the explanatory potential of unlimited time (§15.1). Reviewing both Lyell's *Elements* and Murchison's *Silurian System* for the *Quarterly*, Scrope wrote that he was glad to find that Murchison agreed with him "that the entire series of these changes [in geohistory] from first to last were *progressive*, not *cyclical*, as some geologists are inclined to contend—that the dynamical agents affecting the earth's surface have diminished in energy as the organic creation has become more complicated, multiform and perfect". He insisted that the human species was the final crown of this lengthy process; so he concluded jokingly that "when Mr. Lyell finds . . . a *silver-spoon in grauwacke*, or a locomotive in mica-schist, then, but not sooner, shall we enrol ourselves disciples of the Cyclical theory of Geological formations". In *this* sense, Lyell's "absolute uniformity" had found almost no adherents; of the two geological "sects" that Whewell had identified and named (§24.4), Lyell himself remained in splendid isolation as almost the only full-blooded "uniformitarian", a Messiah almost bereft of apostles.[9]

32.3 REFINING TERTIARY GEOHISTORY

Leaving aside Lyell's "controverted positions" on what he called "absolute uniformity", the citation for the Royal Society's award showed that the great respect which other geologists felt for Lyell's inventory of actual causes, and for his deployment of them in causal explanation, was equaled or even exceeded by their admiration for the way in which he had brought order to the rather confusing stratigraphy of the Tertiary formations and for his consequent reconstruction of the era of geohistory that led up to the present. In fact, the committee that recommended Lyell for a Royal Medal also proposed that the next such medal for geology should be awarded for research "towards a System of Geological Chronology, founded on an Examination of Fossil Remains and their attendant Phenomena". A geochronology of this kind was of course just what Lyell himself was trying to achieve.[10]

Lyell's interpretation of Tertiary geohistory had depended on his quantitative method of dating the scattered deposits, which in turn depended on Deshayes's authoritative identifications of large "populations" of fossil molluskan species (§25.4). The relation between the two savants was delicate: Deshayes had been paid by Lyell to identify his shells, and had then seen his results published first in his patron's work. More importantly, Deshayes did not share Lyell's conception of geohistory. He had recently published an illustrated work on "shells characteristic of formations", in which he emphasized that correlations should be based on whole assemblages of fossils, although certain "key" species were "characteristic" at different degrees of resolution, as it were, of the stratigraphy. However, he thought that the "sharp divisions" [*divisions nettes*] established by his Tertiary identifications—and implicit in the tables and illustrations he had sent to Lyell (Fig. 25.4)—corresponded to sharply distinct periods of geohistory, whereas Lyell regarded them as random samples from a continuum of gradual change (§26.1).[11]

Lyell's other Parisian collaborator was more amenable. As he reported to Murchison, Prévost accepted the new names for the periods of Tertiary geohistory, and came round to the viewpoint that underlay them, once he realized that it could incorporate his own earlier conclusion that even the Paris Basin contained deposits from more than one distinct period (§10.3; *BLT* §9.6). Important issues lay, as usual, behind all such apparently recondite arguments, as Lyell explained to Murchison:

> Prévost, although against the 'divisions nettes' which Deshayes' tables seemed to desire to establish, has been quite brought round by my view of the question, especially by what I have said by contemplating an 'upper, middle & lower' in each of the 3 periods (Eocene, Mio- & Plio-cene). This view of it—which Deshayes had not taken [i.e., accepted], or rather which is directly against D.—has persuaded Prévost to adopt my terms. He was trying 'Lutetian' for Eocene, but admits that all geographical terms are bad in as much as you have often *2 or more* epochs in *one and the same* basin.[12]

9. Scrope, "Murchison's *Silurian*" (1839), 115; as usual, the primary meaning of "progressive" was *directional*, progress in the sense of improvement being confined to the organic sphere. This article also reviewed Mantell, *Wonders of geology* (1838), and Phillips, *Treatise on geology* (1837–39), but Scrope focused almost wholly on Murchison's book and De la Beche, *Report on Devon* (1839). In defining Lyell's position the qualification "almost" is needed, because by 1839 he was no longer quite alone (see §33.2). Bartholomew, "Singularity of Lyell" (1979), reviews his intellectual isolation.

10. Council of Royal Society, 1 December 1834 [London-RS: Council minutes 12: 149]. In the event no award was made for this topic, perhaps because it was felt that Lyell had already dealt with it as well as anyone could.

11. Deshayes, *Coquilles caractéristiques* (1831), 15–16, gave the examples of *Lucina divaricata* as a "key" species for all the Tertiaries together, *Cardium porulosum* for the Parisian formations (Lyell's Eocenes) as a whole, and *Cucullaea cravatina* for the lower part of the Coarse Limestone in particular.

12. Lyell to Murchison, 20 June 1833 [London-GS: M/L17/15]; see Prévost, discussion at the Société Géologique on 16 June 1834 [*BSGF* 4: 412–13]. The proposed "Lutetian" (which was adopted by later

Lyell's work in the years that followed showed that the fortunes of his Tertiary geohistory continued to be of crucial importance to him. On his first field trip to Scandinavia in 1834, primarily in pursuit of a quite different problem (see §33.1), he went first to Denmark to check a worrying Tertiary anomaly. Johann Georg Forchhammer (1794–1865), the professor of mineralogy at Copenhagen, had reported earlier that at one Danish locality a limestone with Tertiary fossils was sandwiched *between* beds with Chalk fossils. This "dilemma", as Forchhammer called it, threatened Lyell's claim that the boundary between Cretaceous and Tertiary was everywhere sharp and the faunal contrast absolute, as the result of a huge gap in the fossil record (see Fig. 26.1). However, when he and Forchhammer visited the place, Lyell satisfied both himself and the Dane that the strata were so disturbed at this point that the alternation was only apparent, so his high-level theorizing was saved. Yet in the next edition of the *Principles* he significantly softened his original hard line on this issue, in the face of criticism; he conceded that the period separating the youngest Cretaceous strata from the oldest Tertiaries might have been one of exceptionally rapid faunal turnover, caused by equally rapid geographical and environmental change. Such was the flexibility over "uniformity" that he was obliged to admit: some periods might have been less uniform than others, though he did not think such changes ever deserved to be called "catastrophic".[13]

Lyell's greatest Tertiary problems came, however, from his own country, and involved the "Crag" of East Anglia. His own early collections of fossils from these sandy deposits, when subjected to Deshayes's expert analysis in Paris, had yielded the "Older Pliocene" dating that he recorded in the original *Principles*. But in 1835 the young Suffolk naturalist Edward Charlesworth (1813–93) claimed at the Geological Society that there was field evidence for two distinct deposits in his county, a "Red Crag" of iron-stained sands clearly overlying what he called "Coralline Crag", of contrasting appearance and with quite different fossils. Lyell noted this in his next edition, but denied that the difference was sufficient to warrant assigning them to separate periods of geohistory. So Charlesworth went on the attack again in 1836, this time at the British Association meeting in Bristol, where he described *three* distinct Crag deposits, adding a "Norwich" one that lay above the others and was therefore even younger. More seriously, Charlesworth criticized Lyell's entire percentage method of dating as being founded on "fallacies". This made his attack almost as threatening to Lyell as Murchison's Devon attack at the same meeting was to De la Beche (§31.1). Tellingly, Charlesworth pointed out that whereas Deshayes had identified 40% of extant species in the Crag, another expert, Heinrich Beck (1796–1863) of Copenhagen, who had just been in England, had examined the same collections (among them, Lyell's) and put that figure at zero, judging that *none* of the fossils could be matched exactly with any living species. Charlesworth argued that Lyell's percentage method "must therefore depend upon his [i.e., the geologist's] own estimate of the characters which constitute specific distinctions, and which is evidently liable to the greatest possible amount of variation": depending on the expert witness chosen, the Crag might be judged "*eocene* in Denmark, *miocene* in England, and *pliocene* in France".[14]

This argument brought into the open an issue of which Lyell had been well aware, when he had noted the advantage of having all his collections identified by one and the same conchologist (i.e., Deshayes). For it was well recognized among naturalists that they themselves varied greatly in the breadth of variation that they allowed in their organisms, when defining what should *count* as a species: then as now, there were notorious "splitters" and equally notorious "lumpers" (the informal terms used by modern taxonomists). Lyell the pragmatist had been content to skirt this problem by relying on Deshayes's uniform standards; but Beck's judgment deviated so far from that of the Frenchman that Lyell felt it necessary to beard the Dane in his den on his next foreign trip, to discover how he could have reached such a divergent conclusion. Beck was the curator of the exceptionally fine shell collection of Prince Christian, a keen amateur collector (and one of the Geological Society's three treasured "royal members"); so the visiting savant enjoyed all the facilities that a princely museum could afford (and the hobnobbing with royalty that he relished). Nonetheless, Lyell discovered the hard way that it was time-consuming work to appreciate ranges of variation and to decide on the limits of species, not least because Deshayes's specimens and Beck's could not be studied side by side, except as proxy pictures:

> Even with the aid of Beck & all the books & specimens at hand it often takes us a day to go through the evidence relating to a single species, of which you can rarely form an opinion till you have compared many species of the same genus & many individuals of each, of different ages, sexes & countries, so as to appreciate the true laws of specific distinction in the group, whether it be called genus or sub-genus, to which the species in question belongs.[15]

Lyell and his wife later traveled back across the Continent to Paris, where he hoped to discuss the problem with Deshayes (in the event the Frenchman was away from home). Since he did not plan to attend the British Association meeting in Liverpool, he primed a friend to be ready to report his latest opinion on the Crag question, if Charlesworth were to mount another attack in his absence. Despite Beck's figures, Lyell remained convinced that the Crag—he still treated all the deposits as being of about the same age—were of Older Pliocene age (i.e.,

---

geologists, though with a narrower meaning) was derived from the Roman name for Paris, in the same antiquarian and topographical style that Murchison used for "Silurian" (§30.3).

13. Lyell, *Principles*, 4th ed. (1835), 1: 194–208, an "entirely recast" chapter. Forchhammer, "Chalk formation of Denmark" (1828), had been sent (in English) direct to one of the Edinburgh periodicals: a straw in the wind that was blowing original geological reports from the "minor" European nations increasingly towards Britain as well as to Paris. On Lyell's Danish visit, see Wilson, *Lyell* (1972), 394–96; the locality was on the island of Møn. Subsequent research reinforced the opinion of Lyell's contemporaries that the junction between Cretaceous and Tertiary (in modern jargon, the "K/T boundary") marked a catastrophic mass extinction, though its cause was, and remains, controversial.

14. Charlesworth, "Crag-formation" (1835), read at the Geological Society on 27 May; "Comparative age of Tertiary deposits" (1837), 111, read at the British Association on 26 August 1836; "Observations on the Crag" (1837), 7; also Lyell, *Principles* 3 (1833), 170–82; and 4th ed. (1835), 4: 87–88. A fuller account of the Crag controversy, here much simplified, is in Wilson, *Lyell* (1972), 461–95.

15. Lyell (and Mary Lyell) to her father (Horner), 23 June 1837 [Kirriemuir-KH: quoted in Wilson, *Charles Lyell* (1972), 471]. As an example of their difference of practice, Deshayes's "key" Tertiary species *Lucina divaricata* was treated by Beck as six or eight distinct species (Wilson, 422).

equivalent to Brocchi's Subapennines in Italy: §26.1; *BLT* §9.4). More significantly, however, he retreated tactically from his own percentage method of dating them. He defended his continuing use of his three or four Tertiary subdivisions, but he fell back on the now widespread use of a purely qualitative assessment of the overall similarity between any fossil fauna and that of the present world:

> I am convinced that independently of the relative percentage of recent shells, about which naturalists may differ according to their notions of what constitutes a specific difference, there are other characters in the entire assemblage of forms of shells belonging to each great tertiary epoch, which will enable us to classify the deposits according to the approach which they make to the type of organisation now existing in the neighbouring seas; and that this approach will serve as a chronological test of the eras to which tertiary deposits may respectively belong.[16]

However, Lyell's confidence in his dating method revived later, after he had made no fewer than three field trips to East Anglia to study the Crag in greater detail and to check the observations of Charlesworth and other local geologists. He then agreed with them about the positions and relative ages of the three sets of Crag deposits. The London naturalist George Sowerby identified their respective fossils for him, and Lyell was gratified to hear that the proportions of extant species confirmed the point: by Sowerby's standards the Norwich Crag had 50–60%, the Red Crag 30%, and the Coralline Crag only 19%. So although Lyell had to modify his earlier conclusion that all the Crag deposits dated from his Older Pliocene period, his percentage method was saved for the time being; indeed, having assigned the oldest Crag to the Miocene on that criterion, East Anglia now bore out his earlier assertion that any one Tertiary basin might contain deposits dating from more than one period. On the other hand, his comparisons between the fossils of the oldest Crag and those of Desnoyers's Miocene formations in Touraine (§12.1) showed that even when the percentages were similar the component species might be quite different. But this again was far from fatal: he attributed it to a hypothetical land barrier between France and England in Miocene times, which could have separated distinct faunal "provinces" like those that naturalists recognized in present seas. Finally, having recognized another deposit (at Cromer on the Norfolk coast) as even younger than the Norwich Crag, and having had its fossils identified as 90–95% extant (and thus of about the same putative age as his Sicilian formations), Lyell replaced the "Newer Pliocene" of his original classification with his new term "*Pleistocene*" (i.e., *most* recent); it was a significant improvement and completion of his distinctive terminology for Tertiary geohistory.[17]

In the longer run, however, Lyell's was a Pyrrhic victory. His use of several conchologists with different views about what should count as a species—at the extremes, Deshayes was a lumper, Beck a splitter—showed that in practice his percentage method could not be as objective and rigorously quantitative as he had originally hoped, and as he had presented it in the first edition of his *Principles*. In his subsequent publications it became much less prominent in his description of the Tertiaries. And his early hopes of extending his natural chronometer back into the Secondary era evaporated altogether (which is probably why he abandoned his projected book on the use of fossil shells in geology: §32.1). In effect, he fell back on a much less precise notion: that faunas had indeed changed continuously, by the piecemeal turnover of species, but probably not at any uniform rate. This was still an improvement on Smithian stratigraphy, in that the relative age of a formation would be estimated on the basis of its whole fossil assemblage, and not just by a handful of supposedly "characteristic" forms. But that improved method was already widely in use, for example in Prévost's and Deshayes's work and more recently in Phillips's. And so, in this crucial part of his work, Lyell's only lasting monument was his trio (now enlarged into a quartet) of names for the Tertiary periods. His earlier dream of dating geohistory quantitatively had faded in the hard light of paleontological practice.[18]

32.4 THE "MYSTERY OF MYSTERIES"

What ultimately forced Lyell to downplay and almost abandon his natural chronometer was thus the problem of the variability of species. After the Geological Society awarded Deshayes a grant for his work on Tertiary mollusks, Lyell commented to De la Beche on the issue of "haute philosophie" that underlay it, namely "the additional element in the variability of a species which *time* gives, & the necessity of allowing a greater range to the variations in proportion as there is more time to multiply the differences of local circumstances capable of acting thro' generation on the same species".[19]

Closely linked to this—if, as Lyell claimed, species were indeed real entities in nature—was the problem of their origin. He had already set this in a geological context, when he adopted Candolle's concept of floral "provinces" and gave it a geohistorical dimension; even on his way home from Sicily he had imagined how the often puzzling geographical distribution of terrestrial organisms might be explained in terms of the geohistory of the landmasses they inhabited (§20.1). Whereas the extinction of species could easily be imagined as either piecemeal or catastrophic, the origin of the new species that had evidently taken their places could be inferred only from the fossil record. Lyell had argued that both processes were "gradual" in the original sense of that word: species had died out one by one, probably as a result of changing environmental conditions, and new species had taken their place in an equally piecemeal and localized fashion. But no authentic case of the latter kind of event had ever been observed. Lyell had had to explain this away, arguing that it was improbable that any naturalist would ever be in the right place at the right time to see it (§24.2).

16. Lyell to Darwin, 29 August–5 September 1837 [printed in Burkhardt *et al.*, *Correspondence of Darwin* 2 (1986): 41–43]. On his relation to Darwin, see §33.2.

17. Lyell, "Tertiary deposits commonly called Crag" (1839), read at the Geological Society on 8 May. He had already introduced "Pleistocene" in his *Élémens* (1839), the French edition of *Elements*; the Older Pliocene became henceforth just "Pliocene". Wilson, *Lyell* (1972), 478–89, describes his East Anglian fieldwork.

18. Rudwick, "Lyell's dream" (1978), describes the rise and fall of his percentage method.

19. Lyell to De la Beche, 5 April 1836 [Cardiff-NMW: no. 900].

So the origin of species remained deeply problematic: it was, in Lyell's terms, one notable "cause" of which the effects were manifest, but which could not be proved to be "actual" because it had never been observed. However, it was widely accepted among savants, even in Britain, that some kind of *natural* process, as yet unknown, must be responsible for the origin of new species. Lamarck's transformism might be deeply unacceptable on many levels, and unsatisfactory even in Geoffroy's version (§17.1). But Lyell was pleased to find that Herschel, whose opinions he treated with a degree of respect bordering on deference, agreed with him that what the astronomer called the "mystery of mysteries, the replacement of extinct species by others", was probably due to *some* kind of natural process as yet undiscovered.[20]

The solution of the problem of the origin of species was thus a well-recognized part of Lyell's agenda for the future research that geologists and other naturalists—not only he himself—might undertake. There was no reason to suppose that it could not be solved in due course, within his ambitious overall program for reconstructing the whole of geohistory and explaining it as far as possible in terms of actual causes.[21]

32.5 CONCLUSION

Lyell had from the start wanted his *Principles* to appeal both to his fellow geologists and to the general reading public. Both audiences were treated to a succession of editions, each updated to reinforce his argument that actual causes were wholly adequate to explain everything in the deep past. Lyell then catered specifically to the general public by producing his briefer *Elements of Geology*. However, this presaged a significant split in his work. The new volume was confined to what he (and Scrope) called "geology proper", namely a description of all the formations and their fossils, and their interpretation in terms of geohistory. To avoid duplication, this kind of material was then excised from the *Principles*, which in the next edition became purely a treatise on actual causes, leaving aside their uses in geohistorical explanation. This division of Lyell's work into two separate genres reflected the mixed fortunes of his originally unified geotheory.

Lyell's fellow geologists were of course more knowledgeable than the general public, and therefore more discriminating about his claims. As the citation for his award from the Royal Society showed, they welcomed some components of his work but rejected others. They praised his explanatory use of actual causes, but doubted if they were wholly adequate. In particular, the evidence suggested to them that the elevation of mountain ranges, at infrequent intervals in the vast spans of geohistory, might have involved the sudden and even catastrophic buckling of the earth's crust in certain regions, on a scale that—unsurprisingly—had never been witnessed in the very brief span of human history. Geologists were even more critical of Lyell's continuing claim that the pattern of geohistory had been that of a Huttonian cyclicity or steady state, a claim that Lyell nonetheless repeated in his new *Elements*. His critics, on the contrary, were impressed by the growing evidence that both the history of life and that of its inorganic environment had changed overall in a clearly directional manner. On this point current research seemed strongly in favor of Lyell's critics.

The most valuable part of Lyell's work, in their eyes, was his reconstruction of geohistory in the Tertiary era, the part closest to the present world. But even here Lyell faced mounting difficulties, when it became clear that his quantitative natural chronometer for Tertiary time—based on the rates of change of molluskan faunas—was deeply dependent on the taxonomic standards of the naturalists who identified his fossil shells. In the event, Lyell saved the phenomena, and his underlying conception of gradual faunal change in the course of geohistory, but only at the cost of having almost to abandon his cherished chronometer and fall back on a much less rigorous qualitative method.

Lyell's chronometer was undermined by the naturalists' diverse opinions about what, in the light of the variability of the fossil shells themselves, should *count* as a species. Their practical problems with taxonomy highlighted the theoretical problem of explaining the origin of new species in the course of geohistory. This was an Achilles' heel in Lyell's geotheory: it was an evident and genuine "cause" that could not be proved to be "actual" because it had never been observed. On the other hand, Lyell had more success with actual causes of another kind, namely those that literally underlay the movements of the earth's crust. It was on these that his vision of an ever-changing geography, and its ever-changing faunas and floras, largely depended. This is the subject of the next chapter.

20. Herschel to Lyell, 20 February 1836 [Philadelphia-APS, printed in full in Cannon, "Impact of uniformitarianism" (1961)]; most of the letter was printed at the time, with Lyell's reluctant consent, in Babbage, *Ninth Bridgewater Treatise* (1837), 202–17 (see 203). Lyell to Herschel, 1 June 1836 [*LLJ* 1: 464–69], was his response, with his comment on the origin of species in a long postscript. Herschel's use of "mystery" shows how the original and traditional meaning of that word [*musterion*] had now been reduced to no more than a potentially soluble *puzzle*; but for some savants, such as Whewell (§24.4), the origin of species—and most of all, the origin of Man—might still be a transcendent *musterion* in an otherwise wholly naturalistic interpretation of the natural world.

21. Rupke, "Neither creation nor evolution" (2005), suggests plausibly that Lyell favored something like the "autochthonous generation" theory developed by many German-speaking naturalists in subsequent decades. This "third way"—no less naturalistic than transformism—has been widely overlooked in the predominantly anglophone historiography of the "Darwinian revolution".

CHAPTER 33

Actual causes on trial (1834–39)

33.1 THE QUESTION OF CRUSTAL ELEVATION

In Lyell's (and Prévost's) private and public arguments with von Buch and Élie de Beaumont on volcanic "craters of elevation" (§32.2), Lyell's tactical victory—if it was one—did little to dent the core of his opponents' theory, which concerned not volcanoes but mountain ranges. Sudden large-scale "epochs of elevation", if indeed they had punctuated a long and otherwise relatively peaceful geohistory, were a much more serious threat to Lyell's concept of "absolute uniformity": they were unwitnessed in human history and therefore could not count as an actual cause. His own alternative explanation (§22.1) was that even the elevation of major mountain ranges involved nothing more than a sequence of thousands of small movements, each no greater in magnitude than those recorded in human history, spread over the vast spans of time available in geohistory; his argument was epitomized by the Temple of Serapis, which was displayed in the frontispiece of each successive edition of the *Principles* (Fig. 21.2). As outlined earlier, Lyell imagined the earth's geography as changing ceaselessly—and, in Huttonian fashion, non-directionally—by the elevation of some parts of the crust and the concurrent depression of others. As De la Beche put it scornfully in nursery-rhyme terms, "Here we go up, up, up, / Here we go down, down, down"; and the earth moved, as his draft caricature also suggested, under the eye of a Father Time whose clock was calibrated in "Millions of Centuries" (Fig. 22.6).

This was where Lyell's insistence on the total adequacy of actual causes was most obviously on trial. Crustal movements *recorded in human history* remained for him the most important of all the many and varied actual causes that he had described and analyzed in the *Principles*. On their reality and magnitude depended the plausibility of his claim that similar movements, spread through vast spans of time, were adequate to account for even the highest mountains. The Andes had become a focus of this argument, ever since Maria Graham reported to the Geological Society that the 1822 earthquake in Chile had produced a very small but distinct rise of the land (§8.3). On the other hand, Élie de Beaumont had suggested that the Andes might have arisen suddenly and all at once in the most recent of his putative "epochs of elevation" (Fig. 23.1); and he had hinted that this event in turn might have caused a mega-tsunami capable of producing "diluvial" effects around the world. This possible link with diluvial theory made the Andes a case that urgently demanded Lyell's explanatory attention. He insisted that all the evidence showed that the mountains had risen *gradually*—that is, in small steps, the original meaning of the word—in movements linked causally with the subterranean processes that also produced earthquakes.

Yet in one celebrated case, on the other side of the world from the Andes, it had long been claimed that the land was rising *insensibly* slowly, without any earthquakes at all. The alleged rise of the land around the Baltic was a large fly in Lyell's ointment, not because it was imperceptibly slow but because it lacked any obvious causal agency. Its factual reality had been vouched for by no less an authority than von Buch—a geologist with far longer and more extensive fieldwork experience than Lyell—but in his *Principles* the latter had airily dismissed it as an "extraordinary notion" (§22.1). However, his confidence in rejecting von Buch's testimony was shaken by a report by the Scottish chemist and mineralogist James Finlay Weir Johnston (1796–1855), the reader in chemistry at Durham. Johnston criticized Lyell for his doubts; confirmed, from his own travels in Sweden, the authenticity of what the great Prussian had claimed; and suggested that it might be explained by the earth's "secular refrigeration". So when revising his work Lyell became more cautious: in his third edition he conceded, rather grandly, "I am willing to hope that a personal examination of the country will soon entirely remove from my mind all remaining scepticism". He was all the more concerned to check the facts of the case, because the arch-skeptic Greenough, stung by Lyell's ambitious theorizing, had just used his third presidency of the Geological Society to pour scorn on *all* reports of crustal elevation as an "actual cause", including Maria Graham's on Chile. Lyell had no wish to seem to be on the same side as Greenough on this issue or any other.[1]

As soon as the revisions for his third edition were complete, Lyell therefore left London to devote his now customary summer fieldwork to this worrying problem in Sweden. After resolving en route a quite separate and stratigraphical puzzle in Denmark (§32.3), he traveled on to Stockholm and then northward to the shores of the Gulf of Bothnia. Here certain coastal rocks had been inscribed from time to time to mark the current sea level, ever since Celsius, a century earlier, had reported that local people claimed that it had been changing within their own lifetimes (*BLT* §3.5). Like his compatriot and contemporary Linnaeus, Celsius had assumed it was the sea that was falling worldwide (in modern terms, eustatically), his own country providing uniquely clear evidence of it because the sea level could be recorded with exceptional accuracy in the almost tideless Baltic. Since Celsius's time, however, the discovery that the rate of change was not uniform, but progressively greater towards the north, had convinced

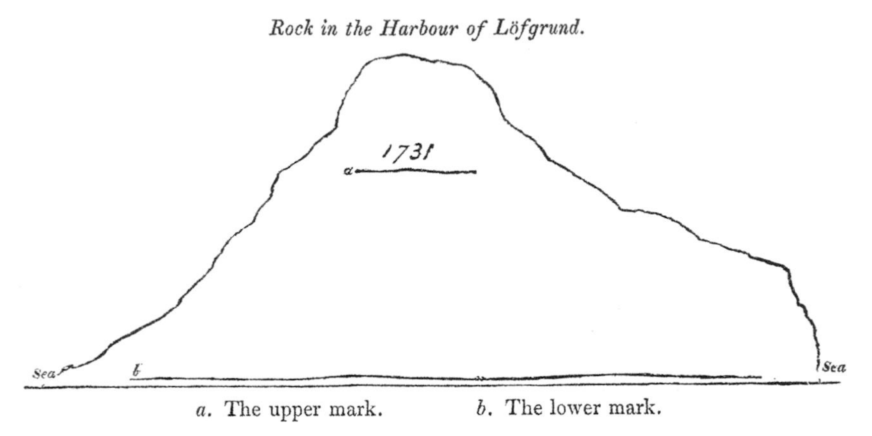

Fig. 33.1 An inscribed rock on a shore in the Gulf of Bothnia: a sketch made by Lyell during his fieldwork in Sweden in 1834, showing that the sea level at that point in the Baltic had fallen by about three feet since 1731, and by a few inches even since the "lower mark" had been made, probably in 1820. Seeing such rocks for himself convinced Lyell that Sweden was indeed rising insensibly slowly, at least in the north, a claim that he had earlier rejected because there were no records of earthquakes in the region. This drawing illustrated the Bakerian Lecture that he gave at the Royal Society after his return to England.

savants such as von Buch that it was the earth that was rising in the Baltic region, although differentially, and that the cause had to be subterranean (§8.1). Lyell too, after seeing some of the inscribed rocks for himself, was convinced that the marks were reliable and the effect genuine (Fig. 33.1).[2]

Later, at Uddevalla on the west coast of Sweden, Lyell also saw for himself, high above sea level, the gravels with marine shells of living species that the elder Brongniart had described some years before (§13.4). Among the fossils were barnacles attached to solid rock on the former shoreline. This eliminated any possibility that the shells had been swept up to that level by some "deluge" or mega-tsunami, and hence dispelled any doubt that the sea itself had been at that height at the time. Lyell took this as evidence that the land had already been rising, long before the times of human history.[3]

1. Lyell, *Principles*, 3rd ed. (1834), 1: 334n; Johnston, "Elevation in Scandinavia" (1833); Greenough, "Address at the anniversary" (1834), 54–69, read on 21 February. Kölbl-Ebert, "Observing orogeny" (1999) and "George Bellas Greenough" (2003), describe his attack on Lyell's theorizing and hence on Graham's competence as an observer. Babbage, "Temple of Serapis" (1834), read on 12 March (i.e., soon after Greenough's address), supported Lyell with a possible *physical* cause of the crustal movements there, as part of a broader "explanation of the vast cycles [*sic*] presented by the phaenomena of geology"; see also Dolan, "Representing novelty" (1998). This, however, was earth-physics rather than geohistory, and is therefore tangential to the present narrative.

2. Fig. 33.1 is reproduced from Lyell, "Rising of the land in Sweden" (1835), 19 (read on 27 November, 18 December 1834). The "lower mark" was said to have been made in 1820, when the Swedish Royal Academy of Sciences had sponsored a thorough study of the phenomenon; Lyell saw it barely above the water, but a local pilot told him that on a calmer day it would have been at least four inches [10cm] up. On Lyell's Scandinavian journey, see Wilson, *Lyell* (1972), 391–407; also Dott, "Ups and downs of eustasy" (1992). The marks are on a vertical face of a large erratic block of gneiss on the northwest shore of the small island of Löfgrund, which lies off the east coast of Sweden about 17km northeast of Gefle [now Gävle]. Bergsten, "Uplift in Sweden" (1954), 95–97, analyzes the historic and continuing elevation of the rock: see the photograph (now already half a century out of date) in fig. 10.

3. Bravais, "Ancien niveau dans le Finmark" (1840), later described measurable gradients in similar raised shorelines in the far north of Norway, sloping upwards away from the outer coast, but this differential elevation was not attributed to post-glacial isostatic rebound until later in the century:

The inscribed rocks and the raised beach—evidence respectively historical and geological—combined to confirm Lyell's new belief that Scandinavia was rising slowly but steadily on a long timescale: "here we go up, up, up". After his return to Britain, the Doubting Thomas announced his conversion at the British Association meeting in Edinburgh. Later that year, having been invited to give the Bakerian Lecture at the Royal Society, he devoted this prestigious occasion to a major paper on "the gradual rising of the land" in Sweden. Lyell's first paper to the Royal Society prompted it to announce a few days later that it had awarded him its Royal Medal (in fact, primarily for his *Principles*: §32.1). In keeping with the Society's rigorous scientific standards, Lyell indulged in no theoretical speculations about the phenomenon he described. But he realized that he could turn it to good account in the service of his geotheory: although its (literally) underlying cause remained obscure, it did allow him to invoke crustal elevation even in regions where there was no record of any earthquakes. In that sense it extended the battery of actual causes that he could deploy in explaining geohistory.

33.2 WITNESSES TO ELEVATION IN SOUTH AMERICA

Early in 1835, with perfect timing for Lyell's argument (though disastrously for those more directly involved), Chile suffered an earthquake even more catastrophic than that of 1822; coastal towns were reduced to rubble and the devastation was made far worse by the twenty-foot tsunami that swept in immediately afterwards. Lyell soon heard from an expatriate British businessman that at Concepcion the land had risen by three or four feet. Later in the year he had even sounder evidence of the reality of that effect. Some time after the earthquake a British naval frigate had been wrecked on the nearby coast. Its captain was court-martialed; but he was acquitted on the testimony of Robert Fitzroy (1805–65), the captain of another British naval ship that was in Chilean waters at the time. Fitzroy stated under oath that one of the coastal islands had risen by ten feet, and that the pattern of currents had changed, implying that his fellow officer should not be blamed. Fitzroy's word could hardly be doubted, for he was not only an officer and a gentleman but also an experienced hydrographer engaged in a coastal survey for the Admiralty. As Lyell commented in jubilation to Sedgwick, "Give me but a *few* thousand centuries, and I will get contorted and fractured beds above water in Chili, horizontal ones in Sweden, &c." Since he had succeeded Greenough as president of the Geological Society, Lyell was able to ensure that these testimonies were reported there effectively. But his opponents, notably Greenough, did doubt the word of even a gentlemanly naval officer. They counter-attacked by marshalling another set of witnesses for the following meeting, denying that in earlier quakes, notably that of 1822, there had been any elevation at all (and thus rejecting Graham's testimony among others). However, on this point it was Lyell's side that was judged to have won.[4]

At the first of these meetings, Sedgwick added to the implications of the latest reported uplift of the Chilean coastline, by reading extracts from letters that had been sent to his Cambridge colleague Henslow by a young naturalist who was on board Fitzroy's little ship in South America; fuller extracts had already been read by Henslow himself at the Cambridge Philosophical Society, where they were considered important enough to be printed for its members and other savants. The writer, Charles Robert Darwin (1809–82), had been known to both Henslow and Sedgwick while he was still an undergraduate at Cambridge. Henslow had encouraged him in botany and zoology, and Sedgwick took him on his own first foray into Wales (§30.3) and, in effect, taught him the practice of field geology. Not long after that trip, Fitzroy, on Henslow's recommendation, had invited Darwin to join his forthcoming hydrographical voyage around the coasts of South America, to alleviate the rigid hierarchy of naval life by keeping him company as a fellow gentleman. In the event, however, the *Beagle*'s unofficial naturalist—"the philosopher", as the crew called him—spent more time away on expeditions on land than he did at sea, because he found the terrestrial natural history of absorbing interest (and he was chronically seasick). Darwin collected plants and animals industriously, and commented on them intelligently; but his frequent letters home, and to Henslow as his contact with the scientific world, made it clear that he was captivated above all by geology, and was increasingly defining himself as a budding *geologist.* He said he was torn, like the proverbial donkey, between two equally fascinating aspects of the science: the Primary rocks and their minerals, and the Secondary and Tertiary formations with their fossils. But he soon chose to focus on the latter.[5]

Fitzroy had given his young companion a copy of the first volume of Lyell's *Principles* before they left England at the end of 1831, and Darwin received the other two in South America soon after they were published. There were many other geological books in the substantial scientific library on board, but in the event it was Lyell's eloquent work that molded Darwin's outlook most profoundly. "I hear that your Theory of the Earth [*sic*] is supposed to be the same as what is contained in Lyell's 3d Vol.", one of his well-informed sisters commented, on the evidence of his letters and even before that volume brought Darwin the Huttonian finale of Lyell's geotheory (§26.2). In the earlier part of Darwin's travels, on the east coast of the continent, his most important geological work was Cuvierian rather than Lyellian: he found large fossil bones (and purchased others), and he knew enough comparative anatomy to identify among them the megatherium and to recognize that others were new, before shipping them all

Wegmann, "Déplacement des lignes de rivage" (1967), summarized in "Moving shorelines" (1969).

4. Caldcleugh, "Earthquake in Chile" (1836), read at the Royal Society, 26 November 1835; Geological Society, 18 November, 2 December 1835 [*PGS* 2: 208–16]; Lyell to Sedgwick, 25 October, 6 December 1835 [*LLJ* 1: 457–61].

5. Darwin, *Letters to Henslow* (1835), marked "For private distribution" but in fact widely available; "Geological notes" (1835), read at the Geological Society on 18 November. Herbert, *Darwin geologist* (2005), is a detailed account of his geological career. In the present context see especially Secord, "Darwin's early geology" (1991); also Barrett, "Sedgwick-Darwin geologic tour" (1974); Rudwick, "Darwin and the world of geology" (1985); and Stoddart, "Darwin and the seeing eye" (1995). Browne, *Darwin: voyaging* (1995), is a superbly readable and reliable account of the part of his life relevant here. Burkhardt *et al.*, *Correspondence of Darwin* (1985-) [hereafter, *CCD*], is an equally superb collection that makes all earlier editions obsolete; vols. 1–2 (1985–86) cover the period (to 1843) relevant to the present narrative. Since letters *to* Darwin are included, this edition is also definitive, for example, for Lyell's letters to him, being far more reliable than the frequently flawed and incomplete transcriptions in the Victorian *LLJ*.

back to England for more expert eyes. However, he was equally impressed to see vast tracts of land covered with deposits containing marine mollusk shells, among which the mussels even retained their color; although earthquakes were unknown in the region, he took this as proof of geologically recent uplift. Having studied the formation underlying these plains, he told Henslow "I conjecture . . . that the main bed is somewhere about the meiocene period (using Mr Lyell's expression); judging from what I have seen of the present shells of Patagonia".[6]

After surveying the complex channels and coasts around Tierra del Fuego (waters of immense commercial and strategic importance before the Panama Canal was cut), the *Beagle* sailed around to the west coast of South America. Here Darwin found another Tertiary formation, which he guessed from its fossil shells might be Eocene in age. Even more exciting for Lyell, however, would have been Darwin's report that he had found an "abundance of recent shells [i.e., of extant species] at an elevation of thirteen hundred feet". Darwin had become a true Lyellian, for he added, "I suppose the thirteen hundred feet elevation must be owing to a succession of small elevations, such as in 1822". When the successor to that earthquake struck Chile, the *Beagle* was in Valdivia, but soon sailed north to its epicenter at Concepcion, where Darwin was shocked at the destruction it had caused. Having, like Fitzroy, seen the evidence with his own eyes, he took the elevation that accompanied both earthquakes so much for granted that he did not mention it to Henslow. Instead he took a longer or more geohistorical view: he reported having found a bed of oysters among the roots of forest trees, 350 feet above sea level on the island of Chiloé, adding that "I can now prove that both sides of the Andes have risen in the [geologically] recent period to a considerable height". And in his own mind he was evidently effacing any radical distinction between equable uplift and movements that might fold and contort strata, for he also told Henslow that "in the modern Tertiary strata I have examined 4 bands of disturbance, which reminded me on a small scale of the famous tract in the Isle of Wight", with its spectacular vertical folding of Tertiary formations (*BLT* §9.4).[7]

Later, in one of his most ambitious expeditions, Darwin was able to cross the Andes by high passes, and to get a rough idea of the internal structure of this major mountain range. "I cannot imagine any part of the world presenting a more extraordinary scene of the breaking up of the crust of the globe", he told Henslow afterwards. As at some famous localities in the Alps (*BLT* §9.5)—though Darwin was apparently unaware of them—he found ammonites and other Secondary fossils as high as 12,000 feet; he inferred in good Lyellian fashion that the granites forming the peaks were "fluid in the tertiary period"; as he put it, "that all this should have been produced in so very recent a period is indeed remarkable". But more important for Lyell was Darwin's earlier evidence that the mountains had not risen all at once, as Élie de Beaumont proposed, but in many successive steps. What Lyell needed most to hear from his young admirer was further evidence for this kind of *gradual* elevation. He had already used one such Chilean locality to good effect in the *Principles*. Inland from Coquimbo, the naval captain Basil Hall (1788–1844), the son of Hutton's friend Sir James, had earlier found three horizontal terraces high above sea level, which Lyell interpreted as ancient sea beaches. Darwin made a point of visiting the place, and duly confirmed at first hand what his hero had inferred.[8]

33.3 DARWIN'S THEORY OF A DYNAMIC EARTH

"How I long for the return of Darwin!", Lyell exclaimed to Sedgwick, shortly after excerpts from Darwin's letters had been read at the Geological Society. The feeling was reciprocated. Even before the *Beagle* returned to England, after a long voyage by way of Tahiti, New Zealand, Australia, and the Cape, Darwin asked Sedgwick to propose him for membership of the Geological Society: clearly he regarded this as the most important affiliation for his prospective new life as, primarily, a *geologist*. Soon after his return he was duly elected, and found Lyell the most helpful of those he called the Society's "great guns". At first he based himself in Cambridge, but he soon moved to London as the center of scientific activity in Britain. Much of his time was spent farming out his collections to specialists. For example, he offered his fossil bones to Clift, who suggested Buckland; Buckland was too busy, and in the end Clift's assistant Owen described and identified them in fine Cuvierian style. In addition to all his work with his specimens, Darwin prepared to write a naturalist's account of the voyage, to complement Fitzroy's nautical one; he also made plans for writing geological books of his own, based on what he had seen in South America.[9]

The first fruits of Darwin's research were reported to the Geological Society in 1837. Not surprisingly, in view of the earlier arguments at the Society, his first paper was on "Proofs of recent elevation on the coast of Chili". He gave details of the evidence for the elevation that had followed both earthquakes, and noted—with a perceptive comment on the social distribution of the knowledge—that he had "met with no intelligent person who doubted the rise of the land [in 1822], or with any of the lower order who doubted that the sea had fallen". But he claimed that the whole coast had been rising, "though insensibly", even in the years between the two quakes. The implication was that the jerky effect associated with earthquakes and the imperceptibly gradual Baltic effect were, as Lyell

6. Darwin to Henslow, March 1834 and 24 July 1834 [*CCD* 1: 368–71, 397–403]; this and later letters cited below were excerpted in Darwin, *Letters to Henslow* (1835); Catherine Darwin to Darwin, 27 November 1833 [*CCD* 1: 356–58]. The library on the *Beagle* (over 200 books) is reconstructed in *CCD* 1: 553–66. See also Herbert, "Darwin as geological author" (1991). Darwin was not the only European naturalist at work in South America. For example the Muséum in Paris sent Alcide d'Orbigny (1802–57) on extensive travels in 1826–33: see the advance summary of his geological results in Cordier *et al.*, "Voyage de M. Alcide d'Orbigny" (1834), and his own full account in vol. 3 (1842) of Orbigny, *Amérique méridionale* (1834–47).

7. Darwin to Henslow, 24 July 1834, March 1835 [*CCD* 1: 397–403, 436–38].

8. Darwin to Henslow, 18 April 1835 [*CCD* 1: 440–45]; Hall (B.), *Chili, Peru, and Mexico* (1824) 2: 6–11; Lyell, *Principles* 3 (1833), 131–32.

9. Lyell to Sedgwick, 6 December 1835 [*LLJ* 1: 460–61]; Darwin to Henslow, 9 July, 30 October 1836, to Caroline Darwin, 9 November 1836, and to Owen, 19 December 1836 [*CCD* 1: 499–501, 512–15, 518–19, 526–27]; Darwin, *Journal and remarks* (1839). Darwin recognized that on purely scientific grounds the Muséum in Paris would have been the most appropriate destination for his vertebrate fossils; but they had been collected on a British naval expedition, and on political grounds could not well be sent abroad, least of all to Britain's ancient rival and former enemy. However, he soon joined the Société Géologique, being sponsored by Murchison and Lyell, and its *Bulletin* thereafter kept him well informed about current Continental research.

already suspected, two manifestations of one and the same kind of crustal movement. Darwin extended this actual cause back into geohistory by mentioning the terraces with marine shells that he had seen, such as those near Coquimbo at up to 250 feet above sea level. And he generalized the effect still further, by relating these features to what he had seen on the east coast. He inferred that the whole of South America was rising on a tilt, more rapidly and jerkily in the west, insensibly slowly in the east, just as Scandinavia, on a much smaller scale, appeared to be rising more rapidly in the Gulf of Bothnia than further south.[10]

Lyell was delighted. Darwin had shown him the paper in advance, asking for his comments. "The idea of the Pampas going up, at the rate of an inch in a century, while the Western Coast and Andes rise many feet and unequally, has long been a dream of mine", Lyell told him; "What a splendid field you have to write upon!" What Darwin next wrote upon, at least for public consumption, were the fossil mammals that Owen had identified. This gave him another opportunity to emphasize the recent elevation of Patagonia: he claimed that "the movements seem to have been so regular, that the amount of elevation becomes a measure of time". His extinct mammals were found in the same formation as living molluskan species; these were, he claimed, facts that "fully confirm the remarkable law,

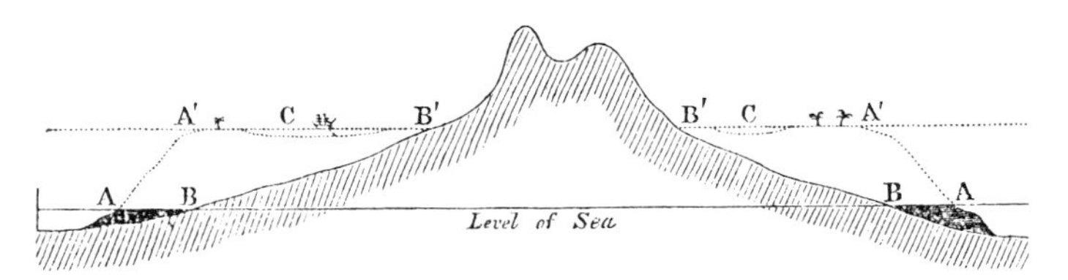

[No. 4]
A A—Outer edge of the reef at the level of the sea.
B B—Shores of the island.
A′ A′—Outer edge of the reef, after its upward growth during a period of subsidence.
C C—The lagoon-channel between the reef and the shores of the now encircled land.
B′ B′—The shores of the encircled island.
N.B. In this, and the following wood-cut, the subsidence of the land could only be represented by an apparent rise in the level of the sea.

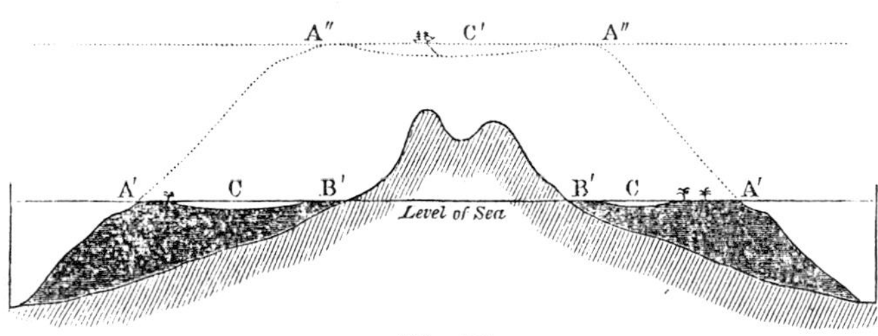

[No. 5.]
A′ A′—Outer edges of the barrier-reef at the level of the sea. The cocoa-nut trees represent coral-islets formed on the reef.
C C—The lagoon-channel.
B′ B′—The shores of the island, generally formed of low alluvial land and of coral detritus from the lagoon-channel.
A″ A″—The outer edges of the reef now forming an atoll.
C′—The lagoon of the newly-formed atoll. According to the scale, the depth of the lagoon and of the lagoon-channel is exaggerated.

Fig. 33.2 Darwin's diagrammatic sections to show how the possible origin of atolls from fringing coral reefs could indicate that the underlying crustal plate is slowly subsiding. The upper section (No. 4) shows how, if the crustal subsidence (or rise in relative sea level) were slow enough, the growth of coral could keep pace with it, so that a fringing reef around the shore of a volcanic island (both shown in solid tones) would become a reef encircling a lagoon (as shown in dotted outline). The lower section (No. 5) shows how the same form of reef (now shown in solid tones) would in turn become an atoll (dotted outline) once the original volcano had sunk below sea level. These wood engravings, published in Darwin's book on coral reefs (1842), illustrated what he first expounded at the Geological Society in 1837. He used this theory about coral reefs to support a more fundamental steady-state theory of oscillating crustal plates, which showed him to be a fully "uniformitarian" Lyellian. (By permission of the Syndics of Cambridge University Library)

often insisted upon by Mr. Lyell, that 'the longevity of the species' among mammalia has been of shorter duration than among molluscs" (§18.3).[11]

Darwin's third paper to the Geological Society, read less than a month later, extended his theory of crustal movement still further, and from a quite unexpected and novel direction. He formulated an ingenious *causal* explanation for the varied forms of coral reefs. He had seen a few with his own eyes in the Pacific and Indian Oceans, but he would have been able to study many more on paper, for example on the nautical charts carried on the *Beagle* and perhaps others at the Admiralty after his return. In the open ocean, reefs were of three forms: fringing reefs close to an island shore; encircling reefs around an island, with a lagoon between; and "lagoon islands" or atolls, encircling a lagoon with no island in the center. Continental coasts might have either fringing reefs close inshore, or barrier reefs offshore with a lagoon on the inner side. Lyell had argued that atolls marked the positions of the rims of volcanic craters, built up by the coral organisms as the volcano sank (§24.3): in effect, they were signs of a sinking former continent. Darwin greatly improved this hypothesis by bringing all the other forms of reef under the same kind of explanation. What traditional physical geography (*BLT* §2.2) had treated as discrete forms now became transient stages in a smooth temporal series, and Darwin claimed to have found all the anticipated intermediate forms. Provided a continent or an ocean floor sank slowly enough, the coral organisms could build the solid structure upwards, keeping pace with the sinking foundations and maintaining themselves in the shallow water they were known to require. A fringing reef along a continental coast would then turn slowly into an offshore barrier reef; a fringing reef around an oceanic volcano would likewise become an encircling reef, and, after the former volcano finally sank out of sight, an atoll (Fig. 33.2).[12]

Darwin's ingenious explanation of coral reefs was for him only a means to an end. What he identified as the "main object" of his paper was to use the geographical distribution of reefs to plot a series of tracts of the earth's crust that were either sinking or rising. Having become convinced that South America was rising, and having adopted Lyell's steady-state "system", Darwin anticipated that other crustal plates must be sinking at the same time, and equally slowly: "here

10. Darwin, "Proofs of recent elevation" (1837), read at the Geological Society on 4 January.

11. Lyell to Darwin, 26 December 1836 [*CCD* 1: 532–33]; Darwin, "Extinct Mammalia" (1837), read at the Geological Society on 3 May. Owen, "Fossil mammalia" (1839–42), was his colleague's later, full report. Around this time, as his notebooks show, Darwin began to differentiate *privately* between his theorizing about crustal elevation and about speciation: see the time chart in Rudwick, "Darwin in London" (1982), 201.

12. Fig. 33.2 is reproduced from Darwin, *Coral reefs* (1842) [Cambridge-UL: S365.c.84.3], 98, 100; his preliminary paper, "Elevation and subsidence" (1837), was read at the Geological Society on 31 May. The crucial evidence that living coral growth was confined to shallow water was well known from Quoy and Gaimard, "Accroissement des polypes lithophytes" (1825). Darwin had seen some of the reefs of the Tuamotus and Tahiti in the Pacific, and at Cocos-Keeling in the Indian Ocean. Stoddart, "Darwin, Lyell, coral reefs" (1976), is a fine account of the theory; "Darwin and the seeing eye" (1995) analyzes the paradox that he was "verbally so visual, visually so limited", and reproduces a selection of those curiously limited images. Darwin's sequence, explaining the origin of one form (the atoll) in terms of a smooth and gradual change from a quite different form (the fringing reef), by way of an *intermediate* form (the reef encircling a lagoon), was closely analogous to, and probably a template for, his simultaneous theorizing about the origin of new species.

we go up, up, up; here we go down, down, down". The coral reefs, he argued, acted as "monuments [*sic*] over subsided land": the classic antiquarian metaphor (*BLT* Chap. 4) here made a rare reappearance, the idea behind it being now taken for granted by geologists. Raised marine terraces or beaches, and fossil coral reefs high above sea level, were corresponding signs of elevated land. Darwin's theoretical ambitions were nothing less than those of global geotheory: as he put it in his conclusion, by plotting the tracts that were currently sinking and those that were rising, "we obtain some insight into the system [*sic*] by which the crust of the globe is modified during the endless cycle [*sic*] of changes". Lyell had gained a true disciple: he himself was no longer the one and only full-blooded "uniformitarian" (§24.4).[13]

Lyell recognized at once that Darwin's theory of coral reefs had trumped his own. As he told Herschel, after reading the draft of Darwin's paper, "I must give up my volcanic crater theory for ever, though it cost me a pang at first, for it accounted for so much . . . Yet in spite of this, the whole theory is knocked on the head". Herschel was at the Cape, mapping the stars of the southern heavens; Lyell urged him to tell any naval commanders who put in at the Cape and called on him (as Fitzroy and Darwin had done) to look out for further evidence to test Darwin's theory. Lyell suffered the loss of his own theory quite cheerfully, for he recognized that Darwin had out-Lyelled him. His disciple's explanation was even more firmly grounded in coral growth as an actual cause; it offered a much better account of the varied forms of reefs; and it provided further evidence that crustal plates were rising or subsiding very gradually. Above all, therefore, it was further ammunition to use against any suggestion that continents or mountain ranges were elevated suddenly, or that geohistory had been marked by any such large-scale violent events.[14]

Darwin himself soon developed his underlying ideas of crustal movement still further in what was, in the event, the last of his major papers to the Geological Society. His recent election as one of its secretaries—Whewell, the current president, had persuaded him to stand—signaled his new status as a highly respected young geologist. His impressive earlier papers gave him, in effect, the license to be openly theoretical and even global in his explanatory ambitions. His own observations in South America were his starting point, but his goal was a general explanation of crustal mobility, an ahistorical theory in earth-physics. He argued that earthquakes, volcanic eruptions, and elevation of the land, occurring more or less simultaneously over vast areas, were "of the greatest importance, as forming parts of one great action, and being the effects of one great cause". In place of the usual focus on mountains, as for example in Élie de Beaumont's theorizing, Darwin argued that "mountain-chains are only subsidiary, and attendant operations on continental elevations". He concluded in Lyellian style that "mountain-chains are formed by a long succession of small movements", and that their elevation, and that of whole continents, "are phaenomena now in progress". And he suggested how all these effects might be caused by movements of "fluid rock" at great depth, and how the highly tilted rocks that he had seen in the Andes could have been displaced by a long succession of small movements rather than violently and all at once.[15]

In short, Darwin proposed an ambitious *causal* explanation of continental elevation, which in turn would fit into an even more ambitious Lyellian—or indeed neo-Huttonian—geotheory of endlessly rising and subsiding crustal plates. As he himself had noted privately, with sublime confidence, while still on his voyage, "geology of whole world will turn out simple". This is just how Darwin's paper was perceived, when its reading was followed by a lively discussion at the Geological Society. De la Beche doubted whether a Lyellian succession of small movements could account for the huge overfolds seen in the Alps (though not by Darwin, and not found by him in the Andes). But Lyell reported that Phillips "pronounced a panegyric" on the *Principles* and its "actual cause doctrine", predicting that in future geologists would be divided on the issue of uniformity only in one sense, "some contending for greater pristine forces, others satisfied like Lyell and Darwin with the same intensity as nature now employs". Darwin was clearly seen as a steady-state "uniformitarian" in as full-blooded a sense as Lyell himself. Conversely, however, Lyell now recognized the extent to which "uniformity" in *other* senses was increasingly accepted by the leading geologists: "I was much struck with the different tone in which my gradual causes were treated by all, including De la Bêche, from that which they [the causes] experienced in the same room four years ago, when Buckland, De la Bêche, Sedgwick, Whewell, and some others treated them with as much ridicule as was consistent with politeness in my presence".[16]

33.4 DARWIN'S TEST CASE IN SCOTLAND

Darwin's most theoretical paper, just summarized, was written in the ahistorical mode of earth-physics, as an explanation that aspired to be valid at all times and in all places (*BLT* §2.4). It would be only marginally relevant to the present narrative, were it not that Darwin himself linked it to geohistory: he extended his explanation back from the present world of observed or recorded effects into the particularities of the geological past. While in Chile he had noted not only the minor elevations associated with the earthquakes of 1835 and 1822, but also the obviously far older raised beaches or terraces, now elevated hundreds of feet

13. Darwin, "Elevation and subsidence" (1837). *Coral reefs* (1842), the first of his planned geological volumes, included a fold-out map of the Indian and Pacific Oceans (plus the Caribbean), distinguishing in blue and red the reefs and other features that he interpreted as signs that the crust was, respectively, sinking or rising; although there were some anomalies, the map suggested a vast series of alternating crustal plates; a lengthy appendix (151–205) documented his evidence in detail.

14. Lyell to Herschel, 24 May 1837 [*LLJ* 2: 11–13]; to be "knocked on the head" was euphemistic slang for an outright killing.

15. Darwin, "Connexion of volcanic phaenomena" (1838), read at the Geological Society on 7 March, published fully as "Volcanic phenomena in South America" (1840). Rhodes, "Darwin's theory of the earth" (1991), gives a superb account of this paper's significance. On his role as a geological theorist, see Herbert, *Darwin geologist* (2005); also Rudwick, "Darwin in London" (1982) and "Darwin and the world of geology" (1985). His better-known theorizing about speciation—at the time, strictly private—had the same ahistorical character, in that he was searching for a solution to Herschel's "mystery of mysteries" (§32.4) in terms of processes that could produce new species at *any* time in past, present, or future.

16. Lyell to Horner, 12 March 1838 [*LLJ* 2: 39–41]; it is not clear which earlier occasion he was referring to. De la Beche's contemporaries often spelled his name with a circonflex, which suggests that he himself pronounced it with a soft "e".

Fig. 33.3 The "Parallel Roads of Glen Roy", as illustrated by John MacCulloch in his detailed study (1817) of the terraces in this and adjacent valleys in the Scottish Highlands. By 1838, when Darwin studied them on his only major field trip after returning to Britain, geologists were agreed that they were ancient shorelines, probably of a former lake. Darwin, on the contrary, interpreted them as the shorelines of a former fiord or deep inlet of the sea, and used them as evidence for his general theory of the slow and equable elevation of a continent on a rising crustal plate. (By permission of the Syndics of Cambridge University Library)

above sea level. Those he saw near Coquimbo had already been described in Lyell's *Principles* as "Parallel Roads". This phrase alluded to some much more famous terraces in Glen Roy, a valley in the remote Lochaber area of the Scottish Highlands (they were called "Roads", because they had been interpreted by local people as man-made forest rides, dating from before the valleys were deforested). So it is not surprising that during the summer following the reading of his paper on global tectonics Darwin visited Lochaber to see the Parallel Roads of Glen Roy for himself. It was the most substantial piece of geological fieldwork that he ever undertook after his return to Britain. Clearly he hoped to use the Roads as further support—all the more persuasive for being already well known to other geologists—for his grand theory of continental elevation. The most spectacular terraces were in Glen Roy itself, where three ran horizontally, high up on the sides of the valley (Fig. 33.3).[17]

Geologists agreed that the Roads were ancient beaches cut by wave action into the steep hillsides; traced laterally they were perfectly horizontal, but in profile they sloped outward, like any modern beach. About twenty years before Darwin arrived, they had been described in detail by John MacCulloch (1773–1835) and Thomas Dick Lauder (1784–1848), two Scottish savants working independently but at almost the same time. Their respective maps were both accurate, displaying the curiously limited extent of the terraces. The two upper ones in Glen Roy stopped abruptly where that valley opened into the much larger Glen Spean, whereas the lowest terrace continued all round Glen Spean, stopping in much the same way where it in turn opened into the Great Glen that cuts right across Scotland. Twenty years later, Darwin provided yet another map, in all essentials the same as those of his predecessors. Many of the relevant "facts" about the terraces were thus not in dispute (Fig. 33.4).[18]

MacCulloch had reviewed several alternative explanations of the terraces, and concluded that the least unsatisfactory was to attribute them to temporary lakes that had been dammed by equally temporary barriers of some kind across the mouths of the valleys, at the points where the terraces end. This accounted for their limited extent, but entailed postulating barriers of which no obvious traces remained. He acknowledged that this was a major flaw, but he judged that his preferred interpretation was nonetheless the best available. He did briefly mention a radical alternative, namely that the terraces were *marine* beaches, but he considered that this was even less satisfactory: it failed to explain why the terraces had only been preserved so locally, and the total absence of seashells among the pebbles on the beaches also told against it. Lauder inferred likewise that the terraces marked the shorelines of former lakes, though he too recognized the problems with that explanation. But he improved it substantially, and explained why the terraces were at those specific heights, when he reported that in three cases (a fourth remained anomalous) there was a col at the head of the relevant valley at exactly the same level as one of the terraces. He interpreted these cols as the overflows from successive lakes, and he drew a set of small maps to illustrate the putative stages by which the various lakes had drained into each other, and later drained away altogether when the mysterious barriers disappeared. This showed clearly that Lauder's goal, like MacCulloch's, was *geohistorical*: he was trying to reconstruct the sequence of concrete events by which the topography of the area in the distant past had eventually become the topography of the present.[19]

Darwin prepared himself for his field trip to Glen Roy by making a careful study of MacCulloch's and Lauder's papers. He also wrote himself an agenda,

17. Fig. 33.3 is reproduced from MacCulloch, "Parallel Roads of Glen Roy" (1817) [Cambridge-UL: Q365.b.12.4], pl. 16, engraved from MacCulloch's own drawing, made with the usual vertical exaggeration (but in reality the topography is nonetheless dramatic). MacCulloch's and other early interpretations are summarized, and Darwin's is analyzed in detail, in Rudwick, "Darwin and Glen Roy" (1974), which also reproduces (98) the similar landscape view published in Darwin's paper (see below); the following brief account is necessarily much simplified.

18. Fig. 33.4 is reproduced from Darwin, "Parallel Roads of Glen Roy" (1839) [Cambridge-UL: T340:1.b.85.130], pl. 1, read at the Royal Society on 7 February. MacCulloch, "Parallel Roads" (1817); Lauder, "Parallel Roads" (1821). Ben Nevis (1344m), the highest hill in Britain, is southwest of the area shown; the town of Fort William at its foot is on the shore of Loch Linnhe.

19. Lauder, "Parallel Roads of Lochaber" (1821), pl. 8 [reproduced in Herbert, *Darwin geologist* (2005), 266], shows his paleogeographical reconstructions of the area. His engraved views of the Roads, based on his own sketches, are much cruder than MacCulloch's; two are reproduced in Herbert, 264, and Rudwick, "Visual language for geological science" (1976), fig. 19. Lauder, significantly, was best known to the reading public as a *historian*, an antiquarian, and the author of historical novels in the manner of Walter Scott.

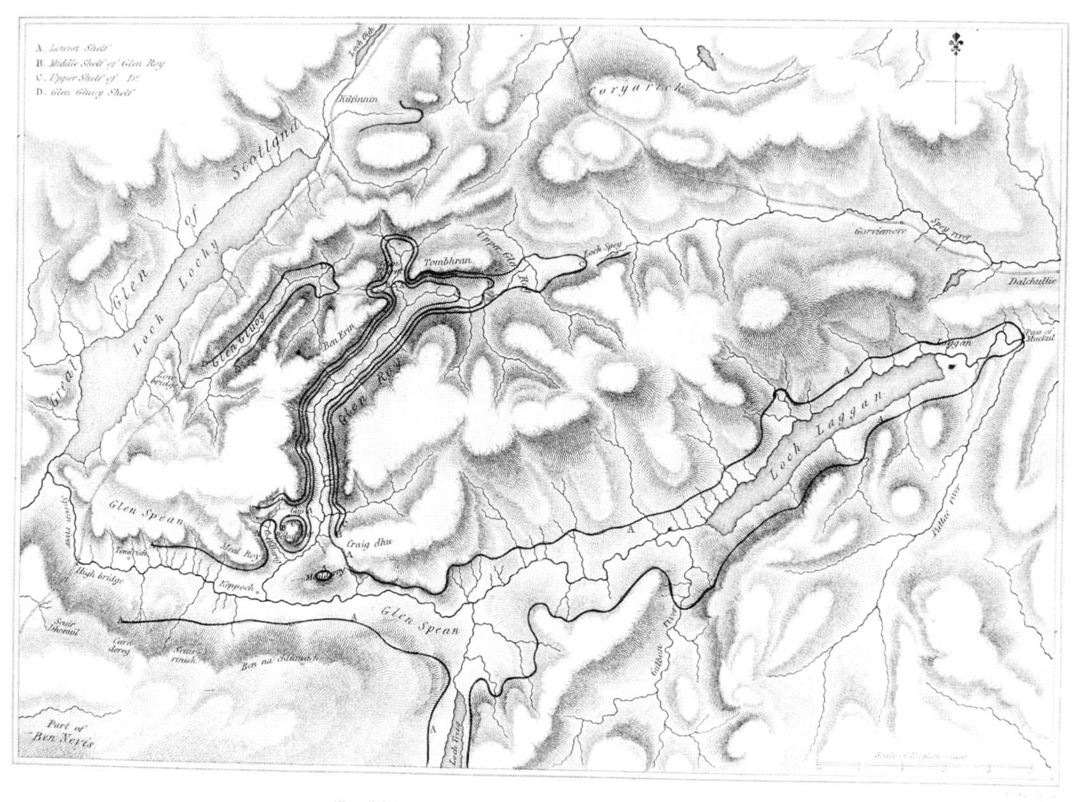

Fig. 33.4 Darwin's map of the "Parallel Roads" in Glen Roy and adjacent valleys, showing how the upper two terraces (*B*, *C*) were confined to Glen Roy itself (left center), whereas the lowest terrace (*A*) also ran all round Glen Spean to the south (and into another tributary valley on its south side, containing Loch Treig); Darwin determined by barometric measurement that the terrace (*D*) in Glen Gluoy, just west of Glen Roy, was slightly higher than the highest one (*C*) in the latter, so that in total there were localized terraces at *four* distinct levels (at 1278, 1266, 1184, and 972 feet above sea level). There are no terraces in the Great Glen (top left), which, however, does contain Loch Lochy and other freshwater lakes such as Loch Ness, and ends (beyond the map to the southwest) in the marine inlet of Loch Linnhe. MacCulloch and Lauder had both interpreted the terraces as the successive shorelines of former lakes, presumably dammed by temporary barriers of which—puzzlingly—no traces remained. In contrast, Darwin argued that they represented successive shorelines of former fiords or inlets of the sea (like the existing Loch Linnhe), marking pauses in the gradual elevation of the whole landmass; but he then had no clear explanation for their localized extent. The scale is of five miles (about 8km). (By permission of the Syndics of Cambridge University Library)

which shows that he intended to look for evidence to support the *marine* interpretation that MacCulloch had rejected, but which Lyell had adopted for the terraces near Coquimbo. For example, he reminded himself to search for seashells among the pebbles on the ancient beaches, and for barnacles wherever there was solid bedrock on the terraces. But in the field he found none of either, and no other clear evidence that the sea had ever been at these great heights; unambiguous raised beaches, like the one on which Serapis stands (Fig. 8.4) and the one that Lyell had studied in Sweden, were well known around Scottish coasts, but only up to about 100 feet (30m) above sea level. Darwin explained this away as a failure of preservation. He also explained away, in the same manner, the even more serious objection that the terraces themselves were preserved only in these specific valleys, and not more generally on hillsides at the same altitudes elsewhere in the Highlands. Imperfect and differential preservation was the panacea for all Darwin's explanatory ailments, just as it had been for Lyell when faced with the almost complete lack of mammalian fossils in the Secondary formations (§17.2, §24.3). Furthermore, he had to give up all the explanatory advantages of Lauder's discovery of cols that could have been the overflows of temporary lakes; Darwin dismissed them as mere "col-coincidences" that had been nothing more than shallows connecting one former arm of the sea with another (without explaining why each terrace was found only on one side). Nonetheless, Darwin found his fieldwork exhilarating: as he told Lyell after he got home, "I enjoyed five days of the most beautiful weather, with gorgeous sunsets, & all nature looking as happy, as I felt". He had found it "far the most remarkable area I ever examined . . . I can assure you Glen Roy has astonished me". He was certainly being frank, however, when he also told his mentor, "I have fully convinced myself, (after some doubting at first) that the shelves are sea-beaches, —although I could not find a trace of a shell, & I think I can explain away [*sic*] most, if not all, the difficulties".[20]

In view of the explanatory problems it posed, it is not surprising that it took Darwin six weeks of hard work to write up his paper on Glen Roy, or that he called it "one of the most difficult and instructive tasks I have ever engaged on". But its explanatory difficulties were matched by its theoretical importance to him; it was no accident that he put the task into top priority. If he could convince other geologists that the terraces were marine beaches, it would show that the Scottish Highlands had been elevated, with pauses, by at least 1300 feet (400m) in the geologically very recent past; and the precise horizontality of the terraces would prove that the underlying crustal plate had risen in an astonishingly equable manner. "I cannot doubt", he told Lyell, "that the molten matter beneath the earth's crust possesses a high degree of fluidity, almost like the sea beneath the Polar ice": Scotland, and perhaps the whole of northern Europe, was floating on "liquid rock", rather like one of the ice floes that he must have seen floating around Tierra del Fuego. The paper was so important to Darwin that he submitted it not to the Geological but to the Royal Society, which elected him a Fellow a week later. The paper was read early in 1839; Sedgwick, acting as the Society's referee, reported that he was inclined to accept its argument, and that anyway the paper deserved publication. Darwin's very first attempt at writing out an extended scientific argument was published later that year in the *Philosophical Transactions*. Its subsequent fate, not anticipated by Darwin, Lyell, Sedgwick, or anyone else, belongs later in this narrative (see §35.3).[21]

20. Darwin to Lyell, 9 August 1838 [*CCD* 2: 95–99]; his brief manuscript agenda [Cambridge-UL: DAR 50] is printed in Rudwick, "Darwin and Glen Roy" (1974), 179–81. In 1845, a Scottish naturalist discovered a similar col (and putative overflow) in an inconspicuous side valley on the level of terrace (*B*), thereby eliminating the anomaly and further strengthening a lake interpretation: Rudwick, 140–45.

21. Darwin, "Parallel Roads" (1839), read on 7 February; Darwin to Lyell, 14 September 1838 [*CCD* 2: 104–8]; Sedgwick to Royal Society, 26 March 1839 [London-RS: RR.1.46, 88–90; printed in Rudwick, "Darwin and Glen Roy" (1974), 181–83, but not in *CCD*]. In his report, Sedgwick urged Darwin—a naturalist with a curiously under-developed visual sense—to include a map (Fig. 33.4), without which (or even with which) his convoluted prose was and is quite difficult to follow.

33.5 CONCLUSION

The question of crustal elevation was a crucial part of Lyell's campaign to show that actual causes were adequate to explain *all* the traces of past geohistory. He might have welcomed the alleged slow rise of the land around the Baltic as evidence for his gradualist view of elevation, but he had doubted its reality because it was not accompanied by any earthquakes. However, once he saw the evidence for himself and became convinced that the effect was genuine, he used it to enlarge his explanatory repertoire. Landmasses could evidently rise insensibly slowly, even without earthquakes; given enough time, continents could rise from the ocean depths. But the elevation of mountain ranges was much more problematic.

The Andes were the focal point of this focal problem. Élie de Beaumont had suggested that the range had been upheaved suddenly and all at once, whereas Lyell asserted that it had risen gradually over a long span of time. The 1835 earthquake in Chile was taken by Lyell to support his interpretation. Its effect—a small rise of the land even better attested than that of 1822—was witnessed by the naval officer Fitzroy and his companion, the ship's unofficial naturalist Darwin. Darwin returned from the voyage of the *Beagle* a keen geologist and a convinced Lyellian. In papers read at the Geological Society he developed an ambitious global geotheory, according to which crustal plates of continental dimensions were in slow and continuous movement, either up or down, as revealed respectively by raised beaches and by coral reefs and atolls. The steady-state character of Darwin's geotheory, and its total dependence on actual causes, marked him out as a true "uniformitarian", Lyell's first and only full-blooded convert.

The crucial test case of Darwin's Lyellian geotheory was his interpretation of the famous Parallel Roads of Glen Roy in Scotland. His predecessors MacCulloch and Lauder had both inferred that these terraces were ancient beaches formed when there were temporary lakes in the relevant valleys, though they had conceded that—puzzlingly—there were no traces of the barriers that must have impounded the lakes. Darwin, on the contrary, took his cue from the similar (but much lower) terraces he had seen in Chile, which Lyell had interpreted as marine beaches dating from a time when South America had not yet risen to its present level. Darwin applied the same interpretation to the Scottish terraces, although he had to explain away the fact that—puzzlingly—they bore no trace of a marine origin and were confined to a few valleys in one part of Scotland. This unresolved puzzle, and therefore unsatisfactory test case, was local, but its implications were global.

Although Darwin's geotheory was in itself causal rather than geohistorical, his concept of crustal plates—moving vertically, unlike the huge horizontal movements of modern plate-tectonic theory—did embody a specific reconstruction of the most recent portion of geohistory. It was therefore relevant to the most awkward problem in Lyell's entire geohistorical project. This did not lie in the deepest past, although his "metamorphic" theory for eliminating any alleged "vestige of a beginning" represented by the Primary rocks was indeed failing to gain much support (§27.4). What gave him much greater cause for concern was the part of geohistory closest to the present, which—paradoxically—was still surprisingly obscure. For all Lyell's efforts to demystify it, the ghost of Buckland's early "diluvial" theory (§6.1; *BLT* §10.5) still haunted the shadowy borderland between present and deep past. Although the biblical Flood had become almost irrelevant to geological debate (§23.1), some kind of far earlier "geological deluge" or mega-tsunami still seemed a highly plausible possibility; anyway, there were still some very puzzling "diluvial" phenomena to be explained. These are the subject of the next chapter.

CHAPTER 34

Explaining erratics (1833–40)

34.1 EXTENDING THE GEOLOGICAL DELUGE

The diluvial theory, like any other important theory in any of the sciences, had not stood still over the years. It had been modified and improved, to take account not only of theoretical objections but also of new empirical evidence. In the 1830s there were still three main classes of evidence for which diluvial explanations continued to be invoked, but they were not of equal significance.

The first comprised the *erosional* features that seemed to many geologists to be inexplicable in terms of ordinary actual causes. The question of the excavation of valleys had pitched "diluvialists" against "fluvialists", but the debate had been inconclusive (§20.3). In fact the most significant outcome had been a tacit agreement—though Sedgwick, for example, had made it explicit (§20.4)—that it was probably futile to look for any singular (or monocausal) explanation, since valleys were so diverse in their topographical forms and in their relation to the structure of the underlying rocks. So, for example, even diluvialists now conceded that some of the valleys in Auvergne might well have been excavated by the long-continued erosive action of the streams and rivers that still flow in them, as Scrope had claimed a decade earlier, and French naturalists half a century before that (§15.1; *BLT* §4.3, §4.4). Yet the puzzling topography of the drainage basin of the Thames, for example, might demand some kind of exceptional "diluvial" cause, as Buckland and Conybeare had argued (§13.2, §20.3; *BLT* §10.5). Above all, violent "aqueous currents" *of some kind* seemed the most plausible explanation for the broad but deep valleys of distinctively U-shaped profile found in the Alps and other less mountainous regions (Fig. 13.1). Erosion by the puny streams that now flow in such valleys—in some cases visibly carving out a small gully of contrasting V-shaped profile along the floor of the valley—was clearly not the answer to the puzzle, however vast the span of time allowed (Fig. 20.1).

The second class of evidence for a diluvial event comprised *depositional* features, notably the peculiar "*boulder clay*" that blanketed large areas of Scandinavia, lowland Britain, and other parts of northern Europe. This had no obvious parallel among deposits visibly being formed in the present world, and its origin remained very puzzling. But the third class of evidence—strictly speaking, it combined subclasses of the other two—was by far the most telling. Large *erratic blocks* (Figs. 6.4, 13.4) were often underlain by distinctive surfaces of *scratched bedrock*, smoothed and sometimes even polished. These were the most difficult features to explain, without invoking some kind of unusual process or exceptional event in recent geohistory, and they therefore played by far the most important roles in the debates about the geologically recent past.

In the 1830s, the general agreement among geologists—including Buckland—that the "diluvium" had not, after all, had anything to do with the *biblical* Flood (§23.1, §29.1) did nothing to lessen the *geological* puzzle posed by the diluvial features, and most of all the erratics and scratched bedrock. Twenty years earlier, James Hall had considered how a submarine mega-earthquake might have produced a mega-tsunami, a body of water massive and violent enough to have moved large boulders past the future site of Edinburgh, scratching the underlying bedrock on the way. At the same time and in much the same way, von Buch in Berlin had considered how a catastrophic Alpine flood might have moved much larger erratic blocks from the Alps all the way to the Jura (Fig. 6.4; *BLT* §10.2). But, as geologists of several nations jointly demonstrated during the 1820s, even this was dwarfed by the scale on which erratics composed of the granite and other Primary rocks of Scandinavia had somehow been carried across the Baltic and dumped all over the north German plain and its extensions westward into the Netherlands and eastward into Russia (§13.3).

A vivid reminder of the sheer scale of the problem, which von Buch would have seen whenever he was back in Berlin, was the vast ornamental granite bowl that had been set up in the center of the Prussian capital: it had been carved from just a part of one of the large erratics strewn across the plains of Brandenburg, and it was composed of a granite that could be traced back to its source on the far side of the Baltic. Geologists living in or visiting St. Petersburg had a similar reminder in the huge granite erratic—derived from somewhere in the direction of Finland—that had been used as a plinth for the famous equestrian statue erected by the empress Catherine, commemorating the founder of the city Peter the Great.[1]

Further descriptions of the northern erratics continued to appear in the 1830s, reinforcing a sense of the astonishing magnitude of the putative diluvial event. For example, in 1836 Georg Gottlieb Pusch (1790–1846), a mining engineer based in Warsaw, described and mapped a spread of erratics stretching from his home city south toward Krakow; he treated this as evidence of "a colossal flood penetrating with immense velocity [*mit ungeheurer Geschwindigkeit*]" even further than previous reports (see Fig. 13.7, no. 9).[2]

A new degree of precision was brought to all these field studies of erratics by the Swedish geologist Nils Gabriel Sefström (1787–1845), who taught at the mining school in Falun. Extending Brongniart's earlier work in Sweden (Fig. 13.6), he gave the Academy of Sciences in Stockholm a detailed description of the bedrock surfaces scratched with "furrows" [*räfflor*] oriented in very specific directions, which he had mapped in detail in his home area and then throughout southern Sweden (Figs. 34.1, 34.2).[3]

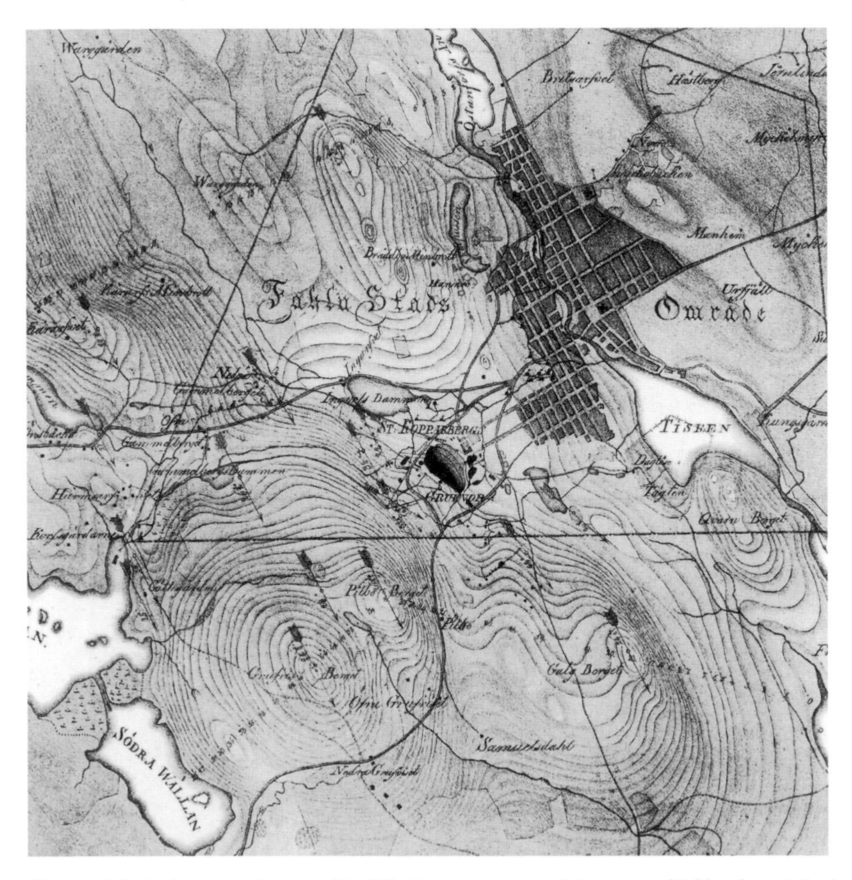

Fig. 34.1 Sefström's large-scale map of the hills above the copper-mining town of Fahlun [now Falun] in central Sweden, showing (by long tailed arrows) the precise orientation, at specific points, of the "furrows" or scratched bedrock surfaces that he attributed to stones carried along in an ancient "flood", which had swept violently across the country, undeflected by the hilly topography. The area shown here is about 6km square. (By permission of the Syndics of Cambridge University Library)

1. The Berlin bowl, produced by a feat of stone turning on an unprecedented scale, was placed in the Lustgarten in the 1820s; happily it survived the devastation of Berlin in the Second World War, though weathering has removed its original polish. The St. Petersburg erratic plinth had been made known to the Geological Society by Strangways (§13.3).

2. Pusch, *Beschreibung von Polen* 2 (1836), 588; his description of the "diluvial" deposits (567–89) came near the end of a comprehensive account of the geology of Poland. The erratics were mapped in his accompanying *Atlas von Polen* (1836), pl. 1, no. 28 [*Urfelsblok Ablagerungen*]: see index on pl. 2.

3. Fig. 34.1 is reproduced from Sefström, "Undersökning af de räfflor" (1838) [Cambridge-UL: T340:7.d.1.94], part of pl. 7 (dated 1836); Fig. 34.2 is reproduced from his "Untersuchung über die Furchen" (1838) [Cambridge-UL: CP352.c.16.131], pl. 5; this was redrawn from pl. 8 (also dated 1836) of the Swedish paper, but made the direction symbols much clearer. See also Frängsmyr, *Upptäckten av istiden* (1976), 73–76, and "Glacial theory" (1985).

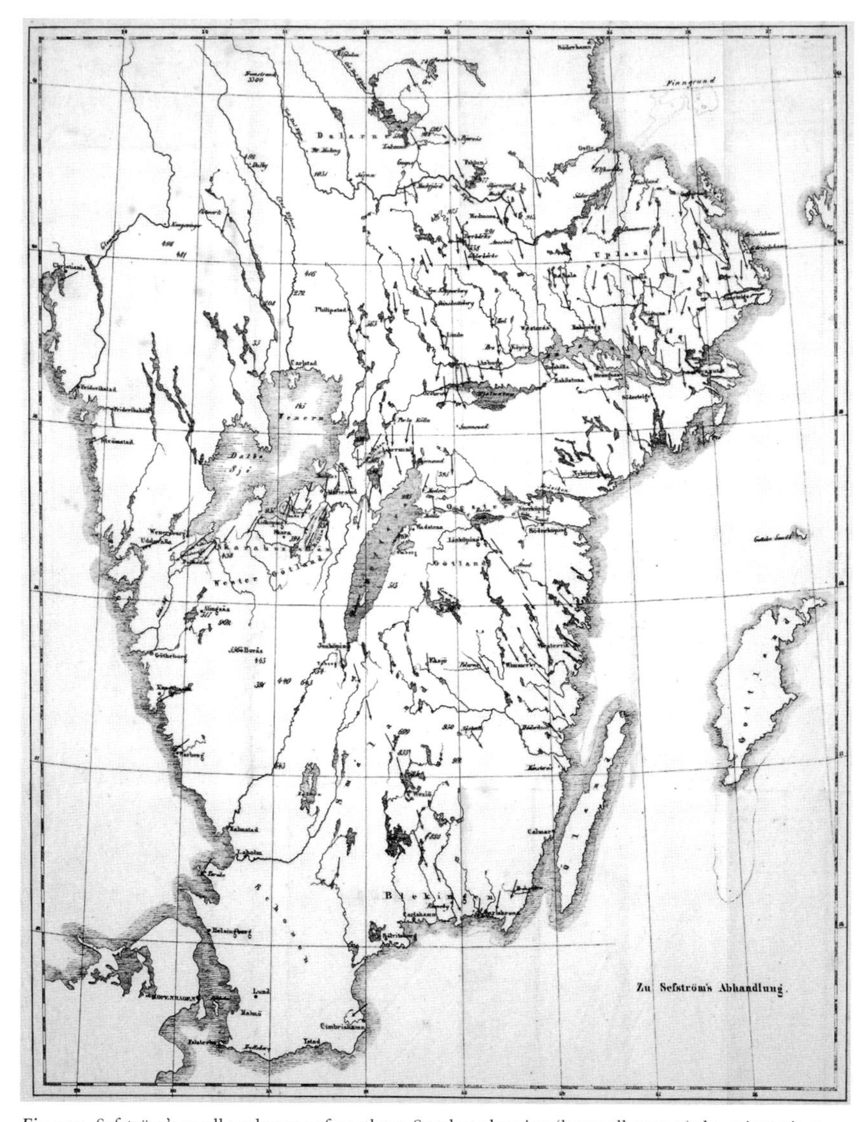

Fig. 34.2 Sefström's small-scale map of southern Sweden, showing (by small arrows) the orientations of scratched bedrock surfaces at many different localities (Fahlun is near the northern edge of the map, west of Gefle [now Gävle] on the east coast). They implied that the inferred southward-flowing "petridelaunic" or stony flood had fanned out from its origin somewhere still further north. On this map the total spread of the arrows, and of the putative deluge, extended about 700km from north to south; but Sefström claimed that it had extended still further south, beyond the Baltic, across the north German plain (see Fig. 13.7). (By permission of the Syndics of Cambridge University Library)

Sefström took it for granted that the event that had scratched the bedrock on such a vast scale must have been an *aqueous* current of some kind, or what other geologists had long called "the geological deluge". He inferred that it had been violent; that it had continued for some considerable time; and that it had been at least 1500 feet (500m) in depth, for it had swept right over the tops of hills up to that height. He gave it the impressively Classical name of "petridelaunic flood" [*petridelauniska floden*], because it had carried the "little rolled stones" that had scratched the bedrock en route before ending up in the diluvial deposits. After presenting his paper in Stockholm he traveled extensively through the German lands to the south of the Baltic, finding what he took to be further traces of the same event. An excerpt from his paper was published promptly in Sweden, and translated into German and English, so his research became known throughout Europe even in advance of its full publication in 1838. It certainly drew attention to the sheer scale of the scratched bedrock in Scandinavia.[4]

Sefström distinguished his "petridelaunic flood" of rolling stones [*rullsten*] from whatever mysterious agency had shifted the larger erratics [*jordsten*]. This separated what other geologists regarded as two closely related phenomena, but Sefström would have realized that large erratics posed far greater explanatory problems than small stones. Von Buch, for example, had recognized the physical difficulties involved in attributing erratics to any merely watery flood, however violent. After the catastrophic natural dam burst in the Val de Bagnes in 1818, he had used that unquestionably actual cause to improve his earlier hypothesis: a similar violent torrent of dense *liquid mud*, but on a far larger scale, might have been able to move far larger blocks of rock through far greater distances (*BLT* §10.5). But although a huge flow of liquid mud—a kind of subaerial turbidity current—might conceivably have swept erratics from the high Alps down the Rhône valley and all the way to the Jura, the same explanation could not easily explain the erratics that had somehow been transported from Scandinavia across the Baltic: the mountains in Scandinavia seemed inadequate to have initiated the flow, and anyway it was not clear how the blocks could have been moved *uphill* out of the Baltic basin and on to the north German plain, by any process powered simply by gravity. There was no mistaking the continuing ambition of geologists, led by von Buch, to find a physically plausible (and of course purely natural) way to account for the erratic distribution of erratic blocks. But it remained quite unclear whether any kind of huge and violent "flood" or "deluge", even one of dense mud rather than pure water, was an adequate explanation.[5]

34.2 ERRATICS AND ICEBERGS

The only moderately plausible alternative to *some* kind of violent "aqueous current" (muddy or not) was derived from what had been suggested some thirty years earlier by the Prussian naturalist Wrede, namely that the German erratics might have *floated* across the Baltic, embedded in and given buoyancy by the ice floes that were found drifting there when the frozen winter sea began to break up in spring (*BLT* §10.2). Wrede had chosen as his best example a huge erratic that was still lapped by the waters of the Baltic, and that might have drifted in on a large ice floe and been dropped or grounded there at a time of higher sea level. But he had explained the latter in terms of a change in the earth's axis of rotation, a style of geotheory that was already going out of favor. And anyway, as de

4. Sefström, "Grossen urweltlicher Fluth" (1836), was a summary of his full account in "Undersökning af de räfflor" (1838); the latter (with illustrations) was followed immediately by his "Sednare observationer" (1838), describing his German travels in the summer of 1836.

5. Gard, *Débâcle du Giétro* (1988), includes much documentation on the Val de Bagnes disaster, with superb reproductions of contemporary illustrations and modern photographs of the site; I regret that I did not know of this work (for which I am indebted to Jean-Paul Schaer) in time to cite it in *BLT*.

BOULDERS DRIFTED BY ICE ON SHORES OF THE ST. LAWRENCE.

Fig. 34.3 "Boulders drifted by ice on shores of the St. Lawrence" at latitude 46°, looking upstream near Richelieu Rapid between Montreal and Quebec: an engraving published in Lyell's *Principles* in 1840, based on a sketch made in the spring of 1836 by a British naval officer serving in Canada. The pool *a* on the beach (behind the boat) had been occupied the previous winter by the seventy-ton block *b* (to the right of the pool), which had moved several feet when the ice broke up. Lyell inferred that all these granite blocks had been shifted over time by the same means; the underlying bedrock was limestone and slate, so they had evidently been moved from elsewhere and were true erratics. He claimed that such examples proved that even erratics now far from any water could have been transported by the same actual cause, if the geography in the past had been different.

Luc had argued, there was no independent evidence that the north German plain had been submerged below sea level in the geologically recent past. So for many years geologists had given little attention to the possible role of floating ice—in the form of either floes or bergs—in the transport of erratic blocks (§13.4).

However, in this respect as in so many others, Lyell emerged as the exception. In his original *Principles*, he had imagined the Alpine glaciers loaded with future erratics, breaking off into icebergs floating on any temporary lake dammed by glacial ice (as in the Val de Bagnes); when the dam eventually burst (again as in the historic Alpine disaster), "the flood would instantly carry down the icebergs, with their burden, to the low country at the base of the Alps". He repeated this rather ad hoc thought experiment in later editions, but must have had a growing sense of its inadequacy. For in his sixth edition (1840) he responded to the increasing attention that other geologists were giving to the phenomena of glaciers (see below) by adding a new chapter. Here he invoked flotation by ice floes or icebergs as a fully adequate explanation for the strange distribution of erratics, not only in the Alps but everywhere, provided that adequate amounts of subsequent crustal elevation were conceded. He presented this as an impeccably actual cause, by describing numerous examples of large blocks of rock being seen to move as winter ice covering frozen rivers or even coastal water broke up in spring. One case was particularly significant, in that the St. Lawrence in Canada

Fig. 34.4 A sailor's sketch of a twelve-foot block of rock embedded twenty feet above sea level in an Antarctic iceberg. This wood engraving was published in 1839, with a comment by Darwin noting its importance for the debate about the origin of the large "erratic blocks" scattered on land around many parts of Europe and the Americas. Darwin, following his mentor Lyell, inferred that many erratics now found on land could have been transported from the glaciers of polar or mountain regions on icebergs such as this, at an earlier time of higher relative sea level. (By permission of the Syndics of Cambridge University Library)

was on the same latitude as central France, yet had a present climate that was quite different (Fig. 34.3).[6]

Invoking a formerly higher relative sea level—to account for major differences of geography, and hence of ocean currents and local climates—was for Lyell no problem at all, since he was convinced on other grounds that segments of the earth's crust, whether continents or ocean floors, were forever either sinking or rising (§22.1). And drifting icebergs, no less than coastal ice floes, were for him an admirably actual cause, since sailors reported that from time to time they saw icebergs carrying large blocks of rock, which would obviously drop to the sea floor and become the erratics of the future, as soon as the icebergs melted (Fig. 34.4).[7]

Lyell noted how, in the north Atlantic, icebergs from the Greenland glaciers (Fig. 7.4) were sometimes sighted by—and were of course a serious hazard to—transatlantic shipping as far south as the Azores, and in the southern hemisphere they were seen as far north as the Cape: hence "the area over which the effects

6. Fig. 34.3 is reproduced from Lyell, *Principles*, 6th ed. (1840), 1: pl. 4, drawn by Lieut. Bowen and explained on 372–73; the chapter is at 369–84. See also *Principles* 3 (1833), 148–50; and for example 4th ed. (1835) 4: 59–61. He may have been sent the picture after another naval officer described similar cases from the same area, and also rock-bearing icebergs probably originating in the Greenland glaciers: Bayfield, "Transportation of rocks" (1836), read at the Geological Society on 6 January.

7. Fig. 34.4 is reproduced from Enderby, "Discoveries in the Antarctic Ocean" (1839) [CambridgeUL: Q690.c.8.9], 526, printed immediately before Darwin, "Rock seen on an iceberg" (1839); the latter estimated that the berg was about 300 feet (90m) in height (mostly underwater, of course). It was sketched (very crudely) by the mate of a schooner that was then some 1400 miles from the eponymous Enderby Land, the nearest known coast of the still barely known Antarctic continent; but Enderby guessed that there might be land no more than 300 miles (480km) away. (His reported coordinates show that in fact the ship was about 750km north of the then-undiscovered Knox Coast in Wilkes Land.) See also Mills, "Darwin and the iceberg theory" (1983).

of moving ice may be experienced, comprehends a large portion of the globe". He had already argued that quite minor differences in the elevation of parts of the earth's crust, in the geologically recent past, could have generated substantial differences in the distribution of land and sea. For example, much of low-lying northern Europe might have been a shallow sea; and this in turn could have generated substantial differences in *local* climates and ocean currents, without any change in the condition of the earth as a whole. So he now pointed out that under such a former climatic regime in the north Atlantic region, icebergs might have drifted in abundance even further south than at present, and might, for example, have dropped their cargo of erratics all over what are now low-lying land areas such as southern Sweden, northern Germany, the Netherlands, and much of Britain. Lyell's own research on the continuing slow rise of the land in Sweden (§33.1) fitted neatly into this drift theory; the North Sea and the Baltic, as shrunken remnants of these more extensive former seas, would have been no obstacle at all. At a pinch and with some awkwardness, Lyell even included the Alpine erratics under the same explanation, although this entailed postulating a much greater degree of subsequent crustal elevation in that part of Europe, to account for the erratics found high up on the Jura. In a paper on eastern England, Lyell also suggested that Superficial deposits such as boulder clay (for which he proposed using the Scottish farmers' term "*till*") should therefore be collectively called "*drift*". This word was, of course, no less theory-loaded than Buckland's "diluvium", which it was tacitly designed to replace, thereby eliminating any hint of the dreaded Deluge.[8]

Darwin's work in the Scottish Highlands showed likewise the lengths (or, more literally, the heights) to which Lyell's explanation had to be taken, in order to explain the distribution of erratics. Darwin used it to account for the emplacement of those he saw in Lochaber at altitudes up to 2200 feet (670m), well above even the highest of the terraces in Glen Roy (§33.4). That was the minimum depth to which Scotland would have had to be submerged in the geologically recent past—and the height to which it must have been elevated subsequently—if the erratic phenomena were to be fully explained by Lyell's theory of drifting ice floes and icebergs. While he was in Chile, Darwin had seen less ambiguous evidence of Tertiary elevation up to 1300 feet (§33.2), and evidently found it no great strain to infer that another crustal block might have risen even further during a much briefer span of time. Other geologists were more skeptical, above all because there was no obvious independent evidence for such dramatic changes of sea level in Europe during Pleistocene (post-Pliocene) time alone. So there were plenty of problems about Lyell's theory; it was an explanation that most other geologists found less plausible than that of violent "aqueous currents", mega-tsunamis, or mud flows, though both theories were acknowledged to be less than fully satisfactory.[9]

34.3 THE RECONSTRUCTION OF MEGA-GLACIERS

A third way of explaining erratics began as a theory of local scope and modest pretensions. Back in 1817 the then newly founded and peripatetic Sociéte

Helvétique (soon to be the model for the German Association and thence also for the British) had offered a prize for research on climate change in Switzerland, prompted by civic concern at recent climatic deterioration. After the Val de Bagnes disaster of 1818 (*BLT* §10.5) the topic was redefined to deal specifically with the recent extension of the Alpine glaciers and the impact of the worsening climate on pastures and forests. In 1822, the prize had been awarded to Ignaz Venetz (1788–1859), the civil engineer responsible for public works in the canton of Valais (in which the Val de Bagnes lies). From his fieldwork in the Val de Bagnes and other valleys, Venetz had concluded that the extent of the Alpine glaciers had fluctuated within historical times, and that they were currently near their maximum on the latest cycle; he estimated that the "*moraines*" or ridges of rocky débris found at the snouts of glaciers—obviously derived from rocks embedded in the ice and dropped where it melted—probably dated only from the past few centuries. But an observant local huntsman, Jean-Pierre Perraudin (1767–1858), pointed out to him that there were similar moraines much further down the valleys, and higher up on their flanks, often now concealed in forest. Echoing Buffon's classic phrase, Venetz inferred that these dated "from an epoch *that is lost in the night of time*"; but he assured his compatriots that the glaciers were very unlikely to return in the foreseeable future to "the gigantic height at which we find so many vestiges".[10]

Venetz's prizewinning work had remained unpublished until, in 1829, he summarized it at the annual meeting of the Société Helvétique, held that year, appropriately, at the famous monastic refuge at the top of the Great Saint-Bernard pass across the Alps, with glacial scenery all around it. The published report of the meeting showed that he now set his own local observations in the much wider context of what had been reported as "diluvial" traces elsewhere in Europe:

> Mr Venetz read a paper on the extension that he has inferred that glaciers formerly had, and on their retreat to their present limits. He attributes the masses of blocks of Alpine rocks that are scattered at various points in the Alps and the Jura, as well as in several regions in northern Europe, to the existence of immense glaciers that have since disappeared, and of which these blocks formed the moraines. He supports this hypothesis by citing several facts that he has observed in the neighborhood of glaciers in the Alps of the Valais.[11]

8. Lyell, *Principles*, 6th ed. (1840), 1: 377–84; "Boulder formation" (1840), read on 22 January. The term "Drift" was later adopted by the increasingly powerful Geological Survey of Great Britain, headed (after De la Beche's death) by Murchison, and it continued to be used on all the Survey's geological maps long after Lyell's theory of drifting icebergs was abandoned.

9. Rudwick, "Darwin and Glen Roy" (1974), fig. 6 (123), gives a graphical representation of the way that Darwin subsumed the erratics at high altitude, the famous "Roads" somewhat lower, and the uncontroversial "raised beaches" close to sea level, all under the explanatory model of a slowly rising crustal block.

10. Venetz, "Température dans les Alpes" (1833), 38, revised from his paper "written in 1821"; see Weidmann, "Ignace Venetz" (1972), Forel, "Jean-Pierre Perraudin" (1899), and Schaer, "Agassiz et les glaciers" (2000), 234–35.

11. Report of 15th meeting (1829) of Société Helvétique [*ASHSN* 1830: 31], quoted here in full. Discussions of the papers read were not recorded, but Weidmann, "Ignace Venetz" (1972), infers that this one was greeted with embarrassed silence.

In the much briefer report on the meeting in the Genevan *Bibliothèque Universelle*, a passing mention of Venetz's paper may have alerted geologists beyond Switzerland that he had attributed the Alpine erratics to "glaciers that have disappeared"; but at least his ideas became well known to his Swiss colleagues. Among those who had been at the meeting was the geologist Jean de Charpentier (1786–1855), the director of the salt mines at Bex, further down the Rhône valley from the Val de Bagnes. Charpentier was at first skeptical, but after some fieldwork with Venetz he became convinced that his friend was right. So, three years later, when Lyell visited Charpentier during his geological honeymoon (§25.2), the latter explained that he was now convinced that the distribution of the Alpine erratics was too orderly for von Buch's kind of "*grand débâcle*" of turbulent water or mud. According to Lyell, Charpentier said:

> Now many observers, & I among others, have found [that] the rocks along the levels where the great [erratic] blocks lie are worn smooth, just as the manner in which glaciers, descending with blocks upon them, wear the rocks. *Above* the lines of these rolled blocks the rocks are never so worn. If, therefore, it were only granted that the Alps were once *higher* than they now are, the rolled & angular huge blocks may have travelled on ancient glaciers.[12]

Fig. 34.5 The "Pierre des Marmettes" [Marmots' Stone], an erratic block of Alpine granite about ten meters in height, perched high up on the side of the Rhône valley, roughly opposite Charpentier's home at Bex. This was his prime example among the erratics that he described and pictured in *Essay on Glaciers* (1841), the book in which he later expounded in full his theory of the former vast extension of the Alpine glaciers. He identified its distinctive granite as having come from an area some 30km further up the valley. The huge size of the block combined with the view from it—out across the broad valley and up towards the high Alps in the far distance—was (and is) almost unsurpassed for impressing on any visitor the sheer scale of what needed explanation.

Charpentier was here adding another telltale feature to the moraines and erratics that Venetz had described. At the snouts of present glaciers, where they were melting, smoothed surfaces of bedrock could often be seen beneath the ice, and they were scored with scratches that were obviously due to the blocks of rock embedded in the ice at the base of the glacier: clearly it acted on the underlying bedrock much like a sheet of coarse glasspaper on a surface of wood in a carpenter's workshop (see Fig. 35.4). But Charpentier had found similar scratched rock surfaces in his own part of the Rhône valley, far from any of the present glaciers and high above the valley floor, though only up to the height at which morainic débris and erratic blocks were also found. If his inference—and Venetz's—was correct, an enormous glacier must once have filled the whole of the upper Rhône valley. It was a suggestion as spectacular as the erratics themselves (Fig. 34.5).[13]

Charpentier was probably uneasy about supporting Venetz in public unless he could also suggest a causal explanation for such a vast former extension of the glaciers. Significantly, his remarks to Lyell imply that he did not even consider the possibility that the climate might have been far colder than in the present world. That idea was, almost literally, inconceivable to him: like most other geologists, he took it for granted that on a steadily cooling earth the climate in the geologically recent past might have been slightly warmer than at present, but certainly not far colder. With a major change of *global* climate tacitly ruled out, Charpentier conjectured instead that the Alps themselves might have been much higher at the time; combined with a much heavier snowfall, they could then have generated far larger and longer glaciers. This explanatory deus ex machina drew Lyell's scorn, though he was probably too polite to disclose his skepticism to his host. But writing to Horner, his father-in-law, he commented, "you see they are beginning as usual, *in the last resort*, to seek a natural explanation". Lyell, as usual, could not see—or would not concede—that "catastrophist" explanations such as Buckland's original diluvial theory, no less than Charpentier's new idea, employed impeccably natural and indeed "actual" processes, although on a much larger scale than any witnessed in human history.[14]

12. Lyell (and Mary Lyell) to her father (Horner), 5 August 1832 [Kirriemuir-KH; quoted in Wilson, *Lyell* (1972), 369]; *BU* 41 (1829): Sci. et Arts, 263–64. Charpentier was the son of Werner's colleague Johann von Charpentier (1738–1805) and had been born in Freiberg; but he had migrated to Switzerland, was employed in the francophone canton of Vaud (adjacent to Valais), and used the French form of his name.

13. Fig. 34.5 is reproduced from Charpentier, *Essai sur les glaciers* (1841), pl. 1, explained on 126, 360, lithographed from a drawing by Christian Gottlieb Steinlen (1779–1847), an art teacher at Vevey well known for his landscapes around Lac Léman. The source of the erratic was located in Val Ferret at the eastern end of the Mont Blanc massif. By the time Charpentier published this image, a "*petit pavillon*" had been built on top, to enhance its attractions as a tourist viewpoint. The erratic and its pavilion are now, less romantically, in the parking lot of the local hospital, but it remains a superb viewpoint despite the modern sprawl of the town of Monthey far below. For any historical comprehension of the problem of erratics, the Pierre des Marmettes is a site of outstanding importance; surpassed only, perhaps, by the similar-sized Pierre à Bot above Neuchâtel (Fig. 6.4; *BLT* §10.2), which is far further from the same granitic source (see Fig. 34.6).

14. Lyell (and Mary Lyell) to Horner, 5 August 1832 [Kirriemuir-KH: quoted in Wilson, *Lyell* (1972), 369]. Charpentier, *Essai sur les glaciers* (1841), 243–44, makes it explicit that his initial skepticism about Venetz's ideas was due to the more obvious attractions of the rehabilitated theory of a cooling earth (§9.2). Rudwick, "Glacial theory" (1970), first suggested more generally that this was by far the most important reason for the reluctance of geologists in the 1830s to abandon the diluvial theory and invoke major glacial extension or even some kind of circumpolar or global "Ice Age" (see Chap. 35).

In 1833 Venetz's prizewinning paper was at last published in full by the Société Helvétique. This must have strengthened Charpentier's confidence in glacial extension; for the next year, when the Société met in Lucerne, he read a paper that generously promoted Venetz's ideas "on the present and past state of the Valais glaciers". However, there was still surprisingly little interest, so in 1835 Charpentier took a decisive step to propagate the theory beyond Switzerland. He sent his Lucerne paper to Paris, where it was published in the *Annales des Mines* with the title "On the probable cause of the transport of the erratic blocks in Switzerland"; German and English translations soon ensured that his ideas were known to geologists throughout Europe. Charpentier summarized Venetz's earlier work and added the grounds for his own endorsement of it. He argued that an ancient "*glacier-monstre*" (or "*mega-glacier*", as it will be called here) must have extended as far as the "*terrain erratique*" itself: that is, along the entire length of the upper Rhône valley and its many side valleys, over the site of the present Lac Léman (Lake of Geneva), right across the Swiss plain, and up onto the lower slopes of

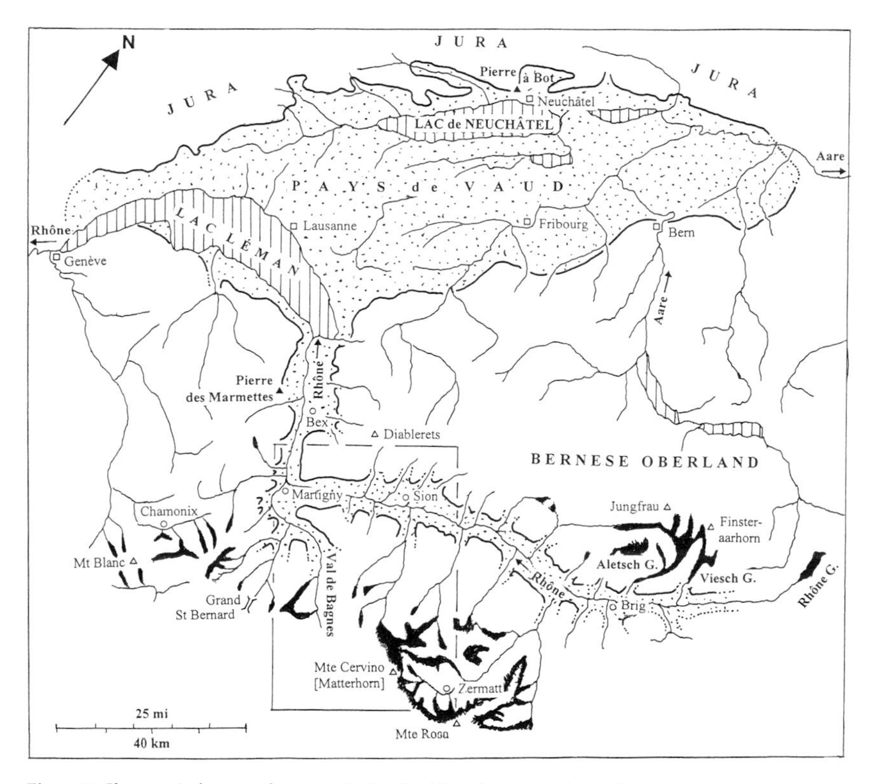

Fig. 34.6 Charpentier's map of western Switzerland (1841), redrawn in modern style (and greatly reduced in size) in order to clarify its main features; the small rectangle marks the part reproduced in Fig. 34.7. Existing glaciers are shown in solid black: even the Rhône and Aletsch glaciers (bottom right)—the latter the longest in the Alps—are dwarfed by the *mega-glacier* (stippled) reconstructed by Charpentier from the evidence of the lines of lateral moraine (thick lines) on both sides of the Rhône valley and beyond it. He inferred that the Rhône mega-glacier had flowed past Charpentier's home at Bex and the Pierre des Marmettes (Fig. 34.5), fanned out over the site of Lac Léman (the Lake of Geneva) and across the Pays de Vaud or Swiss plain, and lapped up on to the foothills of the Jura hills, where among the erratics was the famous Pierre à Bot (Fig. 6.4) above Neuchâtel. Charpentier's explanation of the Rhône erratics was far more satisfactory than von Buch's earlier one in terms of a huge and violent aqueous current or mud flow (*BLT* §10.2), though it left the *cause* of the vast glacial extension uncertain and controversial.

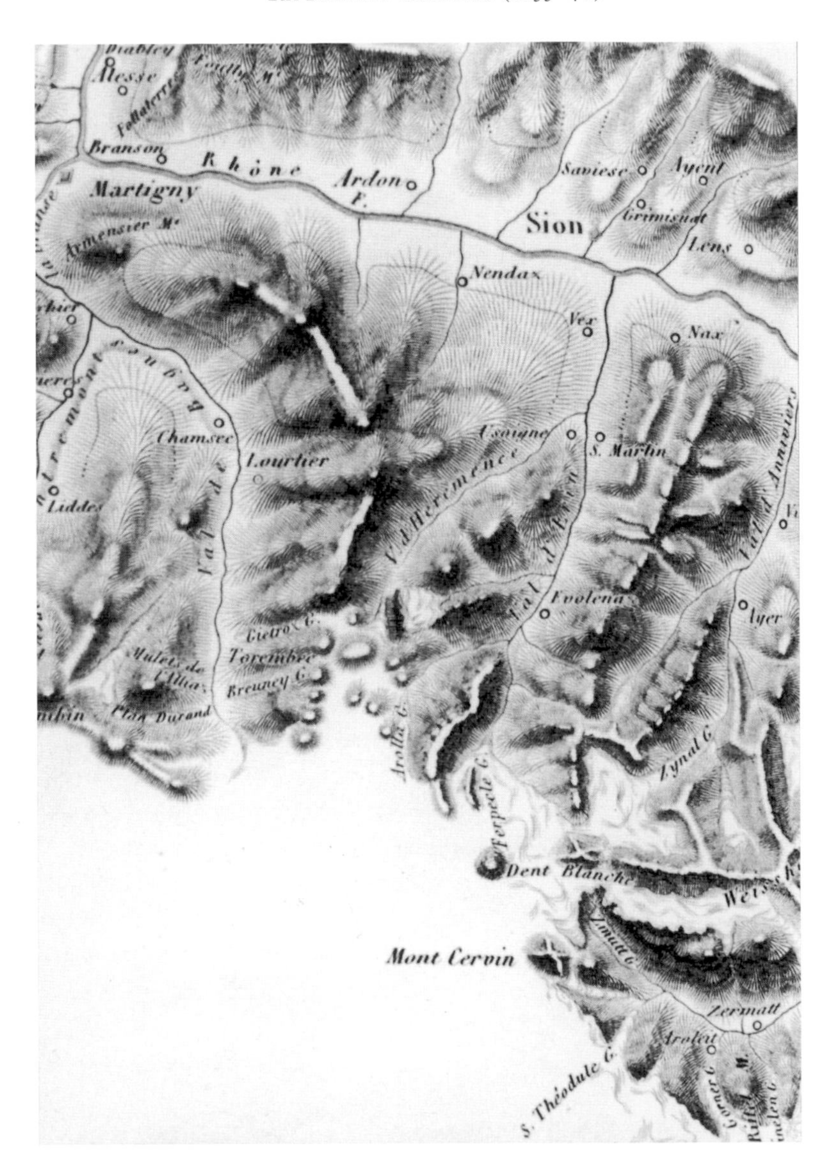

Fig. 34.7 A small part of Charpentier's map of western Switzerland (1841), illustrating the theory of glacial extension that he first expounded in public in 1834. It shows the style in which he depicted the mountains and existing snowfields and glaciers around the Matterhorn [*Mont Cervin*], and the moraines (marked as thin lines) along the flanks of a part of the upper Rhône valley and its side valleys (the Val de Bagnes, notorious for its catastrophic natural dam burst in 1818, is on the left).

the Jura hills. In effect, therefore, it was a tacit replacement of von Buch's earlier "diluvial" explanation of the same widespread phenomena (*BLT* §10.2). Charpentier conceded that "such a hypothesis seems at first sight implausible, shocking, even extravagant", but it was strongly supported by observational evidence (Figs. 34.6, 34.7).[15]

15. Fig. 34.6 is traced from, and Fig. 34.7 reproduced from a part of, the map in Charpentier, *Essai sur les glaciers* (1841). He mapped the *terrain erratique* only in the Rhône basin, claiming that any mega-glacier in the Aare basin (to the northeast) was likely to have been much less extensive (281–83); he treated the Italian territory to the south of the main Alpine watershed as terra incognita. The positions of the two famous erratics mentioned by name in the text are also marked on this redrawn version of his map (though not on the original). Charpentier's theory was first published (without illustrations) in "Transport des blocs erratiques" (1835), read at Lucerne on 29 July 1834 (quotation on 226). Venetz, "Température dans les Alpes" (1833), was his long-delayed paper, written in 1821. I once had the great good luck, while flying high over the Alps in exceptionally clear winter weather, to see the entire extent of Charpentier's mega-glacier accurately replicated or simulated by a layer of low cloud, with the snow-covered Alps and Jura rising above it, and the existing real glaciers plunging into it.

Charpentier distinguished clearly between his geohistorical reconstruction of the vanished mega-glacier that had filled the upper Rhône valley and the Swiss plain, and any causal or earth-physics explanation of this astonishing glacial extension: the former was based on detailed field evidence of the earlier presence of glacial ice, whereas the latter was necessarily speculative. Charpentier himself sought the cause in what he called "one of the best proved and demonstrated facts [*sic*] in geology". Following Élie de Beaumont's ideas (§9.4), a buckling of the earth's crust—the result of slow cooling and shrinkage at great depth—could have caused a sudden "epoch of elevation", which might have raised the Alps to Andean or Himalayan heights, generating the mega-glacier that had left such widespread traces. At a later time, however, the cracks and cavities left in the crust by the upheaval would have caused the range to collapse to its present height in "a kind of contraction [*tassement*]", which in turn would have caused the glaciers to shrink to their present modest size. Charpentier's theory was based impeccably on the actual causes of present glaciers, and their evident ability to move large blocks of rock and scratch the underlying rock surfaces; but the cause of their former huge extension and later shrinkage remained hypothetical.[16]

Nonetheless, Charpentier's sensational reconstruction of a mega-glacier in the upper Rhône valley could potentially be extended without difficulty to the rest of the Alps. Von Buch had briefly noted several other areas on the north side of the main watershed where similar spreads of erratics were known (*BLT* §10.2), and these could now be converted in the mind's eye from "aqueous currents" or mud flows into further mega-glaciers; one of them, for example, would have extended from Mont Blanc down the Arve valley almost to Geneva, judging by the trail of erratics that Saussure had described half a century earlier. The similar erratics on the south side of the Alps, some of which De la Beche had mapped and illustrated near Como (Figs. 13.4, 13.5), could be reinterpreted in the same way, replacing his putative "violent rush" of water with another mega-glacier, this one flowing south. Following Charpentier, all these immense glaciers could be explained as the products of a mountain range that might have stood at that time at Andean or Himalayan heights. However, it remained to be seen whether geologists would find these startling inferences plausible.

In principle, the similar erratics and scratched bedrock surfaces that were already known across vastly more extensive low-lying regions of northern Europe (Fig. 13.7)—Sefström's study of the Swedish case (§34.1) was first published only a year after Charpentier's—could have been attributed, following Esmark's earlier speculation (§13.4), to further mega-glaciers or ice-sheets. But Charpentier's causal explanation of glacial extension could hardly be applied to them unless, most implausibly, the Scandinavian range had stood at that time even higher than the Alps; and no mountains of any kind were known to lie to the north of the erratics strewn across northern North America, such as those that Bigsby had reported on the shores of Lake Huron (§13.3). So the manifestly similar features shared by the Alpine and northern cases seemed difficult to bring together under any single explanation. The reconstruction of the most recent phase of geohistory therefore remained almost as problematic as ever.[17]

34.4 CONCLUSION

The theory of a geological "deluge", far from being in retreat during the 1830s, was being improved in the light of new ideas and new evidence. It was now fully uncoupled from its earlier association with the biblical Flood (and the similar events recorded in other ancient traditions), because geologists now recognized that the "diluvial" features were far too old to be the traces of any event in the early history of literate humankind. Yet, equally clearly, they dated from a *geologically* very recent time, near the boundary between the present world and the vast expanses of prehuman geohistory. This was the period that Lyell named "Pleistocene" (§32.3), precisely in order to blend it into the rest of the Tertiary era (his "Pliocene", etc.) and to efface any sharp disjunction at the borderline with the present. But most other geologists insisted that this period, whatever it was to be called, had been marked by events of a far from ordinary kind.

Erosional features such as valleys remained ambiguous, since their very diversity of form suggested that no single causal explanation would be applicable to all: some might be "fluvial" in origin and others truly "diluvial". Depositional features such as boulder clay or "till" remained problematic, and not obviously explicable by any known actual cause. But those boulders large and exotic enough to deserve the name "erratics", together with the strange surfaces of scratched or polished bedrock often found underlying them, focused the diluvial debates more sharply than any other features. Any adequate interpretation of erratic blocks and scratched bedrock had to account not only for those found in and around the Alps, but also for those spread even more widely across northern Europe and northern North America. There were three alternative explanations, but each had its weak points.

The diluvial option, often linked with Hall's earlier hypothesis about the potential power of a mega-tsunami, had been improved by von Buch's suggestion that such currents might have been composed of liquid mud, which would be better able to move large blocks of rock. Sefström's careful survey of the scratched bedrock surfaces across much of Sweden then demonstrated the even larger scale of the phenomena in northern Europe: he attributed them to a sudden, deep, and violent "petridelaunic flood" bearing masses of small rocks from north to south, though the cause of the flood itself remained enigmatic. It was, of course, a geological deluge totally different—in both physical character and apparent age—from the biblical Flood or any other event recorded in human history.

Lyell's option, designed to avoid having to invoke any such catastrophic event, revived and developed earlier claims that drifting ice floes and icebergs could

16. Charpentier's earlier papers, cited above, were amplified in this respect in his "Conjectures sur les grandes révolutions" (1836), probably a lecture for the general public in Lausanne. The moraines and erratics were far too well preserved for him to attribute the geologically very recent reduction in the height of the Alps to the very slow processes of ordinary erosion.

17. A manuscript dated 1836 shows that Venetz was continuing to work privately on his conjecture about a glaciation covering both the Alps and northern Europe: Weidmann, "Ignace Venetz" (1972). But he had no causal explanation for it.

have provided the necessary flotation to shift even large erratics for long distances from their areas of origin. He argued that this impeccably actual cause was adequate to explain the observed distribution of erratics, both in northern Europe and in the Alps (and indeed elsewhere in the world), provided only that it was conceded that crustal movements in this geologically recent period had been substantial enough to have generated *local* geographies (and hence local climates) quite different from those of the present. This, he argued, would save the phenomena satisfactorily, without having to concede that the *global* climate had been any different: Lyellian uniformity would be safely preserved.

The third option was that the erratics had been shifted by vastly extended glaciers. This too was based on an actual cause, since existing glaciers were observably capable of moving blocks of rock to form moraines where the ice melted, and rocks embedded in the ice were observably capable of scratching and polishing the underlying bedrock. But Venetz noted that there were moraines and erratics far beyond the existing Alpine glaciers, and he inferred that the glaciers themselves must once have been similarly extended. After initial skepticism, Charpentier agreed; he mapped the "erratic terrain" in his part of Switzerland, and argued sensationally that a mega-glacier must have not only filled the entire upper Rhône valley, but also spread out across the Swiss plain and up onto the Jura hills beyond it. Like other geologists, he probably found it literally inconceivable that this vast extension of the Alpine glaciers, and their subsequent shrinkage, could have been due to a huge climatic fluctuation—first far colder, then warming up again—in the geologically very recent past. Instead, he invoked Élie de Beaumont's widely supported theory of the origin of mountain ranges by "epochs of elevation" (§9.4), and he attributed the extension of the glaciers to the recent upheaval of the Alps themselves, probably to Andean or Himalayan heights. The subsequent retreat of the glaciers, attributed to a more recent reduction in the height of the mountains, remained more problematic. However, this kind of explanation was difficult to extend to the northern erratics, for there were no comparably high mountain ranges and yet the erratics were spread even more widely.

The explanatory problems raised by erratics and scratched bedrock were therefore unresolved. Yet they remained right at the heart of debates about geohistory as a whole, because they impinged crucially on the problems of reconstructing the part of geohistory closest to the present world. The next chapter describes how these problems came much closer to being resolved.

CHAPTER 35

Snowball earth? (1835–40)

35.1 AGASSIZ'S "ICE AGE" FOR THE ALPS

The vast dispersion of erratics in the Alpine region and much further north, and of the scratched or polished bedrock often found underlying them, was explained in the 1830s by invoking one of three alternative causal agencies: either a sudden violent mega-tsunami, perhaps muddy and stony rather than just watery; or the calm drifting, grounding, and melting of erratic-bearing icebergs or ice floes, during a geologically recent period of higher relative sea level; or—most sensationally—by vastly extended glaciers spreading from an Alpine range that had been, at the time, much higher than at present, creating a *locally* colder climate. However, the continuing arguments among geologists reflected the inadequacy of each of these three explanations. There was a fourth alternative, but to most geologists it seemed even less satisfactory. To suggest that *global* climate had been far colder in the geologically very recent past flew in the face of their expectations. Their growing consensus that the long-term trend in geohistory had been one of global cooling implied that the climate might have been slightly warmer, but certainly not far colder, during the "diluvial" or Pleistocene period. Even Lyell, the most prominent geologist to stand outside this consensus, found it an unacceptably catastrophic deviation from the "uniformity" of an earth in a steady state or at least in an extremely slow or long-wave cyclicity (§21.4).

This possibility of explaining the puzzling features in terms of vastly more extensive glaciation on a global scale had been raised a decade earlier by Esmark (§13.4). His paper could not easily have been overlooked, for although it was published in Danish it was translated for Jameson's Edinburgh periodical and summarized in a Parisian one. Yet it had attracted little attention, probably because Esmark embedded his idea in an unacceptably speculative style of geotheory: it was explicitly inspired by Whiston's seventeenth-century cosmogony, complete with changes in the earth's orbit.

More recently, Esmark's idea had been revived by Reinhard Bernhardi (1797–1849), the professor of natural sciences at the Dreissigacker forestry school near Meiningen in Saxony. Bernhardi had studied at Göttingen under Hausmann, and must have known of the latter's work on the German and Dutch erratics (§13.3); in 1832, he too won a prize at Haarlem, for a general review of geology in which he too abstained from offering any causal explanation of the science's most puzzling phenomenon. But he may have felt he could afford to be more openly speculative in the article that he published the same year in the new and specifically geological periodical edited by Leonhard and Bronn (§25.2). There it was much less likely to be overlooked by other geologists than Esmark's had been. However, his interpretation of the north German erratics was explicitly offered as "a further step" beyond Esmark's, for it invoked an even greater former extension of the polar ice cap, as a "colossal sea of ice [*Eismeer*]" right across northern Germany. This must have made it even more incredible than Esmark's speculation, and it may be no accident that none of the editors of other scientific periodicals around Europe saw fit to reprint or translate it, even in summary. Nonetheless, geologists who were familiar with Bernhardi's German evidence, and with the supposedly diluvial evidence elsewhere, might have recognized the explanatory potential of a radically glacial interpretation, if only the period of great cold could be accounted for in some way less speculative than Esmark's.[1]

In the event, the further development of such theorizing came from an unexpected direction. During the winter of 1835–36, shortly after Charpentier had publicized Venetz's ideas on glacial extension (§34.3), the problem of the Alpine erratics was reviewed in lectures given by Agassiz, the young professor of natural history at Neuchâtel (§30.1). He could hardly ignore it, for it was posed most acutely by the erratics furthest from their source, namely those on the Jura, and most vividly by the huge and famous Pierre à Bot stranded on a hillside high above his home town (Fig. 6.4). As possible explanations he mentioned von Buch's transport in violent currents of mud and Lyell's flotation on icebergs; but he preferred a new and surprising idea of his own, namely that the erratics had slid down an icy surface during the sudden upheaval of the Alps. Almost as an afterthought he also mentioned Venetz's ideas about glacial extension, but dismissed them as "bizarre".[2]

However, when Agassiz and his wife needed a vacation the following spring, they chose to spend it at Bex, staying in a house close to Charpentier's. Two years earlier, the latter had invited Agassiz to visit him, but at that time erratics had not been one of the obvious attractions of Bex. Now they certainly were, and Agassiz was probably taken by Charpentier to see the supposed glacial phenomena, or at least told where to go, just as Lyell had been (§34.3). Whether Agassiz then found Venetz's (and Charpentier's) ideas any less bizarre is not clear: he may just have shelved the problem, since he was much preoccupied with his intensive research on fossil fish (§30.1).[3]

A year later, however, Agassiz's interest in the problem of erratics was expressed for the first time in a more public arena. At a meeting of Neuchâtel's local scientific society, the visiting botanist Karl Friedrich Schimper (1803–67) of Heidelberg, a friend of Agassiz's from their student days, read a botanical paper and also handed out a scientific poem that he had written earlier (in German) about a hypothetical global "Ice Age" [*Eiszeit*] in the distant past. At the next meeting of the society one of its members, prompted by Schimper's poetic speculation, described the erratics he had seen in 1833 in Scania, a part of southern Sweden that Sefstöm had since mapped for its scratched bedrock (Fig. 34.2). Commenting on this, Agassiz adopted Schimper's "geopoetic" concept: he attributed the Alpine erratics to "the action and movement of immense sheets [*nappes*] of ice", and therefore thought it unsurprising that they were found on an even larger scale in northern Europe. He dated these effects before "the present creation" (that is, to prehuman geohistory), and predicted that they would "give the key to the solution of many phenomena on which the science still has only more or less satisfactory hypotheses".[4]

What Agassiz meant by these hints became clearer a few months later, when the Société Helvétique held its 1837 meeting in his home town. The ambitious thirty-year-old, its president for the occasion, gave the opening "discourse", but abandoned the customary review of recent work by its members in favor of expounding his own sensational theory of a geologically recent "Ice Age". This explicitly combined his own ideas with Schimper's, while crediting Venetz and Charpentier only with the "facts" about ancient moraines. Indeed, he criticized and rejected the latter's claim that a mega-glacier had ever crossed the Swiss plain or reached the Jura. Instead, recruiting Sefström's Swedish evidence (and tacitly Bernhardi's too), he claimed that *before* the upheaval of the Alps the earth was already "covered with ice at least from the North Pole as far as the shores of the Mediterranean and the Caspian Sea"; he later extended the ice cover "as far as the Atlas" in North Africa, and probably envisaged a literally global "Snowball Earth" (see below). The sudden upheaval of the Alps—like Charpentier, he took its geologically recent date from Élie de Beaumont (§23.1)—had then locally tilted the surface of this *static* ice-sheet, and blocks of Alpine rock had slid down its sloping surface until they came to rest as erratics on the slopes of the Jura. Only later still, as the earth began to warm up again, had these static ice-sheets melted away,

1. Bernhardi, "Felsbruchstücke und Geschiebe" (1832). His prize essay at Teyler's, "Abhandlung über Geologie" (1832), answered a question set in 1827 on "What is now known in geology": see his brief review of "opinions on the catastrophe that ended this [most recent] period" (315–17); also Himbergen, "Prijsvragen" (1978). Mielecke, "Reinhard Bernhardi" (1973), gives biographical information.

2. An auditor's notes on Agassiz's lectures [Neuchâtel-IG: V.F.19.3] are quoted in Schaer, "Agassiz et les glaciers" (2000), 233–34. The present narrative is much indebted to this, by far the best account of Agassiz's role in the glacial debate, and to Professor Schaer's guidance in the field during the INHIGEO meeting at Neuchâtel and in the Swiss Alps in 1998.

3. Charpentier to Agassiz, 31 December 1833 [Cambridge/MA-HLH: bMS.Am.1419/243], had mentioned that his local area was interesting for "soulèvement" theory and plutonic effects, but that his visitor would find few fossils to interest him.

4. Société des Sciences Naturelles de Neuchâtel, meetings on 15 February, 1 March 1837 [*MSSNN* 2: 13]. Charles Godet (the author of the paper on Scania) had been present at the 1829 meeting of the Société Helvétique, when Venetz read his paper suggesting a vast glacial extension (§34.3); he and Schimper may well have known the work of Esmark and Bernhardi. (In 1960 Harry Hess famously described his own seminal suggestion of sea-floor spreading—a decisive moment in the development of the modern theory of plate tectonics—as an "essay in geopoetry".)

their remnants transformed into moving glaciers that had gradually shrunk to their present small size (see Fig. 34.6). So the putative "Ice Age" had *preceded*, and was quite distinct from, the time at which the glaciers of the upper Rhône had been more extensive than at present: as he put it, "it would thus be a grave error to confuse the glaciers [*glaciers*] that descend from the Alpine summits with the phenomena of the epoch of the great ice-sheets [*glaces*] that preceded their existence".[5]

Agassiz's reasons for proposing this theory—itself rather bizarre—became apparent in the later part of his speech. He used it to explain the most recent of what he claimed had been a sequence of mass extinctions in the course of geohistory. This was the crucial link between his Ice Age theory and his otherwise unrelated research on fossils. He tacitly rejected the growing trend among other geologists, such as Phillips, toward interpreting the history of life in less catastrophic terms (§30.4); in effect he retained Cuvier's early view that it had been sharply divided into discrete periods separated by sudden "révolutions" (*BLT* §9.3). For if all fossil organisms had been well adapted to their respective environments—as, following Cuvier, he certainly believed they had—it followed, in his view, that only a drastic catastrophe could have wiped them out. And only a succession of such events could account for the sequence of fossil faunas, including the fossil fish faunas that he himself was currently describing. But then another major problem with his theory became apparent: "it is a supposition that appears to be in direct contradiction to the well-known facts that demonstrate a substantial cooling of the earth since the most remote times". A conflict with the well-established theory of a cooling earth here became explicit (it was already clear in the case of Charpentier: §34.3).

Agassiz solved this puzzle with an ingenious hypothesis, which fitted neatly into "the progressive development of the earth since its origin" that he had outlined in his lecture course the previous year. In that synopsis, ten "epochs", from the "*actuelle*" back to that of the "primitive schists", were separated by the kind of successive "upheavals" [*soulèvements*] proposed by Élie de Beaumont (§9.4); the penultimate epoch (just before the present world) was that of the diluvium, the erratic blocks, and the upheaval of both the Alps and the Himalaya. Onto this quite conventional geohistory (described, as usual, retrospectively), Agassiz now projected a huge enlargement of the minor climatic oscillations that Venetz had inferred for the recent centuries of human history in Switzerland. As Agassiz put it, "nothing has hitherto proved to us that this [geohistorical] cooling has been continuous, and that it has operated without oscillations". If in fact it had been stepwise, and if in addition each step had been separated from the next by a severe drop in temperature followed by a partial recovery (both somehow related, he thought, to a mass extinction and subsequent renewal of life), the most recent of these catastrophes might have been unique and unprecedented, simply because the global temperature had for the first time dropped temporarily below freezing point and therefore generated a global Ice Age or Snowball Earth (Fig. 35.1).[6]

Agassiz's speech certainly had the desired effect of creating a sensation, and putting its author into the spotlight that he craved, but the assembled savants

qu'elle a diminué subitement et considérablement à la fin de chaque époque, avec la disparition des êtres organisés qui la caractérisent, pour se relever avec l'apparition d'une nouvelle création au commencement de l'époque suivante, bien qu'à un degré inférieur à la température moyenne de l'époque précédente; en sorte que la diminution de la température du globe pourrait être exprimée par la ligne suivante: —ʟ—ʟ—

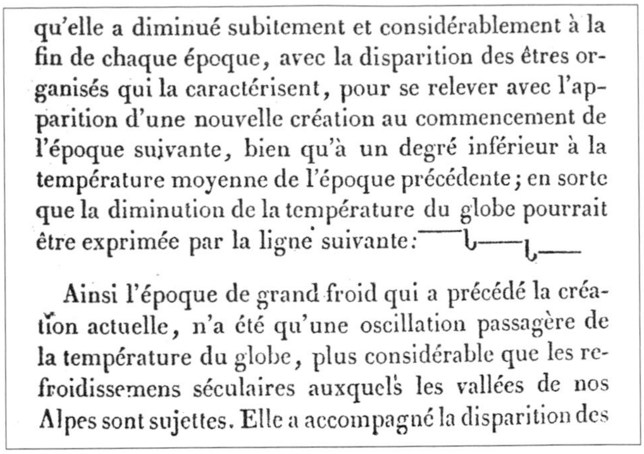

Ainsi l'époque de grand froid qui a précédé la création actuelle, n'a été qu'une oscillation passagère de la température du globe, plus considérable que les refroidissemens séculaires auxquels les vallées de nos Alpes sont sujettes. Elle a accompagné la disparition des

Fig. 35.1 Agassiz's tiny graph—improvised from the printer's stock typographical symbols—illustrating his theory of an intensely cold "Ice Age", first published in his "Neuchâtel discourse" (1837). In the course of geohistorical time (horizontal axis, left to right), global temperature (vertical axis) has fallen, not smoothly and continuously but in discrete steps, separated by sudden episodes of severe cooling followed by partial warming to a new stable level. This model could explain how each successive set of organisms had enjoyed a stable environment for a finite period, followed by its mass extinction and its replacement (somehow!) by new organisms equally well adapted to a different environment. By implication, it was only in the most recent of these catastrophic events that global climates had cooled sufficiently to generate a temporary but drastic Ice Age, at least down to what are now temperate latitudes but probably on a truly global scale (not unlike the modern theory of a "Snowball Earth" episode or episodes at a far earlier time in geohistory).

were not convinced. Among them was von Buch, who was revisiting the scene of some of his early research on erratics (*BLT* §10.2); he drew on the authority of his wide field experience to reject altogether what he called "the glacial theory". Others present, such as Charpentier, Élie de Beaumont, and Studer, rejected Agassiz's specific version, with its putative Snowball Earth or global Ice Age, on the grounds that it was incompatible with known actual causes; and the published account of the meeting tacitly censured Agassiz for having surrendered to "the seductions of brilliant hypotheses" in the outmoded style of speculative geotheory. Still, his ideas received plenty of publicity: his speech was distributed as a preprint before it appeared in the Société's own periodical; the version published in Geneva in the widely read *Bibliothèque Universelle* was translated into English for Jameson's periodical in Edinburgh; and a brief report in German appeared later at Heidelberg in Leonhard and Bronn's periodical. He also sent a letter to the Académie in Paris (also subsequently translated for Jameson), specifically on the Jura erratics, urging that a forthcoming French expedition should look out

5. Agassiz, "Discours à Neuchâtel" (1837), xxv, xxix, read on 24 July. The phrase "bis zum Atlas" is in his German translation of his full work, *Untersuchungen über die Gletscher* (1841), 284. Later in his life he even interpreted landforms in tropical Brazil as glacial in origin, so it is likely that from the start he envisaged the Ice Age as a geologically recent "Snowball Earth" (the modern phrase refers to a putative episode or episodes of a similar kind in the far deeper past of the Proterozoic or late Precambrian).

6. Fig. 35.1 is reproduced from Agassiz, "Discours à Neuchâtel" (1837), xxx; it was reprinted, in a less crude form, in "Des glaciers, des moraines" (1837), 392, and in *Études sur les glaciers* (1840), 328 [trans. Carozzi, *Glaciers by Agassiz* (1967), lviii, 176]. Agassiz, "Cours de géologie" for 1836 [Cambridge/MA-HLH: *pAB85.Ag167.836c]; see also his and Schimper's joint "New ideas on the development of the animal kingdom", presented at Neuchâtel on 3 May 1837 [summary in *MSSNN* 2: 15]. As usual in naturalists' discourse at this time, the word "*création*" did not necessarily denote immediate divine agency, though it could.

for similar features in the polar regions. In each case, Agassiz dissociated himself from Charpentier and Venetz, relegating all their evidence for glacial extension to a subordinate role in his theory and minimizing his indebtedness to them.[7]

35.2 EXTENDING THE ICE AGE

In the summer of 1838, a year after he gave the speech at Neuchâtel, Agassiz at last remedied his total lack of firsthand knowledge of existing glaciers by basing himself at the refuge on the Grimsel pass, near the Rhône glacier. The field trip was innovative in that he took six collaborators with him, among them an artist. Although he was primarily concerned to work out the physics of glacial movement, he did also see for himself the kind of evidence on which Charpentier's (and Venetz's) argument for glacial extension had been based (Fig. 35.2).[8]

After a quick visit to Chamonix to see the Mont Blanc glaciers, Agassiz—the stereotypical Young Man in a Hurry—hurried away to attend the annual field meeting of the Société Géologique de France, held this year at Porrentruy (across the Jura from Neuchâtel and just short of the frontier with France). He and

Fig. 35.2 The lower end of the Viesch glacier in the Swiss Alps and its arc of terminal and lateral moraines, as depicted (rather crudely and inadequately) in Agassiz's *Studies on Glaciers* (1840). He inferred that the valley in the foreground had been carved out when, as Venetz and Charpentier had already claimed (§34.3), glaciers such as this extended much further than at present. The broad valley of U-shaped profile—with its smooth steep sides, its floor of "rounded polished and scratched" bedrock surfaces, and its "blocks scattered on the polished rock"—contrasts with the narrow "erosions of the torrent" carved within it, thus allowing "a direct comparison between the action of ice and water on rocks". This glacial interpretation of characteristic Alpine topography was one of the explanatory benefits of the theory of glacial extension, whether Charpentier's version or Agassiz's.

Fig. 35.3 Agassiz's picture of a surface of scratched and polished bedrock at the foot of the Jura hills near Neuchâtel, published in his *Studies on Glaciers* (1840) but already mentioned in his "Neuchâtel discourse" in 1837—in which he first expounded his sensational theory of an "Ice Age"—and demonstrated to other geologists on a field trip in 1838. Such features were closely similar to the scratched rock surfaces underlying existing glaciers in the Alps (see Fig. 35.4) and those traced by Charpentier throughout the upper Rhône valley.

several other geologists then went straight on to Basel for the annual meeting of the Société Helvétique. On both occasions Agassiz continued his campaign for his full-blown Ice Age theory, explicitly recruiting Sefström's "diluvial" evidence in Sweden (§34.1) to his glacialist cause. *Before* the upheaval of the Alps, he argued, "the whole of Europe was covered with ice [*glaces*]", wiping out the mammalian megafauna; the erratic blocks, moraines, and polished rocks all dated from a distinctly *later* period. His auditors, who included not only Charpentier and Studer but also Buckland, remained unconvinced, though the criticism was less intense than it had been at Neuchâtel; and some of those present said they would suspend judgment until they had studied their home areas for relevant

7. Agassiz, "Des glaciers, des moraines" (1837); "Blocs erratiques du Jura" (1837), read on 2 October; Agassiz to Leonhard, 9 March 1838 [printed in *NJM* 1838: 304–5]. He described his theory more clearly in a statement printed within Macaire, "Blocs du Jura" (1837), which corrected Macaire, "Compte rendu" (1837), the official report of the meeting. See also Schaer, "Rôle d'Agassiz" (2002), on his relentless self-promotion.

8. Fig. 35.2 is reproduced from Agassiz, *Études sur les glaciers* (1840), pl. 9, drawn and lithographed by Joseph Bettanier, and explained on 336. The quoted phrases are printed on the translucent overlay (pl. 9a) that acts as a key to the plate. Schaer, "Agassiz et les glaciers" (2000), 242, reproduces a proof copy of this plate, on which Agassiz wrote instructions to the artist to improve it; but he was in such a hurry to get his work published (see §36.1) that it remained unaltered. Viesch [now Fiesch] is in the upper Rhône valley 15 km northeast of Brig; the glacier and its meltwater stream flow down a side valley from the north (see Fig. 34.6).

evidence. In any case, what they saw under Agassiz's guidance on the field trip after the Porrentruy meeting was reported as having softened their skepticism (Fig. 35.3).[9]

The most important and heavyweight convert—though not immediately—was Studer, the highly respected professor of geology at Bern (§25.2). After the meeting in Basel, he wrote a shrewd assessment of "the more recent explanations of the phenomena of erratic blocks" for the *Neues Jahrbuch*. He criticized Charpentier's, because it made the extension of the glaciers dependent on the upheaval of the Alps, whereas the present distribution of glaciers worldwide implied that other factors beside mere altitude (and latitude) must be involved. And the Alpine phenomena were so similar to those in Scandinavia that it seemed highly implausible to propose "two wholly different theories" to explain them. Studer conceded that some of the difficulties might be solved by Agassiz's hypothesis of a "general ice-covering [*Eisbedeckung*] of the earth . . . from the pole to the equator"—he certainly took his compatriot to be invoking a full-blown Snowball Earth—but in the end he concluded that torrential currents of some kind still remained a better explanation.[10]

However, when in the following year (1839) the Société Helvétique met in Studer's home city of Bern, Agassiz suggested a joint trip to the glaciers around Zermatt. On the spot, Studer was duly convinced that the scratched bedrock surfaces had indeed been produced by the rocks embedded in glaciers. He then continued his fieldwork alone, over the watershed and into the Aosta valley on the Italian side of the Alps. Here he found clear traces of an ancient mega-glacier extending southward right down to Ivrea on the plain of the Po, matching Charpentier's mega-glacier flowing northward from the Rhône valley across the Swiss plain. On his return to Bern, Studer reported his fieldwork in a letter to the *Neues Jahrbuch* and a paper to the Société Géologique (the latter eventually translated into English for Jameson's periodical). By the same criteria of erratics and scratched bedrock, he was now convinced that it could "scarcely be doubted" that Charpentier's mega-glacier had indeed reached the Jura during "the diluvial period", and that the covering of ice must also have affected Sefström's Sweden and the parts of Germany and Britain in which similar features were reported.

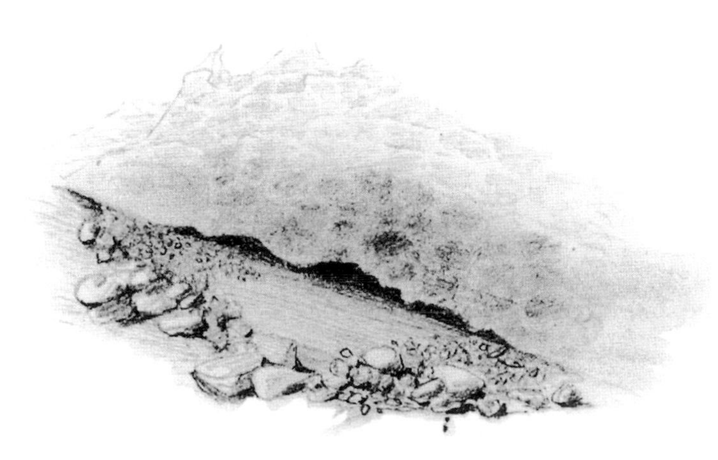

Fig. 35.4 The melting edge of an Alpine glacier, showing the stones and boulders which, while still embedded in the slowly moving ice, would have scratched the underlying bedrock. This sketch was made during one of Agassiz's annual field trips to study existing glaciers; possibly on the second (1839), when his colleague Studer, on seeing these features, agreed that they were valid criteria for tracing the vast former extension of the Alpine glaciers. (By permission of the Institut de Géologie, Universite de Neuchâtel)

Although, as he put it, "such colossal results must make us cautious", Studer now considered them plausible. The key to his change of mind was, of course, his firsthand study of the "actual cause" of glacial agency (Fig. 35.4).[11]

Bringing the Alpine and northern phenomena under the same explanatory umbrella—as Studer did—implied that the notion of glacial extension should be detached from any connection with the upheaval of the Alps, and suggested that the common cause might lie in a recent period of *general* cold throughout Europe, however unexpected and unexplained that might be. This interpretation was promptly and perhaps independently reinforced by the Alsatian naturalist Édouard Renoir, who taught physics in Belfort and was also a good geologist. Like Studer, he was among those who heard Agassiz at Porrentruy but was unconvinced. However, he then traveled to the Alps; studied the Viesch glacier (Fig. 35.2) and others, to learn how to recognize the traces of former glaciers; and visited Charpentier, who took him to see the nearby moraines and the huge erratic blocks (Fig. 34.5). Renoir was promptly converted, at least to the idea of major glacial extension. After returning home, he looked for glacial traces in the valleys of the nearby Vosges hills, and found clear signs of several small glaciers, which had evidently been quite separate from Charpentier's mega-glacier in the Swiss plain on the other side of the Jura. Renoir's paper on the Vosges was read at the Société Géologique in Paris at the same meeting as Studer's on the Alps, and immediately after it; it was of course taken to support Studer's conclusion. A few months later the civil engineer Henri-Joseph Hogard (1776–1873), who had also been at the Porrentruy meeting, presented his local scientific society in Épinal with a paper that confirmed Renoir's reconstruction. Yet another local amateur, Édouard Collomb (1801–75), later published in Paris a detailed survey that made the vanished Vosges glaciers vividly convincing to geologists everywhere (Fig. 35.5).[12]

9. Fig. 35.3 is reproduced from Agassiz, *Études sur les glaciers* (1840), pl. 17, a version of which may have been one of the "demonstrations graphiques" that accompanied his speech: "Discours à Neuchâtel" (1837), xii–xiii, xxxii. See also his Porrentruy paper, "Observations sur les glaciers" (1838), and the account of the field trip in Anonymous, "Réunion extraordinaire" (1838), 417–18. This spot above the village of Landeron, 12km northeast of Neuchâtel, was later quarried away, but similar scratched and polished surfaces are still exposed at Le Mail, in the suburbs of Neuchâtel itself.

10. Studer, "Erratischer Blöcke" (1838); an English translation (a rather poor one) was published in Silliman's American periodical, not in Jameson's Scottish one.

11. Fig. 35.4 is reproduced from an undated drawing, probably by Jacques Bourckhardt [Neuchâtel-IG: 35/3.1, no. 15]; see Schaer, "Agassiz et les glaciers" (2000), 239–40. Studer to Leonhard, 8 October 1839 [printed in *NJM* 1840: 208–11], and Studer, "Phénomènes de l'époque diluvienne" (1840), read on 2 December 1839. Like many other geologists (and notably the German-speaking), Studer continued to use "diluvial" as a descriptive term for the period in which the Superficial deposits had been formed, even after abandoning any strictly diluvial explanation of them.

12. Fig. 35.5 is reproduced from Collomb, *Anciens glaciers des Vosges* (1847) [London-BL: 7105.c.5], pl. opp. 38 (the snout of the glacier, immediately behind *A*, is drawn unrealistically distinct from the rest of it); see also his preliminary report, "Phénomène erratique dans les Vosges" (1845), read at the Société Géologique on 16 June. He worked in the famous textile factory in the town of Wesserling (sited at *A*), 11km northwest of Thann (Haut-Rhin): see Fig. 35.6. Renoir, "Glaciers des Vosges" (1840), read on 2 December 1839, and soon translated for Jameson's periodical, was the *first* public report on the Vosges glaciers; his "Traces des anciens glaciers" (1841), read on 21 December 1840, reinforced his argument by relating his own Alpine fieldwork to a recent expedition to Arctic Russia. The moraine (now largely in the wooded Parc de Wesserling) that straddles the Thur valley was Renoir's prime example and remains one of the most striking signs of the former glaciation of the Vosges. Hogard, "Traces de glaciers" (1840), dated 10 March, and "Glaciers dans les Vosges" (1845), read at the Société Géologique on 20

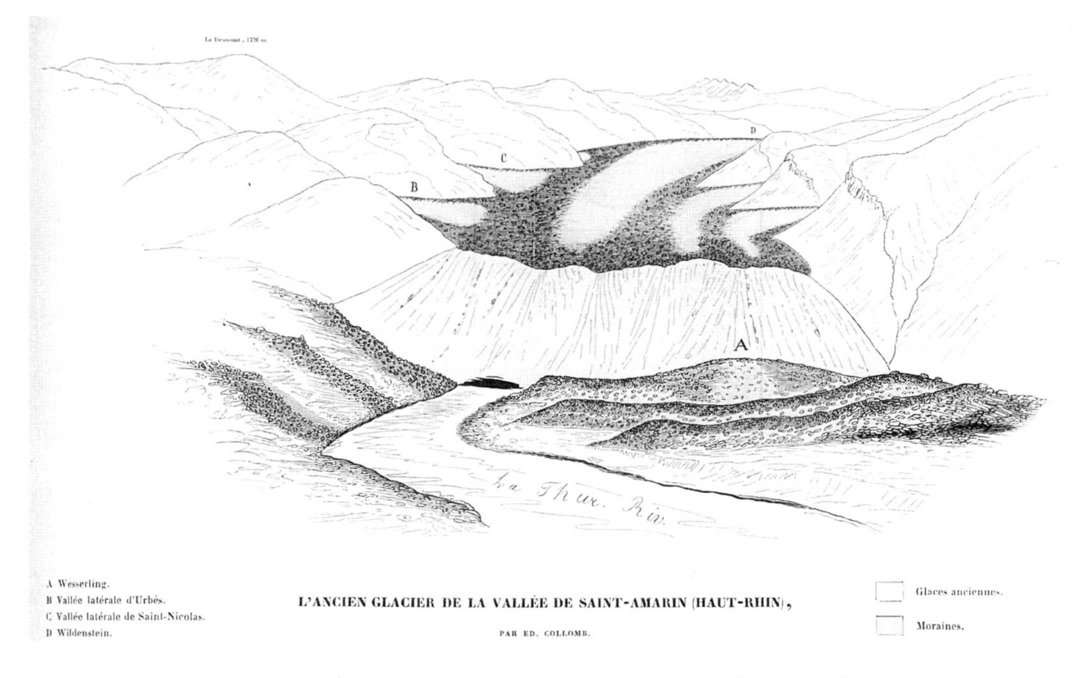

Fig. 35.5 Édouard Collomb's semi-aerial landscape view of the valley of the Thur, in the Vosges hills, with its former glacier (and tributary glaciers in the side valleys *B* and *C*), as reconstructed by him and other local geologists; the terminal moraine blocking the main valley (in front of *A*) marked its maximum extension (see Fig. 35.6). The evidence of erratics and scratched bedrock implied that the ice had been up to 500m thick; this strongly suggested that there had been a very cold climate in Europe at a geologically recent period, unconnected with the allegedly recent upheaval of the Alps further south. (By permission of the British Library)

In one respect these reports from Alsace were even more significant than Studer's. The vanished mega-glaciers that Studer now acknowledged on both sides of the Alps could still have been explained, following Charpentier, as dating from a time when the mountains were much higher than at present, or at least as having been causally connected with their upheaval. In contrast, these modest reports of modest little glaciers in the modest hills of the Vosges implied that, as Studer suggested, the climate of *the whole of Europe* must have been much colder in the geologically recent past, without any comparable major change in the general topography; for according to Élie de Beaumont's tectonic research (§9.4) the Vosges had *not* been upheaved in the recent past, but far further back in geohistory. In other words, the local fieldwork in the Vosges supported Studer's version of the glacial theory—with glacial conditions far more widespread than Charpentier proposed, but not universal—rather than Agassiz's bizarre notion of a static global ice-sheet or Snowball Earth episode *preceding* the period of widespread glaciers (Fig. 35.6).[13]

Yet a glacial theory along the lines of Studer's could be established only if the kind of evidence found in the Vosges could also be recognized in other regions even further from existing glaciers. Similarly detailed fieldwork was imperative, to discover how widespread in Europe the former glacial conditions had been. Buckland was in fact already planning just such research in his home country. After attending the earlier meetings in Porrentruy and Basel he had made a trip

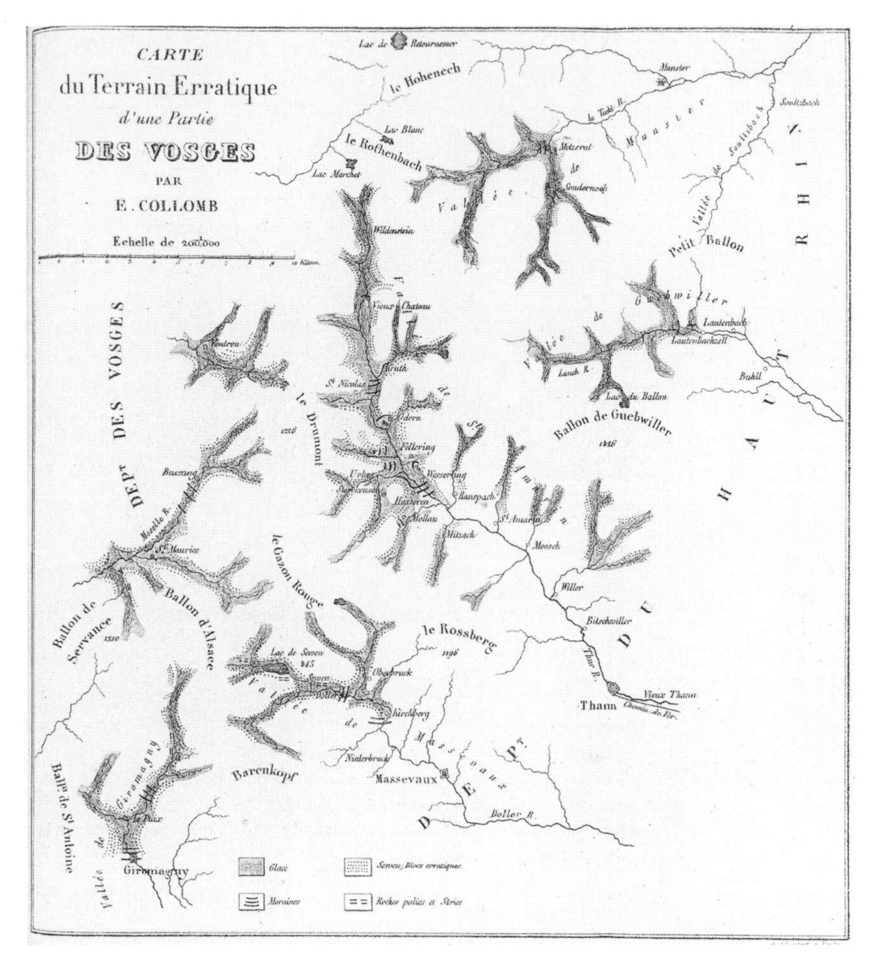

Fig. 35.6 Collomb's map of the southern part of the Vosges hills in Alsace, showing his reconstruction of the many small valley glaciers that had left their traces in terminal moraines (arcuate lines across the valleys), scratched bedrock (short straight lines along the valley sides), and erratics (stippled). The scale is of 11km. Collomb's reconstructed glacial landscape (Fig. 35.5) looks up the Thur valley past the terminal moraine at Wesserling (dead center). Although this map was not published until 1847, fieldwork by the local geologists Renoir and Hogard had made the evidence for valley glaciers in the Vosges well known since 1839. (By permission of the British Library)

to the Alps to see some of its glaciers for himself. His wife told Agassiz that "Dr Buckland is as far as ever from agreeing with you"; but Buckland himself conceded that a more limited glacial theory might indeed explain the Swiss phenomena, though "it would not apply to the granite blocks and transported [i.e., "diluvial"] gravel of England, which I can only explain by referring to currents of water".

January. These three local geologists may have been rivals rather than collaborators, but in either case their joint involvement reflected the perceived importance of the Vosges. Renoir mentioned that yet another local geologist had found similar glacial traces in the similar hilly massif of the Schwarzwald [Black Forest] on the other (German) side of the Rhine valley.

13. Fig. 35.6 is reproduced from Collomb, *Anciens glaciers des Vosges* (1847) [London-BL: 7105.c.5], pl. opp. 180. Thann, the largest town on the map (lower right), is 20km west-northwest of Mulhouse (Haut-Rhin). To reconstruct former glaciers in the Vosges was much more striking than to do so in Scotland (see below), because many of the warm Alsatian hillsides were (and are) covered by vineyards, not by cold peaty moorland.

Some such *combination* of ice and water was compatible with his earlier conjecture about the extinct megafauna found in Alaska (§14.2). But it remained to be seen whether he and Agassiz's other English friends could be brought round to accept the Swiss naturalist's highly speculative geotheory, or only some more modest version of the Ice Age.[14]

35.3 THE ICE AGE IN BRITAIN

In 1840 Agassiz left his research team on the Aar glacier to continue studying the movement of the ice and other physical problems. He himself traveled to Britain, primarily for further research on fossil fish (§30.1). But he also hoped to convert the British geologists to his Ice Age theory and to find further evidence for it on their home ground. In advance of his guest's arrival, Buckland used his position as the current president of the Geological Society to get a paper by Agassiz squeezed into the last meeting before the summer recess. It argued that the scratched surfaces of bedrock in the Alps were demonstrably produced not by any kind of aqueous current but by stones embedded in glacial ice, thereby implying that such features could be used as reliable criteria for glacial extension. But these ideas were already well known, and most of the English geologists found Buckland's theory of violent currents, or Lyell's theory of flotation on icebergs, more plausible explanations of the erratics in Britain.[15]

Agassiz arrived in Britain in time to attend the British Association meeting in Glasgow, where he reported on his latest research on fossil fish (and confirmed that those that Murchison had just brought back from Russia were of Old Red Sandstone age, thereby helping finally to resolve the "great Devonian controversy": §31.1). Equally important was his summary (in French) of the Swiss evidence for his glacial theory, aided by a display of the plates for his forthcoming book (Fig. 35.2). Lyell, De la Beche, Murchison, and others took part in a lively discussion, which Murchison summarized—presciently—by telling the absent Sedgwick that "I think we shall end in having a compromise between himself and us of the floating icebergs". Agassiz said that his glacial theory would be strengthened if he could find traces of vanished glaciers in Scotland, and that he himself intended to look for them.[16]

When the meeting was over, Agassiz toured the Highlands with Buckland as his guide. They visited fossil fish localities in the Old Red Sandstone; but first, and more importantly, many well-known "diluvial" localities, which Buckland already suspected were ripe for reinterpretation. Like Renoir and Hogard in the Vosges the previous year, they duly found all the distinctive features—erratics, moraines, scratched and polished bedrock—that could be explained far more satisfactorily as the traces of extensive valley glaciers than as diluvial effects. Agassiz immediately sent a report to Jameson, who judged the "Discovery of the former existence of glaciers in Scotland" so important that he had it published in the *Scotsman* newspaper, making it widely known without waiting for the next issue of his own periodical.[17]

In view of Darwin's recent weighty paper on Lochaber (§33.4), the Parallel Roads of Glen Roy were a particularly significant test case for the new understanding of the recent geohistory of Scotland. In the light of their brief fieldwork there, Agassiz and Buckland rejected Darwin's interpretation and adopted Lauder's and MacCulloch's earlier inference that the terraces marked the successive shorelines of a series of lakes. But the enigmatic barriers that had impounded the water—which for the earlier authors had then vanished so completely and mysteriously—were a problem no more: they had been composed of glacial ice, which had simply melted away. Agassiz claimed in fine Lyellian style that "the phenomenon must have been precisely analogous to the glacier-lakes of the Tyrol" and to the one in the Val de Bagnes that had notoriously burst its icy dam in 1818. A simple thought experiment based on Mont Blanc—which for British savants was a more familiar part of the Alps—later made the analogy even clearer. As Agassiz recalled, "I shall never forget the impression I experienced at the sight of the terraced mounds of [erratic] blocks which occur at the mouth of the valley of Loch Treig, where it joins Glen Spean; it seemed to me as if I were looking at the numerous moraines of the neighbourhood of Tines, in the valley of Chamonix" (Fig. 35.7).[18]

After the two geologists parted, Agassiz crossed to Ireland, primarily to see further fossil fish collections but also to look for glacial traces. He reported to Buckland that "there is no doubt that glaciers also covered Ireland", and later he told Humboldt that "I've seen my [*sic*] polished surfaces and scratches *down to sea level* all over the plain that slopes from Enniskillen toward Dublin". Crossing back to Scotland, he traveled to Edinburgh and was shown the spectacular scratched bedrock surfaces on Corstorphine Hill. James Hall, some thirty years earlier, had famously attributed them to a violent mega-tsunami (*BLT* §10.2), but now they could be reinterpreted as marking the passage of a slow-grinding ice-sheet. And Agassiz was delighted to be shown further scratched surfaces on Blackford Hill, which were so undercut as to be almost impossible to attribute to drifting icebergs: "that is the work of the ice", he exclaimed to his guide Charles Maclaren (1782–1866), the editor of the *Scotsman* and a good amateur geologist.

14. Agassiz to Buckland, [autumn 1838] [copy in Cambridge/MA-HLH: bMS.Am.1419/107], the first of his extant letters to be written in (slightly faulty) English, summarizes his full-blown Ice Age theory in informal style; Buckland and Mary Buckland to Agassiz, [1838] [both quoted briefly, but without dates, in Agassiz (E. C.), *Louis Agassiz* (1885) 1: 289–90]. Agassiz was already in close contact with Buckland through his work on fossil fish (§30.1), and was also translating and editing a German edition of Buckland's Bridgewater Treatise (§29.2), *Geologie und Mineralogie* (1838–39).

15. Agassiz, "Polished and striated surfaces" (1840), read (at least in summary: there were ten other papers!) on 10 June.

16. Agassiz, "Glaciers and boulders" (1841), read at BAAS on 22 September 1840; Murchison to Sedgwick, 26 September 1840 [quoted in Geikie, *Murchison* (1875) 1: 306–8]; the Glasgow discussion was reported in *Athenaeum*, 17 October 1840, 824.

17. Boylan, "Lyell and glaciation" (1998); also Herries Davies, "Tour of the British Isles" (1968) and *Earth in decay* (1969), 173–87.

18. Fig. 35.7 is reproduced from Agassiz, "Glacial theory" (1842) [Cambridge-UL: Q340:1.c.6.47], pl. 4, quotation on 222. Agassiz marked the higher terraces as extending on to the south side of Glen Spean, where they were required on his reconstruction but where no one else before (or since) had seen or mapped them (compare with Darwin's map: Fig. 33.4). See Rudwick, "Darwin and Glen Roy" (1974), 130–36 and map (fig. 7), which also describes in detail how Agassiz's rather crude reconstruction was progressively amended and improved through further fieldwork by other geologists. Among the existing Alpine analogues with which Agassiz was familiar was the little Märjelensee, dammed by the great Aletsch glacier, about 18km northeast of Brig.

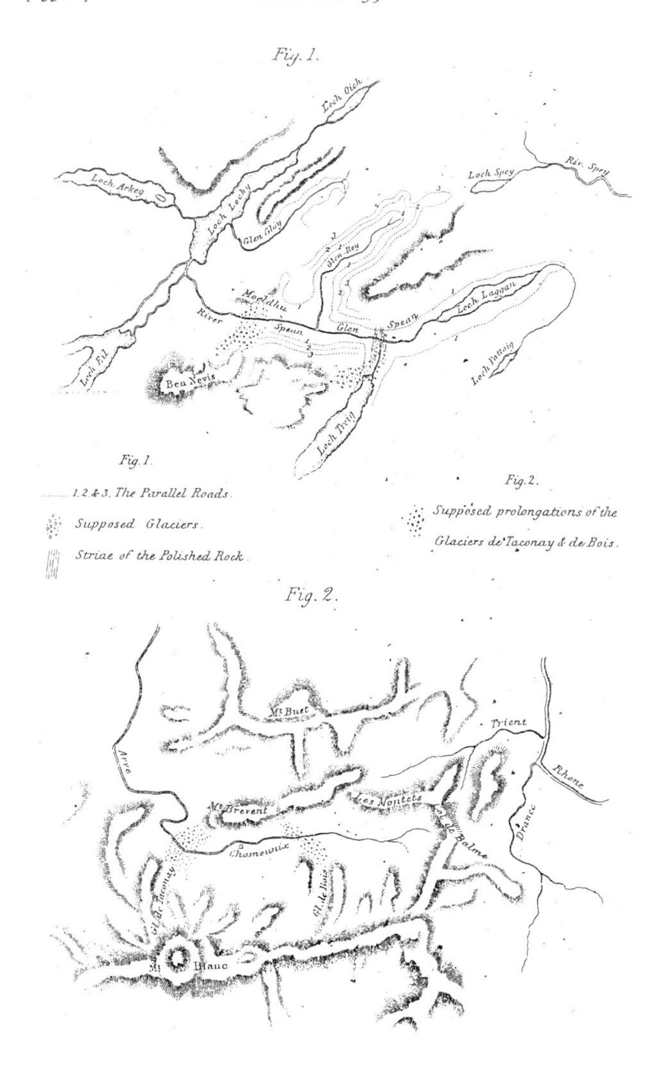

Fig. 35.7 The Parallel Roads of Glen Roy in the Scottish Highlands (see Figs. 33.3, 33.4), reinterpreted by Agassiz in 1840 as the traces of vanished glacial lakes. His sketch map of Lochaber (fig. 1) showed how two former glaciers (stippled), flowing north from Ben Nevis and from the valley now occupied by Loch Trieg, could have pushed out across Glen Spean, impounding between them a lake that also extended up Glen Roy; later, after the lake fell from level 3 to level 2, a retreat of the Trieg glacier could have allowed the glacial lake to extend at level 1 right up to the head of Glen Spean (right); a still later retreat of the Ben Nevis glacier would have drained the lake almost entirely, leaving Lochs Trieg and Laggan as its remnants. The sketch map of the Alps around Chamonix (fig. 2) showed how analogous former extensions (stippled) of two of the existing Mont Blanc glaciers could—hypothetically—have impounded a similar temporary lake on the site of Chamonix, thus demonstrating that his interpretation of Glen Roy was impeccably grounded in actual causes. (By permission of the Syndics of Cambridge University Library)

All such cases implied that ice-sheets had covered a vast expanse of low-lying Ireland and Scotland, and had not been confined to hilly regions such as the Highlands.[19]

Meanwhile, Buckland went straight to meet Lyell at the latter's family home on the edge of the Highlands. There he recognized glacial features in abundance, and the leading exponent of the best alternative interpretation was duly impressed. "Lyell has adopted your theory *in toto*!!!", Buckland told Agassiz in triumph; "On my showing him a beautiful cluster of moraines within two miles of his father's house, he instantly accepted it, as solving a host of difficulties that have all his life embarrassed him". After Buckland left, Lyell stayed on for three weeks of intensive fieldwork, reinterpreting familiar features of his home territory in the light of his new insight. It was for him a spectacular conversion.[20]

Buckland now designed a grand presentation of the glacial theory, to take place as soon as the Geological Society reconvened after its summer recess. At the first meeting Agassiz restated (in French) his own distinctive theory of an Ice Age, insisting that erratics "were not pushed forward by the glaciers, as conjectured by M. de Charpentier" and that "the glaciers did not advance from the

Alps into the plains, but that they gradually withdrew towards the mountains from the plains that they once covered"; covered, that is, with a *static* sheet of ice. Likewise it followed, from his recent fieldwork in Scotland and Ireland, "not only [that] glaciers once existed in the British Islands, but that large sheets (*nappes*) of ice covered all the surface", that is, at an *earlier* "long period of intense cold". But he conceded ancillary roles for both the rival explanations: "the disappearance of great bodies of ice produced enormous debacles and considerable currents", he argued, which could account for those "diluvial" gravels that did seem to have been deposited in water; and "masses of ice were set afloat and conveyed, in diverging directions, the blocks with which they were charged", so that Lyell's earlier iceberg explanation might have some partial validity.[21]

Buckland's paper came next, and described the evidence for extensive vanished glaciers in Scotland, based on all his recent fieldwork; that Castle Hill at the heart of Edinburgh had once been (in modern terms) a nunatak completely surrounded by a vast ice-sheet was a sensational inference for the many in his audience who would have been familiar with the city. A second part of the paper, which was deferred until after Lyell had had his say on further Scottish evidence, described similar signs of extensive former glaciation in northern England. The problems with any diluvial explanation of the famous erratics of Shap granite, for example, were "entirely removed by the application of the glacial theory": they could now be understood as the products of ice-sheets flowing in three directions away from the granitic outcrop. It was a telling example of the explanatory power of the new theory.[22]

Lyell, the third member of this powerful troika, confined himself to Forfarshire, but his analysis was far-reaching. "He is a complete convert", Buckland confidently told the Society's secretary beforehand, and so he proved to be. He readily conceded that "Professor Agassiz's extension to Scotland of the glacial theory" had made him change his mind on the familiar features around his family home, and he now gave them a glacial interpretation. He also admitted that he "had been diverted from the consideration of a long-continued covering of snow on the Scottish mountains, by the knowledge that the climate of Great Britain

19. Agassiz to Humboldt, 27 December 1840 [printed in Marcou, *Agassiz* (1896) 1: 169–74]; Maclaren, "Glacial theory of Agassiz" (1842), 351. The locality on Blackford Hill (south of the center of Edinburgh) later became known as the Agassiz Rock and is now marked by a commemorative plaque. Hall's scratched rock surfaces are still spectacularly exposed on Corstorphine Hill (west of the center of Edinburgh). On the social context of Agassiz's visit to Edinburgh, see Finnegan, "Work of ice" (2004).

20. Buckland to Agassiz, 15 October 1840, and Agassiz to Buckland, 17 October 1840 [quoted in Agassiz (E. C.), *Louis Agassiz* (1885) 1: 309]. Lyell's notebooks for 13 October to 6 November 1840 [Kirriemuir-KH: notebooks 84, 85] leave no doubt about his full conversion to glacialism at this point: see Boylan, "Lyell and glaciation" (1998), 148–53, which reproduces some of his field sketches. The use of the term "conversion" in the present narrative is justified by the geologists' reports of their experience of suddenly seeing familiar things in an unexpectedly new way, and in some cases by their own use of this and similar words.

21. Agassiz, "On glaciers" (1840), read at the Geological Society on 4 November 1840 and translated in summary for its *Proceedings*. On this and the subsequent papers, see Boylan, "Lyell and glaciation" (1998).

22. Buckland, "Glaciers in Scotland" (1840), read on 4, 18 November, 2 December. Around this time he planned a revised edition of his *Reliquiae diluvianae* (§6.3); its title was to be enlarged "*et glaciales*", but it was never published.

during the several tertiary epochs was warmer than it is at present". In other words, although he remained adamantly opposed to the theory of a cooling earth, he too had never until recently imagined that the climate of Europe in the geologically recent past could have been much *colder* than at present. However, he now accepted that "several oscillations of temperature" had affected northern latitudes; he had already suspected this when, the previous year, mollusk shells of Arctic species were sent to him from Pleistocene deposits in Canada, matching similar reports from Scotland.[23]

Lyell restated the glacial problem in geohistorical terms of "three distinct phases of action". What needed explaining causally were "1st, the coming of the [glacial] epoch; 2nd, its continuance in full intensity; and 3rd, its gradual retreat". Unlike Agassiz, he envisaged the first phase as one of slowly deteriorating climate, during which glaciers advanced "progressively, century by century, gaining ground" and eventually reaching the sea. The relevant analogy in the present world, he argued, was not the Alps but icy Antarctic islands such as South Georgia. In the second phase, at the height of the glacial period, Lyell assumed that there would have been "vast accumulations of snow filling the plains and valleys to a great height", with only the highest peaks protruding as nunataks; and that "the erratic blocks were detached and conveyed almost imperceptibly along the surface of the frozen snow to great distances". In the third and final phase, as the ice melted, "the boulders were deposited in the various situations in which they are now found".

There was a long and heated discussion after the reading of Buckland's paper, and another after Lyell's: Murchison, Greenough, and Whewell were all in different ways among the skeptics. At one point Greenough, as ever the arch-skeptic, objected that in Switzerland the glaciers "must have crossed Lake Geneva, and ascended very high mountains [i.e., the Jura]" (Fig. 34.6). "Does Professor Agassiz suppose", he asked derisively, "that the lake of Geneva was occupied by a glacier 3000 feet thick?" Agassiz at once replied, "At least!" To anyone who had traveled in the region and was familiar with its topography, nothing could have illustrated more clearly the radical character of what the Swiss naturalist was claiming for Europe as a whole.[24]

35.4 CONCLUSION

The three rival explanations of erratic blocks and scratched bedrock surfaces—that they were traces of a "diluvial" mega-tsunami, or of submergence in a sea with drifting and melting icebergs, or of vast glaciers extending from a newly elevated mountain range—were joined in the later 1830s by a fourth and even more sensational possibility. Agassiz, previously known for his work on fossil fish, unexpectedly joined the fray when he argued that the whole earth—or, at least, the northern hemisphere as far south as north Africa—had been covered by a static sheet of ice during an "Ice Age", *before* the sudden and geologically recent upheaval of the Alps; the latter had merely produced a tilted surface of ice, down which the erratics could slide from Alps to Jura. Only after this had a warming climate melted the static ice-sheet and turned its remnants into slowly retreating valley glaciers. Agassiz integrated this surprising "glacial theory" into his larger picture of geohistory, by suggesting that the Ice Age had been just the most recent in a succession of catastrophic climatic crises in the history of life.

After belatedly going to see for himself some of the existing Alpine glaciers and their effects, Agassiz promoted his new theory at meetings attended by many of the leading European geologists, but he failed to convince them. However, the highly respected Studer later became a heavyweight convert, both to Charpentier's notion of a formerly vast extension of the Alpine glaciers, and to the idea of even more widespread glaciation in northern Europe. When the local geologists Renoir and Hogard found clear traces of small glaciers in the hills of the Vosges in Alsace, Studer's idea of a recent episode of *widespread* glacial conditions became even more plausible. In effect, his modified glacial theory was detached *both* from Charpentier's more local explanation in terms of a sudden upheaval of the Alps, *and* from Agassiz's more global notion of a "Snowball Earth" largely or wholly covered with static ice.

Like the Alsatian geologists, Buckland was among those who returned to their own countries determined to test Agassiz's ideas by searching for traces of former glaciers in regions far from any existing ones. When Agassiz next visited Britain, primarily to continue his research on fossil fish, he also took the opportunity to promote his glacial theory. In Buckland's company he toured the Scottish Highlands, where they found abundant evidence of recent glaciation. Most notably, Agassiz proposed a new and glacial explanation of the famous Parallel Roads of Glen Roy, which was far more convincing than Darwin's had been. Even Lyell was converted—despite the strikingly "catastrophic" character of Agassiz's theory—to the more modest notion of the former existence of glaciers in Britain, because it explained the phenomena more satisfactorily than either a diluvial theory or his own earlier theory based on floating icebergs. However, many other leading geologists were unconvinced, and the fate of the glacial theory remained uncertain.

∾

23. Lyell, "Shells collected in Canada" (1839), and Smith (J.), "Climate of the newer pliocene" (1839), both read on 24 April; Lyell, "Glaciers in Forfarshire" (1840), read on 18 November, 2 December. Buckland to Lonsdale, 12 October 1840 [London-GS: LR6/32].

24. Notes on these famous discussions, made at the time, were published in 1883 but reprinted more accessibly in Woodward, *Geological Society* (1907), 138–44, and in Thackray, *Fellows fight* (2003), 95–101. See also Boylan, "Lyell and glaciation" (1998), 153–56.

Fig. 36.1 "Zermatt glacier: upper part", as portrayed in Agassiz's *Studies on Glaciers* (1840); the bands of debris on the glacier were explained as the products of the lateral moraines of tributary glaciers further upstream. Agassiz is one of the trio relaxing in the foreground, while another of his colleagues studies the rocks behind him. Although crude by contemporary artistic standards, this and his other landscape views (see Fig. 35.2) were useful proxies for the experience of seeing the existing Alpine glaciers at first hand.

CHAPTER 36

Taking stock for the future (1840–45)

36.1 THE PLEISTOCENE ICE AGE

Agassiz had known that Charpentier was preparing a full account of his own glacial research, but he himself gained all the credit by rushing his similar book into print before his rival's; it went to press even before he had found traces of former glaciers far away in Scotland (§35.3). His *Studies on Glaciers* (1840) was the fruit of intensive but hasty work by himself and his team of artists and other collaborators; he added insult to injury by dedicating it to Venetz and Charpentier while hardly mentioning them in the rest of the book. Most of the text simply described the existing Alpine glaciers; a penultimate chapter dealt with the evidence for their former extension; and the volume concluded with a restatement of the full-blown Ice Age theory that Agassiz had first sketched in his "Neuchâtel discourse" (§35.2). The accompanying atlas included many large landscape views of the Alpine glaciers, but compared with the superb topographical art being produced elsewhere at this time they were crude in style (Fig. 36.1).[1]

Charpentier's *Essay on glaciers* (1841), published just a few months later, was more modest in scope. Its illustrations were confined to his pictures of the more

1. Fig. 36.1 is reproduced from Agassiz, *Études sur les glaciers* (1840), pl. 3, explained on 333–34; the peak on the central skyline is the Breithorn (4165m). On the translucent overlay (pl. 3a) the figures are identified by letters, "A" (on the left of the trio) probably denoting Agassiz himself. His field trips to the glaciers, which were repeated annually until he emigrated to the United States in 1846, included a growing entourage of assistants and admiring collaborators, giving his research project a modern flavor of multidisciplinary teamwork. His preface was dated, somewhat dramatically, "At the Grimsel hospice, 20 August 1840", shortly before he suspended his glacial fieldwork and traveled for the third time to Britain. His own German translation, *Untersuchungen über die Gletscher* (1841), made his work even more widely accessible; there was no English translation until the modern edition (with fine large reproductions of the plates) by Carozzi, *Glaciers by Agassiz* (1967).

spectacular erratics (Fig. 34.5) and his map of the Rhône basin and its inferred mega-glacier (Figs. 34.6, 34.7). But his detailed descriptions of the Alpine evidence for glacial extension made his book just as valuable as Agassiz's, if not more so, to the many geologists who literally followed in his footsteps in subsequent years. However, it appeared in print after Agassiz had published his volume and after news of his sensational new discoveries in Britain had become well known to geologists throughout Europe. As Charpentier's reviewer in the *Bibliothèque Universelle* commented wryly, "in science, as at table, *Tarde venientibus ossa* [latecomers get only the bones]". Agassiz the relentlessly self-promoting publicist gained all the glory, although the glacial theory that was eventually accepted by all geologists was in most respects closer to Charpentier's version than to what Conybeare, writing to Lyell, dubbed derisively "the Bucklando-Agassizean Universal Glacier".[2]

Darwin was among those who, like Lyell, soon learned to see familiar topography with fresh eyes. In 1842, during what turned out to be his very last piece of serious geological fieldwork, he revisited the hills of north Wales that he had first seen in Sedgwick's company before he sailed to South America. Following where Buckland had been the previous year, Darwin found his own perceptions of Snowdonia transformed as radically as Studer's had been in the Alps, Renoir's in the Vosges, and most recently Lyell's on the edge of the Highlands. "Eleven years ago", he told Fitton, "I spent a whole day in the valley [Cwm Idwal], where yesterday every thing but the Ice of the Glacier was palpably clear to me, and then [in 1831] I saw nothing but plain water, and bare Rock". Yet Darwin remained extremely reluctant to relinquish his own totally non-glacial interpretation of the Glen Roy terraces in the similarly hilly region of Lochaber. The reason is not hard to find. Not only did he have an understandable attachment to his first substantial piece of published research. More fundamentally, he believed that Glen Roy had given him crucial evidence for the massive crustal elevation required by his Lyellian theory of global tectonics (§33.3, §33.4), which in turn was intimately linked with his nascent private theorizing about species origins. So in his published account of his Welsh fieldwork he combined a recent deep submergence of the whole of lowland Britain with some *local* glaciation in its more hilly regions, "for on that point I cannot *of course* doubt Agassiz and Buckland". As a Lyellian analogue in the present world, he cited a recent account of the Arctic glaciers of Spitzbergen (now Svalbard), where valley glaciers descend to sea level and turn into floating ice-sheets, which then break up into drifting icebergs.[3]

This was in fact just the middle ground that Murchison had foreseen as a likely outcome (§35.3), and that Darwin's hero had already adopted. Soon after the famous discussions of the glacial theory at the Geological Society, Lyell retracted his adherence to anything like Agassiz's Ice Age, withdrew his paper from consideration for full publication in the Society's *Transactions*, and reverted to his earlier iceberg or "drift" explanation for erratic blocks. He did still accept the former existence of glaciers in the Highlands and other hilly regions, because they could be explained—on his Humboldtian climate theory (§21.4)—as the result of a changed configuration of land, sea, and ocean currents in the part of the world that is now Europe, without any major *global* change of climate. But the theory of an Ice Age with sheets of *land-ice* covering vast areas of lowland Europe and North America seemed to him, on reflection, to be a different matter altogether. Having campaigned tirelessly against other catastrophist theories, notably Élie de Beaumont's notion of the sudden upheaval of volcanoes and mountain ranges (§26.2), Lyell could not in the end bring himself to abandon the principles of his *Principles* by adopting a still more blatantly catastrophist hypothesis.

The state of play at this point was shrewdly summarized by James David Forbes (1809–68), the professor of "natural philosophy" (i.e., physics) at Edinburgh, in a long article in the *Edinburgh Review.* Forbes, like his exact contemporary Darwin, was a rising star in the geological world. He had already spent much time at Agassiz's field station on the Aar glacier and elsewhere in the Alps, making accurate measurements in order to work out the physics of glacial movement. But he rightly treated this as the first of "two very distinct questions" within the glacial theory. The second was not a matter of physics but of geohistory. Had there been huge and widespread glaciers during the Pleistocene period, where now there were only small ones or none at all? He cited the erratic blocks on the Jura, such as the famous Pierre à Bot (Fig. 6.4), as the "stumbling blocks" of current geology; and Lyell was his obvious target when he commented that erratics were "enough, alone and all at once, to overthrow any hypothesis as to the omnipotence of causes now in action, unmodified in intensity, however long continued". Likewise he cited the surfaces of scratched and polished bedrock as the glacial theory's "*experimentum crucis*", because they could be seen beneath glacial ice (Fig. 35.4) but could not be shown to be produced under any circumstances by material suspended in running water. Nonetheless, Forbes noted that the glacial theory still had many powerful critics or skeptics; he conceded that it was as yet imperfect, but concluded that it was probably on the right lines.[4]

What eventually tipped the balance in favor of an Ice Age more extensive than Charpentier's but less sensational than Agassiz's—in effect, a theory close to what Studer had suggested—was the product of exploration in the polar regions during just the years in which the glacial controversy was at its height. Further expeditions in search of the elusive Northwest Passage and the north magnetic

2. Charpentier, *Essai sur les glaciers* (1841); Macaire, "*Glaciers* par Charpentier" (1842), 390; Conybeare to Lyell, February 1841 [Philadelphia-APS; printed in Rudwick, "Critique of uniformitarian geology" (1967)]. The attention given to Charpentier's book was also diluted by several other important works on the Alpine glaciers published around the same time: Godefroy, *Notice sur les glaciers* (1840), and Rendu, *Théorie des glaciers* (1840), for example, had already been reviewed by Macaire.

3. Darwin to Fitton, 23 June 1842 [*CCD* 2: 321–22]; Darwin, "Ancient glaciers of Caernarvonshire" (1842); Buckland, "Glacia-diluvial phaenomena" (1842), read at the Geological Society on 15 December 1841; Martins, "Glaciers du Spitzberg" (1840), read by Darwin in translation; Mills, "Darwin and the iceberg theory" (1983). Darwin clung to his interpretation of Glen Roy for another twenty years, while further fieldwork by Scottish geologists progressively improved and reinforced Agassiz's sketchy glacial-lakes theory; only in 1861 did he reluctantly abandon his own alternative, calling his 1839 paper in retrospect "a great failure" and even "one long gigantic blunder": see Rudwick, "Darwin and Glen Roy" (1974), 130–53.

4. [Forbes], "Glacier theory" (1842), quotation on 81; it was nominally a review of nine publications, by Venetz, Charpentier, Agassiz, and others. Forbes, "Observations on glaciers" (1842) and "Leading phenomena of glaciers" (1843), expounded his own concept of glacial movement as *viscous* flow; see also his travel book *Alps of Savoy* (1843), Cunningham, *Forbes* (1990), and Hevly, "Heroic science of glacier motion" (1996).

Fig. 36.2 Part of the icy coast of Antarctica discovered and mapped in 1841 by the British expedition of 1839-43. Its commander James Clark Ross (1800-62) named the peak in this view Mount Sabine (3850m), the mountains the Admiralty Range, and the whole region Victoria Land, in honor of, respectively, the expedition's chief scientific instigator, its governmental sponsors, and his young sovereign. From the 1840s onwards, reports from both polar regions, and images such as this, helped to give much greater plausibility to claims that there had been far more extensive glaciation—though not Agassiz's global "Snowball Earth"—in the Pleistocene period of geohistory. (By permission of the Syndics of Cambridge University Library)

pole had the incidental effect of confirming the sheer scale of the *ice-sheets* covering most of Greenland and parts of other Arctic landmasses (§7.4). This kind of large-scale glaciation, rather than the limited valley glaciers of the Alps, then seemed the most directly relevant analogue for the apparently even more extensive glaciation implied by the scratched bedrock and erratic blocks found so widely in northern Europe and North America. It became evident that sheets of *land-ice* (bearing rock debris) could move, however slowly, even where there was no steep gradient to help, although the physics involved remained controversial. At much the same time, an expedition to locate the south magnetic pole had the similarly incidental effect that it mapped much more of the coasts of Antarctica, which was then confirmed as being a huge continent almost wholly covered in a sheet of land-ice. This again suggested, far more vividly than any little Alpine glaciers, what even larger areas of the northern hemisphere might have looked like, in the geohistorically very recent past (Fig. 36.2).[5]

In the long run, then, a solid synthesis—more substantial than any mere compromise—was constructed more or less along the lines of Studer's early assessment of the problem; that is, somewhere between Agassiz's theory of a catastrophic global Ice Age—a precursor of the modern theory that infers a "Snowball Earth" in the far deeper past—and Lyell's (and Darwin's) concept of relatively minor and local glaciation caused by equally local changes in geography and climate. The consensual outcome was a reconstruction of geohistory in the Pleistocene period as one in which climatic conditions had been far colder than at the present time, not just in northern and central Europe but around the world (and notably in North America); and it was envisaged as a period that had indeed had a drastic effect on terrestrial faunas, though not the kind of total mass extinction that Agassiz (and Cuvier before him) had imagined. What could have caused such a dramatic departure from the generally equable trends of earlier geohistory remained mysterious and controversial.

The remainder of this chapter summarizes how some of the other focal problems that have been described earlier in this narrative were being perceived by the 1840s. This will show how all this research contributed to the growing confidence of geologists that a coherent narrative of the history of the earth and its life, all the way from the earliest decipherable periods up to the present, was being pieced together with substantial reliability.

36.2 PHILLIPS AND GLOBAL GEOHISTORY

The Pleistocene was now claimed—though not without counter-claims—as an exceptionally cold period on the threshold of the present world. The rest of the Tertiary era, before that enigmatic episode, seemed less problematic. Early in the century the Tertiary had been treated as a relatively brief penultimate phase of geohistory, recorded more or less fully in the classic sequence of formations that Brongniart and Cuvier had described in the Paris Basin (*BLT* Chap. 9). But further fieldwork in many other parts of Europe had then shown that some of the other Tertiary "basins" were of different ages, as judged by the assemblages of fossils found in their formations, and as confirmed occasionally by the more direct evidence of superposition (§12.1). This had led to a vast expansion of the perceived magnitude of the Tertiary era, though it still seemed to be dwarfed by the earlier eras of geohistory. Lyell had attempted to calibrate Tertiary geohistory quantitatively, by constructing a natural chronometer based on the apparently gradual changes in its molluskan fauna (§23.3, §26.1). But this ambitious project had broken down in the face of chronic problems with the identification and taxonomy of the shells (§32.3). What survived, and passed into general use by geologists, was Deshayes's qualitative divisions of the Tertiary faunas, as expanded by Lyell and given memorable names by Whewell (*Eocene*, etc.). These enabled

5. Fig. 36.2 is reproduced from Ross (J. C.), *Antarctic regions* (1847) 1 [Cambridge-UL: Zz.1.26], pl. opp. 183, drawn by J. L. Davis on the west side of the Ross Sea at about 72° South; the edge of the vast Ross ice shelf was found (and depicted, pl. at 232–33) at about 78° South; the geography of Antarctica as a whole, as then known, is shown on the "South Polar Chart" at the end of vol. 2. Cawood, "Terrestrial magnetism" (1977) and "Magnetic Crusade" (1979), describe the worldwide geomagnetic research to which Ross's expedition was intended to contribute, and for which Edward Sabine (1788–1883) was a leading adviser to the British government. On the Arctic, see for example Ross (J.), *Narrative of a second voyage* [1829–33] (1835), and Levere, *Science and the Canadian Arctic* (1993). On the further course of the long arguments over the glacial theory, see Davies, *Earth in decay* (1969), chap. 8, and Oldroyd, "Glaciation in the English Lake District" (1999); and, for the world beyond Britain, Schaer, "Agassiz et les glaciers" (2000).

the course of Tertiary geohistory—its organisms, sedimentary deposits, volcanic eruptions, crustal upheavals, and so on—to be reconstructed with a degree of precision that became a shining exemplar for the rest of geohistory.

The Tertiary era, then, had fulfilled its potential—foreseen with characteristic insight by Cuvier (*BLT* §9.3)—to become the cognitive gateway, as it were, to still earlier periods of geohistory. By 1840, the stratigraphy of the younger of these *pre*-Tertiary (Secondary) formations had been resolved in broad outline on a pan-European level, and tentatively even on a global scale, by the recognition of "groups" or "systems" named (in downward or retrospective order) as *Cretaceous*, *Jurassic*, and *Trias*. These names were derived respectively from the distinctive Chalk of northwest Europe, the formations in the Jura hills on the Franco-Swiss border, and a strikingly tripartite sequence of formations in Germany. But the names were beginning to transcend any such regional connotations, and were being used increasingly—by geologists of all European nations—to denote distinctive *periods* of geohistory, regardless of the kinds of rocks involved.[6]

Beneath the Triassic formations of Germany and other parts of western Europe were further limestones and sandstones, some of the latter with major salt deposits; this group was therefore often referred to as "Saliferous". During Murchison's extensive fieldwork in Russia in 1840, he was shown apparently equivalent but much thicker formations in the region around the city of Perm to the west of the Urals. After his return to Britain he proposed that they be termed *Permian*. In effect, they spanned the gap between the Triassic formations and what Conybeare had named the Carboniferous (including the coal deposits that were literally fueling the burgeoning Industrial Revolution). At much the same time, the almost consensual resolution of the "great" and long-running Devonian controversy—"Devonian" formations being recognized as unexpectedly equivalent to the Old Red Sandstone—spanned the similar gap between Phillips's Carboniferous faunas and Murchison's Silurian formations (§31.1).[7]

Finally, below the Silurian, Sedgwick's Cambrian was the lowest and oldest of all the "groups" or "systems" of formations with any fossils at all (§30.3). Unfortunately for Sedgwick, his Cambrian fossils were rare and poorly preserved, and he was unable to define a Cambrian *fauna* clearly distinct from the Silurian. So Murchison, in characteristically imperious style, annexed most of Sedgwick's sequence to his own stratigraphical territory. This soon led to vehement arguments and mutual recriminations, and ultimately to their notorious long-term estrangement. However, in the 1840s the French geologist Joachim Barrande (1799–1883)—a royalist who, after the July Revolution, had exiled himself to Bohemia (now the Czech Republic) and worked there as a civil engineer—reported finding a diverse and distinctive set of very well-preserved trilobites in a formation *underlying* those with Murchison's typical Silurian forms. He called this oldest Bohemian fauna "*Primordial*"; but Murchison's campaign had been so effective that Barrande included his "Primordial" within the Silurian (later it became part of the basis for the modern concept of a distinctive Cambrian period). Whatever the formation was to be called, however, there was no mistaking

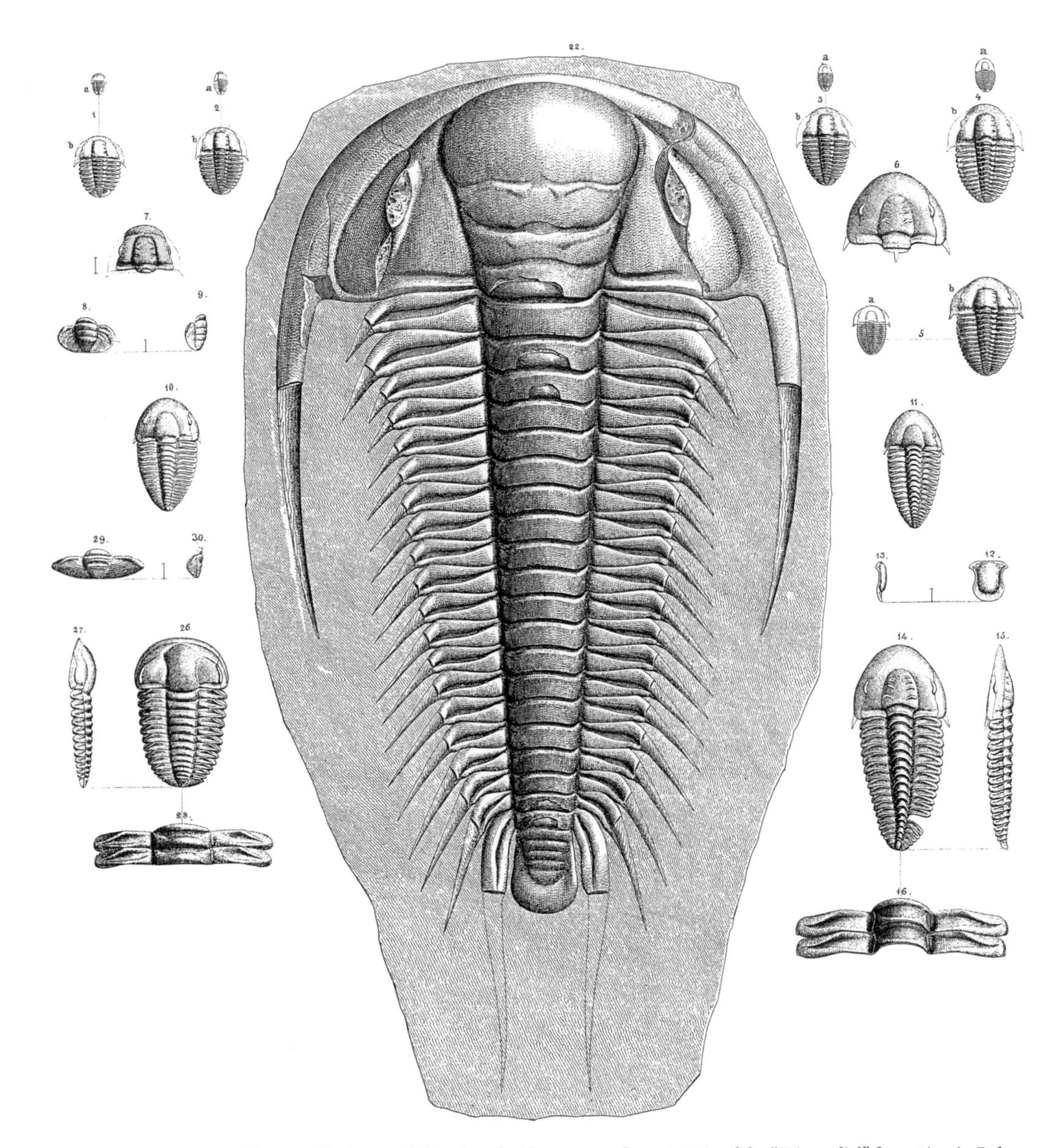

Fig. 36.3 The large trilobite *Paradoxides*, a genus characteristic of the "Primordial" formation in Bohemia (later identified as Cambrian), which Barrande first reported in the 1840s. This superbly preserved specimen (18cm in length) exemplified the large size and complex anatomy of these earliest of all known organisms; this was generally taken to be utterly incompatible with any transformist (evolutionary) interpretation of the history of life. On this lithographed plate, the large *Paradoxides* specimen is flanked by drawings of two other "Primordial" trilobite genera, showing some of the tiny early growth stages (top rows, drawn at natural size and enlarged) that enabled Barrande to reconstruct in detail the life-history (ontogeny) of these extremely ancient animals. (By permission of the Syndics of Cambridge University Library)

6. In Germany, Friedrich August von Alberti (1795–1878), a geologist responsible for salt mines, had defined the Trias as comprising the distinctive *Muschelkalk* (a limestone with marine fossils somewhat like those of the overlying Jurassic), sandwiched between two sets of sandstones and other rocks (*Keuper* and *Bunter*): Alberti, *Monographie des Bunten Sandsteins* (1834), 323–24; Nitsch, "Keuper, 1820–34" (1996). In Britain the limestone was missing and the whole "group" was a part of the New Red Sandstone.

7. Murchison, "Observations géologiques sur la Russie" (1841), 902, first proposed "Système Permien", not only for the formations in the Perm region but also for a well-known German limestone, the

the anatomical complexity of the "primordial" forms of life that it contained (Fig. 36.3).[8]

Significantly, it was Phillips who suggested how the entire record of geohistory could be characterized on the basis of the history of life that was emerging with increasing clarity in these years. Back in 1834, when he first lectured in London as Lyell's successor at King's College, he had broken with the convention established in "Conybeare and Phillips" (the latter no relation: §3.2) and many other works, including Lyell's. He had summarized current knowledge of stratigraphy in true or *geohistorical* order, forwards from the oldest rocks with "*the most ancient reliquiae of living beings*". In 1838 he had expanded these lectures into a long article on geology for the *Encyclopaedia Britannica* (it was also published as a book). Being a work for the general reading public, he explained his ideas by invoking the analogies with *human* history that had been decisive in establishing the practice of geohistory long before (*BLT* Chap. 4). Although "not at present reducible to the scale of historical time" in terms of years, he explained that "the history of the successive formation of the crust of the globe is so far like a narration of human events, that it admits of being placed in chronological order, and classed in periods more or less characteristic".[9]

In the 1840s, Phillips, who by then had about as wide and varied an experience of fossils as anyone in the scientific world, suggested how the whole known global history of the earth could be summarized, on the basis of the fossil record, in a way that would transcend any local or regional stratigraphy. His first step in this direction was a brief article that he published in 1840 in the *Penny Cyclopaedia*,

Fig. 36.4 John Phillips in 1840, painted around the time that he first proposed his tripartite division of the whole of geohistory. The curtain behind him is drawn back to disclose in iconic form the basis of his growing scientific reputation: books, an ammonite, and quartz crystals of museum quality. (By permission of the National Portrait Gallery, London)

one of the early products of the new steam-press technology that catered to the rapidly growing mass market for inexpensive reading matter of an "improving" kind. His starting point was Sedgwick's earlier proposal that the Cambrian and Silurian should jointly be termed "*Palaeozoic*", to denote an era of "ancient life" (§30.3). But Phillips suggested, on the basis of the general character of the fossil record, that this term should be extended upwards in the stratigraphical pile (or forwards in geohistory) to embrace at least the Old Red Sandstone (or Devonian) and perhaps the whole of the Carboniferous and even the overlying formations. At the same time, he coined two new words to match: "*Mesozoic*" (i.e., middling life) and "Kainozoic" (i.e., recent life, later amended to "*Cenozoic*"). These, he implied, might prove useful to denote the rest of the history of life. Whewell's earlier suggestions to Lyell about the naming of the Tertiary periods (§23.3) were his obvious precedent for these respectably Classical neologisms (Fig. 36.4).[10]

The following year these proposals were given a more permanent form, and reached a far more knowledgeable set of readers, in Phillips's book on *The Palaeozoic Fossils of Cornwall, Devon and West Somerset.* This described the fossils that he (and many fossilists and other collectors) had found in southwest England while he was working for De la Beche's new state-supported Geological Survey (§30.2); the book was generously illustrated with Phillips's own fine lithographed drawings of the more significant specimens. However, in addition to detailed descriptions of the fossils and stratigraphy of the "Palaeozoic" rocks of southwest England, Phillips included an important essay in which he summarized his

---

Zechstein (and its modest English equivalent the Magnesian Limestone), and the lower (pre-Trias) part of the New Red Sandstone. Murchison *et al.*, *Geology of Russia* (1845), contained his (and his collaborators') full descriptions.

8. Fig. 36.3 is reproduced from Barrande, *Systême Silurien* 1 (1852) [Cambridge-UL: 376.a.90.5], part of pl. 10. This was published in his full description of the Bohemian fossils, but his striking discoveries became known to geologists throughout Europe much earlier, even in advance of his preliminary report, *Silurien de Bohême* (1846). His "*Étage C*" containing "Primordial" trilobites (in modern terms, of Middle Cambrian age) was sandwiched between unfossiliferous "Azoic" sedimentary and volcanic rocks (*étages* A and B), and four *étages* (D through G) with trilobites that he identified as Silurian. The sequence was unambiguous, the rocks being gently folded in a simple syncline; this *Paradoxides* came from its south limb at Ginetz [now Jince], 10km north of Pribram; the other specimens from its north limb at Skrey [now Skrye], 15km south of Rakovnik (both localities being to the west of Prague). Trilobites of these genera were already known from Sweden but their relation to the typical Silurian fauna had been unclear; the first British specimens were found by Phillips in 1848: Barrande, *Systême Silurien* 1 (1852), 63–66d. See Horný, "Joachim Barrande" (1980), and Secord, *Victorian geology* (1986), 276–81.

9. Phillips, *Syllabus on geology* (1834); *Geology* (1838), 67. In the latter he also outlined (291–95) several methods that might one day lead to a *quantified* geochronology, though he regarded it as a very distant goal. Two decades later, however, he hazarded an estimate of about 96 million years since the start of the Cambrian, based on the total thickness of the sedimentary rocks and their present rates of accumulation: Phillips, *Life on the earth* (1860), 122–37; see Morrell, "Genesis and geochronology" (2001), and *Phillips* (2005), 352–58. This was compatible with the estimates put forward around the same time by William Thomson (1824–1907), later Lord Kelvin, on independent and strictly physical grounds: Burchfield, *Kelvin* (1975) and "Age of the earth" (1998). Such estimates jointly lie at the conceptual roots of modern geochronologies based on post-Kelvin radiometric methods: Rudwick, "Geologists' time" (1999).

10. Fig. 36.4 is reproduced from a portrait by Alexander Craig [London-NPG: no. 23461 (whereabouts of original unknown)], probably painted in 1840 during the British Association meeting in Glasgow. Phillips, "Palaeozoic series" (1840); he amended "Kainozoic" to "Cenozoic" (its modern spelling) to conform with the standard rules for transliterating Greek into English (Whewell's analogous "Eocene" etc. were derived from the same word *kainos*).

proposals for a globally valid classification of sedimentary formations (and implicitly of geohistory). Each of his three major eras (Palaeozoic, Mesozoic, and Cenozoic) was now subdivided into three parts, Lower, Middle, and Upper (implicitly: early, middle, and late). Together they spanned the entire range of geohistory, from the most recent of Lyell's Tertiaries downward, or back in time, to the Transition formations (and even the Primaries, in which in fact no fossils had yet been found). Phillips pointedly declined to use the new names Cambrian, Silurian, and Devonian, treating them as terms of only local application (and thereby effectively snubbing Murchison, the relentlessly hostile critic of his friend and colleague De la Beche). His own terms, in contrast, were to be of global validity, because they were based not on rock formations but on fossils, on the record of a worldwide history of life itself. A small table, tacked onto the end of a *Penny Cyclopedia* article mainly about the formations that Murchison was about to name "Permian", encapsulated Phillips's summary of geohistory (Fig. 36.5).[11]

It appears then that the rocks of this great and varied Saliferous System may be best placed in relation to the other systems of strata by help of a further analysis, as in the scheme subjoined, in which we have added all the other groups of strata to illustrate our general view:—

Proposed titles depending on the series of organic affinities.		Ordinary titles.
Cainozoic strata	upper	Pleiocene Tertiaries.
	middle	Meiocene Tertiaries.
	lower	Eocene Tertiaries.
Mesozoic strata	upper	Cretaceous System.
	middle	Oolitic System.
	lower	New Red Formation.
Palæozoic strata	upper	Magn. Limestone Formation. Carboniferous System.
	middle	Eifel and South Devon.
	lower	Transition strata. Primary strata.

The determination of two of the types of the Palæozoic strata is yet imperfect, because the series of forms intermediate between the transition and carboniferous periods is not fully investigated. It is probable that the Eifel and South Devon fossils make the middle Palæozoic type.

Fig. 36.5 Phillips's tripartite classification of stratigraphy and geohistory, as he first defined it in print in 1841. The three major eras were based on sharply distinct "systems of life"; but their subdivisions expressed the relatively gradual way in which the faunas and floras had changed *within* each era. Sedgwick's Cambrian and Murchison's Silurian were included in "Transition strata", and their recently contentious Devonian (§31.1) was referred to only as "Eifel and South Devon", because Phillips regarded such "Ordinary titles" as sets of rock formations that were probably merely local. In contrast, his own "Proposed titles" such as "Middle Palaeozoic" were designed to be of global validity because they were based on the worldwide history of life. Phillips published this table in an encyclopedia article on the "Saliferous System" (for which Murchison the same year proposed the name "Permian"), but it paralleled the more detailed account in his great monograph on the *Palaeozoic Fossils* of southwest England (also published in 1841). (By permission of the Syndics of Cambridge University Library)

In summary, Phillips's Palaeozoic, the first and oldest major era, was represented by the Transition and older Secondary formations, or everything from Sedgwick's and Murchison's ancient rocks (and even the underlying Primaries) through the Carboniferous and on into the still younger "Saliferous" or Permian. Phillips claimed that the fossils of all these jointly constituted "one system of life". The Mesozoic was represented by all the younger Secondary formations, from the widespread New Red Sandstone (which, with the Muschelkalk of Germany, had now been defined as the Trias), through the "Oolitic" or Jurassic, to those already being termed Cretaceous (up to and including the Chalk itself). The Cenozoic, the third and last major era, was represented by Lyell's three Tertiary groups, including the puzzling Superficial or diluvial deposits (just then in the throes of being considered as possibly glacial), and the present world itself. Phillips's tripartite classification was soon recognized by geologists as fulfilling a long felt need. It made good sense of the broader shape of geohistory, and his terms were soon adopted internationally (and continue to be used by modern geologists, with their meaning almost unchanged).

When Phillips explained his tripartite division of geohistory, he did not deploy the customary analogy with human history that he had invoked earlier. But

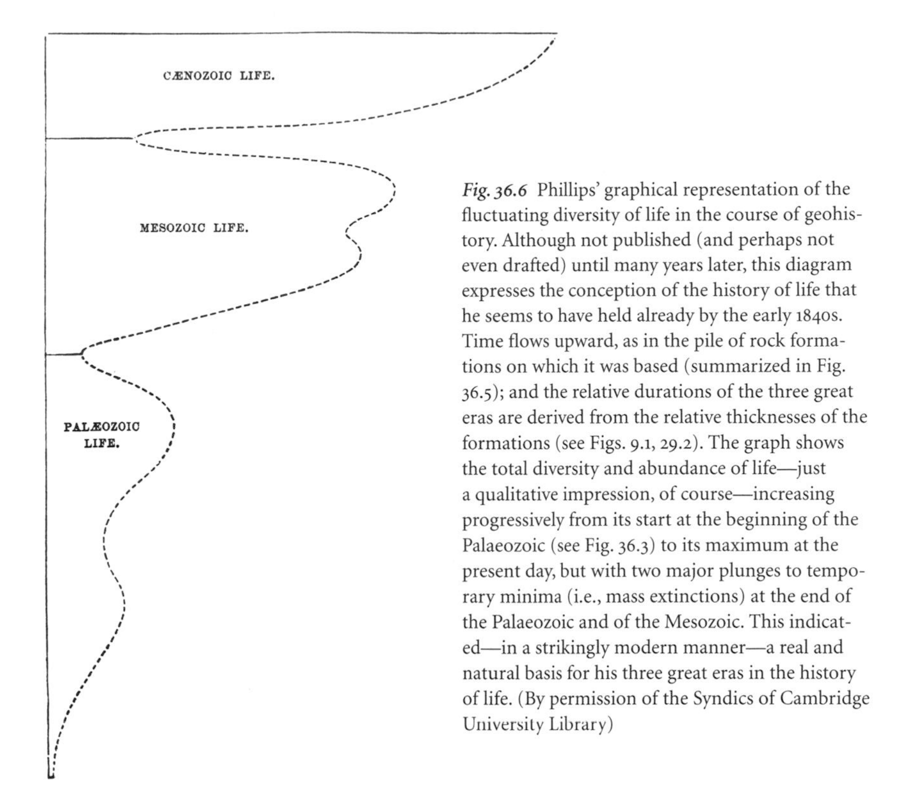

Fig. 36.6 Phillips' graphical representation of the fluctuating diversity of life in the course of geohistory. Although not published (and perhaps not even drafted) until many years later, this diagram expresses the conception of the history of life that he seems to have held already by the early 1840s. Time flows upward, as in the pile of rock formations on which it was based (summarized in Fig. 36.5); and the relative durations of the three great eras are derived from the relative thicknesses of the formations (see Figs. 9.1, 29.2). The graph shows the total diversity and abundance of life—just a qualitative impression, of course—increasing progressively from its start at the beginning of the Palaeozoic (see Fig. 36.3) to its maximum at the present day, but with two major plunges to temporary minima (i.e., mass extinctions) at the end of the Palaeozoic and of the Mesozoic. This indicated—in a strikingly modern manner—a real and natural basis for his three great eras in the history of life. (By permission of the Syndics of Cambridge University Library)

those of his readers who had enjoyed (or endured) a Classical education would have noticed that his Palaeozoic, Mesozoic, and Cenozoic paralleled the traditional tripartite classification of human history. The "Antiquity" of ancient Greek and Roman civilization had been followed, after its collapse, by the "Middle Ages" of the great cathedrals but also (in the eyes of Protestants) of a corrupt Catholicism, before the dramatic new light of Renaissance and Reformation ushered in the era of "Modern" history. Likewise, in Phillips's outline of geohistory, the Palaeozoic era of trilobites and tree-ferns had been followed, after a sharp break, by the Mesozoic era of ammonites, belemnites, and bizarre reptilian monsters. After another sharp break above the Chalk came the start of the Cenozoic era, with its quite modern Tertiary faunas and floras leading without any comparable major break to those of the present world. Whether or not the geologists who adopted Phillips's outline of geohistory appreciated the parallel with human history, they were certainly well aware that what his names denoted was a *history* of life on earth, which marked the major eras in the history of the earth itself. Phillips's tripartite history of life epitomized the way that the fossil record was now established as the principal means of plotting geohistory (Fig. 36.6).[12]

11. Fig. 36.5 is reproduced from Phillips, "Saliferous system" (1841) [Cambridge-UL: XXVI.2.22], 355. See Phillips, *Palaeozoic fossils* (1841), esp. "Notices and inferences" (155–82); also Morrell, *Phillips* (2005), 166–70, and Rudwick, *Devonian controversy* (1985), 387–89.

12. Fig. 36.6 is reproduced from Phillips, *Life on the earth* (1860) [Cambridge-UL: MF.20.47], 66. This is the only image in the present book that dates from well beyond its chronological limit; but the exception is justified, because it gives graphical expression (in both senses) to what was evidently in Phillips's mind even in the early 1840s. The two major plunges in diversity that separate the three eras,

36.3 AGASSIZ AND THE "GENEALOGY" OF LIFE

The most striking feature of geohistory disclosed by all this research was, of course, the directional and even "progressive" character of the history of life. Not only had organisms apparently become more abundant and diverse; the fossil record also suggested that "higher" or more complex groups had been added successively to the world's faunas and floras in the course of time (Chaps. 4, 12). By the 1840s, it was becoming almost impossible to doubt that the fossil record was a broadly reliable trace of the true history of life. Lyell did continue to doubt

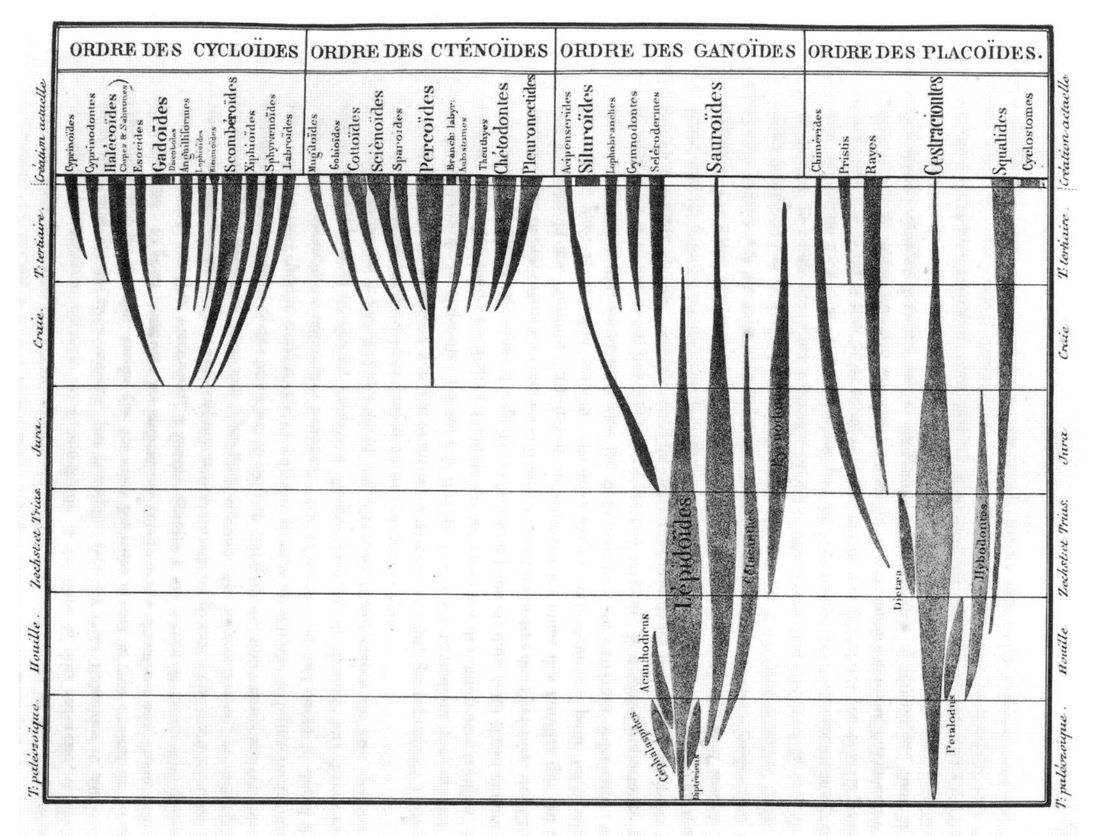

Fig. 36.7 The "Genealogy of the Class of Fish": Agassiz's diagrammatic summary of their history, published in 1843 in his *Researches on Fossil Fish.* Geological time flows upward (as in the stratigraphical pile in which it is recorded); diversity is expressed on the horizontal dimension. Each taxonomic family is represented by a "spindle" shaped on the basis of an impression of its relative abundance and diversity through time; the families are grouped in turn in Agassiz's four "orders". The placoids and ganoids are found even in the "*T[errain] paléozoique*" or pre-Carboniferous formations and are abundant through the rest of the Secondaries, but less so in the "*T[errain] tertiaire*"; the ctenoids and cycloids appear only in the Cretaceous [*Craie*] and are abundant in the Tertiary and in the present world [*Création actuelle*]. The earliest two periods constitute Agassiz's Age of Fish, before there were any "higher" vertebrates; the overall effect is one of increasing diversity through time. Although Agassiz called the diagram a "genealogy", he rejected Lamarckian (and, later, Darwinian) transformism; but this kind of "spindle diagram" was later used by others—as it still is by modern paleontologists—to depict an *evolutionary* history of life. Rather surprisingly, the diagram does not depict the repeated drastic episodes of extinction, particularly at the boundaries between the successive periods, that Agassiz stressed in the context of his glacial theory (§35.1). (By permission of the Syndics of Cambridge University Library)

it, claiming that the record was so deeply and systematically imperfect that any appearance of "progression" was an illusion; but he found himself stranded in a minority of one. At least in relation to the history of life, Lyell's concept of "uniformity" was held by no one but himself (and to some extent by his disciple Darwin).[13]

For example, Agassiz's painstaking research on fossil fish—for which he was praised as much as he was criticized for his speculative Ice Age—enabled him to plot their expanding diversity in detail against the dimension of time represented by the stratigraphical pile. What he had expressed verbally in his early papers (§30.1) was represented visually in a striking diagram published in one of the final fascicles of his great monograph on fossil fish (Fig. 36.7).[14]

The causal processes underlying this directional and even "progressive" history of life were obscure and highly contentious. The theory of a cooling earth (§9.2) remained attractive, and indeed it was reinforced—and set in a cosmic context—by the renewed plausibility of Laplace's earlier "nebular hypothesis" for the origin of the whole solar system. A steadily cooling earth could at least account for a sequence of environments to which the successive faunas and floras might have been well adapted. But it could not readily explain the organic changes themselves.[15]

On the other hand, the detailed history of life that a quarter-century's research had disclosed did not fit easily into the mold of Lamarckian transformism (or, as it was later termed, *evolution*). The progression appeared to be neither linear nor insensibly gradual. New and arguably "higher" groups such as mammals and flowering plants did indeed appear relatively late in geohistory, long after their "lower" relatives such as reptiles and ferns; but the latter continued to flourish in the present world alongside the former, which had not wholly replaced them. And even the earliest known invertebrates were astonishingly complex animals. Buckland had emphasized that they were by no means the crude and poorly adapted "rough drafts" [*ébauches*] that Lamarckian transformism was taken to entail (§29.3); even Barrande's "Primordial" trilobites (Fig. 36.3)—in modern terms, of Cambrian age—were not in any obvious way simpler or more "primitive" than Murchison's Silurian ones, though there were fewer of them. And in the underlying Primaries (largely, in modern terms, *Precambrian* formations) no

---

and make their faunas so sharply distinct, are those defined by modern geologists as the mass extinctions at the Permo-Triassic and Cretaceous-Tertiary ("K/T") boundaries, both now regarded as even more drastic or "catastrophic" than Phillips depicted them.

13. Lyell continued to deny the directionality of the fossil record until after Darwin published his *Origin of species* (1859), and then conceded it only with reluctance. Benton, "Progressionism in the 1850s" (1982), describes a celebrated episode during Lyell's rearguard campaign: an anatomically "advanced" fossil reptile was found in Scotland in what was claimed to be a formation of Old Red Sandstone (Devonian) age, i.e., back in Agassiz's "age of fish"; but the rock was later redated—on good field evidence—as New Red Sandstone (Triassic), whereupon the anomaly vanished and Lyell's case collapsed. Bartholomew, "Singularity of Lyell" (1979), is an important review of the long history of Lyell's intellectual isolation.

14. Fig. 36.7 is reproduced from Agassiz, *Poissons fossiles* (1833–43) 1 [Cambridge-UL: MA.62.17], pl. at 170 (in a fascicle issued in 1843). Agassiz used "*paléozoique*" in the sense proposed by Sedgwick (§30.3), not (yet) in Phillips's expanded sense.

15. See, for example, Schaffer, "Nebular hypothesis" (1989), and, more generally, Bowler, *Fossils and progress* (1976).

fossil remains had yet been found, and certainly not the still simpler organisms that were expected on any transformist theory. At the other end of geohistory, the human species—taken to be self-evidently the culmination of any "progress" in the history of life—was unquestionably a very recent arrival, even if it were to be found to date from before the diluvial or glacial disruption rather than after it (§28.1, §28.2); in any case it was quite distinct from the recently discovered fossil primates that were its closest analogues (§28.3).[16]

It was therefore no surprise that Sedgwick, for example, reacted with such vehemence to the notorious *Vestiges of the Natural History of Creation* (1844), which became a *cause célèbre* in Britain (though not in the rest of Europe) in subsequent years. For the Scottish popular science writer Robert Chambers (1802–71) had the effrontery—in the eyes of many British savants—to use the latest geological research to give the history of life a Lamarckian interpretation. Sedgwick's intemperate public attack on the tantalizingly anonymous *Vestiges* was fueled by his fully justified fears about its sensational impact on the popular apprehension of the "place of Man in Nature". But he was also indignant at, even angered by, the way it distorted and misused all the careful and painstaking work of geologists and paleontologists in the preceding decades.[17]

Darwin, on the other hand, with a strategic sense that matched Cuvier's a generation earlier, had already recognized that the problem of organic change was best tackled from the other end. After his return from South America, and in parallel with his public work on geology (§33.3), he had begun exploring privately the possible causes of the origin of species in general (with, in modern terms, a tentative theory of speciation by micro-evolution), setting aside the more refractory problems of the origin of major groups (in modern terms, macro-evolution) and of life itself, not to mention the thorny issue of the origin of the human species. In 1842, even before Chambers's bestseller appeared, Darwin sketched an outline of his radically novel theory of the origin of species by natural selection, and two years later he even wrote out a publishable "essay" in case he died before completing the much fuller version (on the scale of Lyell's three-volume *Principles*) that he was planning. He was deliberately pursuing an agenda set by Lyell: not only to look for a small-scale process that could cumulatively generate large-scale results, but more specifically to find an observable "actual cause" for the piecemeal origins of species to match that of their equally piecemeal extinctions (§24.2), and thereby to solve Herschel's "mystery of mysteries" (§32.4). Darwin's innovative research on species-origins was radically different from the transformist conceptions of Lamarck, his followers, and (in Chambers's case) his vulgarizers; Darwin—"I, the geologist", as he had called himself—was transposing Lyell's *geohistorical* approach into zoology and botany, even more profoundly than his mentor.[18]

36.4 WHEWELL'S HISTORICAL-CAUSAL SCIENCE

However, by the 1840s the work of Lyell himself was experiencing an ever more mixed reception. He continued to be greatly admired and respected by geologists for his research on the Tertiary formations and his reconstruction of Cenozoic geohistory, and for his persuasive advocacy of actual causes (§32.1). But, as the present narrative has emphasized, he had been by no means the first to demonstrate the explanatory potential of actual causes, and he was widely criticized for claiming more for them than the evidence seemed to warrant. All geologists agreed that much—perhaps most—of what had happened in the deep past could indeed be explained in terms of actual causes; but they continued to suspect that some causal processes might no longer operate at their former intensity, or do so only at long intervals, and that some might even have ceased to operate altogether. None of this affected in any way the geologists' assumption that all such processes, in the past as in the present, were based on physico-chemical foundations that had remained at all times unchanged: even the most violent "catastrophes" would have broken no "laws of nature". Yet catastrophes there might have been, in the opinion of most geologists, disrupting occasionally the vast spans of more tranquil geohistory. The upheaval of mountain ranges, even if no longer envisaged as all-at-once events, remained an obvious candidate for this kind of interpretation; but the geohistorically recent disruption of some kind of Ice Age was a new example that put all others into the shade. All in all, Lyell's rigorous concept of "absolute uniformity" was looking tattered by the 1840s, as he had tacitly conceded in the successive transformations of his *Principles* and *Elements* (Chap. 32).

It took a relative outsider, but a polymath, to define in fundamental terms what the still quite young science of geology had now become. If Conybeare was arguably the most intelligent of all its leaders in Britain—and recognized internationally as such, for example by his earlier election to the Académie des Sciences in Paris—then Sedgwick's Cambridge colleague Whewell was an even more formidable intellectual, who was analyzing the science from an even wider perspective. In 1837 Lyell had persuaded Whewell to succeed him as president of the Geological Society, flattering him as a "Universal" among "men of science".

16. What has since been termed the "Cambrian explosion" of animal life became, during the rest of the century, increasingly embarrassing to Darwinian evolutionists who insisted on an extreme gradualism, because further fieldwork around the world showed that many of the underlying Precambrian formations were as unaltered as Paleozoic ones, and equally suitable for the preservation of fossils. One candidate Precambrian fossil, optimistically named *Eozoon* [dawn animal], was soon debunked as an inorganic metamorphic structure. Only in the twentieth century were genuine traces of Precambrian life discovered (mostly microfossils, but also a few larger "soft-bodied" or shell-less animals); this has allowed the puzzling Cambrian explosion to be to some extent explained, and not merely explained away as an artifact of an imperfect record.

17. Chambers, *Vestiges* (1844); [Sedgwick], "[Review of] *Vestiges*" (1845). Secord, *Victorian sensation* (2000), is a superbly comprehensive account of the secretive production and manifold kinds of reception of Chambers's book. *Vestiges* also appropriated the nebular hypothesis from the astronomers, who were equally incensed. It was, in short, a Victorian forerunner of those popular books and television programs that misrepresent current research and raise the blood pressure of modern scientists in much the same way.

18. Darwin's manuscript "Sketch of 1842" and "Essay of 1844" are printed in De Beer, *Darwin and Wallace* (1958), 41–153; see Browne, *Darwin: voyaging* (1995), 436–47, for their place in his ongoing work. Rudwick, "Darwin in London" (1982), analyzes his work in these early years in relation to the continuum between private and public. The present volume comes to an end, not coincidentally, around the start of the period covered by the superb scholarly research (by the so-called "Darwin industry") on the development of his evolutionary theory, leading toward his *Origin of species* (1859) and still later works, and on its reception and application throughout the biological and human sciences.

During his tenure Whewell had given two fine presidential addresses, bringing his wide-ranging intellect to bear on some of the fundamental problems of the science. But even more significant was his three-volume *History of the Inductive Sciences* (1837), which in turn was designed as the empirical foundation for his two-volume *Philosophy of the Inductive Sciences* (1840). These great works both covered the same vast field of knowledge, namely all the natural sciences. They were to have a profound impact on scientific thinking for the rest of the century, and not only in the anglophone world.[19]

Here, however, attention must focus on the last major division in Whewell's magisterial classification and analysis of the natural sciences, for this is where the science of geology was to be found. Its category came last, not because it was the least important but because it was the most novel, and did not fit into any of Whewell's other and more traditional major groups of sciences. For he recognized that geology had become the sole example (thus far) of a new *kind* of natural science, matched only by the *human* sciences [*Wissenschaften*] of "comparative philology" and "comparative archaeology". Geology was novel, in that it now combined two previously distinct kinds of investigation. One was "*palaeontology*", a term that Whewell used in an unusual, broad, and literal sense, to denote the study not just of fossils but of *all* entities of which the relics survived from the unobservable deep past. The other part—its analogue already well established in medicine, for example—was "*aetiology*", the analysis of the causes of natural entities and events of all kinds. So Whewell invented another neologism, "*palaetiology*", that incorporated both terms. Geology was a "palaetiological" science because its practitioners had now learned not only how to reconstruct the deep past of the earth's *history* from its fragmentary traces in the present—"bursting the limits of time", in Cuvier's famous phrase (*BLT* §9.3)—but also how to infer the *causes* of the events so reconstructed (or at least how to search for plausible causes). Causal earth-physics (*BLT* §2.4)—now broadened in practice to include, for example, the enigmatic cause of new species—was combined with the much newer practice of geohistorical reconstruction, which had originally taken its cue from human historiography (*BLT* Chap. 4). Although Whewell's neologism was never widely adopted by geologists—their way of working was too well established and too much taken for granted to need it—it did express what their practice had now become.[20]

36.5 CONCLUSION

To avoid straining the patience of readers (and the endurance of the author) any further, the narrative in this book has been brought to an end in the mid-1840s, with the arguments over glacial extension and the putative Ice Age still in full swing, as they would continue to be for many years to come. But the glacial debate has been traced far enough to illustrate its decisive role in the broader story of the transformation of the earth sciences by the penetration of *geohistorical* ways of practicing them.

The most important point about the controversy over the Ice Age was that any such episode in the geologically recent past was *totally unexpected* by leading geologists of all stripes: by Buckland no less than by Lyell, to mention just two representative figures. It was drastic enough to count as a catastrophe, yet this particular catastrophe was the very last kind of event that might have been anticipated, on an earth that most geologists believed was cooling very slowly from its unimaginably remote origin as a fiery ball in space. On the other hand it was too drastic, and in geological terms too sudden and catastrophic, to have been anticipated, on an earth that a very few geologists (notably Lyell and his disciple Darwin) believed to be in an endless steady state of dynamic equilibrium, oscillating on a vast and stately cycle of gradual but directionless change. In other words, it did not fit into *either* of the current large-scale geotheories that geologists were cautiously bringing back into play.

Equally unexpected, in another way, was the complex story of the history of life that had been pieced together during the quarter-century covered by the present volume. Based initially on Smithian stratigraphy, vastly enlarged and deepened by the geohistorical perspective inspired by Cuvier, and brought to maturity by geologists such as Smith's nephew Phillips, this research disclosed a history of life that, while certainly directional and even "progressive", could not have been predicted (or rather, retrodicted) in any detail from grand theoretical schemes such as Lamarck's transformism. Phillips's definition of the eras of ancient, middling, and modern kinds of life (Paleozoic, Mesozoic, and Cenozoic) made sense of the broad outlines of geohistory. But the earliest organisms known in the fossil record, those of the early Paleozoic, were not crude or simple or imperfect, as Lamarck's theory was taken to predict, but complex and evidently well adapted to their original environments. Progressively "higher" forms of life did not simply replace earlier or "lower" groups of organisms but were successively added to them, culminating in the appearance of the human species in the geologically recent past. The *cause* of this contingent sequence of organic change and enrichment remained profoundly mysterious, but the *reality* of some such progressiveness was accepted by almost all geologists; it was denied almost uniquely by Lyell, holding out heroically for his concept of "absolute uniformity". Yet behind the scenes his disciple Darwin was busy working on a possible causal explanation of the origin of species, which might plug a glaring gap in Lyell's account of geohistory in terms of actual causes, and thereby save his "uniformity" in another of its manifold meanings.

However, Lyell was also on the defensive about yet another meaning of "uniformity". This was his rejection of any kind of physical "catastrophe" in the deep past—however natural its causation—if it was unmatched *in magnitude* in the present world. On the possibly abrupt upheaval of mountain ranges, for ex-

~ 19. For an introduction to an extensive scholarly literature, see Fisch, *Whewell philosopher* (1991); Yeo, *Defining science* (1993); and the essays in Fisch and Schaffer, *William Whewell* (1991).

20. Whewell, *History of the sciences* (1837), 3: 481–624; *Philosophy of the sciences* (1840) 2: 95–165. At least one scholar did promptly recognize the potential value of Whewell's term in his own field: "May Palaetiology, on the higher theme of Man, obtain as numerous and scientific inquirers as she already possesses on the subject of the earth!": Winning, *Comparative philology* (1838), 15. Constraints of space here make it impracticable to do more than touch on Whewell's concept, but see the important essay by Hodge, "Whewell and palaetiological science" (1991).

ample, and their possible causal connection with occasional episodes of mass extinction, the jury was still out. But on the "catastrophe" of a geologically recent Pleistocene period of globally cold climate—even if it had not been Agassiz's Ice Age of "Snowball Earth" intensity—the evidence against Lyellian "uniformity" was mounting steadily.

Even more clearly than any of the other developments that have been described in this lengthy narrative, the widespread acceptance of at least a moderate version of the glacial theory signaled the far more profound if tacit recognition by geologists that their science should properly be grounded in a *geohistorical* approach; and that such a practice, which had first been developed by deliberately exploring and exploiting a set of analogies with *human* history, entailed a view of geohistory as equally unpredictable (even in retrospect) and ineluctably *contingent*. Whewell, a polymath with a keen and well-informed interest in geology, recognized that this still youthful natural science was uniquely "palaetiological", in that it was successfully combining the reconstruction of the earth's deep *history* with its *causal* explanation. In conjunction with the physical sciences as a resource for the latter, the most fundamental conceptual model for the future of the science of geology, as for its past half-century, would therefore be that of *history*. This is what modern earth-scientists—whether they know it or not—have inherited from their forerunners in the astonishingly fruitful decades covered by this book and its predecessor. This is the human achievement that they have absorbed so thoroughly into their everyday practice that they can, and rightly do, take it completely for granted.

CONCLUDING (UN)SCIENTIFIC POSTSCRIPT

This is, obviously, the concluding part of the present volume. It is also a postscript: literally, in that I drafted it some time after completing the rest of the book; and structurally, in that it comments on the narrative as it were from outside. It is scientific in that it deals with the observations, debates, and theories with which the narrative has been concerned; but also in that it considers even more briefly the implications of this work for understanding the ways in which the practice of the sciences generates reliable knowledge of the natural world. And since I have borrowed this title from a writer beside whom all of us who work in such "science studies" are mere pygmies, it is also perhaps unscientific—in a sense faintly echoing Kierkegaard's—in that I make my own viewpoint somewhat more explicit here than in the body of the volume. Conversely, I have left these comments without footnoted references: those already initiated into the mysteries of science studies will have no difficulty seeing where I position myself in recent and current debates.

I introduced *Bursting the Limits of Time* (*BLT*) by promising that it would trace a revolution in human thought no less profound, but much less widely recognized, than those associated with Copernicus, Darwin, and Freud. The revolution comprised the realization that our human species is a recent newcomer on our planet, confined to a mere sliver of time at the tail end of an immensely long, diverse, and eventful history of the earth: what earth scientists now call *geohistory*. It was not just that the timescale was vastly enlarged, since there had always been the possibility that the universe was eternal and the timescale therefore infinite. It was rather that the "place of Man in nature"—as our politically incorrect ancestors expressed it—was radically reshaped, in such a way that *non-human* nature was now seen to be as significant in the dimension of time as it had long been recognized to be in the dimension of space. Of course, this was often construed as meaning that in both respects humanity itself was *not* significant, but that existential inference did not (and does not) follow from the science.

I introduced the present volume, *Worlds Before Adam* (*WBA*), as the sequel to *BLT*. It has continued the narrative far enough to show how the modern scientific conception of geohistory was well established—at least in outline and in general character—even before the middle of the nineteenth century. *BLT* showed how, back in the late eighteenth, *historical* ways of exploring and thinking about the earth had been marginal and not widely valued by savants (among whom were those who would now be called "scientists"). Yet, within the first few decades of the political "age of revolution", this historical approach to the natural world came to be seen as the cutting edge of a newly named and exciting science of "geology". However, at the point at which the narrative in *BLT* was suspended, the value of treating the natural world as *having its own history* was still more a matter of promise than of achievement. For example, Buckland's reconstruction of Kirkdale Cave, as a den of extinct hyenas and a part of an "antediluvial" ecosystem, was rightly treated by his contemporaries as a sensational vindication of Cuvier's programmatic aspiration to "burst the limits of time" by making *prehuman* history reliably knowable to human beings confined to the present. But it remained no more than a promising exemplar, until the same methods were applied much more broadly and consistently to the relics of *all* periods of geohistory. This widening use of a geohistorical approach, and its spectacular fruitfulness in scientific practice, is what the narrative in *WBA* has traced through the subsequent decades of the political "age of reform".

I have borrowed these convenient labels from mainstream historians, for the subtitles of my volumes, not only to signal whereabouts in human history my story lies, but also to indicate that these scientific developments were not insulated from the wider human world in which they took place. However, they were not necessarily influenced in any crude sense, let alone determined, by the circumstances of their societal setting. The latter remained mainly a *context*, the stage on which the scientific action was played out: sometimes constraining the plot, but also often enabling it. For example, British commercial and strategic expansion provided the *occasions* for Buckland to study fossil bones from Burma and Alaska, which he then used to argue for the worldwide reach of his putative "diluvial" event. But the specimens themselves, and his use of them, were in no way tainted with imperialism, any more than the earlier arguments about extinction, based on fossil bones from the wilds of Siberia and the banks of the Ohio, were either soiled with Russian autocracy or perfumed with American democracy. Likewise, some of the first fossil primate bones were found by British expatriates in India; but the fossil record was not thereby smeared with colonialism or orientalism, and the geohistory based on them would have been no different if the bones had been found by Bengalis, or if the only specimens known had been those that were in fact found at almost the same time in the heart of the French countryside.

Similarly, the course of geological debate was greatly constrained by the long Revolutionary and Napoleonic wars, which restricted the movements of geologists and limited their personal exchanges and particularly their opportunities for fieldwork; and at the coming of peace they lost no time putting this right. But the military antagonism between the warring nations, and the political rivalries of those enjoying the subsequent peace, had little if any substantive impact on the scientific arguments that continued unabated among savants. For example, they themselves showed no correlation between political conservatism and an insistence on slow and gradual causal processes in geology, or between political radicalism and an emphasis on geological "catastrophes" and "revolutions". To claim that the informal and international "Republic of Letters" (which included the natural sciences) usually stood above the political fray is therefore no self-deceiving idealization, but reflects a social reality. So, in a quite different way, does my geographical and cultural focus on Europe (which of course included my native offshore island), rather than more distant parts of the world or non-European cultures. In the late eighteenth and early nineteenth centuries, even North America was as yet peripheral to the scientific world, and its contributions were valued primarily as empirical grist for theoretical or interpretative mills that were grinding away in Europe. This was even more the case with non-European or non-Western cultures; their bodies of indigenous knowledge, however significant they may have been in other contexts, had only the most marginal input into what became the science of geology.

I must emphasize that *BLT* and *WBA* have not set out to describe the history of the earth sciences as a whole, even during the few decades that the volumes jointly cover. Nor do I claim to trace the origins of "geology" as a new science, except insofar as a historical (or rather, geohistorical) perspective came to be seen as one of its defining features. So I have given little attention to many important parts of the earth sciences, such as mineralogy and petrology, structural and tectonic geology, igneous and metamorphic geology, and still less to economic geology and the technologies of mineral exploration and mining, except when and where any of these impinged directly on my main theme of the reconstruction of geohistory. Paleontology, in contrast, has been given a lot of attention: not because it was my own professional pursuit before I reinvented myself as a historian (and it remains my favorite science), but because in the decades covered by my volumes fossils turned out to provide some of the most reliable evidence for building up a picture not only of the history of life itself but also of its environment and the physical state of the earth's surface.

Both my volumes have traced the debates about geohistory primarily among those leading savants who had earned for themselves an international reputation in the relevant sciences, and who were active in an international network of scholarly exchange. My volumes comprise, without apology, an elitist account of certain aspects of the formative decades of the science of geology; but the elite was one based not on birth or wealth or social class but on intellectual originality and sheer hard work. Many of the most prominent figures were indeed relatively wealthy—whether by inheritance (von Buch, Boué) or through marriage (Prévost, Scrope) or by exceptionally well-paid employment (Cuvier, Brongniart)—because in an age with few public sources to pay research costs some private wealth was a necessary condition for productive work. But readers of my volumes will have noticed that those who would now be called "professionals"— many of them, such as university professors, being poorly paid, then as now—were far more prominent on the scientific stage at this period than is commonly supposed. In this respect, as in many others, Great Britain was the great exception; but I have tried throughout these volumes to transcend the overwhelmingly anglophone and anglocentric bias of much of the historiography of the earth sciences, and instead to replicate in my narrative the international and multilingual world of the leading scientific actors themselves.

However, I have not in fact limited my account to the activities and ideas of those who comprised the scientific elite. As my narrative should clearly illustrate, their work was at every point sustained and made possible by a gradient of lesser figures, ranging from geologists with substantial but strictly local expertise down to fossil collectors with great skill in finding valuable specimens but little ability to interpret what they found; and this gradient of competence cut right across the social spectrum, since for example collectors ranged from lower-class "fossilists" who relied on their finds to make a living, to aristocrats with private "cabinets" of greater scientific value than many public museums. Nonetheless, my focus has been on the intellectual elite, because it was through their exchanges and arguments that claims to new scientific knowledge were tested most rigorously and effectively; and their work was as necessarily international as that of their successors in the modern scientific world.

The outlook of the leading geologists was sustained and made possible by a well-developed infrastructure of scientific communication, as should have been apparent throughout my narrative. It has been realistic to assume—and this can usually be confirmed from the documentary evidence—that any significant new idea or discovery reported in one part of the scientific world became known elsewhere within a year or two at the most. Leading geologists visited each other and attended each others' meetings as frequently as they could, often across national frontiers and linguistic boundaries, and corresponded when they could not. Rapidly improving systems of roads and public coaches on land, fast sailing ships and then the new steamships at sea, and above all a rapid and reliable mail system, international as well as internal to each country, all contributed to the efficient exchange of ideas and evidence.

One important feature of this infrastructure was the rapidly expanding network of scientific periodicals, each of which routinely reprinted, translated, or at least summarized the more important contents of others. But men of science—the gendered phrase matched the social reality—were not wholly dependent on translations, since those with any ambition would make sure they could read the international scientific language of the time (French), just as modern scientists of all nationalities must do the same with its present successor (American English). Furthermore, books, periodicals, and offprints of articles were routinely exchanged not only between their authors or editors, but also between the various state-supported "academies of the sciences" (or, in Britain, the private Royal Societies in London and Edinburgh), and between the new specialist societies (for example, the Geological Society in London and the Société Géologique in Paris). The annual meetings of the peripatetic scientific societies established successively in Switzerland, the German states, and Britain (and later in the Italian states, the United States, and elsewhere), were likewise in practice as international as circumstances allowed.

The narrative in my volumes has followed the actors not only in being highly international but also in being strongly oriented toward the *particular*, toward the importance of seeing particular specimens in museums and particular localities in the field *with one's own eyes*. This should not be dismissed as a sign of crude positivism, naive Baconian inductivism, or any other philosophical ism. Again and again, the opinions of geologists were decisively changed by seeing crucial specimens and localities *for themselves*: the famous extinct volcanoes of the Massif Central in France, for example, and the equally famous huge Alpine erratic block stranded high above Neuchâtel, remained powerful in their impact decade after decade. I have therefore tried to follow the actors in their emphasis on the importance of *fieldwork* (and also to follow them, literally, by "re-treading" or replicating their fieldwork as far as my own circumstances have allowed). Fieldwork was not just a symbolic activity establishing a geologist's stamina or virility. The hammer rightly became the science's icon, because fieldwork was the arena in which he (very rarely she) encountered the primary evidence that demanded interpretation and understanding. This evidence often ran counter to earlier expectations and required awkward or even painful adjustments of belief. But such adjustments were in fact made; geologists did often change their minds in the face of what they saw with their own eyes, *in the field*. And they often had analogous experiences in public museums and private cabinets, when they saw important or even unique specimens for the first time.

Of course, geologists were not always able to see the primary evidence for themselves, or not yet. So those who *had* seen it, and had found a particular interpretation persuasive on the spot, would then try to convey the same experience to others, through the use of *proxy* specimens, landscape views, and so on (in the form of drawings, engravings, and lithographs), in their papers at meetings and in their publications. Although they often presented this as the disinterested laying out of the "facts", their intensive and subtle use of proxy pictures—which I have replicated and exemplified as far as possible in these volumes—was as deeply *rhetorical* (in the proper non-pejorative sense) as their accompanying texts. Jointly, words and images were designed to present the *evidence* for a particular interpretation as persuasively as possible, to make them in effect eloquent "*witnesses*", so as to make as strong a *case* as possible. Such pervasive legal metaphors, and the previous legal training of many prominent geologists (such as Lyell), were neither trivial nor coincidental.

Another set of metaphors held center stage in *BLT* but has been much less prominent in my narrative in *WBA*, for good and important reasons. The primary theme of both volumes has been the transformation of the sciences of the earth—or at least, of major parts of them—by the infusion of a sense of *the earth's own history*. My claim has been that this took place by the deliberate and conscious transposition of a powerful set of analogies and metaphors from *human* historiography into the world of nature; and that this was achieved, not way back in the seventeenth century in the age of Hooke and Steno, nor as late as the mid-nineteenth in the age of von Ranke and Lyell, but quite specifically in the late eighteenth in the work of naturalists such as Desmarest and Soulavie, during the few decades covered by my narrative in *BLT*. It was then that the vocabulary of human history was knowingly transposed into the study of the earth. Nature's own history, it was claimed, could be reconstructed by treating rocks and fossils, mountains and volcanoes, as *nature's* documents and monuments, *nature's* coins and inscriptions, *nature's* archives and annals: Soulavie called himself nature's archivist; Cuvier, nature's antiquarian or historian.

This strategy proved so fruitful and successful that within a generation or two it was taken for granted among those who now called themselves geologists. The explicit historical metaphors therefore sank largely out of sight, and only occasionally resurfaced in the expert debates that have been my primary focus in *WBA*. Historical analogies remained valuable literary tools when geologists wanted to communicate to a wider public what their science had now become (as in the popular writing of Mantell and Phillips); but it was no longer necessary in what they wrote for each other. Its crucial conceptual role in the science of geology has therefore been almost forgotten by their modern successors (who may, for example, be unaware of the human-historical origins of terms such as "epoch", "period", "era", and "chronology" itself).

The main theme of my volumes, then, has been the progressive transformation of the scientific study of the earth by the injection of historical ways of interpreting what can be observed: the earth, and by extension the natural world as a whole, came to be seen as having their own *histories*. Initially, this appeared difficult to combine with older and better established ways of studying nature: reconstructing a unique sequence of distinctive events in geohistory was quite different from detecting repetitive patterns that might reveal underlying "laws of nature", unchanging and ahistorical. So in the period covered in *BLT* the new *geohistorical* practice developed alongside the older *causal* practice of earth physics [*physique de la terre*], and independently of it. Priority was given to establishing *what* in fact had happened in the deep past, and (in relative terms) *when*, shelving the often much more difficult problems of determining *how* the events might have been caused. But during the subsequent period covered in *WBA* the geohistorical and the causal were increasingly brought together and integrated, and this integration came to be regarded as definitive for the science of "geology".

The chief tool for integrating history and causation was the "actualistic" method of using the observable present as the key to the unobservable deep past. The study of the earth's "actual causes", or processes *now* ("actually") at work in the world, could lead to the conclusion that they could *or could not* have generated the effects observed in the traces of the past. Even a small stream, for example, might conceivably be capable of eroding the deep valley in which it flows, given enough time; but the same cause could not conceivably have shifted a huge erratic block tens or hundreds of miles from its point of origin, no matter how much time was allowed. The method obviously depended on the underlying uniformity, through time, of the *kinds* of causal processes involved (and, even more obviously, of the physical "laws of nature" that made them possible), while leaving open the question of the uniformity or otherwise of their magnitude or intensity of action at different times. This use of actual causes to integrate geohistory with a causal understanding of the deep past was first explored during the period covered in *BLT*; by the period covered in *WBA* it was being deployed *routinely* by geologists on all sides. It was not invented by Lyell, as a persistent historical myth would have it, nor even by his allies and immediate predecessors such as Prévost, von Hoff, and Scrope, though Lyell's insistence on an extreme kind of uniformity (Whewell's "uniformitarianism") did make the issues more explicit than before.

What separated Lyell from almost all his contemporaries was his ambition to follow Hutton in constructing an overarching "theory of the earth" or *geotheory* of a particular kind. He claimed that the earth had been in steady-state or dynamic equilibrium, or at least in a stately cyclicity, as far back in the deep past as any evidence survived, and that it could be expected to continue in the same way into the furthest imaginable future. Back in Hutton's time, at the start of the period covered in *BLT*, geotheory had been a flourishing scientific genre. But such mega-theorizing had then fallen into disrepute for being uncontrollably speculative, and it was displaced by the more modest goals of finding adequate explanations for more limited ranges of features or events. However, during the period covered in *WBA* geotheory was making a comeback, quite independently of Lyell, as evidence accumulated that the earth had had an extremely hot origin in the unimaginably remote past and had been cooling exponentially ever since. So the Lyellian debates saw an updated version of Hutton's geotheory pitched against an updated version of Buffon's: a steady-state or cyclically changing earth pitched against one with an unrepeated and broadly directional *history*. Within the latter, there was ample explanatory space for allowing the possible role of occasional "catastrophes" punctuating far longer stretches of relative calm; this "catastrophism" claimed only that the intensity of geological processes might have been far from uniform through time, never that the ahistorical "laws of nature" had been breached or suspended, let alone that supernatural or divine agency had been involved.

My narrative in *WBA* ends rather abruptly with the early versions of the glacial theory. This is not as arbitrary as it may seem, because the glacial theory, more than any other single development, put a huge question mark against *both* the rival geotheories just mentioned, and therefore against the genre itself. The much earlier eclipse of geotheory (as described in *BLT*) had been due above all to the criticism that too many theories were chasing too few facts; in effect, there was a call for a moratorium on mega-theorizing until the relevant facts were better established. The second time round, however, it was the new facts themselves that made the goals of geotheory questionable. The inference that there had been a drastic "Ice Age" in the (geologically) very recent past was utterly unexpected on both the current rival geotheories, steady-state and directional. Geologists on both sides wriggled hard to avoid the implications of accepting the factuality of a major glacial episode in the Pleistocene period (while generally rejecting Agassiz's extreme "Snowball Earth" version of it). In the long run, they concluded implicitly that the genre of geotheory itself was flawed, insofar as it aspired to provide a predictive model that specified top-down what "must" or "should" have happened in the course of geohistory. Instead, they came to accept that geohistory, like the human history on which it had first been modeled, needed to be reconstructed bottom-up from the detailed evidence of what in fact had happened; like human history it was deeply and ineluctably *contingent* and therefore unpredictable even in retrospect.

Lyell's steady-state geotheory became in the long run untenable, as he himself eventually conceded, at least for the history of life, when, belatedly and half-heartedly, he accepted Darwin's evolutionary theory. Yet he fully deserves his prominent position in my narrative in *WBA*, no less than Cuvier deserves his in *BLT*. If—as I believe—the major "revolution" traced in both volumes merits a place alongside those associated with Copernicus, Darwin, and Freud, and if—as I concede only grudgingly—it too must be given a label based on Dead White Males, then I suggest it should have the somewhat inelegant title of the *Cuviero-Lyellian revolution*. (Historians of the sciences have rightly become cautious about the notion of scientific "revolutions", but it can still help to make sense of the *longue durée* of human understandings of nature.) For it was Cuvier above all who demonstrated how to reconstruct the vanished deep past from its surviving traces, and thereby, in his own phrase, to "burst the limits of time". What he himself achieved mainly with fossil mammals and reptiles became the exemplar for later geologists and paleontologists to extend to other organisms and to their physical environments, and so to build up a cumulatively reliable picture of the whole of geohistory, inorganic as well as organic. And then it was Lyell above all who demonstrated, particularly by his exemplary reconstruction of the Tertiary periods, how much of this geohistory—even if not, as he himself hoped, all of it—could be explained in causal terms by applying the actualistic method more thoroughly than ever before, thereby integrating history and causation in a way that has remained definitive for earth scientists ever since.

I was somewhat surprised to find Lyell looming so large in this story, as I reviewed the sources and constructed my narrative; not least because much of my own earlier research was focused on demythologizing his quasi-heroic status and giving his many critics (the so-called catastrophists) a more sympathetic hearing. So I had expected that Lyell would turn out to be a less central figure in my story than he has been in the traditional historiography. But in writing my narrative, and trying in a sense to re-live the debates of the time, I found the sources compelling me to give Lyell much of his traditional prominence, although no longer in heroic mode, and on strictly geological grounds rather than just those of his influence on Darwin's later evolutionary theorizing. I now judge that his impact on other geologists was profound and largely positive, even though at the same time they rejected some of his most cherished beliefs (and their modern successors, knowingly or not, have adopted much the same dual position).

This summary of my two volumes may give my fellow historians the impression that I am irredeemably "Whiggish" in my interpretation of the history of geology, and that I see the developments described and analyzed in *BLT* and *WBA* as leading ineluctably toward the truths embodied in the present state of the science. This is not so, or not so baldly. Any geologist who has lived through the dramatic revival of catastrophism that started around 1980—most widely known for the theory of a major impact from outer space, which supposedly wiped out the last of the dinosaurs at the Cretaceous-Tertiary ("K/T") boundary in geohistory—should have learned to be cautious about equating the current state of scientific knowledge with ultimate Truth. In earlier decades of the twentieth century any suggestion of major "catastrophic" events in geohistory was roundly condemned by most geologists as utterly "unscientific", on grounds that echoed Lyell's arguments more than a century earlier while generally lacking his subtle quality of thought. (In some of my own detailed work on the evolutionary history of brachiopods, back in the 1960s, I felt I had to insist, against this prejudice, that the evidence for a drastic mass extinction at the Permo-Triassic boundary should be faced and explained, and not merely explained away; and my colleague Brian Harland and I were struggling even more against the tide when we suggested around the same time that there might have been what has more recently been called a "Snowball Earth" episode of drastic and even worldwide glaciation in late Precambrian geohistory.) If the canons of acceptable explanation in geology have changed so dramatically in a mere quarter-century of recent history, it is certainly unwise to treat current scientific opinion as, unproblematically, the standard of truth by which the work of more distant generations, such as those of Cuvier and Lyell, should be judged.

Yet a cautiously skeptical attitude of this kind is quite compatible with believing that the practice of the natural sciences, geology included, has led and continues to lead toward improved understandings of the natural world, in a way that scientists often take too uncritically for granted, and that "science studies" people who analyze their practice are sometimes too reluctant to concede. Scientific knowledge is, of course, endlessly *corrigible.* And scientific opinions are changed, not only by changing scientific fashions but also—as it should be unnecessary to point out—by encounters with new scientific evidence. This can perhaps be seen more clearly and unambiguously in a science like geology, in which the decisive evidence often consists of "huge solid facts" discovered and studied in the field, and in which theoretical stumbling blocks can literally be stumbled over, than it is in sciences where the scientist is separated from the primary evidence by a formidable array of computerized instrumentation, or where an elaborate superstructure of theory is erected on minuscule foundations of empirical evidence. This should suggest the value of attending to the sheer *diversity* of the sciences (which was the underlying theme of my Tarner Lectures at Cambridge many years ago, from which both *BLT* and *WBA* have been derived). The sciences are not all the same, not even all the natural sciences; and we do them no justice and ourselves no favors by continuing to treat physics (or any other single science) as the standard by which all other kinds of knowledge are to be judged either adequate or deficient.

I hope my lengthy narrative in these two volumes has illustrated how the course and character of geohistory did become more reliably and fully knowable, notwithstanding many false leads and dead ends. It is no accident that the outline of geohistory that was becoming clear by the end of my narrative—with, for example, Phillips's definition of Paleozoic, Mesozoic, and Cenozoic eras—remains almost unchanged in the practice of modern geologists. Its contents have of course been hugely refined, in a degree of detail and with a quantitative precision that would have astounded and delighted Phillips and his contemporaries. And they would not, I think, have been disconcerted to find his three great eras now reduced in relative significance (as, jointly, the "Phanerozoic", or those with a clear fossil record), by the charting of the even vaster expanses of still earlier (Precambrian) time, reaching back almost to the birth of our planet. For this unimaginably lengthy and diversely eventful geohistory has been and continues to be reconstructed by deploying essentially the same methods as those they themselves first worked out almost two centuries ago.

There is one specific way in which my narrative may have given some readers the false impression that I see the reconstruction of geohistory in simple linear terms as a story of successive discoveries leading toward modern Truth. I have indeed highlighted the historical importance of the *first* cases of distinctive kinds of interpretation: Desmarest's first reconstruction of the physical geohistory of a specific area; Soulavie's first consistent application of human-historical analogies to geohistory; Cuvier's first "resurrection" of particular extinct animals; Buckland's first reconstruction of a vanished ecosystem; Gressly's first paleogeographical mapping of distinct depositional and ecological "facies"; and so on. But I have described these "firsts" not as celebratory milestones on the road to Truth, but rather as influential *exemplars* that enlarged the interpretative repertoire that was available thereafter to other geologists. I hope my narrative has illustrated my belief in the historical importance of such "paradigms" of practice (the most enduringly useful meaning of Thomas Kuhn's protean and much overworked term). They imprint a genuine linearity, or rather, a cumulative character, on the history of geohistorical research; for once a new kind of interpretation had been devised, and its exemplar had shown it to be valuable and fruitful, it was hardly likely to be disinvented or forgotten, but instead was added to an ever-enlarging repertoire.

Another potentially serious misreading of my narrative would be a result of the necessarily condensed character of my account. It may seem highly inappropriate—or even insulting to my most stalwart readers—to describe two lengthy volumes as condensed; but anyone who has studied in detail any of the episodes I have tried to cover will know that my account is at many points quite sketchy, in relation to the barely explored riches of the available sources. Among such sources, even some major scientific books have as yet been studied with little more than what Kuhn once castigated as mere "preface history"; most of the abundant scientific periodicals published in several languages in the early nineteenth century have only been skimmed by historians (and certainly by me); and some major archives of scientific correspondence and other manuscripts have hardly begun to be exploited. I speak of what I know, having filled another substantial volume with a reconstruction of a single important episode—which here in *WBA* is condensed into a short section of one chapter—simply because I had discovered almost by chance that it could be traced in rich *and revealing* detail from a huge mass of previously untapped sources.

However, I allude here to my own *Great Devonian Controversy* for a much more substantial reason. The chief value of its detailed reconstruction of the course of one intensive debate among geologists was that it exposed the ineluctably "agonistic" or argumentative character of the social and cognitive process by which this—and, I believe, all—scientific knowledge was and is generated. Controversy is not an embarrassing blemish on the fair face of Science, as some scientists seem to feel, but what animates it and gives it life. New interpretations are forged through argument, hammering away on the anvil of what all sides find themselves obliged to accept as relevant evidence. And the Devonian arguments exemplify a pattern that I suspect is widespread if not universal in the sciences. Scientific controversy does not usually result in the triumph of one view over another, truth over error, as the rewriting of history by one side often suggests in retrospect. Instead it results in the construction of an interpretation not anticipated by *either* side: genuinely *new* knowledge, to which both sides have contributed even if one of them fails to admit it. I believe that most of the episodes covered in *BLT* and *WBA* show the same pattern, which I could have demonstrated more clearly if it had been practicable to trace the course of debate in greater detail. For example, Buckland's diluvial theory and Lyell's drift-ice theory were initially opposed, but both were eventually transcended and superseded by the glacial theory. Yet much of Agassiz's apparently triumphant Ice Age theorizing was soon rejected, or never accepted, in a way that modern glaciologists and Quaternary geologists, celebrating him as their founding father, tend to overlook.

An emphasis on controversy has not in fact been lacking in traditional accounts of the history of geology during the period covered in my two volumes. But often it has been assigned a much simpler—indeed simplistic—role. As already hinted, the history has been presented as a series of arguments between Good Guys and Bad Guys: Vulcanists fighting Neptunists, Uniformitarians fighting Catastrophists, Evolutionists fighting Fixists, and so on. My narrative has not confronted these controversies head-on, because I have been determined not to be deflected or distracted from my primary goal of tracing the development of geohistorical ways of understanding the earth. But I hope that my account has suggested, at least in passing, how and why these seductively simple polarities are deeply misleading. No geologist thought that *all* rocks were formed either by igneous or by aqueous processes; the meanings of uniformity and of catastrophe were both highly diverse; so were Lamarck's, Geoffroy's, and Darwin's versions of transformism, which were criticized on correspondingly diverse grounds; and so on.

I have left to the end of this concluding essay any mention of the supposed controversy in the history of geology that now looms largest in the public mind, or at least in the minds of many of my friends and colleagues who over the years have asked me about my long-term research project (of which these two volumes are the product). On hearing that I was exploring the history of a dawning realization that our species is a recent newcomer at the tail end of a lengthy geohistory, their reaction was almost always, "I suppose then that you'll be dealing with the conflict between geology and Genesis" or "between geologists and The Church" or—most sweepingly—"between Science and Religion". In fact I have tried, on the contrary, to follow the historical actors themselves, not only in their accent on fieldwork and their international outlook and so on, but also in the way they treated any such "conflict" as *marginal.*

The great fallacy in the "conflict thesis"—a fallacy sedulously fostered by those modern commentators who can fairly be described as crusading atheistic fundamentalists—is that it treats both sides of the supposed conflict as reified and ahistorical entities: "Science" and "Religion". In reality, everything depended, then as now, on when, where, and who. In the case of the kind of science that eventually became geology, the issues in the late seventeenth century, in the time of Steno and Burnet, were quite different from those in the period covered in *BLT* and *WBA*, which in turn were quite different from those of the late nineteenth century in the time of Huxley and White (the latter the influential propagator of "warfare" imagery). Even in the period covered in my volumes, the issues arising in Britain were recognized at the time as being bizarrely different from those in the rest of Europe. And in early nineteenth-century Britain, the issues that exercised those at the Geological and Royal Societies, for example, were quite different from those that either alarmed or delighted some sections of the wider public.

I have chosen to focus my history of geohistory on the scientific elite; other equally valid and important histories could—and, I hope, will—be written from the perspective of the wider public (or rather, publics). For example, those savants who came to be called geologists, whether or not they regarded themselves as religious believers, treated the question of the *timescale* of geohistory as having been settled long ago and once and for all. In their opinion it had become clear beyond question that the timescale was vast, far beyond the whole of recorded human history and indeed literally beyond human imagination, even though no quantitative figures could be attached to it. Some of them saw a parallel between the emerging scientific picture of geohistory and the poetic imagery of the first Creation story in Genesis, but others rejected any such parallel or simply had no interest in the matter. But all of them repudiated the "scriptural geology" propounded by some of their contemporaries among the general public (almost exclusively in the anglophone world), with its insistence on a "short timescale" of no more than a few millennia for the whole of cosmic history. Their attitude was closely analogous to the reaction of modern earth scientists to the similar "young-earth" theorizing of some religious fundamentalists in modern America.

Though the social settings were quite different, in both these situations the underlying issue was and is that of *literalism* in biblical interpretation. In the period covered in *BLT* and *WBA*, the development of scholarly biblical criticism—motivated as often by the desire to deepen religious belief as to undermine it—had already influenced savant circles, even in benighted Britain, far more deeply than modern secularist myths might suggest (it long antedated Strauss's *Das Leben Jesu* and George Eliot's translation of it, for example). Literalistic readings of biblical texts had already been widely displaced by historically and culturally sensitive interpretations, often with the intention of uncovering deeper *religious* meaning. On the issue of the earth's timescale there was therefore no significant conflict between geology and Genesis, or between geologists and a "Church" that in reality was far from monolithic. The only conflict—sometimes and locally—was between scientific savants (including those who were religious believers) on the one hand, and specific sections of the wider public on the other.

The false polarizations fostered by the conflict thesis have obscured the diversity of the issues involved in the encounter between new natural-scientific knowledge and other bodies of knowledge (some of them equally new). Even within the supposed field of "geology and Genesis", it has obscured the contrast between the question of the earth's origin and prehuman timescale and that of the historicity of the biblical Flood (in effect, between the first chapter of Genesis and the sixth to eighth). For during the period covered in my two volumes, most savants, whether religious or not, treated the Flood story—unlike the Creation story—as clearly *historical* in character, whether the history was regarded as reliable or merely legendary. This brought it into contact with the new geohistory, since it purported to record an event that might have had its place *both* in human history *and* in the physical history of the earth. In effect, therefore, it might represent the boundary zone between recorded human history and prehuman geohistory. However, Genesis was not unique in this respect. As Cuvier recognized, a mass of ancient records in other cultures, ranging as far away as China, also deserved to be mined for their possible evidence for an exceptional watery catastrophe, however obscure and garbled *all* of them (including the Genesis account) might be. His admirer Buckland, adapting the issue to the local circumstances of England (and indeed of Oxford), narrowed down Cuvier's multicultural approach to the single issue of the historicity of the Genesis account. Yet in the long run even Buckland followed his geological colleagues, and notably the pious Sedgwick, in abandoning the equation between the puzzling "diluvial" phenomena and *any* early human records, recognizing that the former were much more ancient than the latter. So Buckland became a leading proponent of the new glacial theory, which turned his "diluvium" into the traces of an Ice Age long antedating any literate human culture. All this was not a triumph of Science over Religion, but simply a case of the usual scientific learning process, in the course of which the chronologies of geohistory and human history came to be more clearly distinguished and differentiated.

Finally, I have suggested in *BLT* that the Judeo-Christian cultural tradition had a far more profound role in the shaping of the new practice of geohistory, and a strongly positive one at that. What was transposed from human history into geohistory, from culture into nature, was not only a fertile set of metaphors and analogies, but also an underlying belief in the *historical* and therefore profoundly *contingent* character of the world, both human and non-human. And this in turn was derived primarily from the Judeo-Christian sense that human lives, and the whole history of the world, were under the sovereignty of God as, ultimately, creator and sustainer of all. The savants who were most effective in constructing a fully historical approach to the study of the earth—who recognized that every feature of the earth had *its own history*—were those (such as de Luc and Buckland) who shared the Judeo-Christian sense of the contingency of the world's history, the sense that the course of past events could at any point have been different, and might be unpredictably different in the future. For this meant that geohistory could be reconstructed only by attending closely to the traces of what *in fact* had happened, which might be utterly unexpected and not what those doing the reconstructing might have anticipated. In contrast, those savants (such as Hutton) who were sustained by a deistic metaphysics, stressing the supremacy of unchanging causation, were antithetical to any true geohistory and played no major part in its reconstruction. Even Lyell, who in certain respects exemplified the fusion of these intellectual traditions, clearly owed the geohistorical component of his synthesis to his Christian cultural environment, as represented for example by his mentor Buckland.

This suggestion that the recognition of the historical character of the natural world, and specifically of the sheer contingency of geohistory, was fostered by the Judeo-Christian understanding of the contingency of human history under divine sovereignty, does not amount to a claim in favor of the validity of that or any other religious tradition. Nor do I propose it primarily as a positive counterweight to the relentlessly negative picture presented by the outworn historiography of intrinsic "conflict", though it is that. What is more important is that it suggests the kind of interaction that there has been—and continues to be in the modern world—between the construction of new scientific knowledge and the lives of the human beings who do the constructing. The savants who came to call themselves geologists learned how to reconstruct the *history* of nature—"bursting the limits of time" to recover an unexpected sequence of "worlds before Adam"—by integrating their exploration of mountains and volcanoes, rocks and fossils, with their understanding of their own lives, lived in social worlds that likewise had their own *histories*. For our modern world, hurtling towards self-induced environmental catastrophe, the cautionary implications of the extremely long, complex, and contingent geohistory that these early geologists first began to reconstruct should need no further emphasis.

SOURCES

1. PLACES AND SPECIMENS

In all the historical research that lies behind the writing of this book (as also of its predecessor *BLT*), I have tried to follow historical actors not only in what they read and wrote, as represented by conventional textual sources, but also in what they *saw*, as represented by features of historical importance that can be studied in the field or in museums (for example, a specific rock outcrop or fossil specimen).

Listed below are the maps I have found helpful in following *in the field* the work of historical figures mentioned in this book, and in trying to understand their interpretations of what they saw, in the light of seeing the same features for myself (unlike the fauna and flora, many geological features have changed relatively little in the past two centuries, once one mentally subtracts the modern overlay of superhighways, power lines, urbanization, etc.). I list topographical rather than geological maps, because the latter are inherently interpretative and their use in the field intrudes modern concepts and conclusions into what should be an exercise in *historical* understanding. Most of the maps listed are on a fairly small scale, which is best suited to a preliminary appreciation of the issues; larger-scale maps are also invaluable while one is in the field. Some localities are listed more than once, because some sheets overlap substantially.

The second category of "material" sources, namely the historically decisive specimens that are still on public display in certain museums, should also logically be listed here. But since they have been cited in the footnotes with the same style of abbreviation as the libraries and archives that contain manuscript sources and pictures, the two lists have been combined, and all the abbreviations of this kind are explained in the "Manuscripts and Pictures" section below.

Like the textual sources listed in the bibliographies that follow, both lists are confined to features and specimens that I myself have been able to study at first hand. Many others, equally relevant and instructive, might have been included, had I had the time and opportunity to see them.

France

Institut Géographique National (1:100000):

31 *St-Dié, Mulhouse, Bâle*: southern Vosges, Thann, Wesserling.

37 *Dijon, Tournus*: Grotte d'Osselle.

38 *Besançon, Lausanne*: Jura, Neuchâtel (Switzerland).

42 *Clermont-Ferrand, Montluçon*: Monts Dore, Puy de Dôme, Olby, Lac Aidat, Lac Chambon, Boulade, etc. (Parc Naturel Régional des Volcans d'Auvergne) [overlaps with sheet 49].

45 *Annecy, Lausanne*: Genève, Arve valley, Mont Blanc massif, Chamonix, upper Rhône valley, Lac Léman.

49 *Clermont-Ferrand, Aurillac*: Monts Dore, Monts du Cantal, Puy de Dôme, etc. (Parc Naturel Régional des Volcans d'Auvergne) [overlaps with sheet 42].

50 *St-Étienne, Le Puy-en-Velay*: Le Puy, upper Loire.

53 *Grenoble, Mont Blanc*: Mont Blanc massif.

59 *Privas, Alès*: Ardèche valley, Aubenas, Jaujac, Thueyts, etc. [Vivarais volcanoes].

63 *Tarbes, Auch*: Sansan.

66 *Avignon, Montpellier*: Rhône delta, Lunel-Viel, Sommières.

72 *Béziers, Perpignan*: Narbonne, Bize-Minervois.

90 *Environs de Paris*: Seine valley, Montmartre, Sèvres, Grignon, Fontainebleau, etc. [Paris Basin].

Germany

Fritsch Wanderkarte (1:35000):

123 *Gössweinstein, Pottenstein*: Muggendorf; Gailenreuth and other caves.

Italy

Istituto Geografico Centrale (1:50000):

4 *Massiccio di Monte Bianco*: Mont Blanc massif, Chamonix.

Touring Club Italiano, Carta turistica (1:50000):

Parco del Etna: Etna and environs.

Kompass Carta turistica (1:50000):

91 *Lago di Como, Lago di Lugano*: Lake Como, Bellagio.

Touring Club Italiano, Grande carta stradale (1:200000):

Campania e Basilicata: Vesuvio, Campi Flegrei, Pompeii, Ercolano [Herculaneum].

Sicilia: Etna, Catania, Siracusa, Val di Noto, Agrigento, Enna.

Toscana: Florence, Siena, north and south flanks of Apennines.

Veneto, Friuli, Venezia, Giulia: Bolca, Euganean hills, Po delta.

Switzerland

Carte Nationale de la Suisse / Landeskarte der Schweiz (1:100000):

35 *Vallorbe*: Pays de Vaud, Jura.

31 *Saane/Sarine*: Neuchâtel.

40 *Lac Léman*: Genève, Lac Léman (Lake of Geneva), Jura.

41 *Col du Pillon*: upper Rhône valley, Monthey, Bex, Diablerets massif.

42 *Oberwallis*: upper Rhône valley, Fiesch glacier, Berner Alpen.

46 *Val de Bagnes*: Mont Blanc massif, Chamonix, Grand St-Bernard, Val de Bagnes.

47 *Monte Rosa*: Zermatt, Pennine Alps.

United Kingdom

Ordnance Survey, Landranger series (1:50000):

34 *Fort Augustus, Glen Albyn & Glen Roy*: Glen Spean, Glen Roy (Parallel Roads).

54 *Dundee & Montrose, Forfar & Arbroath*: Strathmore, Kirriemuir.

66 *Edinburgh, Penicuik & North Berwick*: Edinburgh Castle Rock, Corstorphine Hill.

94 *Whitby & Eskdale, Robin Hood's Bay*: Whitby, Yorkshire coastline.

100 *Malton & Pickering, Helmsley & Easingwold*: Vale of Pickering, Kirkdale.

137 *Ludlow, Wenlock Edge*: south Shropshire [Murchison's "type" Silurian area]

151 *Stratford-upon-Avon, Warwick & Banbury* and 164 *Oxford, Chipping Norton & Bicester*: Cherwell and upper Thames valleys, Stonesfield.

192 *Exeter & Sidmouth* and 193 *Taunton & Lyme Regis, Chard & Bridport*: Lyme Regis, East Devon and Dorset coastline.

196 *The Solent & Isle of Wight, Southampton & Portsmouth*: Isle of Wight, Hampshire coastline.

198 *Brighton & Lewes, Haywards Heath*: Lewes, Cuckfield.

202 *Torbay & South Dartmoor*: Torquay, Kent's Cavern.

Ordnance Survey, Explorer series (1:25000):

164 *Gower, Llanelli*: Gower peninsula, Paviland.

2. MANUSCRIPTS AND PICTURES

Below is a key to the abbreviations used in the footnotes to denote the libraries, archives, and museums in which cited manuscripts and reproduced prints and paintings are held. Also listed here, for convenience, are museums that display historically important specimens mentioned in the text (see "Places and Specimens", above).

Cambridge-ES	Department of Earth Sciences, University of Cambridge
Cambridge-TC	Library of Trinity College, Cambridge [Whewell papers]
Cambridge-UL	University Library, Cambridge [Sedgwick and Darwin papers]
Cambridge/MA-HLH	Houghton Library, Harvard University, Cambridge (Massachusetts) [Agassiz papers]
Cardiff-NMW	Department of Geology, National Museum of Wales, Cardiff [De la Beche papers]
Edinburgh-UL	University Library, Edinburgh [Lyell papers]
Exeter-DRO	Devon Record Office, Exeter [Buckland papers, formerly on loan but now returned to private ownership]
Keyworth-BGS	British Geological Survey, Keyworth (Nottinghamshire) [De la Beche papers]
Kirriemuir-KH	Kinnordy House, Kirriemuir (Tayside) [Lyell papers: private archive]
London-BL	British Library, London
London-BM	Department of Prints and Drawings, British Museum, London
London-GS	Library of Geological Society, London [Society archives and Murchison papers]
London-KC	Department of Geology, King's College, London
London-NHM	Natural History Museum, London
London-NPG	National Portrait Gallery, London
London-RCS	Museum of Royal College of Surgeons, London
London-RS	Library of Royal Society, London
London-UCL	D. M. S. Watson Library, University College London [Greenough papers]
London-WL	Wellcome Trust Library for the History and Understanding of Medicine, London
Madrid-MCN	Museo Nacional de Ciencias Naturales, Madrid
Narbonne-AM	Archives municipales, Narbonne (Aude) [Tournal papers]
Neuchâtel-BV	Bibliothèque de la ville de Neuchâtel
Neuchâtel-IG	Bibliothèque de l'Institut de Géologie, Université de Neuchâtel [Agassiz papers]

Oxford-MNH	Museum of Natural History [formerly University Museum], Oxford [Buckland and Phillips papers]
Paris-IF	Bibliothèque de l'Institut de France, Paris [Cuvier papers]
Paris-MHN	Bibliothèque Centrale, Muséum National d'Histoire Naturelle, Paris [Cuvier and Brongniart papers]
Philadelphia-APS	Library of the American Philosophical Society, Philadelphia [Lyell papers]
Washington-SI	Smithsonian Institution, Washington, D.C.
Wellington-ATL	Alexander Turnbull Library, Wellington (New Zealand) [Mantell papers]

3. PRINTED SOURCES: PRIMARY

In this bibliography, alphabetization of proper names with "de", "von", "van", etc. follows what was customary at the time or is usual in recent historical work: e.g., von Buch and von Hoff are listed under B and H respectively, not V; but the Englishman De la Beche is listed under D, not B. Alphabetization of double-barrelled family names likewise follows contemporary usage: e.g. Léonce Élie de Beaumont and Étienne Geoffroy Saint-Hilaire are listed under E and G respectively, not B and S; but William Fox-Strangways and George Poulett Scrope are both listed under S, not F or P. Contemporary abstracts and translations are not listed, unless they are specifically commented on in the main text. Most of those published in periodicals can be traced easily through the *Royal Society catalogue of scientific papers 1800–1863* (London, 1867–72); those published as books are listed in the catalogues of the British Library, Bibliothèque National, Library of Congress, and other major libraries, and in works such as Louis Agassiz's *Bibliographia zoologiae et geologiae* (London, 1848). Modern reprints, and modern translations of books and articles into English, are noted in brackets at the end of the entries on the originals, where I am aware of them.

The following list is a key to the abbreviations used in this bibliography for items published in periodicals and multi-volume 'dictionaries'. References include not only page numbers but also *plate* numbers; plates are often physically separated from texts and are all too easily missed. Also listed here, for convenience, are a few abbreviations for other periodicals and institutions cited in the footnotes.

AAGC	Atti dell'Accademia Gioenia di Scienze Naturali di Catania
ACP	Annales de chimie et de physique [Paris]
AJS	American Journal of Science [New Haven]
AKPAW	Abhandlungen der Königlichen-Preussischen Akademie der Wissenschaften, physikalische Klasse [Berlin]
AM	Annales des mines [Paris]
AP	Annals of philosophy [London]
APC	Annalen der Physik und Chemie [Leipzig]
ASEV	Annales de la Société d'Émulation, Département des Vosges [Épinal]
ASHSN	Actes de la Société Helvétique des Sciences Naturelles
ASN	Annales des sciences naturelles [Paris]
BAAS	British Association for the Advancement of Science
BC	British Critic [London]
BI	Biblioteca Italiana [Milan]
BSGF	Bulletin de la Société Géologique de France [Paris]
BSNG	Bulletin des sciences naturelles et de géologie [Paris]
BSNM	Bulletin de la Société Impériale des Naturalistes de Moscou
BU	Bibliothèque universelle: Sciences et arts [Geneva]
CO	Christian observer [London]

CRAS Comptes-Rendus de l'Académie des Sciences [Paris]
CSRSG Commentationes Societatis Regiae Scientiarum Goettingensis
DSG Denkschriften der Schweizerischen Gesellschaft für die gesammte Naturwissenschaften / Mémoires de la Société Helvétique des Sciences Naturelles
DSN Dictionnaire des sciences naturelles [Paris]
EJS Edinburgh journal of science
ENPJ Edinburgh new philosophical journal [continuation of *EPJ*]
EPJ Edinburgh philosophical journal
ER Edinburgh review
GDNA Gesellschaft Deutscher Naturforscher und Ärzte
HARST Histoire de l'Académie Royale des Sciences, des Inscriptions, et des Belles-Lettres de Toulouse
JASB Journal of the Asiatic Society of Bengal [Calcutta]
JG Journal de géologie [Paris]
JMGGP Jahrbuch der Mineralogie, Geognosie, Geologie und Petrefaktenkunde [Heidelberg]
JP Journal de physique, de l'histoire naturelle et des arts [full title varies] [Paris]
JRGS Journal of the Royal Geographical Society [London]
KDVA Kongelige Danske Videnskaberne Selskabs naturvidenskabelige og mathematiske afhandlingar [Copenhagen]
KVAH Kongliga Vetenskaps Academiens handlingar [Stockholm]
MARS Mémoires de l'Académie Royale des Sciences [Paris]
MG Mémoires geologiques [Paris]
MIN Mémoires de l'Institut National des Sciences Naturelles et des Arts: Sciences mathématiques et physiques [Paris]
MMHN Mémoires du Muséum d'Histoire Naturelle [Paris]
MN Magazin for Naturvidenskaberne [Christiania]
MNEGN Magazin für den neuesten Entdeckungen in der gesammte Naturkunde [Berlin]
MNH Magazine of natural history [London]
MPCSA Mémoires physiques et chimiques de la Société d'Arceuil [Paris]
MSGF Mémoires de la Société Géologique de France [Paris]
MSHNP Mémoires de la Société d'Histoire Naturelle de Paris
MSLN Mémoires de la Société Linnéenne de Normandie [Caen]
MSSNN Mémoires de la Société des Sciences Naturelles de Neuchâtel
MT Mineralogishes Taschenbuch
MU Magasin universel [Paris]
MWS Memoirs of the Wernerian Society [Edinburgh]
NAALC Nova acta physico-medica Academiae Caesareae Leopoldino-Carolinae [Breslau and Bonn]
NAMHN Nouvelles annales du Muséum d'Histoire Naturelle [Paris] [continuation of *MMHN*]
NBSP Nouvelle bulletin scientifique de la Société Philomathique [Paris]
NDSG Neue Denkschriften der Schweizerischen Gesellschaft für die gesammte Naturwissenschaften / Nouveaux mémoires de la Société Helvétique des Sciences Naturelles [continuation of *DSG*]
NER New Edinburgh review
NJM Neues Jahrbuch für Mineralogie, Geognosie, Geologie und Petrefaktenkunde [Heidelberg] [continuation of *JMGGP*]
NVHMW Natuurkundige verhandlingen van der Hollandsche Maatschappij der Wetenschappen [Haarlem] [continuation of *VHMW*]

PC	Penny cyclopaedia [London]
PGS	Proceedings of the Geological Society [London]
PM	Philosophical magazine . . . [London] [further title varies]
PTRS	Philosophical transactions of the Royal Society [London]
QJS	Quarterly journal of science [London]
QR	Quarterly Review [London]
RB	Revue bibliographique [Paris]
RBAAS	Reports of the British Association for the Advancement of Science
RE	Revue encyclopédique [Paris]
TCPS	Transactions of the Cambridge Philosophical Society
TGS	Transactions of the Geological Society [London]
TRSE	Transactions of the Royal Society of Edinburgh
VHMW	Verhandelingen uitgegeeven door de Hollandsche Maatschappij der Weetenschappen te Haarlem
VTTG	Verhandelingen uitgegeven door Teylers Tweede Genootschap [Haarlem]
ZJ	Zoological journal [London]
ZM	Zeitschrift für Mineralogie [Frankfurt and Heidelberg]
ZNÜ	[Goethe's] Zur Naturwissenschaft überhaupt [Weimar]

Agassiz, Louis. 1833–43. *Recherches sur les poissons fossiles, comprenant une introduction à l'étude de ces animaux; l'anatomie comparée des systèmes organiques qui peuvent contribuer à faciliter la détermination des espèces fossiles; une nouvelle classification des poissons, exprimant leurs rapports avec la série des formations; l'exposition des lois de leur succession et de leur développement durant toutes les metamorphoses du globe terrestre, accompagnée de considérations géologiques générales; enfin, la description d'environ mille espèces qui n'existent plus et dont on a rétabli les caractères d'après les débris qui sont contenus dans les couches de la terre.* 2 vols. and *Atlas* 2 vols. Neuchâtel.

———. 1834. On a new classification of fishes, and on the geological distribution of fossil fishes. *PGS* 2 (37): 99–102.

———. 1837. Discours prononcé à l'ouverture des séances de la Société Helvétique des Sciences Naturelles, à Neuchâtel, le 24 juillet 1837. *ASHSN* 22me session: v–xxxii.

———. 1837. Des glaciers, des moraines et des blocs erratiques: discours prononcé à l'ouverture des séances de la Société Helvétique des Sciences Naturelles, à Neuchâtel, le 24 juillet 1837. *BU* (n.s.) 12: 369–93.

———. 1837. Sur les blocs erratiques du Jura. *CRAS* 5: 506–8.

———. 1838. Observations sur les glaciers. *BSGF* 9: 443–50.

———. 1840. *Études sur les glaciers.* Neuchâtel.

———. 1840. On the polished and striated surfaces of the rocks which form the beds of the glaciers in the Alps. *PGS* 3 (71): 321–22.

———. 1840. On glaciers, and the evidence of their having once existed in Scotland, Ireland and England. *PGS* 3 (72): 327–32.

———. 1841. On glaciers and boulders in Switzerland. *RBAAS* 10th meeting [1840], Trans. secs. 113–14.

———. 1841. *Untersuchungen über die Gletscher.* Solothurn.

———. 1842. The glacial theory and its recent progress. *ENPJ* 33: 217–83.

———. 1842. On the succession and development of organised beings at the surface of the terrestrial globe; being a discourse delivered at the inauguration of the Academy of Neuchâtel. *ENPJ* 33: 388–99 [trans. of *De la succession et du développement des êtres organisés* . . ., Neuchâtel, 1841].

———. 1844–45. *Monographie des poissons fossiles du Vieux Grès Rouge ou Système Dévonien (Old Red Sandstone) des Iles Britanniques et de Russie.* 2 vols. Neuchâtel.

Alberti, Friedrich August von. 1834. *Beiträge zu einer Monographie des Bunten Sandsteins, Muschelkalk und Keupers, und die Verbindung dieser Gebilde zu einer Formation.* Stuttgart and Tübingen.

Anonymous. 1829. Géologie de l'Auvergne [running title of review of Lecoq and Bouillet, *Vues et coupes*, and other works]. *BU* 40: 304–17; 41: 74–82.
Anonymous. ["A layman"]. 1834. On the infidel tendency of certain scientific speculations. *CO* 1834: 199–208.
Anonymous. 1838. Réunion extraordinaire de la Société [Géologique de France] à Porrentruy (Suisse, canton de Berne), du 5 au 12 septembre 1838. *BSGF* 9: 356–450.
Ansted, David Thomas. 1838. On a new genus of fossil multilocular shells, found in the slate-rocks of Cornwall. *TCPS* 6: 415–22, pl. 8.
[Arnold, Thomas]. 1825. [Review of] *Römische Geschichte* von B. G. Niebuhr. *History of Rome. By* B. G. Niebuhr. 2 vols. Berlin. 1811, 1812 [and other works]. *QR* 32: 67–92.
Auldjo, John. 1833. *Sketches of Vesuvius, with short accounts of its principal eruptions, from the commencement of the Christian era to the present time.* London.
Babbage, Charles. 1830. *Reflections on the decline of science in England and on some of its causes.* London.
———. 1834. Observations on the Temple of Serapis at Pozzuoli, near Naples; with remarks on certain causes which may produce geological cycles of great extent. *PGS* 2 (36): 72–76.
———. 1837. *The ninth Bridgewater Treatise: a fragment.* London.
Baker, W. E., and H. M. Durand. 1836. Sub-Himálayan fossil remains of the Dádúpur collection. *JASB* 5: 739–41, pl. 47.
Bacon, Francis. 1620. *Novum organum.* London [ed. Graham Rees and Maria Wakely, *Oxford Francis Bacon* 11 (2004)].
Barrande, Joachim. 1846. *Notice préliminaire sur le Systême Silurien et les trilobites de Bohême.* Leipzig.
———. 1852. *Systême Silurien du centre de la Bohême* 1. Prague and Paris.
Basset, César Auguste. 1815. *Explication de Playfair sur la théorie de la terre par Hutton, et examen comparatif des systèmes géologiques fondés sur le feu et sur l'eau, par M. Murray; en réponse à l'Explication de Playfair.* Paris.
Basterot, Barthélemy de. 1825. Description géologique du bassin tertiaire du sud-ouest de la France. *MSHNP* 2: 1–100, pls. 1–7.
Bayfield, Henry Wolsey. 1836. A notice on the transportation of rocks by ice. *PGS* 2 (43): 223.
Beechey, Frederick William. 1831. *Narrative of a voyage to the Pacific and Beering's Strait, to cooperate with the Polar expeditions: performed in His Majesty's ship Blossom . . . in the years 1825, 26, 27, 28.* London.
Berghaus, Heinrich Karl Wilhelm. 1845–48. *Physikalischer Atlas oder Sammlung von Karten, auf denen die hauptsächlichsten Erscheinungen der anorganischen und organischen Natur nach ihrer geographischen Verbreitung und Vertheilung bildlich dargestellt sind.* 2 vols. Gotha.
Bernhardi, Reinhard. 1832. Abhandlung über das, was man von der Geologie jetzt (1830) weiss. *VTTG* 21: 1–356.
———. 1832. Wie kamen die aus dem Norden stammenden Felsbruchstücke und Geschiebe, welche man in Norddeutschland und den benachbarten Ländern findet, an ihre gegenwärtige Fundorte? *JMGGB* 3: 257–67.
Bertrand-Roux, Jacques-Mathieu. 1823. *Description géognostique des environs du Puy en Velay, et particulièrement du bassin au milieu duquel cette ville est située.* Paris and Le Puy.
Bigsby, John Jeremiah. 1821. Geological and mineralogical observations on the north-west portion of Lake Huron. *AJS* 3: 254–72.
———. 1823. Notes on the geography and geology of Lake Huron. *TGS* (2) 1 (2): 175–209, pl. 31.
Blainville, Henri Marie Ducrotay de. 1818. Resumé des principaux travaux dans les différentes sciences physiques publiés pendant l'année 1817. *JP* 86: 5–97.
———. 1820. Analyse des principaux travaux faits ou publiés dans les différentes sciences physiques dans le cours de l'année 1819. *JP* 90: 5–105.
———. 1822. Analyse des principaux travaux dans les sciences physiques, publiés dans l'année 1821. *JP* 94: v–lxxvii.
———. 1827. *Mémoire sur les bélemnites, considérées zoologiquement et géologiquement.* Paris.
———. 1835. Mémoire sur le dodo, autrement dronte (Didus ineptus L.). *NAMHN* 4: 1–36, pls. 1–4.
———. 1839–64. *Ostéographie, ou description iconographique comparée du squelette et du système dentaire des mammifères récents et fossiles pour servir de base à la zoologie et à la géologie.* 3 vols. and *Atlas.* Paris.

Boitard, Pierre. 1838. L'homme fossile: étude paléontologique. *MU* 5: 209–24.
———. 1861. *Études antédiluviennes: Paris avant les hommes, l'homme fossile, etc., histoire naturelle du globe terrestre.* Paris.
Boué, Ami. 1820. *Essai géologique sur l'Écosse.* Paris.
———. 1820. Short comparison of the volcanic rocks of France with those of a similar nature found in Scotland. *EPJ* 2: 326–32.
———. 1823. Outline of a geological comparative view of the south-west and north of France, and the south of Germany. *EPJ* 9: 128–48.
———. 1824. Mémoire géologique sur le sud-ouest de la France; suivi d'observations comparatives sur le nord du même royaume, et en particulier sur les bords du Rhin. *ASN* 2: 387–423, pl. 19; 3: 55–81, 299–317; 4: 125–74; *Atlas* 2: 67–68, pls. 18–19.
———. 1825. Synoptical table of the formations of the crust of the earth and of the chief subordinate masses. *EPJ* 13: 130–45.
———. 1827. Synoptische Darstellung der die Erdrinde ausmachenden Formazionen, so wie der wichtigsten, ihnen untergeordneten Massen. *ZM* 1827, vol. 2: 1–239, 1 pl.
———. [1831]. *Carte géologique de l'Europe.* Paris.
———. 1831. Compte rendu des progrès de la géologie. *BSGF* 1: 71–75, 94–97, 105–24.
———. 1832. Résumé des progrès de la géologie en 1830 et 1831. *BSGF* 2: 133–218.
———. 1832. Le déluge, le diluvium, et l'époque alluviale ancienne. *MG* 1: 145–64.
———. 1832. Les Principes de la Géologie . . . par M. Ch. Lyell. *MG* 1: 317–56.
———. 1833. Résumé des progrès de la géologie et de quelques unes de ses principales applications, pendant l'année 1832. *BSGF* 3: i–clxxxviii.
———. 1834. Résumé des progrès des sciences géologiques pendant l'année 1833. *BSGF* 5.
Bravais, Auguste. 1840. Sur les lignes d'ancien niveau de la mer dans le Finmark. *CRAS* 10: 691–93.
Breislak, Scipione. 1797. *Topografia fisica della Campania.* Florence.
———. 1818. *Institutions géologiques.* 3 vols. and *Atlas.* Milan.
Brocchi, Giovanni Battista. 1814. *Conchiologia fossile subapennina con osservazioni geologiche sugli Apennini e sul suolo adiacente.* 2 vols. Milan.
———. 1819. Notizia di alcune osservazioni fisiche fatte nel tempio di Serapide a Pozzuoli. *BI* 14: 193–201.
Brochant de Villiers, André Jean Marie. 1835. Notice sur la carte géologique générale de la France. *CRAS* 1835: 423–29.
Broderip, William John. 1828. Observations on the jaw of a fossil mammiferous animal, found in the Stonesfield slate. *ZJ* 3: 408–12, pl. 11.
[———]. 1836. [Review of] 1.*Recherches sur les poissons fossiles*. . . par Louis Agassiz . . . [and] 2. *Rapport sur les poissons fossiles découvertes en Angleterre.* Par Louis Agassiz. Neuchâtel. 1835. *QR* 55: 433–45.
Brongniart, Adolphe. 1822. Sur la classification et la distribution des végétaux fossiles en général, et sur ceux des terrains de sédiment supérieur en particulier. *MMHN* 8: 203–40, 297–348, pls. 12–17.
———. 1828. Considérations générales sur la nature de la végétation qui couvrait la surface de la terre aux diverses époques de formation de son écorce. *ASN* 15: 225–58.
———. 1828. *Prodrome d'une histoire des végétaux fossiles.* Paris.
———. 1828–37. *Histoire des végétaux fossiles, ou recherches botaniques et géologiques sur les végétaux renfermés dans les diverses couches du globe.* 2 vols. Paris.
Brongniart, Alexandre. 1821. Notice sur des végétaux fossiles traversant les couches de terrain. *AM* 6: 359–70, pl. 3.
———. 1828. Notice sur les blocs des roches des terrains de transport en Suède. *ASN* 14: 5–22, pl. 1.
———. 1829. Théorie de la structure de l'écorce du globe. *DSN* 54: 1–256. [also published as] *Tableau des terrains qui composent l'écorce du globe, ou essai sur la structure de la partie connue de la terre.* Paris.
———. 1829. Volcans. *DSN* 58: 334–446. [also published as] *Des volcans et des terrains volcaniques.* Paris.
———. [1829?]. *Tableau théorique de la succession et de la disposition la plus générale en Europe, des terrains et roches qui compose l'écorce de la terre, or exposition graphique du Tableau des terrains* [single sheet]. Paris.

Brongniart, Alexandre, and Anselme-Gaëtan Desmarest. 1822. *Histoire naturelle des crustacés fossiles, sous les rapports zoologiques et géologiques.* Paris.

Bronn, Heinrich Georg. 1828. Ueber noch einige Petrefakten-Sammlungen in Italien. *ZM* 1828, vol. 2: 417–29.

———. 1831. *Italiens Tertiär-Gebilde und deren organische Einschlüsse: vier Abhandlungen.* Heidelberg.

———. 1835–38. *Lethaea geognostica, oder Abbildungen und Beschreibungen der für die Gebirgs-Formationen bezeichnendsten Versteinerungen.* Stuttgart.

Buch, Leopold von. 1810. Ueber dem Gabbro, mit einigen Bemerkungen über den Begriff einer Gebirgsart. *MNEGN* 4e Jg.: 128–149.

———. 1810. *Reisen durch Norwegen und Lappland.* 2 vols. Berlin.

———. 1815. Ueber die Ursachen der Verbreitung grosser Alpengeschiebe. *AKPAW* 1804–11: 161–86, 1 pl.

———. 1820. Ueber die Zusammensetzung der basaltischen Inseln und über Erhebungs-Cratere. *AKPAW* 1818–19: 51–68.

———. 1824. Ueber die geognostischen Systeme von Deutschland. *MT* 1824: 501–6.

———. 1825. *Physikalische Beschreibung der Canarischen Inseln.* 2 vols. Berlin.

[———]. 1826. *Geognostische Karte von Deutschland und den umliegenden Staaten in 42 Blättern, nach der vorzüglichsten mitgetheilten Materialen herausgegeben.* Berlin.

———. 1827. Ueber die Verbreitung grosser Alpengeschiebe. *APC* 9:575–88, pl. 7.

———. 1832. *Über Ammoniten, über ihre Sonderung in Familien, über die Arten, welche in den älteren Gebirgsschichten vorkommen, und über Goniatiten inbesondere.* Berlin.

Buckland, William. 1820. *Vindiciae geologicae; or the connexion of geology with religion explained, in an inaugural lecture delivered before the University of Oxford, May 15, 1819, on the endowment of a Readership in Geology by His Royal Highness the Prince Regent.* Oxford.

———. 1821. Description of the quartz rock of the Lickey Hill in Worcestershire, and of the strata immediately surrounding it; with considerations on the evidence of a recent deluge afforded by the gravel beds of Warwickshire and Oxfordshire, and the valley of the Thames from Oxford downwards to London; and an appendix, containing analogous proofs of diluvial action, collected from various authorities. *TGS* 5 (2): 506–44, pls. 36–37.

———. 1821. Notice of a paper laid before the Geological Society on the structure of the Alps and adjoining parts of the Continent, and their relation to the Secondary and Transition rocks of England. *AP* (n.s.) 1: 450–68.

———. 1823. *Reliquiae diluvianae; or, observations on the organic remains contained in caves, fissures, and diluvial gravel, and on other geological phenomena, attesting the action of an universal deluge.* London.

———. 1824. *Reliquiae diluvianae* . . . 2nd ed. London.

———. 1824. Notice on the Megalosaurus or great fossil lizard of Stonesfield. *TGS* (2) 1 (2): 390–96, pls. 40–44.

———. 1827. Observations on the bones of hyaenas and other animals in the cavern of Lunel near Montpelier, and in the adjacent strata of marine formations. *PGS* 1 (1): 3–6.

———. 1827. Relation d'une découverte récente d'os fossiles faite dans la partie orientale de la France, à la grotte d'Osselles ou Quingey, sur les bords du Doubs, cinq lieues au-dessous de Besançon. *ASN* 10: 306–19.

———. 1827. Letter of Professor Buckland to Professor Jameson, and of Captain Sykes to Professor Buckland, on the interior of the dens of living hyaenas. *ENPJ* 2: 377–80.

———. 1828. Note sur des traces de tortues observées dans le grès rouge. *ASN* 13: 85–86.

———. 1828. Geological account of a series of animal and vegetable remains and of rocks collected by J. Crawfurd, Esq., on a voyage up the Irawadi to Ava in 1826 and 1827. *TGS* (2) 2 (3): 377–92, pls. 36–44.

———. 1828. On the Cycadeoideae, a family of fossil plants found in the Oolite quarries of the Isle of Portland. *TGS* (2) 2 (3): 395–402, pls. 46–49.

———. 1829. On the discovery of a new species of pterodactyle in the Lias at Lyme Regis. *TGS* (2) 3 (1): 217–22, pl. 27.

———. 1829. On the discovery of coprolites, or fossil faeces, in the Lias at Lyme Regis, and in other formations. *TGS* (2) 3 (1): 223–36, pls. 28–31.

———. 1830. Antediluvian human remains. *AJS* 18: 393–94.

———. 1831. On the occurrence of the remains of elephants and other quadrupeds, in the cliffs of frozen mud, in Eschscholtz Bay, within Beering's Strait, and in other distant parts of the shores of the Arctic seas. *In* Beechey, *Voyage to the Pacific* 2: 593–612, pls. 1–3.

———. 1833. On the fossil remains of the megatherium recently imported into England from South America. *RBAAS* 1st and 2nd meetings: 104–7.

———. 1836. *Geology and mineralogy considered with reference to natural theology.* 2 vols. London.

———. 1837. *Geology and mineralogy considered with reference to natural theology.* 2nd ed. 2 vols. London.

———. 1838–39. *Geologie und Mineralogie in Beziehung zur natürlichen Theologie.* 2 vols. Neuchâtel.

———. 1840. Memoir on the evidences of glaciers in Scotland and the north of England. *PGS* 3 (72): 332–37, 345–48.

———. 1842. On the glacia-diluvial phaenomena in Snowdonia and the adjacent parts of north Wales. *PGS* 3 (84): 579–84.

Burtin, François-Xavier. 1784. *Oryctographie de Bruxelles, ou description des fossiles tant naturels qu'accidentels découverts jusqu'à ce jour dans les environs de ce ville.* Brussels.

Caldcleugh, Alexander. 1836. An account of the great earthquake experienced in Chile on the 20th of February, 1835; with a map. *PTRS* 1836: 21–26, pl. 1.

Candolle, Augustin-Pyramus de. 1820. Géographie botanique. *DSN* 18: 359–422.

Cautley, Proby Thomas, and Hugh Falconer. 1840. Notice on the remains of a fossil monkey from the Tertiary strata of the Sewalik Hills in the north of Hindoostan. *TGS* (2) 5 (3): 499–504.

[Chambers, Robert]. 1844. *Vestiges of the natural history of creation.* London [repr., ed. James A. Secord, Chicago (University of Chicago Press), 1994].

Champollion, Jean-François. 1824. *Précis du système hiéroglyphique des anciens Égyptiens, ou recherches sur les élémens premiers de cette écriture sacrée, sur leurs diverses combinaisons, et sur les rapports de ce système avec les autres méthodes graphiques Égyptiennes.* Paris.

Charlesworth, Edward. 1835. Observations on the Crag-formation and its organic remains; with a view to establish a division of the Tertiary strata overlying the London Clay in Suffolk. *PM* (3) 7:81–94.

———. 1837. Observations on the Crag and on the fallacies involved in the present classification of Tertiary deposits. *PM* (3) 10: 1–9.

———. 1837. On some fallacies involved in the results relating to the comparative age of Tertiary deposits obtained from the application of the test recently introduced by Mr Lyell and M. Deshayes. *ENPJ* 22: 110–16.

Charpentier, Jean de. 1835. Notice sur la cause probable de transport des blocs erratiques de la Suisse. *AM* (3) 8: 219–36.

———. 1836. Quelques conjectures sur les grandes révolutions qui ont changés la surface de la Suisse, et particulièrement celle du canton de Vaud, pour l'amener à son état actuel. *BU* 1836 4: 1–12.

———. 1841. *Essai sur les glaciers et sur le terrain erratique du bassin du Rhône.* Lausanne.

Christol, Jules de. 1829. Cavernes à ossements renfermant des débris humains. *BSNG* 18: 101–2.

———. 1829. *Notice sur les ossemens humains fossiles des cavernes du département du Gard.* Montpellier [also published, without pl., in *AM* (2) 5: 517–30].

Christol, Jules de, and A. Bravard. 1828. Mémoire sur de nouvelles espèces d'hyènes fossiles, découvertes dans le caverne de Lunel-Viel, près Montpellier. *MSHNP* 4: 368–78, pl. 23B.

Clarke, Edward Daniel. 1810–23. *Travels in various countries of Europe, Asia and Africa.* 6 vols. London.

Clift, William. 1828. On the fossil remains of two new species of mastodon, and of other vertebrated animals, found on the left bank of the Irawadi. *TGS* (2) 2 (3): 369–75, pls. 36–44.

———. 1835. Some account of the remains of the megatherium sent to England from Buenos Ayres by Woodbine Parish, Jun., Esq., F.G.S., F.R.S. *TGS* (2) 3 (3): 437–50, pls. 43–46.

Collomb, Édouard. 1845. Sur les traces du phénomène erratique dans les Vosges. *BSGF* (2) 2: 506–11, pl. 16.

———. 1847. *Preuves de l'existence d'anciens glaciers dans les vallées des Vosges et du terrain erratique de cette contrée.* Paris.

Conybeare, William Daniel. 1822. Additional notices on the fossil genera Ichthyosaurus and Plesiosaurus. *TGS* (2) 1 (1): 103–23, pls. 15–22.

———. 1823. Memoir illustrative of a general geological map of the principal mountain chains of Europe. *AP* (n.s.) 5: 1–16, 135–49, 210–18, 278–89, 356–59, pl. 19; 6: 214–19.

———. 1824. On the discovery of an almost perfect skeleton of the Plesiosaurus. *TGS* (2) 1 (2): 381–89, pls. 48–49.

———. 1829. On the hydrographical basin of the Thames, with a view more especially to investigate the causes which have operated in the formation of the valleys of that river, and its tributary streams. *PGS* 1 (12): 145–49.

———. 1829. Answer to Dr Fleming's view of the evidence from the animal kingdom, as to the former temperature of the northern regions. *ENPJ* 7: 142–52.

———. 1830. On Mr Lyell's 'Principles of Geology'. *PM* (n.s.) 8: 215–19. [continued as next item]

———. 1830–31. An examination of those phaenomena of geology, which seem to bear most directly on theoretical speculations. *PM* (n.s.) 8: 359–62, 401–6; 9: 19–23, 111–17, 188–97, 258–70.

———. 1831. *Inaugural address on the application of classical and scientific education to theology; and on the evidences of natural and revealed religion.* London.

———. 1832–34. Inquiry how far the theory of M. Elie de Beaumont concerning the parallelism of lines of elevation of the same geological Aera, is agreeable to the phaenomena as exhibited in Great Britain. *PM* (3) 1: 118–26; 4: 404–14.

———. 1833. Report on the progress, actual state, and ulterior prospects of geological science. *RBAAS* 1st and 2nd meetings: 356–414, 1 pl.

———. 1834. *An elementary course of lectures on the criticism, interpretation, and leading doctrines of the Bible; delivered at Bristol College in the years 1832, 1833.* London.

———. 1834. Rev. W. D. Conybeare in reply to a layman, on geology. *CO* 1834: 306–9.

———. 1839. *An analytical examination into the character, value, and just application of the writings of the Christian Fathers during the Ante-Nicene period.* Oxford.

Conybeare, William Daniel, and William Phillips. 1822. *Outlines of the geology of England and Wales, with an introductory compendium of the general principles of that science, and comparative views of the structure of foreign countries. Part I.* London [facsimile repr. Farnborough (Greg), 1969].

[Copleston, Edward]. 1823. [Review of] *Reliquiae Diluvianae*... By the Rev. William Buckland, F.R.S. &c. London. 1823. *QR* 29: 138–65.

Cordier, Louis. 1827. Essai sur la température de l'intérieur de la terre. *MMHN* 15: 161–244 [also *MARS* 7: 473–555].

Cordier, Louis, *et al.* 1834. Rapport sur les résultats scientifiques du voyage de M. Alcide d'Orbigny dans l'Amérique du Sud pendant les années 1826, 1827, 1828, 1829, 1830, 1831, 1832 et 1833. *NAMHN* 3: 84–115.

Costa, Oronzio Gabriele. 1829. *Catalogo sistematico e ragionato de'testacei delle Due Sicilie.* Naples.

Crawfurd, John. 1827. *Brief narrative of an embassy from the Governor-General of India to the King of Ava, in 1826–27.* London.

———. 1829. *Journal of an embassy from the Governor-General of India to the court of Ava, in the year 1827. With an appendix, containing the description of fossil remains, by Professor Buckland and Mr Clift.* London.

Croizet, Jean-Baptiste, and A. C. G. Jobert. 1826–28. *Recherches sur les ossemens fossiles du département du Puy-de-Dôme.* Paris.

Culley, Matthew. 1829. A few facts and observations as to the power which running water exerts in removing heavy bodies. *PGS* 1 (12): 149.

Cuvier, Georges. 1812. *Recherches sur les ossemens fossiles de quadrupèdes, où l'on rétablit les caractères de plusieurs espèces d'animaux que les révolutions du globe paroissent avoir détruites.* 4 vols. Paris.

———. 1812. Sur un nouveau rapprochement à établir entre les classes qui composent le règne animal. *AMHN* 19: 73–84.

———. 1817. *Le règne animal distribué d'après son organisation, pour servir de base à l'histoire naturelle des animaux et d'introduction à l'anatomie comparée.* 4 vols. Paris.

———. 1821–24. *Recherches sur les ossemens fossiles de quadrupèdes, où l'on rétablit les caractères de plusieurs animaux dont les révolutions du globe ont détruites les espèces.* 2nd ed. 5 vols. in 7. Paris.

———. 1825. *Recherches sur les ossemens fossiles de quadrupèdes*... 3rd ed. 5 vols. in 7. Paris [facsimile repr. Paris, 1985].
———. 1826. *Discours sur les révolutions de la surface du globe, et sur les changemens qu'elles ont produits dans le règne animal.* Paris and Amsterdam [facsimile repr. Brussels, 1969].
———. 1826. Rapport verbal sur un ouvrage [by Croizet and Jobert] intitulé: Recherches sur les ossemens fossiles du département du Puy-de-Dôme. *ASN* 9: 273–78.
———. 1828. Rapport verbal fait à l'Académie des Sciences sur un ouvrage de MM. l'abbé Croiset [*sic*] et Jobert ainé, intitulé *Recherches sur les ossemens fossiles du département du Puy-de-Dôme. ASN* 15: 218–24.
———. 1830. Note sur quelques ossemens qui paraissent appartenir au dronte, espèce d'oiseau perdue seulement depuis deux siècles. *BSNG* 22: 122–25.
Darwin, Charles. 1835. *Extracts from letters addressed to Professor Henslow.* Cambridge [repr. Barrett, *Papers of Darwin* (1977) 1, 3–16].
———. 1835. Geological notes made during a survey of the east and west coasts of South America, in the years 1832, 1833, 1834, and 1835, with an account of a traverse section of the Cordilleras of the Andes between Valparaiso and Mendoza. *PGS* 2 (42): 210–12 [repr. Barrett, *Papers of Darwin* (1977) 1, 16–19].
———. 1837. Observations of proofs of recent elevation on the coast of Chili, made during the survey of His Majesty's ship Beagle, commanded by Capt. Fitzroy, R.N. *PGS* 2 (48): 446–49 [repr. Barrett, *Papers of Darwin* (1977) 1, 41–43].
———. 1837. A sketch of the deposits containing extinct Mammalia in the neighbourhood of La Plata. *PGS* 2 (51): 542–44 [repr. Barrett, *Papers of Darwin* (1977) 1, 44–45].
———. 1837. On certain areas of elevation and subsidence in the Pacific and Indian oceans, as deduced from the study of coral formations. *PGS* 2 (51): 552–54 [repr. Barrett, *Papers of Darwin* (1977) 1, 46–49].
———. 1838. On the connexion of certain volcanic phaenomena, and on the formation of mountain chains and volcanos, as the effects of continental elevations. *PGS* 2 (56): 654–60.
———. 1839. Observations on the Parallel Roads of Glen Roy, and of other parts of Lochaber in Scotland, with an attempt to prove that they are of marine origin. *PTRS* 1839: 39–81, pls. 1–2 [repr. Barrett, *Papers of Darwin* (1977) 1, 87–137].
———. 1839. Note on a rock seen on an iceberg in 61° South latitude. *JRGS* 9: 528–29 [repr. Barrett, *Papers of Darwin* (1977) 1, 137–39].
———. 1839. *Journal and remarks 1832–1836.* London [vol. 3 of *Narrative of the surveying voyages of His Majesty's ships Adventure and Beagle, between the years 1826 and 1836, describing their examination of the southern shores of South America, and the Beagle's circumnavigation of the globe*].
——— (ed.). 1839–42. *The zoology of the voyage of H. M. S. Beagle under the command of Captain Fitzroy, R.N., during the years 1832 to 1836.* 4 vols. London.
———. 1840. On the connexion of certain volcanic phaenomena in South America; and on the formation of mountain chains and volcanos, as the effects of the same power by which continents are elevated. *TGS* (2) 5 (3): 601–31 [repr. Barrett, *Papers of Darwin* (1977) 1, 53–86].
———. 1842. *The structure and distribution of coral reefs. Being the first part of the geology of the voyage of the Beagle, under the command of Capt. Fitzroy, R.N., during the years 1832 to 1836.* London.
———. 1842. Notes on the effects produced by the ancient glaciers of Caernarvonshire and on the boulders transported by floating ice. *PM* (3) 21: 180–88 [repr. Barrett, *Papers of Darwin* (1977) 1, 163–71].
———. 1859. *On the origin of species by means of natural selection, or the preservation of favoured races in the struggle for life.* London [facsimile repr. Cambridge, Mass. (Harvard University Press), 1964, and other eds.].
Daubeny, Charles Giles Brindle. 1820–21. On the volcanoes of the Auvergne. *EPJ* 3: 359–67; 4: 89–97, 300–15.
———. 1825. Sketch of the geology of Sicily. *EPJ* 13: 107–18, 254–69, pl. 4.
———. 1826. *A description of active and extinct volcanos; with remarks on their origin, their chemical phaenomena, and the character of their products, as determined by the condition of the earth during the period of their formation.* London.
———. [1826?]. *A tabular view of volcanic phaenomena comprising a list of the burning moun-*

tains that have been noticed at any time since the commencement of historical records together with the dates of their respective eruptions, and of the principal earthquakes connected with them [single sheet]. Oxford and London.

———. 1831. On the diluvial theory, and on the origin of the valleys of Auvergne. *ENPJ* 10: 201–29.

Davy, Humphry. 1830. *Consolations in travel; or, the last days of a philosopher.* London.

De la Beche, Henry Thomas. 1824. *A selection of the geological memoirs contained in the Annales des Mines, together with a synoptical table of equivalent formations and M. Brongniart's table of the classification of mixed rocks.* London.

———. 1825. Notice on the diluvium of Jamaica. *AP* (n.s.) 10: 54–58.

———. 1825. *Notes on the present condition of the negroes in Jamaica.* London.

———. 1827. Remarks on the geology of Jamaica. *TGS* (2) 2 (2): 143–94.

———. 1830. *Sections and views, illustrative of geological phaenomena.* London.

———. 1831. *A geological manual.* London.

———. 1832. *A geological manual.* 2nd ed. London.

———. 1832. *Handbuch der Geognosie.* Berlin.

———. 1833. *Manuel géologique.* Paris.

———. 1834. *Researches in theoretical geology.* London.

———. 1839. *Report on the geology of Cornwall, Devon and West Somerset.* London.

De la Beche, Henry Thomas, and William Daniel Conybeare. 1821. Notice of the discovery of a new fossil animal, forming a link between the ichthyosaurs and the crocodile; together with general remarks on the osteology of the ichthyosaurus. *TGS* 5 (2): 559–94, pls. 40–42.

De la Rive, Auguste. 1829. Notice sur la quinzième session de la Société Helvétique des Sciences Naturelles, réunie à l'Hospice du Grand Saint-Bernard, les 21, 22, et 23 juillet 1829. *BU* 41: 263–64.

Deshayes, Gérard-Paul. 1824–37. *Description des coquilles fossiles des environs de Paris.* 3 vols. Paris.

———. 1831. Tableau comparatif des espèces de coquilles fossiles des terrains tertiaires de l'Europe, et des espèces de fossiles de ces terrains entr'eux. *BSGF* 1: 185–89.

———. 1831. *Description de coquilles caractéristiques des terrains.* Paris.

Desmarest, Nicholas. 1823. *Carte topographique et minéralogique d'une partie du département du Puy-de-Dôme dans le ci-devant province d'Auvergne où sont déterminées la marche et les limites des matières fondues & rejettées par les volcans ainsi que les courants anciens & modernes pour servir aux recherches sur l'histoire naturelle des volcans* [sheet map]. Paris.

Desnoyers, Jules. 1825. Mémoire sur la Craie, et sur les terrains tertiaires du Cotentin. *MSHNP* 2: 176–248, pl. 9.

———. 1829. Observations sur un ensemble de dépôts marins plus récens que les terrains tertiaires du bassin de la Seine, et constituent une *formation géologique* distincte; précédées d'un aperçu *de la non simultanéité des bassins tertiaires. ASN* 16: 171–214, 402–91.

———. 1832. Considérations sur les ossemens humains des cavernes du midi de la France. *BSGF* 2: 126–33.

———. 1832. Rapport sur les travaux de la Société Géologique, pendant l'année 1831. *BSGF* 2: 226–327.

Devèze de Chabriol, J. S., and Jean-Baptiste Bouillet. 1827. *Essai géologique et minéralogique sur les environs d'Issoire, département du Puy-de-Dôme, et principalement sur la montagne de Boulade, avec la description et des figures lithographiées des ossemens fossiles qui ont été recueillis.* Clermont-Ferrand.

Donati, Vitaliano. 1750. *Della storia naturale marina dell'Adriatico.* Venice.

Duméril, A. M. C. *et al.* 1837. Rapport sur la découverte de plusieurs ossements fossiles de quadrumanes, dans le dépôt tertiaire de Sansan, près d'Auch, par M. Lartet. *CRAS* 4: 981–98, 1 pl.

Dumont, André-Hubert. 1832. Mémoire sur la constitution géologique de la Province de Liège. *MASBB* 8.

Duncan, Henry. 1828. An account of the tracks and footmarks of animals found impressed on sandstone in the quarry of Corncockle Muir in Dumfriesshire. *TRSE* 11: 194–209, pl. 8.

Ebel, Johann Gottfried. 1808. *Ueber den Bau der Erde in dem Alpengebirge, zwischen 12 Längen- und 2–4 Breitgraden, nebst einigen Betrachtungen über die Gebirge und den Bau der Erde überhaupt, mit geognostischen Karten.* 2 vols. Zurich.

Élie de Beaumont, Léonce. 1829–30. Recherches sur quelqu'unes des révolutions de la surface du globe, présentant différents exemples de coïncidence entre le redressement des couches de certains systèmes de montagnes, et les changements soudains qui ont produits les lignes de démarcation qu'on observe entre certain étages consecutifs des terrains de sédiment. *ASN* 18: 5–25, 284–416; 19: 5–99, 177–240, pls. 1–3.
———. 1834. Faits pour servir à l'histoire de l'Oisans. *MSHNP* 5: 1–32.
———. 1836. Recherches sur la structure et sur l'origine du Mont Etna. *AM* (3) 9: 175–216, pls. 1–5, 575–630; 10: 351–70, 507–76.
Enderby, Charles. 1839. Discoveries in the Antarctic Ocean, in February 1839. Extracted from the journal of the schooner Eliza Scott, commanded by Mr John Balleny. *JRGS* 9: 517–28.
Esmark, Jens. 1798. *Kurze Beschreibung einer mineralogischen Reise durch Ungarn, Siebenbürgen und das Bannat.* Freiberg.
———. 1816. Description of a new ore of tellurium. *TGS* 3: 413–14.
———. 1824. Bidrag til vor jordklodes historie. *MN* 3: 27–49.
———. 1826. Remarks tending to explain the geological history of the earth. *ENPJ* 2: 107–21.
Falconer, Hugh, and Proby Thomas Cautley. 1836. Sivatherium giganteum, a new fossil ruminant genus, from the valley of the Markanda, in the Siválik branch of the Sub-Himálayan mountains. *JASB* 5: 38–50, pl. 1.
———. 1845–49. *Fauna antiqua Sivalensis, being the fossil zoology of the Siwalik Hills, in the north of India.* 2 vols. London.
Ferrara, Francesco. 1793. *Storia generale dell'Etna, che comprende la descrizione di questa montagna; la storia delle sue eruzioni e dei suoi fenomeni; la descrizione ragionata dei suoi prodotti; e la conoscenza di tutto ciò, che può servire alla storia dei volcani.* Catania.
———. 1810. *I Campi Flegrei della Sicilia e delle isole che le sono intorno o descrizione fisica e mineralogica di queste isole.* Messina.
———. 1818. *Descrizione dell'Etna con la storia delle eruzioni e il catalogo dei prodotti.* Palermo.
[Fitton, William Henry]. 1823. [Review of] *Reliquiae diluvianae* . . . By the Reverend William Buckland . . . London. J. Murray. *ER* 39: 196–234.
———. 1824. Inquiries respecting the geological relations of the beds between the Chalk and the Purbeck Limestone in the south-east of England. *AP* (n.s.) 8: 365–83, pl. 33.
———. 1824. Additions to a paper in the last number of the Annals of Philosophy. *AP* (n.s.) 8: 458–62.
———. 1828. On the strata from whence the fossil described in the preceding notice [Broderip, "Observations on the jaw"] was obtained. *ZJ* 3: 412–18.
———. 1829. [Address to the Geological Society at the Annual General Meeting, 20th February 1829]. *PGS* 1 (10): 112–34.
———. 1830. Observations on part of the Low Countries and the north of France, principally near Maestricht and Aix-la-Chapelle. *PGS* 1 (14): 161–64.
Fleming, John. 1821. Observations on the mineralogy of the neighbourhood of Cork. *MWS* 3: 83–103.
———. 1822. *The philosophy of zoology; or a general view of the structure, functions and classification of animals.* 2 vols. Edinburgh.
[———]. 1823. [Review of] Essay on the theory of the earth, by M. Cuvier, &c. &c. With mineralogical notes, and an account of Cuvier's geological discoveries, by Professor Jameson . . . 4th edition. *NER* 4: 381–98.
———. 1824. Remarks illustrative of the influence of society on the distribution of British animals. *EPJ* 11: 287–305.
———. 1826. The geological deluge, as interpreted by Baron Cuvier and Professor Buckland, inconsistent with the testimony of Moses and the phenomena of nature. *EPJ* 14: 205–39.
———. 1828. *A history of British animals, exhibiting the descriptive characters and systematical arrangement of the genera and species of quadrupeds, birds, reptiles, fishes, mollusca, and radiata of the United Kingdom; including the indigenous, extirpated and extinct kinds, together with periodical and occasional visitants.* Edinburgh.
———. 1829. On the value of the evidence from the animal kingdom, tending to prove that the Arctic regions formerly enjoyed a milder climate than at present. *ENPJ* 6: 277–86.
———. 1829. Additional remarks on the climate of the Arctic regions, in answer to Mr Conybeare. *ENPJ* 8: 65–74.

[Forbes, James David]. 1842. The glacier theory [review of eight publications, 1833–41]. *ER* 75: 49–105.
———. 1842. Professor Forbes' account of his recent observations on glaciers. *ENPJ* 33: 338–52.
———. 1843. An attempt to explain the leading phenomena of glaciers. *ENPJ* 35: 221–52.
———. 1843. *Travels through the Alps of Savoy and other parts of the Pennine chain, with observations on the phenomena of glaciers.* Edinburgh.
Forchhammer, Johann Georg. 1828. On the Chalk formation of Denmark. *EJS* 9: 56–68.
Fourier, Jean-Baptiste Joseph. 1820. Extrait d'une mémoire sur le refroidissement séculaire du globe terrestre. *ACP* 13: 418–38.
———. 1822. *Théorie analytique de la chaleur.* Paris.
———. 1824. Remarques générales sur les températures du globe terrestre et des espaces planétaires. *ACP* 27: 136–67.
Franklin, John. 1828. *Narrative of a second expedition to the shores of the Polar Sea in the years 1825, 1826, and 1827.* London [repr. Rutland, VT, 1971].
[Gay-Lussac, Louis Joseph, and Dominique François Jean Arago (eds.)]. 1821. Nouvelles observations sur la température de la terre à différentes profondeurs. *ACP* 16: 78–85.
Gemmellaro, Carlo. 1831. Sopra la fisionomia delle montagne di Sicilia: cenno geologico. *AAGC* 5: 73–93.
———. 1834. Relazione dei fenomeni del nuovo vulcano, sorte dal mare fra la costa di Sicilia e l'Isola di Pantelleria nel mese di Luglio 1831. *AAGC* 8: 271–98.
Gemmellaro, Giuseppe. 1828. *Quadro istorico delle eruzioni dell'Etna, cominciando dell'epoca dei Sicani sino oggi 1824, formato dietro lo studio de più accreditati scrittori del vulcano e dopo molte accurate osservazioni che ha per oggietto principale mostrare l'origine il corso e l'epoca di ogni eruzione* [sheet map]. London.
Gemmellaro, Mario. [ca. 1820?] *Prospetto meridionale dell'Etna* [sheet print]. [Catania?]
Geoffroy Saint-Hilaire, Étienne. 1818–22. *Philosophie anatomique: [1] Des organes respiratoires sous le rapport de la détermination et de l'identité de leurs pièces osseuses; [2] Des monstruosités humaines.* 2 vols. Paris.
———. 1825. Considérations générales sur la monstruosité; et description d'un genre nouveau observé dans l'espèce humain, et nommé Aspalasome. *ASN* 4: 450–70, pl. 21.
———. 1825. Recherches sur l'organisation des gavials; sur leurs affinités naturelles, desquelles résulte la nécessité d'une autre distribution générique, *Gavialis*, *Teleosaurus* et *Steneosaurus*; et sur cette question, si les gavials (*Gavialis*), aujourd'hui répandus dans les parties orientales de l'Asie, descendent, par voie non interrompue de géneration, des gavials antédiluviens, soit des gavials fossiles, dits crocodiles de Caen (*Teleosaurus*), soit des gavials fossiles du Havre et de Honfleur (*Steneosaurus*). *MMHN* 12: 97–155, pls. 5, 6.
———. 1826. Sur les déviations organiques provoquées et observées dans un établissement d'incubations artificielles. *MMHN* 13: 289–96.
———. 1828. Mémoire où l'on propose de rechercher dans quels rapports de structure organique et de parenté sont entre eux les animaux des âges historiques, et vivant actuellement, et les espèces antédiluviennes et perdues. *MMHN* 17: 209–29.
———. 1831. *Recherches sur de grands sauriens trouvés à l'état fossile vers les confins maritimes de la Basse-Normandie, attribués d'abord au crocodiles, puis déterminés sous les noms de Teléosaurus et Sténéosaurus.* Paris [preprint of "Divers mémoires sur de grands sauriens", *MARS* 12 (1833): 1–138, pl. 1].
———. 1835. *Études progressives d'un naturaliste.* Paris.
———. 1836. Études sur l'orang-outang, et considérations philosophiques au sujet de la race humaine [title of concluding part]. *CRAS* 2: 521–25, 601–3; 3: 27–31.
———. 1837. Sur la singularité et de la haute portée en philosophie naturelle de l'existence d'une espèce de singe trouvée à l'état fossile dans le midi de la France. *CRAS* 5: 35–42.
Godefroy, Charles. 1840. *Notice sur les glaciers, les moraines et les blocs erratiques des Alpes; avec un table analytique.* Paris and Geneva.
Goethe, Johann Wolfgang von. 1823. Architektonisch-naturhistorisches Problem. *ZNÜ* 2 (1): 79–88, 1 pl.
Goldfuss, Georg August. 1826–44. *Petrefacta Germaniae tam ea, quae in Museo Universitatis Regiae Borussicae Fredericiae Wilhelmiae Rhenanae servantur quam alia quaecunque in Museis Hoeninghausiano Muensteriano aliisque extant, iconibus et descriptionibus illustrata.* 3 vols. Düsseldorf.

———. 1831. Beiträge zur Kenntnis verschiedener Reptilien der Vorwelt. *NAALC* 15: 61–128, 7 pls.
Graham, Maria. 1824. An account of some effects of the late earthquakes in Chili. *TGS* (2) 1 (2): 413–15.
———. 1824. *Journal of a residence in Chile, during the year 1822, and a voyage from Chile to Brazil in 1823.* London.
Greenough, George Bellas. 1820. *A geological map of England and Wales.* London.
———. 1834. Address delivered at the anniversary meeting of the Geological Society, on the 21st of February 1834. *PGS* 2 (35): 42–70.
Gressly, Amanz. 1838–41. Observations géologiques sur le Jura Soleurois. *NDSG* 2: 1–112, pls. 1–5; 4: 113–241, pls. 6–12; 5: 245–349, pls. 13–14. [pp. 8–29 trans. Cross and Homewood, "Gressly's role" (1997)]
Hall, Basil. 1824. *Extracts from a journal, written on the coasts of Chili, Peru, and Mexico, in the years 1820, 1821, 1822.* London.
Hamilton, William. 1776. *Campi Phlegraei: Observations on the volcanos of the Two Sicilies, as they have been communicated to the Royal Society of London . . .* Naples [pls. repr. Carlo Knight, *Les fureurs de Vésuve*, Paris (Gallimard), 1992].
Hart, John. 1825. *Description of the skeleton of the fossil deer of Ireland, Cervus Megaceros, drawn up at the instance of the committee of natural philosophy of the Royal Dublin Society.* Dublin.
——— [printed in error as "Part"]. 1826. Description du squelette du daim fossile d'Irlande (Cervus megaceros) du Muséum de la Société Royale de Dublin. *ASN* 8: 389–411, pl. 39.
Hausmann, Johann Friedrich Ludwig. 1828. De origine saxorum, per Germaniae septentrionalis regiones arenosas dispersorum commentatio. *CSRSG* 7: 3–34.
———. 1831. Verhandeling ter beantwoording der vrage: Welke is de oorsprong der graniet- en andere primitieve rotsblokken en stukken van zeer verschillende grootte, die in grote menigte over de vlakten der Nederlanden en van het Noordelijk Duitschland, en in de zandgronden verspreid liggen? *VHMW* 19: 272–400.
Hawkins, Thomas. 1834. *Memoirs of Ichthyosauri and Plesiosauri, extinct monsters of the ancient earth.* London.
———. 1840. *The book of the great sea-dragons, ichthyosauri and plesiosauri, gedolim taninim, of Moses, extinct monsters of the ancient earth.* London.
Herschel, John Frederick William. 1830. *A preliminary discourse on the study of natural philosophy.* London [facsimile repr. New York (Johnson), 1966].
———. 1832. On the astronomical causes which may influence geological phenomena. *TGS* (2) 3 (2): 293–99.
Hibbert, Samuel. 1836. On the fresh-water limestone of Burdiehouse in the neighbourhood of Edinburgh, belonging to the Carboniferous group of rocks; with supplementary notes on other fresh-water limestones. *TRSE* 13: 169–282, pls. 5–12.
Hoff, Karl Ernst Adolf von. 1801–05. *Das Teutsche Reich vor der französischen Revolution und nach dem Frieden zu Luneville. Ein geographisch-statistische Parallele, nebst einigen Urkunden und andere Karte.* 2 vols. Gotha.
———. 1822–41. *Geschichte der durch Überlieferung nachgewiesenen natürlichen Veränderungen der Erdoberfläche: ein Versuch.* 5 vols. Gotha.
[Hoffmann, Friedrich]. 1832. Ueber des in mittelländischen Meere entstandene vulcanische Eiland, genannt Corao, Nerita, Isola Ferdinandea, Graham Island und Julia, nebst einigen Nachrichten über kraterförmige Inseln ähnlichen Ursprungs. *APC* 24: 65–109, pls. 2, 3.
Hogard, Henri. 1840. Observations sur les traces de glaciers, qui, à une époque reculée, paraissent avoir recouvert la chaîne des Vosges, et sur les phénomènes géologiques qu'ils ont pu produire. *ASEV* 4: 91–112.
———. 1845. Note sur les traces d'anciens glaciers dans les Vosges. *BSGF* (2) 2: 249–55, pl. 6.
Hombres-Firmas. L. A. d'. 1821. Notice sur les ossemens humains fossiles. *JP* 92: 227–33.
Home, Everard. 1814. Some account of the fossil remains of an animal more nearly allied to the fishes than to any of the other classes of animals. *PTRS* 1814: 571–77, pls. 17–20.
———. 1816. Some farther account of the fossil remains of an animal, of which a description was given to the [Royal] Society in 1814. *PTRS* 1816: 318–21, pls. 13–16.
———. 1818. Additional facts respecting the fossil remains of an animal, on the subject of which two papers have been printed in the Philosophical Transactions, showing that the bones of the sternum resemble those of the ornithorhynchus paradoxus. *PTRS* 1818: 24–32, pls. 2, 3.

———. 1819. An account of the fossil skeleton of the Proteo-saurus. Reasons for giving the name Proteo-saurus to the fossil skeleton which has been described. *PTRS* 1819: 209–16, pls. 13–15.

Horner, Leonard. 1831. On the new volcanic island in the Mediterranean, and its connection with the extinct volcanic island of Pantellaria, and the hot springs of Sciacca on the coast of Sicily. *PGS* 1 (23): 338–39.

Hugi, Franz Joseph. 1830. *Naturhistorische Alpenreise*. Solothurn.

Humboldt, Alexander von. 1817. Des lignes isothermes et de la distribution de la chaleur sur le globe. *MPCSA* 3: 462–602 ["Extrait" in *ACP* 5: 102–11 (1817), with 1 pl. not in full paper].

———. 1822. Indépendance des formations. *DSN* 23: 56–385.

———. 1823. *Essai géogno stique sur le gisement des roches dans les deux hémisphères*. Paris.

Huot, Jean-Jacques Nicolas. 1824. Observations sur le banc de Grignon, sur le calcaire renfermant les végétaux, et sur les couches supérieures de cette localité. *ASN* 3: 5–15, and *Atlas* 3: 68–69, pl. 1.

———. 1824. Notice géologique sur le prétendu fossile humain trouvé près de Moret, département de Seine-et-Marne. *ASN* 3: 138–48.

[Jameson, Robert?]. 1826. Observations on the nature and importance of geology. *ENPJ* 1: 293–302.

———. 1826. General observations on the former and present geological condition of the countries discovered by Captains Parry and Ross. *ENPJ* 2: 104–6.

Johnston, James Finlay Weir. 1833. On the gradual elevation of the land in Scandinavia. *ENPJ* 15: 34–48.

Jorio, Andrea di. 1817. *Guida di Pozzuoli e contorni*. Naples.

———. 1820. *Ricerche sul Tempio di Serapide in Pozzuoli*. Naples.

Keferstein, Christian. 1821. *General Charte von Teutschland auf der vom Hauptmann Weiland gezeichneten Charte geognostisch begraenzt*. Weimar.

———. 1821–24. *Teutschland geognostisch-geologisch dargestellt und mit Charten und Durchschnittszeichnungen erläutert*. 3 vols. Weimar.

———. 1825. *Tabellen über die vergleichende Geognosie. Ein Versuch*. Halle.

Klipstein, August Wilhelm von, and Johann Jakob Kaup. 1836. *Beschreibung und Abbildung von dem in Rheinhessen aufgefundenen colossalen Schedel des* Dinotherii gigantei, *mit geognostischen Mittheilungen über die knochenführenden Bildungen des mittelrheinischen Tertiärbeckens*. Darmstadt.

König, Charles. 1823. Account of the rock specimens collected by Captain Parry, during the northern voyage of discovery, performed in the years 1819 and 1820. *QJS* 15: 11–22.

Koninck, Laurent Guillaume. 1842–44. *Description des animaux fossiles qui se trouvent dans le terrain carbonifère de Belgique*. 2 vols. Liège.

Laplace, Pierre Simon. 1796. *Exposition du systême du monde*. Paris.

———. 1820. Sur la diminution de la durée du jour par le refroidissement de la terre. *ACP* 13: 410–17.

Lartet, Édouard. 1837. Note sur les ossements fossiles des terrains tertiaires de Simorre, de Sansan, etc., dans le département de Gers, et sur la découverte récente d'une machoire de singe fossile. *CRAS* 4: 85–93, 1 pl.

———. 1837. Nouvelles observations sur une machoire inférieure fossile, crue d'un singe voisin de gibbon, et sur quelques dents et ossements attribués à d'autres quadrumanes. *CRAS* 4: 583–84.

Lauder, Thomas Dick. 1821. On the Parallel Roads of Lochaber. *TRSE* 9: 1–64, pls. 1–9.

Lecoq, Henri, and Jean-Baptiste Bouillet. 1830. *Vues et coupes des principales formations géologiques du département du Puy-de-Dôme, accompagnées de la description et des échantillons des roches qui les composent*. 2 vols. Paris.

Lindley, John, and William Hutton. 1831–37. *The fossil flora of Great Britain; or, figures and descriptions of the vegetable remains found in a fossil state in this country*. 3 vols. London.

Luc, Jean-André de. 1809. *Traité élémentaire de géologie*. Paris.

———. 1810–11. *Geological travels*. 3 vols. London.

———. 1813. *Geological travels in some parts of France, Switzerland and Germany*. 2 vols. London.

Lund, Peter Wilhelm. 1838. Blik paa Brasiliens dyreverden for sidste jordomvaeltning: anden afhandling: Pattedyrene. *KDVA* 8: 61–144.

———. 1839. Coup d'oeil sur les espèces éteintes de mammifères du Brésil; extrait de quelques mémoires présentés à l'Académie des Sciences de Copenhague. *ASN* (*Zool.*) 11: 214–34.

Lyell, Charles. 1825. On a dike of serpentine, cutting through sandstone in the county of Forfar. *EJS* 3: 112–26.

———. 1826. On a recent formation of freshwater limestone in Forfarshire, and on some recent deposits of freshwater marl. *TGS* (2) 2 (1): 72–96, pls. 10–13.

[———]. 1826. [Review of the publications of six provincial English scientific institutions]. *QR* 34: 153–79.

[———]. 1826. [Review of] *Transactions of the Geological Society of London.* Vol. 1, 2d series. London. 1824. *QR* 34: 507–40.

[———]. 1827. [Review of] *Memoir on the geology of central France, including the volcanic formations of Auvergne, the Velay, and the Vivarais, with a volume of maps and plates.* By G. P. Scrope, F.R.S, F.G.S. London. 1827. *QR* 36: 437–83.

———. 1830–33. *Principles of geology, being an attempt to explain the former changes of the earth's surface, by reference to causes now in operation.* 3 vols. London [facsimile repr. Chicago (University of Chicago Press), 1990–91].

———. 1832–34. *Lehrbuch der Geologie: ein Versuch, die früheren Veränderungen der Erdoberfläche durch noch jetzt wirksame Ursachen zu erklären.* 3 vols. Leipzig.

———. 1834–35. *Principles of geology, being an inquiry how far the former changes of the earth's surface are referable to causes now in operation.* 3rd ed. 4 vols. London.

———. 1835. On the proofs of a gradual rising of the land in certain parts of Sweden. *PTRS* 1835: 1–38, pls. 1, 2.

———. 1835. *Principles of geology . . .* 4th ed. 4 vols. London.

———. 1837. Address to the Geological Society, delivered at the anniversary, on the 17th of February, 1837. *PGS* 2 (49): 479–523.

———. 1838. *Elements of geology.* London.

———. 1839. *Élémens de géologie.* Paris.

———. 1839. On the relative ages of the tertiary deposits commonly called 'Crag' in Norfolk and Suffolk. *MNH* (n.s.) 3: 313–30.

———. 1839. Remarks on some fossil and recent shells, collected by Capt. Bayfield, R.N., in Canada. *PGS* 3 (63): 119–20.

———. 1840. *Principles of geology. Or, the modern changes of the earth and its inhabitants considered as illustrative of geology.* 6th ed. 3 vols. London.

———. 1840. On the boulder formation or Drift, and associated freshwater deposits composing the mud cliffs of eastern Norfolk. *PGS* 3 (67): 171–79.

———. 1840. On the geological evidence of the former existence of glaciers in Forfarshire. *PGS* 3 (72): 337–45.

———. 1841. *Elements of geology.* 2nd ed. 2 vols. London.

———. 1841–42. *Grundsätze der Geologie oder die neuen Veränderungen der Erde und ihre Bewohner zu geologischen Erläuterungen. Übersetzt nach der sechsten Originalauflage . . .* 3 vols. Weimar.

———. 1843–48. *Principes de géologie, ou illustrations de cette science empruntées aux changements modernes que la terre et ses habitants ont subis. Ouvrage traduit de l'anglais, sur le sixième édition . . .* 4 vols. Paris.

———. 1845. *Travels in North America; with geological observations on the United States, Canada, and Nova Scotia.* 2 vols. London.

———. 1863. *The geological evidences of the antiquity of man, with remarks on theories of the origin of species by variation.* London.

Lyell, Charles, and Roderick Impey Murchison. 1829. On the excavation of valleys, as illustrated by the volcanic rocks of central France. *ENPJ* 7: 15–48, pls. 1–3.

———. 1829. Sur les dépôts lacustres tertiaires du Cantal, et leurs rapports avec les roches primordiales et volcaniques. *ASN* 18: 173–214, pls. 12, 13.

Macaire, Isaac. 1837. Compte rendu de la session de la Société Helvétique des Sciences Naturelles à Neuchâtel (juillet 1837). *BU* (n.s.) 10: 368–82.

———. 1837. Sur les blocs erratiques du Jura, avec rectification de M. Agassiz. *BU* (n.s.) 11: 416–18.

———. 1842. [Review of] *Essai sur les glaciers et sur les terrains erratiques du bassin du Rhône,* par Jean de Charpentier. Lausanne, 1841. *BU* (n.s.) 37: 390–411.

MacCulloch, John. 1817. On the Parallel Roads of Glen Roy. *TGS* 4 (2): 314–92, pls. 15–22.
McEnery, John. 1859. *Cavern researches, or, discoveries of organic remains, and of British and Roman reliques, in the caves of Kent's Hole, Anstis Cove, Chudleigh, and Berry Head.* London.
Maclaren, Charles. 1842. The glacial theory of Prof. Agassiz. *AJS* 42: 346–65 [repr. from booklet with same title, Edinburgh, 1841].
Maelen, Philippe van den. 1831. *Dictionnaire géographique de la province de Liége, précédé d'un fragment du mémorial de l'Établissement Géographique de Bruxelles.* Brussels.
Mantell, Gideon Algernon. 1822. *The fossils of the South Downs; or illustrations of the geology of Sussex.* London.
———. 1825. Notice on the Iguanodon, a newly discovered fossil reptile, from the sandstone of Tilgate Forest, in Sussex. *PTRS* 1825: 179–86, pl. 14.
———. 1826. On the Iron-Sand formation of Sussex. *TGS* (2) 2 (1): 131–34.
———. 1827. *Illustrations of the geology of Sussex: containing a general view of the geological relations of the south-eastern part of England; with figures and descriptions of the fossils of Tilgate Forest.* London.
———. 1831. The geological age of reptiles. *ENPJ* 11: 181–85.
———. 1838. *The wonders of geology; or, a familiar exposition of geological phenomena; being the substance of a course of lectures delivered at Brighton.* 2 vols. London.
———. 1851. *Petrifactions and their teachings; or, a hand-book to the Gallery of Organic Remains of the British Museum.* London.
Marcet, Jane. 1805. *Conversations on chemistry.* London.
Martins, Charles. 1840. Observations sur les glaciers du Spitzberg, comparés à ceux de la Suisse et de le Norvège. *BU* (n.s.) 28: 139–72.
Miller, John Samuel. 1821. *A natural history of the Crinoidea, or lily-shaped animals; with observations on the genera, Asteria, Euryale, Comatula & Marsupites.* Bristol.
———. 1826. Observations on belemnites. *TGS* (2) 2 (1): 45–62, pls. 7–9.
Mitchell, Thomas Livingstone. 1831. An account of the limestone caves at Wellington Valley, and of the situation, near one of them, where fossil bones have been found. *PGS* 1 (21): 321–22.
Münster, Georg. 1834. Mémoire sur les clymènes et les goniatites du calcaire de transition du Fichtelgebirge. *ASN* (*Zool.*) 2: 65–99, pls. 1–6.
Murchison, Roderick Impey. 1832. An address to the Geological Society, delivered on the evening of the 17th February, 1832. *PGS* 1 (25): 362–86.
———. 1833. Address to the Geological Society, on the evening of the 15th of February 1833. *PGS* 1 (30): 438–64.
———. 1834. On the structure and classification of the Transition rocks of Shropshire, Herefordshire and part of Wales, and on the lines of disturbance which have affected that series of deposits, including the valley of elevation of Woolhope. *PGS* 2 (34): 13–18, table.
———. 1835. On the Silurian system of rocks. *PM* (3) 7: 46–52.
———. 1839. *The Silurian system, founded on geological researches in the counties of Salop, Hereford, Radnor, Montgomery, Caermarthen, Brecon, Pembroke, Monmouth, Gloucester, Worcester, and Stafford; with descriptions of the coal-fields and overlying formation.* London.
———. 1841. Observations géologiques sur la Russie. *BSNM* 1841: 901–9.
Murchison, Roderick Impey, and Charles Lyell. 1829. On the tertiary fresh water formations of Aix in Provence including the coal-field of Fuveau. *ENPJ* 7: 287–98, pls. 5, 6.
Murchison, Roderick Impey, Édouard de Verneuil, and Alexander von Keyserling. 1845. *The geology of Russia in Europe and the Ural mountains / Géologie de la Russie en Europe et des montagnes d'Oural.* 2 vols. London and Paris.
[Murray, John]. 1802. *A comparative view of the Huttonian and Neptunian systems of geology: in answer to the Illustrations of the Huttonian Theory of the Earth, by Professor Playfair.* Edinburgh.
Necker, L. A. 1826. Sur les filons granitiques et porphyritiques de Valorsine et sur le gisement des couches coquillières des montagnes de Sales, des Fizs et de Platet. *BU* 33: 62–92.
Niebuhr, Barthold Georg. 1811–12. *Römische Geschichte.* 2 vols. Berlin.
———. 1828. *The history of Rome.* Cambridge.
Nolan, Frederick. 1833. *The analogy of revelation and science established.* Oxford.
Omalius d'Halloy, Jean-Baptiste d'. 1822. Observations sur un essai de carte géologique de la France, des Pay-Bas, et des contrées voisins. *AM* 7: 353–76.

———. 1831. *Éléments de géologie*. Paris.
Orbigny, Alcide d'. 1834–47. *Voyage dans l'Amérique méridionale executé pendant les années 1826–1833*. 10 vols. Paris.
Owen, Richard. 1834. On the generation of the marsupial animals, with a description of the impregnated uterus of the kangaroo. *PTRS* 1834: 333–64.
———. 1839–42. Fossil mammalia. *In* Darwin, *Zoology of the Beagle*, Part 1.
Paley, William. 1794. *A view of the evidences of Christianity, in three parts*. 3 vols. London.
———. 1802. *Natural theology: or, evidences of the existence and attributes of the Deity, collected from the appearances of nature*. London.
Pander, Christian Heinrich, and E. d'Alton. 1821. *Das Riesen-Faulthier, Bradypus giganteus, abgebildet, beschrieben, und mit den verwandten Geschlechtern verglichen*. Bonn.
Parkinson, James. 1804–11. *Organic remains of a former world. An examination of the mineralized remains of the vegetables and animals of the antediluvian world; generally termed extraneous fossils*. 3 vols. London.
———. 1822. *Outlines of oryctology. An introduction to the study of fossil organic remains, especially those found in the British strata: intended to aid the student in his enquiries respecting the nature of fossils and their connection with the formation of the earth*. London.
Parry, William Edward. 1821–24. *Journal of a voyage for the discovery of a north-west passage from the Atlantic to the Pacific: performed in the years 1819–20*. London.
———. 1826. *Journal of a third voyage for the discovery of a north-west passage from the Atlantic to the Pacific; performed in the years 1824–25, in His Majesty's ships Hecla and Fury*. London.
Penn, Granville. 1822. *A comparative estimate of the mineral and Mosaical geologies*. London.
Phillips, John. 1829. *Illustrations of the geology of Yorkshire; or a description of the strata and organic remains. . . Part I. The Yorkshire coast*. York.
———. 1834. *Syllabus of a course of eight lectures on geology*. London.
———. 1836. *Illustrations of the geology of Yorkshire; or a description of the strata and organic remains. . . Part II. The Mountain Limestone district*. London.
———. 1837–39. *A treatise on geology*. 2 vols. London.
———. 1838. *A treatise on geology, forming the articles under that head in the seventh edition of the Encyclopaedia Britannica*. Edinburgh.
———. 1840. Palaeozoic series. *PC* 17: 153–54.
———. 1841. Saliferous system. *PC* 20: 354–55.
———. 1841. *Figures and descriptions of the Palaeozoic fossils of Cornwall, Devon and west Somerset; observed in the course of the Ordnance Geological Survey of that district*. London.
———. 1860. *Life on the earth: its origin and succession*. Cambridge and London.
Playfair, John. 1802. *Illustrations of the Huttonian theory of the earth*. Edinburgh [facsimile repr. New York (Dover), 1956].
Powell, Baden. 1833. *Revelation and science*. Oxford.
———. 1838. *The connexion of natural and divine truth; or, the study of the inductive philosophy considered as subservient to theology*. London.
Prévost, Constant. 1821. Sur un nouvel exemple de la réunion de coquilles marines et de coquilles fluviatiles dans les mêmes couches. *JP* 92: 418–27.
———. 1822. Observations sur les grès coquillers de Beau-Champ et sur les mélanges de coquilles marines et fluviatiles dans les couches intérieures de la formation de gypse des environs de Paris. *JP* 94: 1–18.
———. 1823. De l'importance de l'étude des corps organisés vivans pour la géologie positive, et description d'une nouvelle espèce de mollusque testacé du genre Melanopsis. *MSHNP* 1: 259–68.
———. 1825. Observations sur le gisement du *Mégalosaure* fossile. *NBSP* 1825: 41–43.
———. 1825. Observations sur les schistes calcaires oolitiques de Stonesfield en Angleterre, dans lesquels ont été trouvés plusieurs ossemens fossiles de mammifères. *ASN* 4: 389–417, pls. 17, 18.
———. 1825. De la formation des terrains des environs de Paris. *NBSP* 1825: 74–77, 88–90 [trans. Bork, "Constant Prévost" (1990), 28–30].
———. 1827. Examen de cette question géologique: les continents que nous habitons ont'ils été à plusieurs reprises submergés par la mer? *ACP* 35: 439–43.
———. 1828. Les continens actuels, ont-ils éte, à plusieurs reprises, submergés par la mer? *MSHNP* 4: 249–346.

———. 1831. Sur le nouvel islot volcanique de la mer de Sicile. *ASN* 24: 103–12, pl. 4.
———. 1831. *L'Ile Julia, le 29 septembre 1831, 2 h[eur]es après-midi* . . . [sheet print]. Paris.
———. 1835. Notes sur l'île Julia, pour servir à l'histoire de la formation des montagnes volcaniques. *MSGF* 2: 91–124, pls. 5, 6.
———. 1835. *Académie des Sciences; Section de Géologie et Minéralogie. Candidature de M. Constant Prévost.* Paris.
———. [1835?]. *Documents pour l'histoire des terrains tertiaires.* [Paris].
Prichard, James Cowles. 1826. *Researches on the physical history of mankind.* 2 vols. London [facsimile repr. Chicago (University of Chicago Press), 1973].
Pusch, Georg Gottlieb. 1833–36. *Geognostische Beschreibung von Polen, so wie der übrigen Nordkarpathen-Länder.* 2 vols. Stuttgart and Tübingen.
———. 1836. *Geognostyscher Atlas von Polen.* Stuttgart.
Quoy, Jean René Constant, and Joseph Paul Gaimard. 1825. Mémoire sur l'accroissement des polypes lithophytes considéré géologiquement. *ASN* 6: 273–90.
Ramond de Carbonnières, Louis. 1818. Nivellement barométrique des Monts-Dores et des Monts-Dômes, disposé par order de terrains. *MIN* 14: 1–138.
Razumovsky, Gregor Kirilovich. 1819. *Coup d'oeil géognostique sur le nord de l'Europe en général, et particulièrement de la Russie . . . seconde édition, fort augmentée.* Berlin.
———. 1829. Des gros blocs de roches que l'on trouve épars ou accumulés sur le terrain de natures très-diverses. *ASN* 18: 133–47.
Reboul, Henri. 1833. *Géologie de la période quaternaire et introduction à l'histoire ancienne.* Paris.
Rendu, [Louis]. 1840. *Théorie des glaciers de la Savoie.* Chambéry.
[Rennie, James]. 1828. *Conversations on geology; comprising a familiar explanation of the Huttonian and Wernerian systems; the Mosaic geology, as explained by Mr Granville Penn; and the late discoveries of Professor Buckland, Humboldt, Dr MacCulloch, and others.* London. [This anonymous work is often attributed, erroneously, to Granville Penn.]
Renoir, E. 1840. Note sur les glaciers qui ont recouvert anciennement la partie méridionale de la chaîne des Vosges. *BSGF* 11: 53–65.
———. 1841. Sur les traces des anciens glaciers qui ont comblé les vallées des Alpes du Dauphiné et sur celles de même nature qui paraissent résulter de quelques-unes des observations fait par M. Robert dans la Russie septentrionale. *BSGF* 12: 68–82.
Richardson, John. 1828. Topographical and geological notes. *In* Franklin, *Second expedition to the Polar Sea*, Appendix I (i–lvii).
Risso, Giovanni Antonio. 1826. *Histoire naturelle des principes productions de l'Europe méridionale et particulièrement de celles des environs de Nice et des Alpes maritimes.* 5 vols. Paris.
Ross, James Clark. 1847. *A voyage of discovery and research in the southern and Antarctic regions, during the years 1839–43.* 2 vols. London.
Ross, John. 1835. *Narrative of a second voyage in search of a north-west passage, and of a residence in the Arctic regions during 1829, 1830, 1831, 1832, 1833.* London.
Schlotheim, Ernst Friedrich von. 1820. *Die Petrefaktenkunde auf ihrem jetzigen Standpunkte durch die Beschreibung seiner Sammlung versteinerter und fossiler Überreste des Thier- und Pflanzenreichs der Vorwelt erläutert.* Gotha.
Schmerling, Philippe-Charles. 1831. Cavernes à ossemens fossiles, découvertes jusqu'à ce jour dans le province de Liége. *In* Maelen, *Province de Liége*, appendice, 3–7.
———. 1833. Sur les cavernes à ossemens de la province de Liége. *BSGF* 3: 217–22.
———. 1833. Ueber die Knochenhöhlen bei Lüttich. *NJM* 1: 38–48, 592–600.
———. 1833–34. *Recherches sur les ossemens fossiles découvertes dans les cavernes de la province de Liége.* 2 vols. Liège.
Scoresby, William, Jr. 1820. *An account of the Arctic regions, with a history and description of the northern whale-fishery.* 2 vols. Edinburgh [facsimile repr. Newton Abbot (David and Charles), 1969].
Scrope, George Poulett. 1825. *Considerations on volcanos, the probable cause of their phenomena, the laws which determine their march, the disposition of their products, and their connexion with the present state and past history of the globe; leading to the establishment of a new theory of the earth.* London.
———. 1827. *Memoir on the geology of central France, including the volcanic formations of Auvergne, the Velay and the Vivarais.* 2 vols. London.

———. 1830. On the gradual excavation of the valleys in which the Meuse, the Moselle, and some other rivers flow. *PGS* 1 (14): 170–71.
[———]. 1830. [Review of] *Principles of geology* by Charles Lyell, F.R.S. 2 vols. Lond. 1830. *QR* 43: 411–69.
———. 1830. *On credit currency and its superiority to coin, in support of a petition for the establishment of a cheap, safe and sufficient circulating medium.* London.
———. 1833. *Principles of political economy, deduced from the natural laws of social welfare and applied to the present state of Britain.* London.
[———]. 1835. [Review of] *Principles of geology* . . . By Charles Lyell, Esq., F.R.S., President of the Geological Society of London. Third edition. In 4 vols. 12mo. 1835. *QR* 53: 406–48.
[———]. 1839. [Review of Murchison's *Silurian*, Lyell's *Elements*, Phillips's *Treatise*, Mantell's *Wonders*, and De la Beche's *Report*]. *QR* 64: 102–20.
———. 1858. *The geology and extinct volcanos of central France.* London.
Sedgwick, Adam. 1820. On the physical structure of those formations which are immediately associated with the primitive ridge of Devonshire and Cornwall. *TCPS* 1: 89–146.
———. 1821. *A syllabus of a course of lectures on geology.* Cambridge.
———. 1825. On the origin of alluvial and diluvial formations. [Continued as] On diluvial formations. *AP* (n.s.) 9: 241–57; 10: 18–37.
———. 1830. Address delivered by the President [of the Geological Society, on 19 February, 1830]. *PGS* 1 (15): 187–212.
———. 1831. Address to the Geological Society, delivered on the evening of the anniversary, Feb. 18, 1831. *PGS* 1 (20): 281–316.
———. 1831. *A discourse on the studies of the university.* Cambridge [facsimile repr. Leicester (Leicester University Press), 1969].
———. 1835. Remarks on the structure of large mineral masses, and especially on the chemical changes produced in the aggregation of stratified rocks during different periods after their deposition. *TGS* (2) 3 (3): 461–86, pl. 47.
———. 1835. Introduction to the general structure of the Cumbrian mountains; with a description of the great dislocations by which they have been separated from the neighbouring Carboniferous chains. *TGS* (2) 4 (1): 47–68, pls. 4, 5.
———. 1838. A synopsis of the English series of stratified rocks inferior to the old red sandstone; with an attempt to determine the successive natural groups and formations. *PGS* 2 (58): 675–85.
[———]. 1845. [Review of] *Vestiges of the natural history of creation.* London. 1845. *ER* 82: 1–85 [facsimile repr. Lynch, *Vestiges* (2000) 1].
Sedgwick, Adam, and Roderick Impey Murchison. 1829. On the Tertiary deposits of the Vale of Gosau in the Salzburg Alps. [Continued as] On the Tertiary formations which range along the flanks of the Salzburg and Bavarian Alps. *PGS* 1 (13): 153–59.
———. 1836. On the *Silurian* and *Cambrian* systems, exhibiting the order in which the older sedimentary strata succeed each other in England and Wales. *RBAAS* 1835, Trans. Sec.: 59–61.
Sefström, Nils Gabriel. 1836. Ueber die Spuren einer sehr grossen urweltlicher Fluth. *APC* 38: 614–18.
———. 1838. Undersökning af de räfflor, hvaraf Skandinaviens berg äro med bestämd riktning fårade, samt om deras sannolika uppkomst. *KVAH* 1836: 141–227, pls. 6–9.
———. 1838. Tillägg till föregående afhandling. Sednare observationer, anställde dels i Sverige dels i andra länder. *KVAH* 1836: 228–55.
———. 1838. Untersuchung über die auf den Felsen Skandinaviens in bestimmter Richtung vorhandenden Furchen und derene wahrscheinliche Entstehung. *APC* 43: 533–67.
Serres, Marcel de. 1818. Mémoire sur les terrains d'eau douce ainsi que sur les animaux et les plantes qui vivent alternativement dans les eaux douces et les eaux salées. *JP* 87: 31–46, 118–35, 161–78.
———. 1823. Observations sur les ossemens humains découverts dans les crevasses des terrains secondaires, et en particulier sur ceux que l'on observe dans le caverne de Durfort, dans le département du Gard. *BU* 23: 277–95; 24: 11–35.
———. 1826. Note sur les cavernes à ossemens et les brèches osseuses du midi de la France. *ASN* 9: 200–213.
———. 1828. Observations générales sur les cavernes à ossements et les brèches osseuses du midi de la France. *MSLN* 1828: 16–58.

———. 1829. *Géognosie des terrains tertiaires, ou tableau des principaux animaux invertébrés des terrains marins tertiaires, du Midi de la France.* Montpellier.

———. 1830. Lettre adressée au Président de l'Académie des Sciences de Paris. *BSNG* 22: 33–36.

———. 1830. Sur les ossemens humains découverts dans certaines cavernes du midi de la France, mêlés et confondus dans les mêmes limons où existent le nombreuses espèces de mammifères terrestres, considérées jusqu'à présent comme fossiles et comme antédiluviennnes. *JG* 2: 184–91.

———. 1832–33. De la contemporanéité de l'homme et des espèces d'animaux perdues. *RE* 55: 48–73; 57: 252–81; 59: 379–414.

———. 1833–34. Mémoire sur la question de savoir si des animaux terrestres ont cessé d'exister depuis l'apparition de l'homme, et si l'homme a été contemporain des espèces perdues, ou du moins qui ne paraissent plus avoir de représentants sur la terre. *BU* 1833 2: 277–314; 1834 1: 160–76, 231–56, 352–84.

———. 1835. Essai sur les cavernes à ossemens et sur les causes qui les y ont accumulés. *NVH-MW* 22: 1–222.

———. 1838. *De la cosmogonie de Moïse, comparée aux faits géologiques.* Paris.

Serres, Marcel de, Joseph-Marie Dubrueil, and Jeanjean. 1839. *Recherches sur les ossemens humatiles des cavernes de Lunel-Viel.* Montpellier.

Smith, James. 1839. On the climate of the newer pliocene tertiary period. *PGS* 3 (63): 118–19.

Smith, William. 1817. *Stratigraphical system of organized fossils, with reference to the specimens of the original geological collection in the British Museum; explaining their state of preservation and their use in identifying the British strata.* London.

Smythe, George Walter. [1832]. *Views and description of the late volcanic island off the coast of Sicily.* London.

Sowerby, James [and James de Carle Sowerby]. 1812–46. *The mineral conchology of Great Britain; or coloured figures and descriptions of those remains of testaceous animals or shells, which have been preserved at various times and depths in the earth.* 7 vols. London.

Spix, J. B. von, and Carl Friedrich Martius. 1823–31. *Reise in Brasilien, auf Befehl S. M. Max. Josephs I, Königs von Baiern, in den Jahren 1817–1820.* 3 vols. and *Atlas.* Munich.

Sternberg, Kaspar Maria von. 1820–38. *Versuch einer geognostisch-botanischen Darstellung der Flora der Vorwelt.* Leipzig.

Strangways, William Thomas Horner Fox-. 1821. Geological sketch of the environs of Petersburg. *TGS* 5: 392–458, pls. 28–31.

———. 1822. An outline of the geology of Russia. *TGS* (2) 1 (1): 1–39, pls. 1, 2.

Studer, Bernhard. 1827. Geognostische Bemerkungen über einige Theile der nördlichen Alpenkette. *ZM* 1827, vol. 1: 1–46.

———. 1838. Über die neueren Erklarungen des Phänomens erratischer Blöcke. *NJM* 1838: 278–87.

———. 1840. Notice sur quelques phénomènes de l'époque diluvienne. *BSGF* 11: 49–52.

[Sumner, John]. 1829. [Review of] *A new system of geology* . . . by Andrew Ure. *BC* 6: 387–412.

Tournal, Paul. 1827. Note sur deux cavernes à ossemens, découvertes à Bire [i.e., Bize], dans les environs de Narbonne. *ASN* 12: 78–82.

———. 1828. Note sur la caverne de Bize près Narbonne. *ASN* 15: 348–51.

———. 1829. Sur les ossemens humains, mêlés, dans les cavernes de Bize, à des débris de mammifères terrestres d'espèces perdues. *AM* 5: 515–17.

———. 1829. Considérations théoriques sur les cavernes à ossemens de Bize, près Narbonne (Aude), et sur les ossemens humains confondus avec des restes d'animaux appartenant à des espèces perdues. *ASN* 18: 242–58.

———. 1831. Observations sur les ossemens humains et les objets de fabrication humaine confondus avec des ossemens de mammifères appartenant à des espèces perdues. *BSGF* 1: 195–200.

———. 1832. Cavernes à ossemens de Bize (Aude). *BSGF* 2: 380–82.

———. 1833. Considérations générales sur les phénomènes des cavernes à ossemens. *ACP* 52: 161–81.

———. 1834. Cavernes à ossemens. *HARST* 3: 52–63.

Ure, Andrew. 1829. *A new system of geology, in which the great revolutions of the earth and animated nature, are reconciled at once to modern science and sacred history.* London.

Venetz, Ignaz. 1833. Mémoire sur les variations de la température dans les Alpes de la Suisse. (Redigé en 1821). *DSG* 1: 1–38.

Webster, Thomas. 1824. A comparison between the beds below the Chalk in the Isle of Wight, and in the counties of Surrey, Kent, and Sussex. *AP* (n.s.) 8: 465–67.

———. 1825. Reply to Dr Fitton's paper in the 'Annals of Philosophy' for November [1824], entitled 'Inquiries respecting the geological relations of the beds between the Chalk and the Purbeck Limestone in the south-east of England'. *AP* (n.s.) 9: 33–50, pl. 35.

———. 1826. Observations on the Purbeck and Portland beds. *TGS* (2) 2 (1): 37–44.

[Whewell, William]. 1831. [Review of] *Principles of geology* . . . By Charles Lyell, Esq., F.R.S., For. Sec. to the Geol. Soc., &c. In 2 vols. Vol. I . . . Murray. 1830. *BC* 9: 180–206.

[———]. 1832. [Review of] *Principles of geology* . . . By Charles Lyell, Esq., F.R.S., Professor of geology in King's College, London. Vol. II. London. 1832. *QR* 47: 103–32.

[———]. 1834. [Review of] *On the connexion of the physical sciences.*By Mrs Somerville . . . *QR* 51: 54–68.

———. 1837. *History of the inductive sciences from the earliest to the present time.* 3 vols. London.

———. 1840. *The philosophy of the inductive sciences, founded upon their history.* 2 vols. London.

Whiston, William. 1696. *A new theory of the earth, from its original, to the consummation of all things. Wherein the creation of the world in six days, the universal deluge, and the general conflagration, as laid down in Holy Scripture, are shewn to be perfectly agreeable to reason and philosophy.* London.

Winning, William Balfour. 1838. *A manual of comparative philology, in which the affinity of the Indo-European languages is illustrated, and applied to the primeval history of Europe, Italy, and Rome.* London.

4. PRINTED SOURCES: SECONDARY

In this bibliography, the quite high proportion of items published more than twenty years ago may surprise or puzzle readers who are more familiar with the practice of the natural sciences than with that of their history. It is not that the pace of change in the history of the sciences is any less rapid than in the sciences themselves; nor that the author has failed to keep up to date with research in his own field. It is rather that historical books and articles tend to enjoy a much longer useful half-life than scientific publications; and that, in the specific area covered by the present volume, much of the finest relevant research was published in the 1970s and 1980s, whereas in more recent times swings of historiographical fashion have, regrettably, reduced the flow.

Sources used and cited in *Bursting the Limits of Time* are not cited again here, unless they have been used substantially for this volume; otherwise they can be found by following the cross-references to *BLT*. Conversely, I make no apology for citing rather fully my own relevant publications; the present volume is intended to integrate much of this earlier research into a unified narrative, but of course the detail has had to be drastically curtailed (some 450 pages of *Great Devonian Controversy* and nearly 100 pages of "Darwin and Glen Roy", for example, are represented here by the brief sections §31.1 and §33.4 respectively).

Biographical sources are listed here only if they add substantially to information in Charles C. Gillispie (ed.), *Dictionary of Scientific Biography*, 16 vols. (New York 1970–80); in J. C. Poggendorff, *Biographisch-litterarisches Handwörterbuch zur Geschichte der exacten Wissenschaften*, 2 vols. (Leipzig, 1863); or in the *Oxford Dictionary of National Biography*, 60 vols. (Oxford, 2004) and equivalent biographical dictionaries of other nations. Not listed here, nor cited in the footnotes, are the many recent "popular" books on the history of the earth sciences, most of which are wholly dependent on more scholarly secondary sources (not always adequately acknowledged).

The following list is a key to the abbreviations used in this bibliography for items published in periodicals (those with single-word titles are not abbreviated, and are not included in the list). For convenience, a few abbreviations used in the footnotes for other frequently cited works are also listed.

ABMNG Abhandlungen und Berichte Museums der Natur Gotha
AHPB Annals of the history and philosophy of biology
AJ Alpine journal
ANH Archives of natural history [continuation of *JSBNH*]
AP Annales de paléontologie
APB Acta psychiatrica Belgica
AS Annals of science
ASGN Annales de la Société Géologique du Nord [Lille]
BBMNH Bulletin of the British Museum (Natural History)
BGSA Bulletin of the Geological Society of America
BJHS British journal for the history of science
BM Bulletin Murithienne
BMHNP Bulletin du Muséum d'Histoire Naturelle de Paris
BMSAP Bulletin et mémoires de la Société d'Anthropologistes de Paris.
BP Biology and philosophy
BSGF Bulletin de la Société Géologique de France
BSNSN Bulletin de la Société Neuchâteloise des Sciences Naturelles
BSPF Bulletin de la Société Préhistorique Française
BSVSN Bulletin de la Société Vaudoise des Sciences Naturelles
CCD Correspondence of Charles Darwin [see Burkhardt *et al.*, 1985–, below]
EGH Eclogae geologicae Helveticae
ESH Earth sciences history
GA Geografiska annaler [Stockholm]
GSAM Geological Society of America, memoirs
GSASP Geological Society of America, special papers
GSLM Geological Society [London], memoirs
GSLSP Geological Society [London], special publications
HPLS History and philosophy of the life sciences
HN Histoire et nature
HS History of science
JGE Journal of geological education
JGS Journal of the Geological Society
JHB Journal of the history of biology
JHI Journal of the history of ideas
JHMAS Journal of the history of medicine and allied sciences
JSBNH Journal of the Society for the Bibliography of Natural History
LLJ Life, letters and journals of Sir Charles Lyell [see Lyell (K.) 1881, below]
McG Mercian geologist
MG Modern geology
MKNAW Mededelingen der Koninklijke Nederlandse Akademie van Wetenschappen: afdeling Letterkunde
MSÉM Mémoires de la Société d'Émulation de Montbéliard
MSGF Mémoires de la Société Géologique de France
MSVSN Mémoires de la Société Vaudoise des Sciences Naturelles
NRRSL Notes and records of the Royal Society of London
PAPS Proceedings of the American Philosophical Society
PGA Proceedings of the Geologists' Association
PPP Palaegeography, palaeoclimatology, palaeoecology
RHS Revue d'histoire des sciences
SC Science in context

SHPS	Studies in the history and philosophy of science
SSS	Social studies of science
ST	Sciences de la terre
TCFHG	Travaux du Comité Français pour l'Histoire de la Géologie [COFRHIGEO]
TWNFC	Transactions of the Woolhope Naturalists' Field Club
VS	Victorian studies
ZJLS	Zoological journal of the Linnean Society

Agassiz, Elizabeth Cary. 1885. *Louis Agassiz; his life and correspondence.* 2 vols. Boston (Houghton, Mifflin).

Aldhouse-Green, Stephen (ed.). 2000. *Paviland Cave and the 'Red Lady': a definitive report.* Bristol (Academic and Specialist Press).

Alfieri, Vittorio Enzo (ed.). 1981. *De Aetna. Il testo di Pietro Bembo tradotto e presentado da Vittorio Enzo Alfieri. . . . Iconographia*, Palermo (Sellerio).

Allen, David E. 1976. *The naturalist in Britain: a social history.* London (Allen Lane).

Altick, Richard D. 1978. *The shows of London.* Cambridge, Mass. (Harvard University Press).

Amsterdamska, Olga. 1987. *Schools of thought: the development of linguistics from Bopp to Saussure.* Dordrecht (Reidel).

Anonymous. (ed.). 1978. *'Teyler' 1778–1978: studies en bijdragen over Teylers Stichting naar aanleiding van het tweede eeuwfeest.* Haarlem (Teylers Museum).

Appel, Toby A. 1987. *The Cuvier-Geoffroy debate: French biology in the decades before Darwin.* New York and London (Oxford University Press).

Backenköhler, Dirk. 2002. Cuviers langer Schatten: 'Il n'y a point d'os humains fossiles'. *In* Hossfeld and Junker, *Entstehung biologischer Disziplinen*, 133–47.

Bardet, N., and J. W. M. Jagt. 1996. *Mosasaurus hoffmani*, le 'Grand animal fossile des carrières de Maestricht', deux siècles d'histoire. *BMHNP* 18: 569–93.

Barrett, Paul H. 1974. The Sedgwick-Darwin geologic tour of North Wales. *PAPS* 118: 146–64.

——— (ed.). 1977. *The collected papers of Charles Darwin.* Chicago (University of Chicago Press).

Bartholomew, Michael. 1973. Lyell and evolution: an account of Lyell's response to the prospect of an evolutionary ancestry for Man. *BJHS* 6: 261–303.

———. 1979. The singularity of Lyell. *HS* 17: 276–93.

Barton, Ruth. 2003. 'Men of science': language, identity and professionalization in the mid-Victorian scientific community. *HS* 41: 73–119.

Benton, Michael. 1982. Progressionism in the 1850s: Lyell, Owen, Mantell and the Elgin fossil reptile *Leptopleuron* (*Telerpeton*). *ANH* 11: 123–36.

Bergsten, Folke. 1954. The land uplift in Sweden from the evidence of the old water marks. *GA* 36: 81–111.

Berkeley, Edmund, and Dorothy Smith Berkeley. 1988. *George William Featherstonhaugh: the first U. S. government geologist.* Tuscaloosa (University of Alabama Press).

Blanckaert, Claude. 1999. Les fossiles de l'imaginaire: temps de la nature et progrès organique. *Romanticisme* no. 104: 85–101.

———. 2000. Avant Adam: les représentations analogiques de l'homme fossile dans la première moitié du XIXe siècle. *In* Ducros and Ducros, *L'homme préhistorique*, 23–61.

Blanckaert, Claude, Claudine Cohen, Pietro Corsi, and Jean-Louis Fischer (eds.). 1997. *Le Muséum au premier siècle de son histoire.* Paris (Muséum National d'Histoire Naturelle).

Blundell, Derek J., and Andrew C. Scott (eds.). 1998. *Lyell: the present is the key to the past.* London (Geological Society) [*GSLSP* 143].

Bork, Kennard B. 1990. Constant Prévost (1787–1856): the life and contributions of a French uniformitarian. *JGE* 38: 1–30.

Bourdier, Franck. 1969. Geoffroy Saint-Hilaire versus Cuvier: the campaign for paleontological evolution. *In* Schneer, *History of geology*, 36–61.

Bowden, A. J., C. V. Burek, and R. Wilding (eds.). 2005. *History of palaeobotany: selected essays.* Bath (Geological Society) [*GSLSP* 241].

Bowler, Peter J. 1976. *Fossils and progress: paleontology and the idea of progressive evolution in the nineteenth century.* New York (Science History).

Boylan, Patrick J. 1970. An unpublished portrait of Dean William Buckland, 1784–1856. *JSBNH* 5: 350–54.

———. 1984. *William Buckland, 1784–1856: scientific institutions, vertebrate palaeontology, and Quaternary geology.* Leicester (University of Leicester [Ph.D. dissertation]) [British Library Document Supply Service microfilm D56595/85 DSC].

———. 1998. Lyell and the dilemma of Quaternary glaciation. *In* Blundell and Scott, *Lyell,* 145–59.

Brock, William H. 1966. The selection of the authors of the Bridgewater Treatises. *NRRSL* 21: 162–79.

Brooke, John H. 1979. The natural theology of the geologists: some theological strata. *In* Jordanova and Porter, *Images of the earth,* 39–64 [2nd ed., 1997, 53–74].

———. 1991. *Science and religion: some historical perspectives.* Cambridge (Cambridge University Press).

Browne, Janet. 1983. *The secular ark: studies in the history of biogeography.* New Haven (Yale University Press).

———. 1995. *Darwin: voyaging.* London (Pimlico).

Bruijn, J.G. de. 1977. *Inventaris van de prijsvragen uitgescheven door de Hollandsche Maatschappij der Wetenschappen, 1753–1917.* Haarlem (Hollandsche Maatschappij der Wetenschappen).

Brush, Stephen G. 1987. The nebular hypothesis and the evolutionary worldview. *HS* 25: 245–78.

———. 1996. *Nebulous earth: the origin of the solar system and the age of the earth from Laplace to Jeffreys.* Cambridge (Cambridge University Press).

Buffetaut, E., J. M. Mazin, and E. Salmon (eds.). 1982. *Actes du symposium paléontologique Georges Cuvier.* Montbéliard (Ville de Montbéliard).

Bultingaire, L. 1932. Iconographie de Georges Cuvier. *AMHN* (6) 9: 1–12, pls. 1–11.

Burchfield, Joe D. 1974. Darwin and the dilemma of geological time. *Isis* 65: 300–321.

———. 1975. *Lord Kelvin and the age of the earth.* New York (Science History).

———. 1998. The age of the earth and the invention of geological time. *In* Blundell and Scott, *Lyell,* 137–43.

Burek, Cynthia, and Bettie Higgs (eds.). 2007. *The role of women in the history of geology.* London (Geological Society) [*GSLSP,* 281].

Burkhardt, Frederick, and Sydney Smith [and, later, others] (eds.). 1985–. *The correspondence of Charles Darwin.* vols. 1–. Cambridge (Cambridge University Press). [abbreviated in footnotes to *CCD.*]

Burns, James. 2007. John Fleming and the geological deluge. *BJHS* 40:205–25.

Calloway, Jack M., and Elizabeth L. Nicholls (eds.). 1997. *Ancient marine reptiles.* San Diego (Academic Press).

Camardi, Giovanni. 1999. Charles Lyell and the uniformity principle. *BP* 14: 537–60.

Cannon, Walter Faye. 1961. The impact of uniformitarianism: two letters from John Herschel to Charles Lyell, 1836–1837. *PAPS* 105: 301–14.

———. 1976. Charles Lyell, radical actualism, and theory. *BJHS* 9: 104–20.

——— [Susan Faye Cannon]. 1978. *Science in culture: the early Victorian period.* New York (Science History).

Carozzi, Albert V. (ed.). 1967. *'Studies on glaciers', by Louis Agassiz.* New York (Hafner).

Cawood, John. 1977. Terrestrial magnetism and the development of international collaboration in the early nineteenth century. *AS* 34: 551–87.

———. 1979. The Magnetic Crusade: science and politics in early Victorian Britain. *Isis* 70: 493–518.

Challinor, John. 1961–64. Some correspondence of Thomas Webster, geologist (1773–1844). *AS* 17: 175–96; 18: 147–75; 19: 49–79, 285–97; 20: 59–80, 143–64.

Chaloner, William G., and Hugh L. Pearson. 2005. John Lindley, the reluctant palaeobotanist. *In* Bowden *et al., History of palaeobotany,* 29–39.

Chester, D. K., A. M. Duncan, J. E. Guest, and C. R. J. Kilburn. 1985. *Mount Etna: the anatomy of a volcano.* London (Chapman and Hall).

Ciancio, Luca. 2005. *Teatro del mutamento: immagini del 'Tempio di Serapide' (1750–1900).* Rovereto (Nicolodi).

Clark, John Willis, and Thomas McKenny Hughes. 1890. *The life and letters of the Reverend Adam Sedgwick . . .* 2 vols. Cambridge (Cambridge University Press).

Clark, Kevin. 1925. A pioneer of prehistory. *Blackfriars* 6: 606–13, 640–48, 726–28.

Cleal, Christopher J., Maureen Lazarus, and Annette Townsend. 2005. Illustrators and illustrations during the 'Golden Age' of palaeobotany. *In* Bowden *et al., History of palaeobotany,* 41–61.

Cleevely, Ron J. 1974. The Sowerbys, the *Mineral conchology,* and their fossil collection. *JSBNH* 6: 418–81.

———. 1974. A provisional bibliography of natural history works by the Sowerby family. *JSBNH* 6: 482–559.

Cleevely, Ron J., and Sandra D. Chapman. 2000. The two states of Mantell's *Illustrations of the geology of Sussex:* 1827 and c. 1829. *ANH* 27: 23–50.

Cohen, I. Bernard. 1985. *Revolution in science.* Cambridge, Mass. (Harvard University Press).

Coleman, William. 1962. Lyell and the 'reality' of species. *Isis* 53: 325–38.

———. 1963. A note on the early relationship between Georges Cuvier and Louis Agassiz. *JHMAS* 18: 51–63.

———. 1964. *Georges Cuvier zoologist: a study in the history of evolution theory.* Cambridge, Mass. (Harvard University Press).

Comment, Bernard. 1999. *The Panorama.* London (Reaktion).

Corsi, Pietro. 1978. The importance of French transformist ideas for the second volume of Lyell's *Principles of Geology. BJHS* 11: 221–44.

———. 1988. *The age of Lamarck: evolutionary theories in France, 1790–1830.* Berkeley (University of California Press) [revised ed. of *Oltre il mito: Lamarck e la scienze naturali del suo tempo,* Bologna (Il Mulino), 1983].

———. 1988. *Science and religion: Baden Powell and the Anglican debate, 1800–1860.* Cambridge (Cambridge University Press).

———. 2001. *Lamarck: genèse et enjeux du transformisme, 1770–1830.* Paris (CNRS) [revised ed. of *Age of Lamarck,* 1988].

———. 2005. Before Darwin: transformist concepts in European natural history. *JHB* 38: 67–83.

Crosland, Maurice P. 1992. *Science under control: the French Academy of Sciences 1795–1914.* Cambridge (Cambridge University Press).

Cross, Timothy A., and Peter W. Homewood. 1997. Amanz Gressly's role in founding modern stratigraphy. *BGSA* 109: 1617–30.

Cunningham, Frank F. 1990. *James David Forbes: pioneer Scottish glaciologist.* Edinburgh (Scottish Academic Press).

Curwen, E. Cecil (ed.). 1940. *The journal of Gideon Mantell, surgeon and geologist.* London (Oxford University Press).

Daudin, Henri. 1926. *Cuvier et Lamarck: les classes zoologiques et l'idée de série animale (1790–1830).* Paris (Félix Alcan) [repr. 1983].

Davies, Gordon L.: *see* Herries Davies, Gordon L.

Dean, Dennis R. 1980. Graham Island, Charles Lyell, and the craters of elevation controversy. *Isis* 71: 571–88.

———. 1985. The rise and fall of the deluge. *JGE* 33: 84–93.

———. 1997. The early career of Charles Lyell. *MG* 21: 215–23.

———. 1999. *Gideon Mantell and the discovery of dinosaurs.* Cambridge (Cambridge University Press).

———. 1999. Lyell and Murchison in France. *MG* 21: 313–34.

———. 1999. Charles Lyell in Italy and Sicily, 1828–1829. *MG* 21: 335–64.

De Beer, Gavin R. (ed.). 1958. *Darwin and Wallace: evolution by natural selection.* Cambridge (Cambridge University Press).

De Beer, Gavin R., and F. J. North. 1950. Sir Henry De la Beche's attempt on Mont Blanc in 1819. *AJ* 57: 493–500, 2 pls.

Dehérain, Henri. 1908. *Catalogue des manuscrits du fonds Cuvier (travaux et correspondence scientifique) conservés à la Bibliothèque de l'Institut de France* [1]. Paris (Honoré Champion).

Delair, Justin B., and William A. S. Sarjeant. 1975. The earliest discoveries of dinosaurs. *Isis* 66: 5–25.

———. 2002. The earliest discoveries of dinosaur bones: the records re-examined. *PGA* 113: 185–97.

Delisle, Richard. 1998. Les origines de la paléontologie humaine: essai de réinterprétation. *L'Anthropologie* 102: 3–19.

Desmond, Adrian. 1979. Designing the dinosaur: Richard Owen's response to Robert Edward Grant. *Isis* 70: 224–34.

———. 1984. Interpreting the origin of mammals: new approaches to the history of palaeontology. *ZJLS* 82: 7–16.

———. 1989. *The politics of evolution: morphology, medicine, and reform in radical London.* Chicago (University of Chicago Press).

Deville, Charles Sainte-Claire. 1878. *Coup-d'oeil historique sur la géologie et sur les travaux d'Élie de Beaumont.* Paris (Masson).

Dolan, Brian P. 1998. Representing novelty: Charles Babbage, Charles Lyell, and experiments in early Victorian geology. *HS* 36: 299–327.

———. 2000. *Exploring European frontiers: British travellers in the age of Enlightenment.* Basingstoke (MacMillan).

Dott, Robert H., Jr. (ed.). 1992. *Eustasy: the historical ups and downs of a major geological concept. GSAM* 180.

———. 1992. An introduction to the ups and downs of eustasy. *In* Dott, *Eustasy*, 1–16.

Ducros, Albert, and Jacqueline Ducros (eds.). 2000. *L'homme préhistorique: images et imaginaire.* Paris (L'Harmattan).

Duvernoy, M. 1939. Sophie Duvaucel d'après des correspondences inédites. *MSÉM* 51: 51–86.

Dvorak, John J., and Giuseppe Mastrolorenzo. 1991. The mechanisms of recent vertical crustal movements in Campi Flegrei caldera, southern Italy. *GSASP* 263.

Edmonds, J. M., and J. A. Douglas. 1976. William Buckland, F.R.S. (1784–1856) and an Oxford geological lecture, 1823. *NRRSL* 30: 141–67.

Elena, Alberto. 1988. The imaginary Lyellian revolution. *ESH* 7: 126–33.

Eyles, Joan M. 1978. G. W. Featherstonhaugh (1780–1866), F.R.S., F.G.S., geologist and traveller. *JSBNH* 8: 381–95.

Farlow, James A., and M. K. Brett-Surman (eds.). 1997. *The complete dinosaur.* Bloomington (Indiana University Press).

Feaver, William. 1975. *The art of John Martin.* Oxford (Clarendon).

Finnegan, Diarmid A. 2004. The work of ice: glacial theory and scientific culture in early Victorian Edinburgh. *BJHS* 37: 29–52.

Fisch, Menachem. 1991. *William Whewell, philosopher of science.* Oxford (Clarendon).

Fisch, Menachem, and Simon Schaffer (eds.). 1991. *William Whewell: a composite portrait.* Oxford (Clarendon).

Fischer, P. 1872. Note sur les travaux scientifiques d'Édouard Lartet. *BSGF* (2) 29: 246–66.

Forel, F.-A. 1899. Jean-Pierre Perraudin de Lourtier. *BSVSN* 35: 104–13 [also, abridged, in *EGH* 6: 169–75, 1899–1900].

Fox, Robert. 1980. The savant confronts his peers: scientific societies in France, 1815–1914. *In* Fox and Weisz, *Organization of science*, 241–82.

Fox, Robert, and George Weisz (eds.). 1980. *The organization of science in France, 1808–1914.* Cambridge (Cambridge University Press); Paris (Maison de l'Homme).

Frängsmyr, Tore. 1976. *Upptäckten av istiden: studier i den moderna geologins framväxt.* Stockholm (Almqvist & Wiksell).

———. 1985. The emergence of the glacial theory: a Scandinavian aspect. *Organon* 21: 231–35.

Freeman, Michael J. 2004. *Victorians and the prehistoric: tracks to a lost world.* New Haven (Yale University Press).

Frigo, Margherita, and Lorenzo Sorbini. 1997. *600 fossili per Napoleone*. Verona (Museo Civico di Storia Naturale di Verona).

Gard, Jean-Michel (ed.). 1988. *16 juin 1818, débâcle du Giétro: exposition thématique sur la géographie, la géologie et la glaciologie de la vallée de Bagnes*. Le Châble [Valais] (Musée de Bagnes).

Gaudant, Jean. 1980. Louis Agassiz (1807–1873), fondateur de la paléoichthyologie. *RHS* 33: 151–62.

———. 1991. Les cent-cinquante ans de la première carte géologique de France. *TCFHG* (3) 5: 79–83.

——— (ed.). 2005. *Dolomieu et la géologie de son temps*. Paris (École des Mines de Paris).

Geikie, Archibald. 1875. *Life of Sir Roderick Murchison . . . based on his journals and letters, with notices of his scientific contemporaries and a sketch of the rise and growth of Palaeozoic geology in Britain*. 2 vols. London (John Murray).

———. 1897. *The founders of geology*. London (Macmillan) [2nd ed., 1905, repr. New York (Dover), 1962].

Geus, Armin. 1997. Specimens and visual representations of animals and plants extinct in historical time. *In* Mazzolini, *Non-verbal communication*, 391–409.

Gohau, Gabriel. 1995. Constant Prévost (1787–1856), géologue critique. *MSGF* 168: 77–82.

———. 1998. Léonce Élie de Beaumont (1798–1874): pour le bicentenaire de sa naissance. *TCFHG* (3) 12: 71–77.

Gordon, [Elizabeth Oke]. 1894. *The life and correspondence of William Buckland, D.D., F.R.S.* London (John Murray).

Gosselet, Jules. 1896. Constant Prévost: coup d'oeil rétrospectif sur la géologie en France pendant la première moitié du XIXe siècle. *ASGN* 25.

Gould, Stephen Jay. 1985. The freezing of Noah. *In* Gould, *The flamingo's smile*, Harmondsworth (Penguin), 114–25.

———. 1987. *Time's arrow, time's cycle: myth and metaphor in the discovery of geological time*. Cambridge, Mass. (Harvard University Press).

———. 1996. The Razumovsky duet. *In* Gould, *Dinosaur in a haystack*. New York (Harmony), 260–71.

———. 2000. Lyell's pillars of wisdom. *In* Gould, *The lying stones of Marrakesh: penultimate reflections on natural history*. London (Jonathan Cape), 147–68.

Grayson, Donald K. 1983. *The establishment of human antiquity*. New York (Academic Press).

———. 1984. Nineteenth-century explanations of Pleistocene extinctions. *In* Martin and Klein, *Quaternary extinctions*, 5–39.

———. 1990. The provision of time depth for paleoanthropology. *In* Laporte, *Framework for paleoanthropology*, 1–13.

Greene, Mott T. 1982. *Geology in the nineteenth century: changing views of a changing world*. Ithaca (Cornell University Press).

Hacking, Ian. 1990. *The taming of chance*. Cambridge (Cambridge University Press).

Haile, N. S. 1997. The 'piddling school' of geology. *Nature* 387: 650.

Hall, A. Rupert. 1969. *The Cambridge Philosophical Society: a history, 1819–1969*. Cambridge (Cambridge Philosophical Society).

Hall, Marie Boas. 1984. *All scientists now: the Royal Society in the nineteenth century*. Cambridge (Cambridge University Press).

Hallam, Anthomy. 1989. *Great geological controversies*. 2nd ed. Oxford (Oxford University Press).

Hamm, Ernst P. 1993. Bureaucratic *Statistik* or actualism? K. E. A. von Hoff's *History* and the history of geology. *HS* 31: 151–76.

Harcourt, E. W. (ed.). 1880–1905. *The Harcourt papers*. 14 vols. Oxford (privately printed).

Harman, Peter (ed.). 1985. *Wranglers and physicists: studies on Cambridge physics in the nineteenth century*. Manchester (Manchester University Press).

Hartwig, Walter Carl. 1995. *Protopithecus*: rediscovering the first fossil primate. *HPLS* 17: 447–60.

Henderickx, Liliane. 1994. Philippe-Charles Schmerling (1790–1836) révèle l'antiquité de l'homme grâce aux dépôts antédiluviens des grottes liégeoises. *APB* 94: 183–212.

Herbert, Sandra. 1991. Charles Darwin as a prospective geological author. *BJHS* 24: 159–92.

———. 2005. *Charles Darwin, geologist.* Ithaca (Cornell University Press).

Herries Davies, Gordon L. 1968. The tour of the British Isles made by Louis Agassiz in 1840. *AS* 24: 131–46.

——— 1969. *The earth in decay: a history of British geomorphology 1578–1878.* London (Macdonald).

———. 2007. *Whatever is under the earth: the Geological Society of London 1807 to 2007.* London (Geological Society).

Hevly, Bruce. 1996. The heroic science of glacier motion. *Osiris* (n.s.) 11: 66–86.

Himbergen, E. J. van. 1978. De prijsvragen van de Tweede Genootschap 1778–1978. *In* Anonymous, *'Teyler'*, 37–55.

Hineline, Mark L. 1993. *The visual culture of the earth sciences 1863–1970.* San Diego (University of California San Diego [Ph.D. dissertation]) [University Microfilms no. 94 20724].

Hodge, M. J. S. 1991. The history of the earth, life, and Man: Whewell and palaetiological science. *In* Fisch and Schaffer, *William Whewell*, 255–88.

Hooykaas, R. 1959. *Natural law and divine miracle: a historical-critical study of the principle of uniformity in geology, biology, and theology.* Leiden (E. J. Brill).

———. 1970. Catastrophism in geology, its scientific character in relation to actualism and uniformitarianism. *MKNAW*, afd. Letterkunde (n.r.) 33: 271–316.

Horný, R. 1980. Joachim Barrande (1799–1883): life, work, collections. *JSBNH* 9: 365–68.

Hossfeld, Uwe, and Thomas Junker (eds.). 2002. *Die Entstehung biologischer Disziplinen II: Beiträge zur 10. Jahrestagung der DGGTB in Berlin 2001.* Berlin (VWB).

Howe, S. R., T. Sharpe, and H. S. Torrens. 1981. *Ichthyosaurs: a history of fossil 'sea-dragons'.* Cardiff (National Museum of Wales).

Inkster, Ian, and Jack Morrell (eds.). 1983. *Metropolis and province: studies in British culture, 1780–1850.* London (Hutchinson).

Jackson, Peter. 1987. *George Scharf's London: sketches and watercolours of a changing city, 1820–1850.* London (John Murray).

Jeannet, Alphonse. 1928–29. Sur l'ordre et la date de publication des ouvrages d'Agassiz sur les poissons fossiles. *BSNSN* 52: 118–24; 53: 197–98.

Jordanova, L. J., and Roy S. Porter (eds.). 1979. *Images of the earth: essays in the history of the environmental sciences.* Chalfont St. Peter (British Society for the History of Science) [2nd ed. 1997].

Klaver, J. M. I. 1997. *Geology and religious sentiment: the effect of geological discoveries on English society and literature between 1829 and 1859.* Leiden (Brill).

Knell, Simon J. 2000. *The culture of English geology, 1815–1851.* Aldershot (Ashgate).

Kohn, David (ed). 1985. *The Darwinian heritage.* Princeton (Princeton University Press).

Kölbl-Ebert, Martina. 1997. Mary Buckland (née Morland), 1797–1857. *ESH* 16: 33–38.

———. 1997. Charlotte Murchison (née Hugonin), 1788–1869. *ESH* 16: 39–43.

———. 1999. Observing orogeny: Maria Graham's account of the earthquake in Chile in 1822. *Episodes* 22: 36–40.

———. 2002. British geology in the early nineteenth century: a conglomerate with a female matrix. *ESH* 21: 3–25.

———. 2003. George Bellas Greenough (1778–1855): a lawyer in geologist's clothes. *PGA* 114: 247–54.

Kuhn, Thomas S. 1970. *The structure of scientific revolutions*, 2nd ed. Chicago (University of Chicago Press)

Langer, Wolfhart. 1990. Frühe Bilder aus der Vorzeit. *Fossilien* 1990: 202–5.

Laporte, Léo F. (ed.). 1990. *Establishment of a geologic framework for paleoanthropology. GSASP* 242.

Lapparent, A. de. 1880. Travaux de la Société Géologique de France depuis sa fondation. *BSGF* (3) 8: xix–lv.

Laudan, Rachel. 1982. The role of methodology in Lyell's science. *SHPS* 13: 215–49.

———. 1987. *From mineralogy to geology: the foundations of a science, 1650–1830.* Chicago (University of Chicago Press).

Launay, Louis de. 1940. *Une grande famille de savants: les Brongniart.* Paris (G. Rapilly).

Laurent, Goulven. 1976. Actualisme et antitransformisme chez Constant Prévost. *HN* 8: 33–51.

———. 1977. Le cheminement d'Étienne Geoffroy Saint-Hilaire vers un transformisme scientifique. *RHS* 30: 43–70.

———. 1987. *Paléontologie et évolution en France de 1800 à 1860: une histoire des idées de Cuvier et Lamarck à Darwin.* Paris (Comité des Travaux Historiques).

———. 1989. Idées sur l'origine de l'homme en France de 1800 à 1871 entre Lamarck et Darwin. *BMSAP* (n.s.) 1: 105–30.

———. 1993. Ami Boué (1794–1881): sa vie et son oeuvre. *TCFHG* (3) 7: 19–30.

———. 2007. Paléontologie et évolution: la Société Géologique de France, "espace de 'liberté" (1830–1860). *TCFHG* (3) 20: 1–11.

Lawrence, Philip J. 1977. Heaven and earth: the relation of the nebular hypothesis to geology. *In* Yourgrau and Breck, *Cosmology*, 253–81.

———. 1978. Charles Lyell versus the theory of central heat: a reappraisal of Lyell's place in the history of geology. *JHB* 11: 101–28.

Le Bas, M. J. (ed.). 1995. *Milestones in geology: reviews to celebrate 150 volumes of the Journal of the Geological Society.* London (Geological Society) [*GSLM* 16].

Le Guyader, Hervé. 1998. *Geoffroy Saint-Hilaire: un naturaliste visionnaire.* Paris (Belin). [trans. Chicago (University of Chicago Press), 2004]

Levere, Trevor H. 1993. *Science and the Canadian Arctic: a century of exploration, 1818–1918.* Cambridge (Cambridge University Press).

Lewis, C. L. E., and S. J. Knell (eds.). 2001. *The age of the earth: from 4004 B.C. to A.D. 2002.* London (Geological Society) [*GSLSP* 190].

Lindberg, David C., and Ronald L. Numbers (eds.). 1986. *God and nature: historical essays on the encounter between Christianity and science.* Berkeley (University of California Press).

Lippincott, Kristen (ed.). 1999. *The story of time.* London (Merrell Holberton).

Lurie, Edward. 1960. *Louis Agassiz: a life in science.* Chicago (University of Chicago Press).

Lyell, Katherine (ed.). 1881. *Life, letters and journals of Sir Charles Lyell, Bart.* 2 vols. London (John Murray) [abbreviated in footnotes to *LLJ*].

Lynch, John M. (ed.). 2000. *'Vestiges' and the debate before Darwin.* 7 vols. Bristol (Thoemmes Press).

———. (ed.). 2002. *Creationism and scriptural geology, 1817–1857.* 7 vols. Bristol (Thoemmes Press).

Lyon, John. 1969. The search for fossil man: cinq personnages à la recherche du temps perdu. *Isis* 61: 68–84.

McCartney, Paul J. 1976. Charles Lyell and G. B. Brocchi: a study in comparative historiography. *BJHS* 9: 177–89.

———. 1977. *Henry De la Beche: observations on an observer.* Cardiff (National Museum of Wales).

MacLeod, Roy. M. 1971. Of medals and men: a reward system in Victorian science, 1826–1914. *NRRSL* 26: 81–105.

Marcou, Jules. 1896. *Life, letters and works of Louis Agassiz.* New York (Macmillan) [repr. Farnborough (Greg), 1972].

Martin, Paul S., and Richard G. Klein (eds.). 1984. *Quaternary extinctions: a prehistoric revolution.* Tucson (University of Arizona Press).

Mathé, Gerhard. 1985. Karl Ernst Adolf von Hoff (1771 bis 1837): Verdienste und Grenzen eines Gothaische Staatsbeamten und die Förderung geologischen Denkens. *In* Prescher, *Deutscher Geologen*, 118–39.

Mathiot, Charles, and M. Duvernoy (eds.). 1940. Lettres inédits de Charles Laurillard à Georges Louis Duvernoy. *MSÉM* 55: 3–48.

Mavor, Elizabeth (ed.). 1993. *The captain's wife: the South American journals of Maria Graham, 1821–1823.* London (Weidenfeld and Nicolson).

Mazzolini, Renato G. (ed.). 1997. *Non-verbal communication in science prior to 1900.* Florence (Olschki).

Mielecke, Walter. 1973. Reinhard Bernhardi, ein fast vergessener Bahnbrecher für die Lehre vom Inlandeis. *ABMNG* 1973: 9–21.

Millhauser, Milton. 1954. The scriptural geologists: an episode in the history of opinion. *Osiris* 11: 65–86.

Mills, William. 1983. Darwin and the iceberg theory. *NRRSL* 38: 109–27.

Moore, James R. 1986. Geologists and interpreters of Genesis in the nineteenth century. *In* Lindberg and Numbers, *God and nature*, 322–50.

———. (ed.). 1989. *History, humanity, and evolution: essays for John C. Greene.* Cambridge (Cambridge University Press).

Morrell, Jack B. 1976. London institutions and Lyell's career. *BJHS* 9: 132–46 [repr. *Science, culture and politics* (1997), art. XII].

———. 1989. The legacy of William Smith: the case of John Phillips in the 1820s. *ANH* 16: 319–35.

———. 1997. *Science, culture and politics in Britain, 1750–1870.* Aldershot (Ashgate).

———. 2001. Genesis and geochronology: the case of John Phillips (1800–1874). *In* Lewis and Knell, *Age of the earth*, 85–90.

———. 2005. *John Phillips and the business of Victorian science.* Aldershot (Ashgate).

Morrell, Jack B., and Arnold Thackray. 1981. *Gentlemen of science: early years of the British Association for the Advancement of Science.* Oxford (Clarendon).

Moser, Stephanie. 1998. *Ancestral images: the iconography of human origins.* Ithaca (Cornell University Press).

Murray, John, IV. 1919. *John Murray III, 1808–1892: a brief memoir.* London (John Murray).

Nair, Savithri Preetha. 2005. 'Eyes and no eyes': Siwalik fossil collecting and the crafting of Indian palaeontology (1830–1847). *SC* 18: 359–92.

Negrin, Howard E. 1977. *Georges Cuvier: administrator and educator.* New York (New York University [Ph.D. dissertation]) [University Microfilms no. 78–3124].

Neve, Michael. 1983. Science in a commercial city: Bristol, 1820–60. *In* Inkster and Morrell, *Metropolis and province*, 179–204.

Nitsch, Edgar. 1996. Keuper, 1820–34: Geburt eines stratigraphischen Begriffes. *AS* 53: 489–500.

Norman, David B. 1993. Gideon Mantell's 'Mantel-piece': the earliest well-preserved Ornithischian dinosaur. *MG* 18: 225–45.

———. 2000. Henry De la Beche and the plesiosaur's neck. *ANH* 27: 137–48.

North, F. J. 1942. Paviland cave, the "Red Lady", the Deluge, and William Buckland. *AS* 5: 91–128.

———. 1943. Centenary of the glacial theory (notes on manuscripts and publications relating to its origin, development, and its introduction into Britain). *PGA* 54: 1–28.

O'Connor, Ralph. 2007. *The earth on show: fossils and the poetics of popular science, 1802–1856.* Chicago (University of Chicago Press).

Oldroyd, David. 1999. Early ideas about glaciation in the English Lake District. *AS* 56: 175–203.

Orr, Mary. 2007. Keeping it in the family: the extraordinary case of Cuvier's daughters. *In* Burek and Higgs, *Role of women*, 277–86.

Ospovat, Dov. 1977. Lyell's theory of climate. *JHB* 10: 317–39.

Outram, Dorinda. 1984. *Georges Cuvier: vocation, science, and authority in post-Revolutionary France.* Manchester (Manchester University Press).

———. 1997. Le Muséum National d'Histoire Naturelle après 1793: institution scientifique ou champ de bataille pour les familles et les groupes d'influence. *In* Blanckaert *et al.*, *Le Muséum*, 25–30.

Page, Leroy E. 1963. *The rise of the diluvial theory in British geological thought.* Norman (University of Oklahoma [Ph.D. dissertation]) [University Microfilms no. 64–215]

———. 1969. Diluvialism and its critics in Great Britain in the early nineteenth century. *In* Schneer, *History of geology*, 257–71.

Parkinson, Richard. 1999. *Cracking codes: the Rosetta Stone and decipherment.* London (British Museum).

Pemberton, S. G., W. A. S. Sarjeant, and H. S. Torrens. 1996. Footsteps before the Flood: the first scientific reports of vertebrate footprints. *Ichnos* 4: 321–24.

Pemberton, S. G., and M. K. Gingras. 2003. The Reverend Henry Duncan (1774–1846) and the discovery of the first fossil footprints. *Ichnos* 10: 69–75.

Pinto, Manuel Serrano (ed.). 2003. *Geological resources and history . . . Proceedings of the 26th INHIGEO symposium.* Aveiro (University of Aveiro).

Porter, Roy. 1976. Charles Lyell and the principles of the history of geology. *BJHS* 9: 91–103.

———. 1978. Philosophy and politics of a geologist: G. H. Toulmin (1754–1817). *JHI* 39: 435–50.

———. 1982. Charles Lyell: the public and private faces of science. *Janus* 69: 29–50.

Prescher, Hans (ed.). 1985. *Leben und Werken Deutscher Geologen im 18. und 19. Jahrhundert.* Leipzig (VEB/Deutscher Verlag für Grundstoffindustrie).

Purcell, Rosamund Wolff, and Stephen Jay Gould. 1992. *Finders, keepers: eight collectors.* London (Hutchinson Radius)

Rehbock, Philip F. 1985. John Fleming (1785–1857) and the economy of nature. *In* Wheeler and Price, *Linnaeus to Darwin*, 129–40.

Reich, Otto. 1905. *Karl Ernst Adolf von Hoff der Bahnbrecher moderner Geologie: ein wissenschaftliche Biographie.* Leipzig (von Veit).

Rhodes, F. H. T. 1991. Darwin's search for a theory of the earth: symmetry, simplicity, and speculation. *BJHS* 24: 193–229.

Rivier, Henri. 1932. La Société Neuchâteloise des sciences Naturelles 1832–1932: mémoire historique publiée à l'occasion de son centenaire. *BSNSN* 56: 5–83.

Roberts, Michael B. 1998. Geology and Genesis unearthed. *Churchman* 112: 225–55.

Robinson, Arthur H. 1982. *Early thematic mapping in the history of cartography.* Chicago (University of Chicago Press).

Roller, Duane H. D. (ed.). 1971. *Perspectives in the history of science and technology.* Norman (University of Oklahoma Press).

Ross, Sydney. 1962. Scientist: the story of a word. *AS* 18: 65–85, pls. 3–5.

Rostand, Jean. 1966. Étienne Geoffroy Saint-Hilaire et la tératogenèse expérimentale. *RHS* 17: 41–50.

Rudwick, Martin J. S. 1967. A critique of uniformitarian geology: a letter from W. D. Conybeare to Charles Lyell, 1841. *PAPS* 111: 272–87.

———. 1969. Lyell on Etna, and the antiquity of the earth. *In* Schneer, *History of geology*, 288–304 [repr. Rudwick, *Lyell and Darwin* (2005), art. IV].

———. 1970. The glacial theory. *HS* 8: 136–57 [repr. Rudwick, *New science of geology* (2004), art. XIV].

———. 1971. Uniformity and progression: reflections on the structure of geological theory in the age of Lyell. *In* Roller, *Perspectives*, 209–27 [repr. Rudwick, *Lyell and Darwin* (2005), art. I].

———. 1972. *The meaning of fossils: episodes in the history of palaeontology.* London (Macdonald) and New York (American Elsevier) [2nd ed. 1976, repr. Chicago (University of Chicago Press), 1985].

———. 1974. Poulett Scrope on the volcanoes of Auvergne: Lyellian time and political economy. *BJHS* 7: 205–42 [repr. Rudwick, *Lyell and Darwin* (2005), art. III].

———. 1974. Darwin and Glen Roy: a 'great failure' in scientific method? *SHPS* 5: 97–185 [repr. Rudwick, *Lyell and Darwin* (2005), art. X].

———. 1975. Caricature as a source for the history of science: De la Beche's anti-Lyellian sketches of 1831. *Isis* 66: 534–60 [repr. Rudwick, *Lyell and Darwin* (2005), art. VII].

———. 1975. Charles Lyell F.R.S. (1797–1875) and his London lectures on geology, 1832–33. *NRRSL* 29: 231–63 [repr. Rudwick, *Lyell and Darwin* (2005), art. VIII].

———. 1976. Charles Lyell speaks in the lecture theatre. *BJHS* 9: 147–55.

———. 1976. The emergence of a visual language for geological science 1760–1840. *HS* 14: 149–95 [repr. Rudwick, *New science of geology* (2004), art. V].

———. 1977. Historical analogies in the early geological work of Charles Lyell. *Janus* 64: 89–107 [repr. Rudwick, *Lyell and Darwin* (2005), art. V].

———. 1978. Charles Lyell's dream of a statistical palaeontology. *Palaeontology* 21: 225–44 [repr. Rudwick, *Lyell and Darwin* (2005), art. VI].

———. 1979. Transposed concepts from the human sciences in the early work of Charles Lyell. *In* Jordanova and Porter, *Images of the earth*, 67–83 [2nd ed., 77–91].

———. 1982. Charles Darwin in London: the integration of public and private science. *Isis* 73: 186–206 [repr. Rudwick, *Lyell and Darwin* (2005), art. IX].

———. 1985. *The great Devonian controversy: the shaping of scientific knowledge among gentlemanly specialists.* Chicago (University of Chicago Press).

———. 1985. Darwin and the world of geology. *In* Kohn, *Darwinian heritage*, 511–18.

———. 1987. International arenas of geological debate in the early nineteenth century. *ESH* 5: 152–58.

———. 1988. A year in the life of Adam Sedgwick and company, geologists. *ANH* 15: 243–68 [repr. Rudwick, *New science of geology* (2004), art. XI].

———. 1989. Encounters with Adam, or at least the hyaenas: nineteenth-century visual representations of the deep past. *In* Moore, *History, humanity, and evolution*, 231–51.

———. 1990. Introduction. *In* [facsimile reprint of] Lyell, *Principles of geology* 1: vii–lviii.

———. 1991. Bibliography of Lyell's sources. *In* [facsimile reprint of] Lyell, *Principles of geology* 3: appendices 113–60.

———. 1992. *Scenes from deep time: early pictorial representations of the prehistoric world.* Chicago (University of Chicago Press).

———. 1995. Historical origins of the Geological Society's *Journal. In* Le Bas, *Milestones*, 3–6.

———. 1996. Geological travel and theoretical innovation: the role of 'liminal' experience. *SSS* 26: 143–59.

———. 1997. *Georges Cuvier, fossil bones, and geological catastrophes.* Chicago (University of Chicago Press).

———. 1997. *Recherches sur les ossements fossiles*: Georges Cuvier et le collecte des alliés internationaux. *In* Blanckaert *et al.*, *Le Muséum*, 591–606 [trans. Rudwick, *New science of geology* (2004), art. VIII].

———. 1998. Lyell and the *Principles of Geology. In* Blundell and Scott, *Lyell*, 3–15 [repr. Rudwick, *Lyell and Darwin* (2005), art. II].

———. 1999. Geologists' time: a brief history. *In* Lippincott, *Story of time*, 250–53 [repr. Rudwick, *New science of geology* (2004), art. I].

———. 2000. Cuvier's paper museum of fossil bones. *ANH* 27: 51–68 [repr. Rudwick, *New science of geology* (2004), art. IX].

———. 2001. Jean-André de Luc and nature's chronology. *In* Lewis and Knell, *Age of the earth*, 51–60 [repr. Rudwick, *New science of geology* (2004), art. VI].

———. 2004. Travel, travel, travel: geological fieldwork in the 1830s. *In* Rudwick, *New science of geology*, art XII.

———. 2004. *The new science of geology: studies in the earth sciences in the age of revolution.* Aldershot (Ashgate).

———. 2005. *Lyell and Darwin, geologists: studies in the earth sciences in the age of reform.* Aldershot (Ashgate).

———. 2005. Picturing nature in the Enlightenment. *PAPS* 149: 279–303.

———. 2005. *Bursting the limits of time: the reconstruction of geohistory in the age of revolution.* Chicago (University of Chicago Press) [abbreviated in text and footnotes to *BLT*].

Rupke, Nicolaas. 1983. *The great chain of history: William Buckland and the English school of geology.* Oxford (Clarendon).

———. 2005. Neither creation nor evolution: the third way in mid-19th-century thinking about the origin of species. *AHPB* 10: 143–72.

Russell, E. S. 1916. *Form and function: a contribution to the history of morphology.* London (Murray) [repr. Chicago (University of Chicago Press), 1982].

Sarjeant, William Anthony S. 1974. A history and bibliography of the study of fossil vertebrate footprints in the British Isles. *PPP* 16: 265–378.

——— (ed.). 1996. *Vertebrate fossils and the evolution of scientific concepts.* New York (Gordon and Breach).

———. 2003. Footprints before the Flood: incidents in the study of fossil vertebrate tracks in 19th century Britain. *In* Pinto, *Geological resources and history*, 63–86.

Sarjeant, William Anthony S., and Justin B. Delair. 1980. An Irish naturalist in Cuvier's laboratory: the letters of Joseph Pentland, 1820–1832. *BBMNH* (hist. ser.) 6: 245–319.

Schaer, Jean-Paul. 1998. *Les géologues et le développement de la géologie en pays de Neuchâtel.* Neuchâtel (Muséum d'Histoire Naturelle de Neuchâtel).

———. 2000. Agassiz et les glaciers: sa conduite de la recherche et ses mérites. *EGH* 93: 231–56.

———. 2002. Le rôle d'Agassiz en glaciologie ou la réussite d'un entrepreneur scientifique ambitieux. *TCFHG* (3) 15: 77–87.

Schaffer, Simon. 1989. The nebular hypothesis and the science of progress. *In* Moore, *History, humanity, and evolution*, 131–64.

Schneer, Cecil J. (ed.). 1969. *Toward a history of geology*. Cambridge, Mass. (MIT Press).

Secord, James A. 1982. King of Siluria: Roderick Murchison and the imperial theme in nineteenth-century British geology. *VS* 25: 413–42.

———. 1986. *Controversy in Victorian geology: the Cambrian-Silurian dispute*. Princeton (Princeton University Press)

———. 1986. The Geological Survey of Great Britain as a research school, 1839–1855. *HS* 24: 223–75.

———. 1991. Edinburgh Lamarckians: Robert Jameson and Robert E. Grant. *JHB* 24: 1–18.

———. 1991. The discovery of a vocation: Darwin's early geology. *BJHS* 24: 133–57.

———. 1997. Introduction. *In* [abridged edition of] *Lyell: Principles of geology*, ix–xliii. London (Penguin).

———. 2000. *Victorian sensation: the extraordinary publication, reception, and secret authorship of Vestiges of the Natural History of Creation*. Chicago (University of Chicago Press).

Smith, Crosbie. 1985. Geologists and mathematicians: the rise of physical geology. *In* Harman, *Wranglers and physicists*, 49–83.

Smith, Jean Chandler. 1993. *Georges Cuvier: annotated bibliography of his published works*. Blue Ridge Summit, Penn. (Smithsonian Institution Press).

Sommer, Marianne. 2003. The Romantic cave? The scientific and poetic quests for subterranean spaces in Britain. *ESH* 22: 172–208.

———. 2004. An amusing account of a cave in Wales: William Buckland (1784–1856) and the Red Lady of Paviland. *BJHS* 37: 53–74.

Stafford, Robert A. 1989. *Scientist of the empire: Sir Roderick Murchison, scientific exploration, and Victorian imperialism*. Cambridge (Cambridge University Press).

Stoczkowski, W. 1993. La préhistoire: les origines du concept. *BSPF* 60: 13–21.

Stoddart, David R. 1976. Darwin, Lyell, and the geological significance of coral reefs. *BJHS* 9: 199–218.

———. 1995. Darwin and the seeing eye: iconography and meaning in the *Beagle* years. *ESH* 14: 3–36.

Sturges, Paul. 1984. *A bibliography of George Poulett Scrope: geologist, economist and local historian*. Boston (Baker Library, Harvard Business School).

Surdez, Maryse. 1973. Catalogue des archives de Louis Agassiz (1807–1873). *BSNSN* 73: 4–202.

Swainston, Stephanie, and Alison Brookes. 2000. Paviland Cave and the 'Red Lady': the history of collection and investigation. *In* Aldhouse-Green, *Paviland Cave*, 19–64.

Taquet, Philippe. 1982. Cuvier, Buckland, Mantell, et les dinosaures. *In* Buffetaut *et al.*, *Symposium Cuvier*, 475–94.

———. 2003. Quand les reptiles marins anglais traversaient la Manche: Mary Anning et Georges Cuvier, deux acteurs de la découverte et de l'étude des ichthyosaures et des plésiosaures. *AP* 89: 37–64.

Taquet, Philippe, and Daniel Contini. 1997. William Buckland et le "mégalosaure" de Franche-Comté: données historiques, stratigraphiques et paléontologiques. *AP* 83: 93–110.

Tasch, Paul. 1985. A matter of priority: Lyell and Deshayes *et al. Compass* 63: 6–17.

Taylor, Michael A. 1994. The plesiosaur's birthplace: the Bristol Institution and its contribution to vertebrate palaeontology. *ZJLS* 112: 179–96.

———. 1997. Before the dinosaur: the historical significance of the fossil marine reptiles. *In* Calloway and Nicholls, *Ancient marine reptiles*, xix–xlvi.

Thackray, John C. 1979. T. T. Lewis and Murchison's *Silurian system*. *TWNFC* 42: 186–93.

———. 2003. *To see the Fellows fight: eye witness accounts of meetings of the Geological Society and its Club, 1822–1868*. Chalfont St. Peter (British Society for the History of Science; monograph 12).

Topham, Jonathan. 1992. Science and popular education in the 1830s: the role of the Bridgewater Treatises. *BJHS* 25: 397–430.

———. 1998. Beyond the 'common context': the production and reading of the Bridgewater Treatises. *Isis* 89: 233–62.

Torrens, Hugh S. 1990. The scientific ancestry and historiography of *The Silurian system*. *JGS* 147: 657–62.

———. 1995. Mary Anning (1799–1847) of Lyme: 'the greatest fossilist the world ever knew'. *BJHS* 28: 257–84.

———. 1996. The dinosaurs and dinomania over 150 years. *In* Sarjeant, *Vertebrate fossils*, 255–84.

———. 1997. Politics and paleontology: Richard Owen and the invention of dinosaurs. *In* Farlow and Brett-Surman, *Complete dinosaur*, 175–90.

Twyman, Michael. 1970. *Lithography 1800–1850: the techniques of drawing on stone in England and France and their application in works of topography*. London (Oxford University Press).

Vaccari, Ezio. 1998. Lyell's reception on the continent of Europe: a contribution to an open historiographical problem. *In* Blundell and Scott, *Lyell*, 39–52.

Van Riper, A. Bowdoin. 1993. *Men among the mammoths: Victorian science and the discovery of human prehistory*. Chicago (University of Chicago Press).

Wegmann, Eugène. 1963. L'exposé original de la notion de facies par A. Gressly (1814–1865). *ST* 9: 83–119.

———. 1967. Évolution des idées sur le déplacement des lignes de rivage: origines en Fennoscandie. *MSVSN* 14: 129–90.

———. 1969. Changing ideas about moving shorelines. *In* Schneer, *History of geology*, 386–414.

Weidmann, Marc. 1972. A propos d'Ignace Venetz (1788–1859). *BM* 89: 5–9.

Wheeler, Alwyne, and James H. Price (eds.). 1985. *From Linnaeus to Darwin: commentaries on the history of biology and geology*. London (Society for the History of Natural History).

Wilson, Leonard G. 1972. *Charles Lyell: the years to 1841. The revolution in geology*. New Haven (Yale University Press).

———. 1980. Geology on the eve of Charles Lyell's first visit to America, 1841. *PAPS* 124: 168–202.

———. 1998. *Lyell in America: transatlantic geology, 1841–1853*. Baltimore (Johns Hopkins University Press).

———. 1998. Lyell: the man and his times. *In* Blundell and Scott, *Lyell*, 21–37.

Woodward, Horace B. 1907. *The history of the Geological Society of London*. London (Geological Society).

Worsley, Peter. 2006. Jens Esmark, 'Vassryggen' and early glacial theory in Britain. *McG* 16: 161–72.

Wyatt, John. 1995. *Wordsworth and the geologists*. Cambridge (Cambridge University Press).

Yeo, Richard R. 1991. William Whewell's philosophy of knowledge and its reception. *In* Fisch and Schaffer, *William Whewell*, 175–99.

———. 1993. *Defining science: William Whewell, natural knowledge, and public debate in early Victorian Britain*. Cambridge (Cambridge University Press).

Yourgrau, Wolfgang, and Allen D. Breck (eds.). 1977. *Cosmology, history, and theology*. New York (Plenum).

INDEX

References to a page number include text in figure captions on that page. Footnotes (shown by "n" after the page number) are given only if they are not mentioned in the text on that page.